## Geometric Gradient

### Geometric Series Present Worth:

To Find $P$    $(P/A, g, i, n)$
Given $A_1, g$    When $i = g$    $P = A_1[n(1+i)^{-1}]$

To Find $P$    $(P/A, g, i, n)$
Given $A_1, g$    When $i \neq g$    $P = A_1 \left[ \dfrac{1 - (1+g)^n(1+i)^{-n}}{i - g} \right]$

$A_j = A_1(1+g)^{j-1}$

## Continuous Compounding at Nominal Rate r

**Single Payment:**    $F = P[e^{rn}]$      $P = F[e^{-rn}]$

**Uniform Series:**    $A = F \left[ \dfrac{e^r - 1}{e^{rn} - 1} \right]$     $A = P \left[ \dfrac{e^{rn}(e^r - 1)}{e^{rn} - 1} \right]$

$F = A \left[ \dfrac{e^{rn} - 1}{e^r - 1} \right]$     $P = A \left[ \dfrac{e^{rn} - 1}{e^{rn}(e^r - 1)} \right]$

## Compound Interest

$i$ = Interest rate per interest period*.

$n$ = Number of interest periods.

$P$ = A present sum of money.

$F$ = A future sum of money. The future sum $F$ is an amount, $n$ interest periods from the present, that is equivalent to $P$ with interest rate $i$.

$A$ = An end-of-period cash receipt or disbursement in a uniform series continuing for $n$ periods, the entire series equivalent to $P$ or $F$ at interest rate $i$.

$G$ = Uniform period-by-period increase or decrease in cash receipts or disbursements; the arithmetic gradient.

$g$ = Uniform *rate* of cash flow increase or decrease from period to period; the geometric gradient.

$r$ = Nominal interest rate per interest period*.

$m$ = Number of compounding subperiods per period*.

---

*Normally the interest period is one year, but it could be something else.

| To Find | Excel |
|---------|-------|
| $P$ | $-\text{PV}(i, n, A, F, \text{Type})$ |
| $A$ | $-\text{PMT}(i, n, P, F, \text{Type})$ |
| $F$ | $-\text{FV}(i, n, A, P, \text{Type})$ |
| $n$ | $\text{NPER}(i, A, P, F, \text{Type})$ |
| $i$ | $\text{RATE}(n, A, P, F, \text{Type, guess})$ |
| $P$ | $\text{NPV}(i, \text{CF}_1 : \text{CF}_n)$ |
| $i$ | $\text{IRR}(\text{CF}_0 : \text{CF}_n)$ |

# ENGINEERING ECONOMIC ANALYSIS

# ENGINEERING ECONOMIC ANALYSIS

ELEVENTH EDITION

Donald G. Newnan
*San Jose State University*

Ted G. Eschenbach
*University of Alaska Anchorage*

Jerome P. Lavelle
*North Carolina State University*

New York   Oxford
OXFORD UNIVERSITY PRESS

Oxford University Press, Inc., publishes works that further Oxford University's objective of excellence in research, scholarship, and education.

Oxford    New York
Auckland    Cape Town    Dar es Salaam    Hong Kong    Karachi
Kuala Lumpur    Madrid    Melbourne    Mexico City    Nairobi
New Delhi    Shanghai    Taipei    Toronto

With offices in
Argentina    Austria    Brazil    Chile    Czech Republic    France    Greece
Guatemala    Hungary    Italy    Japan    Poland    Portugal    Singapore
South Korea    Switzerland    Thailand    Turkey    Ukraine    Vietnam

Published by Oxford University Press, Inc.
198 Madison Avenue, New York, New York 10016
http://www.oup.com

**Library of Congress Cataloging-in-Publication Data**

Newnan, Donald G.
Engineering economic analysis / Donald G. Newnan, Ted G. Eschenbach, Jerome P. Lavelle. — 11th ed.
    p. cm.
Includes bibliographical references and index.
ISBN 978-0-19-977804-1
1. Engineering economy. I. Eschenbach, Ted. II. Lavelle, Jerome P. III. Title.
TA177.4.N48   2012
658.15–dc22

                                                                    2010046035


Printing number: 9 8 7 6 5 4 3 2 1


Printed in the United States of America
on acid-free paper

*Eugene Grant and Dick Bernhard, for leading*
*the field of engineering economic analysis*
*DN*

*Richard Corey Eschenbach, for his lifelong example*
*of engineering leadership and working well with others*
*TE*

*My lovely wife and sweet daughters,*
*who always support all that I do*
*JL*

# CONTENTS IN BRIEF

# CONTENTS

Our goal has been, and still is, to provide an easy-to-understand and up-to-date presentation of engineering economic analysis for today's students. That means the book's writing style must promote the reader's understanding. We humbly note that our approach has been well received by engineering professors—and more importantly—by engineering students through ten previous editions.

## Hallmarks of this Book

Since it was first published in 1976, this text has become the market-leading book for the engineering economy course. It has always been characterized by

- **A focus on practical applications.** One way to encourage students to read the book, and to remember and apply what they have learned in this course, is to make the book interesting. And there is no better way to do that than to infuse the book with real-world examples, problems, and vignettes.
- **Accessibility.** We meet students where they are. Most don't have any expertise in accounting or finance. We take the time to explain concepts carefully while helping them apply them to engineering situations.
- **Superior instructor and student support packages.** To make this course easier to understand, learn, and teach Oxford University Press has produced the best support package available. We offer more for students and instructors than any competing text.

## Changes to the 11<sup>th</sup> Edition

This edition has many significant improvements in coverage. Before going chapter-by-chapter, we'd like to point out a few global changes made to keep this text the most current and most useful in today's courses:

- Additions to the ancillaries for students and faculty include a new set of 36 Excel tutorials, and a test bank approach to replace the Word exam file. Updates to the Student Study Guide, Instructor's Manual, Instructor PowerPoint slides, etc., were also done.
- Six new vignettes have been contributed and another six have been revised to ensure that the set of vignettes is timely, interesting, and motivating to the students. More ethics questions are now part of the *Questions to Consider*.
- Over 270 new problems have been added and another 70 problems have been revised.

- The *Cases in Engineering Economy*, 2$^{nd}$ Edition which was added to the *Enhanced* 10$^{th}$ Edition is still on the student CD, and each chapter now ends with a list of suggested cases.
- Appendix B on TVM Calculators is new. It covers financial calculators and the TVM capabilities of the HP 33s & 35s which can be used on the FE exam. Calculator solutions have been added to existing table-based solutions for representative examples (principally in Chapters 3–7). The approach reinforces the conceptual approach of factor based tables, but adds the computational power that is typically used in the real world. If used by the instructor (and it can easily be skipped), this supports students spending more time on concepts and challenges in engineering economy rather than on financial arithmetic.

Numerous improvements to the content and wording have been made throughout the text. For example:

- Nominal and effective interest has been moved from Chapter 4 to Chapter 3, so that Chapter 3 covers single payment (*P* and *F*) cash flows, and Chapter 4 covers multiple cash flows (*A*, *G*, and *g*). This also supports a better title & focus for Chapter 4 (rather than the old More Interest Formulas).
- Chapter 4 (*Equivalence for Repeated Cash Flows*) now emphasizes that uniform flow assumptions often correspond to the level of detail in feasibility and preliminary analyses, and that cash flow tables and spreadsheets are often matched to greater levels of detail about future cash flows.
- Chapter 5 (*Present Worth Analysis*) now includes a section on bond pricing.
- Chapter 10 (*Uncertainty in Future Events*) has a more complete discussion of potential problems with real option analysis.
- Chapter 11 (*Depreciation*) now covers 50% Bonus Depreciation.
- Chapter 12 (*Income Taxes*) now includes a 1-period example of leverage.
- Chapter 13 (*Replacement Analysis*) now includes a spreadsheet approach for solving for the challenger's economic life with any pattern of O&M costs and of salvage values.
- Chapter 14 (*Inflation and Price Change*) includes an improved explanation of why many problems in the text and the real world are described in constant-$ terms and solved with a real interest rate without explicitly including inflation.
- Chapter 15 (*Selection of a Minimum Attractive Rate of Return*) on selecting the MARR now includes a more complete explanation of the weighted average cost of capital.

There were several significant changes made to the 10$^{th}$ edition and maintained in this one that we want to highlight for faculty who are more familiar with earlier editions of the text.

- A discussion of ethics was included in Chapter 1 (*Making Economic Decisions*) and as part of many vignettes.
- An appendix of multiple choice questions intended to help students prepare for the Fundamentals of Engineering Exam was added (another 24 questions added in 11$^{th}$ edition).

- Chapter 8 (*Choosing the Best Alternative*) was completely rewritten to use the power of spreadsheets to graphically compare alternatives over a full range of possible interest rates. This is also a more intuitive foundation for incremental analysis.
- End-of-chapter problems were grouped under subheadings to make them easier to find and assign.
- "Trust Me, You'll Use This!" testimonials from former engineering economy students were added between chapters 2 and 3.

## The Superior Newnan Support Package

The supplement package for this text was expanded for the 9th and 10th editions, and it has been updated and expanded even further for this edition. No competing text has a more extensive support package.

For student use on the student Companion Website, the in-text CD, or packaged with the text there are the following:

- A set of 36 tutorials on engineering economy applications of Excel by Julie L. Fortune of the University of Alabama in Huntsville.
- A set of 54 cases for more realistic complex problems on the in-text CD. *Cases in Engineering Economy* 2nd by William Peterson and Ted Eschenbach also includes three chapters on case analysis and an example case solution as a fourth chapter.
- A Study Guide by Ed Wheeler of the University of Tennessee, Martin (ISBN 978-0-19-977819-5) is provided free with every new copy of the text.
- Spreadsheet problem modules, written by Thomas Lacksonen of the University of Wisconsin–Stout.
- Interactive multiple-choice problems, written by Paul Schnitzler of the University of South Florida and William Smyer of Mississippi State University.
- Additional Practice FE Exam problems, authored by Karen Thorsett, University of Phoenix.

For instructors there is now an updated and expanded set of supplemental materials:

- An exam file now in test bank format edited by Meenakshi Sundaram of Tennessee Technological University.
- PowerPoint lecture notes for all chapters by Shih Ming Lee of Florida International University and David Mandeville of Oklahoma State University.
- An Instructor's Manual by John M. Usher of Missippi State University and Lawrence Samuelson of Tri-State University and the authors, with complete solutions to all end-of-chapter problems. This is available in print.
- The compound interest tables from the textbook are available for adopting professors who prefer to give closed-book exams. The tables are on the website as PDF files that can be printed in part or in total.

For students and instructors a companion website is available with these supplements and other updates:

- www.oup.com/us/newnan

## Spreadsheets

This edition maintains the approach to spreadsheets that was established in the 8th edition. Rather than relying on spreadsheet templates, the emphasis is on helping students learn to use the enormous capabilities of the spreadsheet software available on nearly every computer. This approach reinforces the traditional engineering economy factor approach, as the equivalent spreadsheet functions (PMT, PV, RATE, etc.) are used frequently.

For students who would benefit from a refresher or introduction on how to write good spreadsheets, there is an appendix to introduce spreadsheets. In Chapter 2, spreadsheets are used to draw cash flow diagrams. Then every chapter from Chapter 4 to Chapter 15 has a concluding section on spreadsheet use. Each section is designed to support the other material in the chapter and to add to the student's knowledge of spreadsheets. If spreadsheets are used, the student will be very well prepared to apply this tool to real-world problems after graduation.

This approach is designed to support a range of approaches to spreadsheets. For example, professors and students can rely on the traditional tools of engineering economy and, without loss of continuity, completely ignore the material on spreadsheets. Or at the other extreme, professors can introduce the concepts and require all computations to be done with spreadsheets. Or a mix of approaches may be taken, depending on the professor, the students, and the particular chapter.

## Acknowledgments

Many people have directly or indirectly contributed to the content of the book. We have been influenced by our Stanford and North Carolina State University educations and our university colleagues, and by students who have provided invaluable feedback on content and form. We are particularly grateful to the following professors and students for their contributions in the form of insights, reviews, contributed problems, and contributed vignettes to this and previous editions.

Kate D. Abel, Stevens Institute of Technology

Mohamed Aboul-Seoud, Rensselaer Polytechnic Institute

V. Dean Adams, University of Nevada Reno

Dick Bernhard, North Carolina State University

James Blackburn, Southern Illinois University

Peter A. Cerenzio, Cerenzio & Panaro Consulting Engineers

Ronald Terry Cutwright, Florida State University

Ronald Deep, University of Dayton

Utpal Dutta, University of Detroit, Mercy

Sandra Duni Eksioglu, University of Florida

David Elizandro, Tennessee Technological University

John Erjavec, University of North Dakota

Peter Flynn, University of Alberta

Herbert Fox, New York Institute of Technology

Ashok Kumar Ghosh, University of New Mexico

Scott E. Grasman, University of Missouri–Rolla

Joseph Hartman, University of Florida

Oliver Hedgepeth, University of Alaska Anchorage

Morgan E. Henrie, University of Alaska Anchorage

Joseph R. Herkert, Arizona State University

Ted Huddleston, University of South Alabama

B. J. Kim, Louisiana Tech University

C. Patrick Koelling, Virginia Polytechnic Institute and State University

Neal Lewis, University of Bridgeport

Hampton Liggett, University of Tennessee

Daniel P. Loucks, Cornell University

Jessica Matson, Tennessee Technological University

Bill McManus, University of Oklahoma

Roger W. Meier, University of Memphis

Brigid Mullany, University of North Carolina–Charlotte

Heather Nachtmann, University of Arkansas

Kim LaScola Needy, University of Arkansas

T. Papagiannakis, Washington State University

Niranjani H. Patel, San Jose State University

William Peterson, Minnesota State University, Mankato

Richard J. Puerzer, Hofstra University

Dariush Rezaei, Kent State University–Trumbull

Duane J. Rosa, West Texas A&M University

John A. Roth, Vanderbilt University

Utpal Roy, Syracuse University

Malgorzata Rys, Kansas State University

C. Thomas Sciance, University of Texas at Austin

Mehdi Semsar, Savannah State University

William F. Shughart II, University of Mississippi

William N. Smyer, Mississippi State University

Meenakshi R. Sundaram, Tennessee Technological University

Arnold L. Sweet, Purdue University

Nicholas Sylvester, University of South Alabama

Kevin Taaffe, University of Florida

Robert E. Taylor, Virginia Polytechnic Institute and State University

William R. Truran, Stevens Institute of Technology

Suleyman Tufekci, University of Florida

John S. Usher, University of Louisville

John Whittaker, University of Alberta

The following individuals are new additions to our list of contributors of vignettes, text or solution reviews, problems, and insight.

Francisco Aguíñiga, Texas A&M University-Kingsville

Magdy Akladios, University of Houston-Clear Lake

Benjamin Armbruster, Northwestern University

Baabak Ashuri, Georgia Institute of Technology

Marsha Berry,U.S. Army Research, Development & Engineering Command

Rebeca Book, Pittsburg State University

William Brown, West Virginia University at Parkersburg

Patrick Brunese, Purdue University

Karen M. Bursic, University of Pittsburgh

Tracy Christofero, Marshall University

Paul Componation, University of Alabama in Huntsville

Gene Dixon, East Carolina University

Colin Drummond, Case Western Reserve University

Alberto Garcia, University of Tennessee

Dolores K. Gooding, University of South Florida

Johnny R. Graham, University of North Carolina at Charlotte

Barry Liner, George Mason University

Hamed Kashani, Georgia Institute of Technology

Paul R. McCright, University of South Florida

Dale McDonald, Midwestern State University

Gillian M. Nicholls, University of Alabama in Huntsville

John O'Haver, University of Mississippi

Renee Petersen, Washington State University

Kevin A. Rider, West Virginia University

Saeid Sadri, Georgia Institute of Technology

Scott Schultz, Mercer University

Don Smith, Texas A&M University

Musaka E. Ssemakula, Wayne State University

Francis M. Vanek, Cornell University

John Whittaker of the University of Alberta wrote a new Chapter 8 for the first Canadian edition and many improvements he made in the second Canadian edition have been incorporated. Neal Lewis of the University of Bridgeport coauthored the appendix on TVM calculators. Finally, we were helped by the professors who participated in the market survey for this book, and whose collective advice helped us shape this new edition.

Finally, our largest thanks go to the professors (and their students) who have developed the products that support this text:

Julie L. Fortune, University of Alabama in Huntsville

Thomas Lacksonen, University of Wisconsin–Stout

Shih Ming Lee, Florida International University

David Mandeville, Oklahoma State University

William Peterson, Minnesota State University, Mankato

Lawrence Samuelson, Tri-State University

Paul Schnitzler, University of South Florida

William Smyer, Mississippi State University

Meenakshi Sundaram, Tennessee Technological University

Karen Thorsett, University of Phoenix

John M. Usher, Mississippi State University

Ed Wheeler, University of Tennessee, Martin

Textbooks are produced through the efforts of many people. Caroline DiTullio, Patrick Lynch, and Rachael Zimmermann have worked with us to make this a timely and improved edition. Jennifer Bossert managed the text's design and production.

This book remains the best text on the market in large part because of feedback from users. We would appreciate hearing comments about the book, including being informed of any errors that have snuck in despite our best attempts to eradicate them. We also look forward to adding problems and vignettes in the next edition that adopters have found effective for their students. Please write us c/o the Engineering Editor at Oxford University Press, 198 Madison Avenue, New York, NY 10016, or email us directly or at highered.us@oup.com. Thanks for using the Newnan book!

Don Newnan
Ted Eschenbach
aftge@uaa.alaska.edu
Jerome Lavelle
jerome_lavelle@ncsu.edu

# ENGINEERING ECONOMIC ANALYSIS

# MAKING ECONOMIC DECISIONS

## Alternative-Fuel Vehicles

Imagine a future where all cars on U.S. roads no longer run on fossil fuel, but instead have alternate power sources. The positive effect of such a scenario on the environment is the motivation behind initiatives such as the Kyoto Protocol, which requires signatory countries to reduce greenhouse gas emissions. As another example, in 2006 California passed the first law in the nation to reduce automobile emissions by 25% by the year 2020. Legal and environmental concerns have increased the visibility of global warming and America's dependence on foreign oil, spurring the drive toward alternative-fuel vehicles.

In 2005, there were only 11 hybrid vehicle models in the U.S, which accounted for 1% of auto sales. The 2010 model year had 29 hybrid models. J.D. Power and Associates predicts an increase to 52 hybrid vehicles by 2012 and 7% of the U.S. market by 2015. By 2020, Toyota expects 30% of its vehicle sales to be hybrids. Although early hybrid-fuel models were small and economical cars, newer models include pickups, crossovers, and luxury cars. Some auto manufacturers, including Renault and Nissan, say the business case for hybrids is weak because they are more expensive to produce. Currently hybrids cost three to four thousand dollars more than traditional vehicles. In 2010 Nissan announced the Leaf, a 5-passenger electric-only car. However, Toyota, the leading manufacturer of alternative-fuel vehicles, makes money selling hybrids such as its Prius model. Consumers will pay more for hybrids because these vehicles have twice the fuel economy of conventional cars.

Engineers who design alternative-fuel vehicles must consider engine power and acceleration, fuel mileage per gallon, distance between refuels, vehicle safety, vehicle maintenance, fuel dispensing and distribution, and environmental and ethical issues. Design decisions influence manufacturing costs and therefore the final sales price. In addition, the entire supply, production, and service chain would be affected by a conversion to alternative-fuel vehicles. Depending on the degree to which alternate sources differ from the current gasoline-based technology, the conversion will be more or less complicated.

The outlook for alternative-fuel vehicles is positive for many reasons. As manufacturing experience increases, alternative-fuel vehicles can be designed to surpass the performance of conventional cars and meet customer expectations. As economies of scale bring the price of alternative vehicles down, more consumers can afford them and more manufacturers can profit from producing them. All of this combined with rising gasoline prices should increase consumers' interest in these vehicles—which could be good news for the environment. ■ ■ ■

*Contributed by Kate D. Abel, Stevens Institute of Technology*

## QUESTIONS TO CONSIDER

1. What marketplace dynamics drive or suppress the development of alternative-fuel vehicles? What role, if any, does government have in these dynamics? What additional charges should government have?

2. Develop a list of concerns and questions that consumers might have regarding the conversion to alternative-fuel vehicles. Which are economic and which are noneconomic factors?

3. From a manufacturer's viewpoint, what are the major concerns potential problems, and overriding goals of producing an alternative-fuel vehicle? How do these affect the vehicle's price?

4. Are there any ethical aspects to the conversion from gasoline power vehicles to alternative-fuel vehicles? List these, and determine how they could be or should be resolved and by whom.

### After Completing This Chapter...

*The student should be able to:*

- Distinguish between simple and complex problems.
- Discuss the role and purpose of engineering economic analysis.
- Describe and give examples of the nine steps in the *economic decision-making process.*
- Select appropriate economic criteria for use with different types of problems.
- Describe common ethical issues in engineering economic decision making.
- Solve simple problems associated with engineering decision making.

This book is about making decisions. **Decision making** is a broad topic, for it is a major aspect of everyday human existence. This book develops the tools to properly analyze and solve the economic problems that are commonly faced by engineers. Even very complex situations can be broken down into components from which sensible solutions are produced. If one understands the decision-making process and has tools for obtaining realistic comparisons between alternatives, one can expect to make better decisions.

Our focus is on solving problems that confront firms in the marketplace, but many examples are problems faced in daily life. Let us start by looking at some of these problems.

## A SEA OF PROBLEMS

A careful look at the world around us clearly demonstrates that we are surrounded by a sea of problems. There does not seem to be any exact way of classifying them, simply because they are so diverse in complexity and "personality." One approach arranges problems by their *difficulty*.

### Simple Problems

Many problems are pretty simple, and good solutions do not require much time or effort.

- Should I pay cash or use my credit card?
- Do I buy a semester parking pass or use the parking meters?
- Shall we replace a burned-out motor?
- If we use three crates of an item a week, how many crates should we buy at a time?

### Intermediate Problems

At a higher level of complexity we find problems that are primarily economic.

- Shall I buy or lease my next car?
- Which equipment should be selected for a new assembly line?
- Which materials should be used as roofing, siding, and structural support for a new building?
- Shall I buy a 1- or 2-semester parking pass?
- What size of transformer or air conditioner is most economical?

### Complex Problems

Complex problems are a mixture of *economic, political,* and *humanistic* elements.

- The decision of Mercedes-Benz to build an automobile assembly plant in Tuscaloosa, Alabama, illustrates a complex problem. Beside the economic aspects, Mercedes-Benz had to consider possible reactions in the U.S. and German auto industries. Would the German government pass legislation to prevent the overseas plant? What about German labor unions?
- The selection of a girlfriend or a boyfriend (who may later become a spouse) is obviously complex. Economic analysis can be of little or no help.

- The annual budget of a corporation is an allocation of resources and all projects are economically evaluated. The budget process is also heavily influenced by noneconomic forces such as power struggles, geographical balancing, and impact on individuals, programs, and profits. For multinational corporations there are even national interests to be considered.

## THE ROLE OF ENGINEERING ECONOMIC ANALYSIS

Engineering economic analysis is most suitable for intermediate problems and the economic aspects of complex problems. They have these qualities:

1. The problem is *important enough* to justify our giving it serious thought and effort.
2. The problem can't be worked in one's head—that is, a careful analysis *requires that we organize* the problem and all the various consequences, and this is just too much to be done all at once.
3. The problem has *economic aspects* important in reaching a decision.

When problems meet these three criteria, engineering economic analysis is useful in seeking a solution. Since vast numbers of problems in the business world (and in one's personal life) meet these criteria, engineering economic analysis is often required.

### Examples of Engineering Economic Analysis

Engineering economic analysis focuses on costs, revenues, and benefits that occur at different times. For example, when a civil engineer designs a road, a dam, or a building, the construction costs occur in the near future; but the benefits to users begin only when construction is finished and then continue for a long time.

In fact nearly everything that engineers design calls for spending money in the design and building stages, and only after completion do revenues or benefits occur—usually for years. Thus the economic analysis of costs, benefits, and revenues occurring over time is called *engineering* economic analysis.

Engineering economic analysis is used to answer many different questions.

- *Which engineering projects are worthwhile?* Has the mining or petroleum engineer shown that the mineral or oil deposit is worth developing?
- *Which engineering projects should have a higher priority?* Has the industrial engineer shown which factory improvement projects should be funded with the available dollars?
- *How should the engineering project be designed?* Has the mechanical or electrical engineer chosen the most economical motor size? Has the civil or mechanical engineer chosen the best thickness for insulation? Has the aeronautical engineer made the best trade-offs between (1) lighter materials that are expensive to buy but cheaper to fly and (2) heavier materials that are cheap to buy and more expensive to fly?

Engineering economic analysis can also be used to answer questions that are personally important.

- *How to achieve long-term financial goals:* How much should you save each month to buy a house, retire, or fund a trip around the world? Is going to graduate school a good investment—will your additional earnings in later years balance your lost income while in graduate school?

* *How to compare different ways to finance purchases:* Is it better to finance your car purchase by using the dealer's low interest rate loan or by taking an available rebate and borrowing money from your bank or credit union?
* *How to make short- and long-term investment decisions:* Should you buy a 1- or 2-semester parking pass? Is a higher salary better than stock options?

## THE DECISION-MAKING PROCESS

Decision making may take place by default; that is, a person may not consciously recognize that an opportunity for decision making exists. This fact leads us to a first element in a definition of decision making. To have a decision-making situation, there must be at least two alternatives available. If only one course of action is available, there is nothing to decide. There is no alternative but to proceed with the single available course of action. (It is rather unusual to find that there are no alternative courses of action. More frequently, alternatives simply are not recognized.)

At this point we might conclude that the decision-making process consists of choosing from among alternative courses of action. But this is an inadequate definition. Consider the following situation.

At a race track, a bettor was uncertain about which of the five horses to bet on in the next race. He closed his eyes and pointed his finger at the list of horses printed in the racing program. Upon opening his eyes, he saw that he was pointing to horse number 4. He hurried off to place his bet on that horse.

Does the racehorse selection represent the process of decision making? Yes, it clearly was a process of choosing among alternatives (assuming the bettor had already ruled out the "do-nothing" alternative of placing no bet). But the particular method of deciding seems inadequate and irrational. We want to deal with rational decision making.

### Rational Decision Making

Rational decision making is a complex process that contains nine essential elements, which are shown in Figure 1-1. Although these nine steps are shown sequentially, it is common for a decision maker to repeat steps, take them out of order, and do steps simultaneously. For example, when a new alternative is identified more data will be required. Or when the outcomes are summarized, it may become clear that the problem needs to be redefined or new goals established.

The value of this sequential diagram is to show all the steps that are usually required, and to show them in a logical order. Occasionally we will skip a step entirely. For example, a new alternative may be so clearly superior that it is immediately adopted at Step 4 without further analysis. The following sections describe the elements listed in Figure 1-1.

#### 1. Recognize the Problem

The starting point in rational decision making is recognizing that a problem exists.

Some years ago, for example, it was discovered that several species of ocean fish contained substantial concentrations of mercury. The decision-making process began with this recognition of a problem, and the rush was on to determine what should be done. Research revealed that fish taken from the ocean decades before and preserved in laboratories also

**FIGURE 1–1** One possible flowchart of the decision process.

1. Recognize problem

2. Define the goal or objective

3. Assemble relevant data

4. Identify feasible alternatives

5. Select the criterion to determine the best alternative

6. Construct a model

7. Predict each alternative's outcomes or consequences

8. Choose the best alternative

9. Audit the result

contained similar concentrations of mercury. Thus, the problem had existed for a long time but had not been recognized.

In typical situations, recognition is obvious and immediate. An auto accident, an overdrawn check, a burned-out motor, an exhausted supply of parts all produce the recognition of a problem. Once we are aware of the problem, we can solve it as best we can. Many firms establish programs for total quality management (TQM) or continuous improvement (CI) that are designed to identify problems so that they can be solved.

### 2. Define the Goal or Objective

The goal or objective can be an overall goal of a person or a firm. For example, a personal goal could be to lead a pleasant and meaningful life, and a firm's goal is usually to operate profitably. The presence of multiple, conflicting goals is often the foundation of complex problems.

But an objective need not be an overall goal of a business or an individual. It may be quite narrow and specific: "I want to pay off the loan on my car by May," or "The plant must produce 300 golf carts in the next 2 weeks," are more limited objectives. Thus, defining the objective is the act of exactly describing the task or goal.

### 3. Assemble Relevant Data

To make a good decision, one must first assemble good information. In addition to all the published information, there is a vast quantity of information that is not written down anywhere but is stored as individuals' knowledge and experience. There is also information that remains ungathered. A question like "How many people in your town would be interested in buying a pair of left-handed scissors?" cannot be answered by examining published data or by asking any one person. Market research or other data gathering would be required to obtain the desired information.

From all this information, what is relevant in a specific decision-making process? Deciding which data are important and which are not may be a complex task. The availability of data further complicates this task. Published data are available immediately at little or no cost; other data are available from specific knowledgeable people; still other data require surveys or research to assemble the information. Some data will be of high quality—that is, precise and accurate, while other data may rely on individual judgment for an estimate.

If there is a published price or a contract, the data may be known exactly. In most cases, the data is uncertain. What will it cost to build the dam? How many vehicles will use the bridge next year and twenty years from now? How fast will a competing firm introduce a competing product? How will demand depend on growth in the economy? Future costs and revenues are uncertain, and the range of likely values should be part of assembling relevant data.

The problem's time horizon is part of the data that must be assembled. How long will the building or equipment last? How long will it be needed? Will it be scrapped, sold, or shifted to another use? In some cases, such as for a road or a tunnel, the life may be centuries with regular maintenance and occasional rebuilding. A shorter time period, such as 50 years, may be chosen as the problem's time horizon, so that decisions can be based on more reliable data.

In engineering decision making, an important source of data is a firm's own accounting system. These data must be examined quite carefully. Accounting data focuses on past information, and engineering judgment must often be applied to estimate current and future values. For example, accounting records can show the past cost of buying computers, but engineering judgment is required to estimate the future cost of buying computers.

Financial and cost accounting are designed to show accounting values and the flow of money—specifically **costs** and **benefits**—in a company's operations. When costs are directly related to specific operations, there is no difficulty; but there are other costs that are not related to specific operations. These indirect costs, or **overhead,** are usually allocated to a company's operations and products by some arbitrary method. The results are generally satisfactory for cost-accounting purposes but may be unreliable for use in economic analysis.

To create a meaningful economic analysis, we must determine the *true* differences between alternatives, which might require some adjustment of cost-accounting data. The following example illustrates this situation.

## EXAMPLE 1–1

The cost-accounting records of a large company show the average monthly costs for the three-person printing department. The wages of the three department members and benefits, such as vacation and sick leave, make up the first category of **direct** labor. The company's indirect or overhead costs—such as heat, electricity, and employee insurance—must be distributed to its various departments in *some* manner and, like many other firms, this one uses *floor space* as the basis for its allocations. Chapter 17 will discuss allocating overhead costs in more detail.

| | |
|---|---|
| Direct labor (including employee benefits) | $ 6,000 |
| Materials and supplies consumed | 7,000 |
| Allocated overhead costs: | |
| 200 m$^2$ of floor area at $25/m$^2$ | 5,000 |
| | $18,000 |

The printing department charges the other departments for its services to recover its $18,000 monthly cost. For example, the charge to run 30,000 copies for the shipping department is:

| | |
|---|---|
| Direct labor | $228 |
| Materials and supplies | 294 |
| Overhead costs | 291 |
| Cost to other departments | $793 |

The shipping department checks with a commercial printer, which would print the same 30,000 copies for $688. The shipping department foreman wants to have the work done externally. The in-house printing department objects to this. As a result, the general manager has asked you to study the situation and recommend what should be done.

## SOLUTION

Much of the printing department's output reveals the company's costs, prices, and other financial information. The company president considers the printing department necessary to prevent disclosing such information to people outside the company. The firm cannot switch to an outside printer for all needs.

A review of the cost-accounting charges reveals nothing unusual. The charges made by the printing department cover direct labor, materials and supplies, and overhead. The allocation of indirect costs is a customary procedure in cost-accounting systems, but it is potentially misleading for decision making, as the following discussion indicates.

The shipping department would reduce its cost by $105 (= $793 – $688) by using the outside printer. In that case, how much would the printing department's costs decline, and which solution is better for the firm? We will examine each of the cost components:

1. *Direct Labor.* If the printing department had been working overtime, then the overtime could be reduced or eliminated. But, assuming no overtime, how much would the saving be? It seems unlikely that a printer could be fired or even put on less than a 40-hour work week. Thus, although there might be a $228 saving, it is much more likely that there will be no reduction in direct labor.

2. *Materials and Supplies.* There would be a $294 saving in materials and supplies.

3. *Allocated Overhead Costs.* There will be no reduction in the printing department's monthly $5000 overhead, and in fact the firm will incur additional expenses in purchasing and accounting to deal with the outside printer.

The firm will save $294 in materials and supplies and may or may not save $228 in direct labor if the printing department no longer does the shipping department work. The maximum saving would be $294 + 228 = $522. Either value of $294 or $522 is less than the $688 the firm would pay the outside printer. For this reason, the shipping department should not be allowed to send its printing to the outside printer.

Gathering cost data presents other difficulties. One way to look at the financial consequences—costs and benefits—of various alternatives is as follows.

- *Market Consequences.* These consequences have an established price in the marketplace. We can quickly determine raw material prices, machinery costs, labor costs, and so forth.
- *Extra-Market Consequences.* There are other items that are not directly priced in the marketplace. But by indirect means, a price may be assigned to these items. (Economists call these prices **shadow prices.**) Examples might be the cost of an employee injury or the value to employees of going from a 5-day to a 4-day, 40-hour week.
- *Intangible Consequences.* Numerical economic analysis probably never fully describes the real differences between alternatives. The tendency to leave out consequences that do not have a significant impact on the analysis itself, or on the conversion of the final decision into actual money, is difficult to resolve or eliminate. How does one evaluate the potential loss of workers' jobs due to automation? What is the value of landscaping around a factory? These and a variety of other consequences may be left out of the numerical calculations, but they must be considered in reaching a decision.

### 4. Identify Feasible Alternatives

One must keep in mind that unless the best alternative is considered, the result will always be suboptimal.[1] Two types of alternatives are sometimes ignored. First, in many situations a do-nothing alternative is feasible. This may be the "Let's keep doing what we are now doing," or the "Let's not spend any money on that problem" alternative. Second, there are often feasible (but unglamorous) alternatives, such as "Patch it up and keep it running for another year before replacing it."

There is no way to ensure that the best alternative *is* among the alternatives being considered. One should try to be certain that all conventional alternatives have been listed and then make a serious effort to suggest innovative solutions. Sometimes a group of people considering alternatives in an innovative atmosphere—**brainstorming**—can be helpful. Even impractical alternatives may lead to a better possibility. The payoff from a new, innovative alternative can far exceed the value of carefully selecting between the existing alternatives.

Any good listing of alternatives will produce both practical and impractical alternatives. It would be of little use, however, to seriously consider an alternative that cannot be adopted. An alternative may be infeasible for a variety of reasons. For example, it might

---

[1]A group of techniques called value analysis or value engineering is used to examine past decisions and current trade-offs in designing alternatives.

violate fundamental laws of science, require resources or materials that cannot be obtained, violate ethics standards, or conflict with the firm's strategy. Only the feasible alternatives are retained for further analysis.

### 5. Select the Criterion to Determine the Best Alternative

The central task of decision making is choosing from among alternatives. How is the choice made? Logically, to choose the best alternative, we must define what we mean by *best*. There must be a **criterion,** or set of **criteria,** to judge which alternative is best. Now, we recognize that *best* is on one end of the following relative subjective judgment:

| Worst | Bad | Fair | Good | Better | Best |
|-------|-----|------|------|--------|------|

*relative subjective judgment spectrum*

Since we are dealing in *relative terms,* rather than *absolute values,* the choice will be the alternative that is relatively the most desirable. Consider a driver found guilty of speeding and given the alternatives of a $175 fine or 3 days in jail. In absolute terms, neither alternative is good. But on a relative basis, one simply makes the best of a bad situation.

There may be an unlimited number of ways that one might judge the various alternatives. Several possible criteria are:

- Create the least disturbance to the environment.
- Improve the distribution of wealth among people.
- Minimize the expenditure of money.
- Ensure that the benefits to those who gain from the decision are greater than the losses of those who are harmed by the decision.[2]
- Minimize the time to accomplish the goal or objective.
- Minimize unemployment.
- Maximize profit.

Selecting the criterion for choosing the best alternative will not be easy if different groups support different criteria and desire different alternatives. The criteria may conflict. For example, minimizing unemployment may require increasing the expenditure of money. Or minimizing environmental disturbance may conflict with minimizing time to complete the project. The disagreement between management and labor in collective bargaining (concerning wages and conditions of employment) reflects a disagreement over the objective and the criterion for selecting the best alternative.

The last criterion—maximize profit—is the one normally selected in engineering decision making. When this criterion is used, all problems fall into one of three categories: neither input nor output fixed, fixed input, or fixed output.

**Neither input nor output fixed.** The first category is the general and most common situation, in which the amount of money or other inputs is not fixed, nor is the amount of benefits or other outputs. For example:

---

[2]This is the Kaldor criterion.

- A consulting engineering firm has more work available than it can handle. It is considering paying the staff for working evenings to increase the amount of design work it can perform.
- One might wish to invest in the stock market, but the total cost of the investment is not fixed, and neither are the benefits.
- A car battery is needed. Batteries are available at different prices, and although each will provide the energy to start the vehicle, the useful lives of the various products are different.

What should be the criterion in this category? Obviously, to be as economically efficient as possible, we must maximize the difference between the return from the investment (benefits) and the cost of the investment. Since the difference between the benefits and the costs is simply profit, a businessperson would define this criterion as **maximizing profit.**

**Fixed input.** The amount of money or other input resources (like labor, materials, or equipment) is fixed. The objective is to effectively utilize them. For economic efficiency, the appropriate criterion is to maximize the benefits or other outputs. For example:

- A project engineer has a budget of $350,000 to overhaul a portion of a petroleum refinery.
- You have $300 to buy clothes for the start of school.

**Fixed output.** There is a fixed task (or other output objectives or results) to be accomplished. The economically efficient criterion for a situation of fixed output is to minimize the costs or other inputs. For example:

- A civil engineering firm has been given the job of surveying a tract of land and preparing a "record of survey" map.
- You must choose the most cost-effective design for a roof, an engine, a circuit, or other component.

For the three categories, the proper economic criteria are:

| Category | Economic Criterion |
|---|---|
| Neither input nor output fixed | Maximize profit = value of outputs − cost of inputs. |
| Fixed input | Maximize the benefits or other outputs. |
| Fixed output | Minimize the costs or other inputs. |

### 6. Constructing the Model

At some point in the decision-making process, the various elements must be brought together. The *objective, relevant data, feasible alternatives,* and *selection criterion* must be merged. For example, if one were considering borrowing money to pay for a car, there is a mathematical relationship between the loan's variables: amount, interest rate, duration, and monthly payment.

Constructing the interrelationships between the decision-making elements is frequently called **model building** or **constructing the model.** To an engineer, modeling may be a scaled *physical representation* of the real thing or system or a *mathematical equation,* or set of equations, describing the desired interrelationships. In economic decision making, the model is usually mathematical.

In modeling, it is helpful to represent only that part of the real system that is important to the problem at hand. Thus, the mathematical model of the student capacity of a classroom might be

$$\text{Capacity} = \frac{lw}{k}$$

where  $l =$ length of classroom, in meters

$w =$ width of classroom, in meters

$k =$ classroom arrangement factor

The equation for student capacity of a classroom is a very simple model; yet it may be adequate for the problem being solved.

### 7. Predicting the Outcomes for Each Alternative

A model and the data are used to predict the outcomes for each feasible alternative. As was suggested earlier, each alternative might produce a variety of outcomes. Selecting a motorcycle, rather than a bicycle, for example, may make the fuel supplier happy, the neighbors unhappy, the environment more polluted, and one's savings account smaller. But, to avoid unnecessary complications, we assume that decision making is based on a single criterion for measuring the relative attractiveness of the various alternatives. If necessary, one could devise a single composite criterion that is the weighted average of several different choice criteria.

To choose the best alternative, the outcomes for each alternative must be stated in a *comparable* way. Usually the consequences of each alternative are stated in terms of money, that is, in the form of costs and benefits. This **resolution of consequences** is done with all monetary and nonmonetary consequences. The consequences can also be categorized as follows:

Market consequences—where there are established market prices available

Extra-market consequences—no direct market prices, so priced indirectly

Intangible consequences—valued by judgment, not monetary prices.

In the initial problems we will examine, the costs and benefits occur over a short time period and can be considered as occurring at the same time. In other situations the various costs and benefits take place in a longer time period. The result may be costs at one point in time followed by periodic benefits. We will resolve these in the next chapter into a *cash flow diagram* to show the timing of the various costs and benefits.

For these longer-term problems, the most common error is to assume that the current situation will be unchanged for the do-nothing alternative. In reality if we do nothing, then current profits will shrink or vanish as a result of the actions of competitors and the expectations of customers; and traffic congestion normally increases over the years as the number of vehicles increases—doing nothing does not imply that the situation will not change.

### 8. Choosing the Best Alternative

Earlier we said that choosing the best alternative may be simply a matter of determining which alternative best meets the selection criterion. But the solutions to most problems in economics have market consequences, extra-market consequences, and intangible consequences. Since the intangible consequences of possible alternatives are left out of the numerical calculations, they should be introduced into the decision-making process at this point. The alternative to be chosen is the one that best meets the choice criterion after considering both the numerical consequences and the consequences not included in the monetary analysis.

During the decision-making process certain feasible alternatives are eliminated because they are dominated by other, better alternatives. For example, shopping for a computer on-line may allow you to buy a custom-configured computer for less money than a stock computer in a local store. Buying at the local store is feasible, but dominated. While eliminating dominated alternatives makes the decision-making process more efficient, there are dangers.

Having examined the structure of the decision-making process, we can ask, When is a decision made, and who makes it? If one person performs *all* the steps in decision making, then she is the decision maker. *When* she makes the decision is less clear. The selection of the feasible alternatives may be the key item, with the rest of the analysis a methodical process leading to the inevitable decision. We can see that the decision may be drastically affected, or even predetermined, by the way in which the decision-making process is carried out. This is illustrated by the following example.

> Liz, a young engineer, was assigned to develop an analysis of additional equipment needed for the machine shop. The single criterion for selection was that the equipment should be the most economical, considering both initial costs and future operating costs. A little investigation by Liz revealed three practical alternatives:
>
> 1. A new specialized lathe
> 2. A new general-purpose lathe
> 3. A rebuilt lathe available from a used-equipment dealer
>
> A preliminary analysis indicated that the rebuilt lathe would be the most economical. Liz did not like the idea of buying a rebuilt lathe, so she decided to discard that alternative. She prepared a two-alternative analysis that showed that the general-purpose lathe was more economical than the specialized lathe. She presented this completed analysis to her manager. The manager assumed that the two alternatives presented were the best of all feasible alternatives, and he approved Liz's recommendation.

At this point we should ask: Who was the decision maker, Liz or her manager? Although the manager signed his name at the bottom of the economic analysis worksheets to authorize purchasing the general-purpose lathe, he was merely authorizing what already had been made inevitable, and thus he was not the decision maker. Rather Liz had made the key decision when she decided to discard the most economical alternative from further consideration. The result was a decision to buy the better of the two *less economically desirable* alternatives.

### 9. Audit the Results

An audit of the results is a comparison of what happened against the predictions. Do the results of a decision analysis reasonably agree with its projections? If a new machine tool was purchased to save labor and improve quality, did it? If so, the economic analysis seems to be accurate. If the savings are not being obtained, what was overlooked? The audit may help ensure that projected operating advantages are ultimately obtained. On the other hand, the economic analysis projections may have been unduly optimistic. We want to know this, too, so that the mistakes that led to the inaccurate projection are not repeated. Finally, an effective way to promote *realistic* economic analysis calculations is for all people involved to know that there *will* be an audit of the results!

# ETHICS

You must be mindful of the ethical dimensions of engineering economic analysis and of your engineering and personal decisions. This text can only introduce the topic, and we hope that you will explore this subject in greater depth.

Ethics can be described variously; however, a common thread is the concept of distinguishing between right and wrong in decision making. Ethics includes establishing systems of beliefs and moral obligations, defining values and fairness, and determining duty and guidelines for conduct. Ethics and ethical behavior are important because when people behave in ethical ways, individuals and society benefit. Usually the ethical choice is reasonably clear, but there are ethical dilemmas with conflicting moral imperatives. Consider an overloaded and sinking lifeboat. If one or more passengers are thrown into the shark-infested waters, the entire lifeboat can be saved. How is the decision made, how is it implemented, and who if anyone goes into the water? Ethical dilemmas also exist in engineering and business contexts. Ethical decision making requires the understanding of problem context, choices, and associated outcomes.

## Ethical Dimensions in Engineering Decision Making

Ethical issues can arise at every stage of the integrated process for engineering decision making described in Figure 1-1. Ethics is such an important part of professional and business decision making that ethical codes or standards of conduct exist for professional engineering societies, small and large organizations, and every individual. Written professional codes are common in the engineering profession, serving as a reference basis for new engineers and a basis for legal action against engineers who violate the code.

One such example is the Code of Ethics of the National Society of Professional Engineers (NSPE). Here is NSPE's fundamental canon of ethical behavior for engineering:
Engineers, in the fulfillment of their professional duties, shall:

- Hold paramount the safety, health and welfare of the public.
- Perform services only in areas of their competence.
- Issue public statements only in an objective and truthful manner.
- Act for each employer or client as faithful agents or trustees.
- Avoid deceptive acts.
- Conduct themselves honorably, responsibly, ethically and lawfully so as to enhance the honor, reputation, and usefulness of the profession.

In addition, NSPE has Rules of Practice and Professional Obligations for its members. Most engineering organizations have similar written standards. For all engineers difficulties arise when they act contrary to these written or internal codes, and opportunities for ethical dilemmas are found throughout the engineering decision-making process. Table 1-1

**TABLE 1–1  Example Ethical Lapses by Decision Process Step**

| Decision Process Step | Example Ethical Lapses |
|---|---|
| 1. Recognize the problem | • "Looking the other way," that is, not to recognize the problem—due to bribes or perhaps fear of retribution for being a "whistle-blower" |
| 2. Define the goal or objective | • Favoring one group of stakeholders by focusing on their objective for a project |
| 3. Assemble relevant data | • Using faulty or inaccurate data |
| 4. Identify feasible alternatives | • Leaving legitimate alternatives out of consideration |
| 5. Select the criterion to determine the best alternative | • Considering only monetary consequences when there are other significant consequences |
| 6. Construct a model | • Using a short horizon that favors one alternative over another. |
| 7. Predict each alternative's outcomes or consequences | • Using optimistic estimates for one alternative and pessimistic ones for the other alternatives |
| 8. Choose the best alternative | • Choosing an inferior alternative, one that is unsafe, adds unnecessary cost for the end user, harms the environment, etc. |
| 9. Audit the result | • Hiding past mistakes |

provides examples of ethical lapses that can occur at each step of the decision-making process.

Ethical dilemmas for engineers may arise in connection with engineering economic analysis in many situations. Following are examples of a few of these.

### *Gaining Knowledge and Building Trust Versus Favors for Influence*

Consider these three situations:

- The salesman for a supplier of HVAC (heating, ventilating, and air conditioning) equipment invites a mechanical engineer and spouse to come along on the company jet for a users' conference at a vacation resort.
- Same salesman and same engineer, but the invitation is for a day of golfing at an exclusive club.
- Same salesman invites the same engineer to lunch.

In each case the salesman is trying to "get the order," and there is likely to be some mix of business—discussing specifications—and pleasure. The first case, which brings up the largest ethical questions, also has the largest business justification. This is the opportunity to meet other users of the products and see displays of the product line. Often, firms and government agencies have strict guidelines that dictate behavior in these situations.

### Cost, Quality, and Functionality

One of the most common conflicts in the conceptual and design phase involves the trade-offs between cost, quality, and required functionality. Most modern products entail many thousands of decisions by designers that ultimately affect the cost and quality for the end user.

- A designer in an engineering consulting firm knows that a "gold-plated" solution would be very profitable for his firm (and for his bonus). This solution may also provide excellent reliability and require little maintenance cost.
- Engineers in the consumer durables division of a multinational company know that by using lower-quality connectors, fasteners, and subcomponents they can lower costs and improve the firm's market position. In addition, they know that these design elements have only a limited usable life, and the firm's most profitable business is repairs and extended warranties.

### The Environment We Live In

Projects for transportation and power generation typically must consider environmental impacts in their design and in deciding whether the project should be done in any form. Who incurs the costs for the project, and who receives the benefits? Many other engineering products are designed to promote recycling, reduce energy usage, and reduce pollution. Ethical issues can be particularly difficult because there are often stakeholders with opposing viewpoints, and some of the data may be uncertain and hard to quantify.

Examples of difficult choices include:

- Protecting the habitat of an endangered species versus flood control projects that protect people, animals, and structures.
- Meeting the needs for electrical power when all choices have some negative environmental impacts:
  - Hydroelectric—reservoir covers land and habitat
  - Coal—underground mining can be dangerous, open-pit mining damages habitat, and burning the coal can cause air pollution
  - Nuclear—disposal of radioactive waste
  - Fuel oil—air pollution and economic dependence
  - Wind—visual pollution of wind farms; birds killed by whirling blades
- Determining standards for pollutants: Is 1 part per million OK, or is 1 part per billion needed?

### Safety and Cost

Some of the most common and most difficult ethical dilemmas involve trade-offs between safety and cost. If a product is "too safe," then it will be too expensive, and it will not be used. Also sometimes the cost is incurred by one party and the risk by another.

- Should the oil platform be designed for the 100-year, 500-year, or 1000-year hurricane?

- Should the auto manufacturer add run-flat tires, stability control, side-cushion airbags, and rear-seat airbags to every car?
- Should a given product design go through another month of testing?
- Are stainless steel valves required, or is it economically better to use less corrosion-resistant valves and replace them more frequently?
- Is it safe to launch the Space Shuttle today?

### Emerging Issues and "Solutions"

Breaches of the law by corporate leaders of Enron, Tyco, and other firms have led to attempts to prevent, limit, and expose financial wrongdoing within corporations. One part of the solution has been the Sarbanes–Oxley Act of 2002, which imposed requirements on firm executives and auditing accounting firms, as well as penalties for violations.

Globalization is another area of increasing importance for ethical considerations. One reason is that different ethical expectations prevail in the world's various countries and regions. A second reason is that jobs may be moved to another country based on differences in cost, productivity, environmental standards, and so on. What may be viewed as a sweatshop from a U.S. perspective may be viewed as a wonderful opportunity to support many families from the perspective of a less developed nation.

## Importance of Ethics in Engineering and Engineering Economy

Many times engineers and firms try to act ethically, but mistakes are made—the data were wrong, the design was changed, or the operating environment was different than expected. In other cases, a choice was made between expediency (profit) and ethics. The firm and management are driven by the need to make a profit, and they expect the engineer to identify when safety will be compromised.

Ethics in engineering economic analysis focuses on how well and how honestly the decision-making process is conducted—the data, method of analysis, recommendations, and follow-up. The first step in avoiding problems is to recognize that ethical issues exist and to make them an explicit part of your decision-making process.

As a student, you've no doubt heard discussions about cheating on exams, plagiarism on written reports, violating university drinking and drug use policies, accepting one job while continuing to interview for others, and selling student sports tickets to nonstudents. You've made your own decisions about your behavior, and you've established patterns of behavior.

You should know that your professors care deeply about the ethical decisions you make at school. Your ethical habits there form a foundation for the character of your work and personal behavior after graduation.

Often recent engineering graduates are asked, "What is the most important thing you want from your supervisor?" The most common response is mentoring and opportunities to learn and progress. When employees with 5, 15, 25, or more years of experience are asked the same question, the most common response at all experience levels is *integrity*. This is what your subordinates, peers, and superiors will expect and value the most from you. Integrity is the foundation for long-term career success.

## ENGINEERING DECISION MAKING FOR CURRENT COSTS

Some of the easiest forms of engineering decision making deal with problems of alternative *designs, methods,* or *materials.* If results of the decision occur in a very short period of time, one can quickly add up the costs and benefits for each alternative. Then, using the suitable economic criterion, the best alternative can be identified. Three example problems illustrate these situations.

### EXAMPLE 1–2

A concrete aggregate mix must contain at least 31% sand by volume for proper batching. One source of material, which has 25% sand and 75% coarse aggregate, sells for $3 per cubic meter ($m^3$). Another source, which has 40% sand and 60% coarse aggregate, sells for $4.40/$m^3$. Determine the least cost per cubic meter of blended aggregates.

### SOLUTION

The least cost of blended aggregates will result from maximum use of the lower-cost material. The higher-cost material will be used to increase the proportion of sand up to the minimum level (31%) specified.

$$\text{Let } x = \text{Portion of blended aggregates from \$3.00/m}^3 \text{ source}$$

$$1 - x = \text{Portion of blended aggregates from \$4.40/m}^3 \text{ source}$$

**Sand Balance**

$$x(0.25) + (1 - x)(0.40) = 0.31$$

$$0.25x + 0.40 - 0.40x = 0.31$$

$$x = \frac{0.31 - 0.40}{0.25 - 0.40} = \frac{-0.09}{-0.15}$$

$$= 0.60$$

Thus the blended aggregates will contain

60% of $3.00/$m^3$ material
40% of $4.40/$m^3$ material

The least cost per cubic meter of blended aggregates is

$$0.60(\$3.00) + 0.40(\$4.40) = 1.80 + 1.76$$

$$= \$3.56/m^3$$

## EXAMPLE 1–3

A machine part is manufactured at a unit cost of 40¢ for material and 15¢ for direct labor. An investment of $500,000 in tooling is required. The order calls for 3 million pieces. Halfway through the order, managers learn that a new method of manufacture can be put into effect that will reduce the unit costs to 34¢ for material and 10¢ for direct labor—but it will require $100,000 for additional tooling. This tooling will not be useful for future orders. Other costs are allocated at 2.5 times the direct labor cost. What, if anything, should be done?

### SOLUTION

Since there is only one way to handle the first 1.5 million pieces, our problem concerns only the second half of the order.

**Alternative A: Continue with Present Method**

| | | |
|---|---|---|
| Material cost | 1,500,000 pieces × 0.40 = | $600,000 |
| Direct labor cost | 1,500,000 pieces × 0.15 = | 225,000 |
| Other costs | 2.50 × direct labor cost = | 562,500 |
| Cost for remaining 1,500,000 pieces | | $1,387,500 |

**Alternative B: Change the Manufacturing Method**

| | | |
|---|---|---|
| Additional tooling cost | | $100,000 |
| Material cost | 1,500,000 pieces × 0.34 = | 510,000 |
| Direct labor cost | 1,500,000 pieces × 0.10 = | 150,000 |
| Other costs | 2.50 × direct labor cost = | 375,000 |
| Cost for remaining 1,500,000 pieces | | $1,135,000 |

Before making a final decision, one should closely examine the *other costs* to see whether they do, in fact, vary as the *direct labor cost* varies. Assuming they do, the decision would be to change the manufacturing method.

## EXAMPLE 1–4

In the design of a cold-storage warehouse, the specifications call for a maximum heat transfer through the warehouse walls of 30,000 joules per hour per square meter of wall when there is a 30°C temperature difference between the inside surface and the outside surface of the insulation. The two insulation materials being considered are as follows:

| Insulation Material | Cost per Cubic Meter | Conductivity $(J\text{-}m/m^2\text{-}{}^\circ C\text{-}hr)$ |
|---|---|---|
| Rock wool | $12.50 | 140 |
| Foamed insulation | 14.00 | 110 |

The basic equation for heat conduction through a wall is

$$Q = \frac{K(\Delta T)}{L}$$

where $Q$ = heat transfer, in J/hr/m$^2$ of wall

$K$ = conductivity, in J-m/m$^2$-$^\circ$C-hr

$\Delta T$ = difference in temperature between the two surfaces, in $^\circ$C

$L$ = thickness of insulating material, in meters

Which insulation material should be selected?

### SOLUTION

Two steps are needed to solve the problem. First, the required thickness of each of the available materials must be calculated. Then, since the problem is one of providing a fixed output (heat transfer through the wall limited to a fixed maximum amount), the criterion is to minimize the input (cost).

**Required Insulation Thickness**

| | | |
|---|---|---|
| Rock wool | $30{,}000 = \dfrac{140(30)}{L}$, | $L = 0.14$ m |
| Foamed insulation | $30{,}000 = \dfrac{110(30)}{L}$, | $L = 0.11$ m |

**Cost of Insulation per Square Meter of Wall**

Unit cost = Cost/m$^3$ × Insulation thickness, in meters

Rock wool        Unit cost = $12.50 × 0.14 m = $1.75/m$^2$

Foamed insulation    Unit cost = $14.00 × 0.11 m = $1.54/m$^2$

The foamed insulation is the lesser-cost alternative. However, there is an intangible constraint that must be considered. How thick is the available wall space? Engineering economy and the time value of money are needed to decide what the maximum heat transfer should be. What is the cost of more insulation versus the cost of cooling the warehouse over its life?

## SUMMARY

### Classifying Problems

Many problems are simple and thus easy to solve. Others are of intermediate difficulty and need considerable thought and/or calculation to properly evaluate. These intermediate problems tend to have a substantial economic component and to require economic analysis. Complex problems, on the other hand, often contain people elements, along with political and economic components. Economic analysis is still very important, but the best alternative must be selected by considering all criteria—not just economics.

## The Decision-Making Process

Rational decision making uses a logical method of analysis to select the best alternative from among the feasible alternatives. The following nine steps can be followed sequentially, but decision makers often repeat some steps, undertake some simultaneously, and skip others altogether.

1. Recognize the problem.
2. Define the goal or objective: What is the task?
3. Assemble relevant data: What are the facts? Is more data needed, and is it worth more than the cost to obtain it?
4. Identify feasible alternatives.
5. Select the criterion for choosing the best alternative: possible criteria include political, economic, environmental, and social. The single criterion may be a composite of several different criteria.
6. *Mathematically model* the various interrelationships.
7. Predict the outcomes for each alternative.
8. Choose the best alternative.
9. Audit the results.

Engineering decision making refers to solving substantial engineering problems in which economic aspects dominate and economic efficiency is the criterion for choosing from among possible alternatives. It is a particular case of the general decision-making process. Some of the unusual aspects of engineering decision making are as follows:

1. Cost-accounting systems, while an important source of cost data, contain allocations of indirect costs that may be inappropriate for use in economic analysis.
2. The various consequences—costs and benefits—of an alternative may be of three types:
    (a) Market consequences—there are established market prices.
    (b) Extra-market consequences—there are no direct market prices, but prices can be assigned by indirect means.
    (c) Intangible consequences—valued by judgment, not by monetary prices.
3. The economic criteria for judging alternatives can be reduced to three cases:
    (a) When neither input nor output is fixed: maximize profit, which equals the difference between benefits and costs.
    (b) For fixed input: maximize benefits or other outputs.
    (c) For fixed output: minimize costs or other inputs.

    The first case states the general rule from which both the second and third cases may be derived.
4. To choose among the alternatives, the market consequences and extra-market consequences are organized into a cash flow diagram. We will see in Chapter 3 that

engineering economic calculations can be used to compare differing cash flows. These outcomes are compared against the selection criterion. From this comparison *plus* the consequences not included in the monetary analysis, the best alternative is selected.

5. An essential part of engineering decision making is the postaudit of results. This step helps to ensure that projected benefits are obtained and to encourage realistic estimates in analyses.

### Importance of Ethics in Engineering and Engineering Economy

One of the gravest responsibilities of an engineer is protecting the safety of the public, clients, and/or employees. In addition, the engineer can be responsible for the economic performance of projects and products on which bonuses and jobs depend. Not surprisingly, in this environment one of the most valued personal characteristics is integrity.

### Decision Making with Current Costs

When all costs and benefits occur within a brief period of time, the time value of money is not a consideration. We still must use the criteria of maximizing profit, minimizing cost, or maximizing benefits.

## PROBLEMS

### Decision Making

**1-1**  Think back to your first hour after awakening this morning. List 15 decision-making opportunities that existed during that hour. After you have done that, mark the decision-making opportunities that you actually recognized this morning and upon which you made a conscious decision.

**1-2**  Some of the following problems would be suitable for solution by engineering economic analysis. Which ones are they?

(*a*) Would it be better to buy a car with a diesel engine or a gasoline engine?

(*b*) Should an automatic machine be purchased to replace three workers now doing a task by hand?

(*c*) Would it be wise to enroll for an early morning class to avoid traveling during the morning traffic rush hours?

(*d*) Would you be better off if you changed your major?

(*e*) One of the people you might marry has a job that pays very little money, while another has a professional job with an excellent salary. Which one should you marry?

**1-3**  Which one of the following problems is *most* suitable for analysis by engineering economic analysis?

(*a*) Some 75¢ candy bars are on sale for 12 bars for $6. Sandy, who eats a couple of candy bars a week, must decide whether to buy a dozen at the lower price.

(*b*) A woman has $150,000 in a bank checking account that pays no interest. She can either invest it immediately at a desirable interest rate or wait a week and know that she will be able to obtain an interest rate that is 0.15% higher.

(*c*) Joe backed his car into a tree, damaging the fender. He has car insurance that will pay for the fender repair. But if he files a claim for payment, they may give him a lower "good driver" rating and charge him more for car insurance in the future.

**1-4**  If you have $300 and could make the right decisions, how long would it take you to become a millionaire? Explain briefly what you would do.

**1-5**  Many people write books explaining how to make money in the stock market. Apparently the authors plan to make *their* money selling books telling other people how to profit from the stock market. Why don't these authors forget about the books and make their money in the stock market?

**1-6**  The owner of a small machine shop has just lost one of his larger customers. The solution to his problem, he says, is to fire three machinists to balance his workforce with his current level of business. The owner says it is a simple problem with a simple solution. The three machinists disagree. Why?

**1-7**  Every college student had the problem of selecting the college or university to attend. Was this a simple, intermediate, or complex problem for you? Explain.

**1-8**  Toward the end of the twentieth century, the U.S. government wanted to save money by closing a small portion of its domestic military installations. While many people agreed that saving money was a desirable goal, people in areas potentially affected by a closing soon reacted negatively. Congress finally selected a panel whose task was to develop a list of installations to close, with the legislation specifying that Congress could not alter the list. Since the goal was to save money, why was this problem so hard to solve?

**1-9**  The college bookstore has put pads of engineering computation paper on sale at half price. What is the minimum and maximum number of pads you might buy during the sale? Explain.

**1-10**  Consider the seven situations described. Which one situation seems most suitable for solution by engineering economic analysis?

(*a*) Jane has met two college students that interest her. Bill is a music major who is lots of fun to be with. Alex is a fellow engineering student, but he does not like to dance. Jane wonders what to do.

(*b*) You drive periodically to the post office to pick up your mail. The parking meters require 25 ¢ for 6 minutes—about twice the time required to get from your car to the post office and back. If parking fines cost $20, do you put money in the meter or not?

(*c*) At the local market, candy bars are 75 ¢ each or three for $2. Should you buy them three at a time?

(*d*) The cost of car insurance varies widely from company to company. Should you check with several insurance companies when your policy comes up for renewal?

(*e*) There is a special local sales tax ("sin tax") on a variety of things that the town council would like to remove from local distribution. As a result, a store has opened up just outside the town and offers an abundance of these specific items at prices about 30% less than is charged in town. Should you shop there?

(*f*) Your mother reminds you that she wants you to attend the annual family picnic. That same Saturday you already have a date with a person you have been trying to date for months.

(*g*) One of your professors mentioned that you have a poor attendance record in her class. You wonder whether to drop the course now or wait to see how you do on the first midterm exam. Unfortunately, the course is required for graduation.

**1-11**  A car manufacturer is considering locating an assembly plant in your region. List two simple, two intermediate, and two complex problems associated with this proposal.

**1-12**  Consider the following situations. Which ones appear to represent rational decision making? Explain.

(*a*) Joe's best friend has decided to become a civil engineer, so Joe has decided that he will also become a civil engineer.

(*b*) Jill needs to get to the university from her home. She bought a car and now drives to the university each day. When Jim asks her why she didn't buy a bicycle instead, she replies, "Gee, I never thought of that."

(*c*) Don needed a wrench to replace the spark plugs in his car. He went to the local automobile supply store and bought the cheapest one they had. It broke before he had finished replacing all the spark plugs in his car.

**1-13**  Identify possible objectives for NASA. For your favorite of these, how should alternative plans to achieve the objective be evaluated?

**1-14**  Suppose you have just 2 hours to determine how many people in your hometown would be interested in buying a pair of left-handed scissors. Give a step-by-step outline of how you would seek to find out within the time allotted.

**1-15**  A college student determines that he will have only $250 per month available for his housing for the coming year. He is determined to continue in the university, so he has decided to list all feasible alternatives for his housing. To help him, list five feasible alternatives.

**1-16**  Describe a situation in which a poor alternative was selected because there was a poor search for better alternatives.

**1-17**  If there are only two alternatives available and both are unpleasant and undesirable, what should you do?

**1-18** The three economic criteria for choosing the best alternative are maximize the difference between output and input, minimize input, and maximize output. For each of the following situations, what is the correct economic criterion?

  (*a*) A manufacturer of plastic drafting triangles can sell all the triangles he can produce at a fixed price. As he increases production, his unit costs increase as a result of overtime pay and so forth. The manufacturer's criterion should be _____.

  (*b*) An architectural and engineering firm has been awarded the contract to design a wharf for a petroleum company for a fixed sum of money. The engineering firm's criterion should be _____.

  (*c*) A book publisher is about to set the list price (retail price) on a textbook. The choice of a low list price would mean less advertising than would be used for a higher list price. The amount of advertising will affect the number of copies sold. The publisher's criterion should be.

  (*d*) At an auction of antiques, a bidder for a particular porcelain statue would be trying to _____.

**1-19** As in Problem 1-18, state the correct economic criterion for each of the following situations.

  (*a*) The engineering school raffled off a car; tickets sold for $2 each or three for $5. When the students were selling tickets, they noted that many people had trouble deciding whether to buy one or three tickets. This indicates the buyers' criterion was _____.

  (*b*) A student organization bought a soft-drink machine for use in a student area. There was considerable discussion over whether they should set the machine to charge 75¢, $1, or $1.25 per drink. The organization recognized that the number of soft drinks sold would depend on the price charged. Eventually the decision was made to charge $1. The criterion was _____.

  (*c*) In many cities, grocery stores find that their sales are much greater on days when they advertise special bargains. However, the advertised special prices do not appear to increase the total physical volume of groceries sold by a store. This leads us to conclude that many shoppers' criterion is _____.

  (*d*) A recently graduated engineer has decided to return to school in the evenings to obtain a master's degree. He feels it should be accomplished in a manner that will allow him the maximum amount of time for his regular day job plus time for recreation. In working for the degree, he will _____.

**1-20** Seven criteria are given in the chapter for judging which is the best alternative. After reviewing the list, devise three additional criteria that might be used.

**1-21** Suppose you are assigned the task of determining the route of a new highway through an older section of town. The highway will require that many older homes be either relocated or torn down. Two possible criteria that might be used in deciding exactly where to locate the highway are:

  (*a*) Ensure that there are benefits to those who gain from the decision and that no one is harmed by the decision.

  (*b*) Ensure that the benefits to those who gain from the decision are greater than the losses of those who are harmed by the decision.

  Which criterion will you select to use in determining the route of the highway? Explain.

**1-22** For the project in Problem 1-21, identify the major costs and benefits. Which are market consequences, which are extra-market consequences, and which are intangible consequences?

**1-23** You must fly to another city for a Friday meeting. If you stay until Sunday morning your ticket will be $200, rather than $800. Hotel costs are $100 per night. Compare the economics with reasonable assumptions for meal expenses. What intangible consequences may dominate the decision?

**1-24** In the fall, Jay Thompson decided to live in a university dormitory. He signed a dorm contract under which he was obligated to pay the room rent for the full college year. One clause stated that if he moved out during the year, he could sell his dorm contract to another student who would move into the dormitory as his replacement. The dorm cost was $5000 for the two semesters, which Jay had already paid.

  A month after he moved into the dorm, he decided he would prefer to live in an apartment. That week, after some searching for a replacement to fulfill his dorm contract, Jay had two offers. One student offered to move in immediately and to pay Jay $300 per month for the eight remaining months of the school year. A second student offered to move in the second semester and pay $2500 to Jay.

Jay estimates his food cost per month is $500 if he lives in the dorm and $450 if he lives in an apartment with three other students. His share of the apartment rent and utilities will be $400 per month. Assume each semester is $4^{1}/_{2}$ months long. Disregard the small differences in the timing of the disbursements or receipts.

(*a*) What are the three alternatives available to Jay?

(*b*) Evaluate the cost for each of the alternatives.

(*c*) What do you recommend that Jay do?

1-25  In decision making we talk about the construction of a model. What kind of model is meant?

1-26  An electric motor on a conveyor burned out. The foreman told the plant manager that the motor had to be replaced. The foreman said that there were no alternatives and asked for authorization to order the replacement. In this situation, is any decision making taking place? If so, who is making the decision(s)?

1-27  Bill Jones's parents insisted that Bill buy himself a new sport shirt. Bill's father gave specific instructions, saying the shirt must be in "good taste," that is, neither too wildly colored nor too extreme in tailoring. Bill found three types of sport shirts in the local department store:

- Rather somber shirts that Bill's father would want him to buy
- Good-looking shirts that appealed to Bill
- Weird shirts that were even too much for Bill

He wanted a good-looking shirt but wondered how to convince his father to let him keep it. The clerk suggested that Bill take home two shirts for his father to see and return the one he did not like. Bill selected a good-looking blue shirt he liked, and also a weird lavender shirt. His father took one look and insisted that Bill keep the blue shirt and return the lavender one. Bill did as his father instructed. What was the key decision in this decision process, and who made it?

1-28  A farmer must decide what combination of seed, water, fertilizer, and pest control will be most profitable for the coming year. The local agricultural college did a study of this farmer's situation and prepared the following table.

| Plan | Cost/acre | Income/acre |
|------|-----------|-------------|
| A    | $ 600     | $ 800       |
| B    | 1500      | 1900        |
| C    | 1800      | 2250        |
| D    | 2100      | 2500        |

The last page of the college's study was torn off, and hence the farmer is not sure which plan the agricultural college recommends. Which plan should the farmer adopt? Explain.

1-29  Identify the alternatives, outcomes, criteria, and process for the selection of your college major. Did you make the best choice for you?

1-30  Describe a major problem you must address in the next two years. Use the techniques of this chapter to structure the problem and recommend a decision.

1-31  Apply the steps of the decision-making process to your situation as you decide what to do after graduation.

1-32  One strategy for solving a complex problem is to break it into a group of less complex problems and then find solutions to the smaller problems. The result is the solution of the complex problem. Give an example in which this strategy will work. Then give another example in which this strategy will not work.

## Ethics

1-33  When you make professional decisions involving investments in engineering projects, what criteria will you use?
*Contributed by D. P. Loucks, Cornell University*

1-34  What are ethics?
*Contributed by D. P. Loucks, Cornell University*

1-35  Analyze the ethical aspects of Problem 1-6.

1-36  Suppose you are a engineer working in a private engineering firm and you are asked to sign documents verifying information that you believe is not true. You like your work and your colleagues in the firm, and your family depends on your income. What criteria can you use to guide your decision regarding this issue?
*Contributed by D. P. Loucks, Cornell University*

1-37  Find the ethics code for the professional society of your major.

(*a*) Summarize its key points.

(*b*) What are its similarities and differences in comparison to NSPE's ethics code?

1-38  Use a personal example or a published source to analyze what went wrong or right with respect to ethics at the assigned stage(s) of the decision-making process.

(*a*) Recognize problem.

(*b*) Define the goal or objective.

(*c*) Assemble relevant data.

(*d*) Identify feasible alternatives.

(*e*) Select the criterion for determining the best alternative.

(*f*) Construct a model.

(*g*) Predict each alternative's outcomes or consequences.

(*h*) Choose the best alternative.

(*i*) Audit the result.

1-39 One of the important responsibilities of municipal assemblies and school boards is public infrastructure, such as roads and schools. Especially for this responsibility, engineers bring skills, knowledge, and perspectives that can improve public decision making. Often the public role is a part-time one, engineers that fulfill it will also have full-time jobs as employees or owners of engineering firms.

(*a*) What are some of the ethical issues that can arise from conflicts between the public- and private-sector roles?

(*b*) What guidelines for process or outcomes are there in your hometown?

(*c*) Summarize and analyze an example situation from the local newspaper, your hometown newspaper, or a news magazine.

1-40 Increasing population and congestion often are addressed through road improvement projects. These may pit the interests of homeowners and business owners in the project area against the interests of people traveling through the improvement project.

(*a*) What are some of the ethical issues that can arise?

(*b*) Summarize and analyze an example situation from the local newspaper, your hometown newspaper, or a news magazine.

(*c*) What guidelines for process or outcomes exist to help address the ethical issues of your example?

1-41 Economic development and redevelopment often require significant acreage that is assembled by acquiring smaller parcels. Sometimes this is done through simple purchase, but the property of an "unwilling seller" can be acquired through the process of eminent domain.

(*a*) What are some of the ethical issues that can arise?

(*b*) Summarize and analyze an example situation from the local newspaper, your hometown newspaper, or a news magazine.

(*c*) What guidelines for process or outcomes exist to help address the ethical issues in your city or state?

1-42 State governments use a variety of advisory and regulatory bodies. Example responsibilities include oversight of professional engineering licensing and the pricing and operation of regulated utilities. Engineers bring skills, knowledge, and perspectives that are useful in this context. Often the public role is a part-time one, and engineers that fulfill it will also have full-time jobs as employees or owners of engineering firms.

(*a*) What are some of the ethical issues that can arise from conflicts between the public- and private-sector roles?

(*b*) What guidelines for process or outcomes are there in your home state?

(*c*) Summarize and analyze an example situation from the local newspaper, your hometown newspaper, or a news magazine.

1-43 Many engineers work in state governments, and some are in high-profile roles as legislators, department commissioners, and so on. Many of these individuals move between working in the private and public sectors.

(*a*) What are some of the ethical issues that can arise as individuals shift between public and private sector roles?

(*b*) What guidelines for process or outcomes are there in your home state?

(*c*) Summarize and analyze an example situation from the local newspaper, your hometown newspaper, or a news magazine.

1-44 In the U.S., regulation of payment for overtime hours is done at the state and federal levels. Because most engineering work is accomplished through projects, it is common for engineers to be asked or required to work overtime as projects near deadlines. Sometimes the overtime is paid at time and a half, sometimes as straight time, and sometimes the engineer's salary is treated as a constant even when overtime occurs. In a particular firm, engineering interns, engineers, and partners may be treated the same or differently.

(*a*) What are some of the legal and ethical issues that can arise?

(*b*) What is the law or regulation in your state or home state?

(*c*) Summarize and analyze an example firm's policies or data on how common the problem is.

**1-45**  At the federal government level, the economic consequences of decisions can be very large. Firms hire lobbyists, legislators may focus on their constituents, and advocacy organizations promote their own agendas. In addition, sometimes some of the players are willing to be unethical.

(*a*) What are some of the ethical issues that can arise?

(*b*) Summarize and analyze an example situation from the local newspaper, your hometown newspaper, or a news magazine.

(*c*) What guidelines for process or outcomes are there for the agency or legislative body involved in your example?

**1-46**  At both state and federal levels, legislators can be involved in "pork barrel" funding of capital projects. These projects may even bypass the economic evaluation using engineering economy that normal projects are subject to.

(*a*) What are some of the ethical issues that can arise?

(*b*) Summarize and analyze an example situation from the local newspaper, your hometown newspaper, or a news magazine.

(*c*) What guidelines for process or outcomes are there for the agency or legislative body involved in your example?

**1-47**  At the international level, a common ethics issue important to engineering and project justification is that of environmental regulation. Often different nations have different environmental standards, and a project or product might be built in either location.

(*a*) What are some of the ethical issues that can arise?

(*b*) Summarize and analyze an example situation from the local newspaper, your hometown newspaper, or a news magazine.

(*c*) What guidelines for process or outcomes exist to help address the ethical issues of your example?

**1-48**  At the international level, a common ethics issue important to engineering and project justification is that of worker health and safety. Often different nations have different standards, and a project or product could be built in either location.

(*a*) What are some of the ethical issues that can arise?

(*b*) Summarize and analyze an example situation from the local newspaper, your hometown newspaper, or a news magazine.

(*c*) What guidelines for process or outcomes exist to help address the ethical issues of your example?

**1-49**  At the international level, engineering decisions are critical in matters of "sustainable development," a common ethics issue.

(*a*) What are some of the ethical issues that can arise?

(*b*) Summarize and analyze an example situation from the local newspaper, your hometown newspaper, or a news magazine.

(*c*) What guidelines for process or outcomes exist to help address the ethical issues of your example?

**1-50**  At the international level, questions arise about whether the U.S. ban on bribery is practical or appropriate. In some countries government workers are very poorly paid, and they can support their families only by accepting money to "grease" a process.

(*a*) What are some of the ethical issues that can arise?

(*b*) Summarize and analyze an example situation from the local newspaper, your hometown newspaper, or a news magazine.

(*c*) What guidelines for process or outcomes exist to help address the ethical issues of your example?

**1-51**  In the 1970s the Ford Motor Company sold its subcompact Pinto model with known design defects. In particular, the gas tank's design and location led to rupture, leaks, and explosion in low-speed, rear-impact collisions. Fifty-nine people burned to death in Pinto accidents. In a cost–benefit analysis weighing the cost of fixing the defects ($11 per vehicle) versus the firm's potential liability for lawsuits on behalf of accident victims, Ford had placed the value of a human life at $200,000. Ford eventually recalled 1.4 million Pintos to fix the gas tank problem for a cost of $30 million to $40 million. In addition the automaker ultimately paid out millions more in liability settlements and incurred substantial damage to its reputation.

(*a*) Critique Ford's actions from the perspective of the NSPE Code of Ethics.

(b) One well-known ethical theory, utilitarianism, suggests that an act is ethically justified if it results in the "greatest good for the greatest number" when all relevant stakeholders are considered. Did Ford's cost–benefit analysis validly apply this theory?

(c) What should engineers do when the product they are designing has a known safety defect with an inexpensive remedy?

*Contributed by Joseph R. Herkert, Arizona State University*

**1-52** The decision-making process used to launch the Challenger shuttle has been extensively analyzed. Briefly summarize the key institutional groups, how the decision was made, and the ethical principles that may have been compromised.

**1-53** One of the elements in the flooding of New Orleans during Hurricane Katrina was the failure of some of the levees that protected the city. Outline the role that ethical failures by engineers may have played in this situation. How could society structure decision making to minimize such failures?

**1-54** Toyota has a long history of engineering-driven, high-quality production processes. However, more recently a series of problems with acceleration and braking has raised questions about quality and ethics. Summarize what seems to have produced the problems, how they have been addressed, and how successful Toyota has been in restoring its reputation.

## Current Costs

**1-55** If you rent a car, you can (1) return it with a full gas tank, (2) return it without filling it and pay $3.45/gallon, or (3) accept a fixed price of $35 for gas. The local price is $3.15/gallon for gasoline, and you expect this car to get 28 miles per gallon. The car has a 20-gallon tank. What choice should you make if you expect to drive:

(a) 150 miles?

(b) 350 miles?

(c) 500 miles?

(d) How do your answers change if stopping at the filling station takes 15 minutes and your time is worth $12/hr?

**1-56** Your car gets 24 miles per gallon (mpg) at 60 miles per hour (mph) and 20 mpg at 70 mph. At what speed should you make a 500-mile trip:

(a) If gas costs $3 per gallon and your time is worth $18/hr?

(b) If gas costs $4 per gallon and your time is worth $12/hr?

(c) If gas costs $5 per gallon and your time is worth $9/hr?

(d) Build a spreadsheet (see Appendix A) to calculate the total trip cost for gas costs of $2, $3, $4, and $5 and values of time of $6, $9, $12, $15, and $18 per hour. Do two tables: one at 60 mph and one at 70 mph.

**1-57** Willie Lohmann travels from city to city in the conduct of his business. Every other year he buys a used car for about $15,000. The dealer allows about $8000 as a trade-in allowance, with the result that the salesman spends $7000 every other year for a car. Willie keeps accurate records, which show that all other expenses on his car amount to 32.3¢ per mile for each mile he drives. Willie's employer has two plans by which salesmen are reimbursed for their car expenses:

A. Willie will receive all his operating expenses, and in addition will receive $3500 each year for the decline in value of the automobile.

B. Willie will receive 52¢ per mile but no operating expenses and no depreciation allowance.

If Willie travels 18,000 miles per year, which method of computation gives him the larger reimbursement? At what annual mileage do the two methods give the same reimbursement?

**1-58** Maria, a college student, is getting ready for three final examinations at the end of the school year. Between now and the start of exams, she has 15 hours of study time available. She would like to get as high a grade average as possible in her math, physics, and engineering economy classes. She feels she must study at least 2 hours for each course and, if necessary, will settle for the low grade that the limited study would yield. If Maria estimates her grade in each subject as follows, how much time should she devote to each class?

| Study Hours | Mathematics Grade | Physics Grade | Engineering Economy Grade |
|---|---|---|---|
| 2 | 25 | 35 | 50 |
| 3 | 35 | 41 | 61 |
| 4 | 44 | 49 | 71 |
| 5 | 52 | 59 | 79 |
| 6 | 59 | 68 | 86 |
| 7 | 65 | 77 | 92 |
| 8 | 70 | 85 | 96 |

**1-59** Two manufacturing companies, located in cities 90 miles apart, have discovered that they both send their trucks four times a week to the other city full of cargo and return empty. Each company pays its driver $185 a day (the round trip takes all day) and have truck operating costs (excluding the driver) of 60¢ a mile. How much could each company save each week if they shared the task, with each sending its truck twice a week and hauling the other company's cargo on the return trip?

**1-60** A city needs to increase its rubbish disposal facilities. There is a choice of two rubbish disposal areas, as follows.

*Area A:* A gravel pit with a capacity of 16 million cubic meters. Owing to the possibility of high groundwater, however, the Regional Water Pollution Control Board has restricted the lower 2 million cubic meters of fill to inert material only (earth, concrete, asphalt, paving, brick, etc.). The inert material, principally clean earth, must be purchased and hauled to this area for the bottom fill.

*Area B:* Capacity is 14 million cubic meters. The entire capacity may be used for general rubbish disposal. This area will require an average increase in a round-trip haul of 5 miles for 60% of the city, a decreased haul of 2 miles for 20% of the city. For the remaining 20% of the city, the haul is the same distance as for Area A.

Assume the following conditions:

- Cost of inert material placed in Area A will be $9.40/m$^3$.
- Average speed of trucks from last pickup to disposal site is 15 miles per hour.
- The rubbish truck and a two-man crew will cost $140 per hour.

- Truck capacity of $4^1/2$ tons per load or 20 m$^3$.
- Sufficient cover material is available at all areas; however, inert material for the bottom fill in Area A must be hauled in.

Which of the sites do you recommend?
(*Answer:* Area B)

**1-61** An oil company is considering adding an additional grade of fuel at its service stations. To do this, an additional 3000-gallon tank must be buried at each station. Discussions with tank fabricators indicate that the least expensive tank would be cylindrical with minimum surface area. What size tank should be ordered?
(*Answer:* 8-ft diameter by 8-ft length)

**1-62** The vegetable buyer for a group of grocery stores has decided to sell packages of sprouted grain in the vegetable section of the stores. The product is perishable, and any remaining unsold after one week in the store is discarded. The supplier will deliver the packages to the stores, arrange them in the display space, and remove and dispose of any old packages. The price the supplier will charge the stores depends on the size of the total weekly order for all the stores.

| Weekly Order | Price per Package |
|---|---|
| Less than 1000 packages | 70¢ |
| 1000–1499 | 56 |
| 1500–1999 | 50 |
| 2000 or more | 40 |

The vegetable buyer estimates the quantity that can be sold per week, at various selling prices, as follows:

| Selling Price | Packages Sold per Week |
|---|---|
| $1.20 | 300 |
| .90 | 600 |
| .80 | 1200 |
| .66 | 1700 |
| .52 | 2300 |

The sprouted grain will be sold at the same price in all the grocery stores.

(a) How many packages should be purchased per week, and at which of the five prices listed above should they be sold?

(b) Build a spreadsheet (see Appendix A) to calculate the profit for every combination of selling price and weekly order size.

**1-63** Cathy Gwynn, a recently graduated engineer, decided to invest some of her money in a "Quick Shop" grocery store. The store emphasizes quick service, a limited assortment of grocery items, and rather high prices. Cathy wants to study the business to see if the store hours (currently 0600 to 0100) can be changed to make the store more profitable. Cathy assembled the following information.

| Time Period | Daily Sales in the Time Period |
|---|---|
| 0600–0700 | $ 40 |
| 0700–0800 | 80 |
| 0800–0900 | 120 |
| 0900–1200 | 400 |
| 1200–1500 | 500 |
| 1500–1800 | 600 |
| 1800–2100 | 800 |
| 2100–2200 | 200 |
| 2200–2300 | 60 |
| 2300–2400 | 120 |
| 2400–0100 | 40 |

The cost of the groceries sold averages 70% of sales. The incremental cost to keep the store open, including the clerk's wage and other incremental operating costs, is $20 per hour. To maximize profit, when should the store be opened, and when should it be closed?

**1-64** Jim Jones, a motel owner, noticed that just down the street the "Motel 36" advertises a $36-per-night room rental rate on its sign. As a result, this competitor has rented all 80 rooms every day by late afternoon. Jim, on the other hand, does not advertise his rate, which is

$54 per night, and he averages only a 68% occupancy of his 50 rooms.

There are a lot of other motels nearby, but only Motel 36 advertises its rate on its sign. (Rates at the other motels vary from $48 to $80 per night.) Jim estimates that his actual incremental cost per night for each room rented, rather than remaining vacant, is $12. This $12 pays for all the cleaning, laundering, maintenance, utilities, and so on. Jim believes his eight alternatives are:

| Alternative | | Resulting Occupancy Rate |
|---|---|---|
| | **Advertise and Charge** | |
| 1 | $35 per night | 100% |
| 2 | 42 per night | 94 |
| 3 | 48 per night | 80 |
| 4 | 54 per night | 66 |
| | **Do Not Advertise and Charge** | |
| 5 | $48 per night | 70% |
| 6 | 54 per night | 68 |
| 7 | 62 per night | 66 |
| 8 | 68 per night | 56 |

What should Jim do? Show how you reached your conclusion.

**1-65** A firm is planning to manufacture a new product. The sales department estimates that the quantity that can be sold depends on the selling price. As the selling price is increased, the quantity that can be sold decreases. Numerically they estimate:

$$P = \$35.00 - 0.02Q$$

where  $P =$ selling price per unit

$Q =$ quantity sold per year

On the other hand, management estimates that the average cost of manufacturing and selling the product will decrease as the quantity sold increases. They estimate

$$C = \$4.00Q + \$8000$$

where $C$ = cost to produce and sell $Q$ per year

The firm's management wishes to produce and sell the product at the rate that will maximize profit, that is, income minus cost will be a maximum. What quantity should the decision makers plan to produce and sell each year?

(*Answer:* 775 units)

**1-66** A manufacturing firm has received a contract to assemble 1000 units of test equipment in the next year. The firm must decide how to organize its assembly operation. Skilled workers, at $33 per hour each, could be assigned to individually assemble the test equipment. Each worker would do all the assembly steps, and it would take 2.6 hours to complete one unit. An alternate approach would be to set up teams of four less skilled workers (at $19 per hour each) and organize the assembly tasks so that each worker does part of the assembly. The four-man team would be able to assemble a unit in one hour. Which approach would result in more economical assembly?

**1-67** A grower estimates that if he picks his apple crop now, he will obtain 1000 boxes of apples, which he can sell at $3 per box. However, he thinks his crop will increase by 120 boxes of apples for each week he delays picking, but that the price will drop at a rate of 15¢ per box per week; in addition, he estimates that approximately 20 boxes per week will spoil for each week he delays picking.

(*a*) When should he pick his crop to obtain the largest total cash return? How much will he receive for his crop at that time?

(*b*) Build a spreadsheet (see Appendix A) to calculate the profit for 0, 1, 2, ..., 6 weeks.

**1-68** On her first engineering job, Joy Hayes was given the responsibility of determining the production rate for

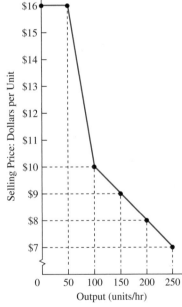

a new product. She has assembled the data presented in the following graphs.

(a) Select an appropriate economic criterion and estimate the production rate based upon it.

(b) Joy's boss told Joy: "I want you to maximize output with minimum input." Joy wonders if it is possible to meet her boss's criterion. She asks for your advice. What would you tell her?

# CASES

**The following cases from *Cases in Engineering Economy* 2 $^{nd}$ are suggested as matched with this chapter.**

CASE 1   **New Office Equipment**

Student develops requirements for new copier machines with little given data. Organizational thinking required.

CASE 2   **Budgeting Issues**

Strategies for a group's operating budget request. One focus is ethics.

# ENGINEERING COSTS AND COST ESTIMATING

## North Slope Natural Gas Pipeline

**C**rude oil from Alaska's North Slope has long been important to the U.S. Now a major project to bring Alaska North Slope natural gas to the lower 48 is under active consideration. The numbers associated with this private sector endeavor are: 35 trillion cubic feet (TCF) of producible gas reserves, $20 billion of estimated infrastructure costs needed to access and deliver the gas, and a 9-year engineer, design, and build process to make it all happen.

A significant driver for this massive project is consumer demand and supply capabilities. By the end of the decade the annual U.S. new sources of natural gas demand will increase to 18 TCF, and experts estimate that the consumption rate will continue to increase at a 2 to 3% annual rate. All this adds up to a case of demand soon exceeding domestic supply capabilities. That is where the North Slope project comes into play. At a design capacity of 4.5 billion cubic feet per day, the project would not completely fill the projected shortfall, but it would reduce reliance on foreign suppliers.

While the pipeline design configuration is not complete, the leading candidate is the Prudhoe Bay to Canada to Chicago route—approximately 2140 miles. An alternate route is the Prudhoe Bay to Valdez pipeline (approximately 800 miles) to a liquefaction plant for shipping on liquefied natural gas (LNG) oceangoing tankers. Under either route, the proposed pipeline design daily capacity is 4.5 billion cubic feet of gas. Delivering 4.5 billion cubic feet daily puts the estimated revenue increase for the state of Alaska between $2 billion and $3 billion annually.

As is often the case with a private-sector, large-scale project, this one faces many risks. One major risk is the development of accurate cost estimates for engineering and building the pipeline. Another risk is providing the project owners an adequate return on investment when energy prices are volatile and demands are uncertain.

The natural gas resource supply is available, and the projected consumer needs are real. Yet, this project must compete for private investors' capital with other global investment projects. Final decisions on project viability will be based on cost estimates and calculations of rate of return on investment in comparison to other investment opportunities. That is why accurate estimates are essential in the decision process, and ultimately will determine whether North Slope gas finds its way into U.S. markets. ■ ■ ■

*Contributed by Morgan E. Henrie, University of Alaska Anchorage*

## QUESTIONS TO CONSIDER

1. When large-scale projects compete for limited corporate capital funds, what type of cost estimating should be utilized? Can you develop a philosophical statement, from the firm's perspective, regarding the detail needed and confidence justifiable in a cost estimate?

2. When a project is estimated to take 5 to 10 (or more) years from concept to completion, should cost estimates be adjusted for the greater influences and impacts of inflation, adverse government regulatory changes, and changing local economic environment? Why or why not?

3. Should large-scale infrastructure projects be required to meet the same internal rate of return requirements as smaller, shorter-completion-time projects? Why or why not?

4. Are there any ethical issues related to economics, the environment, safety, and so on that project sponsors, consumers, and/or the U.S. government should consider?

## After Completing This Chapter...

### The student should be able to:

- Define various cost concepts.
- Provide specific examples of how and why these engineering cost concepts are important.
- Define engineering cost estimating.
- Explain the three types of engineering estimate, as well as common difficulties encountered in making engineering cost estimates.
- Use several common mathematical estimating models in cost estimating.
- Discuss the impact of the *learning curve* on cost estimates.
- State the relationship between cost estimating and estimating project benefits.
- Draw *cash flow diagrams* to show project costs and benefits.

This chapter defines fundamental cost concepts. These include fixed and variable costs, marginal and average costs, sunk and opportunity costs, recurring and nonrecurring costs, incremental cash costs, book costs, and life-cycle costs. We then describe the various types of estimates and difficulties sometimes encountered. The models that are described include unit factor, segmenting, cost indexes, power sizing, triangulation, and learning curves. The chapter discusses estimating benefits, developing cash flow diagrams, and drawing these diagrams with spreadsheets.

Understanding engineering costs is fundamental to the engineering economic analysis process, and therefore this chapter addresses an important question: Where do the numbers come from?

## ENGINEERING COSTS

Evaluating a set of feasible alternatives requires that many costs be analyzed. Examples include costs for initial investment, new construction, facility modification, general labor, parts and materials, inspection and quality, contractor and subcontractor labor, training, computer hardware and software, material handling, fixtures and tooling, data management, and technical support, as well as general support costs (overhead). In this section we describe several concepts for classifying and understanding these costs.

### Fixed, Variable, Marginal, and Average Costs

**Fixed** costs are constant or unchanging regardless of the level of output or activity. In contrast, **variable** costs depend on the level of output or activity. A **marginal** cost is the variable cost for one more unit, while the **average** cost is the total cost divided by the number of units.

In a production environment, for example, fixed costs, such as those for factory floor space and equipment, remain the same even though production quantity, number of employees, and level of work-in-process may vary. Labor costs are classified as a *variable* cost because they depend on the number of employees and the number of hours they work. Thus *fixed* costs are level or constant regardless of output or activity, and *variable* costs are changing and related to the level of output or activity.

As another example, many universities charge full-time students a fixed cost for 12 to 18 hours and a cost per credit hour for each credit hour over 18. Thus for full-time students who are taking an overload (>18 hours), there is a variable cost that depends on the level of activity, but for most full-time students tuition is a fixed cost.

This example can also be used to distinguish between *marginal* and *average* costs. A marginal cost is the cost of one more unit. This will depend on how many credit hours the student is taking. If currently enrolled for 12 to 17 hours, adding one more is free. The marginal cost of an additional credit hour is $0. However, for a student taking 18 or more hours, the marginal cost equals the variable cost of one more hour.

To illustrate average costs, the fixed and variable costs need to be specified. Suppose the cost of 12 to 18 hours is $1800 per term and overload credits are $120/hour. If a student takes 12 hours, the *average* cost is $1800/12 = $150 per credit hour. If the student were to take 18 hours, the *average* cost would decrease to $1800/18 = $100 per credit hour. If the student takes 21 hours, the *average* cost is $102.86 per credit hour [$1800 + (3 × $120)]/21. Average cost is thus calculated by dividing the total cost for all units by the total number of units. Decision makers use **average** cost to attain an overall cost picture of the investment on a per unit basis.

**Marginal** cost is used to decide whether an additional unit should be made, purchased, or enrolled in. For our example full-time student, the marginal cost of another credit is $0 or $120 depending on how many credits the student has already signed up for.

## EXAMPLE 2-1

The Federation of Student Societies of Engineering (FeSSE) wants to offer a one-day training course to help students in job hunting and to raise funds. The organizing committee is sure that they can find alumni, local business people, and faculty to provide the training at no charge. Thus the main costs will be for space, meals, handouts, and advertising.

The organizers have classified the costs for room rental, room setup, and advertising as fixed costs. They also have included the meals for the speakers as a fixed cost. Their total of $225 is pegged to a room that will hold 40 people. So if demand is higher, the fixed costs will also increase.

The variable costs for food and bound handouts will be $20 per student. The organizing committee believes that $35 is about the right price to match value to students with their budgets. Since FeSSE has not offered training courses before, they are unsure how many students will reserve seats.

Develop equations for FeSSE's total cost and total revenue, and determine the number of registrations that would be needed for revenue to equal cost.

### SOLUTION

Let $x$ equal the number of students who sign up. Then,

$$\text{Total cost} = \$225 + \$20x$$

$$\text{Total revenue} = \$35x$$

To find the number of student registrants for revenue to equal cost, we set the equations equal to each other and solve.

$$\text{Total cost} = \text{Total revenue}$$

$$\$225 + \$20x = \$35x$$

$$\$225 = (\$35 - \$20)x$$

$$x = 225/15 = 15 \text{ students}$$

While this example has been defined for a student engineering society, we could just as easily have described this as a training course to be sponsored by a local chapter of a professional technical society. The fixed cost and the revenue would increase by a factor of about 10, while the variable cost would probably double or triple.

If a firm were considering an in-house short course, the cost of the in-house course would be compared with the cost per employee (a variable cost) for enrolling employees in external training.

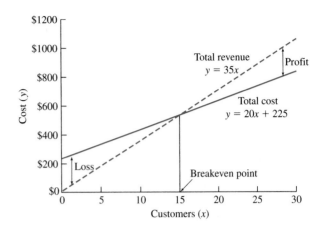

**FIGURE 2–1** Profit–loss breakeven chart for Example 2-1.

From Example 2-1 we see how it is possible to calculate total fixed and total variable costs. Furthermore, these values can be combined into a single **total** cost equation as follows:

$$\text{Total cost} = \text{Total fixed cost} + \text{Total variable cost} \qquad (2\text{-}1)$$

Example 2-1 developed *total cost* and *total revenue* equations to describe a training course proposal. These equations can be used to create what is called a *profit–loss breakeven chart* (see Figure 2-1). A plot of revenues against costs for various levels of output (activity) allows one to illustrate a *breakeven point* (in terms of costs and revenue) and regions of *profit* and *loss*. These terms can be defined as follows.

> *Breakeven point:*  The level of activity at which total costs for the product, good, or service are *equal to* the revenue (or savings) generated. This is the level at which one "just breaks even."

> *Profit region:*  Values of the variable $x$ greater than the breakeven point, where total revenue is greater than total costs.

> *Loss region:*  Values of the variable $x$ less than the breakeven point, where total revenue is less than total costs.

Notice in Figure 2-1 that the *breakeven* point for the number of persons in the training course is 15 people. For more than 15 people, FeSSE will make a profit. If fewer than 15 sign up, there will be a net loss. At the breakeven level, the total cost to provide the course equals the revenue received from the 15 students.

The fixed costs of our simple model are in reality *fixed* over a range of values for $x$. In Example 2-1, that range was 1 to 40 students. If *zero* students signed up, then the course could be canceled and many of the fixed costs would not be incurred. Some costs such as advertising might already have been spent, and there might be cancellation fees. If more than 40 students signed up, then greater costs for larger rooms or multiple sessions would be incurred. The model is valid only within the range named.

When modeling a specific situation, we often use *linear* variable costs and revenues. However, sometimes the relationship may be nonlinear. For example, employees are often paid at 150% of their hourly rate for overtime hours, so that production levels requiring overtime have higher variable costs. Total cost in Figure 2-2 is a fixed cost of $3000 plus

**FIGURE 2–2** Nonlinear variable costs.

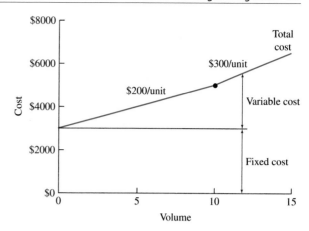

a variable cost of $200 per unit for straight-time production of up to 10 units and $300 per unit for overtime production of up to 5 more units.

Figure 2-2 can also be used to illustrate marginal and average costs. At a volume of 5 units the marginal cost is $200 per unit, while at a volume of 12 units the marginal cost is $300 per unit. At 5 units the average cost is $800 per unit, or (3000 + 200 × 5)/5. At 12 units the average cost is $467 per unit, or (3000 + 200 × 10 + 300 × 2)/12.

## Sunk Costs

A **sunk cost** is money already spent as a result of a *past* decision. If only 5 students signed up for the training course in Example 2-1, the advertising costs would be a *sunk cost*.

Sunk costs must be ignored in engineering economic analysis because current decisions cannot change the past. For example, dollars spent last year to purchase new production machinery is money that is *sunk:* the money has already been spent—there is nothing that can be done now to change that action. As engineering economists we deal with present and future opportunities.

Many times it is difficult not to be influenced by sunk costs. Consider 100 shares of stock in XYZ, Inc., purchased for $15 per share last year. The share price has steadily declined over the past 12 months to a price of $10 per share today. Current decisions must focus on the $10 per share that could be attained today (as well as future price potential), not the $15 per share that was paid last year. The $15 per share paid last year is a *sunk cost* and has no influence on present opportunities.

As another example, when Regina was a sophomore, she purchased a newest-generation laptop from the college bookstore for $2000. By the time she graduated, the most anyone would pay her for the computer was $400 because the newest models were faster and cheaper and had more capabilities. For Regina, the original purchase price was a *sunk cost* that has no influence on her present opportunity to sell the laptop at its current market value of $400.

When we get to Chapters 11 and 12 on depreciation and income taxes, we will find an exception to the rule of *ignore sunk costs*. When an asset is sold or disposed of, then the sunk cost of what was paid for it is important in figuring out how much is owed in taxes. This exception applies only to the after-tax analysis of capital assets.

## Opportunity Costs

An **opportunity cost** is associated with using a resource in one activity instead of another. Every time we use a business resource (equipment, dollars, manpower, etc.) in one activity, we give up the opportunity to use the same resources at that time in some other activity.

Every day businesses use resources to accomplish various tasks—forklifts are used to transport materials, engineers are used to design products and processes, assembly lines are used to make a product, and parking lots are used to provide parking for employees' vehicles. Each of these resources costs the company money to maintain for those intended purposes. However, the amount is not just made up of the dollar cost; it also includes the opportunity cost. Each resource that a firm owns can feasibly be used in several alternative ways. For instance, the assembly line could produce a different product, and the parking lot could be rented out, used as a building site, or converted into a small airstrip. Each of these alternative uses would provide some benefit to the company.

A firm that chooses to use the resource in one way is giving up the benefits that would be derived from using it in those other ways. The benefit that would be derived by using the resource in this "other activity" is the **opportunity cost** of using it in the chosen activity. Opportunity cost may also be considered a **forgone opportunity cost** because we are forgoing the benefit that could have been realized. A formal definition of opportunity cost might be:

> An *opportunity cost* is the benefit that is *forgone* by engaging a business resource in a chosen activity instead of engaging that same resource in the forgone activity.

As an example, suppose that friends invite a college student to travel through Europe over the summer break. In considering the offer, the student might calculate all the *out-of-pocket* cash costs that would be incurred. Cost estimates might be made for items such as air travel, lodging, meals, entertainment, and train passes. Suppose this amounts to $3000 for a 10-week period. After checking his bank account, the student reports that indeed he can afford the $3000 trip. However, the *true* cost to the student includes not only his *out-of-pocket* cash costs but also his *opportunity cost*. By taking the trip, the student is giving up the *opportunity* to earn $5000 as a summer intern at a local business. The student's total cost will comprise the $3000 cash cost as well as the $5000 opportunity cost (wages forgone)—the total cost to our traveler is thus $8000.

### EXAMPLE 2–2

A distributor of electric pumps must decide what to do with a "lot" of old electric pumps purchased 3 years ago. Soon after the distributor purchased the lot, technology advances made the old pumps less desirable to customers. The pumps are becoming obsolescent as they sit in inventory. The pricing manager has the following information.

| | |
|---|---|
| Distributor's purchase price 3 years ago | $ 7,000 |
| Distributor's storage costs to date | 1,000 |
| Distributor's list price 3 years ago | 9,500 |
| Current list price of the same number of new pumps | 12,000 |
| Amount offered for the old pumps from a buyer 2 years ago | 5,000 |
| Current price the lot of old pumps would bring | 3,000 |

Looking at the data, the pricing manager has concluded that the price should be set at $8000. This is the money that the firm has "tied up" in the lot of old pumps ($7000 purchase and $1000 storage), and it was reasoned that the company should at least recover this cost. Furthermore, the pricing manager has argued that an $8000 price would be $1500 less than the list price from 3 years ago, and it would be $4000 less than what a lot of new pumps would cost ($12,000 − $8000). What would be your advice on price?

### SOLUTION

Let's look more closely at each of the data items.

*Distributor's purchase price 3 years ago:* This is a sunk cost that should not be considered in setting the price today.

*Distributor's storage costs to date:* The storage costs for keeping the pumps in inventory are sunk costs; that is, they have been paid. Hence they should not influence the pricing decision.

*Distributor's list price 3 years ago:* If there have been no willing buyers in the past 3 years at this price, it is unlikely that a buyer will emerge in the future. This past list price should have no influence on the current pricing decision.

*Current list price of newer pumps:* Newer pumps now include technology and features that have made the older pumps less valuable. Directly comparing the older pumps to those with new technology is misleading. However, the price of the new pumps and the value of the new features help determine the market value of the old pumps.

*Amount offered by a buyer 2 years ago:* This is a forgone opportunity. At the time of the offer, the company chose to keep the lot and thus the $5000 offered became an opportunity cost for keeping the pumps. This amount should not influence the current pricing decision.

*Current price the lot could bring:* The price a willing buyer in the marketplace offers is called the asset's *market value.* The lot of old pumps in question is believed to have a current market value of $3000.

From this analysis, it is easy to see the flaw in the pricing manager's reasoning. In an engineering economist analysis we deal only with *today's* and prospective *future* opportunities. It is impossible to go back in time and change decisions that have been made. Thus, the pricing manager should recommend to the distributor that the price be set at the current value that a buyer assigns to the item: $3000.

### Recurring and Nonrecurring Costs

**Recurring costs** refer to any expense that is known and anticipated, and that occurs at regular intervals. **Nonrecurring costs** are one-of-a-kind expenses that occur at irregular intervals and thus are sometimes difficult to plan for or anticipate from a budgeting perspective.

Examples of recurring costs include those for resurfacing a highway and reshingling a roof. Annual expenses for maintenance and operation are also recurring expenses. Examples of nonrecurring costs include the cost of installing a new machine (including any facility modifications required), the cost of augmenting equipment based on older technology to restore its usefulness, emergency maintenance expenses, and the disposal or close-down costs associated with ending operations.

In engineering economic analyses *recurring costs* are modeled as cash flows that occur at regular intervals (such as every year or every 5 years). Their magnitude can be estimated, and they can be included in the overall analysis. *Nonrecurring costs* can be handled easily in our analysis if we are able to anticipate their timing and size. However, this is not always so easy to do.

### Incremental Costs

One of the fundamental principles in engineering economic analysis is that in choosing between competing alternatives, the focus is on the *differences* between those alternatives. This is the concept of **incremental costs.** For instance, one may be interested in comparing two options to lease a vehicle for personal use. The two lease options may have several specifics for which costs are the same. However, there may be incremental costs associated with one option but not with the other. In comparing the two leases, the focus should be on the differences between the alternatives, not on the costs that are the same.

### EXAMPLE 2–3

Philip is choosing between model *A* (a budget model) and model *B* (with more features and a higher purchase price). What *incremental costs* would Philip incur if he chose model *B* instead of the less expensive model *A*?

| Cost Items | Model A | Model B |
|---|---|---|
| Purchase price | $10,000 | $17,500 |
| Installation costs | 3,500 | 5,000 |
| Annual maintenance costs | 2,500 | 750 |
| Annual utility expenses | 1,200 | 2,000 |
| Disposal costs after useful life | 700 | 500 |

### SOLUTION

We are interested in the incremental or *extra* costs that are associated with choosing model *B* instead of model *A*. To obtain these we subtract model *A* costs from model *B* costs for each category (cost item) with the following results.

| Cost Items | (Model B Cost – A Cost) | Incremental Cost of B |
|---|---|---|
| Purchase price | 17,500 – 10,000 | $7500 |
| Installation costs | 5,000 – 3,500 | 1500 |
| Annual maintenance costs | 750 – 2,500 | –1750/yr |
| Annual utility expenses | 2,000 – 1,200 | 800/yr |
| Disposal costs after useful life | 500 – 700 | –200 |

Notice that for the cost categories given, the incremental costs of model *B* are both positive and negative. Positive incremental costs mean that model *B* costs more than model *A*, and negative incremental costs mean that there would be a *savings* (reduction in cost) if model *B* were chosen instead.

Because model *B* has more features, a decision would also have to consider the incremental benefits offered by those features.

## Cash Costs Versus Book Costs

A *cash cost* requires the cash transaction of dollars "out of one person's pocket" into "the pocket of someone else." When you buy dinner for your friends or make your monthly car payment you are incurring a **cash cost** or **cash flow.** Cash costs and cash flows are the basis for engineering economic analysis.

*Book costs* do not require the transaction of dollars "from one pocket to another." Rather, **book costs** are cost effects from past decisions that are recorded "in the books" (accounting books) of a firm. In one common book cost, asset depreciation (which we discuss in Chapter 11), the expense paid for a particular business asset is "written off" on a company's accounting system over a number of periods. Book costs do not ordinarily represent cash flows and thus are not included in engineering economic analysis. One exception to this is the impact of asset depreciation on tax payments—which are cash flows and are included in after-tax analyses.

## Life-Cycle Costs

The products, goods, and services designed by engineers all progress through a **life cycle** very much like the human life cycle. People are conceived, go through a growth phase, reach their peak during maturity, and then gradually decline and expire. The same general pattern holds for products, goods, and services. As with humans, the duration of the different phases, the height of the peak at maturity, and the time of the onset of decline and termination all vary depending on the individual product, good, or service. Figure 2-3 illustrates the typical phases that a product, good, or service progresses through over its life cycle.

**Life-cycle costing** refers to the concept of designing products, goods, and services with a full and explicit recognition of the associated costs over the various phases of their life cycles. Two key concepts in life-cycle costing are that the later design changes are made, the higher the costs, and that decisions made early in the life cycle tend to "lock in" costs that will be incurred later. Figure 2-4 illustrates how costs are committed early in the product life cycle—nearly 70–90% of all costs are set during the design phases. At the same time, as the figure shows, only 10–30% of cumulative life-cycle costs have been spent.

Figure 2-5 reinforces these concepts by illustrating that later product changes are more costly and that earlier changes are easier (and less costly) to make. When planners try to save money at an early design stage, the result is often a poor design that results in change orders during construction and prototype development. These changes, in turn, are more costly than working out a better design would have been.

From Figures 2-4 and 2-5 we see that the time to consider all life-cycle effects and make design changes is during the needs and conceptual/preliminary design phases—before a lot of dollars are committed. Some of the life-cycle effects that engineers should consider

Beginning ──────────────── Time ──────────────→ End

| Needs Assessment and Justification Phase | Conceptual or Preliminary Design Phase | Detailed Design Phase | Production or Construction Phase | Operational Use Phase | Decline and Retirement Phase |
|---|---|---|---|---|---|
| Requirements | Impact Analysis | Allocation of Resources | Product, Goods, & Services Built | Operational Use | Declining Use |
| Overall Feasibility | Proof of Concept | Detailed Specifications | All Supporting Facilities Built | Use by Ultimate Customer | Phase Out |
| Conceptual Design Planning | Prototype/ Breadboard | Component and Supplier Selection | Operational Use Planning | Maintenance and Support | Retirement |
|  | Development and Testing | Production or Construction Phase |  | Processes, Materials, and Methods Used | Responsible Disposal |
|  | Detailed Design Planning |  |  | Decline and Retirement Planning |  |

**FIGURE 2–3** Typical life cycle for products, goods, and services.

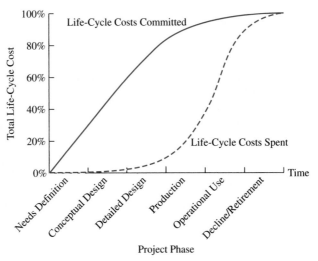

**FIGURE 2–4** Cumulative life-cycle costs committed and dollars spent.

at design time include product costs for liability, production, material, testing and quality assurance, and maintenance and warranty. Other life-cycle effects include product features based on customer input and product disposal effects on the environment. The key point is that engineers who design products and the systems that produce them should consider all life-cycle costs.

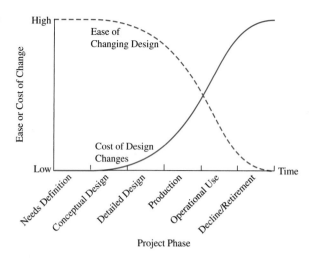

**FIGURE 2-5** Ease of life-cycle design change and costs of change.

## COST ESTIMATING

Engineering economic analysis focuses on the future consequences of current decisions. Because these consequences are in the future, usually they must be estimated and cannot be known with certainty. Estimates that may be needed in engineering economic analysis include purchase costs, annual revenue, yearly maintenance, interest rates for investments, annual labor and insurance costs, equipment salvage values, and tax rates.

Estimating is the foundation of economic analysis. As in any analysis procedure, the outcome is only as good as the numbers used to reach the decision. For example, a person who wants to estimate her federal income taxes for a given year could do a very detailed analysis, including social security deductions, retirement savings deductions, itemized personal deductions, exemption calculations, and estimates of likely changes to the tax code. However, this very technical and detailed analysis will be grossly inaccurate if poor data are used to predict the next year's income. Thus, to ensure that an analysis is a reasonable evaluation of future events, it is very important to make careful estimates.

### Types of Estimate

The American poet and novelist Gertrude Stein wrote in *The Making of Americans* in 1925 that "a rose is a rose is a rose is a rose." However, what holds for roses does not necessarily hold for estimates because "an estimate is not an estimate." Rather, we can define three general types of estimate whose purposes, accuracies, and underlying methods are quite different.

*Rough estimates:*   Order-of-magnitude estimates used for high-level planning, for determining project feasibility, and in a project's initial planning and evaluation phases. Rough estimates tend to involve back-of-the-envelope numbers with little detail or accuracy. The intent is to quantify and consider the order of magnitude of the numbers involved. These estimates require minimum resources to develop, and their accuracy is generally −30 to +60%.

*Notice the nonsymmetry in the estimating error. This is because decision makers tend to underestimate the magnitude of costs (negative economic effects). Also, as*

*Murphy's law predicts, there seem to be more ways for results to be worse than expected than there are for the results to be better than expected.*

*Semidetailed estimates:*   Used for budgeting purposes at a project's conceptual or preliminary design stages. These estimates are more detailed, and they require additional time and resources. Greater sophistication is used in developing semidetailed estimates than the rough-order type, and their accuracy is generally −15 to +20%.

*Detailed estimates:*   Used during a project's detailed design and contract bidding phases. These estimates are made from detailed quantitative models, blueprints, product specification sheets, and vendor quotes. Detailed estimates involve the most time and resources to develop and thus are much more accurate than rough or semidetailed estimates. The accuracy of these estimates is generally −3 to +5%.

*The upper limits of +60% for rough order, +20% for semidetailed, and +5% for detailed estimates are based on construction data for plants and infrastructure. Final costs for software, research and development, and new military weapons often have much higher corresponding percentages.*

In considering the three types of estimate it is important to recognize that each has its unique purpose, place, and function in a project's life. Rough estimates are used for general feasibility activities; semidetailed estimates support budgeting and preliminary design decisions; and detailed estimates are used for establishing design details and contracts. As one moves from rough to detailed design, one moves from less to much more accurate estimates.

However, this increased accuracy requires added time and resources. Figure 2-6 illustrates the trade-off between accuracy and cost. In engineering economic analysis, the resources spent must be justified by the need for detail in the estimate. As an illustration, during the project feasibility stages we would not want to use our people, time, and money to develop detailed estimates for infeasible alternatives that will be quickly eliminated from further consideration. However, regardless of how accurate an estimate is assumed to be, it is only an estimate of what the future will be. There will be some error even if ample resources and sophisticated methods are used.

FIGURE 2–6 Accuracy versus cost trade-off in estimating.

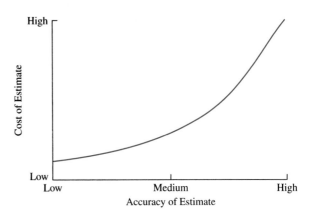

## Difficulties in Estimation

Estimating is difficult because the future is unknown. With few exceptions (such as with legal contracts) it is difficult to foresee future economic consequences exactly. In this section we discuss several aspects of estimating that make it a difficult task.

### One-of-a-Kind Estimates

Estimated parameters can be for one-of-a-kind or first-run projects. The first time something is done, it is difficult to estimate costs required to design, produce, and maintain a product over its life cycle. Consider the projected cost estimates that were developed for the first NASA missions. The U.S. space program initially had no experience with human flight in outer space; thus the development of the cost estimates for design, production, launch, and recovery of the astronauts, flight hardware, and payloads was a "first-time experience." The same is true for any endeavor lacking local or global historical cost data. New products or processes that are unique and fundamentally different make estimating costs difficult.

The good news is that there are very few one-of-a-kind estimates to be made in engineering design and analysis. Nearly all new technologies, products, and processes have "close cousins" that have led to their development. The concept of **estimation by analogy** allows one to use knowledge about well-understood activities to anticipate costs for new activities. In the 1950s, at the start of the military missile program, aircraft companies drew on their in-depth knowledge of designing and producing aircraft when they bid on missile contracts. As another example, consider the problem of estimating the production labor requirements for a brand new product, $X$. A company may use its labor knowledge about Product $Y$, a similar type of product, to build up the estimate for $X$. Thus, although "first-run" estimates are difficult to make, estimation by analogy can be an effective tool.

### Time and Effort Available

Our ability to develop engineering estimates is constrained by time and person-power availability. In an ideal world, it would *cost nothing* to use *unlimited resources* over an *extended period of time*. However, reality requires the use of limited resources in fixed intervals of time. Thus for a rough estimate only limited effort is used.

Constraints on time and person-power can make the overall estimating task more difficult. If the estimate does not require as much detail (such as when a rough estimate is the goal), then time and personnel constraints may not be a factor. When detail is necessary and critical (such as in legal contracts), however, requirements must be anticipated and resource use planned.

### Estimator Expertise

Consider two common expressions: *The past is our greatest teacher*, and *Knowledge is power*. These simple axioms hold true for much of what we encounter during life, and they are true in engineering estimating as well. The more experienced and knowledgeable the engineering estimator is, the less difficult the estimating process will be, the more accurate the estimate will be, the less likely it is that a major error will occur, and the more likely it is that the estimate will be of high quality.

How is experience acquired in industry? One approach is to assign inexperienced engineers relatively small jobs, to create expertise and build familiarity with products and processes. Another strategy used is to pair inexperienced engineers with mentors who have vast technical experience. Technical boards and review meetings conducted to "justify the numbers" also are used to build knowledge and experience. Finally, many firms maintain databases of their past estimates and the costs that were actually incurred.

## ESTIMATING MODELS

This section develops several estimating models that can be used at the rough, semidetailed, or detailed design levels. For rough estimates the models are used with rough data, likewise for detailed design estimates they are used with detailed data.

### Per-Unit Model

The **per-unit model** uses a "per unit" factor, such as cost per square foot, to develop the estimate desired. This is a very simplistic yet useful technique, especially for developing estimates of the rough or order-of-magnitude type. The per-unit model is commonly used in the construction industry. As an example, you may be interested in a new home that is constructed with a certain type of material and has a specific construction style. Based on this information a contractor may quote a cost of $65 per square foot for your home. If you are interested in a 2000-square-foot floor plan, your cost would thus be: 2000 × 65 = $130,000. Other examples where per unit factors are used include

- Service cost per customer
- Safety cost per employee
- Gasoline cost per mile
- Cost of defects per batch
- Maintenance cost per window
- Mileage cost per vehicle
- Utility cost per square foot of floor space
- Housing cost per student

It is important to note that the per-unit model does not make allowances for economies of scale (the fact that higher quantities usually cost less on a per-unit basis). In most cases, however, the model can be effective at getting the decision maker "in the ballpark" of likely costs, and it can be very accurate if accurate data are used.

## EXAMPLE 2–4

Use the per-unit model to estimate the cost per student that you will incur for hosting 24 foreign exchange students at a local island campground for 10 days. During camp you are planning the following activities:

- 2 days of canoeing
- 3 campsite-sponsored day hikes

- 3 days at the lake beach (swimming, volleyball, etc.)
- Nightly entertainment

After calling the campground and collecting other information, you have accumulated the following data:

- Van rental from your city to the camp (one way) is $50 per 15-person van plus gas.
- Camp is 50 miles away, the van gets 10 miles per gallon, and gas is $3 per gallon.
- Each cabin at the camp holds 4 campers, and rent is $10 per day per cabin.
- Meals are $10 per day per camper; no outside food is allowed.
- Boat transportation to the island is $2 per camper (one way).
- Insurance/grounds fee/overhead is $1 per day per camper.
- Canoe rentals are $5 per day per canoe; a canoe holds 3 campers.
- Day hikes are $2.50 per camper (plus the cost for meals).
- Beach rental is $25 per group per half-day.
- Nightly entertainment is free.

## SOLUTION

You are asked to use the per-unit factor to estimate the cost per student on this trip. For planning purposes we assume that there will be 100% participation in all activities. We will break the total cost down into categories of transportation, living, and entertainment.

### Transportation Costs

*Van travel to and from camp:*  2 vans/trip × 2 trips × ($50/van + 50 miles/van × 1 gal/10 miles × $3/gal) = $260

*Boat travel to and from island:*  2 trips × $2/camper × 24 campers = $96

**Transportation costs = 260 + 96 = $356**

### Living Costs

*Meals for the 10-day period:*  24 campers × $10/camper-day × 10 days = $2400

*Cabin rental for the 10-day period:*  24 campers × 1 cabin/4 campers × $10/day/cabin × 10 days = $600

*Insurance/Overhead expense for the 10-day period:*  24 campers × $1/camper-day × 10 days = $240

**Living costs = 2400 + 600 + 240 = $3240**

### Entertainment Costs

*Canoe rental costs:*  2 canoe days × 24 campers × 1 canoe/3 campers × $5/day/canoe = $80

*Beach rental costs:*  3 days × 2 half-days/day × $25/half-day = $150

*Day hike costs:*   24 campers × 3 day hikes × $2.50/camper-day hike = $180

*Nightly entertainment:*   This is free! Can you believe it?

$$\text{Entertainment costs} = 80 + 150 + 180 + 0 = \$410$$

**Total Cost**

$$\text{Total cost for 10-day period} = \text{Transportation costs} + \text{Living costs} + \text{Entertainment costs}$$
$$= 356 + 3240 + 410 = \$4006$$

Thus, the cost per student would be $4006/24 = $166.92.
Thus, it would cost you $166.92 per student to host the students at the island campground for the 10-day period. In this case the per-unit model gives you a very detailed cost estimate (although its accuracy depends on the accuracy of your data and assumptions you've made).

### Segmenting Model

The **segmenting model** can be described as "divide and conquer." An estimate is decomposed into its individual components, estimates are made at those lower levels, and then the estimates are aggregated (added) back together. It is much easier to estimate at the lower levels because they are more readily understood. This approach is common in engineering estimating in many applications and for any level of accuracy needed. In planning the camp trip of Example 2-4, the overall estimate was **segmented** into the costs for travel, living, and entertainment. The example illustrated the segmenting model (division of the overall estimate into the various categories) together with the unit factor model to make the subestimates for each category. Example 2-5 provides another example of the segmenting approach.

### EXAMPLE 2–5

Clean Lawn Corp., a manufacturer of yard equipment, is planning to introduce a new high-end industrial-use lawn mower called the Grass Grabber. The Grass Grabber is designed as a walk-behind, self-propelled mower. Clean Lawn engineers have been asked by the accounting department to estimate the material costs for the new mower. The material cost estimate will be used, along with estimates for labor and overhead to evaluate the potential of this new model.

### SOLUTION

The engineers decide to decompose the design specifications for the Grass Grabber into its components, estimate the material costs for each of the components, and then sum these costs up to obtain the overall estimate. The engineers are using a segmenting approach to build up their estimate. After careful consideration, the engineers have divided the mower into the following major subsystems: chassis, drive train, controls, and cutting/collection system. Each of these was further divided as appropriate, and unit material costs were estimated at this lowest of levels as follows:

| Cost Item | Unit Material Cost Estimate | Cost Item | Unit Material Cost Estimate |
|---|---|---|---|
| **A.  Chassis** | | **C.  Controls** | |
| A.1  Deck | $ 7.40 | C.1  Handle assembly | $ 3.85 |
| A.2  Wheels | 10.20 | C.2  Engine linkage | 8.55 |
| A.3  Axles | 4.85 | C.3  Blade linkage | 4.70 |
| | $22.45 | C.4  Speed control linkage | 21.50 |
| **B.  Drive train** | | C.5  Drive control assembly | 6.70 |
| B.1  Engine | $38.50 | C.6  Cutting height adjuster | 7.40 |
| B.2  Starter assembly | 5.90 | | $52.70 |
| B.3  Transmission | 5.45 | **D.  Cutting/Collection system** | |
| B.4  Drive disc assembly | 10.00 | D.1  Blade assembly | $10.80 |
| B.5  Clutch linkage | 5.15 | D.2  Side chute | 7.05 |
| B.6  Belt assemblies | 7.70 | D.3  Grass bag and adapter | 7.75 |
| | $72.70 | | $25.60 |

The total material cost estimate of $173.45 was calculated by summing up the estimates for each of the four major subsystem levels (chassis, drive train, controls, and cutting/collection system). It should be noted that this cost represents only the material portion of the overall cost to produce the mowers. Other costs would include labor and overhead items.

In Example 2-5 the engineers at Clean Lawn Corp. decomposed the cost estimation problem into logical elements. The scheme they used of decomposing cost items and numbering the material components (A.1, A.1, A.2, etc.) is known as a **work breakdown structure.** This technique is commonly used in engineering cost estimating and project management of large products, processes, or projects. A work breakdown structure decomposes a large "work package" into its constituent parts, which can then be estimated or managed individually. In Example 2-5 the work breakdown structure of the Grass Grabber has three levels. At the top level is the product itself, at the second level are the four major subsystems, and at the third level are the individual cost items. Imagine what the product work breakdown structure for a Boeing 787 looks like. Then imagine trying to manage the 787's design, engineering, construction, and costing without a tool like the work breakdown structure.

### Cost Indexes

Cost indexes are numerical values that reflect historical change in engineering (and other) costs. The cost index numbers are dimensionless, and they reflect relative price change in either individual cost items (labor, material, utilities) or groups of costs (consumer prices, producer prices). Indexes can be used to update historical costs with the basic ratio relationship given in Equation 2-2.

$$\frac{\text{Cost at time } A}{\text{Cost at time } B} = \frac{\text{Index value at time } A}{\text{Index value at time } B} \tag{2-2}$$

Equation 2-2 states that the ratio of the cost index numbers at two points in time ($A$ and $B$) is equivalent to the dollar cost ratio of the item at the same times (see Example 2-6).

## EXAMPLE 2–6

Miriam is interested in estimating the annual labor and material costs for a new production facility. She was able to obtain the following cost data:

**Labor Costs**

- Labor cost index value was at 124 ten years ago and is 188 today.
- Annual labor costs for a similar facility were $575,500 ten years ago.

**Material Costs**

- Material cost index value was at 544 three years ago and is 715 today.
- Annual material costs for a similar facility were $2,455,000 three years ago.

### SOLUTION

Miriam will use Equation 2-2 to develop her cost estimates for annual labor and material costs.

**Labor**

$$\frac{\text{Annual cost today}}{\text{Annual cost 10 years ago}} = \frac{\text{Index value today}}{\text{Index value 10 years ago}}$$

$$\text{Annual cost today} = \frac{188}{124} \times \$575,500 = \$871,800$$

**Materials**

$$\frac{\text{Annual cost today}}{\text{Annual cost 3 years ago}} = \frac{\text{Index value today}}{\text{Index value 3 years ago}}$$

$$\text{Annual cost today} = \frac{715}{544} \times \$2,455,000 = \$3,227,000$$

Cost index data are collected and published by several private and public sources in the U.S. (and world). The U.S. government publishes data through the Bureau of Labor Statistics of the Department of Commerce. The *Statistical Abstract of the United States* publishes cost indexes for labor, construction, and materials. Another useful source for engineering cost index data is the *Engineering News Record*.

### Power-Sizing Model

The **power-sizing model** is used to estimate the costs of industrial plants and equipment. The model "scales up" or "scales down" known costs, thereby accounting for economies of scale that are common in industrial plant and equipment costs. Consider the cost to build a refinery. Would it cost twice as much to build the same facility with double the capacity? It is unlikely. The *power-sizing model* uses the exponent $(x)$, called the *power-sizing exponent,* to reflect economies of scale in the size or capacity:

$$\frac{\text{Cost of equipment } A}{\text{Cost of equipment } B} = \left(\frac{\text{Size (capacity) of equipment } A}{\text{Size (capacity) of } B}\right)^x \qquad (2\text{-}3)$$

where $x$ is the power-sizing exponent, costs of $A$ and $B$ are at the same point in time (same dollar basis), and size or capacity is in the same physical units for both $A$ and $B$.

The power-sizing exponent $(x)$ can be 1.0 (indicating a linear cost-versus-size/capacity relationship) or greater than 1.0 (indicating *dis*economies of scale), but it is usually less than 1.0 (indicating economies of scale). Generally the size ratio should be less than 2, and it should never exceed 5. This model works best in a "middle" range—not very small or very large.

Exponent values for plants and equipment of many types may be found in several sources, including industry reference books, research reports, and technical journals. Such exponent values may be found in *Perry's Chemical Engineers' Handbook, Plant Design and Economics for Chemical Engineers,* and *Preliminary Plant Design in Chemical Engineering.* Table 2-1 gives power-sizing exponent values for several types of industrial facilities and equipment. The exponent given applies only to equipment within the size range specified.

In Equation 2-3 equipment costs for both $A$ and $B$ occur at the same point in time. This equation is useful for scaling equipment costs but *not* for updating those costs. When the time of the desired cost estimate is different from the time in which the scaling occurs (per Equation 2-3) cost indexes accomplish the time updating. Thus, in cases like Example 2-7 involving both scaling and updating, we use the power-sizing model together with cost indexes.

## TABLE 2–1  Example Power-Sizing Exponent Values

| Equipment or Facility | Size Range | Power-Sizing Exponent |
|---|---|---|
| Blower, centrifugal | 10,000–100,000 ft$^3$/min | 0.59 |
| Compressor | 200–2100 hp | 0.32 |
| Crystallizer, vacuum batch | 500–7000 ft$^2$ | 0.37 |
| Dryer, drum, single atmospheric | 10–100 ft$^2$ | 0.40 |
| Fan, centrifugal | 20,000–70,000 ft$^3$/min | 1.17 |
| Filter, vacuum rotary drum | 10–1500 ft$^2$ | 0.48 |
| Lagoon, aerated | 0.05–20 million gal/day | 1.13 |
| Motor | 5–20 hp | 0.69 |
| Reactor, 300 psi | 100–1000 gal | 0.56 |
| Tank, atmospheric, horizontal | 100–40,000 gal | 0.57 |

### EXAMPLE 2–7

Based on her work in Example 2-6, Miriam has been asked to estimate the cost today of a 2500-ft$^2$ heat exchange system for the new plant being analyzed. She has the following data.

- Her company paid $50,000 for a 1000-ft$^2$ heat exchanger 5 years ago.
- Heat exchangers within this range of capacity have a power-sizing exponent ($x$) of 0.55.
- Five years ago the Heat Exchanger Cost Index (HECI) was 1306; it is 1487 today.

### SOLUTION

Miriam will first use Equation 2-3 to scale up the cost of the 1000-ft$^2$ exchanger to one that is 2500 ft$^2$ using the 0.55 power-sizing exponent.

$$\frac{\text{Cost of 2500-ft}^2 \text{ equipment}}{\text{Cost of 1000-ft}^2 \text{ equipment}} = \left(\frac{2500\text{-ft}^2 \text{ equipment}}{1000\text{-ft}^2 \text{ equipment}}\right)^{0.55}$$

$$\text{Cost of 2500-ft}^2 \text{ equipment} = \left(\frac{2500}{1000}\right)^{0.55} \times 50{,}000 = \$82{,}800$$

Miriam knows that the $82,800 reflects only the scaling up of the cost of the 1000-ft$^2$ model to a 2500-ft$^2$ model. Now she will use Equation 2-2 and the HECI data to estimate the cost of a 2500-ft$^2$ exchanger today. Miriam's cost estimate would be

$$\frac{\text{Equipment cost today}}{\text{Equipment cost 5 years ago}} = \frac{\text{Index value today}}{\text{Index value 5 years ago}}$$

$$\text{Equipment cost today} = \frac{1487}{1306} \times \$82{,}800 = \$94{,}300$$

### Triangulation

Triangulation is used in engineering surveying. A geographical area is divided into triangles from which the surveyor is able to map points within that region by using three fixed points and horizontal angular distances to locate fixed points of interest (e.g., property line reference points). Since any point can be located with two lines, the third line represents an extra perspective and check. We will not use trigonometry to arrive at our cost estimates, but we can use the concept of triangulation. We should approach our economic estimate from different perspectives because such varied perspectives add richness, confidence, and quality to the estimate. **Triangulation** in cost estimating might involve using different sources of data or using different quantitative models. As decision makers, we should always seek out varied perspectives.

### Improvement and the Learning Curve

One common phenomenon observed, regardless of the task being performed, is that as the number of repetitions increases, performance becomes faster and more accurate. This is the concept of learning and improvement in the activities that people perform. From our own experience we all know that our fiftieth repetition is done in much less time than we needed to do the task the first time.

The **learning curve** captures the relationship between task performance and task repetition. In general, as output *doubles*, the unit production time will be reduced to some fixed percentage, the **learning-curve percentage** or **learning-curve rate.** For example, it may take 300 minutes to produce the third unit in a production run involving a task with a 95% learning time curve. In this case the sixth unit doubles the output, so it will take 300(0.95) = 285 minutes to produce. Sometimes the learning curve is also known as the progress curve, improvement curve, experience curve, or manufacturing progress function.

Equation 2-4 gives an expression that can be used for time estimating for repetitive tasks.

$$T_N = T_{\text{initial}} \times N^b \tag{2-4}$$

where $T_N$ = time required for the $N^{\text{th}}$ unit of production

$T_{\text{initial}}$ = time required for the first (initial) unit of production

$N$ = number of completed units (cumulative production)

$b$ = learning curve exponent (slope of the learning curve on a log–log plot)

As just given, a learning curve is often referred to by its percentage learning slope. Thus, a curve with $b = -0.074$ is a 95% learning curve because $2^{-0.074} = 0.95$. This equation uses 2 because the learning-curve percentage applies for doubling cumulative production. The learning-curve exponent is calculated by using Equation 2-5.

$$b = \frac{\log(\text{learning curve expressed as a decimal})}{\log 2.0} \tag{2-5}$$

## EXAMPLE 2–8

Calculate the time required to produce the hundredth unit of a production run if the first unit took 32.0 minutes to produce and the learning-curve rate for production is 80%.

### SOLUTION

Using Equation 2-4, we write

$$T_{100} = T_1 \times 100^b$$

$$T_{100} = T_1 \times 100^{\log 0.80/\log 2.0}$$

$$T_{100} = 32.0 \times 100^{-0.3219}$$

$$T_{100} = 7.27 \text{ minutes}$$

It is particularly important to account for the learning-curve effect if the production run involves a small number of units instead of a large number. When thousands or even millions of units are being produced, early inefficiencies tend to be "averaged out" because of the larger batch sizes. However, in the short run, inefficiencies of the same magnitude can lead to rather poor estimates of production time require-ments, and thus production cost estimates may be understated. Consider Example 2-9 and the results that might be observed if the learning-curve effect is ignored. Notice in this example that a "steady-state" time is given. Steady state is the time at which the phys-ical constraints of performing the task prevent the achievement of any more learning or improvement.

## EXAMPLE 2–9

Estimate the overall labor cost portion due to a task that has a learning-curve rate of 85% and reaches a steady-state value after 16 units of 5.0 minutes per unit. Labor and benefits are $22 per hour, and the task requires two skilled workers. The overall production run is 20 units.

### SOLUTION

Because we know the time required for the 16$^{th}$ unit, we can use Equation 2-4 to calculate the time required to produce the first unit.

$$T_{16} = T_1 \times 16^{\log 0.85 / \log 2.0}$$

$$5.0 = T_1 \times 16^{-0.2345}$$

$$T_1 = 5.0 \times 16^{0.2345}$$

$$T_1 = 9.6 \text{ minutes}$$

Now we use Equation 2-4 to calculate the time requirements for the first 16 units in the production run, and we sum for the total production time required.

$$T_N = 9.6 \times N^{-0.2345}$$

| Unit Number, $N$ | Time (min) to Produce $N^{th}$ Unit | Cumulative Time from 1 to $N$ | Unit Number, $N$ | Time (min) to Produce $N^{th}$ Unit | Cumulative Time from 1 to $N$ |
|---|---|---|---|---|---|
| 1 | 9.6 | 9.6 | 11 | 5.5 | 74.0 |
| 2 | 8.2 | 17.8 | 12 | 5.4 | 79.2 |
| 3 | 7.4 | 24.2 | 13 | 5.3 | 84.5 |
| 4 | 6.9 | 32.1 | 14 | 5.2 | 89.7 |
| 5 | 6.6 | 38.7 | 15 | 5.1 | 94.8 |
| 6 | 6.3 | 45.0 | 16 | 5.0 | 99.8 |
| 7 | 6.1 | 51.1 | 17 | 5.0 | 104.8 |
| 8 | 5.9 | 57.0 | 18 | 5.0 | 109.8 |
| 9 | 5.7 | 62.7 | 19 | 5.0 | 114.8 |
| 10 | 5.6 | 68.3 | 20 | 5.0 | 119.8 |

The total cumulative time of the production run is 119.8 minutes (2.0 hours). Thus the total labor cost estimate would be

$$(2.0 \text{ hours})(\$22/\text{hour per worker})(2 \text{ workers}) = \$88$$

If we ignore the learning-curve effect and calculate the labor cost portion based only on the steady-state labor rate, the estimate would be

$$(0.083 \text{ hour/unit})(20 \text{ units})(\$22/\text{hour per worker})(2 \text{ workers}) = \$73.04$$

This estimate is understated by about 20% from what the true cost would be.

Figures 2-7 and 2-8 illustrate learning curves plotted on linear scales and as a *log–log* plot. On the log–log plot the relationship is a straight line, since doubling the number of units on the *x* axis reduces the time required by a constant percentage.

**FIGURE 2-7** Learning curve of time vs. number of units.

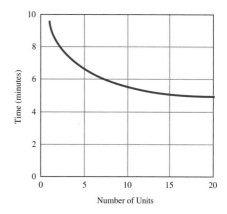

**FIGURE 2-8** Learning curve on log–log scale.

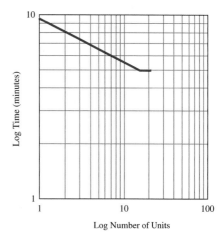

## ESTIMATING BENEFITS

This chapter has focused on cost terms and cost estimating. However, engineering economists must often also estimate benefits. Benefits include sales of products, revenues from bridge tolls and electric power sales, cost reductions from reduced material or labor costs, less time spent in traffic jams, and reduced risk of flooding. Many engineering projects are undertaken precisely to secure these benefits.

The cost concepts and cost-estimating models can also be applied to economic benefits. Fixed and variable benefits, recurring and nonrecurring benefits, incremental benefits, and life-cycle benefits all have meaning. Also, issues regarding the type of estimate (rough, semidetailed, and detailed), as well as difficulties in estimation (one of a kind, time and effort, and estimator expertise), all apply directly to estimating benefits. Last, per-unit, segmented, and indexed models are used to estimate benefits. The concept of triangulation is particularly important for estimating benefits.

The uncertainty in benefit estimates is also typically asymmetric, with a broader limit for negative outcomes. Benefits are more likely to be overestimated than underestimated, so an example set of limits might be (−50%, +20%). One difference between cost and benefit estimation is that many costs of engineering projects occur in the near future (for design and construction), but the benefits are further in the future. Because benefits are often further in the future, they are more difficult to estimate accurately, and more uncertainty is typical.

The estimation of economic benefits for inclusion in our analysis is an important step that should not be overlooked. Many of the models, concepts, and issues that apply in the estimation of costs also apply in the estimation of economic benefits.

## CASH FLOW DIAGRAMS

The costs and benefits of engineering projects occur over time and are summarized on a cash flow diagram (CFD). Specifically, a CFD illustrates the size, sign, and timing of individual cash flows. In this way the CFD is the basis for engineering economic analysis.

A **cash flow diagram** is created by first drawing a segmented time-based horizontal line, divided into time units. The time units on the CFD can be years, months, quarters, or any other consistent time unit. Then at each time at which a cash flow will occur, a vertical arrow is added—pointing down for costs and up for revenues or benefits. These cash flows are drawn to scale.

Unless otherwise stated, cash flows are **assumed** to occur at time 0 or at the **end** of each period. Consider Figure 2-9, the CFD for a specific investment opportunity whose cash flows are described as follows:

| Timing of Cash Flow | Size of Cash Flow |
|---|---|
| At time zero (now or today) | A positive cash flow of $100 |
| 1 time period from today | A negative cash flow of $100 |
| 2 time periods from today | A positive cash flow of $100 |
| 3 time periods from today | A negative cash flow of $150 |
| 4 time periods from today | A negative cash flow of $150 |
| 5 time periods from today | A positive cash flow of $50 |

**FIGURE 2-9** An example cash flow diagram (CFD).

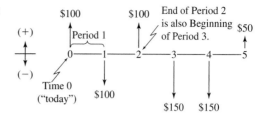

## Categories of Cash Flows

The expenses and receipts due to engineering projects usually fall into one of the following categories:

First cost ≡ expense to build or to buy and install

Operating and maintenance (O&M) ≡ annual expense, such as electricity, labor, and minor repairs

Salvage value ≡ receipt at project termination for sale or transfer of the equipment (can be a salvage cost)

Revenues ≡ annual receipts due to sale of products or services

Overhaul ≡ major capital expenditure that occurs during the asset's life

Individual projects will often have specific costs, revenues, or user benefits. For example, annual operating and maintenance (O&M) expenses on an assembly line might be divided into direct labor, power, and other. Similarly, a public-sector dam project might have its annual benefits divided into flood control, agricultural irrigation, and recreation.

## Drawing a Cash Flow Diagram

The cash flow diagram shows when all cash flows occur. Look at Figure 2-9 and the $100 positive cash flow at the end of period 2. From the time line one can see that this cash flow can also be described as occurring at the *beginning* of period 3. Thus, in a CFD the end of *period t* is the same time as the beginning of *period t* + 1. Beginning-of-period cash flows (such as rent, lease, and insurance payments) are thus easy to handle: just draw your CFD and put them in where they occur. Thus O&M, salvages, revenues, and overhauls are assumed to be end-of-period cash flows.

The choice of time 0 is arbitrary. For example, it can be when a project is analyzed, when funding is approved, or when construction begins. When construction periods are assumed to be short, first costs are assumed to occur at time 0, and the first annual revenues and costs start at the end of the first period. When construction periods are long, time 0 is usually the date of commissioning—when the facility comes on stream.

Perspective is also important when one is drawing a CFD. Consider the simple transaction of paying $5000 for some equipment. To the firm buying the equipment, the cash flow is a cost and hence negative in sign. To the firm selling the equipment, the cash flow is a revenue and positive in sign. This simple example shows that a consistent perspective is required when one is using a CFD to model the cash flows of a problem. One person's cash outflow is another person's inflow.

Often two or more cash flows occur in the same year, such as an overhaul and an O&M expense or the salvage value and the last year's O&M expense. Combining these into one total cash flow per year would simplify the cash flow diagram. However, it is better to show each individually, to ensure a clear connection from the problem statement to each cash flow in the diagram.

### Drawing Cash Flow Diagrams with a Spreadsheet

One simple way to draw cash flow diagrams with "arrows" proportional to the size of the cash flows is to use a spreadsheet to draw a stacked bar chart. The data for the cash flows are entered, as shown in the table part of Figure 2-10. To make a quick graph, select cells B1 to D8, which are the three columns of the cash flow. Then select the graph menu and choose column chart and select the stack option. Except for labeling axes (using the cells for year 0 to year 6), choosing the scale for the *y* axis, and adding titles, the cash flow diagram is done. Refer to the appendix for a review of basic spreadsheet use. (*Note*: A bar chart labels periods rather than using an *x* axis with arrows at times 0, 1, 2, ...)

|   | A | B | C | D | E | F |
|---|---|---|---|---|---|---|
| 1 | Year | Capital Costs | O&M | Overhaul | | |
| 2 | 0 | −80000 | | | | |
| 3 | 1 | | −12000 | | | |
| 4 | 2 | | −12000 | | | |
| 5 | 3 | | −12000 | −25000 | | |
| 6 | 4 | | −12000 | | | |
| 7 | 5 | | −12000 | | | |
| 8 | 6 | 10000 | −12000 | | | |
| 9 | | | | | | |

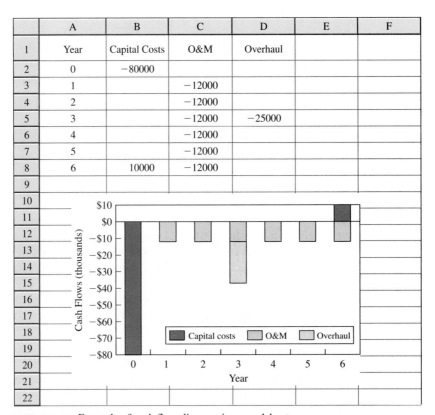

FIGURE 2–10 Example of cash flow diagram in spreadsheets.

## SUMMARY

This chapter has introduced the following cost concepts: fixed and variable, marginal and average, sunk, opportunity, recurring and nonrecurring, incremental, cash and book, and life-cycle. **Fixed costs** are constant and unchanging as volumes change, while **variable costs** change as output changes. Fixed and variable costs are used to find a breakeven value between costs and revenues, as well as the regions of net profit and loss. A **marginal cost** is for one more unit, while the **average cost** is the total cost divided by the number of units.

**Sunk costs** result from past decisions; they should not influence our attitude toward current and future opportunities. Remember, "sunk costs are sunk." **Opportunity costs** involve the benefit that is forgone when we choose to use a resource in one activity instead of another. **Recurring costs** can be planned and anticipated expenses; **nonrecurring costs** are one-of-a-kind costs that are often more difficult to anticipate.

**Incremental costs** are economic consequences associated with the differences between two choices of action. **Cash costs** are also known as **out-of-pocket costs** that represent actual cash flows. **Book costs** do not result in the exchange of money, but rather are costs listed in a firm's accounting books. **Life-cycle costs** are all costs that are incurred over the life of a product, process, or service. Thus engineering designers must consider life-cycle costs when choosing materials and components, tolerances, processes, testing, safety, service and warranty, and disposal.

Cost estimating is the process of "developing the numbers" for engineering economic analysis. Unlike a textbook, the real world does not present its challenges with neat problem statements that provide all the data. **Rough estimates** give us order-of-magnitude numbers and are useful for high-level and initial planning as well as judging the feasibility of alternatives. **Semidetailed estimates** are more accurate than rough-order estimates, thus requiring more resources (people, time, and money) to develop. These estimates are used in preliminary design and budgeting. **Detailed estimates** generally have an accuracy of −3% to 5%. They are used during the detailed design and contract bidding phases of a project.

Difficulties are common in developing estimates. **One-of-a-kind estimates** will have no basis in earlier work, but this disadvantage can be addressed through **estimation by analogy.** Lack of time is best addressed by planning and by matching the estimate's detail to the purpose—one should not spend money developing a detailed estimate when only a rough estimate is needed. **Estimator expertise** must be developed through work experiences and mentors.

Several general models and techniques for developing cost estimates were discussed. The **per-unit** and **segmenting models** use different levels of detail and costs per square foot or other unit. **Cost index data** are useful for updating historical costs to formulate current estimates. The **power-sizing model** is useful for scaling up or down a known cost quantity to account for economies of scale, with different power-sizing exponents for industrial plants and equipment of different types. **Triangulation** suggests that one should seek varying perspectives when developing cost estimates. Different information sources, databases, and analytical models can all be used to create unique perspectives. As the number of task repetitions increases, efficiency improves because of learning or improvement. This is summarized in the **learning-curve percentage,** where doubling the cumulative production reduces the time to complete the task, which equals the learning-curve percentage times the current production time.

Cash flow estimation must include project benefits. These include labor cost savings, avoided quality costs, direct revenue from sales, reduced catastrophic risks, improved traffic flow, and cheaper power supplies. **Cash flow diagrams** are used to model the positive and negative cash flows of potential investment opportunities. These diagrams provide a consistent view of the problem (and the alternatives) to support economic analysis.

# PROBLEMS

## Fixed, Variable, Average, and Marginal Costs

**2-1** Electricity is sold for $0.12 per kilowatt-hour (kWh) for the first 10,000 units each month and $0.09/kWh for all remaining units. If a firm uses 14,000 kWh/month, what is its average and marginal cost?

(*Answer*: average cost is $0.111/kWh)

**2-2** One of your firm's suppliers discounts prices for larger quantities. The first 1000 parts are $13 each. The next 2000 are $12 each. All parts in excess of 3000 cost $11 each. What are the average cost and marginal cost per part for the following quantities?

(*a*) 500

(*b*) 1500

(*c*) 2500

(*d*) 3500

**2-3** A new machine comes with 100 free service hours over the first year. Additional time costs $75 per hour. What are the average and marginal costs per hour for the following quantities?

(*a*) 75

(*b*) 125

(*c*) 250

**2-4** Venus Computer can produce 23,000 personal computers a year on its daytime shift. The fixed manufacturing costs per year are $2 million and the total labor cost is $9,109,000. To increase its production to 46,000 computers per year, Venus is considering adding a second shift. The unit labor cost for the second shift would be 25% higher than the day shift, but the total fixed manufacturing costs would increase only to $2.4 million from $2 million.

(*a*) Compute the unit manufacturing cost for the daytime shift.

(*b*) Would adding a second shift increase or decrease the unit manufacturing cost at the plant?

**2-5** A small machine shop, with 30 hp of connected load, purchases electricity at the following monthly rates (assume any demand charge is included in this schedule) per hp of connected load:

First 50 kWh at 12.6¢ per kWh

Next 50 kWh at 10.6¢ per kWh

Next 150 kWh at 6.0¢ per kWh

Over 250 kWh at 5.7¢ per kWh

The shop uses 2800 kWh per month.

(*a*) Calculate the monthly bill for this shop. What are the marginal and average costs per kilowatt-hour?

(*b*) Suppose Jennifer, the proprietor of the shop, has the chance to secure additional business that will require her to operate her existing equipment more hours per day. This will use an extra 1200 kWh per month. What is the lowest figure that she might reasonably consider to be the "cost" of this additional energy? What is this per kilowatt-hour?

(*c*) Jennifer contemplates installing certain new machines that will reduce the labor time required on certain operations. These will increase the connected load by 10 hp, but since they will operate only on certain special jobs, will add only 100 kWh per month. In a study to determine the economy of installing these new machines, what should be considered as the "cost" of this energy? What is this per kilowatt-hour?

**2-6** Two automatic systems for dispensing maps are being compared by the state highway department. The accompanying breakeven chart of the comparison of these systems (System I vs. System II) shows total yearly costs for the number of maps dispensed per year for both alternatives. Answer the following questions.

(*a*) What is the fixed cost for System I?

(*b*) What is the fixed cost for System II?

(*c*) What is the variable cost per map dispensed for System I?

(*d*) What is the variable cost per map dispensed for System II?

(*e*) What is the breakeven point in terms of maps dispensed at which the two systems have equal annual costs?

(*f*) For what range of annual number of maps dispensed is System I recommended?

(*g*) For what range of annual number of maps dispensed is System II recommended?

(*h*) At 3000 maps per year, what are the marginal and average map costs for each system?

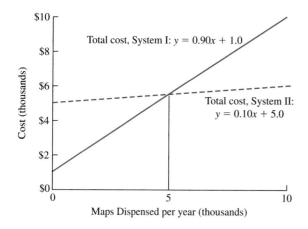

Total cost, System I: $y = 0.90x + 1.0$

Total cost, System II: $y = 0.10x + 5.0$

Cost (thousands)

Maps Dispensed per year (thousands)

**2-7** A privately owned summer camp for youngsters has the following data for a 12-week session:

| | |
|---|---|
| Charge per camper | $480 per week |
| Fixed costs | $192,000 per session |
| Variable cost per camper | $320 per week |
| Capacity | 200 campers |

(a) Develop the mathematical relationships for total cost and total revenue.

(b) What is the total number of campers that will allow the camp to *just break even*?

(c) What is the profit or loss for the 12-week session if the camp operates at 80% capacity?

(d) What are marginal and average costs per camper at 80% capacity?

**2-8** Two new rides are being compared by a local amusement park in terms of their annual operating costs. The two rides are assumed to be able to generate the same level of revenue (thus the focus on costs). The Tummy Tugger has fixed costs of $10,000 per year and variable costs of $2.50 per visitor. The Head Buzzer has fixed costs of $4000 per year and variable costs of $4 per visitor. Provide answers to the following questions so the amusement park can make the needed comparison.

(a) Mathematically determine the breakeven number of visitors per year for the two rides to have equal annual costs.

(b) Develop a breakeven graph that illustrates the following:

• Accurate total cost lines for the two alternatives (show line, slopes, and equations).

• The breakeven point for the two rides in terms of number of visitors.

• The ranges of visitors per year where each alternative is preferred.

**2-9** Consider the accompanying breakeven graph for an investment, and answer the following questions.

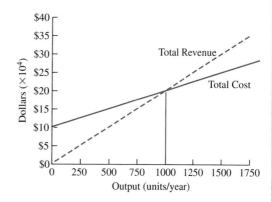

Total Revenue

Total Cost

Dollars ($\times 10^4$)

Output (units/year)

(a) Give the equation for total revenue for $x$ units per year.

(b) Give the equation for total costs for $x$ units per year.

(c) What is the "breakeven" level of $x$?

(d) If you sell 1500 units this year, will you have a profit or loss? How much?

(e) At 1500 units, what are your marginal and average costs?

**2-10** Quatro Hermanas, Inc. is investigating implementing some new production machinery as part of its operations. Three alternatives have been identified, and they have the following fixed and variable costs:

| Alternative | Annual Fixed Costs | Annual Variable Costs per Unit |
|---|---|---|
| A | $100,000 | $20.00 |
| B | 200,000 | 5.00 |
| C | 150,000 | 7.50 |

Determine the ranges of production (units produced per year) over which each alternative would be recommended up to 30,000 units per year.

**2-11** Three alternative designs have been created by Snakisco engineers for a new machine that spreads cheese between the crackers in a Snakisco snack. The costs for the three designs (where $x$ is the annual production rate of boxes of cheese crackers) follow:

| Design | Fixed Cost | Variable Cost ($/x) |
|---|---|---|
| A | $100,000 | 20.5x |
| B | 350,000 | 10.5x |
| C | 600,000 | 8.0x |

(*a*) Management is interested in the production interval of 0–150,000 boxes of crackers per year. Mathematically determine the production volume over which each design (*A* or *B* or *C*) would be chosen.

(*b*) Depict your solution from part (*a*) graphically, clearly labeling your axes and including a *title* for the graph, so that management can easily see the following:

    i. Accurate total cost lines for each alternative (show line, slopes, and line equations).

    ii. Any relevant breakeven, or crossover points.

    iii. Ranges of annual production where each alternative is preferred.

**2-12** A painting operation is performed by a production worker at a labor cost of $1.40 per unit. A robot spray-painting machine, costing $15,000, would reduce the labor cost to $0.20 per unit. If the device would be valueless at the end of 3 years, what is the minimum number of units that would have to be painted each year to justify the purchase of the robot machine?

**2-13** Company *A* has fixed expenses of $15,000 per year and each unit of product has a $0.20 variable cost. Company *B* has fixed expenses of $5000 per year and can produce the same product at a $0.50 variable cost. At what number of units of annual production will Company *A* have the same overall cost as Company *B*?

**2-14** Mr. Sam Spade, the president of Ajax, recently read in a report that a competitor named Bendix has the following relationship between cost and production quantity:

$$C = \$3,000,000 - \$18,000Q + \$75Q^2$$

where $C =$ total manufacturing cost per year and $Q =$ number of units produced per year.

    A newly hired employee, who previously worked for Bendix, tells Mr. Spade that Bendix is now producing 110 units per year. If the selling price remains unchanged, Sam wonders if Bendix is likely to increase the number of units produced per year in the near future. He asks you to look at the information and tell him what you are able to deduce from it.

**2-15** A small company manufactures a certain product. Variable costs are $20 per unit and fixed costs are $10,875. The price–demand relationship for this product is $P = -0.25D + 250$, where $P$ is the unit sales price of the product and $D$ is the annual demand. Use the data (and helpful hints) that follow to work out answers to parts (*a*)–(*e*).

- Total cost = Fixed cost + Variable cost
- Revenue = Demand × Price
- Profit = Revenue − Total cost

Set up your graph with dollars on the *y* axis (between 0 and $70,000) and, on the *x* axis, demand *D*: (units produced or sold), between 0 and 1000 units.

(*a*) Develop the equations for total cost and total revenue.

(*b*) Find the breakeven quantity (in terms of profit and loss) for the product.

(*c*) What profit would the company obtain by maximizing its total revenue?

(*d*) What is the company's maximum possible profit?

(*e*) Neatly graph the solutions from parts (*a*), (*b*), (*c*), and (*d*).

**2-16** A firm believes a product's sales volume (*S*) depends on its unit selling price (*P*) as $S = \$100 - P$. The production cost (*C*) is $1000 + 10S$.

(*a*) Draw a graph with the sales volume (*S*) from 0 to 100 on the *x* axis, and total cost and total income from $0 to $2500 on the *y* axis. On the graph draw the line $C = \$1000 + 10S$. Then plot the curve of total income. Mark the breakeven points on the graph.

(*b*) Determine the breakeven point (lowest sales volume at which total sales income just equals total production cost). (*Hint:* This may be done by trial and error or by using the quadratic equation to locate the point at which profit is zero.)

(*c*) Determine the sales volume (*S*) at which the firm's profit is a maximum. (*Hint:* Write an equation for profit and solve it by trial and error, or as a minima–maxima calculus problem.)

**2-17** A labor-intensive process has a fixed cost of $338,000 and a variable cost of $143 per unit. A capital-intensive (automated) process for the same product has a fixed cost of $1,244,000 and a variable cost of $92.50 per unit. How many units must be produced and sold at $197 each for the automated process to be preferred to the labor-intensive process.
*Contributed by Paul R. McCright, University of South Florida*

**2-18** Heinrich is a manufacturing engineer with the Miller Company. He has determined the costs of producing a new product to be as follows:

Equipment cost: $288,000/year
Equipment salvage value at EOY5 = $41,000
Variable cost per unit of production: $14.55
Overhead cost per year: $48,300

If the Miller Company uses a 5-year planning horizon and the product can be sold for a unit price of $39.75, how many units must be produced and sold each year to break even?

*Contributed by Paul R. McCright, University of South Florida*

**2-19** An assembly line can produce 90 units per hour. The line's hourly cost is $4500 on straight time (the first 8 hours). Workers are guaranteed a minimum of 6 hours. There is a 50% premium for overtime, however, productivity for overtime drops by 5%. What are the average and marginal costs per unit for the following daily quantities?

(*a*) 450

(*b*) 600

(*c*) 720

(*d*) 900

## Sunk and Other Costs

**2-20** A pump has failed in a facility that will be completely replaced in 3 years. A brass pump costing $6000 installed will last 3 years. However, a used stainless steel pump that should last 3 more years has been sitting in the maintenance shop for a year. The pump cost $13,000 new. The accountants say the pump is worth $7000 now. The maintenance supervisor says that it will cost an extra $500 to reconfigure the pump for the new use and that he could sell it used (as is) for $4000.

(*a*) What is the book cost of the stainless steel pump?

(*b*) What is the opportunity cost of the stainless steel pump?

(*c*) How much cheaper or more expensive would it be to use the stainless steel pump rather than a new brass pump?

**2-21** Last year to help with your New Year's resolutions you purchased a $500 piece of fitness equipment. However, you use it only once a week on average. It is December, and you can sell the equipment for $200 (to someone with a New Year's resolution) and rely on the university gym until you graduate in May. If you don't sell until May, you will get only $100. If you keep the heavy piece, you'll have to pay $25 to move it to the city of your new job (where you interned last summer). There is no convenient

gym at the new location. What costs and intangible consequences are relevant to your decision? What should you do?

**2-22** Bob Johnson decided to buy a new home. After looking at tracts of new homes, he decided that a custom-built home was preferable. He hired an architect to prepare the drawings. In due time, the architect completed the drawings and submitted them. Bob liked the plans; he was less pleased that he had to pay the architect a fee of $7000 to design the house. Bob asked a building contractor to provide a bid to construct the home on a lot Bob already owned. While the contractor was working to assemble the bid, Bob came across a book of standard house plans. In the book was a home that he and his wife liked better than the one designed for them by the architect. Bob paid $200 and obtained a complete set of plans for this other house. Bob then asked the contractor to provide a bid to construct this "stock plan" home. In this way Bob felt he could compare the costs and make a decision. The building contractor submitted the following bids:

| | |
|---|---|
| Custom-designed home | $258,000 |
| Stock-plan home | 261,000 |

Bob was willing to pay the extra $3000 for it. Bob's wife, however, felt they should go ahead with the custom-designed home, for, as she put it, "We can't afford to throw away a set of plans that cost $7000." Bob agreed, but he disliked the thought of building a home that is less desirable than the stock plan home. Then he asked your advice. Which house would you advise him to build? Explain.

**2-23** You are reevaluating the pump choice that was made last year by your boss. The expected cost savings have not occurred because the pump is too small. Choice *A*, at $90,000, is to replace the pump with one that is the right size and sell the old one. Choice *B*, at $100,000, is to buy a small pump to use in tandem with the existing pump. The two-pump solution has slightly higher maintenance costs, but it is more flexible and reliable. What criteria should you use? Which choice would you recommend, and why?

**2-24** Consider the situation of owning rental properties that local university students rent from you for the academic year. Develop a set of costs that you could classify as recurring and others that could be classified as nonrecurring.

**2-25** List and classify your costs in this academic year as recurring or nonrecurring.

2-26   Define the difference between a "cash cost" and a "book cost." Is engineering economic analysis concerned with both types of cost? Give an example of each, and provide the context in which it is important.

2-27   In your own words, develop a statement of what the authors mean by "life-cycle costs." It is important for a firm to be aware of life-cycle costs. Can you explain why?

2-28   Most engineering students own a computer. What costs have you incurred at each stage of your computer's life cycle? Estimate the total cost of ownership. Estimate the benefits of ownership. Has it been worth it?

2-29   Estimating benefits is often more difficult than cost estimation. Use the example of car ownership to describe the complicating factors in estimating the costs and benefits.

2-30   In looking at Figures 2-4 and 2-5, restate in your own words what the authors are trying to get across with these graphs. Do you agree that this is an important effect for companies? Explain.

2-31   In the text we describe three effects that complicate the process of making estimates to be used in engineering economy analyses. List these three effects and comment on which might be the most influential.

2-32   Develop an estimate for each of the following situations.

(a) The cost of a 500-mile automobile trip, if gasoline is $3 per gallon, vehicle wear and tear is $0.50 per mile, and our vehicle gets 20 miles per gallon.

(b) The total number of hours in the average human life, if the average life is 75 years.

(c) The number of days it takes to travel around the equator using a hot air balloon, if the balloon averages 100 miles per day, the diameter of the earth is ~4000 miles.

(d) The total area in square miles of the United States of America, if Kansas is an average-sized state. Kansas has an area of 390 miles × 200 miles.

## Segmenting Models

2-33   Northern Tundra Telephone (NTT) has received a contract to install emergency phones along a new 100-mile section of the Snow-Moose Turnpike. Fifty emergency phone systems will be installed about 2 miles apart. The material cost of a unit is $125. NTT will need to run underground communication lines that cost NTT $7500 per mile (including labor) to install. There will also be a one-time cost of $10,000 to network these phones into NTT's current communication system.

(a) Develop a cost estimate of the project from NTT's perspective.

(b) If NTT adds a profit margin of 35% to its costs, how much will it cost the state to fund the project?

2-34   You and your spouse are planning a second honeymoon to the Cayman Islands this summer and would like to have your house painted while you are away. Estimate the total cost of the paint job from the information given below, where:

$$Cost_{total} = Cost_{paint} + Cost_{labor} + Cost_{fixed}$$

*Paint information:* Your house has a surface area of 6000 ft$^2$. One can of paint can cover 300 ft$^2$. You are estimating the cost to put on *two coats* of paint for the entire house.

| Number of Cans Purchased | Cost per Can |
|---|---|
| First 10 cans purchased | $15.00 |
| Second 15 cans purchased | $10.00 |
| Up to next 50 cans purchased | $7.50 |

*Labor information:* You plan to hire five painters who will paint for 10 hours per day each. You estimate that the job will require 4.5 days of their painting time. The painter's rate is $8.75 per hour.

*Fixed cost information:* The painting company charges $1200 per job to cover travel expenses, clothing, cloths, thinner, administration, and so on.

2-35   You want a mountain cabin built for weekend trips, vacations, to host family, and perhaps eventually to retire in. After discussing the project with a local contractor, you receive an estimate that the total construction cost of your 2000-ft$^2$ lodge will be $250,000. The percentage of costs is broken down as follows:

| Cost Items | Percentage of Total Costs |
|---|---|
| Construction permits, legal and title fees | 8 |
| Roadway, site clearing, preparation | 15 |
| Foundation, concrete, masonry | 13 |
| Wallboard, flooring, carpentry | 12 |
| Heating, ventilation, air conditioning (HVAC) | 13 |
| Electric, plumbing, communications | 10 |
| Roofing, flooring | 12 |
| Painting, finishing | 17 |
| | 100 |

(*a*) What is the cost per square foot of the 2000-ft$^2$ lodge?

(*b*) If you are also considering a 4000-ft$^2$ layout option, estimate your construction costs if:

  i. All cost items (in the table) change proportionately to the size increase.

  ii. The first two cost items do not change at all; HVAC changes by 50%; and all others are proportionate.

2-36 SungSam, Inc. is designing a new digital camcorder that is projected to have the following per-unit costs to manufacture:

| Cost Categories | Unit Costs |
|---|---|
| Materials costs | $ 63 |
| Labor costs | 24 |
| Overhead costs | 110 |
| Total unit cost | $197 |

SungSam adds 30% to its manufacturing cost for corporate profit.

(*a*) What unit profit would SungSam realize on each camcorder?

(*b*) What is the overall cost to produce a batch of 10,000 camcorders?

(*c*) What would SungSam's profit be on the batch of 10,000 if historical data show that 1% of product will be scrapped in manufacturing, 3% of finished product will go unsold, and 2% of *sold* product will be returned for refund?

(*d*) How much can SungSam afford to pay for a contract that would lock in a 50% reduction in the unit material cost previously given? If SungSam does sign the contract, the sales price will not change.

## Indexes and Sizing Models

2-37 Estimate the cost of expanding a planned new clinic by 20,000 ft$^2$. The appropriate capacity exponent is 0.66, and the budget estimate for 200,000 ft$^2$ was $15 million. (*Answer*: $1 million)

2-38 Fifty years ago, Grandma Bell purchased a set of gold-plated dinnerware for $55, and last year you inherited it. Unfortunately a fire at your home destroyed the set. Your insurance company is at a loss to define the replacement cost and has asked your help. You do some research and find that the Aurum Flatware Cost Index (AFCI) for gold-plated dinnerware, which was 112 when Grandma Bell bought her set, is at 2050 today. Use the AFCI to update the cost of Bell's set to today's cost to show to the insurance company.

2-39 Your boss is the director of reporting for the Athens County Construction Agency (ACCA). It has been his job to track the cost of construction in Athens County. Twenty-five years ago he created the ACCA Cost Index to track these costs. Costs during the first year of the index were $12 per square foot of constructed space (the index value was set at 100 for that first year). This past year a survey of contractors revealed that costs were $72 per square foot. What index number will your boss publish in his report for this year? If the index value was 525 last year, what was the cost per square foot last year?

2-40 A new aerated sewage lagoon is required in a small town. Earlier this year one was built on a similar site in an adjacent city for $2.3 million. The new lagoon will be 60% larger. Use the data in Table 2-1 to estimate the cost of the new lagoon.

2-41 A refinisher of antiques named Constance has been so successful with her small business that she is planning to expand her shop. She is going to start enlarging her shop by purchasing the following equipment.

| Equipment | Original Size | Cost of Original Equipment | Power-Sizing Exponent | New Equipment Size |
|---|---|---|---|---|
| Varnish bath | 50 gal | $3500 | 0.80 | 75 gal |
| Power scraper | 3/4 hp | $250 | 0.22 | 1.5 hp |
| Paint booth | 3 ft$^3$ | $3000 | 0.6 | 12 ft$^3$ |

What would be the *net* cost to Constance to obtain this equipment—assume that she can trade the old equipment in for 15% of its original cost. Assume there has been no inflation in equipment prices.

**2-42** Refer to Problem 2-41 and now assume the prices for the equipment that Constance wants to replace have not been constant. Use the cost index data for each piece of equipment to update the costs to the price that would be paid today. Develop the overall cost for Constance, again assuming the 15% trade-in allowance for the old equipment. Trade-in value is based on original cost.

| Original Equipment | Cost Index When Originally Purchased | Cost Index Today |
|---|---|---|
| Varnish bath | 154 | 171 |
| Power scraper | 780 | 900 |
| Paint booth | 49 | 76 |

**2-43** Five years ago, when the relevant cost index was 120, a nuclear centrifuge cost $40,000. The centrifuge had a capacity of separating 1500 gallons of ionized solution per hour. Today, it is desired to build a centrifuge with capacity of 4500 gallons per hour, but the cost index now is 300. Assuming a power-sizing exponent to reflect economies of scale, $x$, of 0.75, use the power-sizing model to determine the approximate cost (expressed in today's dollars) of the new reactor.

**2-44** Padre works for a trade magazine that publishes lists of power-sizing exponents (PSE) that reflect economies of scale for developing engineering estimates of various types of equipment. Padre has been unable to find any published data on the VMIC machine and wants to list its *PSE* value in his next issue. Given the following data (your staff was able to find data regarding costs and sizes of the VMIC machine) calculate the *PSE* value that Padre should publish. (*Note:* The VMIC-100 can handle twice the volume of a VMIC-50.)

Cost of VMIC-100 today $100,000
Cost of VMIC-50 5 years ago $45,000
VMIC equipment index today $= 214$
VMIC equipment index 5 years ago $= 151$

## Learning-Curve Models

**2-45** Determine the time required to produce the 2000th item if the first item requires 180 minutes to produce and the learning-curve percentage is 92%. (*Answer:* 72.1 minutes)

**2-46** If 200 labor hours were required to produce the 1st unit in a production run and 60 labor hours were required to produce the 7th unit, what was the *learning-curve rate* during production?

**2-47** Rose is a project manager at the civil engineering consulting firm of Sands, Gravel, Concrete, and Waters, Inc. She has been collecting data on a project in which concrete pillars were being constructed; however not all the data are available. She has been able to find out that the 10th pillar required 260 person-hours to construct, and that a 75% learning curve applied. She is interested in calculating the time required to construct the 1st and 20th pillars. Compute the values for her.

**2-48** Sally Statistics is implementing a system of statistical process control (SPC) charts in her factory in an effort to reduce the overall cost of scrapped product. The current cost of scrap is $X$ per month. If an 80% learning curve is expected in the use of the SPC charts to reduce the cost of scrap, what would the *percentage reduction* in monthly scrap cost be after the charts have been used for 12 months? (*Hint:* Model each month as a unit of production.)

**2-49** Randy Duckout has been asked to develop an estimate of the *per-unit selling price* (the price that each unit will be sold for) of a new line of handcrafted booklets that offer excuses for missed appointments. His assistant Doc Duckout has collected information that Randy will need in developing his estimate:

| | |
|---|---|
| Cost of direct labor | $20 per hour |
| Cost of materials | $43.75 per batch of 25 booklets |
| Cost of overhead items | 50% of direct labor cost |
| Desired profit | 20% of total manufacturing cost |

Doc also finds out that (1) they should use a 75% learning curve for estimating the cost of direct labor, (2) the time to complete the 1st booklet is estimated at 0.60 hour, and (3) the estimated time to complete the 25th booklet should be used as their standard time for the purpose of determining the *unit selling price*. What would Randy and Doc's estimate be for the *unit selling price*?

## Benefit Estimation

**2-50** Develop a statement that expresses the extent to which cost-estimating topics also apply to estimating benefits. Provide examples to illustrate.

**2-51** Large projects, such as a new tunnel under the Hudson, the Big Dig in Boston, the Denver airport, a new military jet, and a natural gas pipeline from Alaska to the Midwest, often take 5 to 15 years from concept to completion.

(*a*) Should benefit and cost estimates be adjusted for the greater influences and impacts of inflation, government regulatory changes, and changing local economic environments? Why or why not?.

(*b*) How does the public budget-making process interact with the goal of accurate benefit and cost estimating for these large projects?

*Contributed by Morgan Henrie, University of Alaska Anchorage*

### Cash Flow Diagrams

**2-52** On December 1, Al Smith purchased a car for $18,500. He paid $5000 immediately and agreed to pay three additional payments of $6000 each (which includes principal and interest) at the end of 1, 2, and 3 years. Maintenance for the car is projected at $1000 at the end of the first year and $2000 at the end of each subsequent year. Al expects to sell the car at the end of the fourth year (after paying for the maintenance work) for $7000. Using these facts, prepare a table of cash flows.

**2-53** Bonka Toys is considering a robot that will cost $20,000 to buy. After 7 years its salvage value will be $2000. An overhaul costing $5000 will be needed in Year 4. O&M costs will be $2500 per year. Draw the cash flow diagram.

**2-54** Pine Village needs some additional recreation fields. Construction will cost $225,000, and annual O&M expenses are $85,000. The city council estimates that the value of added youth leagues is about $190,000 annually. In year 6 another $75,000 will be needed to refurbish the fields. The salvage value is estimated to be $100,000 after 10 years. Draw the cash flow diagram.

**2-55** Identify your major cash flows for the current school term as first costs, O&M expenses, salvage values, revenues, overhauls, and so on. Using a week as the time period, draw the cash flow diagram.

## CASES

**The following cases from *Cases in Engineering Economy* 2^(nd) are suggested as matched with this chapter.**

CASE 3 **Wildcat Oil in Kasakstan**

Rough order-of-magnitude estimation of total facility cost and annual revenue. Option for NPV and different size facility.

CASE 25 **Raster Blaster**

Questions to guide students. Includes breakeven analysis.

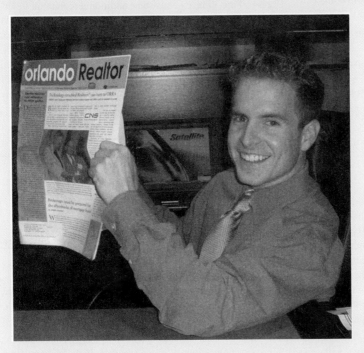

**A**s an industrial engineer for a high-profile corporation, Nick Cotton makes a good living, but he's figured out a way to make his skills work double-time.

Cotton's engineering economics class taught him how to turn his passion for math into profit. At 19, he began to invest in stocks and mutual funds, making use of principles in his undergraduate text. "I was able to calculate the future value of stocks and so could predict how well each would do," he says. He now owns two houses-one of which is still being built and has already appreciated by 15%.

Being the enterprising type, Cotton realized that he could market his solid investment skills. He started a real estate business that has turned out to be as lucrative as his day job. When he breaks down potential and interest for properties up for sale, he demonstrates to his clients how a seemingly small percentage increase in worth each year can compound to become a significant amount of money because of compounding.

Cotton is in prime position to become a tycoon, and he has engineering economics to thank. "No other class has contributed so much to my life," he says. "Who doesn't like making money?" ■■■

**W**hen Kim Brock enrolled in Engineering Economy, she had no idea she would ever use the skills taught in the course, let alone apply them to a career in aeronautics.

"I was just flipping through a magazine and saw an ad for Boeing-thought I might as well send them a résumé," she says.

As it turned out, Brock's understanding of big-picture problem solving made her an ideal job candidate. She now works as an industrial engineer for the Boeing 767 line, designing, improving, and controlling systems of people, materials, and information.

She says that Engineering Economy was a particularly valuable course because it taught her how to project the favorable and unfavorable consequences of a situation. When she conducts design meetings and process improvement workshops, she helps determine the safest and most cost-effective options for Boeing mechanics, right down to the way they hold a drill. "Recently, I had to offer advice on building a fixture to check that a part was working," she says. "After running some analyses, I realized that the fixture would cost more than the part itself."

It's this kind of common sense know-how that Engineering Economy imparts. Brock still owns her copy of an earlier edition of *Engineering Economic Analysis* and says that it's been helpful as she considers buying a condo: "It's a book about the reality of making decisions." ■ ■ ■

# INTEREST AND EQUIVALENCE

### A Prescription for Success

**A** pharmaceutical company manufactured a prescription drug that contained a tablet inside another tablet. This inner tablet, called a "core," needed to be in the exact center of the larger tablet, and tolerances were measured in tenths of millimeters. The process was not robust, and the placement of the inner tablet sometimes drifted, requiring the scrapping of off-quality product and the adjustment of the tablet press. This resulted in significant scrap and tablet press downtime.

A process change was invented in Japan to correct the problem, using a new process to place the inner tablet in the die of the press that made the outer tablet. There were three of these tablet presses in use in the U.S., but modifications were made to one tablet press as a test. The modification to the first press cost $27,000. During the first batch the modified press ran the entire batch without a quality problem and without quality losses. The batch finished compressing in 16 hours, which was considerably faster than the typical time of 24 hours (however, core centering problems could cause a delay of several days).

Additional test batches were run, all with excellent results. A detailed quality examination proved that the modification performed as desired, reducing variation and nearly eliminating product quality scrap. The other two tablet presses were later modified. The total cost for all modifications, including spare parts, was $90,000.

Over the next year, the results of the change were analyzed. Product yield increased from 92.4% to 96.6%, with each 1% yield increase worth $2.4 million per year. Operating efficiency improved, resulting in higher output because of less scrap and less downtime due to quality problems. Production plans called for 240 batches to be processed over the year after the tablet press modification was made. This product was produced daily, but production was reduced from three shifts to two because of the improved efficiency. Production planning could now plan effectively; they knew that a batch could be processed in two shifts, not one to five days.

Year-end accounting showed $10 million saved in the first year. Because the product's patent was about to expire, production was expected to be greatly reduced beyond this time. ■ ■ ■

*Contributed by Neal A. Lewis, University of Bridgeport*

## QUESTIONS TO CONSIDER

1. One year of production had a value of $240 million. What is the value of one batch of product?

2. How many batches needed to be produced to break even on the initial $27,000 investment? (Assume all batches improved the yield by 4.2%. Do not consider the time value of money.)

3. If the first-year savings is considered to be a single end-of-year cash flow, and the entire $90,000 investment is considered to occur at time 0, what is the present value of the project? Assume an interest rate of 15%.

4. If one batch is produced per day, how often are the savings actually compounded?

5. Does a company face any ethical considerations when it improves process efficiency resulting in lost labor hours for employees? Discuss and explain.

### After Completing This Chapter...

#### The student should be able to:

- Define and provide examples of the *time value of money*.
- Distinguish between *simple* and *compound interest*, and use compound interest in engineering economic analysis.
- Explain *equivalence* of cash flows.
- Solve problems using the single payment compound interest formulas.
- Distinguish and apply *nominal* and *effective interest rates*.
- Use *continuously compounded interest* with single payment interest factors.

In the first chapter, we discussed situations where the economic consequences of an alternative were immediate or took place in a very short period of time, as in Example 1-2 (design of a concrete aggregate mix) or Example 1-3 (change of manufacturing method). We totaled the various positive and negative aspects, compared our results, and could quickly reach a decision. But can we do the same if the economic consequences occur over a considerable period of time?

No we cannot, because *money has value over time.* Would you rather (1) receive $1000 today or (2) receive $1000 ten years from today? Obviously, the $1000 today has more value. Money's value over time is expressed by an interest rate. This chapter describes two introductory concepts involving the *time value of money:* interest and cash flow equivalence. Later in the chapter, nominal and effective interest are discussed. Finally, equations are derived for situations where interest is compounded continuously.

## COMPUTING CASH FLOWS

Installing expensive machinery in a plant obviously has economic consequences that occur over an extended period of time. If the machinery was bought on credit, the simple process of paying for it could take several years. What about the usefulness of the machinery? Certainly it was purchased because it would be a beneficial addition to the plant. These favorable consequences may last as long as the equipment performs its useful function. In these circumstances, we describe each alternative as cash **receipts** or **disbursements** at different points in **time.** Since earlier cash flows are more valuable than later cash flows, we cannot just add them together. Instead, each alternative is resolved into a set of **cash flows,** as in Examples 3-1 and 3-2.

### EXAMPLE 3–1

The manager has decided to purchase a new $30,000 mixing machine. The machine may be paid for in one of two ways:

1. Pay the full price now *minus* a 3% discount.
2. Pay $5000 now; at the end of one year, pay $8000; at the end of each of the next four years, pay $6000.

List the alternatives in the form of a table of cash flows.

### SOLUTION

The first plan represents a lump sum of $29,100 now, the second one calls for payments continuing until the end of the fifth year. Those payments total $37,000, but the $37,000 value has no clear meaning, since it has added together cash flows occurring at different times. The second alternative can only be described as a set of *cash flows*. Both alternatives are shown in the cash flow table, with disbursements given negative signs.

| End of Year | Pay in Full Now | Pay Over 5 Years |
|---|---|---|
| 0 (now) | −$29,100 | −$5000 |
| 1 | 0 | −8000 |
| 2 | 0 | −6000 |
| 3 | 0 | −6000 |
| 4 | 0 | −6000 |
| 5 | 0 | −6000 |

This can also be presented as a cash flow diagram:

---

## EXAMPLE 3–2

A man borrowed $1000 from a bank at 8% interest. He agreed to repay the loan in two end-of-year payments. At the end of each year, he will repay half of the $1000 principal amount plus the interest that is due. Compute the borrower's cash flow.

### SOLUTION

In engineering economic analysis, we normally refer to the beginning of the first year as "time 0." At this point the man receives $1000 from the bank. (A positive sign represents a receipt of money and a negative sign, a disbursement.) The time 0 cash flow is +$1000.

At the end of the first year, the man pays 8% interest for the use of $1000 for that year. The interest is $0.08 \times \$1000 = \$80$. In addition, he repays half the $1000 loan, or $500. Therefore, the end-of-year-1 cash flow is −$580.

At the end of the second year, the payment is 8% for the use of the balance of the principal ($500) for the one-year period, or $0.08 \times 500 = \$40$. The $500 principal is also repaid for a total end-of-year-2 cash flow of −$540. The cash flow is:

| End of Year | Cash Flow |
|---|---|
| 0 (now) | +$1000 |
| 1 | −580 |
| 2 | −540 |

In this chapter, we will demonstrate techniques for comparing the value of money at different dates, which is the foundation of engineering economic analysis. We must be able to compare, for example, a low-cost pump with a higher-cost pump. If there were no other consequences, we would obviously prefer the low-cost one. But what if the higher-cost pump is more durable? Then we must consider whether to spend more money now to postpone the future cost of a replacement pump. This chapter will provide the methods for comparing the alternatives to determine which pump is preferred.

## TIME VALUE OF MONEY

We often find that the monetary consequences of any alternative occur over a substantial period of time—say, a year or more. When monetary consequences occur in a short period of time, we simply add up the various sums of money and obtain a net result. But can we treat money this way when the time span is greater?

Which would you prefer, $100 cash today or the assurance of receiving $100 a year from now? You might decide you would prefer the $100 now because that is one way to be certain of receiving it. But suppose you were convinced that you would receive the $100 one year hence. Now what would be your answer? A little thought should convince you that it *still* would be more desirable to receive the $100 now. If you had the money now, rather than a year hence, you would have the use of it for an extra year. And if you had no current use for $100, you could let someone else use it.

Money is quite a valuable asset—so valuable that people are willing to pay to have money available for their use. Money can be rented in roughly the same way one rents an apartment; only with money, the charge is called **interest** instead of rent. The importance of interest is demonstrated by banks and savings institutions continuously offering to pay for the use of people's money, to pay interest.

If the current interest rate is 9% per year and you put $100 into the bank for one year, how much will you receive back at the end of the year? You will receive your original $100 together with $9 interest, for a total of $109. This example demonstrates the time preference for money: we would rather have $100 today than the assured promise of $100 one year hence; but we might well consider leaving the $100 in a bank if we knew it would be worth $109 one year hence. This is because there is a **time value of money** in the form of the willingness of banks, businesses, and people to pay interest for the use of various sums.

### Simple Interest

**Simple interest** is interest that is computed only on the original sum, not on accrued interest. Thus if you were to loan a present sum of money $P$ to someone at a simple annual interest rate $i$ (stated as a decimal) for a period of $n$ years, the amount of interest you would receive from the loan would be

$$\text{Total interest earned} = P \times i \times n = Pin \tag{3-1}$$

At the end of $n$ years the amount of money due you, $F$, would equal the amount of the loan $P$ plus the total interest earned. That is, the amount of money due at the end of the loan would be

$$F = P + Pin \tag{3-2}$$

or $F = P(1 + in)$.

### EXAMPLE 3–3

You have agreed to loan a friend $5000 for 5 years at a simple interest rate of 8% per year. How much interest will you receive from the loan? How much will your friend pay you at the end of 5 years?

### SOLUTION

$$\text{Total interest earned} = Pin = (\$5000)(0.08)(5 \text{ yr}) = \$2000$$

$$\text{Amount due at end of loan} = P + Pin = 5000 + 2000 = \$7000$$

In Example 3-3 the interest earned at the end of the first year is $(5000)(0.08)(1) = \$400$, but this money is not paid to the lender until the end of the fifth year. As a result, the borrower has the use of the $400 for 4 years without paying any interest on it. This is how simple interest works, and it is easy to see why lenders seldom agree to make simple interest loans.

### Compound Interest

With simple interest, the amount earned (for invested money) or due (for borrowed money) in one period does not affect the principal for interest calculations in later periods. However, this is not how interest is normally calculated. In practice, interest is computed by the **compound interest** method. For a loan, any interest owed but not paid at the end of the year is added to the balance due. Then the next year's interest is calculated on the unpaid balance due, which includes the unpaid interest from the preceding period. In this way, compound interest can be thought of as *interest on top of interest*. This distinguishes compound interest from simple interest. In this section, the remainder of the book, and in practice you should **assume that the rate is a compound interest rate**. The very few exceptions will clearly state "simple interest."

### EXAMPLE 3–4

To highlight the difference between simple and compound interest, rework Example 3-3 using an interest rate of 8% per year compound interest. How will this change affect the amount that your friend pays you at the end of 5 years?

### SOLUTION

$$\text{Original loan amount (original principal)} = \$5000$$

$$\text{Loan term} = 5 \text{ years}$$

$$\text{Interest rate charged} = 8\% \text{ per year compound interest}$$

In the following table we calculate on a year-to-year basis the total dollar amount due at the end of each year. Notice that this amount becomes the principal upon which interest is calculated in the next year (this is the compounding effect).

| Year | Total Principal (P) on Which Interest Is Calculated in Year n | Interest (I) Owed at the End of Year n from Year n's Unpaid Total Principal | Total Amount Due at the End of Year n, New Total Principal for Year n+1 |
|------|------|------|------|
| 1 | $5000 | $5000 × 0.08 = 400 | $5000 + 400 = 5400 |
| 2 | 5400 | 5400 × 0.08 = 432 | 5400 + 432 = 5832 |
| 3 | 5832 | 5832 × 0.08 = 467 | 5832 + 467 = 6299 |
| 4 | 6299 | 6299 × 0.08 = 504 | 6299 + 504 = 6803 |
| 5 | 6803 | 6803 × 0.08 = 544 | 6803 + 544 = 7347 |

The total amount due at the end of the fifth year is $7347. This is $347 more than you received for loaning the same amount, for the same period, at simple interest. This is because of the effect of interest being earned on top of interest.

### Repaying a Debt

To better understand the mechanics of interest, let us say that $5000 is owed and is to be repaid in 5 years, together with 8% annual interest. There are a great many ways in which debts are repaid; for simplicity, we have selected four specific ways for our example. Table 3-1 tabulates the four plans.

*Plan 1 (Constant Principal)*, like Example 3-2, repays $1/n^{th}$ of the principal each year. So in Plan 1, $1000 will be paid at the end of each year plus the interest due at the end of the year for the use of money to that point. Thus, at the end of Year 1, we will have had the use of $5000. The interest owed is 8% × $5000 = $400. The end-of-year payment is $1000 principal *plus* $400 interest, for a total payment of $1400. At the end of Year 2, another $1000 principal plus interest will be repaid on the money owed during the year. This time the amount owed has declined from $5000 to $4000 because of the Year 1 $1000 principal payment. The interest payment is 8% × $4000 = $320, making the end-of-year payment a total of $1320. As indicated in Table 3-1, the series of payments continues each year until the loan is fully repaid at the end of the Year 5.

In *Plan 2 (Interest Only)* only the interest due is paid each year, with no principal payment. Instead, the $5000 owed is repaid in a lump sum at the end of the fifth year. The end-of-year payment in each of the first 4 years of Plan 2 is 8% × $5000 = $400. The fifth year, the payment is $400 interest *plus* the $5000 principal, for a total of $5400.

*Plan 3 (Constant Payment)* calls for five equal end-of-year payments of $1252 each. In Example 4-3, we will show how the figure of $1252 is computed. By following the computations in Table 3-1, we see that a series of five payments of $1252 repays a $5000 debt in 5 years with interest at 8%.

*Plan 4 (All at Maturity)* repays the $5000 debt like Example 3-4. In this plan, no payment is made until the end of Year 5, when the loan is completely repaid. Note what happens at the end of Year 1: the interest due for the first year—8% × $5000 = $400—is not paid; instead, it is added to the debt. At the second year then, the debt has increased to $5400. The Year 2 interest is thus 8% × $5400 = $432. This amount, again unpaid, is added to the debt, increasing it further to $5832. At the end of Year 5, the total sum due has grown to $7347 and is paid at that time.

### TABLE 3-1  Four Plans for Repayment of $5000 in 5 Years with Interest at 8%

| (a) Year | (b) Amount Owed at Beginning of Year | (c) Interest Owed for That Year, 8% × (b) | (d) Total Owed at End of Year, (b) + (c) | (e) Principal Payment | (f) Total End-of-Year Payment |
|---|---|---|---|---|---|
| **Plan 1:** Constant principal payment *plus* interest due. | | | | | |
| 1 | $5000 | $ 400 | $5400 | $1000 | $1400 |
| 2 | 4000 | 320 | 4320 | 1000 | 1320 |
| 3 | 3000 | 240 | 3240 | 1000 | 1240 |
| 4 | 2000 | 160 | 2160 | 1000 | 1160 |
| 5 | 1000 | 80 | 1080 | 1000 | 1080 |
| | | $1200 | | $5000 | $6200 |
| **Plan 2:** Annual interest payment and principal payment at end of 5 years. | | | | | |
| 1 | $5000 | $ 400 | $5400 | $   0 | $ 400 |
| 2 | 5000 | 400 | 5400 | 0 | 400 |
| 3 | 5000 | 400 | 5400 | 0 | 400 |
| 4 | 5000 | 400 | 5400 | 0 | 400 |
| 5 | 5000 | 400 | 5400 | 5000 | 5400 |
| | | $2000 | | $5000 | $7000 |
| **Plan 3:** Constant annual payments. | | | | | |
| 1 | $5000 | $ 400 | $5400 | $ 852 | $1252* |
| 2 | 4148 | 331 | 4479 | 921 | 1252 |
| 3 | 3227 | 258 | 3485 | 994 | 1252 |
| 4 | 2233 | 178 | 2411 | 1074 | 1252 |
| 5 | 1159 | 93 | 1252 | 1159 | 1252 |
| | | $1260 | | $5000 | $6260 |
| **Plan 4:** All payment at end of 5 years. | | | | | |
| 1 | $5000 | $ 400 | $5400 | $   0 | $   0 |
| 2 | 5400 | 432 | 5832 | 0 | 0 |
| 3 | 5832 | 467 | 6299 | 0 | 0 |
| 4 | 6299 | 504 | 6803 | 0 | 0 |
| 5 | 6803 | 544 | 7347 | 5000 | 7347 |
| | | $2347 | | $5000 | $7347 |

*The exact value is $1252.28, which has been rounded to an even dollar amount.

Note that when the $400 interest was not paid at the end of Year 1, it was added to the debt and, in Year 2 there was interest charged on this unpaid interest. That is, the $400 of unpaid interest resulted in 8% × $400 = $32 of additional interest charge in Year 2. That $32, together with 8% × $5000 = $400 interest on the $5000 original debt, brought the total interest charge at the end of the Year 2 to $432. Charging interest on unpaid interest is called **compound interest**.

With Table 3-1 we have illustrated four different ways of accomplishing the same task, that is, to repay a debt of $5000 in 5 years with interest at 8%. Having described the alternatives, we will now use them to present the important concept of *equivalence*.

## EQUIVALENCE

When we are indifferent as to whether we have a quantity of money now or the assurance of some other sum of money in the future, or series of future sums of money, we say that the present sum of money is **equivalent** to the future sum or series of future sums.

If an industrial firm believed 8% was a reasonable interest rate, it would have no particular preference about whether it received $5000 now or was repaid by Plan 1 of Table 3-1. Thus $5000 today is equivalent to the series of five end-of-year payments. In fact, *all four repayment plans must be equivalent to each other and to $5000 now at 8% interest.*

Equivalence is an essential factor in engineering economic analysis. In Chapter 2, we saw how an alternative could be represented by a cash flow table. How might two alternatives with different cash flows be compared? For example, consider the cash flows for Plans 1 and 2 from Table 3-1:

| Year | Plan 1 | Plan 2 |
|------|--------|--------|
| 1 | −$1400 | −$400 |
| 2 | −1320 | −400 |
| 3 | −1240 | −400 |
| 4 | −1160 | −400 |
| 5 | −1080 | −5400 |
| | −$6200 | −$7000 |

If you were given your choice between the two alternatives, which one would you choose? Obviously the two plans have cash flows that are different, and you cannot compare the −$6200 and −$7000. Plan 1 requires that there be larger payments in the first 4 years, but the total payments are smaller. To make a decision, we must use the **technique of equivalence**.

We can determine an **equivalent value** at some point in time for Plan 1 and a **comparable equivalent value** for Plan 2, based on a selected interest rate. Then we can judge the relative attractiveness of the two alternatives, not from their cash flows, but from comparable equivalent values. Since Plan 1, like Plan 2, repays a *present* sum of $5000 with interest at 8%, the plans are equivalent to $5000 *now;* therefore, the alternatives are equally attractive at an interest rate of 8%. This cannot be deduced from the given cash flows alone. It is necessary to learn this by determining the equivalent values for each alternative at some point in time, which in this case is "the present."

### Difference in Repayment Plans

The four plans computed in Table 3-1 are equivalent in nature but different in structure. Table 3-2 repeats the end-of-year payment schedules from Table 3-1 and also graphs each plan to show the debt still owed at any point in time. Since $5000 was borrowed at the beginning of the first year, all the graphs begin at that point. We see, however, that the four plans result in quite different amounts of money owed at any other point in time. In Plans 1 and 3, the money owed declines as time passes. With Plan 2 the debt remains constant, while Plan 4 increases the debt until the end of the fifth year. These graphs show an important difference among the repayment plans—the areas under the curves differ greatly. Since the axes are *Money Owed* and *Time,* the area is their product: Money owed × Time, in years.

In the discussion of the time value of money, we saw that the use of money over a time period was valuable, that people are willing to pay interest to have the use of money for periods of time. When people borrow money, they are acquiring the use of money as represented by the area under the curve for Money owed versus Time. It follows that, at a given interest rate, the amount of interest to be paid will be proportional to the area under the curve. Since in each case the $5000 loan is repaid, the interest for each plan is the total *minus* the $5000 principal:

| Plan | Total Interest Paid |
|------|---------------------|
| 1 | $1200 |
| 2 | 2000 |
| 3 | 1260 |
| 4 | 2347 |

We can use Table 3-2 and the data from Table 3-1, to compute the area under each of the four curves, that is, the area bounded by the abscissa, the ordinate, and the curve itself. We multiply the ordinate (Money owed) by the abscissa (1 year) for each of the five years, then *add:*

$$\text{Shaded area} = (\text{Money owed in Year 1})(1 \text{ year})$$

$$+ (\text{Money owed in Year 2})(1 \text{ year})$$

$$+ \cdots$$

$$+ (\text{Money owed in Year 5})(1 \text{ year})$$

or

$$\text{Shaded area } [(\text{Money owed})(\text{Time})] = \textbf{dollar-years}$$

The dollar-years for the four plans would be as follows:

| | Dollar-Years | | | |
| | Plan 1 | Plan 2 | Plan 3 | Plan 4 |
|---|---|---|---|---|
| (Money owed in Year 1)(1 year) | $ 5,000 | $ 5,000 | $ 5,000 | $ 5,000 |
| (Money owed in Year 2)(1 year) | 4,000 | 5,000 | 4,148 | 5,400 |
| (Money owed in Year 3)(1 year) | 3,000 | 5,000 | 3,227 | 5,832 |
| (Money owed in Year 4)(1 year) | 2,000 | 5,000 | 2,233 | 6,299 |
| (Money owed in Year 5)(1 year) | 1,000 | 5,000 | 1,159 | 6,803 |
| Total dollar-years | $15,000 | $25,000 | $15,767 | $29,334 |

**TABLE 3-2 End-of-Year Payment Schedules and Their Graphs**

**Plan 1 (Constant Principal):** At end of each year pay $1000 principal *plus* interest due.

| Year | End-of-Year Payment |
|------|---------------------|
| 1 | $1400 |
| 2 | 1320 |
| 3 | 1240 |
| 4 | 1160 |
| 5 | 1080 |
| | ——— |
| | $6200 |

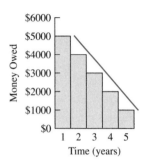

**Plan 2 (Interest Only):** Pay interest due at end of each year and principal at end of 5 years.

| Year | End-of-Year Payment |
|------|---------------------|
| 1 | $ 400 |
| 2 | 400 |
| 3 | 400 |
| 4 | 400 |
| 5 | 5400 |
| | ——— |
| | $7000 |

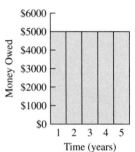

**Plan 3 (Constant Payment):** Pay in five equal end-of-year payments.

| Year | End-of-Year Payment |
|------|---------------------|
| 1 | $1252 |
| 2 | 1252 |
| 3 | 1252 |
| 4 | 1252 |
| 5 | 1252 |
| | ——— |
| | $6260 |

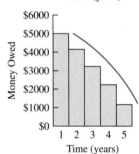

**Plan 4 (All at Maturity):** Pay principal and interest in one payment at end of 5 years.

| Year | End-of-Year Payment |
|------|---------------------|
| 1 | $ 0 |
| 2 | 0 |
| 3 | 0 |
| 4 | 0 |
| 5 | 7347 |
| | ——— |
| | $7347 |

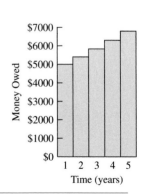

With the area under each curve computed in dollar-years, the ratio of total interest paid to area under the curve may be obtained:

| Plan | Total Interest Paid | Area Under Curve (dollar-years) | Ratio of Total Interest Paid to Area Under Curve |
|------|------|------|------|
| 1 | $1200 | 15,000 | 0.08 |
| 2 | 2000 | 25,000 | 0.08 |
| 3 | 1260 | 15,767 | 0.08 |
| 4 | 2347 | 29,334 | 0.08 |

We see that the ratio of total interest paid to the area under the curve is constant and equal to 8%. Stated another way, the total interest paid equals the interest rate *times* the area under the curve.

From our calculations, we more easily see why the repayment plans require the payment of different total sums of money, yet are actually equivalent. The key factor is that the four repayment plans provide the borrower with different quantities of dollar-years. Since dollar-years times interest rate equals the interest charge, the four plans result in different total interest charges.

## Equivalence Is Dependent on Interest Rate

In the example of Plans 1 to 4, all calculations were made at an 8% interest rate. At this interest rate, it has been shown that all four plans are equivalent to a present sum of $5000. But what would happen if we were to change the problem by changing the interest rate?

If the interest rate were increased to 9%, we know that the interest payment for each plan would increase, and the calculated repayment schedules (Table 3-1, column **f**) could no longer repay the $5000 debt with the higher interest. Instead, each plan would repay a sum *less* than the principal of $5000, because more money would have to be used to repay the higher interest rate. By some calculations (to be explained later in this chapter and in Chapter 4), the equivalent present sum that each plan will repay at 9% interest is:

| Plan | Repay a Present Sum of |
|------|------|
| 1 | $4877 |
| 2 | 4806 |
| 3 | 4870 |
| 4 | 4775 |

At the higher 9% interest, each of the repayment plans of Table 3-1 repays a present sum less than $5000. But they do not repay the *same* present sum. Plan 1 would repay $4877 with 9% interest, while Plan 2 would repay $4806. Thus, with interest at 9%, Plans 1 and 2 are no longer equivalent, for they will not repay the same present sum. The two series of payments (Plan 1 and Plan 2) were equivalent at 8%, but not at 9%. This leads to the conclusion **that equivalence is dependent on the interest rate.** Changing the interest rate destroys the equivalence between two series of payments.

Could we create revised repayment schemes that would be equivalent to $5000 now with interest at 9%? Yes, of course we could: to revise Plan 1 of Table 3-1, we need to increase the total end-of-year payment in order to pay 9% interest on the outstanding debt.

| Year | Amount Owed at Beginning of Year | 9% Interest for Year | Total End-of-Year Payment ($1000 plus interest) |
|------|------|------|------|
| 1 | $5000 | $450 | $1450 |
| 2 | 4000 | 360 | 1360 |
| 3 | 3000 | 270 | 1270 |
| 4 | 2000 | 180 | 1180 |
| 5 | 1000 | 90 | 1090 |

Plan 2 of Table 3-1 is revised for 9% interest by increasing the first four payments to 9% × $5000 = $450 and the final payment to $5450. Two plans that repay $5000 in 5 years with interest at 9% are:

| Revised Year | End-of-Year Payments Plan 1 | Plan 2 |
|------|------|------|
| 1 | $1450 | $ 450 |
| 2 | 1360 | 450 |
| 3 | 1270 | 450 |
| 4 | 1180 | 450 |
| 5 | 1090 | 5450 |

We have determined that Revised Plan 1 is equivalent to a present sum of $5000 and Revised Plan 2 is equivalent to $5000 now; it follows that at 9% interest, Revised Plan 1 is equivalent to Revised Plan 2.

Thus far we have discussed computing equivalent present sums for a cash flow. But the technique of equivalence is not limited to a present computation. Instead, we could compute the equivalent sum for a cash flow at any point in time. We could compare alternatives in "Equivalent Year 10" dollars rather than "now" (Year 0) dollars. Furthermore, the equivalence need not be a single sum; it could be a series of payments or receipts. In Plan 3 of Table 3-1, the series of equal payments was equivalent to $5000 now. But the equivalency works both ways. Suppose we ask, What is the equivalent equal annual payment continuing for 5 years, given a present sum of $5000 and interest at 8%? The answer is $1252.

## SINGLE PAYMENT COMPOUND INTEREST FORMULAS

To facilitate equivalence computations, a series of **interest formulas** will be derived. We use the following notation:

$i = $ *interest rate per interest period;* in the equations the interest rate is stated as a decimal (that is, 9% interest is 0.09)

$n = $ *number of interest periods*

$P = $ *a present sum of money*

$F = $ *a future sum of money* at the end of the $n^{th}$ interest period, which is equivalent to $P$ with interest rate $i$

Suppose a present sum of money $P$ is invested for one year[1] at interest rate $i$. At the end of the year, we should receive back our initial investment $P$, together with interest equal to $iP$, or a total amount $P + iP$. Factoring $P$, the sum at the end of one year is $P(1 + i)$.

Let us assume that, instead of removing our investment at the end of one year, we agree to let it remain for another year. How much would our investment be worth at the end of Year 2? The end-of-first-year sum $P(1 + i)$ will draw interest in the second year of $iP(1 + i)$. This means that at the end of Year 2 the total investment will be

$$P(1 + i) + i[P(1 + i)]$$

This may be rearranged by factoring out $P(1 + i)$, which gives

$$P(1 + i)(1 + i) \quad \text{or} \quad P(1 + i)^2$$

If the process is continued for Year 3, the end-of-the-third-year total amount will be $P(1 + i)^3$; at the end of $n$ years, we will have $P(1 + i)^n$. The progression looks like:

| Year | Amount at Beginning of Interest Period | + Interest for Period | = Amount at End of Interest Period |
|------|------------------------------------------|------------------------|-------------------------------------|
| 1 | $P$ | $+iP$ | $= P(1 + i)$ |
| 2 | $P(1 + i)$ | $+iP(1 + i)$ | $= P(1 + i)^2$ |
| 3 | $P(1 + i)^2$ | $+iP(1 + i)^2$ | $= P(1 + i)^3$ |
| $n$ | $P(1 + i)^{n-1}$ | $+iP(1 + i)^{n-1}$ | $= P(1 + i)^n$ |

In other words, a present sum $P$ increases in $n$ periods to $P(1 + i)^n$. We therefore have a relationship between a present sum $P$ and its equivalent future sum, $F$.

$$\text{Future sum} = (\text{Present sum}) (1 + i)^n$$
$$F = P(1 + i)^n \tag{3-3}$$

This is the **single payment compound amount formula** and is written in functional notation as

$$F = P(F/P, i, n) \tag{3-4}$$

The notation in parentheses $(F/P, i, n)$ can be read as follows:

---

[1]A more general statement is to specify "one interest period" rather than "one year." Since it is easier to visualize one year, the derivation uses one year as the interest period.

To find a future sum $F$, given a present sum, $P$, at an interest rate $i$ per interest period, and $n$ interest periods hence.

or

Find $F$, given $P$, at $i$, over $n$.

Functional notation is designed so that the compound interest factors may be written in an equation in an algebraically correct form. In Equation 3-4, for example, the functional notation is interpreted as

$$F = P\left(\frac{F}{P}\right)$$

which is dimensionally correct. Without proceeding further, we can see that when we derive a compound interest factor to find a present sum $P$, given a future sum $F$, the factor will be $(P/F, i, n)$; so, the resulting equation would be

$$P = F(P/F, i, n)$$

which is dimensionally correct.

## EXAMPLE 3–5

If $500 were deposited in a bank savings account, how much would be in the account 3 years from now if the bank paid 6% interest compounded annually?

## SOLUTION

From the viewpoint of the person depositing the $500, the cash flows are:

The present sum $P$ is $500. The interest rate per interest period is 6%, and in 3 years there are three interest periods. The future sum $F$ is found by using Equation 3-3, where $P = \$500$, $i = 0.06$, $n = 3$, and $F$ is unknown:

$$F = P(1 + i)^n = 500(1 + 0.06)^3 = \$595.50$$

Thus if we deposit $500 now at 6% interest, there will be $595.50 in the account in three years.

ALTERNATE SOLUTION

The equation $F = P(1 + i)^n$ need not be solved. Instead, *the single payment compound amount factor,* $(F/P, i, n)$, is readily found in the tables given in Appendix C.[2] In this case the factor is

$$(F/P, 6\%, 3)$$

Knowing $n = 3$, locate the proper row in the 6% table of Appendix C. To find $F$ given $P$, look in the first column, which is headed "Single Payment, Compound Amount Factor": or $F/P$ for $n = 3$, we find 1.191.

Thus,

$$F = 500(F/P, 6\%, 3) = 500(1.191) = \$595.50$$

This can also be solved by using the time value of money (TVM) calculators described in Appendix B. Note that the $A$ variable is the focus of Chapter 4, and it is 0 here.

$$F = FV(i, n, A, P) = FV(6\%, 3, 0, -500) = \$595.51$$

Before leaving this problem, let's draw another diagram of it, this time from the bank's point of view.

This indicates that the bank receives \$500 now and must make a disbursement of $F$ at the end of 3 years. The computation, from the bank's point of view, is

$$F = 500(F/P, 6\%, 3) = 500(1.191) = \$595.50$$

This is exactly the same as what was computed from the depositor's viewpoint, since this is just the other side of the same transaction. The bank's future disbursement equals the depositor's future receipt.

If we take $F = P(1 + i)^n$ and solve for $P$, then

$$P = F\frac{1}{(1 + i)^n} = F(1 + i)^{-n}$$

This is the **single payment present worth formula.** The equation

$$P = F(1 + i)^{-n} \tag{3-5}$$

in our notation becomes

$$P = F(P/F, i, n) \tag{3-6}$$

---

[2]Appendix C contains compound interest tables for rates between 1/4 and 60%.

## EXAMPLE 3–6

If you wish to have \$800 in a savings account at the end of 4 years, and 5% interest will be paid annually, how much should you put into the savings account now?

### FORMULA SOLUTION

$$F = \$800, \quad i = 0.05, \quad n = 4, \quad P = \text{unknown}$$

$$P = F(1+i)^{-n} = 800(1+0.05)^{-4} = 800(0.8227) = \$658.16$$

Thus to have \$800 in the savings account at the end of 4 years, we must deposit \$658.16 now.

### TABLE SOLUTION

$$P = F(P/F, i, n) = \$800(P/F, 5\%, 4)$$

From the compound interest tables,

$$(P/F, 5\%, 4) = 0.8227$$

$$P = \$800(0.8227) = \$658.16$$

### TVM CALCULATOR SOLUTION

This can also be solved using the time value of money (TVM) calculators described in Appendix B. Note that the $A$ variable is the focus of Chapter 4, and it is 0 here.

$$P = PV(i, n, A, F) = PV(5\%, 4, 0, 800) = -\$658.16$$

Here the problem has an exact answer. In many situations, however, the answer is rounded off, since it can be only as accurate as the input information on which it is based.

## EXAMPLE 3–7

Suppose the bank changed the interest policy in Example 3-5 to "6% interest, compounded quarterly." For this situation, how much money would be in the account at the end of 3 years, assuming a \$500 deposit now?

**SOLUTION**

First, we must be certain to understand the meaning of *6% interest, compounded quarterly*. There are two elements:

> *6% interest:*   Unless otherwise described, it is customary to assume that the stated interest is for a one-year period. *If the stated interest is for any other period, the time frame must be clearly stated.*

> *Compounded quarterly:*   This indicates there are four interest periods per year; that is, an interest period is 3 months long.

We know that the 6% interest is an annual rate because if it were for a different period, it would have been stated. Since we are dealing with four interest periods per year, it follows that the interest rate per interest period is $1\frac{1}{2}\%$ ($=6\%/4$). For the total 3-year duration, there are 12 interest periods. Thus

$$P = \$500, \qquad i = 0.06/4 = 0.015, \qquad n = (3 \times 4) = 12, \qquad F = \text{unknown}$$
$$F = P(1+i)^n = P(F/P, i, n)$$
$$= \$500(1+0.015)^{12} = \$500(F/P, 1\frac{1}{2}\%, 12)$$
$$= \$500(1.196) = \$598$$

A \$500 deposit now would yield \$598 in 3 years.

## NOMINAL AND EFFECTIVE INTEREST

**EXAMPLE 3–8**

Consider the situation where a person deposits \$100 into a bank that pays 5% interest, compounded semiannually. How much would be in the savings account at the end of one year?

**SOLUTION**

Five percent interest, compounded semiannually, means that the bank pays $2\frac{1}{2}\%$ every 6 months. Thus, the initial amount $P = \$100$ would be credited with $0.025(100) = \$2.50$ interest at the end of 6 months, or

$$P \rightarrow P + Pi = 100 + 100(0.025) = 100 + 2.50 = \$102.50$$

The \$102.50 is left in the savings account; at the end of the second 6-month period, the interest earned is $0.025(102.50) = \$2.56$, for a total in the account at the end of one year of $102.50 + 2.56 = \$105.06$, or

$$(P + Pi) \rightarrow (P + Pi) + i(P + Pi) = P(1+i)^2 = 100(1+0.025)^2$$
$$= \$105.06$$

**Nominal interest rate** per year, $r$, is the annual interest rate without considering the effect of any compounding.

In Example 3-8, the bank pays $2^1/2\%$ interest every 6 months. The nominal interest rate per year, $r$, therefore, is $2 \times 2^1/2\% = 5\%$. The federal government mandates that lenders provide the **annual percentage rate (APR)** for any loan. This is the nominal interest rate.

**Effective interest rate** per year, $i_a$, is the *annual* interest rate taking into account the effect of any compounding during the year.

In Example 3-8, we saw that $100 left in the savings account for one year increased to $105.06, so the interest paid was $5.06. The effective interest rate per year, $i_a$, is $5.06/$100.00 = 0.0506 = 5.06\%$. The effective annual interest rate earned on savings accounts, certificates of deposit (CD), bonds, and so on is often called the annual percentage yield (**APY**). To calculate the effective annual interest rate $i_a$, we use the following variables:

$r$ = nominal interest rate per interest period (usually one year)

$i$ = effective interest rate per interest period

$m$ = number of compounding subperiods per time period

Using the method presented in Example 3-8, we can derive the equation for the effective interest rate. If a $1 deposit were made to an account that compounded interest $m$ times per year and paid a nominal interest rate per year, $r$, the *interest rate per compounding subperiod* would be $r/m$, and the total in the account at the end of one year would be

$$\$1\left(1 + \frac{r}{m}\right)^m \qquad \text{or simply} \qquad \left(1 + \frac{r}{m}\right)^m$$

If we deduct the $1 principal sum, the expression would be

$$\left(1 + \frac{r}{m}\right)^m - 1$$

Therefore,

**Effective annual interest rate** $\qquad i_a = \left(1 + \frac{r}{m}\right)^m - 1 \qquad$ (3-7)

where   $r$ = nominal interest rate per year

$m$ = number of compounding subperiods per year

Or, substituting the effective interest rate per compounding subperiod, $i = (r/m)$,

**Effective annual interest rate** $\qquad i_a = (1 + i)^m - 1 \qquad$ (3-8)

where   $i$ = effective interest rate per compounding subperiod

$m$ = number of compounding subperiods per year

Either Equation 3-7 or 3-8 may be used to compute an effective interest rate per year. This and many other texts use $i_a$ for the effective annual interest rate. The Fundamentals of Engineering (FE) exam uses $i_e$ for the effective annual interest rate in its supplied material.

One should note that $i$ was described earlier simply as the interest rate per interest period. We were describing the effective interest rate without making any fuss about it. A more precise definition, we now know, is that $i$ is the *effective* interest rate per interest period. Although it seems more complicated, we are describing the same exact situation, but with more care.

The nominal interest rate $r$ is often given for a one-year period (but it could be given for either a shorter or a longer time period). In the special case of a nominal interest rate that is given per compounding subperiod, the effective interest rate per compounding subperiod, $i_m$, equals the nominal interest rate per subperiod, $r$.

In the typical effective interest computation, there are multiple compounding subperiods ($m > 1$). The resulting effective interest rate is either the solution to the problem or an intermediate solution, which allows us to use standard compound interest factors to proceed to solve the problem.

For **continuous compounding** (which is detailed in the next section),

$$\textbf{Effective annual interest rate} \qquad i_a = e^r - 1 \qquad\qquad (3\text{-}9)$$

## EXAMPLE 3–9

If a savings bank pays $1^1/_2$% interest every 3 months, what are the nominal and effective interest rates per year?

### SOLUTION

Nominal interest rate per year $\qquad r = 4 \times 1^1/_2\% = 6\%$

Effective annual interest rate $\qquad i_a = \left(1 + \dfrac{r}{m}\right)^m - 1$

$$= \left(1 + \dfrac{0.06}{4}\right)^4 - 1 = 0.061$$

$$= 6.1\%$$

Alternately,

Effective annual interest rate $\qquad i_a = (1 + i)^m - 1$

$$= (1 + 0.015)^4 - 1 = 0.061$$

$$= 6.1\%$$

Table 3-3 tabulates the effective interest rate for a range of compounding frequencies and nominal interest rates. It should be noted that when a nominal interest rate is compounded annually, the nominal interest rate equals the effective interest rate. Also, it will

TABLE 3-3  **Nominal and Effective Interest**

| Nominal Interest Rate per Year | Effective Annual Interest Rate, $i_a$ When Nominal Rate Is Compounded | | | | |
|---|---|---|---|---|---|
| r (%) | Yearly | Semiannually | Monthly | Daily | Continuously |
| 1 | 1% | 1.0025% | 1.0046% | 1.0050% | 1.0050% |
| 2 | 2 | 2.0100 | 2.0184 | 2.0201 | 2.0201 |
| 3 | 3 | 3.0225 | 3.0416 | 3.0453 | 3.0455 |
| 4 | 4 | 4.0400 | 4.0742 | 4.0809 | 4.0811 |
| 5 | 5 | 5.0625 | 5.1162 | 5.1268 | 5.1271 |
| 6 | 6 | 6.0900 | 6.1678 | 6.1831 | 6.1837 |
| 8 | 8 | 8.1600 | 8.3000 | 8.3278 | 8.3287 |
| 10 | 10 | 10.2500 | 10.4713 | 10.5156 | 10.5171 |
| 15 | 15 | 15.5625 | 16.0755 | 16.1798 | 16.1834 |
| 25 | 25 | 26.5625 | 28.0732 | 28.3916 | 28.4025 |

be noted that increasing the frequency of compounding (for example, from monthly to continuously) has only a small impact on the effective interest rate. But if the amount of money is large, even small differences in the effective interest rate can be significant.

## EXAMPLE 3–10

A loan shark lends money on the following terms: "If I give you $50 on Monday, you owe me $60 on the following Monday."

(a) What nominal interest rate per year ($r$) is the loan shark charging?
(b) What effective interest rate per year ($i_a$) is he charging?
(c) If the loan shark started with $50 and was able to keep it, as well as all the money he received, loaned out at all times, how much money did he have at the end of one year?

### SOLUTION TO PART a

$$F = P(F/P, i, n)$$

$$60 = 50(F/P, i, 1)$$

$$(F/P, i, 1) = 1.2$$

Therefore, $i = 20\%$ per week.

Nominal interest rate per year = 52 weeks × 0.20 = 10.40 = 1040%

**SOLUTION TO PART b**

Effective annual interest rate $\quad i_a = \left(1 + \dfrac{r}{m}\right)^m - 1$

$$= \left(1 + \frac{10.40}{52}\right)^{52} - 1 = 13{,}105 - 1$$

$$= 13{,}104 = 1{,}310{,}400\%$$

Or

Effective annual interest rate $\quad i_a = (1 + i)^m - 1$

$$= (1 + 0.20)^{52} - 1 = 13{,}104$$

$$= 1{,}310{,}400\%$$

**SOLUTION TO PART c**

$$F = P(1 + i)^n = 50(1 + 0.20)^{52}$$

$$= \$655{,}200$$

With a nominal interest rate of 1040% per year and effective interest rate of 1,310,400% per year, the loan shark who started with $50 had $655,200 at the end of one year.

## CONTINUOUS COMPOUNDING

Two variables we have introduced are

$r$ = nominal interest rate per interest period

$m$ = number of compounding subperiods per time period

Since the interest period is normally one year, the definitions become

$r$ = nominal interest rate per year

$m$ = number of compounding subperiods per year

$\dfrac{r}{m}$ = interest rate per interest period

$mn$ = number of compounding subperiods in $n$ years

### Single Payment Interest Factors: Continuous Compounding

The single payment compound amount formula (Equation 3-3)

$$F = P(1 + i)^n$$

may be rewritten as

$$F = P\left(1 + \frac{r}{m}\right)^{mn}$$

If we increase $m$, the number of compounding subperiods per year, without limit, $m$ becomes very large and approaches infinity, and $r/m$ becomes very small and approaches zero.

This is the condition of **continuous compounding:** that is, the duration of the interest period decreases from some finite duration $\Delta t$ to an infinitely small duration $dt$, and the number of interest periods per year becomes infinite. In this situation of continuous compounding:

$$F = P \lim_{m \to \infty} \left(1 + \frac{r}{m}\right)^{mn} \tag{3-10}$$

An important limit in calculus is

$$\lim_{x \to 0} (1 + x)^{1/x} = 2.71828 = e \tag{3-11}$$

If we set $x = r/m$, then $mn$ may be written as $(1/x)(rn)$. As $m$ becomes infinite, $x$ becomes 0. Equation 3-10 becomes

$$F = P\left[\lim_{x \to 0} (1 + x)^{1/x}\right]^{rn}$$

Equation 3-11 tells us the quantity inside the brackets equals $e$. So returning to Equations 3-3 and 3-5, we find that

$$F = P(1 + i)^n \qquad \text{becomes} \qquad F = Pe^{rn} \tag{3-12}$$

and

$$P = F(1 + i)^{-n} \qquad \text{becomes} \qquad P = Fe^{-rn} \tag{3-13}$$

We see that for continuous compounding,

$$(1 + i) = e^r$$

or, as stated earlier,

**Effective interest rate per year** $\qquad i_a = e^r - 1 \tag{3-14}$

To find compound amount and present worth for continuous compounding and a single payment, we write:

**Compound amount** $\qquad F = P(e^{rn}) = P[F/P, r, n] \tag{3-15}$

**Present worth** $\qquad P = F(e^{-rn}) = F[P/F, r, n] \tag{3-16}$

Square brackets around the factors distinguish continuous compounding. If your calculator does not have $e^x$, use the table of $e^{rn}$ and $e^{-rn}$, provided at the end of Appendix C.

### EXAMPLE 3–11

If you were to deposit $2000 in a bank that pays 5% nominal interest, compounded continuously, how much would be in the account at the end of 2 years?

#### SOLUTION

The single payment compound amount equation for continuous compounding is

$$F = Pe^{rn}$$

where  $r =$ nominal interest rate $= 0.05$

$n =$ number of years $= 2$

$$F = 2000e^{(0.05)(2)} = 2000(1.1052) = \$2210.40$$

There would be $2210.40 in the account at the end of 2 years.

### EXAMPLE 3–12

A bank offers to sell savings certificates that will pay the purchaser $5000 at the end of 10 years but will pay nothing to the purchaser in the meantime. If interest is computed at 6%, compounded continuously, at what price is the bank selling the certificates?

#### SOLUTION

$$P = Fe^{-rn}, \qquad F = \$5000, \qquad r = 0.06, \qquad n = 10 \text{ years}$$

$$P = 5000e^{-0.06 \times 10} = 5000(0.5488) = \$2744$$

Therefore, the bank is selling the $5000 certificates for $2744.

### EXAMPLE 3–13

How long will it take for money to double at 10% nominal interest, compounded continuously?

$$F = Pe^{rn}$$

$$2 = 1e^{0.10n}$$

$$e^{0.10n} = 2$$

or

$$0.10n = \ln 2 = 0.693$$

$$n = 6.93 \text{ years}$$

It will take 6.93 years for money to double at 10% nominal interest, compounded continuously.

### EXAMPLE 3-14

If the savings bank in Example 3-10 changes its interest policy to 6% interest, compounded continuously, what are the nominal and the effective interest rates?

### SOLUTION

The nominal interest rate remains at 6% per year.

$$\text{Effective interest rate} = e^r - 1$$
$$= e^{0.06} - 1 = 0.0618$$
$$= 6.18\%$$

## SUMMARY

This chapter describes cash flow tables, the time value of money, and equivalence. The single payment compound interest formulas were derived. It is essential that these concepts and the use of the interest formulas be understood, since the remainder of this book and the practice of engineering economy are based on them.

*Time value of money:*  The continuing offer of banks to pay interest for the temporary use of other people's money is ample proof that there is a time value of money. Thus, we would always choose to receive $100 today rather than the promise of $100 to be paid at a future date.

*Equivalence:*  What sum would a person be willing to accept a year hence instead of $100 today? If a 9% interest rate is considered to be appropriate, she would require $109 a year hence. If $100 today and $109 a year from today are considered equally desirable, we say the two sums of money are equivalent. But, if on further consideration, we decided that a 12% interest rate is applicable, then $109 a year from today would no longer be equivalent to $100 today. This illustrates that equivalence is dependent on the interest rate.

### Single Payment Formulas

These formulas are for compound interest, which is used in engineering economy.

$$\textbf{Compound amount} \quad F = P(1+i)^n = P(F/P, i, n)$$
$$\textbf{Present worth} \quad P = F(1+i)^{-n} = F(P/F, i, n)$$

where  $i$ = interest rate per interest period (stated as a decimal)

$n$ = number of interest periods

$P$ = a present sum of money

$F$ = a future sum of money at the end of the $n^{th}$ interest period that is equivalent to $P$ with interest rate $i$

This chapter also defined simple interest, where interest does not carry over and become part of the principal in subsequent periods. Unless otherwise specified, all interest rates in this text are compound rates.

### Single Payment Formulas: Continuous Compounding at Nominal Rate *r* per Period

Compound amount:
$$F = P(e^{rn}) = P[F/P, r, n]$$

Present worth:
$$P = F(e^{-rn}) = F[P/F, r, n]$$

Note that square brackets around the factors are used to distinguish continuous compounding.

## PROBLEMS

### Equivalence

**3-1** Explain the difference between simple and compound interest. Which is more common?

**3-2** A woman borrowed $2000 and agreed to repay it at the end of 3 years, together with 10% simple interest per year. How much will she pay 3 years hence?

**3-3** A $5000 loan was to be repaid with 8% simple annual interest. A total of $5350 was paid. How long had the loan been outstanding?

*Problems 3-4 to 3-6 contributed by Hamed Kashani, Saeid Sadri, and Baabak Ashuri, Georgia Institute of Technology*

**3-4** At an interest rate of 10% per year, $100,000 today is equivalent to how much a year from now?

**3-5** A company invested $450,000 ten years ago in a new technology that is now worth $1,000,000. What rate of interest did the company earn on a simple interest basis?

**3-6** How long will it take for an investment to double at a 4% per year simple interest rate?

**3-7** In your own words explain the *time value of money*. From your own life (either now or in a situation that might occur in your future), provide an example in which the time value of money would be important.

**3-8** Which is more valuable, $20,000 received now or $5000 per year for 4 years? Why?

**3-9** Magdalen, Miriam, and Mary June were asked to consider two different cash flows: $500 that they could receive today and $1000 that would be received 3 years from today. Magdalen wanted the $500 dollars today, Miriam chose to collect $1000 in 3 years, and Mary June was indifferent between these two options. Can you offer an explanation of the choice made by each woman?

*Problems 3-10 to 3-13 contributed by Hamed Kashani, Saeid Sadri, and Baabak Ashuri, Georgia Institute of Technology*

**3-10** How long will it take for an investment to double at 4% per year compounding annually?

**3-11** A firm has borrowed $5,000,000 for 5 years at 10% per year compound interest. The firm will make no payments until the loan is due, when it will pay off the interest and principal in one lump sum. What is the total payment?

**3-12** Assume that you save 1 penny a day for 50 years, that you deposit it in the bank at the end of each month, and that there are 30.5 days per month (you save 30.5 cents each month). How much do you have after 50 years, if:

(*a*) The bank does not pay any interest.

(*b*) The bank pays 2% per month interest.

**3-13** A manufacturing company made an investment 10 years ago that is now worth $1,500,000. How much was the initial investment:

(*a*) at a simple interest rate of 10% per year?

(*b*) at an interest rate of 10% per year compounding annually?

### Single Payment Factors

**3-14** Solve the diagram for the unknown $Q$ assuming a 10% interest rate.

$$n = 4$$

(*Answer:* $Q = \$136.60$)

**3-15**  A man borrowed $750 from a bank. He agreed to repay the sum at the end of 3 years, together with the interest at 8% per year. How much will he owe the bank at the end of 3 years?
(*Answer:* $945)

**3-16**  We know that a certain piece of equipment will cost $150,000 in 5 years. How much must be deposited today using 10% interest to pay for it?

**3-17**  Alvin's Uncle Arnold gave him $16,000 from selling the old family farm. Alvin wants to start college and have $12,000 available to buy a used car when he graduates in 4 years. Alvin wants to buy a new computer, software, and peripherals now, and he earns 3% in his savings account. How much can he spend on the computer now and still have enough to grow to the $12,000 he needs when he graduates?
*Contributed by Paul R. McCright, University of South Florida*

**3-18**  Suppose that $2000 is deposited in an account that earns 6% interest. How much is in the account

(*a*)  after 5 years?
(*b*)  after 10 years?
(*c*)  after 20 years?
(*d*)  after 50 years?
(*e*)  after 100 years?

**3-19**  An inheritance will be $20,000. The interest rate for the time value of money is 7%. How much is the inheritance worth now, if it will be received

(*a*)  in 5 years?
(*b*)  in 10 years?
(*c*)  in 20 years?
(*d*)  in 50 years?

**3-20**  Rita borrows $5000 from her parents. She repays them $6000. What is the interest rate if she pays the $6000 at the end of

(*a*)  Year 2?
(*b*)  Year 3?
(*c*)  Year 5?
(*d*)  Year 10?

**3-21**  How long would it take to double your money if you invest it at 4% simple interest? At 4% compound interest?

**3-22**  A savings account earns 8% interest. If $1000 is invested, how many years is it until each of the following amounts is on deposit?

(*a*)  $1360
(*b*)  $2720
(*c*)  $4316
(*d*)  $6848

**3-23**  Mohammed can get a certificate of deposit (CD) at his bank that will pay 3.7% annually for 10 years. If he places $5530 in this CD, how much will it be worth when it matures?

(*a*)  Use the formula.
(*b*)  Use the interest tables and interpolation.
(*c*)  Use a TVM calculator.

*Contributed by Paul R. McCright, University of South Florida*

**3-24**  Ace Manufacturing is building a new facility that will cost $44M. Ace will borrow $40M from First National Bank and pay the remainder immediately as a down payment. Ace will pay 7% interest but will make no payments for 4 years, at which time the entire amount will be due. How large will Ace's payment be?
*Contributed by Paul R. McCright, University of South Florida*

**3-25**  How much must you invest now at 7.9% interest to accumulate $175,000 in 63 years?

**3-26**  (*a*)  If $100 at Time "0" will be worth $110 a year later and was $90 a year ago, compute the interest rate for the past year and the interest rate next year.

(*b*)  Assume that $90 invested a year ago will return $110 a year from now. What is the annual interest rate in this situation?

**3-27**  In 1995 an anonymous private collector purchased a painting by Picasso entitled *Angel Fernandez de Soto* for $29,152,000. The picture depicts Picasso's friend de Soto seated in a Barcelona cafe drinking absinthe. The painting was done in 1903 and was valued then at $600. If the painting was owned by the same family until its sale in 1995, what rate of return did they receive on the $600 investment?

**3-28**  The following series of payments will repay a present sum of $5000 at an 8% interest rate. Use single payment factors to find the present sum that is equivalent to this series of payments at a 10% interest rate.

| Year | End-of-Year Payment |
|------|---------|
| 1 | $1400 |
| 2 | 1320 |
| 3 | 1240 |
| 4 | 1160 |
| 5 | 1080 |

**3-29** What sum of money now is equivalent to $8250 two years later, if interest is 4% per 6-month period? (*Answer:* $7052)

**3-30** A sum of money invested at 2% per 6-month period (semiannually), will double in amount in approximately how many years? (*Answer:* $17\frac{1}{2}$ years)

**3-31** One thousand dollars is borrowed for one year at an interest rate of 1% per month. If the same sum of money could be borrowed for the same period at an interest rate of 12% per year, how much could be saved in interest charges?

**3-32** Sally Stanford is buying a car that costs $12,000. She will pay $2000 immediately and the remaining $10,000 in four annual end-of-year principal payments of $2500 each. In addition to the $2500, she must pay 15% interest on the unpaid balance of the loan each year. Prepare a cash flow table to represent this situation.

**3-33** The local bank offers to pay 5% interest on savings deposits. In a nearby town, the bank pays 1.25% per 3-month period (quarterly). A man who has $3000 to put in a savings account wonders whether the higher interest paid in the nearby town justifies driving there to make the deposit. Assuming he will leave all money in the account for 2 years, how much additional interest would he obtain from the out-of-town bank over the local bank?

**3-34** The following cash flows are equivalent in value if the interest rate is $i$. Which one is more valuable if the interest rate is $2i$?

(i) 0———1———2    (ii) 0———1———2———3

**3-35** The tabulated factors stop at $n = 100$. How can they be used to calculate $(P/F, i, 150)$? $(P/F, i, 200)$?

**3-36** In 1990 Mrs. John Hay Whitney sold her painting by Renoir, *Au Moulin de la Galette,* depicting an open-air Parisian dance hall, for $71 million. The buyer also had to pay the auction house commission of 10%, or a total of $78.1 million. The Whitney family had purchased the painting in 1929 for $165,000.

(*a*) What rate of return did Mrs. Whitney receive on the investment?

(*b*) Was the rate of return really as high as you computed in (*a*)? Explain.

**3-37** A sum of money $Q$ will be received 6 years from now. At 5% annual interest, the present worth of $Q$ is $60. At the same interest rate, what would be the value of $Q$ in 10 years?

### Nominal and Effective Interest Rates

**3-38** A thousand dollars is invested for 7 months at an interest rate of 1% per month. What is the nominal interest rate? What is the effective interest rate? (*Answers:* 12%; 12.68%)

**3-39** A firm charges its credit customers $1\frac{3}{4}\%$ interest per month. What is the effective interest rate?

**3-40** If the nominal annual interest rate is 12% compounded quarterly, what is the effective annual interest rate?

**3-41** A local store charges $1\frac{1}{2}\%$ each month on the unpaid balance for its charge account. What nominal annual interest rate is being charged? What is the effective interest rate?

**3-42** What interest rate, compounded quarterly, is equivalent to a 9.31% effective interest rate?

**3-43** A bank advertises it pays 7% annual interest, compounded daily, on savings accounts, provided the money is left in the account for 4 years. What is the effective annual interest rate?

**3-44** At the Central Furniture Company, customers who buy on credit pay an effective annual interest rate of 16.1%, based on monthly compounding. What is the nominal annual interest rate that they pay?

**3-45** A student bought a $75 used guitar and agreed to pay for it with a single $85 payment at the end of 6 months. Assuming semiannual (every 6 months) compounding, what is the nominal annual interest rate? What is the effective interest rate?

**3-46** A bank is offering to sell 6-month certificates of deposit for $9500. At the end of 6 months, the bank will pay $10,000 to the certificate owner. Based on a 6-month interest period, compute the nominal annual interest rate and the effective annual interest rate.

**3-47** Mr. Sansome withdrew $1000 from a savings account and invested it in common stock. At the end of 5 years, he sold the stock and received a check for $1307. If Mr. Sansome had left his $1000 in the savings account, he would have received an interest rate of 5%, compounded quarterly. Mr. Sansome would like to compute a comparable interest rate on his common stock investment. Based on quarterly compounding, what nominal annual interest rate did Mr. Sansome receive on his investment in stock? What effective annual interest rate did he receive?

**3-48** The treasurer of a firm noted that many invoices were received with the following terms of payment: "2%— 10 days, net 30 days." Thus, if he were to pay the bill within 10 days of its date, he could deduct 2%. On the other hand, if he did not promptly pay the bill, the full amount would be due 30 days from the date of the invoice. Assuming a 20-day compounding period, the 2% deduction for prompt payment is equivalent to what effective annual interest rate?

**3-49** Jim Duggan made an investment of $10,000 in a savings account 10 years ago. This account paid interest of $5\frac{1}{2}\%$ for the first 4 years and $6\frac{1}{2}\%$ interest for the remaining 6 years. The interest charges were compounded quarterly. How much is this investment worth now?

**3-50** First Bank is sending alumni of universities an invitation to obtain a credit card, with the name of their university written on it, for a nominal 9.9% interest per year after 6 months of 0% interest. These interest rates apply to the outstanding debt if not paid by a specified date each month, and hence interest is compounded monthly. If you fail to make the minimum payment in any month, your interest rate could increase (without notice) to a nominal 19.99% per year. Calculate the effective annual interest rates the credit company is charging in both cases
*Contributed by D. P. Loucks, Cornell University*

## Continuous Compounding

**3-51** Select the best of the following five alternatives. Assume the investment is for a period of 4 years and $P = \$10,000$.

    *A.* 11.98% interest rate compounded continuously

    *B.* 12.00% interest rate compounded daily

    *C.* 12.01% interest rate compounded monthly

    *D.* 12.02% interest rate compounded quarterly

    *E.* 12.03% interest rate compounded yearly

**3-52** Traffic at a certain intersection is 2000 cars per day. A consultant has told the city that traffic is expected to grow at a continuous rate of 5% per year for the next 4 years. How much traffic will be expected at the end of 2 years?

**3-53** A bank pays 10% nominal annual interest on special three-year certificates. What is the effective annual interest rate if interest is compounded

    (*a*) Every three months?

    (*b*) Daily?

    (*c*) Continuously?

**3-54** A department store charges $1\frac{3}{4}\%$ interest per month, compounded continuously, on its customer's charge accounts. What is the nominal annual interest rate? What is the effective interest rate? (*Answers:* 21%; 23.4%)

**3-55** If you want a 12% rate of return, continuously compounded, on a project that will yield $6000 at the end of $2\frac{1}{2}$ years, how much must you be willing to invest now? (*Answer:* $4444.80)

**3-56** Bank North advertises, "We pay 6.50%, compounded daily." Bank South says, "We pay 6.50%, compounded continuously." If you deposit $10,000 with Bank South for one year, how much additional interest will you receive?

**3-57** Bart wishes to tour the country with his friends. To do this, he is saving money for a bus.

    (*a*) How much money must Bart deposit in a savings account paying 8% nominal annual interest, compounded continuously, in order to have $8000 in $4\frac{1}{2}$ years?

    (*b*) A friend offers to repay Bart $8000 in $4\frac{1}{2}$ years if Bart gives him $5000 now. Assuming continuous compounding, what is the nominal annual interest rate of this offer?

**3-58** The I've Been Moved Corporation receives a constant flow of funds from its worldwide operations. This money (in the form of checks) is continuously deposited in many banks with the goal of earning as much interest as possible for "IBM." One billion dollars is deposited each month, and the money earns an average of $\frac{1}{2}\%$ interest per month, compounded continuously. Assume all the money remains in the accounts until the end of the month.

    (*a*) How much interest does IBM earn each month?

(b) How much interest would IBM earn each month if it held the checks and made deposits to its bank accounts just four times a month?

**3-59** A forklift truck costs $29,000. A company agrees to purchase such a truck with the understanding that it will make a single payment for the balance due in 3 years. The vendor agrees to the deal and offers two different interest schedules. The first schedule uses an annual effective interest rate of 13%. The second schedule uses 12.75% compounded continuously.

(a) Which schedule should the company accept?

(b) What would be the size of the single payment?

**3-60** How long will it take for $10,000, invested at 5% per year, compounded continuously, to triple in value?

**3-61** A friend was left $50,000 by his uncle. He has decided to put it into a savings account for the next year or so. He finds there are varying interest rates at savings institutions: $4\frac{3}{8}\%$ compounded annually, $4\frac{1}{4}\%$ compounded quarterly, and $4\frac{1}{8}\%$ compounded continuously. He wishes to select the savings institution that will give him the highest return on his money. What interest rate should he select?

**3-62** Jack deposited $500,000 into a bank for 6 months. At the end of that time, he withdrew the money and received $520,000. If the bank paid interest based on continuous compounding:

(a) What was the effective annual interest rate?

(b) What was the nominal annual interest rate?

**3-63** Ace Zenovia Bank and Trust deposits $2,567,223 of excess capital in the Federal Reserve Bank. If the Fed pays 4% interest compounded daily, how much interest will Zenovia earn by leaving the money on deposit for two years? By how much does assuming continuous compounding change the answer?
*Contributed by Paul R. McCright, University of South Florida*

## Minicases

**3-64** The U.S. recently purchased $1 billion of 30-year zero-coupon bonds from a struggling foreign nation. The bonds yield $4\frac{1}{2}\%$ per year interest. The zero-coupon bonds pay no interest during their 30-year life. Instead, at the end of 30 years, the U.S. government is to receive back its $1 billion together with interest at $4\frac{1}{2}\%$ per year.

A U.S. senator objected to the purchase, claiming that the correct interest rate for bonds like this is $5\frac{1}{4}\%$. The result, he said, was a multimillion dollar gift to the foreign country without the approval of Congress. Assuming the senator's math is correct, how much will the foreign country have saved in interest when it repays the bonds at $4\frac{1}{2}\%$ instead of $5\frac{1}{4}\%$ at the end of 30 years?

**3-65** The Apex Company sold a water softener to Marty Smith. The price of the unit was $350. Marty asked for a deferred payment plan, and a contract was written. Under the contract, the buyer could delay paying for the water softener if he purchased the coarse salt for recharging the softener from Apex. At the end of 2 years, the buyer was to pay for the unit in a lump sum, with interest at a rate of 1.5% per quarter-year. According to the contract, if the customer ceased buying salt from Apex at any time prior to 2 years, the full payment due at the end of 2 years would automatically become due.

Six months later, Marty decided to buy salt elsewhere and stopped buying from Apex, whereupon Apex asked for the full payment that was to have been due 18 months hence. Marty was unhappy about this, so Apex offered as an alternative to accept the $350 with interest at 10% per semiannual period for the 6 months that Marty had been buying salt from Apex. Which of these alternatives should Marty accept? Explain.

**3-66** The local garbage company charges $6 a month for garbage collection. It had been their practice to send out bills to their 100,000 customers at the end of each 2-month period. Thus, at the end of February it would send a bill to each customer for $12 for garbage collection during January and February.

Recently the firm changed its billing date: it now sends out the 2-month bills after one month's service has been performed. Bills for January and February, for example, are sent out at the end of January. The local newspaper points out that the firm is receiving half its money before the garbage collection. This unearned money, the newspaper says, could be temporarily invested for one month at 1% per month interest by the garbage company to earn extra income.

Compute how much extra income the garbage company could earn each year if it invests the money as described by the newspaper. (*Answer:* $36,000)

# 4

# EQUIVALENCE FOR REPEATED CASH FLOWS

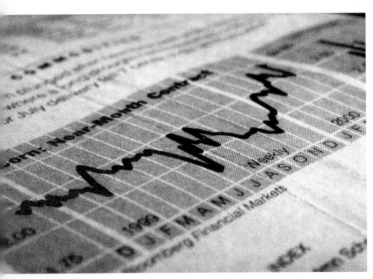

## Anne Scheiber's Bonanza

When Anne Scheiber died in 1995, aged 101, she left an estate worth more than $20 million. It all went to Yeshiva University in New York. The university's officials were grateful. But who, they asked, was Anne Scheiber?

She wasn't a mysterious heiress or a business tycoon, it turned out. She was a retired IRS auditor who had started investing in 1944, putting $5000—her life savings up to that point—into the stock market.

Scheiber's portfolio was not based on get-rich-quick companies. In fact, it contained mostly "garden variety" stocks like Coca-Cola and Exxon. Scheiber typically researched stock purchases carefully and then held onto her shares for years, rather than trading them. She also continually reinvested her dividends.

Despite her increasing wealth, Scheiber never indulged in a lavish lifestyle. Acquaintances described her as extremely frugal and reported that she lived as a near recluse in her small apartment.

What motivated Scheiber to accumulate so much money? In large part, it seems to have been a reaction to her life experiences. Scheiber worked for over 20 years as a tax auditor but failed to receive promotions. At her request, the endowment she left to Yeshiva University was used to fund scholarships and interest-free loans for women students. ▪ ▪ ▪

1. Set aside your calculator and test your intuition for a minute. What annual rate of return do you suppose it would take to turn $5000 into $20 million in 51 years? 20%, 100%?

2. Now get out the calculator and refer to an equation from Chapter 3. What was Anne Scheiber's actual average annual rate of return during the years she was investing? (Assume no added investment from her salary.)

3. How does that rate compare with the overall performance of the stock market from 1944 to 1995?

4. Anne Scheiber's endowment was given to support scholarships and loans for women. Should universities allow donors to place constraints on money they give? What ethical boundaries are appropriate?

### After Completing This Chapter...

*The student should be able to:*

- Solve problems modeled by the uniform series compound interest formulas.
- Use arithmetic and geometric gradients to solve appropriately modeled problems.
- Use *continuously compounded interest* with uniform payment series
- Use spreadsheets and financial functions to model and solve engineering economic analysis problems.

Chapter 3 presented the fundamental components of engineering economic analysis, including formulas to compute equivalent single sums of money at different points in time. Most problems we will encounter are much more complex. This chapter develops formulas for cash flows that are a uniform series or are increasing on an arithmetic or geometric gradient.

## UNIFORM SERIES COMPOUND INTEREST FORMULAS

Many times we will find uniform series of receipts or disbursements. Automobile loans, house payments, and many other loans are based on a **uniform payment series.** Future costs and benefits are often estimated to be the *same* or *uniform* every year. This is the simplest assumption; it is often sufficiently accurate; and often there is no data to support what might be a more accurate model.

Since most engineering economy problems define "a period" as one year, the uniform cash flow is an *annual* cash flow denoted by $A$ for all period lengths. More formally, $A$ is defined as

$A =$ an end-of-period cash receipt or disbursement in a uniform series, continuing for $n$ periods

Engineering economy practice and textbooks (including this one) generally assume that cash flows after Time 0 are end-of-period cash flows. Thus the equations and tabulated values of $A$ and $F$ assume end-of-period timing. Tables for beginning- or middle-of-period assumptions have been built, but they are rarely used.

The horizontal line in Figure 4-1 is a representation of time with four interest periods illustrated. Uniform payments $A$ have been placed at the end of each interest period, and there are as many $A$'s as there are interest periods $n$. (Both these conditions are specified in the definition of $A$.) Figure 4-1 uses January 1 and December 31, but other 1-year or other length periods could be used.

In Chapter 3's section on single payment formulas, we saw that a sum $P$ at one point in time would increase to a sum $F$ in $n$ periods, according to the equation

$$F = P(1+i)^n$$

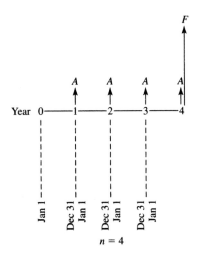

**FIGURE 4-1** The general relationship between $A$ and $F$.

We will use this relationship in our uniform series derivation.

Looking at Figure 4-1, we see that if an amount $A$ is invested at the end of each year for 4 years, the total amount $F$ at the end of 4 years will be the sum of the compound amounts of the individual investments.

In the general case for $n$ years,

$$F = A(1 + i)^{n-1} + \cdots + A(1 + i)^3 + A(1 + i)^2 + A(1 + i) + A \qquad (4\text{-}1)$$

Multiplying Equation 4-1 by $(1 + i)$, we have

$$(1 + i)F = A(1 + i)^n + \cdots + A(1 + i)^4$$
$$+ A(1 + i)^3 + A(1 + i)^2 + A(1 + i) \qquad (4\text{-}2)$$

Factoring out $A$ and subtracting Equation 4-1 gives

$$(1 + i)F = A[(1 + i)^n + \cdots + (1 + i)^4 + (1 + i)^3 + (1 + i)^2 + (1 + i)]$$
$$-F = A[(1 + i)^{n-1} + \cdots + (1 + i)^3 + (1 + i)^2 + (1 + i) + 1]$$
$$\overline{iF = A[(1 + i)^n - 1]} \qquad (4\text{-}3)$$

Solving Equation 4-3 for $F$ gives

$$F = A\left[\frac{(1 + i)^n - 1}{i}\right] = A(F/A, i\%, n) \qquad (4\text{-}4)$$

Thus we have an equation for $F$ when $A$ is known. The term inside the brackets

$$\left[\frac{(1 + i)^n - 1}{i}\right]$$

is called the **uniform series compound amount factor** and has the notation $(F/A, i, n)$.

## EXAMPLE 4-1

You deposit $500 in a credit union at the end of each year for 5 years. The credit union pays 5% interest, compounded annually. Immediately after the fifth deposit, how much can you withdraw from your account?

**SOLUTION**

The diagram on the left shows the situation from your point of view; the one on the right, from the credit union's point of view. Either way, the diagram of the five deposits and the desired computation of the future sum $F$ duplicates the situation for the uniform series compound amount formula

$$F = A\left[\frac{(1+i)^n - 1}{i}\right] = A(F/A, i\%, n)$$

where $A = \$500$, $n = 5$, $i = 0.05$, $F =$ unknown. Filling in the known variables gives

$$F = \$500(F/A, 5\%, 5) = \$500(5.526) = \$2763$$

There will be $2763 in the account following the fifth deposit.

If Equation 4-4 is solved for $A$, we have

$$A = F\left[\frac{i}{(1+i)^n - 1}\right]$$
$$= F(A/F, i\%, n) \tag{4-5}$$

where

$$\left[\frac{i}{(1+i)^n - 1}\right]$$

is called the **uniform series sinking fund**[1] factor and is written as $(A/F, i, n)$.

**EXAMPLE 4–2**

Jim Hayes wants to buy some electronic equipment for $1000. Jim has decided to save a uniform amount at the end of each month so that he will have the required $1000 at the end of one year. The local credit union pays 6% interest, compounded monthly. How much does Jim have to deposit each month?

---

[1] A *sinking fund* is a separate fund into which one makes a uniform series of money deposits ($A$) to accumulate a desired future sum ($F$) by the end of period $n$.

SOLUTION

In this example,

$$F = \$1000, \qquad n = 12, \qquad i = \frac{1}{2}\%, \qquad A = \text{unknown}$$

$$A = 1000(A/F, \frac{1}{2}\%, 12) = 1000(0.0811) = \$81.10$$

Jim would have to deposit $81.10 each month.

If we use the sinking fund formula (Equation 4-5) and substitute for $F$ the single payment compound amount formula (Equation 3-3), we obtain

$$A = F\left[\frac{i}{(1+i)^n - 1}\right] = P(1+i)^n \left[\frac{i}{(1+i)^n - 1}\right]$$

$$A = P\left[\frac{i(1+i)^n}{(1+i)^n - 1}\right] = P(A/P, i\%, n) \tag{4-6}$$

We now have an equation for determining the value of a series of end-of-period payments—or disbursements—$A$ when the present sum $P$ is known.

The portion inside the brackets

$$\left[\frac{i(1+i)^n}{(1+i)^n - 1}\right]$$

is called the **uniform series capital recovery factor** and has the notation $(A/P, i, n)$.

The name *capital recovery factor* comes from asking the question, How large does the annual return, $A$, have to be to "recover" the capital, $P$, that is invested at Time 0? In other words, find $A$ given $P$. This is illustrated in Example 4-3.

## EXAMPLE 4–3

An energy-efficient machine costs $5000 and has a life of 5 years. If the interest rate is 8%, how much must be saved every year to recover the cost of the capital invested in it?

SOLUTION

$$P = \$5000, \qquad n = 5, \qquad i = 8\%, \qquad A = \text{unknown}$$

$$A = P(A/P, 8\%, 5) = 5000(0.2505) = \$1252$$

The required annual savings to recover the capital investment is $1252.

In Example 4-3, with interest at 8%, a present sum of $5000 is equivalent to five equal end-of-period disbursements of $1252. This is another way of stating Plan 3 of Table 3-1. The method for determining the annual payment that would repay $5000 in 5 years with 8% interest has now been explained. The calculation is simply

$$A = 5000(A/P, 8\%, 5) = 5000(0.2505) = \$1252$$

If the capital recovery formula (Equation 4-6) is solved for the present sum $P$, we obtain the uniform series present worth formula

$$P = A\left[\frac{(1+i)^n - 1}{i(1+i)^n}\right] = A(P/A, i\%, n) \tag{4-7}$$

and

$$(P/A, i\%, n) = \left[\frac{(1+i)^n - 1}{i(1+i)^n}\right]$$

which is the **uniform series present worth factor.**

## EXAMPLE 4–4

An investor holds a time payment purchase contract on some machine tools. The investor will receive $140 at the end of each month for a 5-year period. The first payment is due in one month. He offers to sell you the contract for $6800 cash today. If you otherwise can make 1% per month on your money, would you accept or reject the investor's offer?

### SOLUTION

Summarizing the data in a cash flow diagram, we have

Use the uniform series present worth formula to compute the contract's present worth.

$$P = A(P/A, i, n) = 140(P/A, 1\%, 60)$$

$$= 140(44.955)$$

$$= \$6293.70$$

Since you would be paying more than $6293.70, you would receive less than the required 1% per month interest. Reject the investor's offer.

### EXAMPLE 4–5

If $6800 is paid for the time purchase contract in Example 4-4, what is the monthly rate of return?

### SOLUTION

We know $P$, $A$, and $n$, but we do not know $i$. The problem may be solved by using either the uniform series present worth formula

$$0 = -P + A(P/A, i, n)$$

$$P = A(P/A, i, n)$$

or the uniform series capital recovery formula

$$A = P(A/P, i, n)$$

Either way, we have one equation with one unknown.

$$P = \$6800, \qquad A = \$140, \qquad n = 60, \qquad i = \text{unknown}$$

$$P = A(P/A, i, n)$$

$$\$6800 = \$140(P/A, i, 60)$$

$$(P/A, i, 60) = \frac{6800}{140} = 48.571$$

We look through the compound interest tables to find the values of $(P/A, i, 60)$ that are closest to 48.571. Then we compute the rate of return $i$ by interpolation. Entering values from the tables in Appendix C, we find

| **Interest Rate** | **$(P/A, i, 60)$** |
|---|---|
| $\frac{1}{2}\%$ | 51.726 |
| $i$ | 48.571 |
| $\frac{3}{4}\%$ | 48.174 |

The rate of return, which is between $\frac{1}{2}\%$ and $\frac{3}{4}\%$, may indeed be computed by a linear interpolation. The interest formulas are not linear, so a linear interpolation will not give an exact solution. To minimize the error, the interpolation should be computed with interest rates as close to the correct answer as possible. [Since $a/b = c/d$, $a = b(c/d)$], we write

$$\text{Rate of return} \quad i = 0.5\% + a$$

$$= 0.5\% + b(c/d)$$

$$= 0.50\% + 0.25\% \left(\frac{51.726 - 48.571}{51.726 - 48.174}\right)$$

$$= 0.50\% + 0.25\% \left(\frac{3.155}{3.552}\right) = 0.50\% + 0.22\%$$

$$= 0.72\% \text{ per month}$$

The monthly rate of return on our investment would be 0.72% per month.

### Expanded Factors as Calculator Functions

Appendix B describes financial calculators and programmable scientific calculators that can be used instead of or in addition to the tables. Programmable scientific calculators can be used on the FE exam.

These time value of money (TVM) calculators make the same assumptions as the tabulated factors, and the notation is very similar. But fewer numbers have to be entered, and many calculations can be completed much more quickly. Many problems can be solved by entering four values chosen from $i$, $n$, $P$, $A$, and $F$, and solving for the fifth value.

It is much simpler to solve Example 4.5 using

$$i = \text{Rate}(n, A, P, F) = \text{Rate}(60, 140, -6800, 0) = 0.7207\%/\text{month}$$

### Cash Flows That Do Not Match Basic Patterns

## EXAMPLE 4–6

A student is borrowing $100 per year for 3 years. The loan will be repaid 2 years later at a 15% interest rate. Compute how much will be repaid. This is $F$ in the following cash flow table and diagram.

| Year | Cash Flow |
|------|-----------|
| 1 | +100 |
| 2 | +100 |
| 3 | +100 |
| 4 | 0 |
| 5 | −F |

## SOLUTION

We see that the cash flow diagram does not match the sinking fund factor diagram: $F$ occurs two periods later, rather than at the same time as the last $A$. Since the diagrams do not match, the problem is more difficult than those we've discussed so far. The approach to use in this situation is to convert the cash flow from its present form into standard forms, for which we have compound interest factors and compound interest tables.

One way to solve this problem is to consider the cash flow as a series of single payments $P$ and then to compute their sum $F$. In other words, the cash flow is broken into three parts, each one of which we can solve.

$$F = F_1 + F_2 + F_3 = 100(F/P, 15\%, 4) + 100(F/P, 15\%, 3) + 100(F/P, 15\%, 2)$$
$$= 100(1.749) + 100(1.521) + 100(1.322)$$
$$= \$459.20$$

The value of $F$ in the illustrated cash flow is \$459.20.

**ALTERNATE SOLUTION**

A second approach is to calculate an equivalent $F_3$ at the end of Period 3.

Looked at this way, we first solve for $F_3$.

$$F_3 = 100(F/A, 15\%, 3) = 100(3.472) = \$347.20$$

Now $F_3$ can be considered a present sum $P$ in the diagram

and so

$$F = F_3(F/P, 15\%, 2)$$
$$= 347.20(1.322)$$
$$= \$459.00$$

The slightly different value from the preceding computation is due to rounding in the compound interest tables.

This two-step solution can be combined into a single equation:

$$F = 100(F/A, 15\%, 3)(F/P, 15\%, 2)$$
$$= 100(3.472)(1.322)$$
$$= \$459.00$$

### ALTERNATE TVM CALCULATOR SOLUTION

First find $F_3$ as in the previous solution; then that becomes the $P$ for finding $F$.

$$F_3 = FV(i, n, A, P) = FV(15\%, 3, 100, 0) = -347.25, \text{ which is saved as a } P$$
$$F = FV(i, n, A, P) = FV(15\%, 2, 0, -347.25) = \$459.24$$

### EXAMPLE 4–7

Consider the following situation, where $P$ is deposited into a savings account and three withdrawals are made.

The diagram is not in a standard form, indicating that there will be a multiple-step solution. There are at least four different ways of computing the answer. (It is important that you understand how the four computations are made, so please study all four solutions.)

### SOLUTION 1

$$P = P_1 + P_2 + P_3$$
$$= 20(P/F, 15\%, 2) + 30(P/F, 15\%, 3) + 20(P/F, 15\%, 4)$$
$$= 20(0.7561) + 30(0.6575) + 20(0.5718)$$
$$= \$46.28$$

### SOLUTION 2

The second approach converts the three withdrawals into an equivalent value at the end of Period 4.

The relationship between $P$ and $F$ in the diagram is

$$P = F(P/F, 15\%, 4)$$

Next we compute the future sums of the three payments, as follows:

$$F = F_1 + F_2 + 20$$
$$= 20(F/P, 15\%, 2) + 30(F/P, 15\%, 1) + 20$$

Combining the two equations, we have

$$P = [F_1 + F_2 + 20](P/F, 15\%, 4)$$
$$= [20(F/P, 15\%, 2) + 30(F/P, 15\%, 1) + 20](P/F, 15\%, 4)$$
$$= [20(1.322) + 30(1.150) + 20](0.5718)$$
$$= \$46.28$$

### SOLUTION 3

The third approach finds how much would have to be deposited $(P_1)$ at $t = 1$ and then converts that into an equivalent value $(P)$ at $t = 0$.

$$P = P_1(P/F, 15\%, 1) \qquad\qquad P_1 = 20(P/A, 15\%, 3) + 10(P/F, 15\%, 2)$$

Combining, we have

$$P = [20(P/A, 15\%, 3) + 10(P/F, 15\%, 2)] \times (P/F, 15\%, 1)$$

$$= [20(2.283) + 10(0.7561)](0.8696)$$

$$= \$46.28$$

### SOLUTION 4 WITH TVM CALCULATOR

With a TVM calculator, Solution 1 is probably the easiest approach, with changes in $n$ and $F$ as needed.

$$P_1 = PV(i, n, A, F) = PV(15\%, 2, 0, 20) = -15.12$$

$$P_2 = PV(i, n, A, F) = PV(15\%, 3, 0, 30) = -19.73$$

$$P_3 = PV(i, n, A, F) = PV(15\%, 4, 0, 20) = -11.44$$

$$P = -15.12 - 19.73 - 11.44 = -\$46.28$$

## ECONOMIC EQUIVALENCE VIEWED AS A MOMENT DIAGRAM

The similarity between cash flow and free body diagrams allows an analogy that helps some students better understand economic equivalence.[2] Think of the cash flows as forces that are always perpendicular to the axis. Then the time periods become the distances along the axis.

When we are solving for unknown forces in a free body diagram, we know that a moment equation about any point will be in equilibrium. Typically moments are calculated by using the right-hand rule so that counterclockwise moments are positive, but it is also possible to define moments so that a clockwise rotation is positive. We are assuming that **clockwise rotations are positive**. This allows us to make the normal assumptions that positive forces point up and positive distances from the force to the pivot point are measured from left to right. Thus, negative forces point down, and negative distances are measured

---

[2]We thank David Elizandro and Jessica Matson of Tennessee Tech for developing, testing, and describing this approach (*The Engineering Economist*, 2007, Vol. 52, No. 2, "Taking a Moment to Teach Engineering Economy," pp. 97–116).

from the force on the right to the pivot point on the left. These assumptions are summarized in the following diagram.

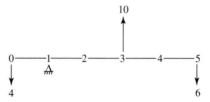

With these assumptions we can write force moment equations at equilibrium for the following diagram.

10

0————1————2————3————4————5

4                                                    6

For example, in the force moment equation about Point 1, the force at 0 is –4, and the distance from the force to the pivot point is 1. Similarly the force at 5 is –6, and the distance from the force on the right to the pivot point on the left is –4. The equilibrium equation for force moments about Point 1 is

$$0 = -4 \times 1 + 10 \times (-2) + -6 \times (-4)$$

To write the cash flow moment equation for the cash flow diagram we need:

1. A sign convention for cash flows, such that positive values point up.
2. A way to measure the moment arm for each cash flow. This moment arm must be measured as $(1 + i)^T$, where $T$ is the number of periods measured *from* the cash flow *to* the pivot point or axis of rotation.
   a. Thus the sign of the distance is moved to the exponent.
   b. For cash flows at the pivot point, $T = 0$, and $(1 + i)^0 = 1$.

We can redraw the simple example with an unknown present cash flow, $P$.

10

0————1————2————3————4————5

P                                                    6

Since $P$ is drawn as a negative cash flow, we put a minus sign in front of it when we write the cash flow moment equation. To rewrite the cash flow moment equation at Year 1, we use the distances from the diagram as the exponents for $(1 + i)^t$:

$$0 = -P \times (1 + i)^1 + 10 \times (1 + i)^{-2} + -6 \times (1 + i)^{-4}$$

If $i = 5\%$, the value of $P$ can be calculated.

$$0 = -P \times 1.05^1 + 10 \times 1.05^{-2} + -6 \times 1.05^{-4}$$
$$P = 10 \times 1.05^{-3} + -6 \times 1.05^{-5} = 8.64 - 4.70 = 3.94$$

## EXAMPLE 4–8

For the cash flow diagram in Example 4-6 (repeated here), write the cash flow moment equations at Years 0, 3, and 5. Solve for $F$ when $i = 15\%$.

### SOLUTION

With Year 0 as the pivot point, the cash flow moment equation is

$$0 = 100 \times (1+i)^{-1} + 100 \times (1+i)^{-2} + 100 \times (1+i)^{-3} - F \times (1+i)^{-5}$$

With Year 3 as the pivot point, the cash flow moment equation is

$$0 = 100 \times (1+i)^2 + 100 \times (1+i)^1 + 100 \times (1+i)^0 - F \times (1+i)^{-2}$$

With Year 5 as the pivot point, the cash flow moment equation is

$$0 = 100 \times (1+i)^4 + 100 \times (1+i)^3 + 100 \times (1+i)^2 - F \times (1+i)^0$$

In each case, the cash flow moment equation simplifies to

$$F = 100 \times (1+i)^4 + 100 \times (1+i)^3 + 100 \times (1+i)^2$$

If $i = 15\%$, then

$$F = 100 \times (1.749 + 1.521 + 1.323) = \$459.23$$

## RELATIONSHIPS BETWEEN COMPOUND INTEREST FACTORS

From the derivations, we see there are several simple relationships between the compound interest factors. They are summarized here.

### Single Payment

$$\text{Compound amount factor} = \frac{1}{\text{Present worth factor}}$$

$$(F/P, i, n) = \frac{1}{(P/F, i, n)} \tag{4-8}$$

### Uniform Series

$$\text{Capital recovery factor} = \frac{1}{\text{Present worth factor}}$$

$$(A/P, i, n) = \frac{1}{(P/A, i, n)} \tag{4-9}$$

$$\text{Compound amount factor} = \frac{1}{\text{Sinking fund factor}}$$

$$(F/A, i, n) = \frac{1}{(A/F, i, n)} \tag{4-10}$$

The uniform series present worth factor is simply the sum of the $n$ terms of the single payment present worth factor

$$(P/A, i, n) = \sum_{t=1}^{n} (P/F, i, t) \tag{4-11}$$

For example:

$$(P/A, 5\%, 4) = (P/F, 5\%, 1) + (P/F, 5\%, 2) + (P/F, 5\%, 3) + (P/F, 5\%, 4)$$
$$3.546 = 0.9524 + 0.9070 + 0.8638 + 0.8227$$

The uniform series compound amount factor equals 1 *plus* the sum of $(n-1)$ terms of the single payment compound amount factor

$$(F/A, i, n) = 1 + \sum_{t=1}^{n-1} (F/P, i, t) \tag{4-12}$$

For example,

$$(F/A, 5\%, 4) = 1 + (F/P, 5\%, 1) + (F/P, 5\%, 2) + (F/P, 5\%, 3)$$
$$4.310 = 1 + 1.050 + 1.102 + 1.158$$

The uniform series capital recovery factor equals the uniform series sinking fund factor *plus i*:

$$(A/P, i, n) = (A/F, i, n) + i \tag{4-13}$$

For example,

$$(A/P, 5\%, 4) = (A/F, 5\%, 4) + 0.05$$
$$0.2820 = 0.2320 + 0.05$$

This may be proved as follows:

$$(A/P, i, n) = (A/F, i, n) + i$$

$$\left[ \frac{i(1 + i)^n}{(1 + i)^n - 1} \right] = \left[ \frac{1}{(1 + i)^n - 1} \right] + i$$

Multiply by $(1 + i)^n - 1$ to get

$$i(1 + i)^n = i + i(1 + i)^n - i = i(1 + i)^n$$

## ARITHMETIC GRADIENT

It frequently happens that the cash flow series is not of constant amount $A$. Instead, there is a uniformly increasing series as shown:

Cash flows of this form may be resolved into two components:

Note that by resolving the problem in this manner, the first cash flow in the arithmetic gradient series becomes zero. This is done so that $G$ is the change from period to period, and because the gradient ($G$) series normally is used along with a uniform series ($A$). To find an equivalent present worth, we already have an equation for $(P/A, i, n)$, and we need to derive an equation for $(P/G, i, n)$. In this way, we will be able to write

$$P = A(P/A, i, n) + G(P/G, i, n)$$

Pause here and look at the tables for $n = 1$. What is the value of $(P/G, i, 1)$ for any $i$?

### Derivation of Arithmetic Gradient Factors

The arithmetic gradient is a series of increasing cash flows as follows:

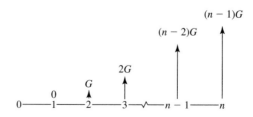

The arithmetic gradient series may be thought of as a series of individual cash flows that can individually be converted to equivalent final cash flows at the end of period $n$.

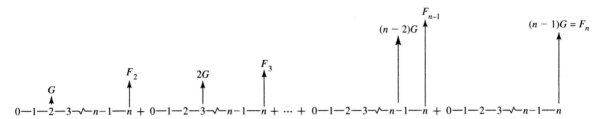

The value of $F$ for the sum of the cash flows $= F_2 + F_3 + \cdots + F_{n-1} + F_n$, or

$$F = G(1+i)^{n-2} + 2G(1+i)^{n-3} + \cdots + (n-2)(G)(1+i)^1 + (n-1)G \quad (4\text{-}14)$$

Multiply Equation 4-14 by $(1+i)$ and factor out $G$, or

$$(1+i)F = G[(1+i)^{n-1} + 2(1+i)^{n-2} + \cdots + (n-2)(1+i)^2 + (n-1)(1+i)^1] \quad (4\text{-}15)$$

Rewrite Equation 4-14 to show other terms in the series,

$$F = G[(1+i)^{n-2} + \cdots + (n-3)(1+i)^2 + (n-2)(1+i)^1 + n - 1] \quad (4\text{-}16)$$

Subtracting Equation 4-16 from Equation 4-15, we obtain

$$F + iF - F = G[(1+i)^{n-1} + (1+i)^{n-2} + \cdots + (1+i)^2 + (1+i)^1 + 1] - nG \quad (4\text{-}17)$$

In the derivation of Equation 4-4, the terms inside the brackets of Equation 4-17 were shown to equal the series compound amount factor:

$$[(1+i)^{n-1} + (1+i)^{n-2} + \cdots + (1+i)^2 + (1+i)^1 + 1] = \frac{(1+i)^n - 1}{i}$$

Thus, Equation 4-17 becomes

$$iF = G\left[\frac{(1+i)^n - 1}{i}\right] - nG$$

Rearranging and solving for $F$, we write

$$F = \frac{G}{i}\left[\frac{(1+i)^n - 1}{i} - n\right] \quad (4\text{-}18)$$

Multiplying Equation 4-18 by the single payment present worth factor gives

$$P = \frac{G}{i}\left[\frac{(1+i)^n - 1}{i} - n\right]\left[\frac{1}{(1+i)^n}\right]$$

$$= G\left[\frac{(1+i)^n - in - 1}{i^2(1+i)^n}\right]$$

$$(P/G, i, n) = \left[\frac{(1+i)^n - in - 1}{i^2(1+i)^n}\right] \quad (4\text{-}19)$$

Equation 4-19 is the **arithmetic gradient present worth factor.** Multiplying Equation 4-18 by the sinking fund factor, we have

$$A = \frac{G}{i}\left[\frac{(1+i)^n - 1}{i} - n\right]\left[\frac{i}{(1+i)^n - 1}\right] = G\left[\frac{(1+i)^n - in - 1}{i(1+i)^n - i}\right]$$

$$(A/G, i, n) = \left[\frac{(1+i)^n - in - 1}{i(1+i)^n - i}\right] = \left[\frac{1}{i} - \frac{n}{(1+i)^n - 1}\right] \tag{4-20}$$

Equation 4-20 is the **arithmetic gradient uniform series factor.**

## EXAMPLE 4–9

Andrew has purchased a new car. He wishes to set aside enough money in a bank account to pay the maintenance for the first 5 years. It has been estimated that the maintenance cost of a car is as follows:

| Year | Maintenance Cost |
|------|------------------|
| 1 | $120 |
| 2 | 150 |
| 3 | 180 |
| 4 | 210 |
| 5 | 240 |

Assume the maintenance costs occur at the end of each year and that the bank pays 5% interest. How much should Andrew deposit in the bank now?

## SOLUTION

The cash flow may be broken into its two components:

Both components represent cash flows for which compound interest factors have been derived. The first is a uniform series present worth, and the second is an arithmetic gradient series present worth:

$$P = A(P/A, 5\%, 5) + G(P/G, 5\%, 5)$$

Note that the value of $n$ in the gradient factor is 5, not 4. In deriving the gradient factor, the cash flow in the first period is zero followed by $(n - 1)$ terms containing $G$. Here there are four terms containing $G$, and it is a 5-period gradient.

$$P = 120(P/A, 5\%, 5) + 30(P/G, 5\%, 5)$$
$$= 120(4.329) + 30(8.237)$$
$$= 519 + 247$$
$$= \$766$$

Andrew should deposit $766 in the bank now.

## EXAMPLE 4-10

On a certain piece of machinery, it is estimated that the maintenance expense will be as follows:

| Year | Maintenance |
|------|-------------|
| 1 | $100 |
| 2 | 200 |
| 3 | 300 |
| 4 | 400 |

What is the equivalent uniform annual maintenance cost for the machinery if 6% interest is used?

## SOLUTION

The first cash flow in the arithmetic gradient series is zero, hence the diagram is *not* in proper form for the arithmetic gradient equation. As in Example 4-9, the cash flow must be resolved into two components:

$$A = 100 + 100(A/G, 6\%, 4) = 100 + 100(1.427) = \$242.70$$

The equivalent uniform annual maintenance cost is $242.70.

### EXAMPLE 4–11

Demand for a new product will decline as competitors enter the market. If interest is 10%, what is an equivalent uniform value?

| Year | Revenue |
|------|---------|
| 1 | $24,000 |
| 2 | 18,000 |
| 3 | 12,000 |
| 4 | 6,000 |

### SOLUTION

The projected cash flow is still a cash flow in Period 1 ($24,000) that defines the uniform series. However, now the gradient or change each year is −$6000.

$$A' = 24,000 - 6000(A/G, 10\%, 4)$$

$$= 24,000 - 6000(1.381)$$

$$= \$15,714$$

The projected equivalent uniform value is $15,714 per year.

### EXAMPLE 4–12

A car's warranty is 3 years. Upon expiration, annual maintenance starts at $150 and then climbs $25 per year until the car is sold at the end of Year 7. Use a 10% interest rate and find the present worth of these expenses.

**SOLUTION**

With the uniform series and arithmetic gradient series present worth factors, we can compute a present sum $P'$.

It is important that you closely examine the location of $P'$. Based on the way the factor was derived, there will be one zero value in the gradient series to the right of $P'$. (If this seems strange or incorrect, review the beginning of this section on arithmetic gradients.)

$$P' = A(P/A, i, n) + G(P/G, i, n)$$
$$= 150(P/A, 10\%, 4) + 25(P/G, 10\%, 4)$$
$$= 150(3.170) + 25(4.378) = 475.50 + 109.45 = 584.95$$

Then

$$P = P'(P/F, 10\%, 3) = 584.95(0.7513) = \$439.47$$

### Reality and the Assumed Uniformity of $A$, $G$, and $g$

The reality of engineering projects is that the annual revenues from selling a new product or annual benefits from using a new highway change each year as demand and traffic levels change. Most annual cash flows are not really uniform.

Thus, why do we define and start with an $A$ that is a uniform annual cost, a $G$ that is a uniform annual gradient, and a $g$ (next section) that is a uniform annual rate of increase?

1. It is easier to start with simpler models. We use cash flow tables and spreadsheets when needed for more complex models.

2. These model cash flows are the basis of the formulas and tabulated factors that are often used in engineering economic analysis.

3. Often in the real world, engineering economy is applied in a feasibility or preliminary analysis. At this stage, annual cash flows for costs and revenues are typically estimated using $A$, $G$, and/or $g$. Not enough is known about the problem for more detailed estimates.

## GEOMETRIC GRADIENT

Earlier, we saw that the arithmetic gradient is applicable where the period-by-period change in a cash receipt or payment is a *constant* amount. There are other situations where the period-by-period change is a **uniform rate, $g$.** Often geometric gradients can be traced to population levels or other levels of activity where changes over time are best modeled as a percentage of the previous year. The percentage or *rate* is constant over time, rather than the *amount* of the change as in the arithmetic gradient.

For example, if the maintenance costs for a car are $100 the first year and they increase at a uniform rate, $g$, of 10% per year, the cash flow for the first 5 years would be as follows:

| Year | | | | Cash Flow |
|---|---|---|---|---|
| 1 | $100.00$ | | $=$ | $\$100.00$ |
| 2 | $100.00 + 10\%(100.00) = 100(1 + 0.10)^1$ | | $=$ | $110.00$ |
| 3 | $110.00 + 10\%(110.00) = 100(1 + 0.10)^2$ | | $=$ | $121.00$ |
| 4 | $121.00 + 10\%(121.00) = 100(1 + 0.10)^3$ | | $=$ | $133.10$ |
| 5 | $133.10 + 10\%(133.10) = 100(1 + 0.10)^4$ | | $=$ | $146.41$ |

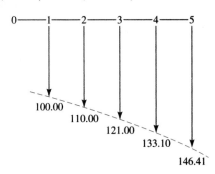

From the table, we can see that the maintenance cost in any year is

$$\$100(1 + g)^{t-1}$$

Stated in a more general form,

$$A_t = A_1(1 + g)^{t-1} \tag{4-21}$$

where    $g =$ uniform *rate* of cash flow increase/decrease from period
to period, that is, the geometric gradient
$A_1 =$ value of cash flow at Year 1 ($100 in the example)
$A_t =$ value of cash flow at any year $t$

Since the present worth $P_t$ of any cash flow $A_t$ at interest rate $i$ is

$$P_t = A_t(1+i)^{-t} \qquad (4\text{-}22)$$

we can substitute Equation 4-21 into Equation 4-22 to get

$$P_t = A_1(1+g)^{t-1}(1+i)^{-t}$$

This may be rewritten as

$$P_t = A_1(1+i)^{-1}\left(\frac{1+g}{1+i}\right)^{t-1} \qquad (4\text{-}23)$$

The present worth of the entire gradient series of cash flows may be obtained by expanding Equation 4-23:

$$P = A_1(1+i)^{-1}\sum_{t=1}^{n}\left(\frac{1+g}{1+i}\right)^{t-1} \qquad (4\text{-}24)$$

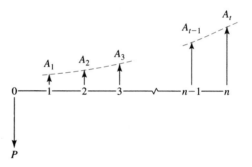

In the general case, where $i \neq g$, Equation 4-23 may be written out as follows:

$$P = A_1(1+i)^{-1} + A_1(1+i)^{-1}\left(\frac{1+g}{1+i}\right) + A_1(1+i)^{-1}\left(\frac{1+g}{1+i}\right)^{2}$$

$$+ \cdots + A_1(1+i)^{-1}\left(\frac{1+g}{1+i}\right)^{n-1} \qquad (4\text{-}25)$$

Let $a = A_1(1+i)^{-1}$ and $b = (1+g)/(1+i)$. Equation 4-25 becomes

$$P = a + ab + ab^2 + \cdots + ab^{n-1} \qquad (4\text{-}26)$$

Multiply Equation 4-26 by $b$:

$$bP = ab + ab^2 + ab^3 + \cdots + ab^{n-1} + ab^n \qquad (4\text{-}27)$$

Subtract Equation 4-27 from Equation 4-26:

$$P - bP = a - ab^n$$
$$P(1-b) = a(1-b^n)$$
$$P = \frac{a(1-b^n)}{1-b}$$

Replacing the original values for $a$ and $b$, we obtain

$$P = A_1(1+i)^{-1}\left[\frac{1-\left(\frac{1+g}{1+i}\right)^n}{1-\left(\frac{1+g}{1+i}\right)}\right] = A_1\left[\frac{1-\left(\frac{1+g}{1+i}\right)^n}{(1+i)-\left(\frac{1+g}{1+i}\right)(1+i)}\right]$$

$$= A_1\left[\frac{1-(1+g)^n(1+i)^{-n}}{1+i-1-g}\right]$$

$$P = A_1\left[\frac{1-(1+g)^n(1+i)^{-n}}{i-g}\right] \qquad (4\text{-}28)$$

where $i \neq g$.

The expression in the brackets of Equation 4-28 is the **geometric series present worth factor**

$$(P/A, g, i, n) = \left[\frac{1-(1+g)^n(1+i)^{-n}}{i-g}\right], \qquad \text{where } i \neq g \qquad (4\text{-}29)$$

In the special case of $i = g$, Equation 4-28 becomes

$$P = A_1 n(1+i)^{-1}$$

$$(P/A, g, i, n) = [n(1+i)^{-1}], \qquad \text{where } i = g \qquad (4\text{-}30)$$

## EXAMPLE 4–13

The first-year maintenance cost for a new car is estimated to be $100, and it increases at a uniform rate of 10% per year. Using an 8% interest rate, calculate the present worth (PW) of the cost of the first 5 years of maintenance.

### STEP-BY-STEP SOLUTION

| Year $n$ | | Maintenance Cost | | $(P/F, 8\%, n)$ | PW of Maintenance |
|---|---|---|---|---|---|
| 1 | 100.00 | = 100.00 | × | 0.9259 = | $ 92.59 |
| 2 | 100.00 + 10%(100.00) | = 110.00 | × | 0.8573 = | 94.30 |
| 3 | 110.00 + 10%(110.00) | = 121.00 | × | 0.7938 = | 96.05 |
| 4 | 121.00 + 10%(121.00) | = 133.10 | × | 0.7350 = | 97.83 |
| 5 | 133.10 + 10%(133.10) | = 146.41 | × | 0.6806 = | 99.65 |
| | | | | | $480.42 |

## SOLUTION USING GEOMETRIC SERIES PRESENT WORTH FACTOR

$$P = A_1 \left[ \frac{1 - (1 + g)^n (1 + i)^{-n}}{i - g} \right], \qquad \text{where } i \neq g$$

$$= 100.00 \left[ \frac{1 - (1.10)^5 (1.08)^{-5}}{-0.02} \right] = \$480.42$$

The present worth of cost of maintenance for the first 5 years is $480.42.

## COMPOUNDING PERIOD AND PAYMENT PERIOD DIFFER

When the various time periods in a problem match, we generally can solve the problem by simple calculations. Thus in Example 4-3, where we had $5000 in an account paying 8% interest, compounded annually, the five equal end-of-year withdrawals are simply computed as follows:

$$A = P(A/P, 8\%, 5) = 5000(0.2505) = \$1252$$

Consider how this simple problem becomes more difficult if the compounding period is changed so that it no longer matches the annual withdrawals.

## EXAMPLE 4-14

On January 1, a woman deposits $5000 in a credit union that pays 8% nominal annual interest, compounded quarterly. She wishes to withdraw all the money in five equal yearly sums, beginning December 31 of the first year. How much should she withdraw each year?

### SOLUTION

Since the 8% nominal annual interest rate $r$ is compounded quarterly, we know that the effective interest rate per interest period, $i$, is 2%; and there are a total of $4 \times 5 = 20$ interest periods in 5 years. For the equation $A = P(A/P, i, n)$ to be used, there must be as many periodic withdrawals as there are interest periods, $n$. In this example we have 5 withdrawals and 20 interest periods.

To solve the problem, we must adjust it so that it is in one of the standard forms for which we have compound interest factors. This means we must first compute either an equivalent $A$ for each 3-month interest period or an effective $i$ for each time period between withdrawals. Let's solve the problem both ways.

**SOLUTION 1**

**Compute an equivalent $A$ for each 3-month time period.**
If we had been required to compute the amount that could be withdrawn quarterly, the diagram would have been as follows:

$$A = P(A/P, i, n) = 5000(A/P, 2\%, 20) = 5000(0.0612) = \$306$$

Now, since we know $A$, we can construct the diagram that relates it to our desired equivalent annual withdrawal, $W$.

Looking at a one-year period,

$$W = A(F/A, i, n) = 306(F/A, 2\%, 4) = 306(4.122)$$

$$= \$1260$$

## SOLUTION 2

**Compute an effective $i$ for the time period between withdrawals.**

Between withdrawals, $W$, there are four interest periods, hence $m = 4$ compounding subperiods per year. Since the nominal interest rate per year, $r$, is 8%, we can proceed to compute the effective annual interest rate.

$$\text{Effective annual interest rate} \quad i_a = \left(1 + \frac{r}{m}\right)^m - 1 = \left(1 + \frac{0.08}{4}\right)^4 - 1$$

$$= 0.0824 = 8.24\% \text{ per year}$$

Now the problem may be redrawn as follows:

$$i = 8.24\% \text{ per year}$$
$$n = 5 \text{ years}$$

This diagram may be directly solved to determine the annual withdrawal $W$ with the capital recovery factor:

$$W = P(A/P, i, n) = 5000(A/P, 8.24\%, 5)$$

$$= P\left[\frac{i(1+i)^n}{(1+i)^n - 1}\right] = 5000\left[\frac{0.0824(1 + 0.0824)^5}{(1 + 0.0824)^5 - 1}\right]$$

$$= 5000(0.2520) = \$1260$$

The depositor should withdraw \$1260 per year.

### Uniform Payment Series: Continuous Compounding at Nominal Rate $r$ per Period

Let us now substitute the equation $i = e^r - 1$ into the equations for end-of-period compounding.

*Continuous Compounding Sinking Fund*

$$[A/F, r, n] = \frac{e^r - 1}{e^{rn} - 1} \tag{4-31}$$

*Continuous Compounding Capital Recovery*

$$[A/P, r, n] = \frac{e^{rn}(e^r - 1)}{e^{rn} - 1} \tag{4-32}$$

*Continuous Compounding Series Compound Amount*

$$[F/A, r, n] = \frac{e^{rn} - 1}{e^r - 1} \qquad (4\text{-}33)$$

*Continuous Compounding Series Present Worth*

$$[P/A, r, n] = \frac{e^{rn} - 1}{e^{rn}(e^r - 1)} \qquad (4\text{-}34)$$

## EXAMPLE 4–15

In Example 4-1, $500 per year was deposited into a credit union that paid 5% interest, compounded annually. At the end of 5 years, $2763 was in the credit union account. How much would there have been if the institution paid 5% nominal interest, compounded continuously?

### SOLUTION

$$A = \$500, \qquad r = 0.05, \qquad n = 5 \text{ years}$$

$$F = A[F/A, r, n] = A\left(\frac{e^{rn} - 1}{e^r - 1}\right) = 500\left(\frac{e^{0.05(5)} - 1}{e^{0.05} - 1}\right)$$

$$= \$2769.84$$

## EXAMPLE 4–16

In Example 4-2, Jim Hayes wished to save a uniform amount each month so he would have $1000 at the end of a year. Based on 6% nominal interest, compounded monthly, he had to deposit $81.10 per month. How much would he have to deposit if his credit union paid 6% nominal interest, compounded continuously?

### SOLUTION

The deposits are made monthly; hence, there are 12 compounding subperiods in the one-year time period.

$$F = \$1000, \qquad r = \text{nominal interest rate/interest period} = \frac{0.06}{12} = 0.005$$

$n = 12$ compounding subperiods in the one-year period of the problem

$$A = F[A/F, r, n] = F\left(\frac{e^r - 1}{e^{rn} - 1}\right) = 1000\left(\frac{e^{0.005} - 1}{e^{0.005(12)} - 1}\right)$$

$$= 1000\left(\frac{0.005013}{0.061837}\right) = \$81.07$$

He would have to deposit $81.07 per month. Note that the difference between monthly and continuous compounding is just 3 cents per month.

## SPREADSHEETS FOR ECONOMIC ANALYSIS

The reality of engineering projects is that the annual revenues from selling a new product or annual benefits from using a new highway change each year as demand and traffic levels change. Often this follows a geometric gradient, which spreadsheets can easily include; but more complex patterns are common. For example, cumulative sales and market penetration of a new product often follow a technology growth curve or *S-curve*, where the volume in each year must be individually estimated.

Similarly, if a monthly model of energy use is built, then fluctuations in air-conditioning and heating costs can be accurately modeled in a spreadsheet. As a final example, consider the construction cost for a large project such as new tunnels under the Hudson River, the new airport that Denver built, a new nuclear power plant, or a new professional sports stadium. The construction period may be 3 to 10 years long, and the costs will be spread in a project-unique way over that time span. Spreadsheets are a powerful tool to model and analyze complex projects.

Spreadsheets are used in most real-world applications of engineering economy. Common tasks include the following:

1. Constructing tables of cash flows.
2. Using annuity functions to calculate a *P*, *F*, *A*, *n*, or *i*.
3. Using a block function to find the present worth or internal rate of return for a table of cash flows.
4. Making graphs for analysis and convincing presentations.
5. Calculating "what-if" for different assumed values of problem variables.

Constructing tables of cash flows relies mainly on spreadsheet basics that are covered in Appendix A. These basics include using and naming spreadsheet variables, understanding the difference between absolute and relative addresses when copying a formula, and formatting a cell. Appendix A uses the example of the amortization schedule shown in Table 3-1, Plan 3. This amortization schedule divides each scheduled loan payment into principal and interest portions and includes the outstanding balance for each period.

Because spreadsheet functions can be found by pointing and clicking on menus, those steps are not detailed. In Excel the starting point is the $f_x$ button. Excel functions are used here; syntax differences for most other spreadsheet programs are minor.

### Spreadsheet Annuity Functions

In tables of engineering economy factors, *i* is the table header, *n* is the row, and two of *P*, *F*, *A*, and *G* define a column. The spreadsheet annuity functions list four arguments chosen from *n*, *A*, *P*, *F*, and *i*, and solve for the fifth argument. The *Type* argument is optional. If it is omitted or 0, then the *A* value is assumed to be the end-of-period cash flow. If the *A* value represents the beginning-of-period cash flow, then a value of 1 can be entered for the Type variable.

| | | |
|---|---|---|
| To find the equivalent *P* | −PV  (*i*, *n*, *A*, *F*, Type) | −PV  (*i*, *n*, *A*,    *F*, Type) |
| To find the equivalent *A* | −PMT(*i*, *n*, *P*, *F*, Type) | −PMT (*i*, *n*,    *P*, *F*, Type) |
| To find the equivalent *F* | −FV  (*i*, *n*, *A*, *P*, Type) | −FV  (*i*, *n*, *A*, *P*,    Type) |
| To find *n* | NPER (*i*, *A*, *P*, *F*, Type) | NPER (*i*,    *A*, *P*, *F*, Type) |
| To find *i* | Rate  (*n*, *A*, *P*, *F*, Type, guess) | Rate  (  *n*, *A*, *P*, *F*, Type, guess) |

The sign convention for the first three functions seems odd to some students. The PV of $200 per period for 10 periods is negative, and the PV of −$200 per period is positive. So a minus sign is inserted to find the equivalent $P$, $A$, or $F$. Without this minus sign, the calculated value is not equivalent to the four given values.

If you have been using a TVM calculator (see Appendix B), then these spreadsheet annuity functions will look very familiar. The most prominent difference is that in the spreadsheet functions $i$ is entered as a decimal or a percentage, rather than an integer value that is assumed to represent a percentage as in the TVM calculators.

## EXAMPLE 4–17

A new engineer wants to save money for down payment on a house. The initial deposit is $685, and $375 is deposited at the end of each month. The savings account earns interest at an annual nominal rate of 6% with monthly compounding. How much is on deposit after 48 months?

### SOLUTION

Because deposits are made monthly, the nominal annual interest rate of 6% must be converted to $1/2\%$ per month for the 48 months. Thus we must find $F$ if $i = 0.5\% = 0.005, n = 48, A = 375$, and $P = 685$. Note both the initial and periodic deposits are positive cash flows for the savings account. The Excel function is multiplied by −1 or −FV(0.005,48,375,685,0), and the result is $21,156.97.

## EXAMPLE 4–18

A new engineer buys a car with 0% down financing from the dealer. The cost with all taxes, registration, and license fees is $15,732. If each of the 48 monthly payments is $398, what is the monthly interest rate? What is the effective annual interest rate?

### SOLUTION

The RATE function can be used to find the monthly interest rate, given that $n = 48, A = -398, P = 15,732$, and $F = 0$. The Excel function is RATE(48,−398,15732,0) and the result is 0.822%. The effective annual interest rate is $1.00822^{12} - 1 = 10.33\%$.

### Spreadsheet Block Functions for Cash Flow Tables

Cash flows can be specified period by period as a block of values. These cash flows are analyzed by **block functions** that identify the row or column entries for which a present worth or an internal rate of return should be calculated. In Excel the two functions are NPV($i$,values) and IRR(values,guess).

| Economic Criterion | Excel Function | Values for Periods |
|---|---|---|
| Net present value | NPV($i$,values) | 1 to $n$ |
| Internal rate of return | IRR(values,guess); guess argument is optional | 0 to $n$ |

Excel's IRR function can be used to find the interest rate for a loan with irregular payments (other applications are covered in Chapter 7).

These block functions make different assumptions about the range of years included. For example, NPV($i$,values) assumes that Year 0 is **not** included, while IRR(values,guess) assumes that Year 0 is included. These functions require that a cash flow be identified for each period. You cannot leave cells blank even if the cash flow is $0. The cash flows for 1 to $n$ are assumed to be end-of-period flows. All periods are assumed to be the same length of time.

Also, the NPV function returns the present worth equivalent to the cash flows, unlike the PV annuity function, which returns the negative of the equivalent value.

For cash flows involving only constant values of $P$, $F$, and $A$ this block approach seems to be inferior to the annuity functions. However, this is a conceptually easy approach for more complicated cash flows, such as arithmetic or geometric gradients. Suppose the years (row 1) and the cash flows (row 2) are specified in columns B through E.

|   | A | B | C | D | E | F |
|---|---|---|---|---|---|---|
| 1 | Year | 0 | 1 | 2 | 3 | 4 |
| 2 | Cash flow | −25,000 | 6000 | 8000 | 10,000 | 12,000 |

NPV(.08,C2:F2)

IRR(B2:F2)

If an interest rate of 8% is assumed, then the present worth of the cash flows can be calculated as = B2+NPV(.08,C2:F2), which equals $4172.95. This is the present worth equivalent to the five cash flows, rather than the negative of the present worth equivalent returned by the PV annuity function. The internal rate of return calculated using IRR(B2:F2) is 14.5%. Notice how the NPV function does not include the Year 0 cash flow in B2 but the IRR function does.

- PW = B2+NPV(.08,C2:F2)   NPV range without Year 0
- IRR = IRR(B2:F2)   IRR range with Year 0

## Using Spreadsheets for Basic Graphing

Often we are interested in the relationship between two variables. Examples include the number and size of payments to repay a loan, the present worth of an MS degree and how long until we retire, and the interest rate and present worth for a new machine. This kind of two-variable relationship is best shown with a graph.

As we will show in Example 4-19, the goal is to place one variable on each axis of the graph and then plot the relationship. Spreadsheets automate most steps of drawing a graph, so that it is quite easy. However, there are two very similar chart types, and we must be careful to choose the *xy* chart, not the *line* chart. Both charts measure the *y* variable, but they treat the *x* variable differently. The *xy* chart measures the *x* variable; thus its *x* value is measured along the *x* axis. For the *line* chart, each *x* value is placed an equal distance along the *x* axis. Thus *x* values of 1, 2, 4, 8 would be spaced evenly, rather than doubling each distance. The *line* chart is really designed to plot *y* values for different categories, such as prices for models of cars or enrollments for different universities.

Drawing an *xy* plot with Excel is easiest if the table of data lists the *x* values before the *y* values. This convention makes it easy for Excel to specify one set of *x* values and several sets of *y* values. The block of *xy* values is selected, and then the chart tool is selected. Then the spreadsheet guides the user through the rest of the steps.

## EXAMPLE 4–19

Graph the loan payment as a function of the number of payments for a possible auto loan. Let the number of monthly payments vary between 36 and 60. The nominal annual interest rate is 12%, and the amount borrowed is $18,000.

## SOLUTION

FIGURE 4–2 Example spreadsheet graph.

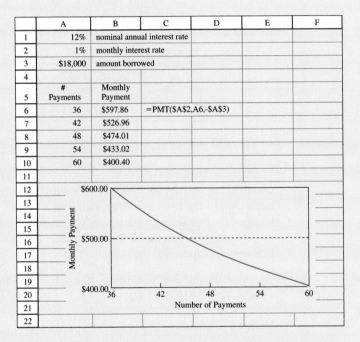

| | A | B | C | D | E | F |
|---|---|---|---|---|---|---|
| 1 | 12% | nominal annual interest rate | | | | |
| 2 | 1% | monthly interest rate | | | | |
| 3 | $18,000 | amount borrowed | | | | |
| 4 | | | | | | |
| 5 | # Payments | Monthly Payment | | | | |
| 6 | 36 | $597.86 | =PMT($A$2,A6,-$A$3) | | | |
| 7 | 42 | $526.96 | | | | |
| 8 | 48 | $474.01 | | | | |
| 9 | 54 | $433.02 | | | | |
| 10 | 60 | $400.40 | | | | |

The spreadsheet table shown in Figure 4-2 is constructed first. Cells A5:B10 are selected, and then the Chartwizard icon is selected. The first step is to select an *xy (scatter)* plot with smoothed lines and without markers as the chart type. (The other choices are no lines, straight lines, and adding markers for the data points.) The second step shows us the graph and allows the option of changing the data cells selected. The third step is for chart options. Here we add titles for the two axes and turn off showing the legend. (Since we have only one line in our graph, the legend is not needed. Deleting it leaves more room for the graph.) In the fourth step we choose where the chart is placed.

Because *xy* plots are normally graphed with the origin set to (0, 0), an attractive plot is best obtained if the minimum and maximum values are changed for each axis. This is done by placing the mouse cursor over the axis and left-clicking. Handles or small black boxes should appear on the axis to show that it has been selected. Right-clicking brings up a menu to select format axis. The scale tab allows us to change the minimum and maximum values.

# SUMMARY

The compound interest formulas described in this chapter, along with those in Chapter 3, will be referred to throughout the rest of the book. It is very important that the reader understand the concepts presented and how these formulas are used. The following notation is used consistently:

$i$ = effective interest rate per interest period[3] (stated as a decimal)

$n$ = number of interest periods

$P$ = a present sum of money

$F$ = a future sum of money at the end of the $n^{\text{th}}$ interest period, which is equivalent to $P$ with interest rate $i$

$A$ = an end-of-period cash receipt or disbursement in a uniform series continuing for $n$ periods; the entire series equivalent to $P$ or $F$ at interest rate $i$

$G$ = uniform period-by-period increase or decrease in cash receipts or disbursements; the arithmetic gradient

$g$ = uniform rate of cash flow increase or decrease from period to period; the geometric gradient

$r$ = nominal interest rate per interest period (see footnote 3)

$i_a$ = effective annual interest rate

$m \doteq$ number of compounding subperiods per period (see footnote 3)

### Single Payment Formulas (Derived in Chapter 3)

Compound amount:

$$F = P(1 + i)^n = P(F/P, i, n)$$

Present worth:

$$P = F(1 + i)^{-n} = F(P/F, i, n)$$

### Uniform Series Formulas

Compound amount:

$$F = A\left[\frac{(1 + i)^n - 1}{i}\right] = A(F/A, i, n)$$

Sinking fund:

$$A = F\left[\frac{i}{(1 + i)^n - 1}\right] = F(A/F, i, n)$$

Capital recovery:

$$A = P\left[\frac{i(1 + i)^n}{(1 + i)^n - 1}\right] = P(A/P, i, n)$$

Present worth:

$$P = A\left[\frac{(1 + i)^n - 1}{i(1 + i)^n}\right] = A(P/A, i, n)$$

---

[3]Normally the interest period is one year, but it could be some other period (e.g., quarter, month, half-year).

## Arithmetic Gradient Formulas

Arithmetic gradient present worth:

$$P = G\left[\frac{(1+i)^n - in - 1}{i^2(1+i)^n}\right] = G(P/G, i, n)$$

Arithmetic gradient uniform series:

$$A = G\left[\frac{(1+i)^n - in - 1}{i(1+i)^n - i}\right] = G\left[\frac{1}{i} - \frac{n}{(1+i)^n - 1}\right] = G(A/G, i, n)$$

## Geometric Gradient Formulas

Geometric series present worth, where $i \neq g$:

$$P = A_1\left[\frac{1 - (1+g)^n(1+i)^{-n}}{i - g}\right] = A_1(P/A, g, i, n)$$

Geometric series present worth, where $i = g$:

$$P = A_1[n(1+i)^{-1}] = A_1(P/A, g, i, n) = A_1(P/A, i, i, n)$$

## Single Payment Formulas: Continuous Compounding at Nominal Rate $r$ per Period

Compound amount:

$$F = P(e^{rn}) = P[F/P, r, n]$$

Present worth:

$$P = F(e^{-rn}) = F[P/F, r, n]$$

Note that square brackets around the factors are used to distinguish continuous compounding.

## Uniform Payment Series: Continuous Compounding at Nominal Rate $r$ per Period

Continuous compounding sinking fund:

$$A = F\left[\frac{e^r - 1}{e^{rn} - 1}\right] = F[A/F, r, n]$$

Continuous compounding capital recovery:

$$A = P\left[\frac{e^{rn}(e^r - 1)}{e^{rn} - 1}\right] = P[A/P, r, n]$$

Continuous compounding series compound amount:

$$F = A\left[\frac{e^{rn} - 1}{e^r - 1}\right] = A[F/A, r, n]$$

Continuous compounding series present worth:

$$P = A\left[\frac{e^{rn} - 1}{e^{rn}(e^r - 1)}\right] = A[P/A, r, n]$$

### Nominal Annual Interest Rate, *r*

The annual interest rate without considering the effect of any compounding. Also called the annual percentage rate (APR).

### Effective Annual Interest Rate, $i_a$

The annual interest rate taking into account the effect of any compounding during the year. Effective annual interest rate (periodic compounding):

$$i_a = \left(1 + \frac{r}{m}\right)^m - 1$$

or

$$i_a = (1 + i)^m - 1$$

Effective annual interest rate (continuous compounding):

$$i_a = e^r - 1$$

## PROBLEMS

Most of these problems could be solved with a spreadsheet, but calculators and tabulated factors are often easier. The icon indicates that a spreadsheet is recommended.

### Uniform Annual Cash Flows

**4-1** For diagrams (*a*) to (*c*), compute the unknown values *B, C, V*, using the minimum number of compound interest factors.

*i* = 10%

(*a*) *C* = value at *t* = 0

*i* = 10%

(*b*) *V* = value at *t* = 5

*i* = 10%

*n* = 5

(*c*) *B* = value at *t* = 0

(*Answers: B* = $228.13; *C* = $634; *V* = $51.05)

**4-2** The cash flows have a present value of 0. Compute the value of *J*, assuming a 10% interest rate.

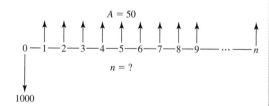

**4-3** For what value of *n*, based on a $3^1/2$% interest rate, do these cash flows have a present value of 0?

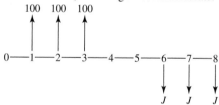

**4-4** The cash flows have a present value of 0. Compute the value of *n*, assuming a 10% interest rate.

**4-5** How much must be deposited now at $5^1/4$% interest to produce $300 at the end of every year for 10 years?

**4-6**   A man buys a car for $18,000 with no money down. He pays for the car in 30 equal monthly payments with interest at 12% per annum, compounded monthly. What is his monthly loan payment?
(*Answer:* $697.47)

**4-7**   A car may be purchased with a $3000 down payment now and 60 monthly payments of $480. If the interest rate is 12% compounded monthly, what is the price of the car?

**4-8**   A company deposits $2000 in a bank at the end of every year for 10 years. The company makes no deposits during the subsequent 5 years. If the bank pays 8% interest, how much would be in the account at the end of 15 years?

**4-9**   A city engineer knows that she will need $25 million in 3 years to replace toll booths on a toll road in the city. Traffic on the road is estimated to be 20 million vehicles per year. How much per vehicle should the toll be to cover the cost of the toll booth replacement project? Interest is 10%. (Simplify your analysis by assuming that the toll receipts are received at the end of each year in a lump sum.)

**4-10**   A student wants to have $30,000 at graduation 4 years from now to buy a new car. His grandfather gave him $10,000 as a high school graduation present. How much must the student save each year if he deposits the $10,000 today and can earn 12% on both the $10,000 and his earnings in a mutual fund his grandfather recommends?

**4-11**   Using linear interpolation, determine the value of $(P/A, 6^{1}/2\%, 10)$ from the compound interest tables. Compute this same value using the equation or a TVM calculator. Why do the values differ?

**4-12**   How many months will it take to pay off a $525 debt, with monthly payments of $15 at the end of each month, if the interest rate is 18%, compounded monthly?
(*Answer:* 50 months)

**4-13**   Tori is planning to buy a car. The maximum payment she can make is $3400 per year, and she can get a car loan at her credit union for 7.3% interest. Assume her payments will be made at the end of each year 1–4. If Tori's old car can be traded in for $3325, which is her down payment, what is the most expensive car she can purchase?
*Contributed by Paul R. McCright, University of South Florida*
*Problems 4-14 to 4-16 contributed by Hamed Kashani, Saeid Sadri, and Baabak Ashuri, Georgia Institute of Technology*

**4-14**   A manufacturing firm spends $500,000 annually for a required safety inspection procedure on its production lines. A new monitoring technology would enable the company to eliminate the need for such inspection. If the interest rate is 10% per year, how much can the company afford to spend on this new technology? The company wants to recover its investment in 15 years.

**4-15**   A student is buying a new car. The car's price is $16,500, the sales tax is 8%, and the title, license, and registration fee is $450 to be paid in cash. The dealer offers to finance 90% of the car's price for 48 months at a nominal interest rate of 9% per year, compounded monthly.

  (*a*) How much cash is paid when the car is purchased?

  (*b*) How much is the monthly payment?

**4-16**   A student is buying a new car. The car's price is $19,500, the sales tax is 8%, and the title, license, and registration fee is $650 to be paid in cash. Instead of buying the car now, the student has decided to save money in equal monthly amounts for 48 months and then pay cash. If the student earns 0.75% per month interest on the money she saves, how much money is the monthly savings?

**4-17**   Rose recently graduated in engineering. Her employer will give her a raise of $3500 per year if she passes the FE exam (Fundamentals of Engineering). Over a career of 40 years, what is the present worth of the raise if the interest rate is 7%?

**4-18**   Jose graduated in engineering 5 years ago. His employer will give him a raise of $10,000 per year if he passes the PE exam (Professional Engineer). Over a career of 35 years, what is the present worth of the raise if the interest rate is 8%?

**4-19**   Brad will graduate next year. When he begins working, he plans to deposit $3000 at the end of each year into an IRA account. If the retirement account pays 4% interest, how much will be in his account after 40 deposits?
*Contributed by Paul R. McCright, University of South Florida*

**4-20**   If the university's College of Engineering can earn 4% on its investments, how much should be in its savings account to fund one $5000 scholarship each year for 10 years?
*Contributed by Paul R. McCright, University of South Florida*

**4-21** Elias makes 7 annual end-of-year deposits of $350 into an account earning 2.5% interest. How much is in the account at the end of Year 11, if he makes no additional deposits and makes no withdrawals?
*Contributed by Paul R. McCright, University of South Florida*

**4-22** Liam dreams of starting his own business to import consumer electronic products to his home country. He estimates he can earn 5% on his investments and that he will need to have $300,000 at the end of year 10 if he wants to give his business a good solid foundation. He now has $28,850 in his account, and he believes he can save $12,000 each year from his income, beginning now. He plans to marry at about the end of year 6 and will skip the investment contribution that year. How far below or above his $300,000 goal will he be?
*Contributed by Paul R. McCright, University of South Florida*

**4-23** Martin pays rent of $500 per month for the 9-month academic year. He is going to travel the world this summer and won't be working. How much must he set aside in his savings account for the 3-month summer to cover his rent for next year? The savings account earns 3% with monthly compounding.

**4-24** How much money should Timothy and Tiffany deposit annually for 20 years in order to provide an income of $30,000 per year for the next 10 years? Assume the interest rate is a constant 4%.
*Contributed by Paul R. McCright, University of South Florida*

**4-25** LaQuita deposits $3500 in her retirement account every year. If her account pays an average of 6% interest and she makes 38 deposits before she retires, how much money can she withdraw in 20 equal payments beginning one year after her last deposit?
*Contributed by Paul R. McCright, University of South Florida*

**4-26** Determine the breakeven resale price 10 years from now of an apartment house that can be bought today for $449,000. Its annual net income is $54,000. The owner wants a 10% annual return on her investment.
*Contributed by D. P. Loucks, Cornell University*

**4-27** Kelsey Construction has purchased a crane that comes with a 3-year warranty. Repair costs are expected to average $3500 per year beginning in Year 4 when the warranty expires. Determine the present worth of the crane's repair costs over its 15-year life. The interest rate is 10%.

**4-28** A young engineer wishes to become a millionaire by the time he is 60 years old. He believes that by careful investment he can obtain a 15% rate of return. He plans to add a uniform sum of money to his investment program each year, beginning on his 20th birthday and continuing through his 59th birthday. How much money must the engineer set aside in this project each year?

**4-29** What amount will be required to purchase, on an engineers's 40th birthday, an annuity to provide him with 30 equal semiannual payments of $1000 each, the first to be received on his 50th birthday, if nominal interest is 4% compounded semiannually?

**4-30** The first of a series of equal semiannual cash flows occurs on July 1, 2011, and the last occurs on January 1, 2024. Each cash flow is equal to $128,000. The nominal interest rate is 12% compounded semiannually. What single amount on July 1, 2015 is equivalent to this cash flow system?

**4-31** On January 1, Frank Jenson bought a used car for $7200 and agreed to pay for it as follows: 1/3 down payment; the balance to be paid in 36 equal monthly payments; the first payment due February 1; an annual interest rate of 9%, compounded monthly.

(*a*) What is the amount of Frank's monthly payment?

(*b*) During the summer, Frank made enough money to pay off the entire balance due on the car as of October 1. How much did Frank owe on October 1?

**4-32** If $i = 12\%$, for what value of $B$ is the present value = 0.

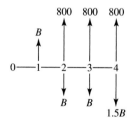

**4-33** Compute $E$ so that the cash flows have a present value of 0.

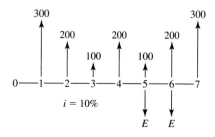

**4-34** A man borrowed $500 from a bank on October 15$^{th}$. He must repay the loan in 16 equal monthly payments, due on the 15$^{th}$ of each month, beginning November 15$^{th}$. If interest is computed at 1% per month, how much must he pay each month? (*Answer:* $33.95)

**4-35** Jerry bought a house for $500,000 and made a $100,000 down payment. He obtained a 30-year loan for the remaining amount. Payments were made monthly. The nominal annual interest rate was 9%. After 10 years (120 payments) he decided to pay the remaining balance on the loan.

(*a*) What was his monthly loan payment?

(*b*) What must he have paid (in addition to his regular 120$^{th}$ monthly payment) to pay the remaining balance of his loan?

**4-36** A man wants to help provide a college education for his young daughter. He can afford to invest $600/yr for the next 4 years, beginning on the girl's 4$^{th}$ birthday. He wishes to give his daughter $4000 on her 18$^{th}$, 19$^{th}$, 20$^{th}$, and 21$^{st}$ birthdays, for a total of $16,000. Assuming 5% interest, what uniform annual investment will he have to make on the girl's 8$^{th}$ through 17$^{th}$ birthdays? (*Answer:* $792.73)

**4-37** Table 3-1 presented four plans for the repayment of $5000 in 5 years with interest at 8%. Still another way to repay the $5000 would be to make four annual end-of-year payments of $1000 each, followed by a final payment at the end of the fifth year. How much would the final payment be?

**4-38** A $150 bicycle was purchased on December 1 with a $15 down payment. The balance is to be paid at the rate of $10 at the end of each month, with the first payment due on December 31. The last payment may be some amount less than $10. If interest on the unpaid balance is computed at 1$^{1}$/$_{2}$% per month, how many payments will there be, and what is the amount of the final payment? (*Answers:* 16 payments; final payment: $1.99)

**4-39** A company buys a machine for $12,000, which it agrees to pay for in five equal annual payments, beginning one year after the date of purchase, at an annual interest rate of 4%. Immediately after the second payment, the terms of the agreement are changed to allow the balance due to be paid off in a single payment the next year. What is the final single payment? (*Answer:* $7778)

**4-40** An engineering student bought a car at a local used car lot. Including tax and insurance, the total price was $6000. He is to pay for the car in 12 equal monthly payments, beginning with the first payment immediately (in other words, the first payment was the down payment). Nominal interest on the loan is 12%, compounded monthly. After six payments (the down payment plus five additional payments), he decides to sell the car. A buyer agrees to pay a cash amount to pay off the loan in full at the time the next payment is due and also to pay the engineering student $2000. If there are no penalty charges for this early payment of the loan, how much will the car cost the new buyer?

**4-41** A realtor sold a house on August 31, 2011, for $150,000 to a buyer in which a 20% down payment was made. The buyer took a 15-year mortgage on the property with an effective interest rate of 8% per annum. The buyer intends to pay off the mortgage owed in yearly payments starting on August 31, 2012.

(*a*) How much of the mortgage will still be owed after the payment due on August 31, 2018, has been made?

(*b*) Solve the same problem by separating the interest and the principal amounts.

**4-42** To provide for a college education for her son, a woman opened an escrow account in which equal deposits were made. The first deposit was made on January 1, 1995, and the last deposit was made on January 1, 2012. The yearly college expenses including tuition were estimated to be $8000, for each of the 4 years. Assuming the interest rate to be 5.75%, how much did the mother have to deposit each year in the escrow account for the son to draw $8000 per year for 4 years beginning January 1, 2012?

**4-43** A bank recently announced an "instant cash" plan for holders of its bank credit cards. A cardholder may receive cash from the bank up to a preset limit (about $500). There is a special charge of 4% made at the time the "instant cash" is sent to the cardholders. The debt may be repaid in monthly installments. Each month the bank charges 1$^{1}$/$_{2}$% on the unpaid balance. The monthly payment, including interest, may be as little as $10. Thus, for $150 of "instant cash," an initial charge of $6 is made and added to the balance due. Assume the cardholder makes a monthly payment of $10 (this includes both principal and interest). How many months are required to repay the debt? If your answer includes a fraction of a month, round up to the next month.

**4-44** An engineer borrowed $3000 from the bank, payable in six equal end-of-year payments at 8%. The bank agreed to reduce the interest on the loan if interest rates declined in the U.S. before the loan was fully repaid. At the end of 3 years, at the time of the third payment, the bank agreed to reduce the interest rate on the remaining debt from 8% to 7%. What was the amount of the equal annual end-of-year payments for each of the first 3 years? What was the amount of the equal annual end-of-year payments for each of the last 3 years?

**4-45** A local finance company will loan $10,000 to a homeowner. It is to be repaid in 24 monthly payments of $499 each. The first payment is due 30 days after the $10,000 is received. What interest rate per month are they charging?   (*Answer: 1½%*)

**4-46** A woman made 10 annual end-of-year purchases of $1000 worth of common stock. The stock paid no dividends. Then for 4 years she held the stock. At the end of the 4 years she sold all the stock for $28,000. What interest rate did she obtain on her investment?

## Relationships Between Factors

**4-47** For some interest rate $i$ and some number of interest periods $n$, the uniform series capital recovery factor is 0.1408 and the sinking fund factor is 0.0408. What is the interest rate? What is $n$?

**4-48** For some interest rate $i$ and some number of interest periods $n$, the uniform series capital recovery factor is 0.1728 and the sinking fund factor is 0.0378. What is the interest rate?

**4-49** Derive an equation to find the end-of-year future sum $F$ that is equivalent to a series of $n$ beginning-of-year payments $B$ at interest rate $i$. Then use the equation to determine the future sum $F$ equivalent to six $B$ payments of $100 at 8% interest. (*Answer: F = $792.28*)

**4-50** If $200 is deposited in a savings account at the beginning of each of 15 years, and the account draws interest at 7% per year, how much will be in the account at the end of 15 years?

**4-51** How can the tables be used to compute $(P/A, 5\%, 150)$? $(P/A, 7\%, 200)$?

**4-52** Prove the following relationships algebraically

(a) $(A/F, i, n) = (A/P, i, n) - i$

(b) $(P/F, i, n) = (P/A, i, n) - (P/A, i, n-1)$

(c) $(P/A, i, n) = (P/F, i, 1) + (P/F, i, 2) + \cdots + (P/F, i, n)$

(d) $(F/A, i, n) = [(F/P, i, n) - 1]/i$

## Arithmetic Gradients

**4-53** Assume a 10% interest rate and find $S$, $T$, and $x$.

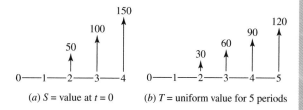

(a) $S$ = value at $t = 0$          (b) $T$ = uniform value for 5 periods

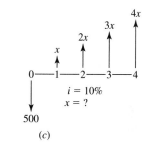

(c)

(*Answers: S = $218.90; T = $54.30; x = $66.24*)

**4-54** Compute the unknown values.

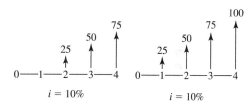

(a) $C$ = value at $t = 0$          (b) $F$ = value at $t = 4$

(c) $A$ = uniform value from $t = 1$ to 4

(*Answers: C = $109.45; F = $276.37; A = $60.78*)

**4-55**  For diagrams (*a*) to (*d*), compute the present values of the cash flows.

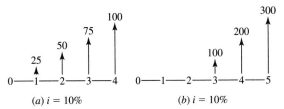

(*a*) *i* = 10%        (*b*) *i* = 10%

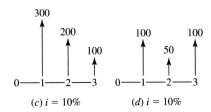

(*c*) *i* = 10%        (*d*) *i* = 10%

**4-56**  Compute the present value of the cash flows.

*i* = 10%

(*Answer:* $589.50)

**4-57**  Use a 15% interest rate to compute the present value of the cash flows.

**4-58**  The cash flows have a present value of 0. Compute the value of *C*, assuming a 10% interest rate.

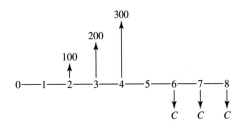

**4-59**  If *i* = 12%, for what value of *G* does the present value equal 0?

**4-60**  The cash flows have a present value equal 0. Compute the value of *D* in the diagram.

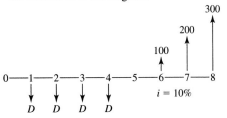

**4-61**  Using a 10% interest rate, for what value of *B* does the present value equal 0?

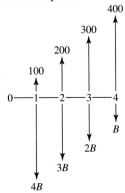

**4-62**  These cash flow transactions are said to be equivalent in terms of economic desirability at an interest rate of 12% compounded annually. Determine the unknown value *A*.

Cash Flow 1        Cash Flow 2

**4-63** For what value of $P$ in the cash flow diagram does the present value equal 0?

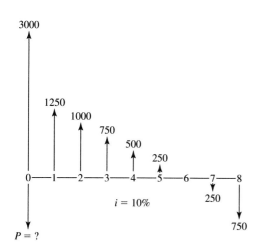

**4-64** Use a 10% interest rate to compute the present value of the cash flows.

**4-65** Consider the following cash flow:

| Year | Cash Flow |
|------|-----------|
| 0 | −$100 |
| 1 | 50 |
| 2 | 60 |
| 3 | 70 |
| 4 | 80 |
| 5 | 140 |

Which one of the following is correct for this cash flow?

A. $100 = 50 + 10(A/G, i, 5) + 50(P/F, i, 5)$

B. $1 = \dfrac{50(P/A, i, 5) + 10(P/G, i, 5) + 50(P/F, i, 5)}{100}$

C. $100(A/P, i, 5) = 50 + 10(A/G, i, 5)$

D. None of the equations are correct.

**4-66** Consider the following cash flow:

| Year | Cash Flow |
|------|-----------|
| 0 | −$P |
| 1 | 1000 |
| 2 | 850 |
| 3 | 700 |
| 4 | 550 |
| 5 | 400 |
| 6 | 400 |
| 7 | 400 |
| 8 | 400 |

Alice was asked to compute the value of $P$ for the cash flow at 8% interest. She wrote three equations:

A. $P = 1000(P/A, 8\%, 8) - 150(P/G, 8\%, 8) + 150(P/G, 8\%, 4)(P/F, 8\%, 4)$

B. $P = 400(P/A, 8\%, 8) + 600(P/A, 8\%, 5) - 150(P/G, 8\%, 4)$

C. $P = 150(P/G, 8\%, 4) + 850(P/A, 8\%, 4) + 400(P/A, 8\%, 4)(P/F, 8\%, 4)$

Which of the equations is correct?

**4-67** It is estimated that the maintenance cost on a new car will be $400 the first year. Each subsequent year, this cost is expected to increase by $100. How much would you need to set aside when you bought a new car to pay all future maintenance costs if you planned to keep the vehicle for 7 years? Assume interest is 5% per annum. (*Answer*: $3937.6)

**4-68** The council members of a small town have decided that the earth levee that protects the town flooding should be rebuilt and strengthened. The town engineer estimates that the cost of the work at the end of the first year will be $85,000. He estimates that in subsequent years the annual repair costs will decline by $10,000, making the second-year cost $75,000; the third-year $65,000, and so forth. The council members want to know what the equivalent present cost is for the first 5 years of repair work if interest is 4%. (*Answer*: $292,870)

**4-69** A firm expects to install smog control equipment on the exhaust of a gasoline engine. The local smog control district has agreed to pay to the firm a lump sum of money to provide for the first cost of the equipment and maintenance during its 10-year useful life. At the end of 10 years the equipment, which initially cost $10,000, is valueless. The firm and the smog control district have agreed that the following are reasonable estimates of the end-of-year maintenance costs:

| Year 1 | $500 | Year 6 | $200 |
|---|---|---|---|
| 2 | 100 | 7 | 225 |
| 3 | 125 | 8 | 250 |
| 4 | 150 | 9 | 275 |
| 5 | 175 | 10 | 300 |

Assuming interest at 6% per year, how much should the smog control district pay to the firm now to provide for the first cost of the equipment and its maintenance for 10 years? *(Answer: $11,693)*

**4-70** A debt of $5000 can be repaid, with interest at 8%, by the following payments.

| Year | Payment |
|---|---|
| 1 | $ 500 |
| 2 | 1000 |
| 3 | 1500 |
| 4 | 2000 |
| 5 | X |

The payment at the end of the fifth year is shown as $X$. How much is $X$?

**4-71** A man is buying a small garden tractor. There will be no maintenance cost during the first 2 years because the tractor is sold with 2 years free maintenance. For the third year, the maintenance is estimated at $80. In subsequent years the maintenance cost will increase by $40 per year. How much would need to be set aside now at 8% interest to pay the maintenance costs on the tractor for the first 6 years of ownership?

**4-72** A woman makes an investment every 3 months at a nominal annual interest rate of 28%, compounded quarterly. Her first investment was $100, followed by investments *increasing* $20 each 3 months. Thus, the second investment was $120, the third investment $140, and so on. If she continues to make this series of investments for a total of 20 years, what will be the value of the investments at the end of that time?

**4-73** A sports star can sign a 6-year contract that starts at $1.2M with increases of $0.3M each year for his expected playing career of 6 years. It is also possible to sign a contract that starts at $0.8M for the first year and then increases at $0.2M each year for 10 years (note some income is deferred until after he retires). If his interest rate for the time value of money is 8%, what is the value of each choice?

*Problems 4-74 to 4-76 contributed by Hamed Kashani, Saeid Sadri, and Baabak Ashuri, Georgia Institute of Technology*

**4-74** A college student is buying a new car, which costs $16,500 plus 8% sales tax. The title, license, and registration fees are $650. The dealer offers her a financing program that starts with a small monthly payment, and the payments will gradually increase. The dealer offers to finance 80% of the car's price for 48 months at a nominal interest rate of 9% per year, compounded monthly. The first payment is $300 and each successive payment will increase by a constant dollar amount $x$.

(*a*) How much is the constant amount $x$?

(*b*) How much is the $48^{\text{th}}$ payment?

**4-75** A firm has $500,000 per year to pay for replacing machinery over the next 5 years. What is the expected cost in Year 1 if the firm has projected that the machinery cost will increase by $15,000 per year? The interest rate is 10% per year.

**4-76** A construction firm can achieve a $15,000 cost savings in Year 1 and increasing by $2000 each year for the next 5 years by upgrading some equipment. At an interest rate of 15%, what is the equivalent annual worth of the savings?

**4-77** Helen can earn 3% interest in her savings account. Her daughter Roberta is 11 years old today. Suppose Helen deposits $4000 today, and one year from today she deposits another $500. Each year she increases her deposit by $500 until she makes her last deposit on Roberta's $18^{\text{th}}$ birthday. What is the annual equivalent of her deposits, and how much is on deposit after the $18^{\text{th}}$ birthday? *Contributed by Paul R. McCright, University of South Florida*

**4-78** Francisco believes in planning ahead. So he decides to find out how much he can draw from a retirement account each year for 30 years if he invests exactly $1000 each year in the account. Suppose that he can earn 5% on this long-term account, that he will make 40 deposits, and that he will make the first withdrawal one year after the last deposit. How big can Francisco's withdrawals be? *Contributed by Paul R. McCright, University of South Florida*

**4-79** Perry is a freshman. He estimates that the cost of tuition, books, room and board, transportation, and other incidentals will be $13,000 this year. He expects these costs to rise about $1500 each year while he is in college. If it will take him 5 years to earn his BS, what is the present cost of his degree at an interest rate of 6%? If he earns an extra $10,000 annually for 40 years, what is the present worth of his degree? *Contributed by Paul R. McCright, University of South Florida*

**4-80**   The following cash flows are equivalent in value if the interest rate is $i$. Which one is more valuable if the interest rate is $2i$?

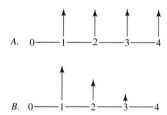

**4-81**   The following cash flows are equivalent in value if the interest rate is $i$. Which one is more valuable if the interest rate is $2i$?

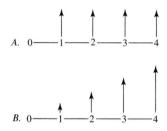

### Geometric Gradients

**4-82**   A set of cash flows begins at $20,000 the first year, with an increase each year until $n = 10$ years. If the interest rate is 8%, what is the present value when

(a)   the annual increase is $2000?

(b)   the annual increase is 10%?

(Answer: (b) $201,405)

**4-83**   A set of cash flows begins at $50,000 the first year, with an increase each year until $n = 15$ years. If the interest rate is 7%, what is the present value when

(a)   the annual increase is $5000?

(b)   the annual increase is 10%?

**4-84**   A set of cash flows begins at $20,000 the first year, with an increase each year until $n = 10$ years. If the interest rate is 10%, what is the present value when

(a)   the annual increase is $2000?

(b)   the annual increase is 10%?

**4-85**   A set of cash flows begins at $20,000 the first year, with a decrease each year until $n = 10$ years. If the interest rate is 8%, what is the present value when

(a)   the annual decrease is $2000?

(b)   the annual decrease is 10%?

**4-86**   The market for a product is expected to increase at an annual rate of 8%. First-year sales are estimated at $60,000, the horizon is 15 years, and the interest rate is 10%. What is the present value?

**4-87**   Fred is evaluating whether a more efficient motor with a life of 5 years should be installed on an assembly line. Energy savings are estimated at $400 for the first year, then increasing by 6% annually. If the interest rate is 10%, what is the present value of the energy savings?

**4-88**   In Problem 4-87, what if the energy savings are increasing by 15% annually?

**4-89**   Suzanne is a recent chemical engineering graduate who has been offered a 5-year contract at a remote location. She has been offered two choices. The first is a fixed salary of $75,000 per year. The second has a starting salary of $65,000 with annual raises of 5% starting in Year 2. (For simplicity, assume that her salary is paid at the end of the year, just before her annual vacation.) If her interest rate is 9%, which should she take?

**4-90**   Mark Johnson saves a fixed percentage of his salary at the end of each year. This year he saved $1500. For the next 5 years, he expects his salary to increase at an 8% annual rate, and he plans to increase his savings at the same 8% annual rate. He invests his money in the stock market. Thus there will be six end-of-year investments (the initial $1500 plus five more). Solve the problem using the geometric gradient factor.

(a)   How much will the investments be worth at the end of 6 years if they increase in the stock market at a 10% annual rate?

(b)   How much will Mark have at the end of 6 years if his stock market investments increase only at 8% annually?

**4-91**   The Macintosh Company has an employee savings plan that allows every employee to invest up to 5% of his or her annual salary. The money is invested in company common stock with the company guaranteeing that the annual return will never be less than 8%. Jill was hired at an annual salary of $52,000. She immediately joined the savings plan investing the full 5% of her salary each year. If Jill's salary increases at an 8% uniform rate, and she continues to invest 5% of it each year, what amount of money is she guaranteed to have at the end of 20 years?

**4-92**   The football coach at a midwestern university was given a 5-year employment contract that paid $225,000 the first year, and increased at an 8% uniform rate in each subsequent year. At the end of the first year's football season, the alumni demanded that the coach be fired. The alumni agreed to buy his remaining years on the contract by paying him the equivalent present sum, computed using a 12% interest rate. How much will the coach receive?

**4-93** A 25-year-old engineer is opening an individual retirement account (IRA) at a bank. Her goal is to accumulate $1 million in the account by the time she retires from work in 40 years. The bank manager estimates she may expect to receive 8% nominal annual interest, compounded quarterly, throughout the 40 years. The engineer believes her income will increase at a 7% annual rate during her career. She wishes to start her IRA with as low a deposit as possible and increase it at a 7% rate each year. Assuming end-of-year deposits, how much should she deposit the first year?

**4-94** A contractor estimates maintenance costs for a new backhoe to be $275 for the first month with a monthly increase of 0.5%. The contractor can buy a 4-year maintenance contract for $18,500 at any point. If the contract is purchased at the same time as the backhoe is purchased, the dealer has offered a 10% discount. Use $i = 0.75\%$ per month. What should the contractor do? *Contributed by Hamed Kashani, Saeid Sadri, and Baabak Ashuri, Georgia Institute of Technology*

**4-95** An engineer will deposit 15% of her salary each year into a retirement fund. If her current annual salary is $80,000 and she expects that it will increase by 5% each year, what will be the present worth of the fund after 35 years if it earns 5% per year? *Contributed by Hamed Kashani, Saeid Sadri, and Baabak Ashuri, Georgia Institute of Technology*

**4-96** Zachary has opened a retirement account that will pay 5% interest each year. He plans to deposit 10% of his annual salary into the account for 39 years before he retires. His first year's salary is $52,000, and he expects the salary to grow 4% each year. How much will be in his account after he makes the last deposit? What uniform amount can he withdraw from the account for 25 years beginning one year after his last deposit? *Contributed by Paul R. McCright, University of South Florida*

**4-97** Eddie is a production engineer for a major supplier of component parts for cars. He has determined that a robot can be installed on the production line to replace one employee. The employee earns $20 per hour and benefits worth $8 per hour for a total annual cost of $58,240 this year. Eddie estimates this cost will increase 6% each year. The robot will cost $16,500 to operate for the first year with costs increasing by $1500 each year. The firm uses an interest rate of 15% and a 10-year planning horizon. The robot costs $75,000 installed and will have a salvage value of $5000 after 10 years. Should Eddie recommend that purchase of the robot? *Contributed by Paul R. McCright, University of South Florida*

**Nominal and Effective Interest Rates**

**4-98** Quentin has been using his credit card too much. His plan is to use only cash until the balance of $8574 is paid off. The credit card company charges 18% interest, compounded monthly. What is the effective interest rate? How much interest will he owe in the first month's payment? If he makes monthly payments of $225, how long until it is paid off? *Contributed by Paul R. McCright, University of South Florida*

**4-99** A bank is offering a loan of $25,000 with a nominal interest rate of 18% compounded monthly, payable in 60 months. (*Hint:* The loan origination fee of 2% will be taken out from the loan amount.)

(*a*) What is the monthly payment?

(*b*) If a loan origination fee of 2% is charged at the time of the loan, what is the effective interest rate?

**4-100** One of the largest car dealers in the city advertises a 3-year-old car for sale as follows:

> Cash price $13,750, or a down payment of $1375 with 45 monthly payments of $361.23.

Susan DeVaux bought the car and made a down payment of $2000. The dealer charged her the same interest rate used in his advertised offer. What is the monthly interest rate? How much will Susan pay each month for 45 months? What effective interest rate is being charged?

**4-101** Pete borrows $10,000 to purchase a used car. He must repay the loan in 48 equal end-of-period monthly payments. Interest is calculated at $1^1/4\%$ per month. Determine the following:

(*a*) The nominal annual interest rate

(*b*) The effective annual interest rate

(*c*) The amount of the monthly payment

**4-102** Picabo borrows $1000. To repay the amount, she makes 12 equal monthly payments of $90.30. Determine the following:

(*a*) The effective monthly interest rate

(*b*) The nominal annual interest rate

(*c*) The effective annual interest rate

**4-103** The *Bawl Street Journal* costs $580, payable now, for a 2-year subscription. The newspaper is published 252 days per year (5 days per week, except holidays). If a 10% nominal annual interest rate, compounded quarterly, is used:

(*a*) What is the effective annual interest rate in this problem?

(b) Compute the equivalent interest rate per $1/252$ of a year.

(c) What is a subscriber's cost per copy of the newspaper, taking interest into account?

**4-104** You are taking a $2000 loan. You will pay it back in four equal amounts, paid every 6 months starting 3 years from now. The interest rate is 6% compounded semiannually. Calculate:

(a) The effective interest rate, based on both semiannual and continuous compounding

(b) The amount of each semiannual payment

(c) The total interest paid

**4-105** In 1555 King Henry borrowed money from his bankers on the condition that he pay 5% of the loan at each fair (there were four fairs per year) until he had made 40 payments. At that time the loan would be considered repaid. What effective annual interest did King Henry pay?

**4-106** A local bank offers a customer a 2-year car loan of $10,000 as follows:

| | |
|---|---:|
| Money to pay for car | $10,000 |
| Two years' interest at 7%: $2 \times 0.07 \times 10,000$ | 1400 |
| | $11,400 |

$$24 \text{ monthly payments} = \frac{11,400}{24} = \$475.00$$

The first payment must be made in 30 days. What is the nominal annual interest rate the bank is receiving?

**4-107** A local lending institution advertises the "51–50 Club." A person may borrow $2000 and repay $51 for the next 50 months, beginning 30 days after receiving the money. Compute the nominal annual interest rate for this loan. What is the effective interest rate?

**4-108** The **Rule of 78s** is a commonly used method of computing the amount of interest when the balance of a loan is repaid in advance.

Adding the numbers representing 12 months gives

$$1 + 2 + 3 + 4 + 5 + \cdots + 11 + 12 = 78$$

If a 12-month loan is repaid at the end of one month, for example, the interest the borrower would be charged is $12/78$ of the year's interest. If the loan is repaid at the end of 2 months, the total interest charged would be $(12+11)/78$, or $23/78$ of the year's interest. After 11 months the interest charge would therefore be $77/78$ of the total year's interest.

Helen borrowed $10,000 on January 1 at 9% annual interest, compounded monthly. The loan was to be repaid in 12 equal end-of-period payments. Helen

made the first two payments and then decided to repay the balance of the loan with the third payment. Thus she will pay the third payment plus an additional sum.

You are to calculate the amount of this additional sum

(a) based on the rule of 78s.

(b) based on exact economic analysis methods.

## Compounding and Payment Periods Differ

**4-109** Upon the birth of his first child, Dick Jones decided to establish a savings account to partly pay for his son's education. He plans to deposit $20 per month in the account, beginning when the boy is 13 months old. The savings and loan association has a current interest policy of 6% per annum, compounded monthly, paid quarterly. Assuming no change in the interest rate, how much will be in the savings account when Dick's son becomes 16 years old?

**4-110** Ann deposits $100 at the end of each month into her bank savings account. The bank paid 6% nominal interest, compounded and paid quarterly. No interest was paid on money not in the account for the full 3-month period. How much was in Ann's account at the end of 3 years?   (*Answer:* $3912.30)

**4-111** What is the present worth of a series of equal quarterly payments of $3000 that extends over a period of 8 years if the interest rate is 10% compounded monthly?

**4-112** What single amount on April 1, 2012, is equivalent to a series of equal, semiannual cash flows of $1000 that starts with a cash flow on January 1, 2010, and ends with a cash flow on January 1, 2019? The interest rate is 14% and compounding is quarterly.

**4-113** A contractor wishes to set up a special fund by making uniform semiannual end-of-period deposits for 20 years. The fund is to provide $10,000 at the end of each of the last 5 years of the 20-year period. If interest is 8%, compounded semiannually, what is the required semiannual deposit?

**4-114** Paco's saving account earns 13% compounded weekly and receives quarterly deposits of $38,000. His first deposit occurred on October 1, 2006, and the last deposit is scheduled for April 1, 2022. Tisha's account earns 13% compounded weekly. Semiannual deposits of $18,000 are made into her account, with the first one occurring on July 1, 2016, and the last one occurring on January 1, 2025. What single amount on January 1, 2017, is equivalent to the sum of both cash flow series?

**4-115** The first of a series of equal monthly cash flows of $2000 occurred on April 1, 2008, and the last of the monthly cash flows occurred on February 1, 2010.

This series of monthly cash flows is equivalent to a series of semiannual cash flows. The first semiannual cash flow occurred on July 1, 2011, and the last semiannual cash flow will occur on January 1, 2020. What is the amount of each semiannual cash flow? Use a nominal interest rate of 12% with monthly compounding on all accounts.

**4-116** A series of monthly cash flows is deposited into an account that earns 12% nominal interest compounded monthly. Each monthly deposit is equal to $2100. The first monthly deposit occurred on June 1, 2008 and the last monthly deposit will be on January 1, 2015. The account (the series of monthly deposits, 12% nominal interest, and monthly compounding) also has equivalent quarterly withdrawals from it. The first quarterly withdrawal is equal to $5000 and occurred on October 1, 2008. The last $5000 withdrawal will occur on January 1, 2015. How much remains in the account after the last withdrawal?

## Continuous Compounding

**4-117** PARC Company has money to invest in an employee benefit plan, and you have been chosen as the plan's trustee. As an employee yourself, you want to maximize the interest earned on this investment and have found an account that pays 14% compounded continuously. PARC is providing you $1200 per month to put into your account for 7 years. What will be the balance in this account at the end of the 7-year period?

**4-118** Barry, a recent engineering graduate, never took engineering economics. When he graduated, he was hired by a prominent architectural firm. The earnings from this job allowed him to deposit $750 each quarter into a savings account. There were two banks that offered savings accounts in his town (a small town!). The first bank was offering 4.5% interest compounded continuously. The second bank offered 4.6% compounded monthly. Barry decided to deposit in the first bank because it offered continuous compounding. Did he make the right decision?

**4-119** What single amount on October 1, 2012, is equal to a series of $1000 quarterly deposits made into an account? The first deposit occurs on October 1, 2012, and the last deposit occurs on January 1, 2026. The account earns 13% compounded continuously.

## Spreadsheets for Economic Analysis

**4-120** Develop a complete amortization table for a loan of $4500, to be paid back in 24 uniform monthly installments, based on an interest rate of 6%.

The amortization table must include the following column headings:

Payment Number, Principal Owed (beginning of period), Interest Owed in Each Period, Total Owed (end of each period), Principal Paid in Each Payment, Uniform Monthly Payment Amount

You must also show the equations used to calculate each column of the table. You are encouraged to use spreadsheets. *The entire table must be shown.*

**4-121** Using the loan and payment plan developed in Problem 4-120, determine the month that the final payment is due, and the amount of the final payment, if $500 is paid for Payment 8 and $280 is paid for Payment 10. This problem requires a separate amortization table giving the balance due, principal payment, and interest payment for each period of the loan.

**4-122** The following beginning-of-month (BOM) and end-of-month (EOM) amounts are to be deposited in a savings account that pays interest at 9%, compounded monthly:

| | |
|---|---|
| Today (BOM 1) | $400 |
| EOM 2 | 270 |
| EOM 6 | 100 |
| EOM 7 | 180 |
| BOM 10 | 200 |

Set up a spreadsheet to calculate the account balance at the end of the first year (EOM12). The spreadsheet must include the following column headings: Month Number, Deposit BOM, Account Balance at BOM, Interest Earned in Each Month, Deposit EOM, Account Balance at EOM. Also, use the compound interest tables to draw a cash flow diagram of this problem and solve for the account balance at the EOM 12.

**4-123** What is the present worth of cash flows that begin at $20,000 and increase at 7% per year for 10 years? The interest rate is 9%. (*Answer:* $169,054)

**4-124** What is the present worth of cash flows that begin at $50,000 and decrease at 12% per year for 10 years? The interest rate is 8%. (*Answer:* $217,750)

**4-125** Net revenues at an older manufacturing plant will be $2 million for this year. The net revenue will decrease 15% per year for 5 years, when the assembly plant will be closed (at the end of Year 6). If the firm's interest rate is 10%, calculate the PW of the revenue stream.

**4-126** What is the present worth of cash flows that begin at $10,000 and increase at 8% per year for 4 years? The interest rate is 6%.

**4-127** What is the present worth of cash flows that begin at $30,000 and decrease at 15% per year for 6 years? The interest rate is 10%.

**4-128** Five annual payments at an interest rate of 9% are made to repay a loan of $6000. Build the table that shows the balance due, principal payment, and interest payment for each payment. What is the annual payment? (Use a spreadsheet function, not the tables.) What interest is paid in the last year? (*Answer:* payment = $1542.55, interest = $127.37)

**4-129** A newly graduated engineer bought furniture for $900 from a local store. Monthly payments for 1 year will be made. Interest is computed at a nominal rate of 6%. Build the table that shows the balance due, principal payment, and interest payment for each payment. What is the monthly payment? (Use a spreadsheet function, not the tables.) What interest is paid in the last month? (*Answer:* payment = $77.46, interest = $0.39)

**4-130** Calculate and print out the balance due, principal payment, and interest payment for each period of a used-car loan. The nominal interest is 12% per year, compounded monthly. Payments are made monthly for 3 years. The original loan is for $11,000.

**4-131** Calculate and print out the balance due, principal payment, and interest payment for each period of a new car loan. The nominal interest is 9% per year, compounded monthly. Payments are made monthly for 5 years. The original loan is for $17,000.

**4-132** For the used car loan of Problem 4-130, graph the monthly payment.

(*a*) As a function of the interest rate (5–15%).
(*b*) As a function of the number of payments (24–48).

**4-133** For the new car loan of Problem 4-131, graph the monthly payment.

(*a*) As a function of the interest rate (4–14%).
(*b*) As a function of the number of payments (36–84).

**4-134** Your beginning salary is $50,000. You deposit 10% at the end of each year in a savings account that earns 6% interest. Your salary increases by 5% per year. What value does your savings book show after 40 years?

**4-135** The market volume for widgets is increasing by 15% per year from current profits of $200,000. Investing in a design change will allow the profit per widget to stay steady; otherwise profits will drop 3% per year. What is the present worth of the savings over the next 5 years? Ten years? The interest rate is 10%.

**4-136** A 30-year mortgage for $120,000 has been issued. The interest rate is 10% and payments are made monthly. Print out the balance due, principal payment, and interest payment for each period.

**4-137** A homeowner may upgrade a furnace that runs on fuel oil to a natural gas unit. The investment will be $2500 installed. The cost of the natural gas will average $60 per month over the year, instead of the $145 per month that the fuel oil costs. If the interest rate is 9% per year, how long will it take to recover the initial investment?

**4-138** Develop a general-purpose spreadsheet to calculate the balance due, principal payment, and interest payment for each period of a loan. The user inputs to the spreadsheet will be the loan amount, the number of payments per year, the number of years payments are made, and the nominal interest rate. Submit printouts of your analysis of a loan in the amount of $15,000 at 8.9% nominal rate for 36 months and for 60 months of payments.

**4-139** Use the spreadsheet developed for Problem 4-138 to analyze 180-month and 360-month house loan payments. Analyze a $100,000 mortgage loan at a nominal interest rate of 7.5% and submit a graph of the interest and principal paid over time. You need not submit the printout of the 360 payments because it will not fit on one page.

**Minicases**

**4-140** Assume that you plan to retire 40 years from now and that you expect to need $2M to support the lifestyle that you want.

(*a*) If the interest rate is 10%, is the following statement approximately true? "Waiting 5 years to start saving doubles what you must deposit each year."

(*b*) If the interest rate is 12%, is the required multiplier higher or lower than for the 5% rate in (*a*)?

(*c*) At what interest rate is the following statement exactly true? "Waiting 5 years to start saving doubles what you must deposit each year."

**4-141** For winners of the California SuperLotto Plus, the choice is between a lump sum and annual payments that increase from 2.5% for the first year to 2.7% for the second year and then increase by 0.1% per year to 5.1% for the 26th payment. The lump sum is equal to the net proceeds of bonds purchased to fund the 26 payments. This is estimated at 45% to 55% of the lump sum amount. At what interest rate is the present worth of the two payment plans equivalent if the lump sum is 45%? If it is 55%?

**C A S E S** 💿

**The following case from *Cases in Engineering Economy* 2<sup>nd</sup> is suggested as matched with this chapter.**

CASE 27 **Can Crusher**

Basic time value of money, costs, and breakeven analysis.

# PRESENT WORTH ANALYSIS

## The Present Value of 30 Years of Benefits

The Columbia River, from its headwaters west of Banff in the Canadian Rockies (Columbia Lake, elevation 820 m) flows 2000 km through British Columbia and the state of Washington before entering the Pacific Ocean at Astoria, Oregon. Measured by the volume of its flow, the Columbia is the largest river flowing into the Pacific from North America. Measured by elevation drop (0.41 m/km compared to the Mississippi's 0.12 m/km), the Columbia alone possesses one-third of the hydroelectric potential of the U.S. Because of the steep mountain trenches and high snowfall in its catchment area, the Columbia's water levels used to fluctuate wildly, and vunerable areas along its banks were subject to seasonal flooding.

Anxious to exploit that hydroelectric power and control flooding, the U.S. and Canadian governments negotiated the Columbia River Treaty. It was signed in 1961 and ratified by the U.S. in 1961 and by Canada in 1964.

> The Treaty requires Canada to store 15.5 Million Acre Feet [the volume of water that would cover an acre of land one foot deep], for flood control in perpetuity. This storage was accomplished with the construction of the Duncan, Hugh Keenleyside, and Mica Dams in Canada. In return for constructing the dams and regulating the water levels, the Province of British Columbia is entitled to half of the electrical downstream power benefits that the water generates on the dams located in the U.S. Canada was also entitled to half of the estimated value of the future flood control benefits in the U.S. in a one-time payment of $69.9 million. (*Virtual Museum of Canada*)

It took Canada several years to ratify the treaty because British Columbia in the 1950s was sparsely populated and not very industrialized, and the prospects of turning a large part of the interior of the province into a reservoir to provide benefits that the residents could not use was not appealing to the premier, W. A. C. Bennett. The impasse was resolved when the U.S. agreed to pay in cash to British Columbia the discounted present value of the first 30 years of the province's benefit entitlement. The treaty said:

> The purchase price of the entitlement shall be $254,400,000, in United States funds as of October 1, 1964, subject to adjustment, in the event of an earlier payment of all or part thereof, to the then present worth, at a discount rate of 4.5% per annum.

The 30 years was up in 1994, and British Columbia then started receiving annual benefits. Today, Canada's 50% share in the downstream benefits is worth annually about $250 to 350 million, which is paid to the government to British Columbia. ■ ■ ■

*Contributed by John Whittaker, University of Alberta*

## QUESTIONS TO CONSIDER

1. Canadian Premier Bennett used the $254 million to finance the Portage Mountain Dam (later renamed the W. A. C. Bennett Dam) and an associated 2730-megawatt power station in northern British Columbia. Was this a good use for the money from the Canadian perspective?

2. Negotiating downstream rights can be a curious tangle. The people at the bottom of the hill can argue that the water is going to flow there anyway and they are free to use it as they want. The people upstream can argue that while the water is on their land, they should be free to use or divert it as they please. Who is right- and wrong-minded in this situation? What are the ethical considerations from both perspectives?

3. Was 4.5%, the discount rate used in 1964, a reasonable one for the governments to choose? Explain and justify your conclusion.

### After Completing This Chapter...

#### The student should be able to:

- Define and apply the *present worth criteria*.
- Compare two competing alternative choices using present worth (PW).
- Apply the PW model in cases with equal, unequal, and infinite project lives.
- Compare multiple alternatives using the PW criteria.
- Develop and use spreadsheets to make *present worth* calculations.

In Chapters 3 and 4 we accomplished two important tasks. First, we presented the concept of equivalence. We can compare cash flows only if we can resolve them into equivalent values. Second, we derived a series of compound interest factors to find those equivalent values.

## ASSUMPTIONS IN SOLVING ECONOMIC ANALYSIS PROBLEMS

One of the difficulties of problem solving is that most problems tend to be very complicated. It becomes apparent that *some* simplifying assumptions are needed to make complex problems manageable. The trick, of course, is to solve the simplified problem and still be satisfied that the solution is applicable to the *real* problem! In the subsections that follow, we will consider six different items and explain the customary assumptions that are made. These assumptions apply to all problems and examples, unless other assumptions are given.

### End-of-Year Convention

As we indicated in Chapter 4, economic analysis textbooks and practice follow the end-of-period convention. This makes "*A*" a series of end-of-period receipts or disbursements.

A cash flow diagram of *P*, *A*, and *F* for the end-of-period convention is as follows:

If one were to adopt a middle-of-period convention, the diagram would be:

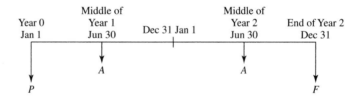

As the diagrams illustrate, only *A* shifts; *P* remains at the beginning of the period and *F* at the end of period, regardless of the convention. The compound interest tables in Appendix C are based on the end-of-period convention.

### Viewpoint of Economic Analysis Studies

When we make economic analysis calculations, we must proceed from a point of reference. Generally, we will want to take the point of view of a total firm when doing industrial economic analyses. Example 1-1 vividly illustrated the problem: a firm's shipping department decided it could save money by outsourcing its printing work rather than by using the in-house printing department. An analysis from the viewpoint of the shipping department supported this, as it could get for $688.50 the same printing it was paying $793.50 for in-house. Further analysis showed, however, that its printing department costs would decline *less* than using the commercial printer would save. From the viewpoint of the firm, the net result would be an increase in total cost.

From Example 1-1 we see it *is* important that the **viewpoint of the study** be carefully considered. Selecting a narrow viewpoint, like that of the shipping department, may result in a suboptimal decision from the firm's viewpoint. It is the total firm's viewpoint that is used in industrial economic analyses. For public-sector problems, the combined viewpoint of the government and the citizenry is chosen, since for many public projects the benefits of faster commuting, newer schools, and so on are received by individuals and the costs are paid by the government.

## Sunk Costs

We know that it is the **differences between alternatives** that are relevant in economic analysis. Events that have occurred in the past really have no bearing on what we should do in the future. When the judge says, "$200 fine or 3 days in jail," the events that led to these unhappy alternatives really are unimportant. It is the current and future differences between the alternatives that *are* important. Past costs, like past events, have no bearing on deciding between alternatives unless the past costs somehow affect the present or future costs. In general, past costs do not affect the present or the future, so we refer to them as *sunk costs* and disregard them.

## Borrowed Money Viewpoint

In most economic analyses, the proposed alternatives inevitably require money to be spent, and so it is natural to ask the source of that money. Thus, each problem has two monetary aspects: one is the **financing**—the obtaining of money; the other is the **investment**—the spending of money. Experience has shown that these two concerns should be distinguished. When separated, the problems of obtaining money and of spending it are both logical and straightforward. Failure to separate financing and investment sometimes produces confusing results and poor decision making.

The conventional assumption in economic analysis is that the money required to finance alternatives/solutions in problem solving is considered to be **obtained at interest rate $i$**.

## Effect of Inflation and Deflation

For the present we will assume that prices are stable or stated in constant-value dollars. This means that a machine that costs $5000 today can be expected to cost the same amount several years hence. Inflation and deflation can be serious problems for after-tax analysis and for cost and revenues whose inflation rates differ from the economy's inflation rates, but we assume stable prices for now.

## Income Taxes

Income taxes, like inflation and deflation, must be considered to find the real payoff of a project. However, taxes will often affect alternatives similarly, allowing us to compare our choices without considering income taxes. We will introduce income taxes into economic analyses in a later chapter.

## ECONOMIC CRITERIA

We have shown how to manipulate cash flows in a variety of ways, and we can now solve many kinds of compound interest problems. But engineering economic analysis is more than simply solving interest problems. The decision process (see Figure 1-1) requires that the outcomes of feasible alternatives be arranged so that they may be judged for **economic efficiency** in terms of the selection criterion. The economic criterion will be one of the following, depending on the situation:

| Situation | Criterion |
|---|---|
| Neither input nor output fixed | Maximize (output–input) |
| For fixed input | Maximize output |
| For fixed output | Minimize input |

We will now examine ways to resolve engineering problems, so that criteria for economic efficiency can be applied.

Equivalence provides the logic by which we may adjust the cash flow for a given alternative into some equivalent sum or series. We must still choose which comparable units to use. In this chapter we will learn how analysis can resolve alternatives into *equivalent present consequences*, referred to simply as **present worth analysis.** Chapter 6 will show how given alternatives are converted into an *equivalent uniform annual cash flow*, and Chapter 7 solves for the interest rate at which favorable consequences—that is, *benefits*—are equivalent to unfavorable consequences—or *costs*.

As a general rule, any economic analysis problem may be solved by any of these three methods. This is true because *present worth*, *annual cash flow*, and *rate of return* are exact methods that will always yield the same recommendation for selecting the best alternative from among a set of mutually exclusive alternatives. Remember that "mutually exclusive" means that selecting one alternative precludes selecting any other alternative. For example, constructing a gas station and constructing a drive-in restaurant on a particular piece of vacant land are mutually exclusive alternatives.

Some problems, however, may be more easily solved by one method. We now focus on problems that are most readily solved by present worth analysis.

## APPLYING PRESENT WORTH TECHNIQUES

One of the easiest ways to compare mutually exclusive alternatives is to resolve their consequences to the present time. The three criteria for economic efficiency are restated in terms of present worth analysis in Table 5-1.

Present worth analysis is most frequently used to determine the present value of future money receipts and disbursements. It would help us, for example, to determine the present worth of an income-producing property, like an oil well or an apartment house. If the future income and costs are known, then we can use a suitable interest rate to calculate the property's present worth. This should provide a good estimate of the price at which the property could be bought or sold. Another application is valuing stocks or bonds based on the anticipated future benefits of ownership.

In present worth analysis, careful consideration must be given to the time period covered by the analysis. Usually the task to be accomplished has a time period associated

### TABLE 5–1  Present Worth Analysis

|  | Situation | Criterion |
|---|---|---|
| Neither input nor output is fixed | Typical, general case | Maximize (present worth of benefits *minus* present worth of costs), that is, maximize net present worth |
| Fixed input | Amount of money or other input resources are fixed | Maximize present worth of benefits or other outputs |
| Fixed output | There is a fixed task, benefit, or other output to be accomplished | Minimize present worth of costs or other inputs |

with it. The consequences of each alternative must be considered for this period of time, which is usually called the **analysis period, planning horizon,** or **project life.**

The analysis period for an economy study should be determined from the situation. In some industries with rapidly changing technologies, a rather short analysis period or planning horizon might be in order. Industries with more stable technologies (like steelmaking) might use a longer period (say, 10–20 years), while government agencies frequently use analysis periods extending to 50 years or more.

Three different analysis-period situations are encountered in economic analysis problems with multiple alternatives:

1. The useful life of each alternative equals the analysis period.
2. The alternatives have useful lives different from the analysis period.
3. There is an infinite analysis period, $n = \infty$.

### Useful Lives Equal the Analysis Period

Since different lives and an infinite analysis period present some complications, we will begin with four examples in which the useful life of each alternative equals the analysis period.

## EXAMPLE 5–1

A firm is considering which of two mechanical devices to install to reduce costs. Both devices have useful lives of 5 years and no salvage value. Device *A* costs $1000 and can be expected to result in $300 savings annually. Device *B* costs $1350 and will provide cost savings of $300 the first year but will increase $50 annually, making the second-year savings $350, the third-year savings $400, and so forth. With interest at 7%, which device should the firm purchase?

### SOLUTION

The analysis period can conveniently be selected as the useful life of the devices, or 5 years. The appropriate decision criterion is to choose the alternative that maximizes the net present worth of benefits minus costs.

$$\text{PW}_A = 300(P/A, 7\%, 5) = -1000 + 300(4.100) = \$230$$

$$\text{PW}_B = -1350 + 300(P/A, 7\%, 5) + 50(P/G, 7\%, 5)$$

$$= -1350 + 300(4.100) + 50(7.647) = \$262.4$$

Device $B$ has the larger present worth and is the preferred alternative.

## EXAMPLE 5–2

Wayne County will build an aqueduct to bring water in from the upper part of the state. It can be built at a reduced size now for $300 million and be enlarged 25 years hence for an additional $350 million. An alternative is to construct the full-sized aqueduct now for $400 million.

Both alternatives would provide the needed capacity for the 50-year analysis period. Maintenance costs are small and may be ignored. At 6% interest, which alternative should be selected?

### SOLUTION

This problem illustrates staged construction. The aqueduct may be built in a single stage, or in a smaller first stage followed many years later by a second stage to provide the additional capacity when needed.

**For the Two-Stage Construction**

$$\text{PW of cost} = \$300 \text{ million} + 350 \text{ million}(P/F, 6\%, 25)$$

$$= \$300 \text{ million} + 81.6 \text{ million}$$

$$= \$381.6 \text{ million}$$

**For the Single-Stage Construction**

$$\text{PW of cost} = \$400 \text{ million}$$

The two-stage construction has a smaller present worth of cost and is the preferred construction plan.

## EXAMPLE 5–3

A purchasing agent plans to buy some new equipment for the mailroom. Two manufacturers have provided bids. An analysis shows the following:

| Manufacturer | Cost | Useful Life (years) | End-of-Useful-Life Salvage Value |
|---|---|---|---|
| Speedy | $1500 | 5 | $200 |
| Allied | 1600 | 5 | 325 |

The equipment of both manufacturers is expected to perform at the desired level of (fixed) output. For a 5-year analysis period, which manufacturer's equipment should be selected? Assume 7% interest and equal maintenance costs.

### SOLUTION

For fixed output, the criterion is to minimize the present worth of cost.

Speedy

$$\text{PW of cost} = 1500 - 200(P/F, 7\%, 5)$$
$$= 1500 - 200(0.7130)$$
$$= 1500 - 143 = \$1357$$

Allied

$$\text{PW of cost} = 1600 - 325(P/F, 7\%, 5) = 1600 - 325(0.7130)$$
$$= 1600 - 232 = \$1368$$

Since it is only the *differences between alternatives* that are relevant, maintenance costs may be left out of the economic analysis. Although the PWs of cost for the two alternatives are nearly identical, we would still choose the one with minimum present worth of cost unless there were other tangible or intangible differences that would change the decision. Buy the Speedy equipment.

## EXAMPLE 5–4

A firm is trying to decide which of two weighing scales it should install to check a package-filling operation in the plant. The ideal scale would allow better control of the filling operation, hence less overfilling. If both scales have lives equal to the 6-year analysis period, which one should be selected? Assume an 8% interest rate.

| Alternatives | Cost | Uniform Annual Benefit | End-of-Useful-Life Salvage Value |
|---|---|---|---|
| Atlas scale | $2000 | $450 | $100 |
| Tom Thumb scale | 3000 | 600 | 700 |

**Atlas Scale**

$$\text{PW of benefits} - \text{PW of cost} = 450(P/A, 8\%, 6) + 100(P/F, 8\%, 6) - 2000$$
$$= 450(4.623) + 100(0.6302) - 2000$$
$$= 2080 + 63 - 2000 = \$143$$

**Tom Thumb Scale**

$$\text{PW of benefits} - \text{PW of cost} = 600(P/A, 8\%, 6) + 700(P/F, 8\%, 6) - 3000$$
$$= 600(4.623) + 700(0.6302) - 3000$$
$$= 2774 + 441 - 3000 = \$215$$

The salvage value of each scale, it should be noted, is simply treated as another positive cash flow. Since the criterion is to maximize the present worth of benefits minus the present worth of cost, the preferred alternative is the Tom Thumb scale.

In Examples 5-1 and 5-4, we compared two alternatives and selected the one in which present worth of benefits *minus* present worth of cost was a maximum. The criterion is called the **net present worth criterion** and written simply as **NPW:**

$$\text{Net present worth} = \text{Present worth of benefits} - \text{Present worth of cost}$$

$$\text{NPW} = \text{PW of benefits} - \text{PW of cost} \tag{5-1}$$

The field of engineering economy and this text often use present worth (PW), present value (PV), net present worth (NPW), and net present value (NPV) as synonyms. Sometimes, as in the foregoing definition, *net* is included to emphasize that both costs and benefits have been considered.

## Useful Lives Different from the Analysis Period

In present worth analysis, there always must be an identified analysis period. It follows, then, that each alternative must be considered for the entire period. In Examples 5-1 to 5-4, the useful life of each alternative was equal to the analysis period. While often this is true, in many situations at least one alternative will have a useful life different from the analysis period. This section describes one way to evaluate alternatives with lives different from the study period.

In Example 5-3, suppose that the Allied equipment was expected to have a 10-year useful life, or twice that of the Speedy equipment. Assuming the Allied salvage value would still be $325 in 10 years, which equipment should now be purchased? We can recompute the present worth of cost of the Allied equipment, starting as follows:

$$\text{PW of cost} = 1600 - 325(P/F, 7\%, 10)$$
$$= 1600 - 325(0.5083)$$
$$= 1600 - 165 = \$1435$$

The present worth of cost has increased. This, of course, is because of the more distant recovery of the salvage value. More importantly, we now find ourselves attempting to compare Speedy equipment, with its 5-year life, against the Allied equipment with a 10-year life. Because of the variation in the useful life of the equipment, we no longer have a situation of *fixed output*. Speedy equipment in the mailroom for 5 years is certainly not the same as 10 years of service with Allied equipment.

For present worth calculations, it is important that we select an analysis period and judge the consequences of each of the alternatives during that period. As such, it is not a fair comparison to compare the NPW of the Allied equipment over its 10-year life against the NPW of the Speedy equipment over its 5-year life.

The firm, its economic environment, and the specific situation are important in selecting an analysis period. If the Allied equipment (Example 5-3) has a useful life of 10 years, and the Speedy equipment will last 5 years, one method is to select an analysis period that is the **least common multiple** of their useful lives. Thus we would compare the 10-year life of Allied equipment against an initial purchase of Speedy equipment *plus* its replacement with new Speedy equipment in 5 years. The result is to judge the alternatives on the basis of a 10-year requirement in the mailroom. On this basis the economic analysis is as follows.

Assuming the replacement Speedy equipment 5 years hence will also cost $1500,

$$PW \text{ of cost} = 1500 + (1500 - 200)(P/F, 7\%, 5) - 200(P/F, 7\%, 10)$$

$$= 1500 + 1300(0.7130) - 200(0.5083)$$

$$= 1500 + 927 - 102 = \$2325$$

For the Allied equipment, on the other hand, we have the following results:

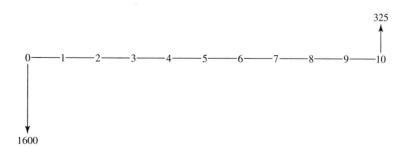

$$\text{PW of cost} = 1600 - 325(P/F, 7\%, 10) = 1600 - 325(0.5083) = \$1435$$

For the fixed output of 10 years of service in the mailroom, the Allied equipment, with its smaller present worth of cost, is preferred.

We have seen that setting the analysis period equal to the least common multiple of the lives of the two alternatives seems reasonable in the revised Example 5-3. However, what if the alternatives had useful lives of 7 and 13 years? Here the least common multiple of lives is 91 years. An analysis period of 91 years hardly seems realistic. Instead, a suitable analysis period should be based on how long the equipment is likely to be needed. This may require that terminal values be estimated for the alternatives at some point prior to the end of their useful lives.

As Figure 5-1 shows, it is not necessary for the analysis period to equal the useful life of an alternative or some multiple of the useful life. To properly reflect the situation at the end of the analysis period, an estimate is required of the market value of the equipment at that time. The calculations might be easier if everything came out even, but this is not essential.

**FIGURE 5-1** Superimposing a 10-year analysis period on 7- and 13-year alternatives.

## EXAMPLE 5-5

A diesel manufacturer is considering the two alternative production machines graphically depicted in Figure 5-1. Specific data are as follows:

|                                               | Alt. 1   | Alt. 2   |
|-----------------------------------------------|----------|----------|
| Initial cost                                  | $50,000  | $75,000  |
| Estimated salvage value at end of useful life | $10,000  | $12,000  |
| Useful life of equipment, in years            | 7        | 13       |

The manufacturer uses an interest rate of 8% and wants to use the PW method to compare these alternatives over an analysis period of 10 years.

|                                                        | Alt. 1   | Alt. 2   |
|--------------------------------------------------------|----------|----------|
| Estimated market value, end of 10-year analysis period | $20,000  | $15,000  |

In this case, the decision maker is setting the analysis period at 10 years rather than accepting a common multiple of the lives of the alternatives, or assuming that the period of needed service is infinite (to be discussed in the next section). This is a legitimate approach—perhaps the diesel manufacturer will be phasing out this model at the end of the 10-year period. In any event, we need to compare the alternatives over 10 years.

As illustrated in Figure 5-1, we may assume that Alternative 1 will be replaced by an identical machine after its 7-year useful life. Alternative 2 has a 13-year useful life. The diesel manufacturer has provided an estimated market value of the equipment at the time of the analysis period. We can compare the two choices over 10 years as follows:

$$PW \text{ (Alt. 1)} = -50{,}000 + (10{,}000 - 50{,}000)(P/F, 8\%, 7) + 20{,}000(P/F, 8\%, 10)$$
$$= -50{,}000 - 40{,}000(0.5835) + 20{,}000(0.4632)$$
$$= -\$64{,}076$$

$$PW \text{ (Alt. 2)} = -75{,}000 + 15{,}000(P/F, 8\%, 10)$$
$$= -75{,}000 + 15{,}000(0.4632)$$
$$= -\$69{,}442$$

To minimize PW of costs the diesel manufacturer should select Alt. 1.

## Infinite Analysis Period: Capitalized Cost

Another difficulty in present worth analysis arises when we encounter an infinite analysis period ($n = \infty$). In governmental analyses, a service or condition sometimes must be maintained for an infinite period. The need for roads, dams, pipelines, and so on, is sometimes considered to be permanent. In these situations a present worth of cost analysis would have an infinite analysis period. We call this particular analysis **capitalized cost.**

Infinite lives are rare in the private sector, but a similar assumption of "indefinitely long" horizons is sometimes made. This assumes that the facility will need electric motors, mechanical HVAC equipment, and forklifts as long as it operates, and that the facility will last far longer than any individual unit of equipment. So the equipment can be analyzed as though the problem horizon is *infinite* or indefinitely long.

Capitalized cost is the present sum of money that would need to be set aside now, at some interest rate, to yield the funds required to provide the service (or whatever) indefinitely. To accomplish this, the money set aside for future expenditures must not decline. The interest received on the money set aside can be spent, but not the principal. When one stops to think about an infinite analysis period (as opposed to something relatively short, like a hundred years), we see that an undiminished principal sum is essential; otherwise one will of necessity run out of money prior to infinity.

In Chapter 3 we saw that

$$\text{Principal sum} + \text{Interest for the period} = \text{Amount at end of period, or}$$

$$P \quad + \quad iP \quad = \quad P + iP$$

If we spend $iP$, then in the next interest period the principal sum $P$ will again increase to $P + iP$. Thus, we can again spend $iP$.

This concept may be illustrated by a numerical example. Suppose you deposited $200 in a bank that paid 4% interest annually. How much money could be withdrawn each year without reducing the balance in the account below the initial $200? At the end of the first year, the $200 would have earned 4%($200) = $8 interest. If this interest were withdrawn, the $200 would remain in the account. At the end of the second year, the $200 balance would again earn 4%($200) = $8. This $8 could also be withdrawn and the account would still have $200. This procedure could be continued indefinitely and the bank account would always contain $200. If more or less than $8 is withdrawn, the account will either increase to ∞ or decrease to 0.

The year-by-year situation would be depicted like this:

$$\textit{Year 1}: \$200 \text{ initial } P \rightarrow 200 + 8 = 208$$

$$\text{Withdrawal } iP = -\underline{\ 8\ }$$

$$\textit{Year 2}: \$200 \rightarrow 200 + 8 = 208$$

$$\text{Withdrawal } iP = -\underline{\ 8\ }$$

$$\$200$$

and so on

Thus, for any initial present sum $P$, there can be an end-of-period withdrawal of $A$ equal to $iP$ each period, and these withdrawals can continue forever without diminishing the initial sum $P$. This gives us the basic relationship:

$$\text{For} \quad n = \infty, \quad A = Pi$$

This relationship is the key to capitalized cost calculations. Earlier we defined capitalized cost as the present sum of money that would need to be set aside at some interest rate to yield the funds to provide the desired task or service forever. Capitalized cost is therefore the $P$ in the equation $A = iP$. It follows that:

$$\textbf{Capitalized cost} \qquad P = \frac{A}{i} \qquad\qquad (5\text{-}2)$$

If we can resolve the desired task or service into an equivalent $A$, the capitalized cost can be computed. The following examples illustrate such computations.

## EXAMPLE 5–6

How much should one set aside to pay $50 per year for maintenance on a gravesite if interest is assumed to be 4%? For perpetual maintenance, the principal sum must remain undiminished after the annual disbursement is made.

### SOLUTION

$$\text{Capitalized cost } P = \frac{\text{Annual disbursement } A}{\text{Interest rate } i}$$

$$P = \frac{50}{0.04} = \$1250$$

One should set aside $1250.

## EXAMPLE 5–7

A city plans a pipeline to transport water from a distant watershed area to the city. The pipeline will cost $8 million and will have an expected life of 70 years. The city expects to keep the water line in service indefinitely. Compute the capitalized cost, assuming 7% interest.

### SOLUTION

The capitalized cost equation

$$P = \frac{A}{i}$$

is simple to apply when there are end-of-period disbursements $A$. Here we have renewals of the pipeline every 70 years. To compute the capitalized cost, it is necessary to first compute an end-of-period disbursement $A$ that is equivalent to $8 million every 70 years.

The $8 million disbursement at the end of each 70-year period may be resolved into an equivalent $A$.

$$A = F(A/F, i, n) = \$8 \text{ million}(A/F, 7\%, 70)$$

$$= \$8 \text{ million}(0.00062)$$

$$= \$4960$$

Each 70-year period is identical to this one, and the infinite series is shown in Figure 5-2.

$$\text{Capitalized cost } P = 8 \text{ million} + \frac{A}{i} = 8 \text{ million} + \frac{4960}{0.07}$$

$$= \$8,071,000$$

**FIGURE 5–2** Using the sinking fund factor to compute an infinite series.

Capitalized Cost   $8 million   $A = 4960$   $n = \infty$

## ALTERNATE SOLUTION 1

Instead of solving for an equivalent end-of-period payment $A$ based on a *future* $8 million disbursement, we could find $A$, given a *present* $8 million disbursement.

$$A = P(A/P, i, n) = 8 \text{ million}(A/P, 7\%, 70)$$

$$= 8 \text{ million}(0.0706) = \$565,000$$

On this basis, the infinite series is shown in Figure 5-3. Carefully note the difference between this and Figure 5-2. Now:

$$\text{Capitalized cost } P = \frac{A}{i} = \frac{565,000}{0.07} = \$8,071,000$$

**FIGURE 5–3** Using the capital recovery factor to compute an infinite series.

$P$   $A = 565,000$   $n = \infty$

## ALTERNATE SOLUTION 2

Another way of solving the problem is to assume the interest period is 70 years long and compute an equivalent interest rate for the 70-year period. Then the capitalized cost may be computed by using Equation 3–8 for $m = 70$

$$i_{70\,\text{yr}} = (1 + i_{1\,\text{yr}})^{70} - 1 = (1 + 0.07)^{70} - 1 = 112.989$$

$$\text{Capitalized cost} = 8 \text{ million} + \frac{8 \text{ million}}{112.989} = \$8,070,803$$

## Multiple Alternatives

So far the discussion has been based on examples with only two alternatives. But multiple-alternative problems may be solved by exactly the same methods. (The only reason for avoiding multiple alternatives was to simplify the examples.) Examples 5-8 and 5-9 have multiple alternatives.

## EXAMPLE 5–8

A contractor has been awarded the contract to construct a 6-mile-long tunnel in the mountains. During the 5-year construction period, the contractor will need water from a nearby stream. She will construct a pipeline to carry the water to the main construction yard. An analysis of costs for various pipe sizes is as follows:

| | Pipe Sizes (in.) | | | |
|---|---|---|---|---|
| | 2 | 3 | 4 | 6 |
| Installed cost of pipeline and pump | $22,000 | $23,000 | $25,000 | $30,000 |
| Cost per hour for pumping | $1.20 | $0.65 | $0.50 | $0.40 |

At the end of 5 years, the pipe and pump will have a salvage value equal to the cost of removing them. The pump will operate 2000 hours per year. The lowest interest rate at which the contractor is willing to invest money is 7%. (The minimum required interest rate for invested money is called the **minimum attractive rate of return,** or MARR.) Select the alternative with the least present worth of cost.

### SOLUTION

We can compute the present worth of cost for each alternative. For each pipe size, the present worth of cost is equal to the installed cost of the pipeline and pump plus the present worth of 5 years of pumping costs.

| | Pipe Size (in.) | | | |
|---|---|---|---|---|
| | 2 | 3 | 4 | 6 |
| Installed cost of pipeline and pump | $22,000 | $23,000 | $25,000 | $30,000 |
| $1.20 \times 2000$ hr $\times (P/A, 7\%, 5)$ | 9,840 | | | |
| $0.65 \times 2000$ hr $\times 4.100$ | | 5,330 | | |
| $0.50 \times 2000$ hr $\times 4.100$ | | | 4,100 | |
| $0.40 \times 2000$ hr $\times 4.100$ | | | | 3,280 |
| Present worth of cost | $31,840 | $28,330 | $29,100 | $33,280 |

Select the 3-in. pipe, since the lowest present worth of cost.

## EXAMPLE 5–9

An investor paid $8000 to a consulting firm to analyze possible uses for a small parcel of land on the edge of town that can be bought for $30,000. In their report, the consultants suggested four alternatives:

| Alternatives | | Total Investment Including Land* | Uniform Net Annual Benefit | Terminal Value at End of 20 yr |
|---|---|---|---|---|
| A | Do nothing | $      0 | $      0 | $      0 |
| B | Vegetable market | 50,000 | 5,100 | 30,000 |
| C | Gas station | 95,000 | 10,500 | 30,000 |
| D | Small motel | 350,000 | 36,000 | 150,000 |

*Includes the land and structures but does not include the $8000 fee to the consulting firm.

Assuming 10% is the minimum attractive rate of return, what should the investor do?

## SOLUTION

Alternative *A* is the "do-nothing" alternative. Generally, one feasible alternative in any situation is to remain in the present status and do nothing. In this problem, the investor could decide that the most attractive alternative is not to purchase the property and develop it. This is clearly a decision to do nothing.

We note, however, that if he does nothing, the total venture would not be a very satisfactory one. This is because the investor spent $8000 for professional advice on the possible uses of the property. But because the $8000 is a past cost, it is a **sunk cost.** The only relevant costs in an economic analysis are *present* and *future* costs. Past events and past, or sunk, costs are gone and cannot be allowed to affect future planning. (Past costs may be relevant in computing depreciation charges and income taxes, but nowhere else.) The past should not deter the investor from making the best decision now.

This problem is one of neither fixed input nor fixed output, so our criterion will be to maximize the present worth of benefits *minus* the present worth of cost; that is, to maximize net present worth.

**Alternative *A*, Do Nothing**

$$NPW = 0$$

**Alternative *B*, Vegetable Market**

$$NPW = -50,000 + 5100(P/A, 10\%, 20) + 30,000(P/F, 10\%, 20)$$

$$= -50,000 + 5100(8.514) + 30,000(0.1486)$$

$$= -50,000 + 43,420 + 4460$$

$$= -\$2120$$

Using a TVM calculator with $\times(-1)$ to change sign, we obtain

$$NPW = -50,000 + PV(10\%, 20, 5100, 30000) \times (-1)$$

$$= -50,000 + 47,878.5 = -\$2121.5$$

**Alternative C, Gas Station**

$$NPW = -95{,}000 + 10{,}500(P/A, 10\%, 20) + 30{,}000(P/F, 10\%, 20)$$

$$= -95{,}000 + 89{,}400 + 4460 = -\$1140$$

or

$$NPW = -95{,}000 + PV(10\%, 20, 10500, 3000) \times (-1)$$

$$= -95{,}000 + 93{,}851.7 = -\$1148.3$$

**Alternative D, Small Motel**

$$NPW = -350{,}000 + 36{,}000(P/A, 10\%, 20) + 150{,}000(P/F, 10\%, 20)$$

$$= -350{,}000 + 306{,}500 + 22{,}290 = -\$21{,}210$$

or

$$NPW = -350{,}000 + PV(10\%, 20, 36000, 150000) \times (-1)$$

$$= -350{,}000 + 328{,}784.8 = -\$21{,}215.2$$

The criterion is to maximize net present worth. In this situation, one alternative has NPW equal to zero, and three alternatives have negative values for NPW. We will select the best of the four alternatives, namely, the do-nothing Alt. *A*, with NPW equal to zero.

## EXAMPLE 5–10

A piece of land may be purchased for $610,000 to be strip-mined for the underlying coal. Annual net income will be $200,000 for 10 years. At the end of the 10 years, the surface of the land will be restored as required by a federal law on strip mining. The reclamation will cost $1.5 million more than the resale value of the land after it is restored. Using a 10% interest rate, determine whether the project is desirable.

## SOLUTION

The investment opportunity may be described by the following cash flow:

| Year | Cash Flow (thousands) |
|------|-----------------------|
| 0    | −$610                 |
| 1–10 | +200 (per year)       |
| 10   | −1500                 |

$$NPW = -610 + 200(P/A, 10\%, 10) - 1500(P/F, 10\%, 10)$$

$$= -610 + 200(6.145) - 1500(0.3855)$$

$$= -610 + 1229 - 578$$

$$= +\$41$$

Since NPW is positive, the project is desirable. (See Appendix 7A for a more complete analysis of this type of problem. At interest rates of 4.07% and 18.29%, NPW = 0.) In this example the unlisted alternative was the do-nothing alternative with NPW = 0.

## EXAMPLE 5–11

Two pieces of construction equipment are being analyzed:

| Year | Alt. A | Alt. B |
|------|--------|--------|
| 0 | −$2000 | −$1500 |
| 1 | 1000 | 700 |
| 2 | 850 | 300 |
| 3 | 700 | 300 |
| 4 | 550 | 300 |
| 5 | 400 | 300 |
| 6 | 400 | 400 |
| 7 | 400 | 500 |
| 8 | 400 | 600 |

At an 8% interest rate, which alternative should be selected?

### SOLUTION

Alternative A

$$\text{PW of benefits} = 400(P/A, 8\%, 8) + 600(P/A, 8\%, 4) - 150(P/G, 8\%, 4)$$

$$= 400(5.747) + 600(3.312) - 150(4.650) = 3588.50$$

$$\text{PW of cost} = 2000$$

$$\text{Net present worth} = 3588.50 - 2000 = +\$1588.50$$

**Alternative B**

$$\text{PW of benefits} = 300(P/A, 8\%, 8) + (700 - 300)(P/F, 8\%, 1)$$

$$+ 100(P/G, 8\%, 4)(P/F, 8\%, 4)$$

$$= 300(5.747) + 400(0.9259) + 100(4.650)(0.7350)$$

$$= 2436.24$$

$$\text{PW of cost} = 1500$$

$$\text{Net present worth} = 2436.24 - 1500$$

$$= +\$936.24$$

To maximize NPW, choose Alt. *A*.

## Bond Pricing

The calculation in Example 5-12 is done routinely when bonds are bought and sold during their life. Bonds are issued at a face or par value (usually $1000), which is received when the bonds mature. There is a coupon interest rate, which is set when the bond is originally issued or sold. The term *coupon interest* dates from the time when bonds were paper rather than electronic, and a paper coupon was detached from the bond to be redeemed in cash. The annual interest paid equals the coupon rate times the face value. This interest is usually paid in two semiannual payments, although some bonds have other compounding periods. A cash flow diagram for the remaining interest payments and the final face value is used with a current interest rate to calculate a price.

## EXAMPLE 5–12

A 15-year municipal bond was issued 5 years ago. Its coupon interest rate is 8%, interest payments are made semiannually, and its face value is $1000. If the current market interest rate is 12.36%, what should be the bond's price? *Note*: The issuer of the bond (city, state, company) makes interest payments to the bondholder (at the coupon rate), as well as a final value payment.

The first 5 years are past, and there are 20 more semiannual payments. The coupon interest rate is the nominal annual rate or APR. Half that, or 4%, of $1000 or $40 is paid at the end of each 6 month period.

The bond's price is the PW of the cash flows that will be received if the bond is purchased. The cash flows are $40 at the end of each of the 20 semiannual periods and the face value of $1000 at the end of period 20.

Since the $40 in interest is received semiannually, the market or effective annual interest rate ($i_a$) must be converted to a semiannual rate. Using Equation 3-8, we obtain

$$(1+i)^2 = 1 + i_a = 1.1236$$

$$(1+i) = 1.06$$

$$i = 6\% \text{ effective semiannual interest rate}$$

$$PW = 40(P/A, 6\%, 20) + 1000(P/F, 6\%, 20)$$

$$= 40(11.470) + 1000(0.3118) = \$770.6$$

The $770.6 is the discounted price; that is, the PW at 12.36% of the cash flows from the $1000 bond. The $229.4 discount raises the investment's rate of return from a nominal 8% for the face value to 12.36% on an investment of $770.6.

This example also illustrates why it is better to separately state cash flows. At the end of period 20 from now, there are two cash flows, $40 and $1000. The $40 is part of the 20-period uniform series, and the $1000 is a single cash flow. All of these numbers come directly from the problem statement. If the two final cash flows are combined into $1040, then the $40 uniform series has only 19 periods—and it is easy to err and forget that change.

Note that if a TVM calculator is used (see Appendix B), then this can be solved as $PV = PV(i, n, A, F) = PV(6, 20, 40, 1000) = -\$770.6$.

## SPREADSHEETS AND PRESENT WORTH

Spreadsheets make it easy to build more accurate models with shorter time periods. When one is using factors, it is common to assume that costs and revenues are uniform for $n$ years. With spreadsheets it is easy to use 120 months instead of 10 years, and the cash flows can be estimated for each month. For example, energy costs for air conditioning peak in the summer, and in many areas there is little construction during the winter. Cash flows that depend on population often increase at $x\%$ per year, such as for electric power and transportation costs.

In spreadsheets any interest rate is entered exactly—so no interpolation is needed. This makes it easy to calculate the monthly repayment schedule for a car loan or a house mortgage. Examples 5-12 and 5-13 illustrate using spreadsheets to calculate PWs.

## EXAMPLE 5-13

NLE Construction is bidding on a project whose costs are divided into $30,000 for start-up and $240,000 for the first year. If the interest rate is 1% per month or 12.68% per year, what is the present worth with monthly compounding?

### SOLUTION

Figure 5-4 illustrates the spreadsheet solution with the assumption that costs are distributed evenly throughout the year ($-20,000 = -240,000/12$).

FIGURE 5-4 Spreadsheet with monthly cash flows.

| | A | B | C | D |
|---|---|---|---|---|
| 1 | 1% | i | | |
| 2 | −30,000 | initial cash flow | | |
| 3 | −240,000 | annual amount | | |
| 4 | | | | |
| 5 | Month | Cash Flow | | |
| 6 | 0 | −30000 | | |
| 7 | 1 | −20000 | | |
| 8 | 2 | −20000 | | |
| 9 | 3 | −20000 | | |
| 10 | 4 | −20000 | | |
| 11 | 5 | −20000 | | |
| 12 | 6 | −20000 | | |
| 13 | 7 | −20000 | | |
| 14 | 8 | −20000 | | |
| 15 | 9 | −20000 | | |
| 16 | 10 | −20000 | | |
| 17 | 11 | −20000 | | |
| 18 | 12 | −20000 | | |
| 19 | NPV | −$255,102 | =NPV(A1,B7:B18)+B6 | |

Since the costs are uniform, the factor solution is

$$PW_{mon} = -30,000 - 20,000(P/A, 1\%, 12) = -\$255,102$$

The value of monthly periods can be illustrated by computing the PW assuming an annual period. The results differ by more than $12,000, because $20,000 at the end of Months 1 through

12 is not the same as $240,000 at the end of Month 12. The timing of the cash flows makes the difference, even though the effective interest rates are the same.

$$PW_{annual} = -30,000 - 240,000(P/F, 12.68\%, 1) = -30,000 - 240,000/1.1268 = -\$242,993$$

## EXAMPLE 5–14

Regina Industries has a new product whose sales are expected to be 1.2, 3.5, 7, 5, and 3 million units per year over the next 5 years. Production, distribution, and overhead costs are stable at $120 per unit. The price will be $200 per unit for the first 2 years, and then $180, $160, and $140 for the next 3 years. The remaining R&D and production costs are $300 million. If $i$ is 15%, what is the present worth of the new product?

### SOLUTION

It is easiest to calculate the yearly net revenue per unit before building the spreadsheet shown in Figure 5-5. Those values are the yearly price minus the $120 of costs, which equals $80, $80, $60, $40, and $20.

FIGURE 5–5 Present worth of a new product.

|    | A | B | C | D | E |
|----|------|-----------|-------------|----------------|---------|
| 1  | 12% | $i$ | | | |
| 2  | | | Net Unit | Cash | |
| 3  | Year | Sales (M) | Revenue | Flow ($M) | |
| 4  | 0 | | | −300 | |
| 5  | 1 | 1.2 | 80 | 96 | |
| 6  | 2 | 3.5 | 80 | 280 | |
| 7  | 3 | 7 | 60 | 420 | |
| 8  | 4 | 5 | 40 | 200 | |
| 9  | 5 | 3 | 20 | 60 | |
| 10 | | D4+NPV(A1,D5:D9) = | | $469 | Million |

## SUMMARY

Present worth analysis is suitable for almost any economic analysis problem. But it is particularly desirable when we wish to know the present worth of future costs and benefits. We frequently want to know the value today of such things as income-producing assets, stocks, and bonds.

For present worth analysis, the proper economic criteria are:

| | |
|---|---|
| Neither input nor output is fixed | Maximize (PW of benefits − PW of costs) or, more simply stated: Maximize NPW |
| Fixed input | Maximize the PW of benefits |
| Fixed output | Minimize the PW of costs |

To make valid comparisons, we need to analyze each alternative in a problem over the same **analysis period** or **planning horizon.** If the alternatives do not have equal lives, some technique must be used to achieve a common analysis period. One method is to select an analysis period equal to the least common multiple of the alternative lives. Another method is to select an analysis period and estimate end-of-analysis-period salvage values for the alternatives.

Capitalized cost is the present worth of cost for an infinite analysis period ($n = \infty$). When $n = \infty$, the fundamental relationship is $A = iP$. Some form of this equation is used whenever there is a problem with an infinite analysis period.

The numerous assumptions routinely made in solving economic analysis problems include the following.

1. Present sums $P$ are beginning-of-period, and all series receipts or disbursements $A$ and future sums $F$ occur at the end of the interest period. The compound interest tables were derived on this basis.

2. In industrial economic analyses, the point of view for computing the consequences of alternatives is that of the total firm. Narrower views can result in suboptimal solutions.

3. Only the differences between the alternatives are relevant. Past costs are sunk costs and generally do not affect present or future costs. For this reason they are ignored.

4. The investment problem is isolated from the financing problem. We generally assume that all required money is borrowed at interest rate $i$.

5. For now, stable prices are assumed. The inflation–deflation problem is deferred to Chapter 14. Similarly, our discussion of income taxes is deferred to Chapter 12.

6. Often uniform cash flows or arithmetic gradients are reasonable assumptions. However, spreadsheets simplify the finding of PW in more complicated problems.

# PROBLEMS

## Present Value of One Alternative

**5-1**   Compute the present value, $P$, for the following cash flows.

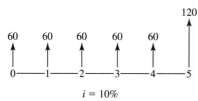

$i = 10\%$

(*Answer:* $P = \$324.71$)

**5-2**   Compute the present value, $P$, for the following cash flows.

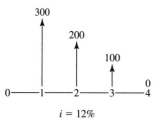

$i = 12\%$

(*Answer:* $P = \$498.50$)

**5-3**   Compute the present value, $P$, for the following cash flows.

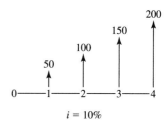

$i = 10\%$

**5-4**   Compute the present value, $P$, for the following cash flows.

$i = 15\%$

**5-5**   Find the value of $Q$ so that the present value is 0.

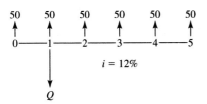

$$i = 12\%$$

**5-6**   Use a geometric gradient formula to compute the present value, $P$, for the following cash flows.

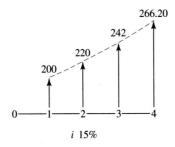

$$i\ 15\%$$

**5-7**   A stonecutter, assigned to carve the headstone for a well-known engineering economist, began with the following design.

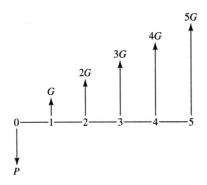

He then started the equation as follows:

$$P = G(P/G, i, 6)$$

He realized he had made a mistake. The equation should have been

$$P = G(P/G, i, 5) + G(P/A, i, 5)$$

The stonecutter does not want to discard the stone and start over. He asks you to help him with his problem. What one compound interest factor can be added to make the equation correct?

$$P = G(P/G, i, 6)(\quad, i, \ )$$

**5-8**   A project has a net present worth of $-\$14,000$ as of January 1, 2015. If a 10% interest rate is used, what was the project NPW as of December 31, 2012?

**5-9**   The annual income from a rented house is $24,000. The annual expenses are $6000. If the house can be sold for $245,000 at the end of 10 years, how much could you afford to pay for it now, if you considered 9% to be a suitable interest rate?
(*Answer:* $219,000)

**5-10**   Consider the following cash flow. At a 6% interest rate, what value of $P$ at the end of Year 1 is equivalent to the benefits at the end of Years 2 through 7?

| Year | Cash Flow |
|------|-----------|
| 1 | $-P$ |
| 2 | $100 |
| 3 | 200 |
| 4 | 300 |
| 5 | 400 |
| 6 | 500 |
| 7 | 600 |

**5-11**   How much would the owner of a building be justified in paying for a sprinkler system that will save $750 a year in insurance premiums if the system has to be replaced every 20 years and has a salvage value equal to 10% of its initial cost? Assume money is worth 7%.
(*Answer:* $8156)

**5-12**   A manufacturer is considering purchasing equipment that will have the following financial effects:

| Year | Disbursements | Receipts |
|------|---------------|----------|
| 0 | $4400 | $   0 |
| 1 | 660 | 880 |
| 2 | 660 | 1980 |
| 3 | 440 | 2420 |
| 4 | 220 | 1760 |

If money is worth 6%, should he invest in the equipment?

5-13 In a present worth analysis of certain equipment, one alternative has a net present worth of +$420, based on a 6-year analysis period that equals the useful life of the alternative. A 10% interest rate was used in the computations.

The alternative device is to be replaced at the end of the 6 years by an identical item with the same cost, benefits, and useful life. Using a 10% interest rate, compute the net present worth of the alternative equipment for the 12-year analysis period.
(*Answer:* NPW = +$657.09)

5-14 On February 1, the Miro Company needs to purchase some office equipment. The company is short of cash and expects to be short for several months. The treasurer has said that he could pay for the equipment as follows:

| Date | Payment |
| --- | --- |
| April 1 | $150 |
| June 1 | 300 |
| Aug. 1 | 450 |
| Oct. 1 | 600 |
| Dec. 1 | 750 |

A local office supply firm will agree to sell the equipment to Miro now and accept payment according to the treasurer's schedule. If interest will be charged at 3% every 2 months, with compounding once every 2 months, how much office equipment can the Miro Company buy now?
(*Answer:* $2020)

5-15 A machine costs $980,000 to purchase and will provide $200,000 a year in benefits. The company plans to use the machine for 13 years and then will sell the machine for scrap, receiving $20,000. The company interest rate is 12%. Should the machine be purchased?

5-16 Annual maintenance costs for a particular section of highway pavement are $2000. The placement of a new surface would reduce the annual maintenance cost to $500 per year for the first 5 years and to $1000 per year for the next 5 years. After 10 years the annual maintenance would again be $2000. If maintenance costs are the only saving,

what investment can be justified for the new surface? Assume interest at 4%.

5-17 A road building contractor has received a major highway construction contract that will require 50,000 m$^3$ of crushed stone each year for 5 years. The stone can be obtained from a quarry for $5.80/m$^3$. As an alternative, the contractor has decided to try to buy the quarry. He believes that if he owned the quarry, the stone would cost him only $4.30/m$^3$. He thinks he could resell the quarry at the end of 5 years for $200,000. If the contractor uses a 10% interest rate, how much would he be willing to pay for the quarry?

5-18 IBP Inc. is considering establishing a new machine to automate a meatpacking process. The machine will save $50,000 in labor annually. The machine can be purchased for $200,000 today and will be used for 10 years. It has a salvage value of $10,000 at the end of its useful life. The new machine will require an annual maintenance cost of $9000. The corporation has a minimum rate of return of 10%. Do you recommend automating the process?

5-19 A firm has installed a manufacturing line for packaging materials. The firm plans to produce 50 tons of packing peanuts at $5000 per ton annually for 5 years, and then 80 tons of packing peanuts per year at $5500 per ton for the next 5 years. What is the present worth of the expected income? The firm's minimum attractive rate of return is 18% per year.?
*Contributed by Hamed Kashani, Saeid Sadri, and Baabak Ashuri, Georgia Institute of Technology*

5-20 A wholesale company has signed a contract with a supplier to purchase goods for $2,000,000 annually. The first purchase will be made now to be followed by 10 more. Determine the contract's present worth at a 7% interest rate.
*Contributed by Hamed Kashani, Saeid Sadri, and Baabak Ashuri, Georgia Institute of Technology*

5-21 A new office building was constructed 5 years ago by a consulting engineering firm. At that time the firm obtained a bank loan for $600,000 with a 12% annual interest rate, compounded quarterly. The terms of the loan call for equal quarterly payments to repay the loan in 10 years. The loan also allows for its prepayment at any time without penalty.

As a result of internal changes in the firm, it is now proposed to refinance the loan through an insurance company. The new loan would be for a 20-year term with an interest rate of 8% per year, compounded quarterly. The new equal quarterly payments would repay the loan in the 20-year period. The insurance company requires the payment of a 5% loan initiation charge (often described as a "5-point loan fee"), which will be added to the new loan.

(*a*) What is the balance due on the original mortgage if 20 payments have been made in the last five years?

(*b*) What is the difference between the equal quarterly payments on the present bank loan and the proposed insurance company loan?

5-22 Argentina is considering constructing a bridge across the Rio de la Plata to connect its northern coast to the southern coast of Uruguay. If this bridge is constructed, it will reduce the travel time from Buenos Aires, Argentina, to São Paulo, Brazil, by over 10 hours, and there is the potential to significantly improve the flow of manufactured goods between the two countries. The cost of the new bridge, which will be the longest bridge in the world, spanning over 50 miles, will be $700 million. The bridge will require an annual maintenance of $10 million for repairs and upgrades and is estimated to last 80 years. It is estimated that 550,000 vehicles will use the bridge during the first year of operation, and an additional 50,000 vehicles per year until the tenth year. These data are based on a toll charge of $90 per vehicle. The annual traffic for the remainder of the life of the bridge will be 1,000,000 vehicles per year. The Argentine government requires a minimum rate of return of 9% to proceed with the project.

(*a*) Does this project provide sufficient revenues to offset its costs?

(*b*) What considerations are there besides economic factors in deciding whether to construct the bridge?

5-23 A student has a job that leaves her with $500 per month in disposable income. She decides that she will use the money to buy a car. Before looking for a car, she arranges a 100% loan whose terms are $500 per month for 36 months at 18% annual interest. What is the maximum car purchase price that she can afford with her loan?

5-24 The student in Problem 5-23 finds a car she likes and the dealer offers to arrange financing. His terms are 12% interest for 60 months and no down payment. The car's sticker price is $24,000. Can she afford to buy this car with her $500 monthly disposable income?

5-25 The student in Problem 5-24 really wants this particular car. She decides to try to negotiate a different interest rate. What is the highest interest rate that she can accept, given a 60-month term and $500 per month payments?

5-26 We know a car can be had for 60 monthly payments of $399. The dealer has set us a nominal interest rate of 4.5% compounded daily. What is the purchase price?

5-27 Compute the present value, $P$, for the following cash flows.

$i = 10\%$

5-28 If $i = 10\%$, compute the present value, $P$, for the following cash flows.

$n = \infty$

5-29 Compute the present value, $P$, for the following cash flows (assume series repeats forever).

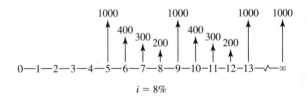

$i = 8\%$

**5-30** By installing some elaborate inspection equipment on its assembly line, the Robot Corp. can avoid hiring an extra worker who would have earned $36,000 a year in wages and an additional $9500 a year in employee benefits. The inspection equipment has a 6-year useful life and no salvage value. Use a nominal 18% interest rate in your calculations. How much can Robot afford to pay for the equipment if the wages and worker benefits were to have been paid

(*a*) At the end of each year

(*b*) Monthly

Explain why the answer in (*b*) is larger.

**5-31** Jerry Stans, a young industrial engineer, prepared an economic analysis for some equipment to replace one production worker. The analysis showed that the present worth of benefits (of employing one less production worker) just equaled the present worth of the equipment costs, assuming a 10-year useful life for the equipment. It was decided not to buy the equipment.

A short time later, the production workers won a new 3-year union contract that granted them an immediate 40¢-per-hour wage increase, plus an additional 25¢-per-hour wage increase in each of the two subsequent years. Assume that in each and every future year, a 25¢-per-hour wage increase will be granted.

Jerry Stans has been asked to revise his earlier economic analysis. The present worth of benefits of replacing one production employee will now increase. Assuming an interest rate of 8%, the justifiable cost of the automation equipment (with a 10-year useful life) will increase by how much? Assume the plant operates a single 8-hour shift, 250 days per year.

**Lives Match**

**5-32** Two alternative courses of action have the following schedules of disbursements:

| Year | A | B |
|---|---|---|
| 0 | −$1300 | |
| 1 | 0 | −$100 |
| 2 | 0 | −200 |
| 3 | 0 | −300 |
| 4 | 0 | −400 |
| 5 | 0 | −500 |
| | −$1300 | −$1500 |

Based on a 6% interest rate, which alternative should be selected?

**5-33** If produced by Method *A*, a product's initial capital cost will be $100,000, its operating cost will be $20,000 per year, and its salvage value after 3 years will be $20,000. With Method *B* there is a first cost of $150,000, an operating cost of $10,000 per year, and a $50,000 salvage value after its 3-year life. Based on a present worth analysis at a 15% interest rate, which method should be used?
*Contributed by Hamed Kashani, Saeid Sadri, and Baabak Ashuri, Georgia Institute of Technology*

**5-34** Quinton's refrigerator has just died. He can get a basic refrigerator or a more efficient refrigerator with an Energy Star designation. Quinton earns 4% compounded annually on his investments, he wants to consider a 10-year planning horizon, and he will use present worth analysis to determine the best alternative. What is your recommendation?
*Contributed by Paul R. McCright, University of South Florida*

| | Basic Unit | Energy Star Unit |
|---|---|---|
| Initial cost | 700 | 800 |
| Delivery and installation | 60 | 60 |
| Professional servicing (year 5) | 100 | 100 |
| Annual energy costs | 120 | 55 |
| Salvage value (year 10) | 150 | 175 |

**5-35** Walt Wallace Construction Enterprises is investigating the purchase of a new dump truck with a 10-year life. Interest is 9%. The cash flows for two likely models are as follows:

| Model | First Cost | Annual Operating Cost | Annual Income | Salvage Value |
|---|---|---|---|---|
| A | $50,000 | $2000 | $9,000 | $10,000 |
| B | 80,000 | 1000 | 12,000 | 30,000 |

(*a*) Using present worth analysis, which truck should the firm buy, and why?

(*b*) Before the construction company can close the deal, the dealer sells out of Model *B* and cannot get any more. What should the firm do now, and why?

**5-36** Two different companies are offering a punch press for sale. Company *A* charges $250,000 to deliver and install the device. Company *A* has estimated that the machine will have maintenance and operating costs of $4000 a year and will provide an annual benefit of $89,000. Company *B* charges $205,000 to deliver and install the device. Company *B* has estimated maintenance and operating costs of the press at $4300 a year, with an annual benefit of $86,000. Both machines will last 5 years and can be sold for $15,000 for the scrap metal. Use an interest rate of 12%. Which machine should your company buy?

5-37 A battery manufacturing plant has been ordered to cease discharging acidic waste liquids containing mercury into the city sewer system. As a result, the firm must now adjust the pH and remove the mercury from its waste liquids. Three firms have provided quotations on the necessary equipment. An analysis of the quotations provided the following table of costs.

| Bidder | Installed Cost | Annual Operating Cost | Annual Income from Mercury Recovery | Salvage Value |
|---|---|---|---|---|
| Foxhill Instrument | $ 35,000 | $8000 | $2000 | $20,000 |
| Quicksilver | 40,000 | 7000 | 2200 | 0 |
| Almaden | 100,000 | 2000 | 3500 | 0 |

If the installation can be expected to last 20 years and money is worth 7%, which equipment should be purchased? (*Answer:* Almaden)

5-38 A new tennis court complex is planned. Each of two alternatives will last 18 years, and the interest rate is 7%. Use present worth analysis to determine which should be selected.
*Contributed by D. P. Loucks, Cornell University*

| | Construction Cost | Annual O&M |
|---|---|---|
| A | $500,000 | $25,000 |
| B | 640,000 | 10,000 |

5-39 In order to improve evacuation routes out of New Orleans in the event of another major disaster such as Hurricane Katrina, the Louisiana Department of Transportation (L-DoT) is planning to construct an additional bridge across the Mississippi River. L-DoT is considering two alternatives: a suspension bridge and a cantilever bridge. The department uses an interest rate of 8% and plans a 50-year life for either bridge. Which design has the most favorable net present worth?
*Contributed by Paul R. McCright, University of South Florida*

| | Suspension Bridge | Cantilever Bridge |
|---|---|---|
| Initial construction costs | $585,000,000 | $470,000,000 |
| Initial land acquisition costs | 120,000,000 | 95,000,000 |
| Annual O&M costs | 1,500,000 | 2,000,000 |
| Annual growth in O&M | 4% | 300,000 |
| Major maintenance (Year 25) | 185,000,000 | 210,000,000 |
| Salvage cost | 30,000,000 | 27,000,000 |

5-40 Teri is an IE at Smith Manufacturing in Sarasota. She has been studying process line G to determine if an automated system would be preferred to the existing labor-intensive system. If Smith wants to earn at least 15% and uses a 15-year planning horizon, which alternative is preferred?
*Contributed by Paul R. McCright, University of South Florida*

| | Labor Intensive | Automated |
|---|---|---|
| Initial cost | $ 0 | $ 55,000 |
| Installation cost | 0 | 15,500 |
| First-year maintenance costs | 1,500 | 4,800 |
| Annual increase in maintenance costs | 200 | 600 |
| First-year labor costs | 116,000 | 41,000 |
| Annual increase in labor costs | 4% | 4% |
| Salvage value (EOY15) | 5,000 | 19,000 |

**5-41** Teléfono Mexico is expanding its facilities to serve a new manufacturing plant. The new plant will require 2000 telephone lines this year, and another 2000 lines after expansion in 10 years. The plant will operate for 30 years.

**Option 1** Provide one cable now with capacity to serve 4000 lines. The cable will cost $200,000 and annual maintenance costs will be $15,000.

**Option 2** Provide a cable with capacity to serve 2000 lines now and a second cable to serve the other 2000 lines in 10 years. Each cable will cost $150,000 and will have an annual maintenance of $10,000.

The telephone cables will last at least 30 years, and the cost of removing the cables is offset by their salvage value.

(a) Which alternative should be selected, assuming a 10% interest rate?

(b) Will your answer to (a) change if the demand for additional lines occurs in 5 years instead of 10 years?

**5-42** A consulting engineer has been hired to advise a town how best to proceed with the construction of a 200,000-m$^3$ water supply reservoir. Since only 120,000 m$^3$ of storage will be required for the next 25 years, an alternative to building the full capacity now is to build the reservoir in two stages. Initially, the reservoir could be built with 120,000 m$^3$ of capacity and then, 25 years hence, the additional 80,000 m$^3$ of capacity could be added by increasing the height of the reservoir. If interest is computed at 4%, which construction plan is preferred?

| | Construction Cost | Annual Maintenance Cost |
|---|---|---|
| **Build in two stages** | | |
| First stage: 120,000-m$^3$ reservoir | $14,200,000 | $ 75,000 |
| Second stage: Add 80,000 m$^3$ of capacity, additional construction and maintenance costs | 12,600,000 | 25,000 |
| **Build full capacity now** 200,000-m$^3$ reservoir | 22,400,000 | 100,000 |

**Lives Differ**

**5-43** Use an 8-year analysis period and a 10% interest rate to determine which alternative should be selected:

| | A | B |
|---|---|---|
| First cost | $5300 | $10,700 |
| Uniform annual benefit | $1800 | $2100 |
| Useful life, in years | 4 | 8 |

**5-44** A man had to have the muffler replaced on his 2-year-old car. The repairman offered two alternatives. For $300 he would install a muffler guaranteed for 2 years. But for $400 he would install a muffler guaranteed "for as long as you own the car." Assuming the present owner expects to keep the car for about 3 more years, which muffler would you advise him to have installed if you thought 20% was a suitable interest rate and the less expensive muffler would only last 2 years?

**5-45** An engineer has received two bids for an elevator to be installed in a new building. Given a 10% interest rate, which bid should be accepted?

| Alternatives | Westinghome | Itis |
|---|---|---|
| Installed cost | $45,000 | $54,000 |
| Annual cost | 2700 | 2850 |
| Salvage value | 3000 | 4500 |
| Life, in years | 10 | 15 |

**5-46** A weekly business magazine offers a 1-year subscription for $58 and a 3-year subscription for $116. If you thought you would read the magazine for at least the next 3 years, and consider 20% as a minimum rate of return, which way would you purchase the magazine: With three 1-year subscriptions or a single 3-year subscription? (*Answer:* Choose the 3-year subscription.)

**5-47** A new alloy can be produced by Process A, which costs $200,000 to implement. The operating cost will be $10,000 per quarter with a salvage value of $25,000 after its 2-year life. Process B will have a first cost of $250,000, an operating cost of $15,000 per quarter, and a $40,000 salvage value after its 4-year life. The interest rate is 8% per year compounded quarterly. Using present value analysis which process should be selected.

*Contributed by Hamed Kashani, Saeid Sadri, and Baabak Ashuri, Georgia Institute of Technology*

**5-48** Which process line should be built for a new chemical? The expected market for the chemical is 20 years. A 9% rate is used to evaluate new process facilities, which are compared with present worth. How much does the better choice save?

| | First Cost | O&M Cost/year | Salvage | Life |
|---|---|---|---|---|
| A | $18M | $5M | $4M | 10 years |
| B | 25M | 3M | 6M | 20 years |

**5-49**  Which equipment is preferred if the firm's interest rate is 9%? In PW terms how great is the difference?

| Alternative | A | B |
|---|---|---|
| First cost | $40,000 | $35,000 |
| Annual O&M | 2,900 | 4,200 |
| Salvage value | 0 | 5,000 |
| Overhaul (Year 6) | 10,000 | Not required |
| Life, in years | 10 | 5 |

**5-50**  North City must choose between two new snow-removal machines. The SuperBlower has a $70,000 first cost, a 20-year life, and an $8000 salvage value. At the end of 9 years, it will need a major overhaul costing $19,000. Annual maintenance and operating costs are $9000. The Sno-Mover will cost $50,000, has an expected life of 10 years, and has no salvage value. The annual maintenance and operating costs are expected to be $12,000. Using a 12% interest rate, which machine should be chosen?

**5-51**  The Crockett Land Winery must replace its present grape-pressing equipment. The two alternatives are the Quik-Skwish and the Stomp-Master. The annual operating costs increase by 12% each year as the machines age. If the interest rate is 9%, which press should be chosen?

| | Quik-Skwish | Stomp-Master |
|---|---|---|
| First cost | $350,000 | $500,000 |
| Annual operating costs | 28,000 | 22,500 |
| Salvage value | 35,000 | 50,000 |
| Useful life, in years | 5 | 10 |

**Perpetual Life**

**5-52**  A small dam was constructed for $2 million. The annual maintenance cost is $15,000. If interest is 5%, compute the capitalized cost of the dam, including maintenance.

**5-53**  A depositor puts $25,000 in a savings account that pays 5% interest, compounded semiannually. Equal annual withdrawals are to be made from the account, beginning one year from now and continuing forever. What is the maximum annual withdrawal?

**5-54**  What amount of money deposited 50 years ago at 8% interest would provide a perpetual payment of $10,000 a year beginning this year?

**5-55**  The president of the E. L. Echo Corporation thought it would be appropriate for his firm to "endow a chair" in the Department of Industrial Engineering of the local university. If the professor holding the chair will receive $134,000 per year, and the interest received on the endowment fund is expected to be 8%, what lump sum of money will the Echo Corporation need to provide to establish the endowment fund?   (*Answer:* $1,675,000)

**5-56**  Dr. Fog E. Professor is retiring and wants to endow a chair of engineering economics at his university. It is expected that he will need to cover an annual cost of $100,000 forever. What lump sum must he donate to the university today if the endowment will earn 10% interest?

**5-57**  The local botanical society wants to ensure that the gardens in the town park are properly cared for. The group recently spent $100,000 to plant the gardens. The members want to set up a perpetual fund to provide $100,000 for future replantings of the gardens every 10 years. If interest is 5%, how much money would be needed to forever pay the cost of replanting?

**5-58**  A home builder must construct a sewage treatment plant and deposit sufficient money in a perpetual trust fund to pay the $5000 per year operating cost and to replace the treatment plant every 40 years. The plant will cost $150,000, and future replacement plants will also cost $150,000 each. If the trust fund earns 8% interest, what is the builder's capitalized cost to construct the plant and future replacements, and to pay the operating costs?

**5-59**  A man who likes cherry blossoms very much wants an urn full of them put on his grave once each year forever after he dies. In his will, he intends to leave a certain sum of money in trust at a local bank to pay the florist's annual bill. How much money should be left for this purpose? Make whatever assumptions you feel are justified by the facts presented. State your assumptions, and compute a solution.

**5-60** An elderly lady decided to distribute most of her considerable wealth to charity and to keep for herself only enough money to provide for her living. She feels that $1000 a month will amply provide for her needs. She will establish a trust fund at a bank that pays 6% interest, compounded monthly. At the end of each month she will withdraw $1000. She has arranged that, upon her death, the balance in the account is to be paid to her niece, Susan. If she opens the trust fund and deposits enough money to pay herself $1000 a month in interest as long as she lives, how much will Susan receive when her aunt dies?

**5-61** A trust fund is to be established for three purposes: (1) to provide $750,000 for the construction and $250,000 for the initial equipment of a small engineering laboratory; (2) to pay the $150,000 per year laboratory operating cost; and (3) to pay for $100,000 of replacement equipment every 4 years, beginning 4 years from now.

At 6% interest, how much money is required in the trust fund to provide for the laboratory and equipment and its perpetual operation and equipment replacement?

**5-62** We want to donate a marble birdbath to the city park as a memorial to cats. We also want to set up a perpetual care fund to cover future expenses "forever." The initial cost of the bath is $5000. Routine annual operating costs are $200 per year, but every fifth year the cost will be $500 to cover major cleaning and maintenance as well as operation.

(a) What is the capitalized cost of this project if the interest rate is 8%?

(b) How much is the present worth of this project if it is to be demolished after 75 years? The final $500 payment in the 75th year will cover the year's operating cost and the site reclamation.

**5-63** A local symphony association offers memberships as follows:

| | |
|---|---|
| Continuing membership, per year | $ 15 |
| Patron lifetime membership | 375 |

The patron membership has been based on the symphony association's belief that it can obtain a 4% rate of return on its investment. If you believed 4% to be an appropriate rate of return, would you be willing to purchase the patron membership? Explain why or why not.

**5-64** A city has developed a plan to provide for future municipal water needs. The plan proposes an aqueduct that passes through 500 feet of tunnel in a nearby mountain. Two alternatives are being considered. The first proposes to build a full-capacity tunnel now for $556,000. The second proposes to build a half-capacity tunnel now (cost = $402,000), which should be adequate for 20 years, and then to build a second parallel half-capacity tunnel. The maintenance cost of the tunnel lining for the full-capacity tunnel is $40,000 every 10 years, and for each half-capacity tunnel it is $32,000 every 10 years.

The friction losses in the half-capacity tunnel will be greater than if the full-capacity tunnel were built. The estimated additional pumping costs in the single half-capacity tunnel will be $2000 per year, and for the two half-capacity tunnels it will be $4000 per year. Based on capitalized cost and a 7% interest rate, which alternative should be selected?

**5-65** Use capitalized cost to determine which type of road surface is preferred on a particular section of highway. Use 12% interest rate.

| | A | B |
|---|---|---|
| Initial cost | $500,000 | $700,000 |
| Annual maintenance | 35,000 | 25,000 |
| Periodic resurfacing | 350,000 | 450,000 |
| Resurfacing interval | 10 years | 15 years |

**5-66** A new bridge project is being evaluated at $i = 5\%$. Recommend an alternative based on the capitalized cost for each.

| | Construction Cost | Annual O&M | Life (years) |
|---|---|---|---|
| Concrete | $50 million | $ 250,000 | 70 |
| Steel | 40 million | 1,000,000 | 50 |

**5-67** A new stadium is being evaluated at $i = 6\%$. Recommend an alternative for the main structural material based on the capitalized cost for each.

| | Construction Cost | Annual O&M | Life (years) |
|---|---|---|---|
| Concrete | $25 million | $ 200,000 | 80 |
| Steel | 21 million | 1,000,000 | 60 |

5-68 A rather wealthy man decides to arrange for his descendants to be well educated. He wants each child to have $60,000 for his or her education. He plans to set up a perpetual trust fund so that six children will receive this assistance in each generation. He estimates that there will be four generations per century, spaced 25 years apart. He expects the trust to be able to obtain a 4% rate of return, and the first recipients to receive the money 10 years hence. How much money should he now set aside in the trust? (*Answer:* $389,150)

5-69 Kansas Public Service Company wishes to determine the capitalized worth of a new windmill at an interest rate of 9% and following costs.
*Contributed by Paul R. McCright, University of South Florida*

| | | | |
|---|---|---|---|
| Initial costs | $ 725,000 | Installation costs | $ 143,000 |
| Annual | 12,000 | Major overhaul | 260,000 |
| O&M costs | | (Year 25) | |
| Expected life | 40 years | Salvage value | 32,000 |

5-70 An open-pit mine must fund an account now to pay for maintenance of a tailing pond in perpetuity (after the mine shuts down in 30 years). The costs until shutdown are part of the mine's operating costs. The maintenance costs begin in 30 years at $300,000 annually. How much must be deposited now if the fund will earn 5% interest?

5-71 A firm wants to sponsor a new engineering lab at a local university. This requires $2.5M to construct the lab, $1.2M to equip it, and $600,000 every 5 years for new equipment. What is the required endowment if the university will earn 6% interest on the funds?

## Multiple Alternatives

5-72 A firm is considering three mutually exclusive alternatives as part of a production improvement program. The alternatives are:

| | A | B | C |
|---|---|---|---|
| Installed cost | $10,000 | $15,000 | $20,000 |
| Uniform annual benefit | 1,625 | 1,530 | 1,890 |
| Useful life, in years | 10 | 20 | 20 |

The salvage value at the end of the useful life of each alternative is zero. At the end of 10 years, Alternative *A* could be replaced with another *A* with identical cost and benefits. The maximum attractive rate of return is 6%. Which alternative should be selected?

5-73 A steam boiler is needed as part of the design of a new plant. The boiler can be fired by natural gas, fuel oil, or coal. A decision must be made on which fuel to use. An analysis of the costs shows that the installed cost, with all controls, would be least for natural gas at $30,000; for fuel oil it would be $55,000; and for coal it would be $180,000. If natural gas is used rather than fuel oil, the annual fuel cost will increase by $7500. If coal is used rather than fuel oil, the annual fuel cost will be $15,000 per year less. Assuming 8% interest, a 20-year analysis period, and no salvage value, which is the most economical installation?

5-74 Austin General Hospital is evaluating new office equipment offered by three companies. In each case the interest rate is 15% and the useful life of the equipment is 4 years. Use NPW analysis to determine the company from which you should purchase the equipment.

| Company | A | B | C |
|---|---|---|---|
| First cost | $15,000 | $25,000 | $20,000 |
| Maintenance and operating costs | 1,600 | 400 | 900 |
| Annual benefit | 8,000 | 13,000 | 11,000 |
| Salvage value | 3,000 | 6,000 | 4,500 |

5-75 The following costs are associated with three tomato-peeling machines being considered for use in a canning plant. If the canning company uses an interest rate of 12%, which is the best alternative? Use NPW to make your decision. (*Note:* Consider the least common multiple as the study period.)

| Machine | A | B | C |
|---|---|---|---|
| First cost | $52,000 | $63,000 | $67,000 |
| Maintenance and operating costs | 15,000 | 9,000 | 12,000 |
| Annual benefit | 38,000 | 31,000 | 37,000 |
| Salvage value | 13,000 | 19,000 | 22,000 |
| Useful life, in years | 4 | 6 | 12 |

5-76 A railroad branch line to a landfill site is to be constructed. It is expected that the railroad line will be used for 15 years, after which the landfill site will be closed and the land turned back to agricultural use. The railroad track and ties will be removed at that time.

In building the railroad line, either treated or untreated wood ties may be used. Treated ties have an installed cost of $6 and a 10-year life; untreated ties are $4.50 with a 6-year life. If at the end of 15 years the ties then in place have a remaining useful life of 4 years or more, they will be used by the railroad elsewhere and have an estimated salvage value of $3 each. Any ties that are removed at the end of their service life, or too close to the end of their service life to be used elsewhere, can be sold for 50¢ each.

Determine the most economical plan for the initial railroad ties and their replacement for the 15-year period. Make a present worth analysis assuming 8% interest.

**5-77** A building contractor obtained bids for some asphalt paving, based on a specification. Three paving subcontractors quoted the following prices and terms of payment:

| Paving Co. | Price | Payment Schedule |
|---|---|---|
| Quick | $85,000 | 50% payable immediately 25% payable in 6 months 25% payable at the end of one year |
| Tartan | 82,000 | Payable immediately |
| Faultless | 84,000 | 25% payable immediately 75% payable in 6 months |

The building contractor uses a 12% nominal interest rate, compounded monthly, in this type of bid analysis. Which paving subcontractor should be awarded the paving job?

**5-78** Given the following data, use present worth analysis to find the best alternative, $A$, $B$, or $C$.

| | A | B | C |
|---|---|---|---|
| Initial cost | $10,000 | 15,000 | $12,000 |
| Annual benefit | 6,000 | 10,000 | 5,000 |
| Salvage value | 1,000 | −2,000 | 3,000 |
| Useful life | 2 years | 3 years | 4 years |

Use an analysis period of 12 years and 10% interest.

**5-79** Consider the following four alternatives. Three are "do something" and one is "do nothing."

| | A | B | C | D |
|---|---|---|---|---|
| Cost | $0 | $50 | $30 | $40 |
| Net annual benefit | 0 | 12 | 4.5 | 6 |
| Useful life, in years | | 5 | 10 | 10 |

At the end of the 5-year useful life of $B$, a replacement is not made. If a 10-year analysis period and a 10% interest rate are selected, which is the preferred alternative?

**5-80** A cost analysis is to be made to determine what, if anything, should be done in a situation offering three "do-something" and one "do-nothing" alternatives. Estimates of the cost and benefits are as follows.

| Alternatives | Cost | Uniform Annual Benefit | End-of- Useful-Life Salvage Value | Useful Life (years) |
|---|---|---|---|---|
| 1 | $500 | $135 | $ 0 | 5 |
| 2 | 600 | 100 | 250 | 5 |
| 3 | 700 | 100 | 180 | 10 |
| 4 | 0 | 0 | 0 | 0 |

Use a 10-year analysis period for the four mutually exclusive alternatives. At the end of 5 years, Alternatives 1 and 2 may be replaced with identical alternatives (with the same cost, benefits, salvage value, and useful life).

(a) If an 8% interest rate is used, which alternative should be selected?

(b) If a 12% interest rate is used, which alternative should be selected?

**5-81** Consider $A–E$, five mutually exclusive alternatives:

| | A | B | C | D | E |
|---|---|---|---|---|---|
| Initial cost | $600 | $600 | $600 | $600 | $600 |
| Uniform annual benefits | | | | | |
| For first 5 years | 100 | 100 | 100 | 150 | 150 |
| For last 5 years | 50 | 100 | 110 | 0 | 50 |

The interest rate is 10%. If all the alternatives have a 10-year useful life, and no salvage value, which alternative should be selected?

**5-82**  An investor has carefully studied a number of companies and their common stock. From his analysis, he has decided that the stocks of six firms are the best of the many he has examined. They represent about the same amount of risk, and so he would like to determine one single stock in which to invest. He plans to keep the stock for 4 years and requires a 10% minimum attractive rate of return.

Which stock if any, should the investor consider buying?   (*Answer:* Spartan Products)

| Common Stock | Price per Share | Annual End-of-Year Dividend per Share | Estimated Price at End of 4 Years |
|---|---|---|---|
| Western House | $23³/₄ | $1.25 | $32 |
| Fine Foods | 45 | 4.50 | 45 |
| Mobile Motors | 30⁵/₈ | 0 | 42 |
| Spartan Products | 12 | 0 | 20 |
| U.S. Tire | 33³/₈ | 2.00 | 40 |
| Wine Products | 52¹/₂ | 3.00 | 60 |

**5-83**  Six mutually exclusive alternatives, *A–F*, are being examined. For an 8% interest rate, which alternative should be selected? Each alternative has a 6-year useful life.

| | Initial Cost | Uniform Annual Benefit |
|---|---|---|
| A | $ 20 | $ 6.00 |
| B | 35 | 9.25 |
| C | 55 | 13.38 |
| D | 60 | 13.78 |
| E | 80 | 24.32 |
| F | 100 | 24.32 |

**5-84**  The management of an electronics manufacturing firm believes it is desirable to automate its production facility. The automated equipment would have a 10-year life with no salvage value at the end of 10 years. The plant engineering department has surveyed the plant and has suggested there are eight mutually exclusive alternatives. If the firm expects a 10% rate of return, which plan, if any, should it adopt?   (*Answer:* Plan 1)

| Plan | Initial Cost (thousands) | Net Annual Benefit (thousands) |
|---|---|---|
| 1 | $265 | $51 |
| 2 | 220 | 39 |
| 3 | 180 | 26 |
| 4 | 100 | 15 |
| 5 | 305 | 57 |
| 6 | 130 | 23 |
| 7 | 245 | 47 |
| 8 | 165 | 33 |

## Bonds

**5-85**  A corporate bond has a face value of $1000 with maturity date 20 years from today. The bond pays interest semiannually at a rate of 8% per year based on the face value. The interest rate paid on similar corporate bonds has decreased to a current rate of 6%. Determine the market value of the bond.

**5-86**  Calculate the present worth of a 4.5%, $5000 bond with interest paid semiannually. The bond matures in 10 years, and the investor wants to make 8% per year compounded quarterly on the investment.

**5-87**  An investor is considering buying a 20-year corporate bond. The bond has a face value of $1000 and pays 6% interest per year in two semiannual payments. Thus the purchaser of the bond will receive $30 every 6 months in addition to $1000 at the end of 20 years, along with the last $30 interest payment. If the investor wants to receive 8% interest, compounded semiannually, how much would he or she be willing to pay for the bond?

**5-88**  You bought a $1000 corporate bond for $900 three years ago. It is paying $30 in interest at the end of every 6 months, and it matures in 4 more years.

(*a*) Compute its coupon rate.

(*b*) Compute its current value, assuming the market interest rate for such investments is 5% per year, compounded semiannually.

*Contributed by D. P. Loucks, Cornell University*

*Problems 5-89 to 5-93 contributed by Meenakshi Sundaram, Tennessee Tech University*

**5-89**  A 6% coupon rate bond has a face value of $1000, pays interest semiannually, and will mature in 10 years. If the current market rate is 8% interest compounded semiannually, what is the bond's price?

**5-90** A Treasury bond with a face value of $5000 and a coupon rate of 6% payable semiannually was bought by Kirt when the market's nominal rate was 8%. The bond matures 20 years from now. What did Kirt pay for the bond?

**5-91** A zero-coupon bond (coupon rate = 0%) with a face value of $10,000 and maturity date in 5 years is being considered for purchase by Pam. The current market interest rate is a nominal 10%, compounded quarterly. How much should she pay for the bond?

**5-92** A city government wants to raise $3 million by issuing bonds. By ballot proposition, the bond's coupon interest rate was set at 8% per year with semiannual payments. However, market interest rates have risen to a nominal 9% interest rate. If the bonds mature in 20 years, how much will the city raise from issuing $3M in bonds.

**5-93** Kal Tech, a manufacturing company, needs to raise $2 million to finance an expansion project. The bonds will have a coupon interest rate of 12%, payable quarterly, and 20 years to maturity. What will the face value of the bonds have to be, if the bonds are to have an interest rate of 12% per year, payable quarterly, and a maturity date of 20 years? The current market interest rate is a nominal semiannual rate of 16% compounded quarterly.

## Spreadsheets

**5-94** Assume monthly car payments of $500 per month for 4 years and an interest rate of 0.5% per month. What initial principal or PW will this repay?

**5-95** Assume annual car payments of $6000 for 4 years and an interest rate of 6% per year. What initial principal or PW will this repay?

**5-96** Assume annual car payments of $6000 for 4 years and an interest rate of 6.168% per year. What initial principal or PW will this repay?

**5-97** Why do the values in Problems 5-94, 5-95, and 5-96 differ?

**5-98** Assume mortgage payments of $1000 per month for 30 years and an interest rate of 0.5% per month. What initial principal or PW will this repay?

**5-99** Assume annual mortgage payments of $12,000 for 30 years and an interest rate of 6% per year. What initial principal or PW will this repay?

**5-100** Assume annual mortgage payments of $12,000 for 30 years and an interest rate of 6.168% per year. What initial principal or PW will this repay?

**5-101** Why do the values in Problems 5-98, 5-99, and 5-100 differ?

**5-102** A construction project has the following end-of month costs. Calculate the PW at a nominal interest rate of 18%.

| January | $ 30,000 | May | $520,000 |
|---------|----------|-----|----------|
| February | 50,000 | June | 460,000 |
| March | 110,000 | July | 275,000 |
| April | 430,000 | August | 95,000 |

**5-103** A factory has averaged the following monthly heating and cooling costs over the last 5 years. Calculate the PW at a nominal interest rate of 12%.

| January | $25,000 | July | $29,000 |
|---------|---------|------|---------|
| February | 19,000 | August | 33,000 |
| March | 15,000 | September | 19,000 |
| April | 9,000 | October | 8,000 |
| May | 12,000 | November | 16,000 |
| June | 18,000 | December | 28,000 |

**5-104** Ding Bell Imports requires a return of 15% on all projects. If Ding is planning an overseas development project with these cash flows, what is the project's net present value?

| Year | 0 | 1 | 2 | 3 | 4 | 5 | 6 | 7 |
|------|---|---|---|---|---|---|---|---|
| Net Cash ($) | 0 | −120,000 | −60,000 | 20,000 | 40,000 | 80,000 | 100,000 | 60,000 |

**5-105** Maverick Enterprises is planning a new product. Annual sales, unit costs, and unit revenues are as tabulated; the first cost of R&D and setting up the assembly line is $42,000. If $i$ is 10%, what is the PW?

| Year | Annual Sales | Cost/unit | Price/unit |
|------|--------------|-----------|------------|
| 1 | $ 5,000 | $3.50 | $6 |
| 2 | 6,000 | 3.25 | 5.75 |
| 3 | 9,000 | 3.00 | 5.50 |
| 4 | 10,000 | 2.75 | 5.25 |
| 5 | 8,000 | 2.5 | 4.5 |
| 6 | 4,000 | 2.25 | 3 |

**5-106** Northern Engineering is analyzing a mining project. Annual production, unit costs, and unit revenues are in the table. The first cost of the mine setup is $8 million. If $i$ is 15%, what is the PW?

| Year | Annual Production (tons) | Cost per ton | Price per ton |
|------|--------------------------|--------------|---------------|
| 1 | 70,000 | $25 | $35 |
| 2 | 90,000 | 20 | 34 |
| 3 | 120,000 | 22 | 33 |
| 4 | 100,000 | 24 | 34 |
| 5 | 80,000 | 26 | 35 |
| 6 | 60,000 | 28 | 36 |
| 7 | 40,000 | 30 | 37 |

## Minicases

**5-107** Bayview's growth is constrained by mountains on one side and the bay on the other. A bridge across the bay is planned, but which plan is best? It can be built with a single deck to meet the needs of the next 20 years, or it can be built with two decks to meet the needs of the next 50 years. The piers can also be built to support two decks, but with only one deck being built now.

Building it all now will cost $160M, and leaving the top deck for later will save $40M. Building that top deck later will cost $70M including the cost of traffic disruption. A single-deck bridge will cost $100M now and $115M in 20 years. Deck maintenance is $1.4M per year per deck. Pier maintenance is $1.2M per year per bridge. If the interest rate is 5%, which design should be built?

If the two-deck bridge is built immediately, then dedicated lanes for buses, carpools, and bicycles can be added. To economically evaluate this use, estimate the cost of the underutilized capacity for the bridge.

**5-108** Florida Power and Light has committed to building a solar power plant. JoAnne, an IE working for FPL, has been tasked with evaluating the three current designs. FPL uses an interest rate of 10% and a 20-year horizon.

### Design 1: Flat Solar Panels

A field of "flat" solar panels angled to best catch the sun will yield 2.6 MW of power and will cost $87 million initially with first-year operating costs at $2 million, growing $250,000 annually. It will produce electricity worth $6.9 million the first year and will increase by 8% each year thereafter.

### Design 2: Mechanized Solar Panels

A field of mechanized solar panels rotates from side to side so that they are always positioned parallel to the sun's rays, maximizing the production of electricity. This design will yield 3.1 MW of power and will cost $101 million initially with first-year operating costs at $2.3 million, growing $300,000 annually. It will produce electricity worth $8.8 million the first year and will increase 8% each year thereafter.

### Design 3: Solar Collector Field

This design uses a field of mirrors to focus the sun's rays onto a boiler mounted in a tower. The boiler then produces steam and generates electricity the same way a coal-fired plant operates. This system will yield 3.3 MW of power and will cost $91 million initially with first-year operating costs at $3 million, growing $350,000 annually. It will produce electricity worth $9.7 million the first year and will increase 8% each year thereafter.

5-109 Your grandparents are asking you for advice on when they should start collecting social security payments. If they wait until age 66, they will collect $2000 per month; but if they start collecting at age 62, they will collect $1600 per month. Assume they live to be 85, and simplify by assuming annual payments.

(a) When do the higher payments catch up in total dollars received with the lower payment that starts earlier?

(b) If their interest rate is 6%, which plan has a higher PW?

# CASES

**The following cases from *Cases in Engineering Economy* $2^{nd}$ are suggested as matched with this chapter.**

CASE 7    **The Board Looks to You**

Bond valuation with realistic business details including early-call premium.

Medium difficulty. Good example of employer taking a small fact and making big assumptions about individual talents which implies why engineering students must keep learning.

CASE 8    **Picking a Price**

Simplified real estate analysis. Demonstrates analysis to screen before acquiring more data and further decision making.

CASE 9    **Recycling?**

Financial analysis of recycling cardboard and selling unusable pallets to recycler.

# ANNUAL CASH FLOW ANALYSIS

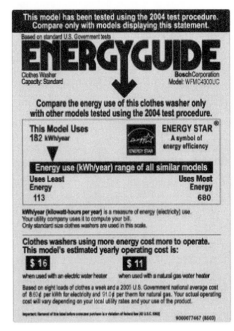

Typical FTC required EnergyGuide label for a clothes washer. According to the California Energy Commission, the typical American household does 400 loads of laundry per year, using 40 gallons of water per load, with a typical non–ENERGY STAR clothes washer. An ENERGY STAR appliance would reduce water and energy consumption by 40%. For more information, see http://www.energystar.gov.

## Are More Efficient Appliances Cost Effective?

With an emphasis on green technologies and energy savings, the U.S. Department of Energy has targeted a reduction of 20% of a home's energy use related to appliances and home electronics. The ENERGY STAR program includes EnergyGuide labels that are required for most major home appliances. New appliances that meet basic energy efficiency criteria are given an ENERGY STAR label.

Whereas a typical residence is used for 75 years, its major appliances last from 10 to 30 years. The long-term trend in energy prices (1–3% increase per year) lulls consumers into ignoring short-term disruptions in energy supply and spikes in energy prices. These spikes are generally accepted by consumers as normal incremental adjustments to their cost of living. The increases in operating costs are not enough to induce consumers to replace existing appliances—except when appliances that have failed must be replaced.

One driver in the slow adoption of energy appliances is the fact that many investment decisions are made by home builders, landlords, and property managers rather than the end-use consumers. To reduce initial costs, construction-level decision makers tend to install less costly, less efficient appliances. Renters have little leverage to push landlords for new, more efficient appliances. Even homeowners may focus on purchase price versus operating costs, since they stay in a location for only 8 years on average and often leave appliances behind. Any energy inefficiencies are assumed to be temporary, since this is not the "dream house" they will live in for an extended period. Although the value of potential energy savings may positively impact resale value, there are no guarantees.

When consumers do decide to replace appliances, EnergyGuide labels are provided to show the energy cost for typical owners so that different models can be compared by all consumers, not just by those who've had engineering economy. Few homeowners know what their average total appliance energy use is or which appliances contribute the most to those costs.

From the EnergyGuide label, the range of energy usage is 113–680 kWh/yr for clothes washers. This energy consumption estimate is based on ratings of electricity and water usage factors. At $0.08/kWh, the estimated electricity costs for comparable models ranges from $9.04/yr to $54.40/yr. An EnergyGuide label for a refrigerator might show a range of energy use from 617 kWh/yr to 698 kWh/yr.

When selecting a clothes washer, wise consumers will consider initial cost and the annual energy savings/costs (is this a benefit or cost?) and will then select a model that suits their needs based on the annual number of washer loads. When people are replacing an existing appliance, changes in life needs (children, empty nest. etc) are important. Choices are based on needs, and making these choices require tools that are part of the skill set provided by engineering economics. ▪ ▪ ▪

*Contributed by Gene Dixon, East Carolina University*

## QUESTIONS TO CONSIDER

1. Should average use be based on family size or lifestyle? How can lifestyle be properly analyzed? For example, a family of two where one or both people work outdoors in construction may have larger laundry demands than a family of four where the parents work in offices. What about the impact of geography and climate: that is, does wearing Bermuda shorts year round versus blue jeans make a significant difference in laundry demands?

2. How would a home builder, landlord, or property manager justify the use of ENERGY STAR appliances in pricing properties?

3. In 2010 some states offered Cash for Appliance Clunkers programs. How would you estimate the impact of this program, anticipating that the program would be paid for with increased taxes (sales, income, or property)?

4. Based on type, major appliances have lifespans of 7 to 15 years, with refrigerators having the longest. If a family stays in a residence on average for 8 years, what considerations should a family use in deciding whether to keep or leave their major appliances when moving?

5. Given the foregoing data, which appliance replacement decision would be more sensitive to changes in the cost of electricity, the clothes washer or the refrigerator? How would you explain your answer to someone not skilled in engineering economics?

6. EnergyGuide labels are not required for televisions, ranges, ovens, clothes dryers, humidifiers, and dehumidifiers. Why would you think some appliances are not required to have EnergyGuide labels?

## After Completing This Chapter...

### *The student should be able to:*

- Define *equivalent uniform annual cost (EUAC)* and *equivalent uniform annual benefits (EUAB)*.
- Resolve an engineering economic analysis problem into its annual cash flow equivalent.
- Conduct an *equivalent uniform annual worth (EUAW) analysis* for a single investment.
- Use EUAW, EUAC, and EUAB to compare alternatives with equal, common multiple, or continuous lives, or over some fixed study period.
- Develop and use spreadsheets to analyze loans for purposes of building an amortization table, calculating interest versus principal, finding the balance due, and determining whether to pay off a loan early.

This chapter is devoted to annual cash flow analysis—the second of the three major analysis techniques. With present worth analysis, we resolved an alternative into an equivalent net present worth, a present worth of cost, or a present worth of benefit. Here we compare alternatives based on their equivalent annual cash flows: the equivalent uniform annual cost (EUAC), the equivalent uniform annual benefit (EUAB), or their difference, the equivalent uniform annual worth: (EUAW) = (EUAB − EUAC).

To prepare for a discussion of annual cash flow analysis, we will review some annual cash flow calculations, then examine annual cash flow criteria.

## ANNUAL CASH FLOW CALCULATIONS

### Resolving a Present Cost to an Annual Cost

In annual cash flow analysis, the goal is to convert money to an equivalent uniform annual cost or benefit. The simplest case is to convert a present sum $P$ to a series of equivalent uniform end-of-period cash flows. This is illustrated in Example 6-1.

### EXAMPLE 6–1

A student bought $1000 worth of home furniture. If the items are expected to last 10 years, what will the equivalent uniform annual cost be if interest is 7%?

$P = 1000$

$n = 10$ years
$i = 7\%$

#### SOLUTION

$$\text{Equivalent uniform annual cost} = P(A/P, i, n)$$
$$= 1000(A/P, 7\%, 10)$$
$$= \$142.40$$

Or with a TVM calculator, we obtain

$$\text{EUAC} = \text{PMT}(i, n, P, F)$$
$$= \text{PMT}(7\%, 10, -1000, 0)$$
$$= \$142.37$$

### Treatment of Salvage Value

When there is a salvage value at the end of an asset's useful life, this decreases the equivalent uniform annual cost.

## EXAMPLE 6–2

The student in Example 6-1 now believes the furniture can be sold at the end of 10 years for $200. Under these circumstances, what is the equivalent uniform annual cost?

### SOLUTIONS

For this situation, the problem may be solved by means of three different calculations.

### SOLUTION 1

$$\text{EUAC} = P(A/P, i, n) - S(A/F, i, n) \tag{6-1}$$
$$= 1000(A/P, 7\%, 10) - 200(A/F, 7\%, 10)$$
$$= 1000(0.1424) - 200(0.0724)$$
$$= 142.40 - 14.48 = \$127.92$$
$$\text{EUAC} = \text{PMT}(i, n, P, F)$$
$$= \text{PMT}(7\%, 10, -1000, 200)$$
$$= \$127.90$$

This method reflects the annual cost of the cash disbursement minus the annual benefit of the future resale value.

### SOLUTION 2

Equation 6-1 describes a relationship that may be modified by an identity presented in Chapter 4:

$$(A/P, i, n) = (A/F, i, n) + i \tag{6-2}$$

Substituting this into Equation 6-1 gives

$$\text{EUAC} = P(A/F, i, n) + Pi - S(A/F, i, n) \tag{6-3}$$
$$= (P - S)(A/F, i, n) + Pi$$
$$= (1000 - 200)(A/F, 7\%, 10) + 1000(0.07)$$
$$= 800(0.0724) + 70 = 57.92 + 70$$
$$= \$127.92$$

This method computes the equivalent annual cost due to the unrecovered $800 when the furniture is sold, and it adds annual interest on the $1000 investment.

**SOLUTION 3**

If the value for $(A/F, i, n)$ from Equation 6-2 is substituted into Equation 6-1, we obtain

$$EUAC = P(A/P, i, n) - S(A/P, i, n) + Si \qquad (6\text{-}4)$$
$$= (P - S)(A/P, i, n) + Si$$
$$= (1000 - 200)(A/P, 7\%, 10) + 200(0.07)$$
$$= 800(0.1424) + 14 = 113.92 + 14 = \$127.92$$

This method computes the annual cost of the $800 decline in value during the 10 years, plus interest on the $200 tied up in the furniture as the salvage value.

When there is an initial disbursement $P$ followed by a salvage value $S$, the annual cost may be computed in the three different ways introduced in Example 6-2.

$$\textbf{EUAC} = P(A/P, i, n) - S(A/F, i, n) \qquad (6\text{-}1)$$

$$\textbf{EUAC} = (P - S)(A/F, i, n) + Pi \qquad (6\text{-}3)$$

$$\textbf{EUAC} = (P - S)(A/P, i, n) + Si \qquad (6\text{-}4)$$

Each of the three calculations gives the same results. In practice, the first method is the most commonly used. The EUAC calculated in Equations 6-1, 6-3, and 6-4 is also known as the *capital recovery cost* of a project.

**EXAMPLE 6–3**

Bill owned a car for 5 years. One day he wondered what his uniform annual cost for maintenance and repairs had been. He assembled the following data:

| Year | Maintenance and Repair Cost for Year |
|------|---------------------------------------|
| 1    | $ 45                                  |
| 2    | 90                                    |
| 3    | 180                                   |
| 4    | 135                                   |
| 5    | 225                                   |

Compute the equivalent uniform annual cost (EUAC) assuming 7% interest and end-of-year disbursements.

**SOLUTION**

The EUAC may be computed for this irregular series of payments in two steps:

1. Use single payment present worth factors to compute the present worth of cost for the 5 years.
2. With the PW of cost known, use the capital recovery factor to compute EUAC.

$$\text{PW of cost} = 45(P/F, 7\%, 1) + 90(P/F, 7\%, 2) + 180(P/F, 7\%, 3)$$
$$+ \ 135(P/F, 7\%, 4) + 225(P/F, 7\%, 5)$$
$$= 45(0.9346) + 90(0.8734) + 180(0.8163) + 135(0.7629) + 225(0.7130)$$
$$= \$531$$
$$\text{EUAC} = 531(A/P, 7\%, 5) = 531(0.2439) = \$130$$

## EXAMPLE 6–4

Bill reexamined his calculations and found that in his table he had reversed the maintenance and repair costs for Years 3 and 4. The correct table is:

| | Maintenance and Repair |
| Year | Cost for Year |
| --- | --- |
| 1 | $ 45 |
| 2 | 90 |
| 3 | 135 |
| 4 | 180 |
| 5 | 225 |

Recompute the EUAC.

**SOLUTION**

This time the schedule of disbursements is an arithmetic gradient series plus a uniform annual cost, as follows:

$$EUAC = 45 + 45(A/G, 7\%, 5)$$

$$= 45 + 45(1.865)$$

$$= \$129$$

Since the timing of the expenditures is different in Examples 6-3 and 6-4, we would not expect to obtain the same EUAC.

The examples have shown four essential points concerning cash flow calculations:

1. There is a direct relationship between the present worth of cost and the equivalent uniform annual cost. It is

$$EUAC = (\text{PW of cost})(A/P, i, n)$$

2. In a problem, expending money increases the EUAC, while receiving money—for example, from an item's salvage value—decreases the EUAC.

3. When there are irregular cash disbursements over the analysis period, a convenient method of solution is to first determine the PW of cost; then use the equation in Item 1 to calculate the EUAC.

4. Where there is an arithmetic gradient, EUAC may be rapidly computed by using the arithmetic gradient uniform series factor, $(A/G, i, n)$.

## ANNUAL CASH FLOW ANALYSIS

The criteria for economic efficiency are presented in Table 6-1. One notices immediately that the table is quite similar to Table 5-1. It is apparent that, if you are maximizing the present worth of benefits, simultaneously you must be maximizing the equivalent uniform annual worth. This is illustrated in Example 6-5.

**TABLE 6–1  Annual Cash Flow Analysis**

| Input/Output | Situation | Criterion |
|---|---|---|
| Neither input nor output is fixed | Typical, general situation | Maximize equivalent uniform annual worth (EUAW = EUAB − EUAC) |
| Fixed input | Amount of money or other input resources is fixed | Maximize equivalent uniform benefits (maximize EUAB) |
| Fixed output | There is a fixed task, benefit, or other output to be accomplished | Minimize equivalent uniform annual cost (minimize EUAC) |

**EXAMPLE 6–5**

A firm is considering which of two devices to install to reduce costs. Both devices have useful lives of 5 years with no salvage value. Device $A$ costs $1000 and can be expected to result in $300 savings annually. Device $B$ costs $1350 and will provide cost savings of $300 the first year; however, savings will increase $50 annually, making the second-year savings $350, the third-year savings $400, and so forth. With interest at 7%, which device should the firm purchase?

**SOLUTION**

**Device A**

$$EUAW = -1000(A/P, 7\%, 5) + 300$$
$$= -1000(0.2439) + 300 = \$56.1$$

**Device B**

$$EUAW = -1350(A/P, 7\%, 5) + 300 + 50(A/G, 7\%, 5)$$
$$= -1350(0.2439) + 300 + 50(1.865)$$
$$= \$64.00$$

To maximize EUAW, select Device $B$.

Example 6-5 was presented earlier, as Example 5-1, where we found:

$$PW_A = -1000 + 300(P/A, 7\%, 5)$$
$$= -1000 + 300(4.100) = \$230$$

This is converted to EUAW by multiplying by the capital recovery factor:

$$EUAW_A = 230(A/P, 7\%, 5) = 230(0.2439) = \$56.1$$

Similarly, for machine $B$

$$PW_B = -1350 + 300(P/A, 7\%, 5) + 50(P/G, 7\%, 5)$$
$$= -1350 + 400(4.100) + 50(7.647) = \$262.4$$

and, hence,

$$EUAW_B = 262.4(A/P, 7\%, 5) = 262.4(0.2439)$$
$$= \$64.0$$

We see, therefore, that it is easy to convert the present worth analysis results into the annual cash flow analysis results. We could go from annual cash flow to present worth just as easily, by using the series present worth factor. And, of course, both methods show that Device $B$ is the preferred alternative.

---

## EXAMPLE 6–6

Three alternatives are being considered to improve an assembly line along with the "do-nothing" alternative. Each of Plans $A$, $B$, and $C$ has a 10-year life and a salvage value equal to 10% of its original cost.

|                                    | Plan A   | Plan B   | Plan C   |
|------------------------------------|----------|----------|----------|
| Installed cost of equipment        | $15,000  | $25,000  | $33,000  |
| Material and labor savings per year | 14,000   | 9,000    | 14,000   |
| Annual operating expenses          | 8,000    | 6,000    | 6,000    |
| End-of-useful life salvage value   | 1,500    | 2,500    | 3,300    |

If interest is 8%, which plan, if any, should be adopted?

### SOLUTION

Since neither installed cost nor output benefits are fixed, the economic criterion is to maximize EUAW = EUAB − EUAC.

|                                         | Plan A    | Plan B   | Plan C    |
|-----------------------------------------|-----------|----------|-----------|
| *Equivalent uniform annual benefit* (EUAB) |           |          |           |
| Material and labor per year             | $14,000   | $9,000   | $14,000   |
| Salvage value $(A/F, 8\%, 10)$          | 104       | 172      | 228       |
| EUAB =                                  | $14,104   | $9,172   | $14,228   |
| *Equivalent uniform annual cost* (EUAC) |           |          |           |
| Installed cost $(A/P, 8\%, 10)$         | $ 2,235   | $3,725   | $ 4,917   |
| Annual operating expenses               | 8,000     | 6,000    | 6,000     |
| EUAC =                                  | $10,235   | $9,725   | $10,917   |
| EUAW = EUAB − EUAC =                     | $ 3,869   | −$ 553   | $ 3,311   |

Based on our criterion of maximizing EUAW, Plan $A$ is the best of the four alternatives. Since the do-nothing alternative has EUAW = 0, it is a more desirable alternative than Plan $B$.

---

## ANALYSIS PERIOD

In Chapter 5, we saw that the analysis period is an important consideration in computing present worth comparisons. In such problems, a common analysis period must be used for all alternatives. In annual cash flow comparisons, we again have the analysis period question. Example 6-7 will help in examining the problem.

## EXAMPLE 6–7

Two pumps are being considered for purchase. If interest is 7%, which pump should be bought?

|  | Pump A | Pump B |
|---|---|---|
| Initial cost | $7000 | $5000 |
| End-of-useful-life salvage value | 1500 | 1000 |
| Useful life, in years | 12 | 6 |

### SOLUTION

The annual cost for 12 years of Pump $A$ can be found by using Equation 6-4:

$$\text{EUAC} = (P - S)(A/P, i, n) + Si$$
$$= (7000 - 1500)(A/P, 7\%, 12) + 1500(0.07)$$
$$= 5500(0.1259) + 105 = \$797$$

Now compute the annual cost for 6 years of Pump $B$:

$$\text{EUAC} = (5000 - 1000)(A/P, 7\%, 6) + 1000(0.07)$$
$$= 4000(0.2098) + 70 = \$909$$

For a common analysis period of 12 years, we need to replace Pump $B$ at the end of its 6-year useful life. If we assume that another pump $B'$ can be obtained, having the same $5000 initial cost, $1000 salvage value, and 6-year life, the cash flow will be as follows:

For the 12-year analysis period, the annual cost for Pump $B$ is

$$\text{EUAC} = [5000 - 1000(P/F, 7\%, 6) + 5000(P/F, 7\%, 6)$$
$$- 1000(P/F, 7\%, 12)] \times (A/P, 7\%, 12)$$
$$= [5000 - 1000(0.6663) + 5000(0.6663) - 1000(0.4440)] \times (0.1259)$$
$$= (5000 - 666 + 3331 - 444)(0.1259)$$
$$= (7211)(0.1259) = \$909$$

The annual cost of $B$ for the 6-year analysis period is the same as the annual cost for the 12-year analysis period. This is not a surprising conclusion when one recognizes that the annual cost of the first 6-year period is repeated in the second 6-year period. Thus the lengthy calculation of

EUAC for 12 years of Pump *B* and *B'* was not needed. By assuming that the shorter-life equipment is replaced by equipment with identical economic consequences, we have avoided a lot of calculations. Select Pump *A*.

### Analysis Period Equal to Alternative Lives

If the analysis period for an economy study coincides with the useful life for each alternative, then the economy study is based on this analysis period.

### Analysis Period a Common Multiple of Alternative Lives

When the analysis period is a common multiple of the alternative lives (for example, in Example 6-7, the analysis period was 12 years with 6- and 12-year alternative lives), a "replacement with an identical item with the same costs, performance, and so forth" is frequently assumed. This means that when an alternative has reached the end of its useful life, we assume that it will be replaced with an identical item. As shown in Example 6-7, the result is that the EUAC for Pump *B* with a 6-year useful life is equal to the EUAC for the entire analysis period based on Pump *B plus* the replacement unit, Pump *B'*.

Under these circumstances of identical replacement, we can compare the annual cash flows computed for alternatives based on their own service lives. In Example 6-7, the annual cost for Pump *A*, based on its 12-year service life, was compared with the annual cost for Pump *B*, based on its 6-year service life.

### Analysis Period for a Continuing Requirement

Many times an economic analysis is undertaken to determine how to provide for a more or less continuing requirement. One might need to pump water from a well as a continuing requirement. There is no distinct analysis period. In this situation, the analysis period is assumed to be long but undefined.

If, for example, we had a continuing requirement to pump water and alternative Pumps *A* and *B* had useful lives of 7 and 11 years, respectively, what should we do? The customary assumption is that Pump *A*'s annual cash flow (based on a 7-year life) may be compared to Pump *B*'s annual cash flow (based on an 11-year life). This is done without much concern that the least common multiple of the 7- and 11-year lives is 77 years. This comparison of "different-life" alternatives assumes identical replacement (with identical costs, performance, etc.) when an alternative reaches the end of its useful life.

This continuing requirement, which can also be described as an *indefinitely long horizon*, is illustrated in Example 6-8. Since this is longer than the lives of the alternatives, we can make the best decision possible given current information by minimizing EUAC or maximizing EUAW or EUAB. At a later time, we will make another replacement and there will be more information on costs *at that time*.

### EXAMPLE 6–8

Pump $B$ in Example 6-7 is now believed to have a 9-year useful life. Assuming the same initial cost and salvage value, compare it with Pump $A$ using the same 7% interest rate.

### SOLUTION

If we assume that the need for $A$ or $B$ will exist for some continuing period, the comparison of costs per year for the unequal lives is an acceptable technique. For 12 years of Pump $A$:

$$\text{EUAC} = (7000 - 1500)(A/P, 7\%, 12) + 1500(0.07) = \$797$$

For 9 years of Pump $B$:

$$\text{EUAC} = (5000 - 1000)(A/P, 7\%, 9) + 1000(0.07) = \$684$$

For minimum EUAC, select Pump $B$.

### Infinite Analysis Period

At times we have an alternative with a limited (finite) useful life in an infinite analysis period situation. The equivalent uniform annual cost may be computed for the limited life. The assumption of identical replacement (replacements have identical costs, performance, etc.) is often appropriate. Based on this assumption, the same EUAC occurs for each replacement of the limited-life alternative. The EUAC for the infinite analysis period is therefore equal to the EUAC computed for the limited life. With identical replacement,

$$\text{EUAC}_{\text{infinite analysis period}} = \text{EUAC}_{\text{for limited life } n}$$

A somewhat different situation occurs when there is an alternative with an infinite life in a problem with an infinite analysis period:

$$\text{EUAC}_{\text{infinite analysis period}} = P(A/P, i, \infty) + \text{Any other annual costs}$$

When $n = \infty$, we have $A = Pi$ and, hence, $(A/P, i, \infty)$ equals $i$.

$$\text{EUAC}_{\text{infinite analysis period}} = Pi + \text{Any other annual costs}$$

### EXAMPLE 6–9

In the construction of an aqueduct to expand the water supply of a city, there are two alternatives for a particular portion of the aqueduct. Either a tunnel can be constructed through a mountain, or a pipeline can be laid to go around the mountain. If there is a permanent need for the aqueduct, should the tunnel or the pipeline be selected for this particular portion of the aqueduct? Assume a 6% interest rate.

**SOLUTION**

|  | Tunnel Through Mountain | Pipeline Around Mountain |
|---|---|---|
| Initial cost | $5.5 million | $5 million |
| Maintenance | 0 | 0 |
| Useful life | Permanent | 50 years |
| Salvage value | 0 | 0 |

**Tunnel**

For the tunnel, with its permanent life, we want $(A/P, 6\%, \infty)$. For an infinite life, the capital recovery is simply interest on the invested capital. So $(A/P, 6\%, \infty) = i$, and we write

$$\text{EUAC} = Pi = \$5.5 \text{ million}(0.06)$$

$$= \$330,000$$

**Pipeline**

$$\text{EUAC} = \$5 \text{ million}(A/P, 6\%, 50)$$

$$= \$5 \text{ million}(0.0634) = \$317,000$$

For fixed output, minimize EUAC. Select the pipeline.

The difference in annual cost between a long life and an infinite life is small unless an unusually low interest rate is used. In Example 6-9 the tunnel is assumed to be permanent. For comparison, compute the annual cost if an 85-year life is assumed for the tunnel.

$$\text{EUAC} = \$5.5 \text{ million}(A/P, 6\%, 85)$$

$$= \$5.5 \text{ million}(0.0604)$$

$$= \$332,000$$

The difference in time between 85 years and infinity is great indeed; yet the difference in annual costs in Example 6-9 is very small.

## Some Other Analysis Period

The analysis period in a particular problem may be something other than one of the four we have so far described. It may be equal to the life of the shorter-life alternative, the longer-life alternative, or something entirely different. One must carefully examine the consequences of each alternative throughout the analysis period and, in addition, see what differences there might be in salvage values, and so forth, at the end of the analysis period.

## EXAMPLE 6–10

Suppose that Alternative 1 has a 7-year life and a salvage value at the end of that time. The replacement cost at the end of 7 years may be more or less than the original cost. If the replacement is retired prior to 7 years, it will have a terminal value that exceeds the end-of-life salvage value. Alternative 2 has a 13-year life and a terminal value whenever it is retired. If the situation indicates that 10 years is the proper analysis period, set up the equations to compute the EUAC for each alternative. Use results from Example 5-5 to compute the results.

### SOLUTION

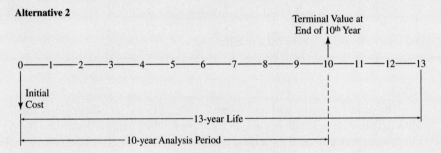

**Alternative 1**

$$\text{EUAC}_1 = [\text{Initial cost} + (\text{Replacement cost} - \text{Salvage value})(P/F, \ i, 7)$$
$$- \ (\text{Terminal value})(P/F, \ i, 10)](A/P, \ i, 10)$$
$$= 64{,}076(A/P, 8\%, 10) \quad \text{using results from Example 5-5}$$
$$= 64{,}076(8.1490) = \$9547$$

**Alternative 2**

$$\text{EUAC}_2 = [\text{Initial cost} - (\text{Terminal value})(P/F, \ i, 10)](A/P, \ i, 10)$$
$$= 69{,}442(A/P, 8\%, 10) \quad \text{using results from Example 5-5}$$
$$= 69{,}442(0.1490) = \$10{,}347$$

Select Alternative 1.

## USING SPREADSHEETS TO ANALYZE LOANS

Loan and bond payments are made by firms, agencies, and individual engineers. Usually, the payments in each period are constant. Spreadsheets make it easy to:

- Calculate the loan's amortization schedule
- Decide how a payment is split between principal and interest
- Find the balance due on a loan
- Calculate the number of payments remaining on a loan.

### Building an Amortization Schedule

As illustrated in previous chapters and Appendix 1, an amortization schedule lists for each payment period: the loan payment, interest paid, principal paid, and remaining balance. For each period, the interest paid equals the interest rate times the balance remaining from the period before. Then the principal payment equals the payment minus the interest paid. Finally, this principal payment is applied to the balance remaining from the preceding period to calculate the new remaining balance. As a basis for comparison with spreadsheet loan functions, Figure 6-1 shows this calculation for Example 6-11.

### EXAMPLE 6–11

An engineer wanted to celebrate graduating and getting a job by spending $2400 on new furniture. Luckily the store was offering 6-month financing at the low interest rate of 6% per year nominal (really $1/2\%$ per month). Calculate the amortization schedule.

### SOLUTION

| | A | B | C | D | E |
|---|---|---|---|---|---|
| 1 | 2400 | Initial balance | | | |
| 2 | 0.50% | $i$ | | | |
| 3 | 6 | $N$ | | | |
| 4 | $407.03 | Payment | $= -PMT(A2,A3,A1)$ | | |
| 5 | | | | | |
| 6 | | | Principal | Ending | |
| 7 | Month | Interest | Payment | Balance | |
| 8 | 0 | | | 2400.00 | =A1 |
| 9 | 1 | 12.00 | 395.03 | 2004.97 | =D8−C9 |
| 10 | 2 | 10.02 | 397.00 | 1607.97 | |
| 11 | 3 | 8.04 | 398.99 | 1208.98 | |
| 12 | 4 | 6.04 | 400.98 | 807.99 | |
| 13 | 5 | 4.04 | 402.99 | 405.00 | |
| 14 | 6 | 2.03 | 405.00 | 0.00 | |
| 15 | | | | =$A$4−B14 | |
| 16 | | | | =Payment−Interest | |
| 17 | | | =$A$2*D13 | | |
| 18 | | | =rate*previous balance | | |

FIGURE 6–1  Amortization schedule for furniture loan.

The first step is to calculate the monthly payment:

$$A = 2400(A/P, \frac{1}{2}\%, 6) = 2400(0.1696) = \$407.0$$
$$= \text{PMT}(0.005, 6, -2400) = \$407.03$$

With this information the engineer can use the spreadsheet of Figure 6–1 to obtain the amortization schedule.

### How Much to Interest? How Much to Principal?

For a loan with constant payments, we can answer these questions for any period without the full amortization schedule. For a loan with constant payments, the functions IPMT and PPMT directly answer these questions. For simple problems, both functions have four arguments $(i, t, n, -P)$, where $t$ is the time period being calculated. Both functions have optional arguments that permit adding a balloon payment (an $F$) and changing from end-of-period payments to beginning-of-period payments.

For example, consider Period 4 of Example 6-11. The spreadsheet formulas give the same answer as shown in Figure 6-1.

$$\text{Interest period } 4 = \text{IPMT}(0.5\%, 4, 6, -2400) = \$6.04$$
$$\text{Principal payment period } 4 = \text{PPMT}(0.5\%, 4, 6, -2400) = \$400.98$$

### Finding the Balance Due on a Loan

An amortization schedule is one used to calculate the balance due on a loan. A second, easier way is to remember that the balance due equals the present worth of the remaining payments. Interest is paid in full after each payment, so later payments are simply based on the balance due.

## EXAMPLE 6–12

A car is purchased with a 48-month, 9% nominal loan with an initial balance of $15,000. What is the balance due halfway through the 4 years?

### SOLUTION

The first step is to calculate the monthly payment, at a monthly interest rate of $3/4\%$. This equals

$$\text{Payment} = 15,000(A/P, 0.75\%, 48) \quad \text{or} \quad = \text{PMT}(0.75\%, 48, -15000)$$
$$= (15,000)(0.0249) = \$373.50 \quad \text{or} \quad = \$373.28$$

The next step will use the spreadsheet answer, because it is more accurate (there are only three significant digits in the tabulated factor).

After 24 payments and with 24 left, the remaining balance equals $(P/A, i, N_{\text{remaining}})$ payment

$$\text{Balance} = (P/A, 0.75\%, 24)\$373.28 \quad \text{or} \quad = \text{PV}(0.75\%, 24, 373.28)$$
$$= (21.889)(373.28) = \$8170.73 \quad \text{or} \quad = \$8170.78$$

Thus halfway through the repayment schedule, 54.5% of the original balance is still owed.

## Pay Off Debt Sooner by Increasing Payments

Paying off debt can be a good investment because the investment earns the rate of interest on the loan. For example, this could be 8% for a mortgage, 10% for a car loan, or 19% for a credit card. When one is making extra payments on a loan, the common question is: How much sooner will the debt be paid off? Until the debt is paid off, any early payments are essentially locked up, since the same payment amount is owed each month.

The first reason that spreadsheets and TVM calculators are convenient relates to fractional interest rates. For example, an auto loan might be at a nominal rate of 13% with monthly compounding or 1.08333% per month. The second reason is that the function NPER or the $n$ key calculates the number of periods remaining on a loan.

NPER can be used to calculate how much difference is made by one extra payment or by increasing all payments by $x\%$. Extra payments are applied entirely to principal, so the interest rate, remaining balance, and payment amounts are all known. $N_{remaining}$ equals NPER($i$, payment, remaining balance) with optional arguments for beginning-of-period cash flows and balloon payments. The signs of the payment and the remaining balance must be different.

## EXAMPLE 6–13

Maria has a 7.5% mortgage with monthly payments for 30 years. Her original balance was $100,000, and she just made her twelfth payment. Each month she also pays into a reserve account, which the bank uses to pay her fire and liability insurance ($900 annually) and property taxes ($1500 annually). By how much does she shorten the loan if she makes an extra *loan* payment today? If she makes an extra *total* payment? If she increases each total payment to 110% of her current total payment?

### SOLUTION

The first step is to calculate Maria's *loan* payment for the 360 months. Rather than calculating a six-significant-digit monthly interest rate, it is easier to use 0.075/12 in the spreadsheet or TVM calculator.

$$\text{Payment} = \text{PMT}(0.075/12, 360, -100000) = \$699.21$$

The remaining balance after 12 such payments is the present worth of the remaining 348 payments.

$$\text{Balance}_{12} = \text{PV}(0.075/12, 348, 699.21) = \$99,077.53$$
$$\text{(after 12 payments, she has paid off \$922!)}$$

If she pays an extra $699.21, then the number of periods remaining is

$$\text{NPER}(0.075/12, -699.21, 99077.53 - 699.21) = 339.5$$

This is 8.5 payments less than the 348 periods left before the extra payment. If she makes an extra total payment, then

$$\text{Total payment} = 699.21 + 900/12 + 1500/12 = \$899.21/\text{month}$$
$$\text{NPER}(0.075/12, -699.21, 99244 - 899.21) = 337.1$$

or 2.4 more payments saved. If she increases each payment by an extra 10% payment on the total payment of $899.21, then

$$\text{NPER}(0.075/12, -(1.1 * 899.21 - 200), 99077.53) = 246.5 \text{ payments}$$

or 101.5 payments saved.

Note that $200 of the total payment goes to pay for insurance and taxes.

## SUMMARY

**Annual cash flow analysis** is the second of the three major methods of resolving alternatives into comparable values. When an alternative has an initial cost $P$ and salvage value $S$, there are three ways of computing the equivalent uniform annual cost:

- EUAC $= P(A/P, i, n) - S(A/F, i, n)$                 (6-1)
- EUAC $= (P - S)(A/F, i, n) + Pi$                    (6-3)
- EUAC $= (P - S)(A/P, i, n) + Si$                   (6-4)

All three equations give the same answer. This quantity is also known as the *capital recovery cost* of the project.

The relationship between the present worth of cost and the equivalent uniform annual cost is

- EUAC $= (\text{PW of cost})(A/P, i, n)$

The three annual cash flow criteria are:

|  |  |
|---|---|
| Neither input nor output fixed | Maximize EUAW = EUAB − EUAC |
| For fixed input | Maximize EUAB |
| For fixed output | Minimize EUAC |

In present worth analysis there must be a common analysis period. Annual cash flow analysis, however, allows some flexibility provided the necessary assumptions are suitable in the situation being studied. The analysis period may be different from the lives of the alternatives, and provided the following criteria are met, a valid cash flow analysis may be made.

1. When an alternative has reached the end of its useful life, it is assumed to be replaced by an identical replacement (with the same costs, performance, etc.).
2. The analysis period is a common multiple of the useful lives of the alternatives, or there is a continuing or perpetual requirement for the selected alternative.

If neither condition applies, it is necessary to make a detailed study of the consequences of the various alternatives over the entire analysis period with particular attention to the difference between the alternatives at the end of the analysis period.

There is very little numerical difference between a long-life alternative and a perpetual alternative. As the value of $n$ increases, the capital recovery factor approaches $i$. At the limit, $(A/P, i, \infty) = i$.

One of the most common uniform payment series is the repayment of loans. Spreadsheets and TVM calculators are useful in analyzing loans (balance due, interest paid, etc.) for several reasons: they have specialized functions, many periods are easy, and any interest rate can be used.

# PROBLEMS

## Annual Calculations

**6-1** Compute the EUAB for these cash flows based on a 10% interest rate.

(*Answer:* EUAB= $35.72)

**6-2** Compute the EUAC for these cash flows.

**6-3** Compute the EUAB for these cash flows.

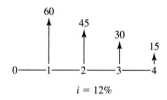

**6-4** If $i = 6\%$, compute the EUAB over 6 years that is equivalent to the two receipts shown.

(*Answer:* EUAB = $52.31)

**6-5** A loan of $1000 is to be repaid in three equal semiannual (every 6 months) payments. If the annual interest rate is 7% compounded semiannually, how much is each payment? (*Answer:* $356.90)

**6-6** When he started work on his twenty-second birthday, D. B. Cooper decided to invest money each month with the objective of becoming a millionaire by the time he reaches age 65. If he expects his investments to yield 18% per annum, compounded monthly, how much should he invest each month? (*Answer:* $6.92 a month.)

**6-7** The average age of engineering students at graduation is a little over 23 years. This means that the working career of most engineers is almost exactly 500 months. How much would an engineer need to save each month to become a millionaire by the end of her working career? Assume a 15% interest rate, compounded monthly.

**6-8** An engineer wishes to have $5 million by the time he retires in 40 years. Assuming 15% nominal interest, compounded continuously, what annual sum must he set aside? (*Answer:* $2011)

**6-9** For the diagram, compute the value of $D$ that results in a net equivalent uniform annual worth (EUAW) of 0.

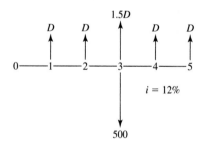

**6-10** An electronics firm invested $60,000 in a precision inspection device. It cost $4000 to operate and maintain in the first year and $3000 in each later year. At the end of 4 years, the firm changed their inspection procedure, eliminating the need for the device. The purchasing agent was very fortunate to sell the inspection device for $60,000, the original price. Compute the equivalent uniform annual cost during the 4 years the device was used. Assume interest at 10% per year. (*Answer:* $9287)

**6-11** A firm is about to begin pilot plant operation, and it could add an optional heat exchanger unit. A unit is now available for $30,000, and it is estimated that the heat exchanger unit will be worth $35,000 after 8 years for use in other company operations. This high salvage value is because the $30,000 purchase price is really a rare bargain. If 15% is an appropriate rate of return, what annual benefit is needed to justify buying the heat exchanger unit? (*Answer:* $4135)

**6-12** A firm purchased some equipment at a very favorable price of $30,000. The equipment resulted in an annual net saving of $1000 per year during the 8 years it was used. At the end of 8 years, the equipment was sold for $40,000. Assuming interest at 8%, did the equipment purchase prove to be desirable?

*Problems 6-13 to 6-16 contributed by Hamed Kashani, Saeid Sadri, and Baabak Ashuri, Georgia Institute of Technology*

**6-13** A couple is saving for their newborn daughter's college education. She will need $25,000 per year for a four-year college program, which she will start when she is 17. What uniform deposits starting 3 years from now and continuing through year 16 are needed, if the account earns 6% interest?

**6-14** How much should a new graduate pay in 10 equal annual payments, starting 2 years from now, in order to repay a $30,000 loan he has received today? The interest rate is 6% per year.

**6-15** A firm is buying an adjacent 1000-acre parcel for a future plant expansion. The price has been set at $30,000 per acre. The payment plan is 25% down, and the balance 2 years from now. If the transaction interest rate is 12% per year, what are the two payments?

**6-16** The manager of a small cleaning company applies for a $25,000 loan at an interest rate of 10% per year. He will repay the loan over 6 years with annual payments. The third through sixth payments are $1500 greater than the first two. Determine the size of the payments.

*Problems 6-17 to 6-20 contributed by Paul R. McCright, University of South Florida*

**6-17** Amanda and Blake have found a house, which owing to a depressed real estate market costs only $201,500. They will put $22,000 down and finance the remainder with a 30-year mortgage loan from Bank of America at 4.65% interest (compounded monthly).

(*a*) How much is their monthly loan payment?

(*b*) How much interest will they pay in the second payment?

(*c*) They will also have the following expenses: property taxes of $2100, homeowners' insurance of $1625, and $290 mortgage insurance (in case one of them dies before the loan is repaid, a requirement of the bank). These annual amounts are paid in 12 installments and added to the loan payment. What will Amanda and Blake's full monthly cost be?

(*d*) If they can afford $1200 per month, can Amanda and Blake afford this house?

**6-18** Helen purchases a new Ford Focus. She negotiates a price of $18,400, trades in her 1993 Contour for $1700, puts down an additional $1000, and borrows the remainder for 3 years at 6% interest. How large will her monthly payments be?

**6-19** To reduce her personal carbon footprint, Zooey is buying a new Ford Escape hybrid. She has negotiated a price of $21,900 and will trade in her old Ford Contour for $2350. She will put another $850 with it and borrow the remainder at 7% interest compounded monthly for 4 years. Prepare a payment schedule for the first three months of payments.

**6-20** Zwango Plus Manufacturing expects that fixed costs of keeping its Zephyr Hills Plant operating will be $1.4 M this year. If the fixed costs increase by $100,000 each year, what is the EUAC for a 10-year period? Assume the interest rate is 12%.

**6-21** For the diagram, compute the value of *C* that results in a net equivalent annual worth (EUAW) of 0.

**6-22** If interest is 10%, what is the EUAB?

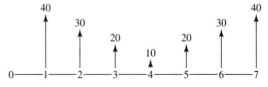

6-23 The maintenance foreman of a plant in reviewing his records found that a large press had the following maintenance cost record:

| | |
|---|---|
| 5 years ago | $ 600 |
| 4 years ago | 700 |
| 3 years ago | 800 |
| 2 years ago | 900 |
| Last year | 1000 |

After consulting with a lubrication specialist, he changed the preventive maintenance schedule. He believes that maintenance will be $900 this year and will decrease by $100 a year in each of the following 4 years. If his estimate of the future is correct, what will be the equivalent uniform annual maintenance cost for the 10-year period? Assume interest at 8%. (*Answer:* $756)

6-24 A motorcycle is for sale for $26,000. The dealer is willing to sell it on the following terms:

No down payment; pay $440 at the end of each of the first 4 months; pay $840 at the end of each month after that until the loan has been paid in full.

At a 12% annual interest rate compounded monthly, how many $840 payments will be required?

6-25 Art Arfons, a K-State-educated engineer, has made a considerable fortune. He wishes to start a perpetual scholarship for engineering students at K-State. The scholarship will provide a student with an annual stipend of $2500 for each of 4 years (freshman through senior), plus an additional $5000 during the senior year to cover job search expenses. Assume that students graduate in 4 years, a new award is given every 4 years, and the money is paid at the beginning of each year with the first award at the beginning of Year 1. The interest rate is 8%.

(*a*) Determine the equivalent uniform annual cost (EUAC) of providing the scholarship.

(*b*) How much money must Art donate to K-State?

6-26 A machine costs $20,000 and has a 5-year useful life. At the end of the 5 years, it can be sold for $4000. If annual interest is 8%, compounded semiannually, what is the equivalent uniform annual cost of the machine?

6-27 Mr. Wiggley wants to buy a new house. It will cost $178,000. The bank will loan 90% of the purchase price at a nominal interest rate of 10.75% compounded weekly, and Mr. Wiggley will make monthly payments. What is the amount of the monthly payments if he intends to pay the house off in 25 years?

6-28 Steve Lowe must pay his property taxes in two equal installments on December 1 and April 1. The two payments are for taxes for the fiscal year that begins on July 1 and ends the following June 30. Steve purchased a home on September 1. Assuming the annual property taxes remain at $850 per year for the next several years, Steve plans to open a savings account and to make uniform monthly deposits the first of each month. The account is to be used to pay the taxes when they are due.

To open the account, Steve deposits a lump sum equivalent to the monthly payments that will not have been made for the first year's taxes. The savings account pays 9% interest, compounded monthly and payable quarterly (March 31, June 30, September 30, and December 31). How much money should Steve put into the account when he opens it on September 1? What uniform monthly deposit should he make from that time on? (*Answers:* Initial deposit $350.28; monthly deposit $69.02)

6-29 Your company must make a $500,000 balloon payment on a lease 2 years and 9 months from today. You have been directed to deposit an amount of money quarterly, beginning today, to provide for the $500,000 payment. The account pays 4% per year, compounded quarterly. What is the required quarterly deposit? *Note:* Lease payments are due at the beginning of the quarter.

6-30 Linda O'Shay deposited $30,000 in a savings account as a perpetual trust. She believes the account will earn 7% annual interest during the first 10 years and 5% interest thereafter. The trust is to provide a uniform end-of-year scholarship at the university. What uniform amount could be used for the student scholarship each year, beginning at the end of the first year and continuing forever?

6-31 An engineer has a fluctuating future budget for the maintenance of a particular machine. During each of the first 5 years, $1000 per year will be budgeted. During the second 5 years, the annual budget will be $1500 per year. In addition, $3500 will be budgeted for an overhaul of the machine at the end of the fourth year, and another $3500 for an overhaul at the end of the eighth year.

The engineer asks you to compute the uniform annual expenditure that would be equivalent to these fluctuating amounts, assuming interest at 6% per year.

6-32   A machine has a first cost of $150,000, an annual operation and maintenance cost of $2500, a life of 10 years, and a salvage value of $30,000. At the end of Years 4 and 8, it requires a major service, which costs $20,000 and $10,000, respectively. At the end of Year 5, it will need to be overhauled at a cost of $45,000. What is the equivalent uniform annual cost of owning and operating this particular machine?

6-33   There is an annual receipt of money that varies from $100 to $300 in a fixed pattern that repeats forever. If interest is 10%, compute the EUAB, also continuing forever, that is equivalent to the fluctuating disbursements.

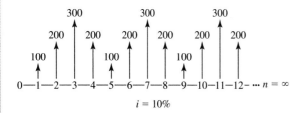

$i = 10\%$

6-34   If the owner earns 5% interest on her investments, determine the equivalent annual cost of owning a car with the following costs (EOY = end of year).

Initial down payment = $2200
Annual payments = $5500, EOY1–EOY4
Prepaid insurance = $1500, growing 8% annually
Gas & oil & minor maintenance = $2000, growing 10% annually
Replacement tires = $650 at EOY4 & $800 at EOY8
Major maintenance = $2400 at EOY5
Salvage value = $3750 at EOY9

*Contributed by Paul R. McCright, University of South Florida*

**Annual Comparisons**

6-35   A construction firm needs a new small loader. It can be leased from the dealer for 3 years for $5500 per year including all maintenance, or it can be purchased for $20,000. The firm expects the loader to have a salvage value of $7000 after 7 years. The maintenance will be $500 the first year and then it will increase by $300 each year. The firm's interest rate is 12% per

year. Compare the EUACs for leasing and buying the loader.

*Contributed by Hamed Kashani, Saeid Sadri, and Baabak Ashuri, Georgia Institute of Technology*

6-36   The Johnson Company pays $2000 a month to a trucker to haul wastepaper and cardboard to the city dump. The material could be recycled if the company were to buy a $60,000 hydraulic press baler and spend $30,000 a year for labor to operate the baler. The baler has an estimated useful life of 30 years and no salvage value. Strapping material would cost $2000 per year for the estimated 500 bales a year that would be produced. A wastepaper company will pick up the bales at the plant and pay Johnson $23 per bale for them. Use an annual cash flow analysis in working this problem.

(a) If interest is 8%, is it economical to install and operate the baler?

(b) Would you recommend that the baler be installed?

6-37   Jenny McCarthy is an engineer for a municipal power plant. The plant uses natural gas, which is currently provided from an existing pipeline at an annual cost of $10,000 per year. Jenny is considering a project to construct a new pipeline. The initial cost of the new pipeline would be $35,000, but it would reduce the annual cost to $5000 per year. Assume an analysis period of 20 years and no salvage value for either the existing or new pipeline. The interest rate is 6%.

(a) Determine the equivalent uniform annual cost (EUAC) for the new pipeline.

(b) Should the new pipeline be built?

6-38   Claude James, a salesman, needs a new car for business use. He expects to be promoted to a supervisory job at the end of 3 years, and he will no longer be "on the road." The company reimburses salesmen each month at the rate of 55¢ per mile driven. Claude finds that there are three different ways of obtaining his chosen car:

A.   Pay cash: the price is $26,000.

B.   Lease the car: the monthly charge is $700 on a 36-month lease, payable at the end of each month; at the end of the 3-year period, the car is returned to the leasing company.

C.   Lease the car with an option to buy at the end of the lease: pay $720 a month for 36 months; at the end of that time, Claude could buy the car, if he chooses, for $7000.

Claude believes he should use a 12% interest rate. If the car could be sold for $7500 at the end of 3 years, which method should he use to obtain it?

**6-39** When he purchased his home, Al Silva borrowed $280,000 at 10% interest to be repaid in 25 equal annual end-of-year payments. After making 10 payments, Al found he could refinance the balance due on his loan at 9% interest for the remaining 15 years.

    To refinance the loan, Al must pay the original lender the balance due on the loan, plus a penalty charge of 2% of the balance due; to the new lender he also must pay a $1000 service charge to obtain the loan. The new loan would be made equal to the balance due on the old loan, plus the 2% penalty charge, and the $1000 service charge. Should Al refinance the loan, assuming that he will keep the house for the next 15 years? Use an annual cash flow analysis in working this problem.

**6-40** A firm must decide whether to provide their salespeople with firm-owned cars or to pay a mileage allowance for their own cars. New cars would cost about $28,000 each and could be resold 4 years later for about $11,000 each. Annual operating costs would be $1200 per year plus 24¢ per mile. If the salespeople drove their own cars, the firm would pay 50¢ per mile. How many miles must each salesperson drive each year for it to be economically practical for the firm to provide the cars? Assume a 10% annual interest rate. Use an annual cash flow analysis.

**6-41** The town of Dry Gulch needs more water from Pine Creek. The town engineer has selected two plans for comparison: a *gravity plan* (divert water at a point 10 miles up Pine Creek and pipe it by gravity to the town) and a *pumping plan* (divert water at a point closer to town). The pumping plant would be built in two stages, with half-capacity installed initially and the other half installed 10 years later.

    The analysis will assume a 40-year life, 10% interest, and no salvage value. Use an annual cash flow analysis to find which plan is more economical.

| | Gravity | Pumping |
|---|---|---|
| Initial investment | $2,800,000 | $1,400,000 |
| Additional investment in 10$^{th}$ year | None | 200,000 |
| Operation and maintenance | 10,000/yr | 25,000/yr |
| Power cost | | |
|   Average first 10 years | None | 50,000/yr |
|   Average next 30 years | None | 100,000/yr |

**6-42** A manufacturer is considering replacing a production machine tool. The new machine, costing $37,000, would have a life of 4 years and no salvage value, but would save the firm $5000 per year in direct labor costs and $2000 per year in indirect labor costs. The existing machine tool was purchased 4 years ago at a cost of $40,000. It will last 4 more years and will have no salvage value. It could be sold now for $10,000 cash. Assume that money is worth 8% and that differences in taxes, insurance, and so forth are negligible. Use an annual cash flow analysis to determine whether the new machine should be purchased.

**6-43** Two possible routes for a power line are under study. Data on the routes are as follows:

| | Around the Lake | Under the Lake |
|---|---|---|
| Length | 15 km | 5 km |
| First cost | $5000/km | $25,000/km |
| Maintenance | $200/km/yr | $400/km/yr |
| Useful life, in years | 15 | 15 |
| Salvage value | $3000/km | $5000/km |
| Yearly power loss | $500/km | $500/km |
| Annual property taxes | 2% of first cost | 2% of first cost |

If 7% interest is used, should the power line be routed around the lake or under the lake? (*Answer:* Around the lake.)

**6-44** An oil refinery must now begin sending its waste liquids through a costly treatment process before discharging them into a nearby stream. The engineering department estimates costs at $300,000 for the first year. It is estimated that if process and plant alterations are made, the waste treatment cost will decline $30,000 each year. As an alternate, a specialized firm, Hydro-Clean, has offered a contract to process the waste liquids for 10 years for a fixed price of $150,000 per year, payable at the end of each year. Either way, there should be no need for waste treatment after 10 years. The refinery manager considers 8% to be a suitable interest rate. Use an annual cash flow analysis to determine whether the Hydro-Clean offer should be accepted.

**6-45** Bill Anderson buys a car every 2 years as follows: initially he makes a down payment of $6000 on a $15,000 car. The balance is paid in 24 equal monthly payments with annual interest at 12%. When he has made the last payment on the loan, he trades in the 2-year-old car for $6000 on a new $15,000 car, and the cycle begins over again.

Doug Jones decided on a different purchase plan. He thought he would be better off if he paid $15,000 cash for a new car. Then he would make a monthly deposit in a savings account so that, at the end of 2 years, he would have $9000 in the account. The $9000 plus the $6000 trade-in value of the car will allow Doug to replace his 2-year-old car by paying $9000 for a new one. The bank pays 6% interest, compounded quarterly.

(*a*) What is Bill Anderson's monthly payment to pay off the loan on the car?

(*b*) After he purchased the new car for cash, how much per month should Doug Jones deposit in his savings account to have enough money for the next car 2 years hence?

(*c*) Why is Doug's monthly savings account deposit smaller than Bill's payment?

**6-46** Two mutually exclusive alternatives are being considered.

| Year | A | B |
|------|------|-------|
| 0 | −$3000 | −$5000 |
| 1 | 845 | 1400 |
| 2 | 845 | 1400 |
| 3 | 845 | 1400 |
| 4 | 845 | 1400 |
| 5 | 845 | 1400 |

One of the alternatives must be selected. Using a 15% nominal interest rate, compounded continuously, determine which one. Solve by annual cash flow analysis.

**6-47** North Plains Biofuels (NPB) has negotiated a contract with an oil firm to sell 150,000 barrels of ethanol per year, beginning in end of year (EOY) 4. The oil firm will pay NPB $10M annually beginning from EOY0 to EOY3 and then $110 per barrel from EOY4 through EOY13.If NPB an interest rate of 15%, which method should be used to produce the biofuels?

*Contributed by Paul R. McCright, University of South Florida*

| | Corn | Algae |
|---|---|---|
| Purchase of land (EOY0) | $1,900,000 | $ 3,800,000 |
| Facility construction (at EOY1) | $5,300,000 | $7,100,000 |
| Annual O&M | $2,450,000 | $2,800,000 |
| Increasing 6% yearly from EOY1 through EOY13 | | |
| Raw materials (corn or algae) | $1,500,000 | $250,000 |
| Annual increase of 8% from EOY4 through EOY13 | | |
| Salvage value (EOY13) | $3,000,000 | $3,600,000 |

**6-48** Which car has a lower EUAC if the owner can earn 5% in his best investment.
*Contributed by Paul R. McCright, University of South Florida*

| | Toyota Corolla | Toyota Prius |
|---|---|---|
| Initial cost | $19,200 | $25,500 |
| Annual maintenance | 1,000 | 1,500 |
| Annual gas & oil (increasing 15% yearly) | 2,500 | 1,200 |
| Salvage value (Year 8) | 8,000 | 10,000 |

## Different Lives

**6-49** A firm is choosing between machines that perform the same task in the same time. Assume the minimum attractive return is 8%. Which machine would you choose? (*Answers: X = $1252; Y = $1106*)

| | Machine X | Machine Y |
|---|---|---|
| First cost | $5000 | $8000 |
| Estimated life, in years | 5 | 12 |
| Salvage value | 0 | $2000 |
| Annual maintenance cost | 0 | 150 |

**6-50** A company must decide whether to buy Machine *A* or Machine *B*:

|  | Machine A | Machine B |
|---|---|---|
| Initial cost | $10,000 | $20,000 |
| Useful life, in years | 4 | 10 |
| End-of-useful-life salvage value | $10,000 | $10,000 |
| Annual maintenance | 1,000 | 0 |

At a 10% interest rate, which machine should be installed? Use an annual cash flow analysis in working this problem. (*Answer:* Machine *A*)

**6-51** Consider the following two mutually exclusive alternatives:

|  | A | B |
|---|---|---|
| Cost | $10,000 | $15,000 |
| Uniform annual benefit | 1,600 | 2,400 |
| Useful life, in years | ∞ | 20 |

Alternative *B* may be replaced with an identical item every 20 years at the same $15,000 cost and will have the same $2400 uniform annual benefit. Using a 10% interest rate and an annual cash flow analysis, which alternative should be selected?

**6-52** A pump is needed for 10 years at a remote location. The pump can be driven by an electric motor if a power line is extended to the site. Otherwise, a gasoline engine will be used. Use an annual cash flow analysis and a 10% interest rate. How should the pump be powered?

|  | Gasoline | Electric |
|---|---|---|
| First cost | $2400 | $6000 |
| Annual operating cost | 1200 | 750 |
| Annual maintenance | 300 | 50 |
| Salvage value | 300 | 600 |
| Life, in years | 5 | 10 |

**6-53** A suburban taxi company is considering buying taxis with diesel engines instead of gasoline engines. The cars average 50,000 km a year.

|  | Diesel | Gasoline |
|---|---|---|
| Vehicle cost | $24,000 | $19,000 |
| Useful life, in years | 5 | 4 |
| Fuel cost per liter | 68¢ | 72¢ |
| Mileage, in km/liter | 35 | 28 |
| Annual repairs | $ 900 | $ 700 |
| Annual insurance premium | 1,000 | 1,000 |
| End-of-useful-life resale value | 4,000 | 6,000 |

Use an annual cash flow analysis to determine the more economical choice if interest is 6%.

**6-54** The manager in a canned food processing plant is trying to decide between two labeling machines.

|  | Machine A | Machine B |
|---|---|---|
| First cost | $15,000 | $25,000 |
| Maintenance and operating costs | 1,600 | 400 |
| Annual benefit | 8,000 | 13,000 |
| Salvage value | 3,000 | 6,000 |
| Useful life, in years | 7 | 10 |

Assume an interest rate of 12%. Use annual cash flow analysis to determine which machine should be chosen.

**6-55** Consider the following three mutually exclusive alternatives:

|  | A | B | C |
|---|---|---|---|
| Cost | $10,000 | $150,000 | $20,000 |
| Uniform annual benefit | 1,000 | 1,762 | 5,548 |
| Useful life, in years | ∞ | 20 | 5 |

Assuming that Alternatives B and C are replaced with identical units at the end of their useful lives, and an 8% interest rate, which alternative should be selected? Use an annual cash flow analysis in working this problem.    (*Answer:* Select C)

6-56    Carp, Inc. wants to evaluate two methods of packaging their products. Use an interest rate of 15% and annual cash flow analysis to decide which is the most desirable alternative.

| | A | B |
|---|---|---|
| First cost | $700,000 | $1,700,000 |
| Maintenance and operating costs | 18,000 | 29,000 |
| + Cost gradient (begin Year 1) | +900/yr | +750/yr |
| Annual benefit | 154,000 | 303,000 |
| Salvage value | 142,000 | 210,000 |
| Useful life, in years | 10 | 20 |

6-57    A college student has been looking for new tires and has found the following alternatives:

| Tire Warranty (months) | Price per Tire |
|---|---|
| 12 | $39.95 |
| 24 | 59.95 |
| 36 | 69.95 |
| 48 | 90.00 |

The student feels that the warranty period is a good estimate of the tire life and that a 10% interest rate is appropriate. Using an annual cash flow analysis, which tire should be purchased?

6-58    Consider the following alternatives:

| | A | B |
|---|---|---|
| Cost | $5000 | $18,000 |
| Uniform annual benefit | 1500 | 6,000 |
| Useful life, in years | 10 | 5 |

The analysis period is 10 years, but there will be no replacement for Alternative B after 5 years. Based on a 15% interest rate, which alternative should be selected? Use an annual cash flow analysis.

6-59    Some equipment will be installed in a warehouse that a firm has leased for 7 years. There are two alternatives:

| | A | B |
|---|---|---|
| Cost | $1000 | $1500 |
| Uniform annual benefit | 550 | 610 |
| Useful life, in years | 3 | 4 |

At any time after the equipment is installed, it has no salvage value. Assume that Alternatives A and B will be replaced at the end of their useful lives by identical equipment with the same costs and benefits. For a 7-year analysis period and a 10% interest rate, use an annual cash flow analysis to determine which alternative should be selected.

6-60    Uncle Elmo needs to replace the family privy. The local sanitary engineering firm has submitted two alternative structural proposals with respective cost estimates as shown. Which construction should Uncle Elmo choose if his minimum attractive rate of return is 6%? Use both a present worth and annual cost approach in your comparison.

| | Masonite | Brick |
|---|---|---|
| First cost | $2500 | $10,000 |
| Annual maintenance | 200 | 100 |
| Salvage value | 100 | 1,000 |
| Service life, in years | 4 | 20 |

6-61    Dick Dickerson Construction, Inc. has asked you to help them select a new backhoe. You have a choice between a wheel-mounted version, which costs $60,000 and has an expected life of 5 years and a salvage value of $2000, and a track-mounted one, which costs $80,000, with a 7-year life and an expected salvage value of $10,000. Both machines will achieve the same productivity. Interest is 8%. Which one will you recommend? Use a annual worth analysis.

**6-62**  A small manufacturing company is evaluating trucks for delivering their products. Truck *A* has a first cost of $22,000, its operating cost will be $5500 per year, and its salvage after 3 years will be $7000. Truck *B* has a first cost of $27,000, an operating cost of $5200, and a resale value of $12,000 after 4 years. At an interest rate of 15% per year, which model should be chosen if an annual worth analysis is performed?
*Contributed by Hamed Kashani, Saeid Sadri, and Baabak Ashuri, Georgia Institute of Technology*

**6-63**  A job can be done with Machine *A* that costs $12,500 and has annual end-of-year maintenance costs of $5000; its salvage value after 3 years is $2000. Or the job can be done with Machine *B*, which costs $15,000 and has end-of-year maintenance costs of $4000 and a salvage value of $1500 at the end of 4 years. These investments can be repeated in the future, and your work is expected to continue indefinitely. Use present worth, annual worth, and capitalized cost to compare the machines. The interest rate is 5%/year.
*Contributed by D. P. Loucks, Cornell University*

## Spreadsheets and Loans

**6-64**  A student loan totals $18,000 at graduation. The interest rate is 6%, and there will be 60 payments beginning 1 month after graduation. What is the monthly payment? What is owed after the first 2 years of payments?  (*Answer*: Payment = $347.99, balance due = $11,439)

**6-65**  The student in Problem 6-64 received $1500 as a graduation present. If an extra $1500 is paid at Month 1, when is the final payment made? How much is it? (*Answer*: $98 in Month 55)

**6-66**  A new car is purchased for $12,000 with a 0% down, 9% loan. The loan is for 4 years. After making 30 payments, the owner wants to pay off the loan's remaining balance. How much is owed?

**6-67**  A year after buying her car, Anita has been offered a job in Europe. Her car loan is for $15,000 at a 9% nominal interest rate for 60 months. If she can sell the car for $12,000, how much does she get to keep after paying off the loan?

**6-68**  (*a*)  You are paying off a debt at a nominal 8% per year by paying $400 at the end of each quarter for the next year. Find the interest paid in the last $400 payment.

(*b*)  If this debt were to be paid off in two equal payments of $1650 at the end of this year and at the end of the next year, find the interest paid in the first $1650 payment. Again the loan rate is a nominal 8% per year compounded quarterly.
*Contributed by D. P. Loucks, Cornell University*

**6-69**  Sam can afford to spend $500 per month on a car. He figures he needs half of it for gas, parking, and insurance. He has been to the bank, and they will loan him 100% of the car's purchase price. (*Note*: If he had a down payment saved, then he could borrow at a lower rate.)

(*a*)  If his loan is at a nominal 12% annual rate over 36 months, what is the most expensive car he can purchase?

(*b*)  The car he likes costs $14,000 and the dealer will finance it over 60 months at 12%. Can he afford it? If not, for how many months will he need to save his $500 per month.

(*c*)  What is the highest interest rate he can pay over 60 months and stay within his budget if he buys the $14,000 car now?

**6-70**  EnergyMax Engineering constructed a small office building for their firm 5 years ago. They financed it with a bank loan for $450,000 over 15 years at 12% interest with quarterly payments and compounding. The loan can be repaid at any time without penalty. The loan can be refinanced through an insurance firm for 8% over 20 years—still with quarterly compounding and payments. The new loan has a 5% loan initiation fee, which will be added to the new loan.

(*a*)  What is the balance due on the original mortgage (20 payments have been made in last 5 years)?

(*b*)  How much will Energy Max's payments drop with the new loan?

(*c*)  How much longer will the proposed loan run?

6-71   Suppose you graduate with a debt of $42,000 that you or someone must repay. One option is to pay off the debt in constant amounts at the beginning of each month over the next 10 years at a nominal annual interest rate of 10%.

   (a) What is the constant beginning-of-month payment?

   (b) Of the first payment, what is the interest and the principal paid?

   (c) Of the last payment, what is the interest and the principal paid?
   *Contributed by D. P. Loucks, Cornell University*

6-72   A $78,000 mortgage has a 30-year term and a 9% nominal interest rate.

   (a) What is the monthly payment?

   (b) After the first year of payments, what is the outstanding balance?

   (c) How much interest is paid in Month 13? How much principal?

6-73   A $92,000 mortgage has a 30-year term and a 9% nominal interest rate.

   (a) What is the monthly payment?

   (b) After the first year of payments, what fraction of the loan has been repaid?

   (c) After the first 10 years of payments, what is the outstanding balance?

   (d) How much interest is paid in Month 25? How much principal?

6-74   A 30-year mortgage for $95,000 is issued at a 9% nominal interest rate.

   (a) What is the monthly payment?

   (b) How long does it take to pay off the mortgage, if $1000 per month is paid?

   (c) How long does it take to pay off the mortgage, if double payments are made?

6-75   A 30-year mortgage for $145,000 is issued at a 6% nominal interest rate.

   (a) What is the monthly payment?

   (b) How long does it take to pay off the mortgage, if $1000 per month is paid?

   (c) How long does it take to pay off the mortgage, if 20% extra is paid each month?

6-76   Solve Problem 6-43 for the breakeven first cost per kilometer of going under the lake.

6-77   Redo Problem 6-53 to calculate the EUAW of the alternatives as a function of miles driven per year to see if there is a crossover point in the decision process. Graph your results.

6-78   Set up Problem 6-36 on a spreadsheet and make all the input data variable. Determine various scenarios which would make the baler economical.

6-79   Develop a spreadsheet to solve Problem 6-41. What is the breakeven cost of the additional pumping investment in Year 10?

## Minicases

6-80   An office building should last 60 years, but this owner will sell it at 20 years for 40% of its construction cost. For the first 20 years it can be leased as Class A space, which is all this owner operates. When the building is sold, the land's cost will be recovered in full.

| | |
|---|---|
| $2.2M | Land |
| $4.1M | Building |
| $640,000 | Annual operating and maintenance |
| 4% | Annual property taxes and insurance |
| | (% of initial investment) |

   (a) If the owner wants a 12% rate of return, what is the required monthly leasing cost?

   (b) Assuming that the building is vacant 5% of the time, what is the required monthly lease?

   (c) What is an example monthly cost per square foot for Class A space in your community?

6-81   A 30-unit apartment building should last 35 years, when it will need to be either replaced or undergo major renovation. Assume the building's value at 35 years will be 10% of its construction cost. Assume it will be sold, and that the land's cost will be recovered in full.

| $3.2M | Land |
|---|---|
| $4.8M | Building |
| $850,000 | Annual operating and maintenance |
| 6% | Annual property taxes and insurance (% of initial investment) |
| 12% | Vacancy rate |

(*a*) If the owner wants a 15% rate of return, what is the required monthly leasing cost for each unit?

(*b*) If turning 2 units into an exercise facility would decrease the vacancy rate by 5%, would that be a good decision?

# CASES

**The following cases from *Cases in Engineering Economy* 2 *nd* are suggested as matched with this chapter.**

CASE 6 **Lease a Lot**

Compares leasing and ownership. Both financing and investing decisions are needed.

CASE 10 **The Cutting Edge**

Make versus buy and machine selection.

CASE 27 **Harbor Delivery Service**

Focus is treatment of sunk costs. More complicated than most. Some discoveries in the data gathering process. Solution uses equation rather than cash flow table.

## CHAPTER 7

# RATE OF RETURN ANALYSIS

## Bar Codes Give a Number; RFID Codes Tell a Story

ISBN 978-0-19-533541-5

**B**oth radio-frequency identification (RFID) and bar code technology encode information, but RFID uses radio waves rather than optical scanning to read that information. This means that the RFID tag can be read without a line of sight. The shift from manual price tags, shipping labels, and baggage tags to bar codes revolutionized operations by reducing costs and errors. Now the shift to RFID is revolutionizing operations again. Today's smart tag is a paper sandwich of a bar code and an RFID tag.

A bar code speaks five or six times in its lifetime; RFID codes can speak 200 to 1000 times each second. Your car keys are an RFID chip allowing you to open or start your car remotely. RFID is in the EZ Pass on your windshield, a rail pass in China, your Smart Passport, a subway pass in Washington, D.C., some employee IDs, and garment labels. RFID chips are used to tightly control the inventory of packs of razor blades—including the control of pilferage. They are used in boxes of Alaska wild seafood to track temperatures from packaging to restaurant delivery. They are required on many deliveries to the military and major retailers.

Because RFID tags carry so much more information, organizations need to measure how they improve performance in many ways. RFID is changing the old ways of operating in logistics and supply chains and business in general. Today's wholesalers, retailers, manufacturers, and military decision makers trust RFID technology to create real-time inventory tracking, while eliminating the jobs of many warehouse workers and cashiers.

As with any technology that allows change, there are some who dislike it. In addition, some raise ethical questions about the ability to track items after purchase and people as they move about a facility. ■ ■ ■

*Contributed by Oliver Hedgepeth, University of Alaska Anchorage, author of* RFID Metrics, *CRC Press (2007)*

## QUESTIONS TO CONSIDER

1. The growth rate for the market in RFID systems from 2010 to 2020 is estimated to be over 50% annually. This is built on many individual projects—each of which must be economically justified. Does the growth rate imply anything about the rate of return on the individual projects?

2. The cost of RFID tags could be as low as 3–5¢ in 2012, much higher than the bar code tag cost of less than a penny. What are the cost and benefit factors that you would use to make a decision to replace bar codes with RFID tags? Or to incur the cost of both in a smart tag?

3. The Smart Passport and its information may not be meaningful without access to a database file on the person being scanned. How would you advise U.S. citizens or Defense Department shippers and logisticians who use Smart Passports when they are traveling outside the country?

4. What would you suggest to management about the 24/7 stream of data each second that could come from an inventory counted by RFID tags?

5. Go to the Internet and find five different RFID applications. List the problems these users think they are solving and why they seem to think the chips will work. Do you think they have thought through the problem sufficiently? Are there ethical or social concerns for each application?

6. The RFID computer chip sewn in the label of your new shirt can still be active or readable when you get home, when you wear the garment to work, and when you return to the store where you bought it. Some consumers have already expressed concern about RFID chips, calling them an invasion of privacy. What ethical and social concerns are raised by RFID chips?

## After Completing This Chapter...

*The student should be able to:*

- Evaluate project cash flows with the *internal rate of return* (IRR) measure.
- Plot a project's present worth (PW) against the interest rate.
- Use an *incremental* rate of return analysis to evaluate competing alternatives.
- Develop and use spreadsheets to make IRR and incremental rate of return calculations.

In this chapter we will examine four aspects of rate of return, the third major analysis method. First, the meaning of "rate of return" is explained; second, calculating the rate of return is illustrated; third, rate of return analysis problems are presented; and fourth, incremental analysis is presented. In an appendix to the chapter, we describe difficulties sometimes encountered when computing an interest rate for cash flow series with multiple sign changes.

**Rate of return** is the most frequently used measure in industry. Problems in computing the rate of return sometimes occur, but its major advantage is that it is a single figure of merit that is readily understood.

Consider these statements:

- The net present worth on a project is $32,000.
- The equivalent uniform annual net benefit is $2800.
- The project will produce a 23% rate of return.

While none of these statements tells the complete story, the third one measures the project's desirability in terms that are widely and easily understood. Thus, this measure is accepted by engineers and business leaders alike.

There is another advantage to rate of return analysis. In both present worth and annual cash flow calculations, one must select an interest rate—and the exact value may be a difficult and controversial item. In rate of return analysis, no interest rate is introduced into the calculations (except as described in Appendix 7A). Instead, we compute a rate of return (more accurately called *internal rate of return*) from the cash flow. To decide how to proceed, the calculated rate of return is compared with a preselected **minimum attractive rate of return,** or simply MARR. This is the same value of *i* used for present worth and annual cash flow analysis.

## INTERNAL RATE OF RETURN

> **Internal rate of return** is the interest rate at which the present worth and equivalent uniform annual worth are equal to 0.

This definition is easy to remember, and it also tells us how to solve for the rate of return. In earlier chapters we did this when we solved for the interest rate on a loan or investment.

Other definitions based on the unpaid balance of a loan or the unrecovered investment can help clarify why this rate of return is also called the *internal rate of return* or IRR. In Chapter 3 we examined four plans to repay $5000 in 5 years with interest at 8% (Table 3-1). In each case the amount loaned ($5000) and the loan duration (5 years) were the same. Yet the total interest paid to the lender varied from $1200 to $2347. In each case the lender received 8% interest each year on the amount of money actually owed. And, at the end of 5 years, the principal and interest payments exactly repaid the $5000 debt with interest at 8%. We say the lender received an "8% rate of return."

> **Internal rate of return** can also be defined as the interest rate paid on the unpaid balance of a *loan* such that the payment schedule makes the unpaid loan balance equal to zero when the final payment is made.

Instead of lending money, we might invest $5000 in a machine tool with a 5-year useful life and an equivalent uniform annual benefit of $1252. The question becomes, What rate of return would we receive on this investment?

We recognize the cash flow as reversed from Plan 3 of Table 3-1. We know that five payments of $1252 are equivalent to a present sum of $5000 when interest is 8%. Therefore, the rate of return on this investment is 8%.

**Internal rate of return** can also be defined as the interest rate earned on the unrecovered *investment* such that the payment schedule makes the unrecovered investment equal to zero at the end of the investment's life.

It must be understood that the 8% rate of return does not mean an annual return of 8% on the $5000 investment, or $400 in each of the 5 years with $5000 returned at the end of Year 5. Instead, each $1252 payment represents an 8% return on the unrecovered investment *plus* a partial return of the investment. This may be tabulated as follows:

| Year | Cash Flow | Unrecovered Investment at Beginning of Year | 8% Return on Unrecovered Investment | Investment Repayment at End of Year | Unrecovered Investment at End of Year |
|------|-----------|---------------------------------------------|-------------------------------------|-------------------------------------|---------------------------------------|
| 0 | −$5000 | | | | |
| 1 | 1252 | $5000 | $ 400 | $ 852 | $4148 |
| 2 | 1252 | 4148 | 331 | 921 | 3227 |
| 3 | 1252 | 3227 | 258 | 994 | 2233 |
| 4 | 1252 | 2233 | 178 | 1074 | 1159 |
| 5 | 1252 | 1159 | 93 | 1159 | 0 |
| | | | $1260 | $5000 | |

This cash flow represents a $5000 investment with benefits that produce an 8% rate of return on the unrecovered investment.

Although the definitions of internal rate of return are stated differently for a loan and for an investment, there is only one fundamental concept. It is that **the internal rate of return is the interest rate at which the benefits are equivalent to the costs**, or the present worth (PW) is 0. Since we are describing the funds that remain within the investment throughout its life, the resulting rate of return is described as the internal rate of return, $i$.

## CALCULATING RATE OF RETURN

To calculate a rate of return on an investment, we must convert the various consequences of the investment into a cash flow series. Then we solve the cash flow series for the unknown value of the internal rate of return (IRR). Five forms of the cash flow equation are as follows:

$$PW \text{ of benefits} - PW \text{ of costs} = 0 \qquad (7\text{-}1)$$

$$\frac{PW \text{ of benefits}}{PW \text{ of costs}} = 1 \qquad (7\text{-}2)$$

$$\text{Present worth} = \text{Net present worth}^1 = 0 \qquad (7\text{-}3)$$

$$\text{EUAW} = \text{EUAB} - \text{EUAC} = 0 \qquad (7\text{-}4)$$

$$PW \text{ of costs} = PW \text{ of benefits} \qquad (7\text{-}5)$$

The five equations represent the same concept in different forms. They relate costs and benefits with the IRR as the only unknown. The calculation of rate of return is illustrated by the following examples.

## EXAMPLE 7–1

An engineer invests $5000 at the end of every year for a 40-year career. If the engineer wants $1 million in savings at retirement, what interest rate must the investment earn?

### SOLUTION

Using Equation 7-3, we write

$$\text{Net PW} = 0 = -\$5000(F/A, i, 40) + \$1,000,000$$

Rewriting, we see that

$$(F/A, i, 40) = \$1,000,000/\$5000 = 200$$

We then look at the compound interest tables for the value of $i$ where $(F/A, i, 40) = 200$. If no tabulated value of $i$ gives this value, then we will interpolate, using the two closest values or solve exactly with a calculator or spreadsheet. In this case $(F/A, 0.07, 40) = 199.636$, which to three significant digits equals 200. Thus the required rate of return for the investment is 7%. (The exact value is 7.007%.)

While this example has been defined in terms of an engineer's personal finances, we could just as easily have said, "a mining firm makes annual deposits of $50,000 into a reclamation fund for 40 years. If the firm must have $10 million when the mine is closed, what interest rate must the investment earn?" Since the annual deposit and required final amount are 10 times larger in each case, the $F/A$ factor and the answer are obviously the same.

---

[1]Remember that present value (PV), present worth (PW), net present value (NPV), and net present worth (NPW) are synonyms in practice and in this text.

### EXAMPLE 7-2

An investment resulted in the following cash flow. Compute the rate of return.

### SOLUTION

$$\text{EUAW} = \text{EUAB} - \text{EUAC} = 0 = 100 + 75(A/G, i, 4) - 700(A/P, i, 4)$$

Here, we have two different interest factors in the equation, and we will solve the equation by trial and error. The EUAW value is a function of $i$. Try $i = 5\%$ first:

$$\begin{aligned}
\text{EUAW}(5\%) &= 100 + 75(A/G, 5\%, 4) - 700(A/P, 5\%, 4) \\
&= 100 + 75(1.439) - 700(0.2820) \\
&= \text{EUAB} - \text{EUAC} = 208 - 197 = +11
\end{aligned}$$

The EUAW is too high. If the interest rate is increased, EUAW will decrease. Try $i = 8\%$:

$$\begin{aligned}
\text{EUAW}(8\%) &= 100 + 75(A/G, 8\%, 4) - 700(A/P, 8\%, 4) \\
&= 100 + 75(1.404) - 700(0.3019) \\
&= \text{EUAB} - \text{EUAC} = 205 - 211 = -6
\end{aligned}$$

This time the EUAW is too low. We see that the true rate of return is between 5% and 8%. Try $i = 7\%$:

$$\begin{aligned}
\text{EUAW}(7\%) &= 100 + 75(A/G, 7\%, 4) - 700(A/P, 7\%, 4) \\
&= 100 + 75(1.416) - 700(0.2952) \\
&= \text{EUAB} - \text{EUAC} = 206 - 206 = 0
\end{aligned}$$

The IRR is 7%.

### EXAMPLE 7-3

A local firm sponsors a student loan program for the children of employees. No interest is charged until graduation, and then the interest rate is 5%. Maria borrows $9000 per year, and she graduates after 4 years. Since tuition must be paid ahead of time, assume that she borrows the money at the start of each year.

  If Maria makes five equal annual payments, what is each payment? Use the cash flow from when she started borrowing the money to when it is all paid back, and then calculate the internal rate of return for Maria's loan. Is this arrangement attractive to Maria?

**SOLUTION**

Maria owes $36,000 at graduation. The first step is to calculate the five equal annual payments to repay this loan at 5%.

$$\text{Loan payment} = \$36,000(A/P, 5\%, 5) = 36,000(0.2310) = \$8316$$

Maria receives $9000 by borrowing at the start of each year. She graduates at the end of Year 4. At the end of Year 4, which is also the beginning of Year 5, interest starts to accrue. She makes her first payment at the end of Year 5, which is one year after graduation.

The next step is to write the present worth equation in factor form, so that we can apply Equation 7-3 and set it equal to 0. This equation has three factors, so we will have to solve the problem by picking interest rates and substituting values. The present worth value is a function of $i$.

$$PW(i) = 9000[1 + (P/A, i, 3)] - 8316(P/A, i, 5)(P/F, i, 4)$$

The first two interest rates used are 0% (because it is easy) and 3% because the subsidized rate will be below the 5% that is charged after graduation.

At 0% any $P/A$ factor equals $n$, and any $P/F$ factor equals 1.

$$PW(0\%) = 9000(4) - 8316(5) = -5180$$

$$PW(3\%) = 9000(1 + 2.829) - 8316(4.580)(0.8885) = 620.5$$

Since $PW(i)$ has opposite signs for 0 and 3, there is a value of $i$ between 0 and 3% which is the IRR. Because the value for 3% is closer to 0, the IRR will be closer to 3%. Try 2% next.

$$PW(2\%) = 9000(1 + 2.884) - 8316(4.713)(0.9238) = -1251$$

As shown in Figure 7-1, interpolating between 2 and 3% leads to

$$IRR = 2\% + (3\% - 2\%)[1251/(1251 + 620.8)] = 2.67\%.$$

This rate is quite low, and it makes the loan look like a good choice.

**FIGURE 7-1** Plot of PW versus interest rate $i$.

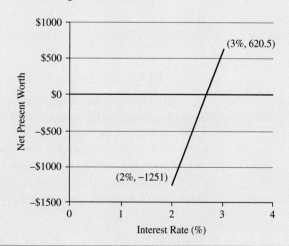

We can prove that the rate of return in Example 7-3 is very close to 2.66% by tabulating how much Maria owes each year, the interest that accrues, and her borrowings and payments.

| Year | Amount Owed at Start of Year | Interest at 2.66% | Cash Flow at End of Year | Amount Owed at End of Year |
|---|---|---|---|---|
| 1 | $ 9,000 | $ 239 | $ 9000 | $ 18,239 |
| 2 | 18,239 | 485 | 9000 | 27,725 |
| 3 | 27,725 | 737 | 9000 | 37,462 |
| 4 | 37,462 | 996 | 0 | 38,459 |
| 5 | 38,459 | 1023 | −8315 | 31,166 |
| 6 | 31,166 | 829 | −8315 | 23,680 |
| 7 | 23,680 | 630 | −8315 | 15,995 |
| 8 | 15,995 | 425 | −8315 | 8,106 |
| 9 | 8,106 | 216 | −8315 | 6* |

*This small amount owed indicates that the interest rate is slightly less than 2.66%.

If in Figure 7-1 net present worth (NPW) had been computed for a broader range of values of $i$, Figure 7-2 would have been obtained. From this figure it is apparent that the error resulting from linear interpolation increases as the interpolation width increases.

**FIGURE 7–2** Replot of NPW versus interest rate $i$ over a larger range of values.

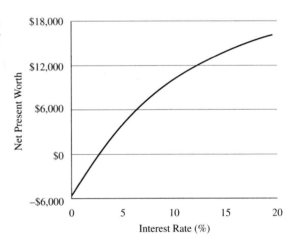

## Plot of NPW Versus Interest Rate $i$

The plot of NPW versus interest rate $i$ is an important source of information. For a cash flow where borrowed money is repaid, the NPW plot would appear as in Figure 7-3. The borrowed money is received early in the time period with a later repayment of an equal sum, plus payment of interest on the borrowed money. In all cases in which interest is charged and the amount borrowed is fully repaid, the NPW at 0% will be negative.

For a cash flow representing an investment followed by benefits from the investment, the plot of NPW versus $i$ (we will call it an **NPW plot** for convenience) would have the form

| Year | Cash Flow |
|------|-----------|
| 0 | $+P$ |
| 1 | $-A$ |
| 2 | $-A$ |
| 3 | $-A$ |
| 4 | $-A$ |
| . | . |
| . | . |
| . | . |

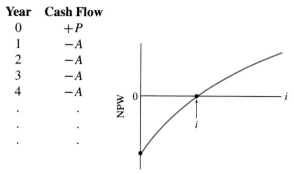

**FIGURE 7-3** Typical NPW plot for borrowed money.

| Year | Cash Flow |
|------|-----------|
| 0 | $-P$ |
| 1 | $+$Benefit $A$ |
| 2 | $+A$ |
| 3 | $+A$ |
| 4 | $+A$ |
| . | . |
| . | . |
| . | . |

**FIGURE 7-4** NPW plot for a typical investment.

of Figure 7-4. As the interest rate increases, future benefits are discounted more heavily and the NPW decreases.

Thus, interest is a charge for the use of someone else's money or a receipt for letting others use our money. The interest rate is almost always positive, but negative interest rates do occur. A loan with a forgiveness provision (not all principal is repaid) can have a negative rate. Some investments perform poorly and have negative rates.

## EXAMPLE 7–4

A new corporate bond was initially sold by a stockbroker to an investor for $1000. The issuing corporation promised to pay the bondholder $40 interest on the $1000 face value of the bond every 6 months, and to repay the $1000 at the end of 10 years. After one year the bond was sold by the original buyer for $950.

(a) What rate of return did the original buyer receive on his investment?

(b) What rate of return can the new buyer (paying $950) expect to receive if he keeps the bond for its remaining 9-year life?

**SOLUTION TO PART a**

The original bondholder sold the bond for less ($950) than the purchase price ($1000). So the semiannual rate of return is less than the 4% semiannual interest rate on the bond.

Since $40 is received each 6 months, we will solve the problem using a 6-month interest period. Let PW of cost = PW of benefits, and write

$$1000 = 40(P/A, i, 2) + 950(P/F, i, 2)$$

Try $i = 1\frac{1}{2}\%$:

$$1000 \doteq 40(1.956) + 950(0.9707) = 78.24 + 922.17 \doteq 1000.41$$

The interest rate per 6 months, $IRR_{6\ mon}$, is very close to $1\frac{1}{2}\%$. This means the nominal (annual) interest rate is $2 \times 1.5\% = 3\%$. The effective (annual) interest rate or IRR is $(1 + 0.015)^2 - 1 = 3.02\%$.

Using a TVM calculator, we can find the exact interest rate for 6 months:

$$i = i(n, A, P, F) = i(2, 40, -1000, 950) = 1.5188\%$$

The effective interest rate is found as

$$IRR = 1.015188^2 - 1 = 3.06\%$$

**SOLUTION TO PART b**

The new buyer will redeem the bond for more ($1000) than the purchase price ($950). So the semiannual rate of return is more than the 4% semiannual interest rate on the bond.

Given the same $40 semiannual interest payments, for 6-month interest periods we write

$$950 = 40(P/A, i, 18) + 1000(P/F, i, 18)$$

Try $i = 5\%$:

$$950 \stackrel{?}{=} 40(11.690) + 1000(0.4155) = 467.60 + 415.50 \stackrel{?}{=} 883.10$$

The PW of benefits is too low. Try a lower interest rate, say, $i = 4\%$:

$$950 \stackrel{?}{=} 40(12.659) + 1000(0.4936) = 506.36 + 493.60$$
$$\stackrel{?}{=} 999.96$$

The value of the 6-month rate $i$, is between 4 and 5%. By interpolation,

$$i = 4\% + (1\%) \left( \frac{999.96 - 950.00}{999.96 - 883.10} \right) = 4.43\%$$

The nominal interest rate is $2 \times 4.43\% = 8.86\%$. The effective interest rate or IRR is $(1 + 0.0443)^2 - 1 = 9.05\%$.

Using a TVM calculator, the 6-month rate is

$$i = i(n, A, P, F) = i(18, 40, -950, 1000) = 4.4082\%$$
$$\text{IRR} = (1.044082)^2 - 1 = 9.011\%$$

## INTEREST RATES WHEN THERE ARE FEES OR DISCOUNTS

Often when firms and individuals borrow money, there are fees charged in addition to the interest. This can be as simple as the underwriting fee that a firm is charged when it sells a bond. In Example 7-5 we add that underwriting fee to the bond in Example 7-4, and look at the bond from the firm's perspective rather than the investor's.

Example 7-6 shows how a cash discount is "unstated interest." When one is buying a new car, this may be stated as a choice between a "cash" rebate and a low-interest loan. Fees and cash discounts raise the interest rate on a loan, and the true cost of the loan is its internal rate of return.

### EXAMPLE 7–5

The corporate bond in Example 7-4 was part of a much larger offering that the firm arranged with the underwriter. Each of the bonds had a face value of $1000 and a life of 10 years. Since $40 or 4% of the face value was paid in interest every 6 months, the bond had a nominal interest rate of 8% per year. If the firm paid the underwriter a 1% fee to sell the bond, what is the effective annual interest rate that the firm is paying on the bond?

## SOLUTION

From the firm's perspective, it receives $1000 minus the fee at time 0, then it pays interest every 6 months for 10 years, and then it pays $1000 to redeem the bond. The 1% fee reduces what the firm receives when the bond is sold to $990. The interest payments are $40 every 6 months. This is easiest to model using twenty 6-month periods.

$$PW(i) = 990 - 40(P/A, i, 20) - 1000(P/F, i, 20)$$

Since the nominal interest rate is 4% every 6 months, we know that the fee will raise this some. So let us use the next higher table of 4.5%.

$$PW(4.5\%) = 990 - 40(P/A, 4.5\%, 20) - 1000(P/F, 4.5\%, 20)$$
$$= 990 - 40(13.008) - 1000(0.4146) = \$55.08$$

We know that the PW of the interest and final bond payoff is $1000 at 4%.

$$PW(4\%) = 990 - 1000 = -10$$

Now we interpolate to find the interest rate for each 6-month period.

$$i = 4\% + (4.5\% - 4\%) \times 10/(10 + 55.08) = 4.077\%$$

The effective annual rate is

$$i_a = 1.04077^2 - 1 = 0.0832 = 8.32\%$$

Using a TVM calculator, we find the 6-month interest rate as

$$i = i(n, A, P, F) = i(20, -40, 990, -1000) = 4.07407\%$$
$$i_a = -1.0407407^2 - 1 = 8.314\%$$

## EXAMPLE 7–6

A manufacturing firm may decide to buy an adjacent property so that it can expand its warehouse. If financed through the seller, the property's price is $300,000, with 20% down and the balance due in five annual payments at 12%. The seller will accept 10% less if cash is paid. The firm does not have $270,000 in cash, but it can borrow this amount from a bank. What is the rate of return or IRR for the loan offered by the seller?

## SOLUTION

One choice is to pay $270,000 in cash. The other choice is to pay $60,000 down and five annual payments computed at 12% on a principal of $240,000. The annual payments equal

$$240,000(A/P, 12\%, 5) = 240,000(0.2774) = \$66,576$$

The two ways of borrowing the money and the *incremental* difference between them can be summarized in a cash flow table. The incremental difference is the result of subtracting the second set of cash flows from the first.

| Year | Pay Cash | Borrow from Property Owner | Incremental Difference |
|---|---|---|---|
| 0 | −$270,000 | −$60,000 | −$210,000 |
| 1 | | −66,576 | 66,576 |
| 2 | | −66,576 | 66,576 |
| 3 | | −66,576 | 66,576 |
| 4 | | −66,576 | 66,576 |
| 5 | | −66,576 | 66,576 |

This choice between (1) −$270,000 now and (2) −$60,000 now and −$66,576 annually for 5 years can be stated by setting them equal in PW terms, as follows:

$$-270,000 = -60,000 - 66,576(P/A, \text{IRR}, 5)$$

The true amount borrowed is $210,000, which is the difference between the $60,000 down payment and the $270,000 cash price. This is the incremental amount paid or invested by the firm instead of borrowing from the seller. Collecting the terms, the PW equation is the same as the PW equation for the incremental difference.

$$0 = -210,000 + 66,576(P/A, \text{IRR}, 5)$$

In either case, the final equation for finding the IRR is

$$\text{PW} = 0 = -210,000 + 66,576(P/A, \text{IRR}, 5)$$
$$(P/A, IRR, 5) = 3.154$$

Looking in the tables we see that the IRR is between 15% and 18%. Interpolating gives us the following:

$$\text{IRR} = 15\% + (18\% - 15\%)[(3.352 - 3.154)/(3.352 - 3.127)] = 17.6\%$$

on a borrowed amount of $210,000. This is a relatively high rate of interest, so that borrowing from a bank and paying cash to the property owner is better. Using a spreadsheet or a TVM calculator gives an exact answer of $17.62\% = i(n, A, P, F) = i(5, \ 66576, \ -210000, \ 0)$

## LOANS AND INVESTMENTS ARE EVERYWHERE

Examples 7-4, 7-5, and 7-6 were about borrowing money through bonds and loans, but many applications for the rate of return are stated in other ways. Example 7-7 is a common problem on university campuses—buying parking permits for an academic year or for a quarter or a semester at a time.

Buying a year's parking permit is investing more money now to avoid paying for another shorter permit later. Choosing to buy a shorter permit is a loan, where the money saved by not buying the annual permit is borrowed to be repaid with the cost of the second semester permit.

## EXAMPLE 7–7

An engineering student is deciding whether to buy two 1-semester parking permits or an annual permit. The annual parking permit costs $180 due August 15th; the semester permits are $130 due August 15th and January 15th. What is the rate of return for buying the annual permit?

### SOLUTION

Before we solve this mathematically, let us describe it in words. We are equating the $50 cost difference now between the two permits with the $130 cost to buy another semester permit in 5 months. Since the $130 is 2.6 times the $50, it is clear that we will get a high interest rate.

This is most easily solved by using monthly periods, and the payment for the second semester is 5 months later. The cash flow table adds a column for the incremental difference to the information in the cash flow diagram.

| Time | Annual Pass | Two Semester Passes | Incremental Difference |
|---|---|---|---|
| Aug. 15th | −$180 | −$ 130 | −$ 50 |
| Jan. 15th | | −130 | 130 |

Setting the two PWs equal to each other we have

$$-\$180 = -\$130[1 + (P/F, i_{mon}, 5)]$$

$$(P/F, i_{mon}, 5) = -\$50/ - \$130 = 5/13 = 0.3846$$

Rather than interpolating, we can use the formula for the P/F factor.

$$1/(1 + i_{mon})^5 = 0.3846$$

$$(1 + i_{mon})^5 = 2.600$$

$$1 + i_{mon} = 1.2106$$

$$i_{mon} = 21.06\%, \text{ which is an extremely high rate per month}$$

On an annual basis, the effective interest rate is $(1.2106^{12} - 1) = 891\%$. Unless the student is planning on graduating in January, it is clearly better to buy the permit a year at a time.

Examples 7-8 and 7-9 reference two common situations faced by firms and by individuals—buying insurance with a choice of payment plans and deciding whether to buy or lease a vehicle or equipment. Like the parking permit case, these situations can be described as investing more now to save money later, or borrowing money now to be paid later.

## EXAMPLE 7–8

An engineering firm can pay for its liability insurance for errors and omissions on an annual or quarterly basis. If paid quarterly, the insurance costs $10,000. If paid annually, the insurance costs $35,000. What is the rate of return for paying on annual basis?

### SOLUTION

This is most easily solved by using quarterly periods. As shown in the cash flow diagram and table, insurance must be paid at the start of each quarter. The cash flow table adds a column for the incremental difference to the information in the cash flow diagram.

| Quarter | Annual Payment | Quarterly Payments | Incremental Difference |
|---------|---------------|--------------------|-----------------------|
| 0 | −$35,000 | −$10,000 | −$25,000 |
| 1 |  | −10,000 | 10,000 |
| 2 |  | −10,000 | 10,000 |
| 3 |  | −10,000 | 10,000 |

Setting the two PWs equal to each other we have

$$-\$35,000 = -\$10,000(1 + (P/A, i_{qtr}, 3))$$

$$(P/A, i_{qtr}, 3) = -\$25,000/-\$10,000 = 2.5$$

$$(P/A, 9\%, 3) = 2.531$$

$$(P/A, 10\%, 3) = 2.487$$

$$i_{qtr} = 0.09 + (0.10 - 0.09)(2.531 - 2.5)/(2.531 - 2.487)$$

$$= 0.09 + 0.01(0.7045) = 9.705\%$$

On an annual basis the effective interest rate is $(1.09705^4 - 1) = 44.8\%$.

Unless the firm is planning on going out of business, it is clearly better to buy the insurance a year at a time. Note that a TVM calculator or spreadsheet gives the exact quarterly interest rate: $i_{qtr} = i(n, A, P, F) = i(3, 10000, -25000, 0) = 9.7010\%$.

## EXAMPLE 7–9

Mountain Environmental Consulting may buy some field equipment for $40,000 or lease it for $2500 per month. In either case, the equipment will be replaced in 2 years. The salvage value of the equipment after 2 years would be $6000. What is the IRR or cost of the lease?

### SOLUTION

The first step is to summarize the cash flows in a table. We must use monthly periods and remember that lease fees are paid at the start of the period.

| Month | Buy | Lease | Incremental Difference |
|---|---|---|---|
| 0 | −$40,000 | −$2500 | −$37,500 |
| 1–23 | | −2500 | 2,500 |
| 24 | 6,000 | 0 | 6,000 |

The easiest way to analyze this example is to simply set the present worths of buying and leasing equal.

$$-\$40,000 + \$6000(P/F, i_{mon}, 24) = -\$2500 - \$2500(P/A, i_{mon}, 23)$$

$$0 = \$37,500 - \$2500(P/A, i_{mon}, 23) - \$6000(P/F, i_{mon}, 24)$$

In other words, if Mountain Environmental leases the equipment, it avoids the expenditure of $37,500 at time 0. It incurs a cost of $2500 at the ends of Months 1 through 23, and it gives up the salvage value of $6000 at the end of Month 24. This cash flow pattern (positive at time 0 and negative in later years) occurs because leasing is a way to borrow money.

The column of incremental difference describes the decision to buy rather than lease, which invests more $37,500 now to avoid the later lease payments.

For TVM calculators and spreadsheets, this equation is easier to solve if $n = 24$ for both factors. There is a total cash flow of −$6000 at the end of Month 24. This cash flow can be split into −$2500 and −$3500. Then both the uniform and single payment periods continue to the end of Month 24.

$$0 = \$37,500 - \$2500(P/A, i_{mon}, 24) - \$3500(P/F, i_{mon}, 24)$$

$$PW(0\%) = \$37,500 - \$2500(24) - \$3500(1) = -\$26,000$$

$$PW(5\%) = \$37,500 - \$2500(13.799) - \$3500(0.3101) = \$1917$$

The answer is clearly between 0% and 5%, and closer to 5%. Try 4% next.

$$PW(4\%) = \$37,500 - \$2500(15.247) - \$3500(0.3901) = -\$1983$$

$$i_{mon} = 4\% + (5\% - 4\%)(1983)/(1983 + 1917) = 4.51\%$$

The effective annual rate is

$$i_a = 1.0451^{12} - 1 = 0.6975 = 69.8\%$$

Using a TVM calculator, we obtain

$$i_{mon} = i(n, A, P, F) = i(24, -2500, 37500, -3500)$$
$$= 4.4886\%$$
$$i_a = 1.044886^{12} - 1 = 69.37\%$$

In this case leasing is an extremely expensive way to obtain the equipment.

## INCREMENTAL ANALYSIS

When there are two alternatives, rate of return analysis is performed by computing the *incremental rate of return*—$\Delta$IRR—on the difference between the alternatives. In Examples 7-6 through 7-9 this was done by setting the present worths of the two alternatives equal.

Now we will calculate the increment more formally. Since we want to look at increments of investment, the cash flow for the difference between the alternatives is computed by taking the higher initial-cost alternative *minus* the lower initial-cost alternative. If $\Delta$IRR is the same or greater than the MARR, choose the higher-cost alternative. If $\Delta$IRR is less than the MARR, choose the lower-cost alternative.

| Two-Alternative Situation | Decision |
| --- | --- |
| $\Delta$IRR $\geq$ MARR | Choose the higher-cost alternative |
| $\Delta$IRR $<$ MARR | Choose the lower-cost alternative |

Rate of return and incremental rate of return analysis are illustrated by Examples 7-10 through 7-13. Example 7-10 illustrates a *very* important point that was not part of our earlier examples of parking permits, insurance, and equipment to be bought or leased. In those cases, the decision that we needed the insurance, permit, or equipment had already been made. We only needed to decide the best way to obtain it. In Example 7-10, there is a *do-nothing* alternative that must be considered.

## EXAMPLE 7–10

If an electromagnet is installed on the input conveyor of a coal-processing plant, it will pick up scrap metal in the coal. Removing this scrap will save an estimated $1200 per year in costs associated with machinery damage. The electromagnetic equipment has an estimated useful life of 5 years and no salvage value. Two suppliers have been contacted: Leaseco will provide the equipment in return for three beginning-of-year annual payments of $1000 each; Saleco will provide the equipment for $2783. If the MARR is 10%, should the project be done, and if so, which supplier should be selected?

### SOLUTION

Before we analyze which supplier should be selected, we must decide whether the do-nothing alternative would be a better choice. Since the first cost of Leaseco is lower than the first cost with Saleco, let us compare Leaseco with doing nothing. This is the same as finding the present worth or IRR of the Leaseco investment.

The cash flow at Time 0 for Leaseco is –$1000. At the end of Years 1 and 2, the firm spends $1000 on the lease and saves $1200 in machinery damage for a net of $200. Then for 3 years the firm saves $1200 annually. Since the firm is investing $1000 to save $4000 spread over 5 years, the arrangement is clearly worthwhile. In fact, at the 10% MARR, the PW is $1813. (Problem 7-23 asks you to find the 49% rate of return.)

To compare Saleco and Leaseco, we first recognize that both will provide equipment with the same useful life and benefits. Thus, the comparison is a fixed-output situation. In rate of return analysis, the method of solution is to examine the differences between the alternatives. We find the incremental cash flows each year to evaluate the additional investment required by Saleco.

| Year | Leaseco | Saleco | Difference Between Alternatives: Saleco – Leaseco |
|------|---------|--------|--------------------------------------------------|
| 0 | −$1000 | −$2783 | −$1783 |
| 1 | $\begin{cases} -1000 \\ 1200 \end{cases}$ | 1200 | +1000 |
| 2 | $\begin{cases} -1000 \\ 1200 \end{cases}$ | 1200 | +1000 |
| 3 | 1200 | 1200 | 0 |
| 4 | 1200 | 1200 | 0 |
| 5 | 1200 | 1200 | 0 |

Compute the NPW at various interest rates on the increment of investment represented by the difference between the alternatives.

| Year n | Cash Flow Saleco − Leaseco | PW* At 0% | At 8% | At 20% | At ∞% |
|--------|----------------------------|-----------|-------|--------|-------|
| 0 | −$1783 | −$1783 | −$1783 | −$1783 | −$1783 |
| 1 | 1000 | 1000 | 926 | 833 | 0 |
| 2 | 1000 | 1000 | 857 | 694 | 0 |
| 3 | 0 | 0 | 0 | 0 | 0 |
| 4 | 0 | 0 | 0 | 0 | 0 |
| 5 | 0 | 0 | 0 | 0 | 0 |
| | NPW = | +217 | 0 | −256 | −1783 |

*Each year the cash flow is multiplied by $(P/F, i, n)$.

At 0%: $(P/F, 0\%, n) = 1$ for all values of $n$
At ∞%: $(P/F, \infty\%, 0) = 1$
$(P/F, \infty\%, n) = 0$ for all other values of $n$

From the plot of these data in Figure 7-5, we see that NPW $= 0$ at $i = 8\%$.

FIGURE 7–5 NPW plot for
Example 7-10.

Thus, the incremental rate of return—ΔIRR—of selecting Saleco rather than Leaseco is 8%. This
is less than the 10% MARR. Thus, select Leaseco.

## EXAMPLE 7–11

You must select one of two mutually exclusive alternatives. (*Note*: Engineering economists
often use the term "mutually exclusive alternatives" to emphasize that selecting one alternative
precludes selecting any other.) The alternatives are as follows:

| Year | Alt. 1 | Alt. 2 |
|------|--------|--------|
| 0 | −$1000 | −$2000 |
| 1 | +1500 | +2800 |

Any money not invested here may be invested elsewhere at the MARR of 6%. If you can choose
only one alternative one time, using the internal rate of return (IRR) analysis method, which one
would you select?

### SOLUTION

Using the IRR analysis method, we will select the lesser-cost alternative (Alt. 1), unless we find
that the additional cost of Alt. 2 produces enough additional benefits to make it preferable instead.
If we consider Alt. 2 in relation to Alt. 1, then

$$\begin{bmatrix} \text{Higher-cost} \\ \text{Alt. 2} \end{bmatrix} = \begin{bmatrix} \text{Lower-cost} \\ \text{Alt. 1} \end{bmatrix} + \begin{bmatrix} \text{Differences between} \\ \text{Alt. 1 and Alt. 2} \end{bmatrix}$$

or

$$\text{Differences between Alt. 1 and Alt. 2} = \begin{bmatrix} \text{Higher-cost} \\ \text{Alt. 2} \end{bmatrix} - \begin{bmatrix} \text{Lower-cost} \\ \text{Alt. 1} \end{bmatrix}$$

The choice between the two alternatives reduces to examining the differences between them. We
can compute the rate of return on the differences between the alternatives.

| Year | Alt. 1 | Alt. 2 | Alt. 2 − Alt. 1 |
|------|--------|--------|-----------------|
| 0 | −$1000 | −$2000 | −$2000 − (−$1000) = −$1000 |
| 1 | 1500 | 2800 | 2800 − 1500 = 1300 |

$$0 = PW_{\text{Alt.2}-\text{Alt.1}} = -1000 + 1300(P/F, i, 1)$$

$$(P/F, i, 1) = \frac{1000}{1300} = 0.7692$$

One can see that if $1000 increases to $1300 in one year, the interest rate must be 30%. The compound interest tables confirm this conclusion. The 30% rate of return on the difference between the alternatives is far higher than the 6% MARR. The additional $1000 investment to obtain Alt. 2 is superior to investing the $1000 elsewhere at 6%. To obtain this desirable increment of investment, with its 30% rate of return, Alt. 2 is selected.

To understand more about Example 7-11, compute the rate of return for each alternative.

*Alternative 1*

$$0 = PW = -\$1000 + \$1500(P/F, i, 1)$$

$$(P/F, i, 1) = \frac{1000}{1500} = 0.6667 \implies i = 50\%$$

*Alternative 2*

$$0 = PW = -\$2000 + \$2800(P/F, i, 1)$$

$$(P/F, i, 1) = \frac{2000}{2800} = 0.7143 \implies i = 40\%$$

The higher rate of return of Alt. 1 is attractive, but Alt. 1 is not the correct solution. Solve the problem again, this time using present worth analysis.

**Present Worth Analysis**

*Alternative 1*

$$\text{NPW} = -1000 + 1500(P/F, 6\%, 1) = -1000 + 1500(0.9434) = +\$415$$

*Alternative 2*

$$\text{NPW} = -2000 + 2800(P/F, 6\%, 1) = -2000 + 2800(0.9434) = +\$642$$

Alternative 1 has a 50% rate of return and an NPW (at the 6% MARR) of +$415. Alternative 2 has a 40% rate of return on a larger investment, with the result that its NPW (at the 6% MARR) is +$642. Our economic criterion is to maximize the return, rather than the rate of return. To maximize NPW, select Alt. 2. This agrees with the incremental rate of return analysis.

## EXAMPLE 7-12

If the computations for Example 7-11 do not convince you, and you still think Alternative 1 would be preferable, try this problem.

You have an extra $20 in your wallet and two alternative ways of lending Bill some money.

(a) Lend Bill $10 with his promise of a 50% return. That is, he will pay you back $15 at the agreed time.

(b) Lend Bill $20 with his promise of a 40% return. He will pay you back $28 at the same agreed time.

You can select whether to lend Bill $10 or $20. This is a one-time situation, and any money not lent to Bill will remain in your wallet. Which alternative do you choose?

### SOLUTION

So you see that a 50% return on the smaller sum is less rewarding to you than 40% on the larger sum? Since you would prefer to have $28 than $25 ($15 from Bill plus $10 remaining in your wallet) after the loan is paid, lend Bill $20.

## EXAMPLE 7-13

Solve Example 7-11 again, but this time compute the interest rate on the increment (Alt. 1 − Alt. 2) instead of (Alt. 2 − Alt. 1). How do you interpret the results?

### SOLUTION

This time the problem is being viewed as follows:

$$\text{Alt. 1} = \text{Alt. 2} + [\text{Alt. 1} - \text{Alt. 2}]$$

| Year | Alt. 1 | Alt. 2 | [Alt.1 − Alt. 2] |
|------|--------|--------|------------------|
| 0 | −$1000 | −$2000 | −$1000 − (−$2000) = $1000 |
| 1 | 1500 | 2800 | 1500 − 2800 = −1300 |

We can write one equation in one unknown:

$$\text{NPW} = 0 = +1000 - 1300(P/F, i, 1)$$

$$(P/F, i, 1) = \frac{1000}{1300} = 0.7692 \Longrightarrow i = 30\%$$

Once again the interest rate is found to be 30%. The critical question is, What does the 30% represent? Looking at the increment again:

| Year | Alt. 1 − Alt. 2 |
|------|-----------------|
| 0 | $1000 |
| 1 | −1300 |

The cash flow does *not* represent an investment; instead, it represents a loan. It is as if we borrowed $1000 in Year 0 ($1000 represents a receipt of money) and repaid it in Year 1 (−$1300 represents a disbursement). The 30% interest rate means this is the amount *we would pay* for the loan.

Is this a desirable borrowing scenario? Since the MARR on investments is 6%, it is reasonable to assume our maximum interest rate on borrowing would also be 6%. Here the interest rate is 30%, which means the borrowing is undesirable. Since Alt. 1 = Alt. 2 plus the undesirable (Alt. 1 − Alt. 2) increment, we should reject Alternative 1. This means we should select Alternative 2—the same conclusion reached in Example 7-11.

Example 7-13 illustrated that one can analyze either **increments of investment** or **increments of borrowing.** When looking at increments of investment, we accept the increment when the incremental rate of return equals or exceeds the minimum attractive rate of return ($\Delta IRR \geq MARR$). When looking at increments of borrowing, we accept the increment when the incremental interest rate is less than or equal to the *minimum* attractive rate of return ($\Delta IRR \leq MARR$). One way to avoid much of the possible confusion is to organize the solution to any problem so that one is examining increments of investment. This is illustrated in the next example.

## EXAMPLE 7–14

A firm is considering which of two devices to install to reduce costs. Both devices have useful lives of 5 years and no salvage value. Device *A* costs $1000 and can be expected to result in $300 savings annually. Device *B* costs $1350 and will provide cost savings of $300 the first year but will increase $50 annually, making the second-year savings $350, the third-year savings $700, and so forth. For a 7% MARR, which device should the firm purchase?

### SOLUTION

This problem has been solved by present worth analysis (Example 5-1) and annual cost analysis (Example 6-5). This time we will use rate of return analysis, which must be done on the incremental investment.

| Year | Device *A* | Device *B* | Difference Between Alternatives: Device *B* − Device *A* |
|------|-----------|-----------|-----------|
| 0 | −$1000 | −$1350 | −$350 |
| 1 | 300 | 300 | 0 |
| 2 | 300 | 350 | 50 |
| 3 | 300 | 400 | 100 |
| 4 | 300 | 450 | 150 |
| 5 | 300 | 500 | 200 |

For the difference between the alternatives, write a single equation with $i$ as the only unknown.

$$PW(i) = 0 = -350 + 50(P/G, i, 5)$$
$$(P/G, i, 5) = 7;  \text{so } i \text{ is between } 9\%, (P/G, 9\%, 5) = 7.111$$
$$\text{and } 10\%, (P/G, 10\%, 5) = 6.862$$
$$i = 9\% + (10\% - 9\%)(7.111 - 7)/(7.111 - 6.862) = 9.45\%$$

The 9.45% IRR is greater than the 7% MARR; therefore, the increment is desirable. Reject Device *A* and choose Device *B*.

### Analysis Period

In discussing present worth analysis and annual cash flow analysis, an important consideration is the analysis period. This is also true in rate of return analysis. The solution method for two alternatives is to examine the differences between the alternatives. Clearly, the examination must cover the selected analysis period. For now, we can only suggest that the assumptions made should reflect one's perception of the future as accurately as possible.

In Example 7-15 the analysis period is a common multiple of the alternative service lives and identical replacement is assumed. This problem illustrates an analysis of the differences between the alternatives over the analysis period.

## EXAMPLE 7–15

Two machines are being considered for purchase. If the MARR is 10%, which machine should be bought? Use an IRR analysis comparison.

|  | Machine X | Machine Y |
|---|---|---|
| Initial cost | $200 | $700 |
| Uniform annual benefit | 95 | 120 |
| End-of-useful-life salvage value | 50 | 150 |
| Useful life, in years | 6 | 12 |

### SOLUTION

The solution is based on a 12-year analysis period and a replacement machine *X* that is identical to the present machine *X*. The cash flow for the differences between the alternatives, is as follows: Difference between Alternatives (or Alts)

| Year | Machine X | Machine Y | Difference Between Alternatives (Alts): Machine Y − Machine X |
|---|---|---|---|
| 0 | −$200 | −$700 | −$500 |
| 1 | 95 | 120 | 25 |
| 2 | 95 | 120 | 25 |
| 3 | 95 | 120 | 25 |
| 4 | 95 | 120 | 25 |
| 5 | 95 | 120 | 25 |

| Year | Machine X | Machine Y | Difference Between Alternatives (Alts): Machine Y − Machine X |
|------|-----------|-----------|------------------------------|
| 6 | 95 / 50 / −200 | 120 | 25 / 150 |
| 7 | 95 | 120 | 25 |
| 8 | 95 | 120 | 25 |
| 9 | 95 | 120 | 25 |
| 10 | 95 | 120 | 25 |
| 11 | 95 | 120 | 25 |
| 12 | 95 / 50 | 120 / 150 | 25 / 100 |

PW of cost (differences) = PW of benefits (differences)

$$500 = 25(P/A, i, 12) + 150(P/F, i, 6) + 100(P/F, i, 12)$$

The sum of the benefits over the 12 years is \$550, which is only a little greater than the \$500 additional cost. This indicates that the rate of return is quite low. Try $i = 1\%$.

$$500 \overset{?}{=} 25(11.255) + 150(0.942) + 100(0.887)$$
$$\overset{?}{=} 281 + 141 + 89 = 511$$

The interest rate is too low. Try $i = 1\frac{1}{2}\%$:

$$500 \overset{?}{=} 25(10.908) + 150(0.914) + 100(0.836)$$
$$\overset{?}{=} 273 + 137 + 84 = 494$$

The internal rate of return on the $Y - X$ increment, $IRR_{Y-X}$, is about 1.3%, far below the 10% minimum attractive rate of return. The additional investment to obtain Machine $Y$ yields an unsatisfactory rate of return, therefore $X$ is the preferred alternative.

## SPREADSHEETS AND RATE OF RETURN ANALYSIS

The spreadsheet functions covered in earlier chapters are particularly useful in calculating internal rates of returns (IRRs). If a cash flow diagram can be reduced to at most one $P$, one $A$, and/or one $F$, then the RATE *investment function* can be used. Otherwise the IRR *block function* is used with a cash flow in each period.

The Excel investment function is RATE($n$, $A$, $P$, $F$, type, guess). The $A$, $P$, and $F$ cannot all be the same sign. The $F$, type, and guess are optional arguments. The "type" is end- or beginning-of-period cash flow (for $A$, but not $F$, which is always end-of-period), and the "guess" is the starting value in the search for the IRR.

Considering Example 7-1, where $P = -8200$, $A = 2000$, and $n = 5$, the RATE function would be:

$$RATE(5, 2000, -8200)$$

which returns an answer of 7.00%, matching Example 7-1.

For Example 7-2, where $P = -700$, $A = 100$, $G = 75$, and $n = 4$, the RATE function cannot be used, since it has no provisions for the arithmetic gradient, $G$. Suppose the years (row 1) and the cash flows (row 2) are specified in columns B through E. The internal rate of return calculated using IRR(B2:F2) is 6.91%.

|   | A | B | C | D | E | F |
|---|---|---|---|---|---|---|
| 1 | Year | 0 | 1 | 2 | 3 | 4 |
| 2 | Cash flow | −700 | 100 | 175 | 250 | 325 |

Figure 7-6 illustrates the use of spreadsheet to graph the present worth of a cash flow series versus the interest rate. The interest rate with present worth equal to 0 is the IRR. The $y$ axis on this graph has been modified so that the $x$ axis intersects at a present worth of −50 rather than at 0. To do this click on the $y$ axis, then right-click to bring up the "format axis" option. Select this and then select the tab for the *scale* of the axis. This has a selection

|   | A | B | C | D | E | F |
|---|---|---|---|---|---|---|
| 1 | Year | Cash flow | | $i$ | PW | |
| 2 | 0 | −700 | | 0% | 150.0 | =$B$2+NPV(D2,$B$3:$B$6) |
| 3 | 1 | 100 | | 2% | 102.1 | |
| 4 | 2 | 175 | | 4% | 58.0 | |
| 5 | 3 | 250 | | 6% | 17.4 | |
| 6 | 4 | 325 | | 8% | −20.0 | |
| 7 | | | | 10% | −54.7 | |
| 8 | | | | | | |
| 9 | | | | | | |
| 10 | | | | | | |
| 11 | | | | | | |
| 12 | | | | | | |
| 13 | | | | | | |
| 14 | | | | | | |
| 15 | | | | | | |
| 16 | | | | | | |
| 17 | | | | | | |
| 18 | | | | | | |
| 19 | | | | | | |

**FIGURE 7–6**  Graphing present worth versus $i$.

for the intersection of the $x$ axis. This process ensures that the $x$-axis labels are outside the graph.

## SUMMARY

Rate of return is the interest rate $i$ at which the benefits are equivalent to the costs or the net present worth equals zero.

There are a variety of ways of writing the cash flow equation in which the rate of return $i$ may be the single unknown. Five of them are as follows:

$$\text{PW of benefits} - \text{PW of costs} = 0$$

$$\frac{\text{PW of benefits}}{\text{PW of costs}} = 1$$

$$\text{NPW} = 0$$

$$\text{EUAB} - \text{EUAC} = 0$$

$$\text{PW of costs} = \text{PW of benefits}$$

*Rate of return analysis:* Rate of return is the most frequently used measure in industry, as the resulting rate of return is readily understood. Also, the difficulties in selecting a suitable interest rate to use in present worth and annual cash flow analysis are avoided.

### Criteria

#### Two Alternatives

Compute the incremental rate of return—$\Delta$IRR—on the increment of *investment* between the alternatives. Then,

- if $\Delta$IRR $\geq$ MARR, choose the higher-cost alternative, or,
- if $\Delta$IRR $<$ MARR, choose the lower-cost alternative

When an increment of *borrowing* is examined, where $\Delta$IRR is the incremental interest rate,

- if $\Delta$IRR $\leq$ MARR, the increment is acceptable, or
- if $\Delta$IRR $>$ MARR, the increment is not acceptable

#### Three or More Alternatives

See Chapter 8.

### Looking Ahead

Rate of return is further described in Appendix 7A. This material concentrates on the difficulties that occur with some cash flows series with multiple sign changes that may yield more than one root for the rate of return equation.

# PROBLEMS

## Rate of Return

**7-1** Compute the rate of return for the following cash flow to within $\frac{1}{2}\%$.

| Year | Cash Flow |
|------|-----------|
| 0 | −$100 |
| 1–10 | +27 |

(*Answer:* 23.9%)

**7-2** The Diagonal Stamp Company, which sells used postage stamps to collectors, advertises that its average price has increased from $1 to $5 in the last 5 years. Thus, management states, investors who had purchased stamps from Diagonal 5 years ago would have received a 100% rate of return each year.

(*a*) To check their calculations, compute the annual rate of return.

(*b*) Why is your computed rate of return less than 100%?

**7-3** A table saw costs $175 at a local store. You may either pay cash for it or pay $35 now and $12.64 a month for 12 months beginning 30 days hence. If you choose the time payment plan, what nominal annual interest rate will you be charged? (*Answer:* 15%)

**7-4** An investment of $5000 in Biotech common stock proved to be very profitable. At the end of 3 years the stock was sold for $25,000. What was the rate of return on the investment?

**7-5** Helen is buying a $12,375 car with a $3000 down payment, followed by 36 monthly payments of $325 each. The down payment is paid immediately, and the monthly payments are due at the end of each month. What nominal annual interest rate is Helen paying? What effective interest rate? (*Answers:* 15%; 16.08%)

**7-6** Peter Minuit bought an island from the Manhattoes Indians in 1626 for $24 worth of glass beads and trinkets. The 1991 estimate of the value of land on this island was $12 billion. What rate of return would the Indians have received if they had retained title to the island rather than selling it for $24?

**7-7** An engineer invests $5800 at the end of every year for a 35-year career. If the engineer wants $1 million in savings at retirement, what interest rate must the investment earn?

**7-8** A mining firm makes annual deposits of $250,000 into a reclamation fund for 20 years. If the firm must have $10 million when the mine is closed, what interest rate must the investment earn?

**7-9** You spend $1000 and in return receive two payments of $1094.60—one at the end of 3 years and the other at the end of 6 years. Calculate the resulting rate of return.

**7-10** Your cat just won the local feline lottery to the tune of 3000 cans of "9-Lives" cat food (assorted flavors). A local grocer offers to take the 3000 cans and in return, supply 30 cans a month for the next 10 years. What rate of return, in terms of nominal annual rate, will you realize on this deal? (Compute to nearest 0.01%.)

**7-11** A woman went to the Beneficial Loan Company and borrowed $3000. She must pay $119.67 at the end of each month for the next 30 months.

(*a*) Calculate the nominal annual interest rate she is paying to within ±0.15%.

(*b*) What effective annual interest rate is she paying?

**7-12** Your cousin Jeremy has asked you to bankroll his proposed business painting houses in the summer. He plans to operate the business for 5 years to pay his way through college. He needs $5000 to purchase an old pickup, some ladders, a paint sprayer, and some other equipment. He is promising to pay you $1500 at the end of each summer (for 5 years) in return for this investment. Calculate your annual rate of return.
*Contributed by Paul R. McCright, University of South Florida*

**7-13** An investor has invested $250,000 in a new rental property. Her estimated annual costs are $6000 and annual revenues are $20,000. What rate of return per year will the investor make over a 30-year period ignoring the salvage value? If the property can be sold for $200,000 what is the rate of return?
*Contributed by Hamed Kashani, Saeid Sadri, and Baabak Ashuri, Georgia Institute of Technology*

**7-14** Installing an automated production system costing $278,000 is initially expected to save Zia Corporation $52,000 in expenses annually. If the system needs $5000 in operating and maintenance costs each year and has a salvage value of $25,000 at Year 10, what is the IRR of this system? If the company wants to earn at least 12% on all investments, should this system be purchased?

*Contributed by Paul R. McCright, University of South Florida*

**7-15** For the following diagram, compute the IRR to within $\frac{1}{2}\%$.

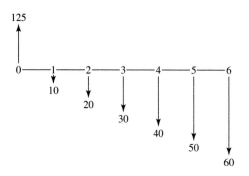

**7-16** For the following diagram, compute the rate of return.

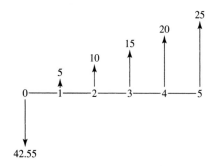

**7-17** Consider the following cash flow:

| Year | Cash Flow |
|------|-----------|
| 0 | −$500 |
| 1 | 200 |
| 2 | 150 |
| 3 | 100 |
| 4 | 50 |

Compute the rate of return represented by the cash flow.

**7-18** Switching to powder coating technology will reduce the emission of volatile organic carbons (VOCs) for a firm's production process. The initial cost is $200,000 with annual costs of $50,000 and revenues of $90,000 in the first year. Revenues are projected

to increase by $2000 annually after Year 1. The salvage value 10 years from now is projected to be $30,000. What rate of return will the firm make on this investment?

*Contributed by Hamed Kashani, Saeid Sadri, and Baabak Ashuri, Georgia Institute of Technology*

**7-19** Consider the following cash flow:

| Year | Cash Flow |
|------|-----------|
| 0 | −$1000 |
| 1 | 0 |
| 2 | 300 |
| 3 | 300 |
| 4 | 300 |
| 5 | 300 |

Compute the rate of return on the $1000 investment to within 0.1%. (*Answer: 5.4%*)

**7-20** A bank proudly announces that it has changed its interest computation method to continuous compounding. Now $2000 left in the bank for 9 years will double to $4000.

(*a*) What nominal interest rate, compounded continuously, is the bank paying?

(*b*) What effective interest rate is it paying?

**7-21** Compute the rate of return for the following cash flow to within 0.5%.

| Year | Cash Flow |
|------|-----------|
| 0 | −$640 |
| 1 | 0 |
| 2 | 100 |
| 3 | 200 |
| 4 | 300 |
| 5 | 300 |

(*Answer: 9.3%*)

**7-22** Compute the rate of return for the following cash flow.

| Year | Cash Flow |
|------|-----------|
| 1–5 | −$233 |
| 6–10 | 1000 |

**7-23** For Example 7-10, find the rate of return on the Leaseco proposal compared with doing nothing. (*Answer: 49%*)

**7-24** For the following diagram, compute the interest rate at which the costs are equivalent to the benefits.

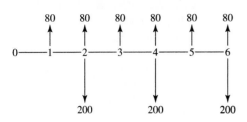

(*Answer:* 50%)

**7-25** For the following diagram, compute the rate of return on the $3810 investment.

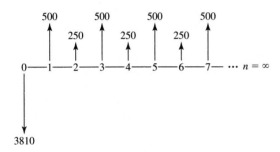

**7-26** For the following diagram, compute the rate of return.

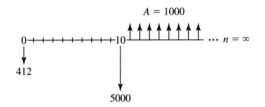

**7-27** Consider the following cash flow:

| Year | Cash Flow |
| --- | --- |
| 0 | −$400 |
| 1 | 0 |
| 2 | 200 |
| 3 | 150 |
| 4 | 100 |
| 5 | 50 |

Write one equation, with $i$ as the only unknown, for the cash flow. In the equation you are not to use more than two single payment compound interest factors. (You may use as many other factors as you wish.) Then solve your equation for $i$.

**7-28** You have just been elected into the "Society of Honorable Engineers." First-year dues are waived in honor of your election. Thus, your first-payment of $200 is due at the end of the year, and annual dues are expected to increase 3% annually. After 40 years you become a life member and no further dues are owed. Instead of paying annual dues, however, you can pay a one-time $2000 life membership fee.

(*a*) Show the equation for determining the rate of return for buying a life membership.
*Contributed by D. P. Loucks, Cornell University*

(*b*) What is the rate of return?

**Net Present Worth vs. *i* and Bonds**

**7-29** For Problem 7-1, graph the PW versus the interest rate for values from 0% to 50%. Is this the typical PW graph for an investment?

**7-30** For Problem 7-21, graph the PW versus the interest rate for values from 0% to 50%. Is this the typical PW graph for an investment?

**7-31** A man buys a corporate bond from a bond brokerage house for $925. The bond has a face value of $1000 and pays 4% of its face value each year. If the bond will be paid off at the end of 10 years, what rate of return will the man receive?
(*Answer:* 4.97%)

**7-32** A well-known industrial firm has issued $1000 bonds that carry 4% nominal annual interest paid semiannually. The bonds mature 20 years from now, at which time the industrial firm will redeem them for $1000 plus the terminal semiannual interest payment. From the financial pages of your newspaper you learn that the bonds may be purchased for $715 each ($710 for the bond plus a $5 sales commission). What nominal annual rate of return would you receive if you purchased the bond now and held it to maturity 20 years from now?
(*Answer:* 6.6%)

**7-33** On April 2, 1998, an engineer bought a $1000 bond of an American airline for $875. The bond paid 6% on its principal amount of $1000, half in each of its April 1 and October 1 semiannual payments; it repaid the $1000 principal sum on October 1, 2011. What nominal rate of return did the engineer receive from the bond if he held it to its maturity (on October 1, 2011)?
(*Answer:* 7.5%)

**7-34** Mildred can purchase a municipal bond with a par (face) value of $5000 that will mature in 3 1/2 years. The bond pays 8% interest compounded quarterly. If she can buy this bond for $4800, what rate of return will she earn?

*Contributed by Paul R. McCright, University of South Florida*

*Problems 7-35 to 7-39 contributed by Meenakshi Sundaram, Tennessee Tech University*

**7-35** Mike buys a corporate bond with a face value of $1000 for $900. The bond matures in 10 years and pays a coupon interest rate of 6%. Interest is paid every quarter.

(*a*) Determine the effective rate of return if Mike holds the bond to maturity.

(*b*) What effective interest rate will Mike get if he keeps the bond for only 5 years and sells it for $950?

**7-36** A 12%, $50,000 bond is offered for sale at $45,000. If the bond interest is payable monthly and the bond matures in 20 years, what nominal and effective rates of return per year will the purchaser make on the investment?

**7-37** An investor purchased a 5%, $5000 bond for $4000. The interest was payable quarterly, and the bond's maturity was 20 years. The bond was kept for only 9 years and sold for $4200 immediately after the 36th interest payment was received. What nominal and effective rates of return per year were made on this investment?

**7-38** At what coupon interest rate will a $20,000 bond yield a nominal 12% interest compounded quarterly if the purchaser pays $18,000 and the bond becomes due in 20 years? Assume the bond interest is payable quarterly.

**7-39** A 9%, $10,000 bond that has interest payable semiannually sells for $8500. Determine what the maturity date should be so that the purchaser may enjoy a 12% nominal rate of return on this investment.

**7-40** ABC Corporation's recently issued bonds paying interest semiannually and maturing in 10 years. The face value of each bond is $1000, and 6.8% is the nominal interest rate.

(*a*) What is the effective interest rate an investor receives?

(*b*) If a 0.75% fee is deducted by the brokerage firm from the initial $1000, what is the effective annual interest rate paid by ABC Corporation?

**7-41** ABC Corporation is issuing some *zero coupon bonds*, which pay no interest. At maturity in 20 years they pay a face value of $10,000. The bonds are expected to sell for $3118 when issued.

(*a*) What is the effective interest rate an investor receives?

(*b*) A 1% fee (based on the face value) is deducted by the brokerage firm from the initial sales revenue. What is the effective annual interest rate paid by ABC Corporation?

## Discounts and Fees

**7-42** The cash price of a machine tool is $3500. The dealer is willing to accept a $1200 down payment and 24 end-of-month monthly payments of $110 each. At what effective interest rate are these terms equivalent? (*Answer:* 14.4%)

**7-43** A local bank makes auto loans. It charges 4% per year in the following manner: if $3600 is borrowed to be repaid over a 3-year period, the bank interest charge is ($3600)(0.04)(3 years) = $432. The bank deducts the $432 of interest from the $3600 loan and gives the customer $3168 in cash. The customer must repay the loan by paying $1/36$ of $3600, or $100, at the end of each month for 36 months. What nominal annual interest rate is the bank actually charging for this loan?

**7-44** Jan purchased 100 shares of Peach Computer stock for $18 per share, plus a $45 brokerage commission. Every 6 months she received a dividend from Peach of 50 cents per share. At the end of 2 years, just after receiving the fourth dividend, she sold the stock for $23 per share and paid a $58 brokerage commission from the proceeds. What annual rate of return did she receive on her investment?

**7-45** A used car dealer advertises financing at 0% interest over 3 years with monthly payments. You must pay a processing fee of $250 at signing. The car you like costs $6000.

(*a*) What is your effective annual interest rate?

(*b*) You believe that the dealer would accept $5200 if you paid cash. What effective annual interest rate would you be paying, if you financed with the dealer?

**7-46** A new car dealer advertises financing at 0% interest over 4 years with monthly payments or a $3000 rebate if you pay cash.

(*a*) The car you like costs $12,000. What effective annual interest rate would you be paying if you financed with the dealer?

(b) The car you like costs $18,000. What effective annual interest rate would you be paying if you financed with the dealer?

(c) The car you like costs $24,000. What effective annual interest rate would you be paying if you financed with the dealer?

**7-47**  A used car dealer advertises financing at 4% interest over 3 years with monthly payments. You must pay a processing fee of $250 at signing. The car you like costs $6000.

(a) What is your effective annual interest rate?

(b) You believe that the dealer would accept $5200 if you paid cash. What effective annual interest rate would you be paying if you financed with the dealer?

(c) Compare these answers with those for Problem 7-45. What can you say about what matters the most for determining the effective interest rate?

**7-48**  Some laboratory equipment sells for $75,000. The manufacturer offers financing at 8% with annual payments for 4 years for up to $50,000 of the cost. The salesman is willing to cut the price by 10% if you pay cash. What is the interest rate you would pay by financing?

**7-49**  A home mortgage with monthly payments for 30 years is available at 6% interest. The home you are buying costs $120,000, and you have saved $12,000 to meet the requirement for a 10% down payment. The lender charges "points" of 2% of the loan value as a loan origination and processing fee. This fee is added to the initial balance of the loan.

(a) What is your monthly payment?

(b) If you keep the mortgage until it is paid off in 30 years, what is your effective annual interest rate?

(c) If you move to a larger house in 10 years and pay off the loan, what is your effective annual interest rate?

(d) If you are transferred in 3 years and pay off the loan, what is your effective annual interest rate?

**TABLE P7–50**

| *Money by Mail* | | Non-Negotiable    INE<br>1/96 | |
|---|---|---|---|
| For the<br>Amount of __$3000 or $2000 or $1000__ **Dollars**<br>Pay to the Order of  **I Feel Rich**<br>**Limited Time Offer** | | | To borrow $3000,<br>$2000, or $1000 |
| For the Amount of $3000 Dollars<br>Pay to the<br>Order of _____**I Feel Rich**_____<br>Total of Payments **$4,280.40**<br>Number of Monthly Payments<br>36 Months<br>Amount Financed $3,246.25 | | APR 18.95%<br>Finance Charge $1034.29<br><br>Monthly Payment $118.90<br>Credit Line Premium<br>$83.46*<br>Credit Disability Premium $162.65* | $3000 loan terms |
| For the Amount of **$2000 Dollars**<br>Pay to the<br>Order of _____**I Feel Rich**_____<br>Total of Payments $2,731.50<br>Number of Monthly Payments<br>30 Months<br>Amount Financed $2,137.25 | | APR 19.95%<br>Finance Charge $594.25<br><br>Monthly Payment $91.05<br>Credit Line Premium<br>$44.38*<br>Credit Disability Premium $92.87* | $2000 loan terms |
| For the Amount of **$1000 Dollars**<br>Pay to the<br>Order of _____**I Feel Rich**_____<br>Total of Payments $1,300.80<br>Number of Monthly Payments<br>24 Months<br>Amount Financed $1055.26 | | APR 20.95%<br>Finance Charge $245.54<br><br>Monthly Payment $54.20<br>Credit Line Premium<br>$16.91*<br>Credit Disability Premium $39.02* | $1000 loan terms |

*Credit insurance. If selected, premium will be paid from amount financed. If not selected,
cash advance is total amount financed.

**7-50** A finance company is using the "Money by Mail" offer shown in Table P7–50. Calculate the yearly nominal IRR received by the company if a customer chooses the loan of $2000 and accepts the credit insurance (life and disability).

**7-51** A finance company is using the "Money by Mail" offer shown in Problem 7-50. Calculate the yearly nominal IRR received by the company if a customer chooses the $3000 loan but declines the credit insurance.

### Investments and Loans

**7-52** An investor bought a one-acre lot on the outskirts of a city for $9000 cash. Each year she paid $80 of property taxes. At the end of 4 years, she sold the lot for a net value of $15,000. What rate of return did she receive on her investment? *(Answer:* 12.92%)

**7-53** A mine is for sale for $240,000. It is believed the mine will produce a profit of $65,000 the first year, but the profit will decline $5000 a year after that, eventually reaching zero, whereupon the mine will be worthless. What rate of return would this $240,000 investment produce for the purchaser of the mine?

**7-54** An apartment building in your neighborhood is for sale for $140,000. The building has four units, which are rented at $500 per month each. The tenants have long-term leases that expire in 5 years. Maintenance and other expenses for care and upkeep are $8000 annually. A new university is being built in the vicinity and it is expected that the building could be sold for $160,000 after 5 years.

   (*a*) What is the internal rate of return for this investment?

   (*b*) Should this investment be accepted if the other options have a rate of return of 12%?

**7-55** A new machine can be purchased today for $300,000. The annual revenue from the machine is calculated to be $67,000, and the equipment will last 10 years. Expect the maintenance and operating costs to be $3000 a year and to increase $600 per year. The salvage value of the machine will be $20,000. What is the rate of return for this machine?

**7-56** An insurance company is offering to sell an annuity for $20,000 cash. In return the firm will guarantee to pay the purchaser 20 annual end-of-year payments, with the first payment amounting to $1100. Subsequent payments will increase at a uniform 10% rate each year (second payment is $1210; third payment

is $1331, etc.). What rate of return would someone who buys the annuity receive?

**7-57** Fifteen families live in Willow Canyon. Although several water wells have been drilled, none has produced water. The residents take turns driving a water truck to a fire hydrant in a nearby town. They fill the truck with water and then haul it to a storage tank in Willow Canyon. Last year truck fuel and maintenance cost $3180. This year the residents are seriously considering spending $100,000 to install a pipeline from the nearby town to their storage tank. What rate of return would the Willow Canyon residents receive on their new water supply pipeline if the pipeline is considered to last

   (*a*) Forever?

   (*b*) 100 years?

   (*c*) 50 years?

   (*d*) Would you recommend that the pipeline be installed? Explain.

**7-58** An investor bought 100 shares of Omega common stock for $9000. He held the stock for 9 years. For the first 4 years he received annual end-of-year dividends of $800. For the next 4 years he received annual dividends of $400. He received no dividend for the ninth year. At the end of the ninth year he sold his stock for $6000. What rate of return did he receive on his investment?

**7-59** One aspect of obtaining a college education is the prospect of improved future earnings in comparison to non-college graduates. Sharon Shay estimates that a college education has a $28,000 equivalent cost at graduation. She believes the benefits of her education will occur throughout 40 years of employment. She thinks that during the first 10 years out of college, her income will be higher than that of a non-college graduate by $3000 per year. During the subsequent 10 years, she projects an annual income that is $6000 per year higher. During the last 20 years of employment, she estimates an annual salary that is $12,000 above the level of the non-college graduate. If her estimates are correct, what rate of return will she receive as a result of her investment in a college education?

**7-60** Upon graduation, every engineer must decide whether to go on to graduate school. Estimate the costs of going full time to the university to obtain a master of science degree. Then estimate the resulting costs and benefits. Combine the various conse- quences into a cash flow table and compute the rate of return. Nonfinancial benefits are probably relevant here too.

**7-61** A popular magazine offers a lifetime subscription for $200. Such a subscription may be given as a gift to an infant at birth (the parents can read it in those early years), or taken out by an individual for him- self. Normally, the magazine costs $12.90 per year. Knowledgeable people say it probably will continue indefinitely at this $12.90 rate. What rate of return would be obtained if a life subscription were pur- chased for an infant, rather than paying $12.90 per year beginning immediately? You may make any reasonable assumptions, but the compound interest factors must be *correctly* used.

**7-62** The following advertisement appeared in the *Wall Street Journal* on Thursday, February 9, 1995:

> There's nothing quite like the Seville SmartLease. Seville SLS $0 down, $599 a month/36 months.

Terms are first month's lease payment of $599 plus $625 refundable security deposit and a consumer down payment of $0 for a total of $1224 due at lease signing. Monthly payment is based on a net capitalized cost of $39,264 for total monthly pay- ments of $21,564. Payment examples based on a 1995 Seville SLS: $43,658 MSRP including desti- nation charge. Tax, license, title fees, and insurance extra. Option to purchase at lease end for $27,854. Mileage charge of $0.15 per mile over 36,000 miles.

(*a*) Set up the cash flows.

(*b*) Determine the interest rate (nominal and effec- tive) for the lease.

**7-63** An engineering student is deciding whether to buy two 1-semester parking permits or an annual permit. The annual parking permit costs $100 due August 15$^{th}$, and the semester permits are $65 due August 15$^{th}$ and January 15$^{th}$. What is the rate of return for buying the annual permit?

**7-64** An engineering student is deciding whether to buy two 1-semester parking permits or an annual per- mit. Using the dates and costs for your university, find the rate of return for the incremental cost of the annual permit.

**7-65** An engineering firm can pay for its liability insurance on an annual or quarterly basis. If paid quarterly, the insurance costs $18,000. If paid annually, the

insurance costs $65,000. What are the quarterly rate of return and the nominal and effective interest rates for paying on an annual basis?

**7-66** An engineering firm can pay for its liability insurance on an annual or quarterly basis. If paid quarterly, the insurance costs $18,000. If paid annually, the insur- ance costs $65,000. Use a spreadsheet to calculate the exact quarterly interest rate for paying on an annual basis.

**7-67** An engineering student must decide whether to pay for auto insurance on a monthly or an annual basis. If paid annually, the cost is $1650. If paid monthly, the cost is $150 at the start of each month. What is the rate of return for buying the insurance on an annual basis?

**7-68** For your auto or home insurance, find out the cost of paying annually or on a shorter term. What is the rate of return for buying the insurance on an annual basis?

**Incremental Analysis**

**7-69** Two alternatives are as follows:

| Year | A | B |
|------|------|------|
| 0 | −$2000 | −$2800 |
| 1 | 800 | 1100 |
| 2 | 800 | 1100 |
| 3 | 800 | 1100 |

If 5% is considered the minimum attractive rate of return, which alternative should be selected?

**7-70** Consider two mutually exclusive alternatives:

| Year | X | Y |
|------|------|------|
| 0 | −$100 | −$50.0 |
| 1 | 35 | 16.5 |
| 2 | 35 | 16.5 |
| 3 | 35 | 16.5 |
| 4 | 35 | 16.5 |

If the minimum attractive rate of return is 10%, which alternative should be selected?

**7-71** Two mutually exclusive alternatives are being considered. Both have a 10-year useful life. If the MARR is 8%, which alternative is preferred?

| | A | B |
|------|------|------|
| Initial cost | $100.00 | $50.00 |
| Uniform annual benefit | 19.93 | 11.93 |

**7-72** Consider two mutually exclusive alternatives:

| Year | X | Y |
|------|------|------|
| 0 | −$5000 | −$5000 |
| 1 | −3000 | 2000 |
| 2 | 4000 | 2000 |
| 3 | 4000 | 2000 |
| 4 | 4000 | 2000 |

If the MARR is 8%, which alternative should be selected?

**7-73** Two mutually exclusive alternatives are being considered. Both have lives of 5 years. Alternative A has a first cost of $2500 and annual benefits of $746. Alternative B costs $6000 and has annual benefits of $1664

If the minimum attractive rate of return is 8%, which alternative should be selected? Solve the problem by

(a) Present worth analysis

(b) Annual cash flow analysis

(c) Rate of return analysis

**7-74** A contractor is considering whether to buy or lease a new machine for her layout site work. Buying a new machine will cost $12,000 with a salvage value of $1200 after the machine's useful life of 8 years. On the other hand, leasing requires an annual lease payment of $3000. Assuming that the MARR is 15% and on the basis of an internal rate of return analysis, which alternative should the contractor be advised to accept? The cash flows are as follows:

| Year (n) | Alt. A (buy) | Alt. B (lease) |
|------|------|------|
| 0 | −$12,000 | −$3000 |
| 1 | | −3000 |
| 2 | | −3000 |
| 3 | | −3000 |
| 4 | | −3000 |
| 5 | | −3000 |
| 6 | | −3000 |
| 7 | | −3000 |
| 8 | +1200 | 0 |

**7-75** Two hazardous environment facilities are being evaluated, with the projected life of each facility being 10 years. The cash flows are as follows:

| | Alt. A | Alt. B |
|------|------|------|
| First cost | $615,000 | $300,000 |
| Maintenance and operating cost | 10,000 | 25,000 |
| Annual benefits | 158,000 | 92,000 |
| Salvage value | 65,000 | −5,000 |

The company uses a MARR of 15%. Using rate of return analysis, which alternative should be selected?

**7-76** The owner of a corner lot wants to find a use that will yield a desirable return on his investment. After much study and calculation, the owner decides that the two best alternatives are:

| | Build Gas Station | Build Soft Ice Cream Stand |
|------|------|------|
| First cost | $80,000 | $120,000 |
| Annual property taxes | 3,000 | 5,000 |
| Annual income | 11,000 | 16,000 |
| Life of building, in years | 20 | 20 |
| Salvage value | 0 | 0 |

If the owner wants a minimum attractive rate of return on his investment of 6%, which of the two alternatives would you recommend?

**7-77** A grocery distribution center is considering whether to invest in RFID or bar code technology to track its inventory within the warehouse and truck loading operations. The useful life of the RFID and bar code devices is projected to be 5 years with minimal or zero salvage value. The bar code investment cost is $100,000 and can be expected to save at least $50,000 in product theft and lost items annually. The RFID system is estimated to cost $200,000 and will save $30,000 the first year, with an increase of $15,000 annually after the first year. For a 5% MARR, should the manager invest in the RFID system or the bar code system? Analyze incrementally using rate of return. *Contributed by Oliver Hedgepeth, University of Alaska Anchorage*

**7-78** A state's department of transportation (DOT) is considering whether to buy or lease an RFID tracking system for asphalt, concrete, and gravel trucks to be used in road paving. Purchasing the RFID system will cost $5000 per truck, with a salvage value of $1500 after the RFID system's useful life of 5-years. However, the DOT considering this purchase is also

looking at leasing this same RFID system for an annual payment of $3500, which includes a full replacement warranty. Assuming that the MARR is 11% and on the basis of an internal rate of return analysis, which alternative would you advise the DOT to consider? Analyze incrementally using rate of return. The number of trucks used in a season varies from 5000 to 7500. Does this matter?
*Contributed by Oliver Hedgepeth, University of Alaska Anchorage*

**7-79** After 15 years of working for one employer, you transfer to a new job. During each of these years your employer contributed (that is, she diverted from your salary) $1500 to an account for your retirement (a fringe benefit), and you contributed a matching amount each year. The whole fund was invested at 5% during that time, and the value of the account now stands at $30,000.

You are now faced with two alternatives. (1) You may leave both contributions in the fund until retirement in 35 years, during which you will get the future value of this amount at 5% interest per year. (2) You may take out the total value of "your" contributions, which is $15,000 (one-half of the total $30,000). You can do as you wish with the money you take out, but the other half will be lost as far as you are concerned. In other words, you can give up $15,000 today for the sake of getting the other $15,000 now. Otherwise, you must wait 35 years to get the accumulated value of the entire fund. Which alternative is more attractive? Explain your choice.

**7-80** In his will, Frank's uncle has given Frank the choice between two alternatives:

Alternative 1   $2000 cash
Alternative 2   $150 cash now plus $100 per month for 20 months beginning the first day of next month

(a) At what rate of return are the two alternatives equivalent?

(b) If Frank thinks the rate of return in (a) is too low, which alternative should he select?

**7-81** A bulldozer can be purchased for $380,000 and used for 6 years, when its salvage value is 15% of the first cost. Alternatively, it can be leased for $60,000 a year. (Remember that lease payments occur at the start of the year.) The firm's interest rate is 12%.

(a) What is the interest rate for buying versus leasing? Which is the better choice?

(b) If the firm will receive $65,000 more each year than it spends on operating and maintenance costs, should the firm obtain the bulldozer? What is the rate of return for the bulldozer using the best financing plan?

**7-82** A diesel generator for electrical power can be purchased by a remote community for $480,000 and used for 10 years, when its salvage value is $50,000. Alternatively, it can be leased for $70,000 a year. (Remember that lease payments occur at the start of the year.) The community's interest rate is 8%.

(a) What is the interest rate for buying versus leasing? Which is the better choice?

(b) The community will spend $80,000 less each year for fuel and maintenance, than it currently spends on buying power. Should it obtain the generator? What is the rate of return for the generator using the best financing plan?

## Analysis Period

**7-83** Two alternatives are being considered:

|  | A | B |
|---|---|---|
| First cost | $9200 | $5000 |
| Uniform annual benefit | 1850 | 1750 |
| Useful life, in years | 8 | 4 |

If the minimum attractive rate of return is 7%, which alternative should be selected?

**7-84** Jean has decided it is time to buy a new battery for her car. Her choices are:

|  | Zappo | Kicko |
|---|---|---|
| First cost | $56 | $90 |
| Guarantee period, in months | 12 | 24 |

Jean believes the batteries can be expected to last only for the guarantee period. She does not want to invest extra money in a battery unless she can expect a 50% rate of return. If she plans to keep her present car another 2 years, which battery should she buy?

**7-85** Two alternatives are being considered:

|  | A | B |
|---|---|---|
| Initial cost | $9200 | $5000 |
| Uniform annual benefit | 1850 | 1750 |
| Useful life, in years | 8 | 4 |

Base your computations on a MARR of 7% and an 8-year analysis period. If identical replacement is assumed, which alternative should be selected?

**7-86**  Two investment opportunities are as follows:

|  | A | B |
|---|---|---|
| First cost | $150 | $100 |
| Uniform annual benefit | 25 | 22.25 |
| End-of-useful-life salvage value | 20 | 0 |
| Useful life, in years | 15 | 10 |

At the end of 10 years, Alt. *B* is not replaced. Thus, the comparison is 15 years of *A* versus 10 years of *B*. If the MARR is 10%, which alternative should be selected?

### Spreadsheets

**7-87**  The Southern Guru Copper Company operates a large mine in a South American country. A legislator in the National Assembly said in a speech that most of the capital for the mining operation was provided by loans from the World Bank; in fact, Southern Guru has only $500,000 of its own money actually invested in the property. The cash flow for the mine is:

| Year | Cash Flow |
|---|---|
| 0 | $−0.5M |
| 1 | 3.5M |
| 2 | 0.9M |
| 3 | 3.9M |
| 4 | 8.6M |
| 5 | 4.3M |
| 6 | 3.1M |
| 7 | 6.1M |

The legislator divided the $30.4 million total profit by the $0.5 million investment. This produced, he said, a 6080% rate of return on the investment. Southern Guru, claiming the actual rate of return is much lower, asks you to compute it.

**7-88**  A young engineer's starting salary is $52,000. The engineer expects annual raises of 3%. The engineer will deposit 10% of the annual salary at the end of each year in a savings account that earns 4%. How much will the engineer have saved for starting a business after 15 years? We suggest that the spreadsheet include at least columns for the year, the year's salary, the year's deposit, and the year's cumulative savings.
(*Answer:* $126,348)

**7-89**  A young engineer's starting salary is $55,000. The engineer expects annual raises of 2%. The engineer will deposit 10% of the annual salary at the end of each year in a savings account that earns 5%. How much will the engineer have saved for retirement after 40 years?

**7-90**  A young engineer's starting salary is $55,000. The engineer expects annual raises of 2%. The engineer will deposit a constant percentage of the annual salary at the end of each year in a savings account that earns 5%. What percentage must be saved so that there will be $1 million in savings for retirement after 40 years?
(*Hint*: Use GOAL SEEK: see last section of Chapter 8.)

**7-91**  Find the average starting engineer's salary for your discipline. Find and reference a source for the average annual raise you can expect. If you deposit 10% of your annual salary at the end of each year in a savings account that earns 4%, how much will you have saved for retirement after 40 years?

### Minicases

**7-92**  Some lenders charge an up-front fee on a loan, which is subtracted from what the borrower receives. This is typically described as "points" (where one point equals 1% of the loan amount). The federal government requires that this be accounted for in the APR that discloses the loan's cost.

(*a*) A 5-year auto loan for $18,000 has monthly payments at a 9% nominal annual rate. If the borrower must pay a loan origination fee of 2 points, what is the true effective cost of the loan? What would the APR be?

(*b*) If the car is sold after 2 years and the loan is paid off, what is the effective interest rate and the APR?

(*c*) Graph the effective interest rate as the time to sell the car and pay off the loan varies from 1 to 5 years.

**7-93**  Some lenders charge an up-front fee on a loan, which is added to what the borrower owes. This is typically described as "points" (where one point equals 1% of the loan amount). The federal government requires that this be accounted for in the APR that discloses the loan's cost.

(*a*) A 30-year mortgage for $220,000 has monthly payments at a 6% nominal annual rate. If a borrower's loan origination fee is 3% (3 points) and it is added to the initial balance, what is the true effective cost of the loan? What would the APR be?

(*b*) If the house is sold after 6 years and the loan is paid off, what is the effective interest rate and the APR?

(*c*) Graph the effective interest rate as the time to sell the house and pay off the loan varies from 1 to 15 years.

CASES ⦿

The following cases from *Cases in Engineering Economy 2nd* are suggested as matched with this chapter.

CASE 12  **Buying a Dream**
Compare three alternative mortgages for a first-time home buyer.

CASE 13  **Guaranteed Return**
Risk versus expected return and choice of interest rate.

CASE 15  **Pave the Stockpile Area?**
Sensitivity analysis for IRR.

CASE 17  **A Free Lunch?**
Is proposal too good to be true from two perspectives? Realistic (unordered) statement of facts.

# APPENDIX 7A

# Difficulties in Solving for an Interest Rate

After completing this chapter appendix, students should be able to:

- Describe why some projects' cash flows cannot be solved for a single positive interest rate.
- Identify when multiple roots can occur.
- Evaluate how many potential roots exist for a particular project.
- Use the *modified internal rate of return (MIRR)* methodology in multiple-root cases.

Example 7A-1 illustrates the situation.

## EXAMPLE 7A–1

The Going Aircraft Company has an opportunity to supply a large airplane to Interair, a foreign airline. Interair will pay $19 million when the contract is signed and $10 million one year later. Going estimates its second- and third-year costs at $50 million each. Interair will take delivery of the airplane during Year 4 and agrees to pay $20 million at the end of that year and the $60 million balance at the end of Year 5. Compute the rate of return on this project.

### SOLUTION

The PW of each cash flow can be computed at various interest rates. For example, for Year 2 and $i = 10\%$: PW $= -50(P/F, 10\%, 2) = -50(0.826) = -41.3$.

| Year | Cash Flow | 0% | 10% | 20% | 40% | 50% |
|------|-----------|------|-------|-------|-------|-------|
| 0 | $19 | $19 | $19 | $19 | $19 | $19 |
| 1 | 10 | 10 | 9.1 | 8.3 | 7.1 | 6.7 |
| 2 | −50 | −50 | −41.3 | −34.7 | −25.5 | −22.2 |
| 3 | −50 | −50 | −37.6 | −28.9 | −18.2 | −14.8 |
| 4 | 20 | 20 | 13.7 | +9.6 | 5.2 | 3.9 |
| 5 | 60 | 60 | 37.3 | +24.1 | 11.2 | 7.9 |
| | PW = | $9 | $0.1 | −$2.6 | −$1.2 | −$0.5 |

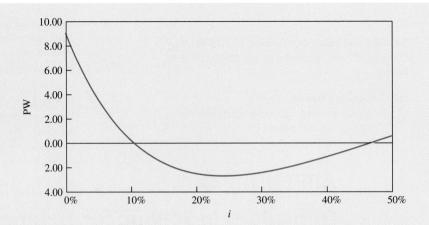

**FIGURE 7A-1** PW plot.

The PW plot for this cash flow is represented in Figure 7A-1. We see that this cash flow produces *two* points at which PW = 0: one at 10.24% and the other at 47.30%.

### Why Multiple Solutions Can Occur

Example 7A-1 produced unexpected results. This happens because there were two changes in the signs of the cash flows. Years 0 and 1 were positive, Years 2 and 3 were negative, and Years 4 and 5 were positive. The cash flow series went from positive cash flows to negative cash flows to positive cash flows.

Most cash flow series have only one change in sign. Investments start with one or more years of negative cash flows followed by many years of positive cash flows. Loans begin with a positive cash flow that is repaid with later negative cash flows. These problems have a unique rate of return because they have a single change in the sign of the cash flows.

Having more than one sign change in the cash flow series can (but might not) produce multiple points or roots at which the PW equals 0. To see how many roots are possible, we link solving an economic analysis problem to solving a mathematical equation.

A project's cash flows are the values from $CF_0$ to $CF_n$.

| Year | Cash Flow |
|------|-----------|
| 0 | $CF_0$ |
| 1 | $CF_1$ |
| 2 | $CF_2$ |
| ⋮ | ⋮ |
| $n$ | $CF_n$ |

The equation to find the internal rate of return, where PW = 0, is written as follows:

$$PW = 0 = CF_0 + CF_1(1+i)^{-1} + CF_2(1+i)^{-2} + \cdots + CF_n(1+i)^{-n} \qquad (7A\text{-}1)$$

If we let $x = (1+i)^{-1}$, then Equation 7A-1 may be written

$$0 = CF_0 + CF_1 x^1 + CF_2 x^2 + \cdots + CF_n x^n \qquad (7A\text{-}2)$$

Equation 7A-2 is an $n$th-order polynomial, and *Descartes' rule* describes the number of positive roots for $x$. The rule is:

If a polynomial with real coefficients has $m$ sign changes, then the number of positive roots will be $m - 2k$, where $k$ is an integer between 0 and $m/2$.

A sign change exists when successive nonzero terms, written according to ascending powers of $x$, have different signs. If $x$ is greater than zero, then the number of sign changes in the cash flows equals the number in the equation. Descartes' rule means that the number of positive roots (values of $x$) of the polynomial cannot exceed $m$, the number of sign changes. The number of positive roots for $x$ must either be equal to $m$ or less by an even integer.

Thus, Descartes' rule for polynomials gives the following:

| Number of Sign Changes, $m$ | Number of Positive Values of $x$ | Number of Positive Values of $i$ |
|:---:|:---:|:---:|
| 0 | 0 | 0 |
| 1 | 1 | 1 or 0 |
| 2 | 2 or 0 | 2, 1, or 0 |
| 3 | 3 or 1 | 3, 2, 1, or 0 |
| 4 | 4, 2, or 0 | 4, 3, 2, 1, or 0 |

If $x$ is greater than 1, the corresponding value of $i$ is negative. If there is only one root and it is negative, then it is a valid IRR. One example of a valid negative IRR comes from a bad outcome, such as the IRR for a failed research and development project. Another example would be a privately sponsored student loan set up so that if a student volunteers for the Peace Corps, only half of the principal need be repaid.

If a project has a negative and a positive root for $i$, then for most projects only the positive root of $i$ is used.

## Projects with Multiple Sign Changes

Example 7A-1 is representative of projects for which there are initial payments (e.g., when the order for the ship, airplane, or building is signed), then the bulk of the costs occur, and then there are more payments on completion.

Projects of other types often have two or more sign changes in their cash flows. Projects with a salvage cost typically have two sign changes. This salvage cost can be large for environmental restoration at termination. Examples include pipelines, open-pit mines, and nuclear power plants. The following cash flow diagram is representative.

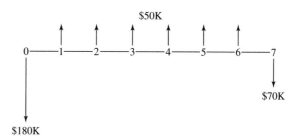

Examples like the cash flow diagram with a first cost, uniform return, and final reme-diation expense have been thoroughly analyzed (Eschenbach, Baker, and Whittaker). Only problems with a very large final cost have double positive roots. More importantly, because the uncertain remediation expenses are much larger than the initial costs, the sensitivity of the results means that the PW values seem also to be unreliable guides for decision making when there are two positive roots for the PW = 0 equation.

Many enhancement projects for existing mines and deposits have a pattern of two sign changes. Example 7A-2 describes an oil well in an existing field. The initial investment recovers more of the resource and speeds recovery of resources that would have been recovered eventually. The resources shifted for earlier recovery can lead to two sign changes.

In Example 7A-3, we consider staged construction, where three sign changes are common.

Other examples can be found in incremental comparisons of alternatives with unequal lives. In the next section we learn how to determine whether each of these examples has multiple roots.

### EXAMPLE 7A–2

Adding an oil well to an existing field costs $4 million (4M). It will increase recovered oil by $3.5M, and it shifts $4.5M worth of production from Years 5, 6, and 7 to earlier years. Thus, the cash flows for Years 1 through 4 total $8M and Years 5 through 7 total −$4.5M. If the well is justified, one reason is that the oil is recovered sooner. How many roots for the PW equation are possible?

### SOLUTION

The first step is to draw the cash flow diagram and count the number of sign changes. The following pattern is representative, although most wells have a longer life.

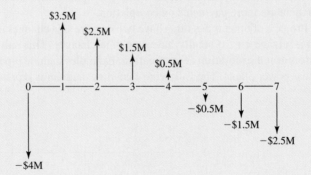

There are two sign changes, thus there may be 0, 1, or 2 positive roots for the PW = 0 equation. The additional recovery corresponds to an investment, and the shifting of recovery to earlier years corresponds to a loan (positive cash flow now and negative later). Thus, the oil wells are neither an investment nor a loan; they are a combination of both.

## EXAMPLE 7A–3

A project has a first cost of $120,000. Net revenues begin at $30,000 in Year 1 and then increase by $2000 per year. In Year 5 the facility is expanded at a cost of $60,000 so that demand can continue to expand at $2000 per year. How many roots for the PW equation are possible?

### SOLUTION

The first step is to draw the cash flow diagram. Then counting the three sign changes is easy.

With three sign changes, there may be 0, 1, 2, or 3 positive roots for the PW = 0 equation.

## Evaluating How Many Roots There Are

The number of sign changes tells us how many roots are possible—not how many roots there are. Rather than covering the many mathematical approaches that *may* tell us if the root will be unique, it is more useful to employ the power of the spreadsheet. A spreadsheet can show us if a root is unique and the value of each root that exists.

The approach is simple. For any set of cash flows, graph the PW as a function of the interest rate. We are interested in positive interest rates, so the graph usually starts at $i = 0$. Since *Descartes' rule* is based on $x > 0$, and $x = (1+i)^{-1}$, sometimes the graph is started close to $i = -1$. This will identify any negative values of $i$ that solve the equation.

If the root is unique, we're done. We've found the internal rate of return. If there are multiple roots, then we know the project's PW at all interest rates—including the one our organization uses. We can also use the approach of the next section to find a *modified internal rate of return*.

The easiest way to build the spreadsheet is to use the NPV($i$,values) function in Excel. Remember that this function applies to cash flows in periods 1 to $n$, so that the cash flow at time 0 must be added in separately. The following examples use a spreadsheet to answer the question of how many roots each cash flow diagram in the last section has.

### EXAMPLE 7A–4

This project is representative of ones with a salvage cost. How many roots for the PW equation exist?

**SOLUTION** Figure 7A-2 shows the spreadsheet calculations and the graph of PW versus $i$.

|    | A    | B         | C | D      | E       | F          | G          |
|----|------|-----------|---|--------|---------|------------|------------|
| 1  | Year | Cash Flow |   | $i$    | PW      |            |            |
| 2  | 0    | −180      |   | −40%   | −126.39 | =$B$2+NPV(D2,$B$3:$B$9) | |
| 3  | 1    | 50        |   | −30%   | 219.99  |            |            |
| 4  | 2    | 50        |   | −20%   | 189.89  |            |            |
| 5  | 3    | 50        |   | −10%   | 114.49  |            |            |
| 6  | 4    | 50        |   | 0%     | 50.00   |            |            |
| 7  | 5    | 50        |   | 10%    | 1.84    |            |            |
| 8  | 6    | 50        |   | 20%    | −33.26  |            |            |
| 9  | 7    | −70       |   | 30%    | −59.02  |            |            |
| 10 |      |           |   | 40%    | −78.24  |            |            |
| 11 | IRR  | 10.45%    |   | 50%    | −92.88  |            |            |
| 12 | root | −38.29%   |   |        |         |            |            |

(Rows 13–22 contain the graph of PW versus $i$, with PW axis from −150.00 to 250.00 and $i$ axis from −40% to 40%.)

**FIGURE 7A–2** PW versus $i$ for project with salvage cost.

In this case, there is one positive root of 10.45%. The value can be used as an IRR. There is also a negative root of $i = −38.29\%$. This root is not useful. With two sign changes, other similar diagrams may have 0, 1, or 2 positive roots for the PW = 0 equation. The larger the value of the final salvage cost, the more likely it is that 0 or 2 positive roots will occur.

## EXAMPLE 7A–5    (7A-2 revisited)

Adding an oil well to an existing field had the following cash flow diagram. How many roots for the PW equation exist and what are they?

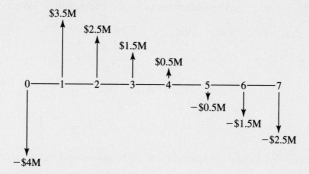

### SOLUTION

Figure 7A-3 shows the spreadsheet calculations and the graph of PW versus $i$.

| | A | B | C | D | E | F | G |
|---|---|---|---|---|---|---|---|
| 1 | Year | Cash Flow | | $i$ | PW | | |
| 2 | 0 | −4.00 | | 0% | −0.50 | =$B$2+NPV(D2,$B$3:$B$9) | |
| 3 | 1 | 3.50 | | 5% | 0.02 | | |
| 4 | 2 | 2.50 | | 10% | 0.28 | | |
| 5 | 3 | 1.50 | | 15% | 0.37 | | |
| 6 | 4 | 0.50 | | 20% | 0.36 | | |
| 7 | 5 | −0.50 | | 25% | 0.29 | | |
| 8 | 6 | −1.50 | | 30% | 0.19 | | |
| 9 | 7 | −2.50 | | 35% | 0.06 | | |
| 10 | | | | 40% | −0.08 | | |
| 11 | root | 4.73% | | 45% | −0.22 | | |
| 12 | root | 37.20% | | 50% | −0.36 | | |
| 13 | | | | | | | |
| 14 | | | | | | | |
| 15 | | | | | | | |
| 16 | | | | | | | |
| 17 | | | | | | | |
| 18 | | | | | | | |
| 19 | | | | | | | |
| 20 | | | | | | | |
| 21 | | | | | | | |
| 22 | | | | | | | |

**FIGURE 7A-3** PW versus $i$ for oil well.

In this case, there are positive roots at 4.73 and 37.20%. These roots are not useful. This project is a combination of an investment and a loan, so we don't even know whether we want a high rate or a low rate. If our interest rate is about 20%, then the project has a positive PW. However, small changes in the data can make for large changes in these results.

It is useful to apply the modified internal rate of return described in the next section.

## EXAMPLE 7A–6    (7A-3 revisited)

A project has a first cost of $120,000. Net revenues begin at $30,000 in Year 1, and then increase by $2000 per year. In Year 5 the facility is expanded at a cost of $60,000 so that demand can continue to expand at $2000 per year. How many roots for the PW equation have a positive value for $i$?

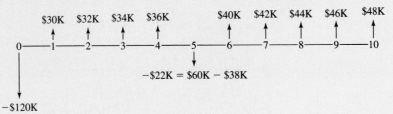

**SOLUTION**  Figure 7A-4 shows the spreadsheet calculations and the graph of PW versus $i$.

|    | A | B | C | D | E | F | G |
|----|------|-----------|---|------|-----------|-------------------------------|---|
| 1  | Year | Cash Flow |   | $i$  | PW        |                               |   |
| 2  | 0    | −120      |   | −80% | 579646330 | =$B$2+NPV(D2,$B$3:$B$12)      |   |
| 3  | 1    | 30        |   | −60% | 735723    |                               |   |
| 4  | 2    | 32        |   | −40% | 17651     |                               |   |
| 5  | 3    | 34        |   | −30% | 1460      |                               |   |
| 6  | 4    | 36        |   | 0%   | 210       |                               |   |
| 7  | 5    | −22       |   | 10%  | 73        |                               |   |
| 8  | 6    | 40        |   | 20%  | 7         |                               |   |
| 9  | 7    | 42        |   | 30%  | −28       |                               |   |
| 10 | 8    | 44        |   | 40%  | −48       |                               |   |
| 11 | 9    | 46        |   | 50%  | −62       |                               |   |
| 12 | 10   | 48        |   | 60%  | −71       |                               |   |
| 13 |      |           |   | 70%  | −78       |                               |   |
| 14 | IRR  | 21.69%    |   | 80%  | −83       |                               |   |
| 15 |      |           |   | 90%  | −87       |                               |   |
| 16 |      |           |   |      |           |                               |   |
| 17 |      |           |   |      |           |                               |   |
| 18 |      |           |   |      |           |                               |   |
| 19 |      |           |   |      |           |                               |   |
| 20 |      |           |   |      |           |                               |   |
| 21 |      |           |   |      |           |                               |   |
| 22 |      |           |   |      |           |                               |   |
| 23 |      |           |   |      |           |                               |   |
| 24 |      |           |   |      |           |                               |   |
| 25 |      |           |   |      |           |                               |   |
| 26 |      |           |   |      |           |                               |   |

**FIGURE 7A-4** PW versus $i$ for project with staged construction.

In this case, there is one positive root of 21.69%. The value can be used as an IRR, and the project is attractive.

When there are two or more sign changes in the cash flow, we know that there are several possibilities concerning the number of positive values of $i$. Probably the greatest danger in this situation is to fail to recognize the multiple possibilities and to solve for a value of $i$. The approach of constructing the PW plot establishes both whether there are multiple roots and what their values are. This may be tedious to do by hand but is very easy with a spreadsheet (or a graphing calculator).

If there is a single positive value of $i$, we have no problem. On the other hand, if there is no positive value of $i$, or if there are multiple positive values, the situation may be attractive, unattractive, or confusing. When there are multiple positive values of $i$, none of them should be considered a suitable measure of the rate of return or attractiveness of the cash flow.

## Modified Internal Rate of Return (MIRR)

Two external rates of return can be used to ensure that the resulting equation is solvable for a unique rate of return—the MIRR. The MIRR is a measure of the attractiveness of the cash flows, but it is also a function of the two external rates of return.

The rates that are *external* to the project's cash flows are (1) the rate at which the organization normally invests and (2) the rate at which it normally borrows. These are external rates for investing, $e_{inv}$, and for financing, $e_{fin}$. Because profitable firms invest at higher rates than they borrow at, the rate for investing is generally higher than the rate for financing. Sometimes a single external rate is used for both, but this requires the questionable assumption that investing and financing happen at the same rate.

The approach is:

1. Combine cash flows in each period ($t$) into a single net receipt, $R_t$, or net expense, $E_t$.
2. Find the present worth of the expenses with the financing rate.
3. Find the future worth of the receipts with the investing rate.
4. Find the MIRR which makes the present and future worths equivalent.

The result is Equation 7A-3. This equation will have a unique root, since it has a single negative present worth and a single positive future worth. There is only one sign change in the resulting series.

$$(F/P, \text{MIRR}, n) \sum_t E_t(P/F, e_{\text{fin}}, t) = \sum_t R_t(F/P, e_{\text{inv}}, n - t) \qquad (7A\text{-}3)$$

There are other external rates of return, but the MIRR has historically been the most clearly defined. All of the external rates of return are affected by the assumed values for the investing and financing rates, so none are a *true* rate of return on the project's cash flow. The MIRR also has an Excel function, so it now can easily be used. Example 7A-7 illustrates the calculation, which is also summarized in Figure 7A-5.

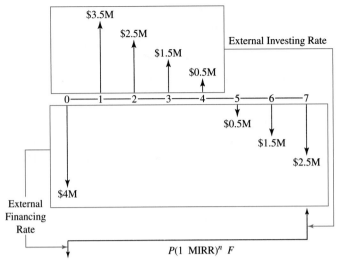

FIGURE 7A-5 MIRR for the oil well.

## EXAMPLE 7A-7 | (7A-2 and 7A-5 revisited)

Adding an oil well to an existing field had the cash flows summarized in Figure 7A-5. If the firm normally borrows money at 8% and invests at 15%, find the modified internal rate of return (MIRR).

### SOLUTION

Figure 7A-6 shows the spreadsheet calculations.

|   | A | B | C | D | E | F | G | H | I |
|---|---|---|---|---|---|---|---|---|---|
| 1 | 8% | external financing rate | | | | | | | |
| 2 | 15% | external investing rate | | | | | | | |
| 3 | Year | 0 | 1 | 2 | 3 | 4 | 5 | 6 | 7 |
| 4 | Cash Flow | −4.00 | 3.50 | 2.50 | 1.50 | 0.50 | −0.50 | −1.50 | −2.50 |
| 5 | 13.64% | | Cell A5 contains = MIRR(A4:I4,A1,A2) | | | | | | |

FIGURE 7A-6 MIRR for oil well.

It is also possible to calculate the MIRR by hand. While more work, the process does clarify what the MIRR function is doing.

1. Each period's cash flow is already a single net receipt or expenditure.
2. Find the present worth of the expenses with the financing rate.

$$PW = -4M - 0.5M(P/F, 8\%, 5) - 1.5M(P/F, 8\%, 6) - 2.5M(P/F, 8\%, 7)$$

$$= -4M - 0.5M(0.6806) - 1.5M(0.6302) - 2.5M(0.5835) = -6.744M$$

3. Find the future worth of the receipts with the investing rate.

$$FW = 3.5M(F/P, 15\%, 6) + 2.5M(F/P, 15\%, 5) + 1.5M(F/P, 15\%, 4) + 0.5M(F/P, 15\%, 3)$$
$$= 3.5M(2.313) + 2.5M(2.011) + 1.5M(1.749) + 0.5M(1.521) = 16.507M$$

4. Find the MIRR that makes the present and future worths equivalent.

$$0 = (1 + MIRR)^n(PW) + FW$$
$$0 = (1 + MIRR)^7(-6.744M) + 16.507M$$
$$(1 + MIRR)^7 = 16.507M/6.744M = 2.448$$
$$(1 + MIRR) = 2.448^{1/7} = 1.1364$$
$$MIRR = 13.64\%$$

The MIRR does allow calculation of a rate of return for *any* set of cash flows. However, the result is only as realistic as the external rates that are used. The MIRR value can depend as much on the external rates that are used, as it does on the cash flows that it is describing.

## SUMMARY

In cash flows with more than one sign change, we find that solving the cash flow equation can result in more than one positive rate of return. Typical situations include a new oil well in an existing field, a project with a significant salvage cost, and staged construction.

In a sign change, successive nonzero values in the cash flow have different signs (that is, they change from $+$ to $-$, or vice versa). Zero sign changes indicates there is no rate of return, as the cash flow is either all disbursements or all receipts.

One sign change is the usual situation, and a single positive rate of return generally results. There will be a negative rate of return whenever loan repayments are less than the loan or an investment fails to return benefits at least equal to the investment.

Multiple sign changes may result in multiple positive roots for $i$. When they occur, none of the positive multiple roots is a suitable measure of the project's economic desirability. If multiple roots are identified by graphing the present worth versus the interest rate, then the modified internal rate of return can be used to evaluate the project.

Graphing the present worth versus the interest rate ensures that the analyst recognizes that the cash flow has multiple sign changes. Otherwise a rate could be found and used that is not in fact a meaningful descriptor of the project.

The modified internal rate of return (MIRR) relies on rates for investing and borrowing that are external to the project. The number of sign changes are reduced to one, ensuring that the MIRR can be found.

# PROBLEMS

Unless the problem asks a different question or provides different data: (1) determine how many roots are possible and (2) graph the PW versus the interest rate to see whether multiple roots occur. If the root is a unique IRR, it is the project's rate of return. If there are multiple roots, then use an *external investing rate* of 12% and an *external borrowing rate* of 6%. Compute and use the MIRR as the project's rate of return.

**7A-1** Find the rate of return for the following cash flow:

| Year | Cash Flow |
|------|-----------|
| 0 | −$15,000 |
| 1 | 10,000 |
| 2 | −8,000 |
| 3 | 11,000 |
| 4 | 13,000 |

(*Answer:* 21.2% IRR)

**7A-2** A group of businessmen formed a partnership to buy and race an Indianapolis-type racing car. They agreed to pay an individual $50,000 for the car and associated equipment. The payment was to be in a lump sum at the end of the year. In what must have been "beginner's luck," the group won a major race the first week and $80,000. The rest of the first year, however, was not as good: at the end of the first year, the group had to pay out $35,000 for expenses plus the $50,000 for the car and equipment. The second year was a poor one: the group had to pay $70,000 just to clear up the racing debts at the end of the year. During the third and fourth years, racing income just equaled costs. When the group was approached by a prospective buyer for the car, they readily accepted $80,000 cash, which was paid at the end of the fourth year. What rate of return did the businessmen obtain from their racing venture? (*Answer:* 9.6% MIRR)

**7A-3** A student organization, at the beginning of the fall quarter, bought and operated a soft-drink vending machine as a means of helping finance its activities. The vending machine cost $75 and was installed at a gasoline station near the university. The student organization pays $75 every 3 months to the station owner for the right to keep the vending machine at the station. During the year the student organization owned the machine, they received the following quarterly income from it, before making the $75 quarterly payment to the station owner:

| Quarter | Income |
|---------|--------|
| Fall | $150 |
| Winter | 25 |
| Spring | 125 |
| Summer | 150 |

At the end of one year, the student group resold the machine for $50. Determine the quarterly cash flow. Then determine a quarterly rate of return, a nominal annual rate, and an effective annual rate.

**7A-4** Given the following cash flow, determine the rate of return on the project.

| Year | Cash Flow |
|------|-----------|
| 0 | −$500 |
| 1 | 2000 |
| 2 | −1200 |
| 3 | −300 |

(*Answer:* 11.3% MIRR)

**7A-5** Given the following cash flow, determine the rate of return on the investment.

| Year | Cash Flow |
|------|-----------|
| 0 | −$500 |
| 1 | 200 |
| 2 | −500 |
| 3 | 1200 |

(*Answer:* IRR = 21.1% IRR)

**7A-6** Given the following cash flow, determine the rate of return on the investment.

| Year | Cash Flow |
|------|-----------|
| 0 | −$100 |
| 1 | 360 |
| 2 | −570 |
| 3 | 360 |

**7A-7** Compute the rate of return on the investment characterized by the following cash flow.

| Year | Cash Flow |
|------|-----------|
| 0 | −$110 |
| 1 | −500 |
| 2 | 300 |
| 3 | −100 |
| 4 | 400 |
| 5 | 500 |

**7A-8** Compute the rate of return on the investment characterized by the following cash flow.

| Year | Cash Flow |
|------|-----------|
| 0 | −$50.0 |
| 1 | 20.0 |
| 2 | −40.0 |
| 3 | 36.8 |
| 4 | 36.8 |
| 5 | 36.8 |

(*Answer:* IRR = 15.4%)

**7A-9** A firm invested $15,000 in a project that appeared to have excellent potential. Unfortunately, a lengthy labor dispute in Year 3 resulted in costs that exceeded benefits by $8000. The cash flow for the project is as follows:

| Year | Cash Flow |
|------|-----------|
| 0 | −$15,000 |
| 1 | 10,000 |
| 2 | 6,000 |
| 3 | −8,000 |
| 4 | 4,000 |
| 5 | 4,000 |
| 6 | 4,000 |

Compute the rate of return for the project.

**7A-10** The following cash flow has no positive interest rate. The project, which had a projected life of 5 years, was terminated early.

| Year | Cash Flow |
|------|-----------|
| 0 | −$50 |
| 1 | 20 |
| 2 | 20 |

There is, however, a negative interest rate. Compute its value.   (*Answer:* −13.7%)

**7A-11** For the following cash flow, compute the rate of return.

| Year | Cash Flow |
|------|-----------|
| 0 | −$20 |
| 1 | 0 |
| 2 | −10 |
| 3 | 20 |
| 4 | −10 |
| 5 | 100 |

**7A-12** Given the following cash flow:

| Year | Cash Flow |
|------|-----------|
| 0 | −$800 |
| 1 | 500 |
| 2 | 500 |
| 3 | −300 |
| 4 | 400 |
| 5 | 275 |

What is the rate of return on the investment? (*Answer:* 26.55% IRR)

**7A-13** Consider the following cash flow.

| Year | Cash Flow |
|------|-----------|
| 0 | −$100 |
| 1 | 240 |
| 2 | −143 |

If the minimum attractive rate of return is 12%, should the project be undertaken?

**7A-14** Refer to the strip-mining project in Example 5-10. Compute the rate of return for the project.

**7A-15** Consider the following cash flow.

| Year | Cash Flow |
|------|-----------|
| 0 | −$500 |
| 1 | 800 |
| 2 | 170 |
| 3 | −550 |

Compute the rate of return on the investment.

**7A-16** Determine the rate of return on the investment for the following cash flow.

| Year | Cash Flow |
|------|-----------|
| 0 | −$100 |
| 1 | 360 |
| 2 | −428 |
| 3 | 168 |

**7A-17** Compute the rate of return on the investment on the following cash flow.

| Year | Cash Flow |
|------|-----------|
| 0 | −$1200 |
| 1 | 358 |
| 2 | 358 |
| 3 | 358 |
| 4 | 358 |
| 5 | 358 |
| 6 | −394 |

**7A-18** Determine the rate of return on the investment on the following cash flow.

| Year | Cash Flow |
|------|-----------|
| 0 | −$3570 |
| 1–3 | 1000 |
| 4 | −3170 |
| 5–8 | 1500 |

**7A-19** Bill bought a vacation lot he saw advertised on television for $800 down and monthly payments of $55. When he visited the lot, he found it was not something he wanted to own. After 40 months he was finally able to sell the lot. The new owner assumed the balance of the loan on the lot and paid Bill $2500. What rate of return did Bill receive on his investment?

**7A-20** Compute the rate of return on an investment having the following cash flow.

| Year | Cash Flow |
|------|-----------|
| 0 | −$ 850 |
| 0 | 600 |
| 2–9 | 200 |
| 10 | −1800 |

**7A-21** Assume that the following cash flows are associated with a project.

| Year | Cash Flow |
|------|-----------|
| 0 | −$16,000 |
| 1 | −8,000 |
| 2 | 11,000 |
| 3 | 13,000 |
| 4 | −7,000 |
| 5 | 8,950 |

Compute the rate of return for this project.

**7A-22** Compute the rate of return for the following cash flow.

| Year | Cash Flow |
|------|-----------|
| 0 | −$200 |
| 1 | 100 |
| 2 | 100 |
| 3 | 100 |
| 4 | −300 |
| 5 | 100 |
| 6 | 200 |

| Year | Cash Flow |
|------|-----------|
| 7 | 200 |
| 8 | −124.5 |

(*Answer:* 20%)

**7A-23** Following are the annual cost data for a tomato press.

| Year | Cash Flow |
|------|-----------|
| 0 | −$210,000 |
| 1 | 88,000 |
| 2 | 68,000 |
| 3 | 62,000 |
| 4 | −31,000 |
| 5 | 30,000 |
| 6 | 55,000 |
| 7 | 65,000 |

What is the rate of return associated with this project?

**7A-24** A project has been in operation for 5 years, yielding the following annual cash flows:

| Year | Cash Flow |
|------|-----------|
| 0 | −$103,000 |
| 1 | 102,700 |
| 2 | −87,000 |
| 3 | 94,500 |
| 4 | −8,300 |
| 5 | 38,500 |

Calculate the rate of return and state whether it has been an acceptable rate of return.

**7A-25** Consider the following situation.

| Year | Cash Flow |
|------|-----------|
| 0 | −$200 |
| 1 | 400 |
| 2 | −100 |

What is the rate of return?

**7A-26** An investor is considering two mutually exclusive projects. She can obtain a 6% before-tax rate of return on external investments, but she requires a minimum attractive rate of return of 7% for these projects. Use a 10-year analysis period to compute the incremental rate of return from investing in Project A rather than Project B.

|  | Project A:<br>**Build Drive-Up<br>Photo Shop** | Project B:<br>**Buy Land<br>in Hawaii** |
|---|---|---|
| Initial capital<br>investment | $58,500 | $ 48,500 |
| Net uniform<br>annual income | 6648 | 0 |
| Salvage value<br>10 years hence | 30,000 | 138,000 |
| Computed rate<br>of return | 8% | 11% |

**7A-27** In January 2003, an investor bought a convertible debenture bond issued by the XLA Corporation. The bond cost $1000 and paid $60 per year interest in annual payments on December 31. Under the convertible feature of the bond, it could be converted into 20 shares of common stock by tendering the bond, together with $400 cash. The next business day after the investor received the December 31, 2005, interest payment, he submitted the bond together with $400 to the XLA Corporation. In return, he received the 20 shares of common stock. The common stock paid no dividends. On December 31, 2007, the investor sold the stock for $1740, terminating his 5-year investment in XLA Corporation. What rate of return did he receive?

**7A-28** A problem often discussed in the engineering economy literature is the "oil-well pump problem." Pump 1 is a small pump; Pump 2 is a larger pump that costs more, will produce slightly more oil, and will produce it more rapidly. If the MARR is 20%, which pump should be selected? Assume that any temporary external investment of money earns 10% per year and that any temporary financing is done at 6%.

| Year | Pump 1<br>($000s) | Pump 2<br>($000s) |
|---|---|---|
| 0 | −$100 | −$110 |
| 1 | 70 | 115 |
| 3 | 70 | 30 |

## CASES

**The following case from *Cases in Engineering Economy* 2$^{nd}$ is suggested as matched with this appendix.**

**CASE 14 Northern Gushers**
Incremental oil production investment with possible double root.

## CHAPTER 8

# CHOOSING THE BEST ALTERNATIVE

## Coauthored with John Whittaker

### The Size of a Pipeline

The U.S. has nearly 200,000 miles of oil pipelines, operating in all 50 states, that transport crude oil and refined liquid petroleum products by using centrifugal pumps to force the fluid through the pipe. When a new pipeline is announced, you may wonder how they decided upon the diameter of pipe to use. Why choose a 20-inch diameter and not a 14- or 30-inch?

Let us say a company needs to transport 70,000 barrels of crude oil a day (bpd) from Cincinnati to Toledo, a distance of 200 miles. The design life of the pipeline is 30 years.

The capital cost of a pipeline is largely dependent upon the weight of steel used. Heavier pipe costs more to buy, haul, handle, place, and weld. The minimum wall thickness is proportional to the diameter because of a requirement to maintain a minimum strength against buckling. The weight, therefore, is approximately proportional to the square of the diameter of the pipe.

Most of the operating costs of a pipeline are related to the electric power consumption required to run the pumps. More flow means more pumps, more power, more pump maintenance, larger electric gear, and costlier operation. Because of the nature of Newtonian fluids, the approximate power consumption for a given flow rate is inversely proportional to the inside pipe diameter to the fifth power. The annual operating costs can be converted to a present value at a given discount rate. The total cost of the pipeline for a given diameter can be found by adding the capital cost to the present value of the 30 years of operating costs. The minimum-cost alternative in this instance might be a pipeline with a 20-inch diameter. ■ ■ ■

*Contributed by John Whittaker, University of Alberta*

1. Sizing and material decisions must be made in all branches of engineering. Should the structure be steel or concrete? Should the transmission be ac or dc? Should the mine be open-pit or underground? Consider a sizing decision from your branch and list the economic and noneconomic factors that influence the decision.

2. Enbridge Energy Partners' Lakehead System, which delivers crude oil across the Midwest via a 1900-mile pipeline system, has been of concern to federal regulators for nearly a decade. Issues related to safety, record keeping, and pipeline operations have resulted in almost 100 citations over that period. The 820,000-gallon oil spill near Marshall, WA in July 2010 was one of the largest pipeline spills in U.S. history and raised concerns related to pipeline corrosion monitoring and spill reporting. What are the ethical, economic, social, environmental, and political issues related to managing the U.S.'s pipeline infrastructure? How do these differ for among the stakeholders?

3. The U.S. government faces a decision regarding approval of the Keystone Pipeline project that will initially deliver 435,000 barrels per day (bpd) of crude oil across the Canadian border (with up to 1.1 million bpd in a few years). What are the environmental, permitting, safety and other aspects of this project that might influence the decision to go forward? What benefits does the project provide the U.S. in terms of energy policy, and what are the implications on reducing other foreign oil imports? Develop a short position statement advising the administration.

## After Completing This Chapter...

*The student should be able to:*

- Use a *graphical technique* to visualize and solve problems involving mutually exclusive choices.
- Define *incremental analysis* and differentiate it from a standard present worth, annual worth, and internal rate of return analyses.
- Use spreadsheets to solve incremental analysis problems.

## INCREMENTAL ANALYSIS

This chapter was once titled *Incremental Analysis,* and it extended the incremental analysis presented in Chapter 7 to multiple alternatives—solely for the rate of return (IRR) measure. The approach relied on a series of numerical comparisons of challengers and defenders.

In this edition we use the more powerful and easier-to-understand approaches of graphical present worth (PW), equivalent uniform annual cost (EUAC), or equivalent uniform annual worth (EUAW) comparisons. These support calculation of incremental rates of return.

The graphical approach has the added benefit of focusing on the difference between alternatives. Often the difference is much smaller than the uncertainty in our estimated data. For ease of grading and instruction, we assume in this text that answers are exact, but in the real world the uncertainty in the data must always be considered.

Engineering design selects one from of a set of feasible alternatives. In engineering economy the words *mutually exclusive* alternatives are often used to emphasize that only one alternative may be implemented. Thus our problem is selecting the best of these mutually exclusive alternatives.

In earlier chapters we did this by maximizing PW, minimizing EUAC, or maximizing EUAW. We do the same here. In Chapter 7 we compared two alternatives incrementally to decide whether the IRR on the increment was acceptable. Any two alternatives can be compared by recognizing that:

[Higher-cost alternative] = [Lower-cost alternative] + [Increment between them]

When there are two alternatives, only a single incremental analysis is required. With more alternatives, a series of comparisons is required. Also, only by doing the analysis step by step can we determine which pairs must be compared. For example, if there are 4 alternatives, then 3 of 6 possible comparisons must be made. For 5 alternatives then 4 of 10 possible comparisons must be made. For $N$ alternatives, $N-1$ comparisons must be made from $N(N-1)/2$ possibilities.

The graphical approach is best implemented with spreadsheets—it provides more information and it is easier to understand and to present to others.

## GRAPHICAL SOLUTIONS

Examples 8-1 and 8-2 illustrate why incremental or graphical analyses for the IRR criterion are required. They show that graphing makes it easy to choose the best alternative. Chapter 4 presented *xy* plots done with spreadsheets, and in Chapter 7 spreadsheets were used to graph the present worth versus the rate of return. In this chapter the present worth of each alternative is one *y* variable for graphs with multiple alternatives (or variables). In Chapter 9, one of the spreadsheet sections will present some of the ways that graphs can be customized for a better appearance.

## EXAMPLE 8-1

The student engineering society is building a snack cart to raise money. Members must decide what capacity the cart should be able to serve. To serve 100 customers per hour costs $10,310, and to serve 150 customers per hour costs $13,400. The 50% increase in capacity is less than 50% of $10,310 because of economies of scale; but the increase in net revenue will be less than 50%, since the cart will not always be serving 150 customers per hour. The estimated net annual income for the lower capacity is $3300, and for the higher capacity it is $4000. After 5 years the cart is expected to have no salvage value. The engineering society is unsure of what interest rate to use to decide on the capacity. Make a recommendation.

### SOLUTION

Since we do not know the interest rate, the easiest way to analyze the problem is to graph the PW of each alternative versus the interest rate, as in Figure 8-1. The Excel function[1]

$$= -\text{cost} + \text{PV(interest rate, life, } -\text{annual benefit)}$$

is used to graph these two equations:

$$PW_{\text{low}} = -\$10,310 + \$3300(P/A, i, 5)$$
$$PW_{\text{high}} = -\$13,400 + \$4000(P/A, i, 5)$$

**FIGURE 8-1** Maximizing PW to choose best alternative.

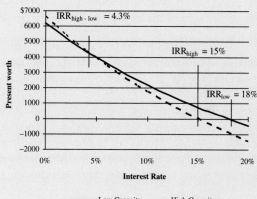

—— Low Capacity   - - - High Capacity

We want to maximize the PW, so the choice between these alternatives is defined by where their curves intersect (the incremental rate of return of 4.3%), not by where those curves intersect the PW = 0 axis. Those internal rates of return of 15 and 18% are irrelevant to the choice between the low- and high-capacity carts.

---

[1]The cost term in the equation has a minus sign because it is a cost and we are calculating a present worth. The annual benefit term has a minus sign because of the sign convention for the PV function.

Figure 8-1 also shows the main advantage of the graphical approach for real-world decision making. We can easily see that for interest rates between 0% and about 8%, there is little difference between the choices. In this region we are better off focusing our decision making on risk, benefits, or costs that we could not quantify, as well as an uncertainties in the data—since the PWs of our two alternatives are nearly the same. These topics are discussed in later chapters. Until then it is clearer if we analyze problems as though all numbers are known precisely.

From Figure 8-1, we see that if the interest rate is below 4.3%, the high-capacity cart has a higher PW. From 4.3 to 18% the lower-capacity cart has a higher PW. Above 18%, the third alternative of doing nothing is a better choice than building a low-capacity cart because the PW < 0. This can be summarized in a choice table:

| Interest Rate | Best Choice |
|---|---|
| 0%   $\leq i \leq 4.3\%$ | High capacity |
| 4.3%   $\leq i \leq 18.0\%$ | Low capacity |
| 18.0% $\leq i$ | Do nothing |

The interest rate where the low- and high-capacity curves intersect can be found by analyzing the *incremental* investment. The following table summarizes the added cost, capacity, and benefit for that incremental investment.

| | Low | High | High – Low |
|---|---|---|---|
| Cost | $10,310 | $13,400 | $3090 |
| Capacity | 100 | 150 | 50 |
| Annual benefit | $3,300 | $4,000 | $700 |
| Life, in years | 5 | 5 | 5 |

If the equation for the PW of the incremental investment is solved for its IRR, the answer is 4.3%. The Excel function is = RATE(life, annual benefit, –first cost) and the TVM calculator function is = $i(n, A, P, F) = i$(life, annual benefit, –first cost, 0).

$$\text{PW}_{\text{high}-\text{low}} = -\$3090 + \$700(P/A, i, 5)$$

This example has been defined in terms of a student engineering society. But very similar problems are faced by a civil engineer sizing the weighing station for trucks carrying fill to a new earth dam, by an industrial engineer designing a package-handling station, and by a mechanical engineer sizing energy conservation equipment.

## EXAMPLE 8–2

Solve Example 7-15 by means of an NPW graph. Two machines are being considered for purchase. If the minimum attractive rate of return (MARR) is 10%, which machine should be bought?

| | Machine X | Machine Y |
|---|---|---|
| Initial cost | $200 | $700 |
| Uniform annual benefit | 95 | 120 |
| End-of-useful-life salvage value | 50 | 150 |
| Useful life, in years | 6 | 12 |

## SOLUTION

Since the useful lives of the two alternatives are different, for an NPW analysis we must adjust them to the same analysis period. If the need seems continuous, then the "replace with an identical machine" assumption is reasonable and a 12-year analysis period can be used. The annual cash flows and the incremental cash flows are as follows:

| End of Year | Machine X | Machine Y | Y − X |
|---|---|---|---|
| | Cash Flows | | |
| 0 | −$200 | −$700 | −$500 |
| 1 | 95 | 120 | 25 |
| 2 | 95 | 120 | 25 |
| 3 | 95 | 120 | 25 |
| 4 | 95 | 120 | 25 |
| 5 | 95 | 120 | 25 |
| 6 | −55 | 120 | 175 |
| 7 | 95 | 120 | 25 |
| 8 | 95 | 120 | 25 |
| 9 | 95 | 120 | 25 |
| 10 | 95 | 120 | 25 |
| 11 | 95 | 120 | 25 |
| 12 | 145 | 270 | 125 |

By means of a spreadsheet program, the NPWs are calculated for a range of interest rates and then plotted on an NPW graph (Figure 8-2).

For a MARR of 10%, Machine $X$ is clearly the superior choice. In fact, as the graph clearly illustrates, Machine $X$ is the correct choice for most values of MARR. The intersection point of the two graphs can be found by calculating $\Delta$IRR, the rate of return on the incremental investment. From the spreadsheet IRR function applied to the incremental cash flows for $Y - X$:

$$\Delta\text{IRR} = 1.3\%$$

FIGURE 8-2 NPW graph.

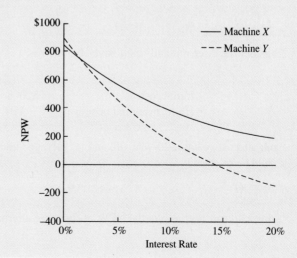

We can see from the graph that for MARR greater than 1.3%, Machine $X$ is the right choice, and for MARR values less than 1.3%, Machine $Y$ is the right choice.

|  | NPW | |
| --- | --- | --- |
| Rate | Machine $X$ | Machine $Y$ |
| 0% | $840.00 | $890.00 |
| 1.322 | 752.24 | 752.24 |
| 2 | 710.89 | 687.31 |
| 4 | 604.26 | 519.90 |
| 6 | 515.57 | 380.61 |
| 8 | 441.26 | 263.90 |
| 10 | 378.56 | 165.44 |
| 12 | 325.30 | 81.83 |
| 14 | 279.77 | 10.37 |
| 16 | 240.58 | −51.08 |
| 18 | 206.65 | −104.23 |
| 20 | 177.10 | −150.47 |

Examples 8-3 through 8-6 increase the number of alternatives being considered, but the same approach is used: graph the PW, EUAW, or EUAC, and for each possible interest rate choose the best alternative. Generally this means maximizing the PW or EUAW, but for Example 8-4 the EUAC is minimized.

Example 8-4 is a typical design problem where the most cost-effective solution must be chosen, but the dollar value of the benefit is not defined. For example, every building must have a roof, but deciding whether it should be metal, shingles, or a built-up membrane is a cost decision. No value is placed on a dry building; it is simply a requirement. In Example 8-4 having the pressure vessel is a requirement.

## EXAMPLE 8–3

Consider the three mutually exclusive alternatives:

|  | A | B | C |
| --- | --- | --- | --- |
| Initial cost | $2000 | $4000 | $5000 |
| Uniform annual benefit | 410 | 639 | 700 |

Each alternative has a 20-year life and no salvage value. If the MARR is 6%, which alternative should be selected?

### SOLUTION

Using a spreadsheet to plot the NPW of the three alternatives produces Figure 8-3.

FIGURE 8–3 NPW graph of Alternatives $A$, $B$, and $C$.

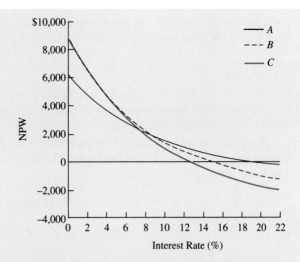

The highest line at any interest rate shows which alternative will provide the maximum NPW at that rate. From the graph we see that Alternative $C$ has the maximum, then $B$, and finally $A$. The precise breakpoints can be found by calculating the $\Delta$IRR of the corresponding increment.

$$\text{NPW}(C - B) = -\$5000 + \$4000 + (\$700 - \$639)(P/A, i, 20) = 0$$

$$\Delta\text{IRR}(C - B) = 2\% \qquad \text{using} = \text{RATE}(20, 700 - 639, 4000 - 5000)$$

$$\text{NPW}(B - A) = -\$4000 + \$2000 + (\$639 - \$410)(P/A, i, 20) = 0$$

$$\Delta\text{IRR}(B - A) = 9.6\% \qquad \text{using} = \text{RATE}(20, 639 - 410, 2000 - 4000)$$

The information from the graph can be presented in a choice table as follows:

| If | | MARR $\geq$ | 9.6% | Choose Alt. $A$ |
|----|------|-------------|------|-----------------|
| If | 9.66% | $\geq$ MARR $\geq$ | 2% | Choose Alt. $B$ |
| If | 2% | $\geq$ MARR $\geq$ | 0% | Choose Alt. $C$ |

and the answer to the original question is select Alternative $B$ if MARR is 6%.

## EXAMPLE 8–4

A pressure vessel can be made out of brass, stainless steel, or titanium. The first cost and expected life for each material are:

|  | Brass | Stainless Steel | Titanium |
|---|---|---|---|
| Cost | $100,000 | $175,000 | $300,000 |
| Life, in years | 4 | 10 | 25 |

The pressure vessel will be in the nonradioactive portion of a nuclear power plant that is expected to have a life of 50 to 75 years. The public utility commission and the power company have not yet agreed on the interest rate to be used for decision making and rate setting. Build a choice table for the interest rates to show where each material is the best.

### SOLUTION

The pressure vessel will be replaced repeatedly during the life of the facility, and each material has a different life. Thus, the best way to compare the materials is using EUAC (see Chapter 6). This assumes identical replacements.

Figure 8-4 graphs the EUAC for each alternative. In this case the best alternative at each interest rate is the material with the *lowest* EUAC. (We maximize worths, but minimize costs.) The Excel function is

$$= \text{PMT}(\text{interest rate}, \text{life}, -\text{first cost})$$

The factor equation is

$$\text{EUAC} = \text{first cost}(A/P, i, \text{life})$$

FIGURE 8–4  EUAC comparison of alternatives.

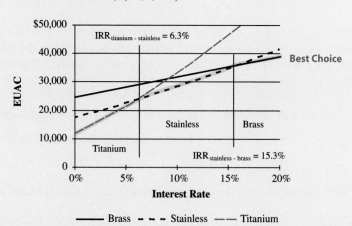

Because these alternatives have different-length lives, calculating the incremental IRRs is best done using the spreadsheet function GOAL SEEK (see the last section of this chapter). The choice table for each material is:

| Interest Rate | Best Choice |
|---|---|
| $0\% \le i \le 6.3\%$ | Titanium |
| $6.3\% \le i \le 15.3\%$ | Stainless steel |
| $15.3\% \le i$ | Brass |

### EXAMPLE 8–5

The following information refers to three mutually exclusive alternatives. The decision maker wishes to choose the right machine but is uncertain what MARR to use. Create a choice table that will help the decision maker to make the correct economic decision.

|  | Machine X | Machine Y | Machine Z |
|---|---|---|---|
| Initial cost | $200 | $700 | $425 |
| Uniform annual benefit | 65 | 110 | 100 |
| Useful life, in years | 6 | 12 | 8 |

## SOLUTION

In this example, the lives of the three alternatives are different. Thus we cannot directly use the present worth criterion but must first make some assumptions about the period of use and the analysis period. As we saw in Chapter 6, when the service period is expected to be continuous and the assumption of identical replacement reasonable, we can assume a series of replacements and compare annual worth values just as we did with present worth values. There is a direct relationship between the present worth and the equivalent uniform annual worth. It is

$$\text{EUAW} = \text{NPW}(A/P, i, n)$$

We will make these assumptions in this instance, and instead of plotting NPW, we shall plot EUAW. This was done on a spreadsheet using the Excel function = benefit + PMT (interest rate, life, initial cost), and the result is Figure 8-5.

The EUAW graph shows that for low values of MARR, $Y$ is the correct choice. Then, as MARR increases, $Z$ and finally $X$ become the preferred machines. This can be expressed in a choice table read directly from the graph. If the "do nothing" alternative is available, the table is the following:

**FIGURE 8–5** EUAW graph.

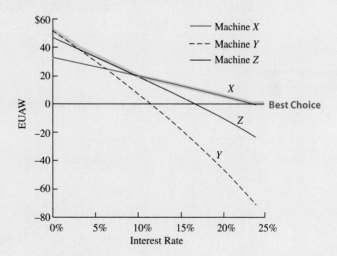

| If | | MARR ≥ 23% | do nothing |
|---|---|---|---|
| If | 23% ≥ MARR ≥ | 10% | choose X |
| If | 10% ≥ MARR ≥ | 3.5% | choose Z |
| If | 3.5% ≥ MARR ≥ | 0% | choose Y |

If the "do nothing" alternative is not available, then the table reads

| | If | MARR ≥ 10% | choose X |
| | If | 10% ≥ MARR ≥ 3.5% | choose Z |
| | If | 3.5% ≥ MARR ≥ 0% | choose Y |

The choice now is back in the decision maker's hands. There is still the need to determine MARR, but if the uncertainty was, for example, that MARR was some value in the range 12 to 18%, it can be seen that for this problem it doesn't matter. The answer is Machine X in any event. If, however, the uncertainty of MARR were in the 7 to 13% range, it is clear the decision maker would have to determine MARR with greater accuracy before solving this problem.

## EXAMPLE 8–6

The following information is for five mutually exclusive alternatives that have 20-year useful lives. The decision maker may choose any one of the options or reject them all. Prepare a choice table.

| | Alternatives | | | | |
|---|---|---|---|---|---|
| | A | B | C | D | E |
| Cost | $4000 | $2000 | $6000 | $1000 | $9000 |
| Uniform annual benefit | 639 | 410 | 761 | 117 | 785 |

## SOLUTION

Figure 8-6 is an NPW graph of the alternatives constructed by means of a spreadsheet.

FIGURE 8-6  NPW graph.

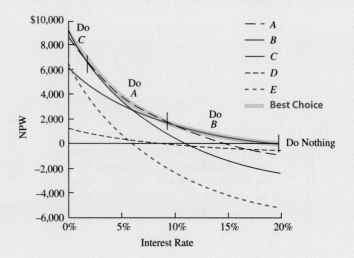

The graph clearly shows that Alternatives $D$ and $E$ are never part of the solution. They are dominated by the other three. The crossover points can either be read from the graph (if you have plotted it at a large enough scale) or found by calculating the $\Delta$IRR of the intersecting graphs. Calculating the incremental interest rates:

$\Delta$IRR$(C - A)$

$\$6000 - \$4000 = (\$761 - \$639)(P/A, i, 20), \quad i = 2\%$

$\Delta$IRR$(A - B)$

$\$4000 - \$2000 = (\$639 - \$410)(P/A, i, 20), \quad i = 9.6\%$

and to find where the NPW of $B$ crosses the 0 axis

IRR$(B)$

$\$2000 = \$410 \quad (P/A, i, 20), \quad i = 20\%$

Placing these numbers in a choice table:

If                  MARR $\geq$ 20%        do nothing
If        20% $\geq$ MARR $\geq$ 9.6%       select $B$
If        9.6% $\geq$ MARR $\geq$  2%       select $A$
If         2% $\geq$ MARR $\geq$  0%       select $C$

A final point to note on this example is that if we view the IRRs of the five alternatives, the only information we can glean from them is that if the "do nothing" alternative is available, it will be chosen if MARR is greater than that of the largest IRR.

|  | A | B | C | D | E |
|---|---|---|---|---|---|
| IRR | 15% | 20% | 11% | 10% | 6% |

There is nothing in the numbers to tell us which alternatives will be in the final solution set; only the NPW graph will show us where the change points will be.[2]

## ELEMENTS IN COMPARING MUTUALLY EXCLUSIVE ALTERNATIVES

1. **Be sure all the alternatives are identified.** In textbook problems the alternatives will be well-defined, but real-life problems may be less clear. Before proceeding, one must have all the mutually exclusive alternatives tabulated, including the "do-nothing" or "keep doing the same thing" alternative, if appropriate.

2. **Construct an NPW or EUAW or EAC graph showing all alternatives plotted on the same axes.** This would be a difficult task were it not for spreadsheets.

---

[2]There are analytical techniques for determining which incremental investments enter the solution set, but they are beyond the scope of this book and, in any case, are redundant in this era of spreadsheets. Interested readers can consult *Economic Analysis for Engineers and Managers* by Sprague and Whittaker (Prentice-Hall, 1987).

3. **Examine the line of best values and determine which alternatives create it, and over what range.**

4. **Determine the changeover points** where the line of best values changes from one alternative to another. These can either be read directly off the graph or calculated, since they are the intersection points of the two curves and, what is more important and meaningful for engineering economy, they are **the $\Delta$IRR of the incremental investment** between the two alternatives.

5. **Create a choice table** to present the information in compact and easily understandable form.

## DOING A SET OF INCREMENTAL CHALLENGER–DEFENDER COMPARISONS

This chapter has focused on doing a graphical comparison of the PW of each alternative over a range of interest rates. We've calculated incremental interest rates, but we've looked at the curves to see which pair of intersecting PW curves we were analyzing. Before spreadsheets, this type of problem was solved by a series of challenger–defender comparisons, where defender $\equiv$ best alternative identified at this stage of the analysis, and challenger $\equiv$ next alternative being evaluated. Example 8-7 illustrates this approach.

The numerically based pairwise incremental comparisons is done at a single interest rate, and it does not show that "near" their intersection, two PW curves have "essentially" the same value. It also does not produce a choice table, since it is at a single interest rate. A choice table can be constructed, but it requires even more incremental comparisons.

### EXAMPLE 8–7

For the alternatives in Example 8-6, conduct a pairwise incremental comparison. Which alternative is the best at an interest rate of 10%? Each of the five alternatives has no salvage value at the end of a 20-year useful life.

| | Alternatives | | | | |
| --- | --- | --- | --- | --- | --- |
| | A | B | C | D | E |
| Cost | $4000 | $2000 | $6000 | $1000 | $9000 |
| Annual benefit | 639 | 410 | 761 | 117 | 785 |

### SOLUTION

1. The first step is to reorder the alternatives in order of increasing cost. This ensures that for each pairwise comparison the increment is an investment. In this example the order becomes $D, B, A, C, E$.

| | Alternatives | | | | |
| --- | --- | --- | --- | --- | --- |
| | D | B | A | C | E |
| Cost | $1000 | $2000 | $4000 | $6000 | $9000 |
| Annual benefit | 117 | 410 | 639 | 761 | 785 |

2. Calculate the IRR of the least expensive alternative to see if it is better than doing nothing at a MARR of 10%.

$$PW_D = 0 = -\$1000 + \$117(P/A, IRR_D, 20)$$

Solving this without a spreadsheet may take several tries and interpolation, but using the Excel function = RATE(life, annual benefit, –first cost) makes it easy and the answer is exact:

$$IRR_D = 9.94\%$$

Because investment $D$ earns less than 10%, doing nothing is preferred to doing $D$.

3. Doing nothing is still our *defender,* and the next Alternative, $B$, becomes the *challenger.* Calculate the IRR of Alternative $B$ to see if it is better than doing nothing at a MARR of 10%.

$$PW_B = 0 = -\$2000 + \$410(P/A, IRR_B, 20)$$

Using the Excel function = RATE(life, annual benefit, –first cost), we find

$$IRR_B = 19.96\%$$

Because investment $B$ earns more than 10%, Alternative $B$ is preferred to doing nothing.

4. Alternative $B$ is now our *defender,* and the next Alternative, $A$, becomes the *challenger.* This comparison must be made incrementally.

$$PW_{A-B} = 0 = -(\$4000 - \$2000) + (\$639 - \$410)(P/A, IRR_{A-B}, 20)$$

Using the Excel function = RATE(life, annual benefit, −first cost), we have

$$IRR_{A-B} = 9.63\%$$

Because the incremental investment $A - B$ earns less than 10%, Alternative $B$ is preferred to doing Alternative $A$.

5. Alternative $B$ is still our *defender,* and the next Alternative, $C$, becomes the *challenger.* This comparison must be made incrementally.

$$PW_{C-B} = 0 = -(\$6000 - \$2000) + (\$761 - \$410)(P/A, IRR_{C-B}, 20)$$

Using the Excel function = RATE(life, annual benefit, −first cost), we have

$$IRR_{C-B} = 6.08\%$$

Because the incremental investment $C - B$ earns less than 10%, Alternative $B$ is preferred to doing Alternative $C$.

6. Alternative $B$ is still our *defender,* and the final Alternative, $E$, becomes the *challenger.* This comparison must be made incrementally.

$$PW_{E-B} = 0 = -(\$9000 - \$2000) + (\$785 - \$410)(P/A, IRR_{E-B}, 20)$$

Using the Excel function = RATE(life, annual benefit, –first cost)

$$IRR_{E-B} = 0.67\%$$

Because the incremental investment $E - B$ earns less than 10%, Alternative $B$ is preferred to doing Alternative $E$.

At an interest rate of 10%, Alternative $B$ is the best choice. But only by referring to Figure 8-6 can we see that Alternative $B$ is the best choice for all interest rates between 9.63% (where Alternative $A$ is equally attractive) and 19.96% (where doing nothing is equally attractive). With other data, the curve for Alternative $B$ might intersect at an incremental comparison we hadn't done. Also this pairwise comparison has not even compared Alternatives $A$ and $C$ so we cannot complete the choice table.

## CHOOSING AN ANALYSIS METHOD

At this point, we have examined in detail the three major economic analysis techniques: present worth analysis, annual cash flow analysis, and rate of return analysis. A practical question is, which method should be used for a particular problem?

While the obvious answer is to use the method requiring the least computations, a number of factors may affect the decision.

1. Unless the MARR—minimum attractive rate of return (or minimum required interest rate for invested money)—is known, neither present worth analysis nor annual cash flow analysis is possible.

2. Present worth analysis and annual cash flow analysis often require far less computation than rate of return analysis.

3. In some situations, a rate of return analysis is easier to explain to people unfamiliar with economic analysis. At other times, an annual cash flow analysis may be easier to explain.

4. Business enterprises generally adopt one, or at most two, analysis techniques for broad categories of problems. If you work for a corporation and the policy manual specifies rate of return analysis, you would appear to have no choice in the matter.

Since one may not always be able to choose the analysis technique computationally best suited to the problem, this book illustrates how to use each of the three methods in all feasible situations. Ironically, the most difficult method to apply when using tabulated factors—rate-of-return analysis—is the one used most frequently by engineers in industry, which is one reason why spreadsheets and TVM calculators are used so much in industry.

## SPREADSHEETS AND GOAL SEEK

An incremental analysis of two alternatives is easily done with the RATE or IRR functions when the lives of the alternatives are the same. The problem is more difficult, however, when the lives are different. As was discussed in Chapters 5 and 6, alternatives having lives of different lengths are usually compared by assuming that the alternatives are repeated until the least common multiple of their lives. This repetition can be done with a spreadsheet, but Excel supports an easier approach.

Excel has a tool called GOAL SEEK that identifies a formula cell, a target value, and a variable cell. This tool causes the variable cell to be changed automatically until the formula cell equals the target value. To find an IRR for an incremental analysis

1. Find alternative $A$'s EUAW or EUAC at its life.
2. Find alternative $B$'s EUAW or EUAC at its life.
3. Compute the difference. This cell is the formula cell for GOAL SEEK.
4. Set the target value to 0.
5. Identify the interest rate cell as the variable cell for GOAL SEEK.

In Excel this tool is accessed by selecting T(ools) on the main toolbar or menu and G(oal seek) on the submenu. As shown in Example 8-8, the variable cell (with the interest rate) must somehow affect the formula cell (difference in equivalent uniform annual worths or costs (EUAWs or EUACs)), although it need not appear directly in the formula cell. In Figures 8-7 and 8-8 the interest rate (cell A1) appears in the EUAC formulas (cells D3 and D4) but not in the formula cell (cell D5).

## EXAMPLE 8–8

Two different asphalt mixes can be used on a highway. The good mix will last 6 years and will cost $600,000 to buy and lay down. The better mix will last 10 years and will cost $800,000 to buy and lay down. Find the incremental IRR for using the more expensive mix.

### SOLUTION

This example would be difficult to solve without the GOAL SEEK tool. The least common multiple of 6 and 10 is 30 years, which is the comparison period. With the GOAL SEEK tool a very simple spreadsheet does the job. In Figure 8-7 the spreadsheet is shown before GOAL SEEK. Figure 8-8 shows the result after the goal (D5) has been set = 0 and A1 has been chosen as the variable cell to change.

|   | A | B | C | D | E | F |
|---|---|---|---|---|---|---|
| 1 | 8.00% | Interest rate | | | | |
| 2 | Alternative | Cost | Life | EUAC | | |
| 3 | Good | 600000 | 6 | 129,789 | $ = -\text{PMT}(\$A\$1,C3,B3)$ | |
| 4 | Better | 800000 | 10 | 119,224 | $ = -\text{PMT}(\$A\$1,C4,B4)$ | |
| 5 | | | difference= | 10,566 | $ = D3 - D4$ | |

**FIGURE 8–7** Spreadsheet before GOAL SEEK.

|   | A | B | C | D | E | F |
|---|---|---|---|---|---|---|
| 1 | 14.52% | Interest rate | | | | |
| 2 | Alternative | Cost | Life | EUAC | | |
| 3 | Good | 600000 | 6 | 156,503 | = −PMT($A$1,C3,B3) | |
| 4 | Better | 800000 | 10 | 156,503 | = −PMT($A$1,C4,B4) | |
| 5 | | | difference= | 0 | = D3−D4 | |

**FIGURE 8–8** Spreadsheet after GOAL SEEK.

The incremental IRR found by GOAL SEEK is 14.52%.

## SUMMARY

For choosing from a set of mutually exclusive alternatives, the rate of return technique is more complex than the present worth or annual cash flow techniques. This results because in the latter two techniques the numbers can be compared directly, whereas with the rate of return it is necessary to consider the *increment of investment*. This is fairly straightforward if there are only two alternatives, but it becomes more and more complex as the number of alternatives increases.

A visual display of the problem can be created by using a spreadsheet to graph the economic value of the alternatives. The steps are as follows:

1. Be sure all the alternatives are identified.
2. Construct a NPW or EUAW (or EUAC) graph showing all alternatives plotted on the same axes.
3. Examine the line of maximum values (or minimum for the EUAC) and determine which alternatives create it, and over what range.
4. Determine the changeover points ($\Delta$IRRs).
5. Create a choice table.

This graphical approach, although more values are calculated, is a more powerful one because, by allowing the decision maker to see the range over which the choices are valid, it provides a form of sensitivity analysis. It also makes it clear that "close" to the changeover point the alternatives are very similar in value.

## PROBLEMS

 These problems are organized such that the (*a*) parts require graphical analysis and as such are much more easily done with spreadsheets, and the (*b*) parts require numerical incremental analysis. Some problems include only one approach.

**Two Action Alternatives**

**8-1** Given the following, construct a choice table for interest rates from 0% to 100%.

| Year | X | Y |
|---|---|---|
| 0 | −$10 | −$20 |
| 1 | 15 | 28 |

**8-2** Consider two mutually exclusive alternatives and the do-nothing alternative:

| Year | Buy X | Buy Y |
|---|---|---|
| 0 | −$100.0 | −$50.0 |
| 1 | 31.5 | 16.5 |
| 2 | 31.5 | 16.5 |
| 3 | 31.5 | 16.5 |
| 4 | 31.5 | 16.5 |

Construct a choice table for interest rates from 0% to 100%.

**8-3** Consider three alternatives *A*, *B*, and "do-nothing." Construct a choice table for interest rates from 0% to 100%.

| Year | A | B |
|------|------|------|
| 0 | −$100 | −$150 |
| 1 | 30 | 43 |
| 2 | 30 | 43 |
| 3 | 30 | 43 |
| 4 | 30 | 43 |
| 5 | 30 | 43 |

**8-4** A firm is considering two alternatives that have no salvage value.

| | A | B |
|------|------|------|
| Initial cost | $10,700 | $5500 |
| Uniform annual benefits | 2,100 | 1800 |
| Useful life, in years | 8 | 4 |

At the end of 4 years, another *B* may be purchased with the same cost, benefits, and so forth.

(*a*) Construct a choice table for interest rates from 0% to 100%.

(*b*) If the MARR is 10%, which alternative should be selected?

**8-5** Your cat's summer kitty-cottage needs a new roof. You are considering the following two proposals and feel a 15-year analysis period is in line with your cat's remaining lives. (There is no salvage value for old roofs.)

| | Thatch | Slate |
|------|------|------|
| First cost | $20 | $40 |
| Annual upkeep | 5 | 2 |
| Service life, in years | 3 | 5 |

(*a*) Construct a choice table for interest rates from 0% to 100%.

(*b*) Which roof should you choose if your MARR is 12%? What is the actual value of the IRR on the incremental cost?

**8-6** Don Garlits is a landscaper. He is considering the purchase of a new commercial lawn mower, either the Atlas or the Zippy. Construct a choice table for interest rates from 0% to 100%.

| | Atlas | Zippy |
|------|------|------|
| Initial cost | $6700 | $16,900 |
| Annual operation and maintenance cost | 1500 | 1,200 |
| Annual benefit | 4000 | 4,500 |
| Salvage value | 1000 | 3,500 |
| Useful life, in years | 3 | 6 |

**8-7** The South End bookstore has an annual profit of $170,000. The owner is considering opening a second bookstore on the north side of the campus. He can lease an existing building for 5 years with an option to continue the lease for a second 5-year period. If he opens the second bookstore, he expects the existing store will lose some business that will he gained by "The North End," the new bookstore. It will take $500,000 of store fixtures and inventory to open The North End. He believes that the two stores will have a combined profit of $260,000 a year after all the expenses of both stores have been paid.

The owner's economic analysis is based on a 5-year period. He will be able to recover this $500,000 investment at the end of 5 years by selling the store fixtures and inventory.

(*a*) Construct a choice table for interest rates from 0% to 100%.

(*b*) The owner will not open The North End unless he can expect a 15% rate of return. What should he do?

**8-8** A paper mill is considering two types of pollution control equipment.

| | Neutralization | Precipitation |
|------|------|------|
| Initial cost | $700,000 | $500,000 |
| Annual chemical cost | 40,000 | 110,000 |
| Salvage value | 175,000 | 125,000 |
| Useful life, in years | 5 | 5 |

(a) Construct a choice table for interest rates from 0% to 100%.

(b) The firm wants a 12% rate of return on any avoidable increments of investment. Which equipment should be purchased?

8-9    A stockbroker has proposed two investments in low-rated corporate bonds paying high interest rates and selling below their stated value (in other words, junk bonds). The bonds are rated as equally risky.

| Bond | Stated Value | Annual Interest Payment | Current Market Price with Commission | Bond Maturity* |
|------|-------|---------|------------|----------|
| Gen Dev | $1000 | $ 94 | $480 | 15 years |
| RJR | 1000 | 140 | 630 | 15 |

*At maturity the bondholder receives the last interest payment plus the bond's stated value.

(a) Construct a choice table for interest rates from 0% to 100%.

(b) Which, if any, of the bonds should you buy if your MARR is 25%?

8-10   Using the current specifications, a new road will cost $1.5M initially, need $120K in annual maintenance, and need to be resurfaced every 10 years for $1.1M. A proposed new specification is expected to be more resistant to wear. The initial cost will be $2.1M with $90K in annual maintenance and resurfacing every 15 years for $1.3M. Use capitalized cost.

(a) Develop a choice table for interest rates from 0% to 25%.

(b) If the highway department's interest rate is 6%, which specification is preferred?

8-11   George is going to replace his car in 3 years when he graduates, but now he needs a radiator repair. The local shop has a used radiator, which will be guaranteed for 2 years, or they can install a new one, which is "guaranteed for as long as you own the car." The used radiator is $250 and the new one is $450. If George assumes the used radiator will last 3 years, but will need to be replaced so he can sell the car, which should he buy?

(a) Develop a choice table for interest rates from 0% to 50%.

(b) George's interest rate on his credit card is 20%. What should he do?

## Multiple Alternatives

8-12   Consider the following alternatives:

|  | A | B | C |
|--|---|---|---|
| Initial cost | $300 | $600 | $200 |
| Uniform annual benefits | 41 | 98 | 35 |

Each alternative has a 10-year useful life and no salvage value.

(a) Construct a choice table for interest rates from 0% to 100%.

(b) If the MARR is 8%, which alternative should be selected?

8-13   The following four mutually exclusive alternatives have no salvage value after 10 years.

|  | A | B | C | D |
|--|---|---|---|---|
| First cost | $75 | $50 | $50 | $85 |
| Uniform annual benefit | 16 | 12 | 10 | 17 |
| Computed rate of return | 16.8% | 20.2% | 15.1% | 15.1% |

(a) Construct a choice table for interest rates from 0% to 100%.

(b) Using 8% for the MARR, which alternative should be selected?   (*Answer: A*)

8-14   The following three mutually exclusive alternatives have no salvage value after 5 years:

|  | A | B | C |
|--|---|---|---|
| First cost | $200 | $300 | $600 |
| Uniform annual benefit | 59.7 | 77.1 | 165.2 |
| Computed rate of return | 15% | 9% | 11.7% |

Construct a choice table for interest rates from 0% to 100%.

8-15   Consider four mutually exclusive alternatives, each having an 8-year useful life:

|  | A | B | C | D |
|--|---|---|---|---|
| First cost | $1000 | $800 | $600 | $500 |
| Uniform annual benefit | 122 | 120 | 97 | 122 |
| Salvage value | 750 | 500 | 500 | 0 |

(a) Construct a choice table for interest rates from 0% to 100%.

(b) If the minimum attractive rate of return is 8%, which alternative should be selected?

**8-16**  Three mutually exclusive projects are being considered:

|  | A | B | C |
|---|---|---|---|
| First cost | $1000 | $2000 | $3000 |
| Uniform annual benefit | 150 | 150 | 0 |
| Salvage value | 1000 | 2700 | 5600 |
| Useful life, in years | 5 | 6 | 7 |

When each project reached the end of its useful life, it would be sold for its salvage value and there would be no replacement.

(a) Construct a choice table for interest rates from 0% to 100%.

(b) If 8% is the desired rate of return, which project should be selected?

**8-17**  Consider three mutually exclusive alternatives and the do-nothing alternative.

|  | A | B | C |
|---|---|---|---|
| Initial cost | $770 | $1406 | $2563 |
| Uniform annual benefit | 420 | 420 | 420 |
| Useful life (years) | 2 | 4 | 8 |
| Rate of return | 6.0% | 7.5% | 6.4% |

The analysis period is 8 years. At the end of 2, 4, and 6 years, Alt. *A* will have an identical replacement. Alternative *B* will have a single identical replacement at the end of 4 years. Construct a choice table for interest rates from 0% to 100%.

**8-18**  Consider the three alternatives:

|  | A | B | C |
|---|---|---|---|
| Initial cost | $1500 | $1000 | $2035 |
| Annual benefit for first 5 years | 250 | 250 | 650 |
| Annual benefit for subsequent 5 years | 450 | 250 | 145 |

Each alternative has a 10-year useful life and no salvage value.
   Construct a choice table for interest rates from 0% to 100%.

**8-19**  Three mutually exclusive alternatives are being considered.

|  | A | B | C |
|---|---|---|---|
| Initial investment | $50,000 | $22,000 | $15,000 |
| Annual net income | 5,093 | 2,077 | 1,643 |
| Rate of return | 8% | 7% | 9% |

Each alternative has a 20-year useful life with no salvage value.

(a) Construct a choice table for interest rates from 0% to 100%.

(b) If the minimum attractive rate of return is 7%, which alternative should be selected?

**8-20**  A business magazine is available for $58 for 1 year, $108 for 2 years, $153 for 3 years, or $230 for 5 years. Assume you will read the magazine for at least the next 5 years. For what interest rates do you prefer each payment plan?

**8-21**  Three office furniture firms that offer different payment plans have responded to a request for bids from a state agency.

|  | Price | Payment Schedule |
|---|---|---|
| OfficeLess | $185,000 | 50% now, 25% in 6 months, 25% in 1 year |
| OfficeMore | 182,000 | 50% now, 50% in 6 months |
| OfficeStation | 198,000 | 100% in 1 year |

(a) Develop a choice table for nominal interest rates from 0% to 50%.

(b) If the agency's MARR is 10%, which vendor's plan is preferred?

**8-22** Consider the following alternatives:

| | A | B | C |
|---|---|---|---|
| Initial cost | $100.00 | $150.00 | $200.00 |
| Uniform annual benefit | 10.00 | 17.62 | 55.48 |
| Useful life (years) | Infinite | 20 | 5 |

Alternatives *B* and *C* are replaced at the end of their useful lives with identical replacements. Use an infinite analysis period.

(*a*) Construct a choice table for interest rates from 0% to 100%.

(*b*) At an 8% interest rate, which alternative is better?

**8-23** Three mutually exclusive alternatives are being studied.

| Year | A | B | C |
|---|---|---|---|
| 0 | −$20,000 | −$20,000 | −$20,000 |
| 1 | 10,000 | 10,000 | 5,000 |
| 2 | 5,000 | 10,000 | 5,000 |
| 3 | 10,000 | 10,000 | 5,000 |
| 4 | 6,000 | 0 | 15,000 |

(*a*) Construct a choice table for interest rates from 0% to 100%.

(*b*) If the MARR is 12%, which alternative should be selected?

**8-24** A firm is considering three mutually exclusive alternatives as part of a production improvement program. The alternatives are as follows:

| | A | B | C |
|---|---|---|---|
| Installed cost | $10,000 | $15,000 | $20,000 |
| Uniform annual benefit | 1,625 | 1,625 | 1,890 |
| Useful life, in years | 10 | 20 | 20 |

For each alternative, the salvage value at the end of useful life is zero. At the end of 10 years, Alt. *A* could be replaced by another *A* with identical cost and benefits.

(*a*) Construct a choice table for interest rates from 0% to 100%.

(*b*) The MARR is 6%. If the analysis period is 20 years, which alternative should be selected?

**8-25** A new 10,000-square-meter warehouse next door to the Tyre Corporation is for sale for $450,000. The terms offered are $100,000 down with the balance being paid in 60 equal monthly payments based on 15% interest. It is estimated that the warehouse would have a resale value of $600,000 at the end of 5 years.

Tyre has the cash and could buy the warehouse but does not need all the warehouse space at this time. The Johnson Company has offered to lease half the new warehouse for $2500 a month.

Tyre presently rents and uses 7000 square meters of warehouse space for $2700 a month. It has the option of reducing the rented space to 2000 square meters, in which case the monthly rent would be $1000 a month. Furthermore, Tyre could cease renting warehouse space entirely. Tom Clay, the Tyre Corp. plant engineer, is considering three alternatives:

1. Buy the new warehouse and lease half the space to the Johnson Company. In turn, the Tyre-rented space would be reduced to 2000 square meters.

2. Buy the new warehouse and cease renting any warehouse space.

3. Continue as is, with 7000 square meters of rented warehouse space.

Construct a choice table for interest rates from 0% to 100%.

**8-26** QZY, Inc. is evaluating new widget machines offered by three companies. The machines have the following characteristics:

| | Company A | Company B | Company C |
|---|---|---|---|
| First cost | $15,000 | $25,000 | $20,000 |
| Maintenance and operating | 1,600 | 400 | 900 |
| Annual benefit | 8,000 | 13,000 | 9,000 |
| Salvage value | 3,000 | 6,000 | 4,500 |
| Useful life, in years | 4 | 4 | 4 |

(*a*) Construct a choice table for interest rates from 0% to 100%.

(*b*) MARR = 15%. From which company, if any, should you buy the widget machine? Use rate of return analysis.

**8-27** The Croc Co. is considering a new milling machine from among three alternatives: each has life of 10 years:

| | Alternative | | |
|---|---|---|---|
| | Deluxe | Regular | Economy |
| First cost | $220,000 | $125,000 | $75,000 |
| Annual benefit | 79,000 | 43,000 | 28,000 |
| Maintenance and operating costs | 38,000 | 13,000 | 8,000 |
| Salvage value | 16,000 | 6,900 | 3,000 |

(a) Construct a choice table for interest rates from 0% to 100%.

(b) MARR = 15%. Using incremental rate of return analysis, which alternative, if any, should the company choose?

8-28  Wayward Airfreight, Inc. has asked you to recommend a new automatic parcel sorter. You have obtained the following bids:

|  | SHIP-R | SORT-Of | U-SORT-M |
|---|---|---|---|
| First cost | $184,000 | $235,000 | $180,000 |
| Salvage value | 38,300 | 44,000 | 14,400 |
| Annual benefit | 75,300 | 89,000 | 68,000 |
| Yearly maintenance and operating cost | 21,000 | 21,000 | 12,000 |
| Useful life, in years | 7 | 7 | 7 |

(a) Construct a choice table for interest rates from 0% to 100%.

(b) Using a MARR of 15% and a rate of return analysis, which alternative, if any, should be selected?

8-29  A firm is considering the following alternatives, as well as a fifth choice: do nothing. Each alternative has a 5-year useful life.

|  | 1 | 2 | 3 | 4 |
|---|---|---|---|---|
| Initial cost | $100,000 | $130,000 | $200,000 | $330,000 |
| Uniform annual net income ($000s) | 26.38 | 38.78 | 47.48 | 91.55 |
| Rate of return | 10% | 15% | 6% | 12% |

(a) Construct a choice table for interest rates from 0% to 100%.

(b) The firm's minimum attractive rate of return is 8%. Which alternative should be selected?

8-30  Construct a choice table for interest rates from 0% to 100%.

|  | Alternatives | | | |
|---|---|---|---|---|
|  | A | B | C | D |
| Initial cost | $2000 | $5000 | $4000 | $3000 |
| Annual benefit | 800 | 500 | 400 | 1300 |
| Salvage value | 2000 | 1500 | 1400 | 3000 |
| Life, in years | 5 | 6 | 7 | 4 |

8-31  In a particular situation, four mutually exclusive alternatives are being considered. Each of the alternatives costs $1300 and has no salvage value after a 10-year life.

| Alternative | Annual Benefit | Rate of Return |
|---|---|---|
| A | $100 at end of first year; increasing $30 per year thereafter | 10.0% |
| B | $10 at end of first year; increasing $50 per year thereafter | 8.8% |
| C | Annual end-of-year benefit = $260 | 15.0% |
| D | $450 at end of first year; declining $50 per year thereafter | 18.1% |

(a) Construct a choice table for interest rates from 0% to 100%.

(b) If the MARR is 8%, which alternative should be selected?    (Answer: Alt. C)

8-32  A more detailed examination of the situation in Problem 8-31 reveals that there are two additional mutually exclusive alternatives to be considered. Both cost more than the $1300 for the four original alternatives. Both have no salvage value after a useful life of 10 years.

| Alternative | Cost | Annual End-of-Years Benefit | Rate of Return |
|---|---|---|---|
| E | $3000 | $ 488 | 10.0% |
| F | 5850 | 1000 | 11.2% |

(a) Construct a choice table for interest rates from 0% to 100%.

(b) If the MARR remains at 8%, which one of the six alternatives should be selected? (Answer: Alt. F)

**8-33** The owner of a downtown parking lot has employed a civil engineering consulting firm to advise him on the economic feasibility of constructing an office building on the site. Betty Samuels, a newly hired civil engineer, has been assigned to make the analysis. She has assembled the following data:

| Alternative | Total Investment* | Total Net Annual Revenue |
|---|---|---|
| Sell parking lot | $ 0 | $ 0 |
| Keep parking lot | 200,000 | 22,000 |
| Build 1-story building | 400,000 | 60,000 |
| Build 2-story building | 555,000 | 72,000 |
| Build 3-story building | 750,000 | 100,000 |
| Build 4-story building | 875,000 | 105,000 |
| Build 5-story building | 1,000,000 | 120,000 |

*Includes the value of the land.

The analysis period is to be 15 years. For all alternatives, the property has an estimated resale (salvage) value at the end of 15 years equal to the present total investment.

(a) Construct a choice table for interest rates from 0% to 100%.

(b) If the MARR is 10%, what recommendation should Betty make?

**8-34** A firm is considering moving its manufacturing plant from Chicago to a new location. The industrial engineering department was asked to identify the various alternatives together with the costs to relocate the plant, and the benefits. The engineers examined six likely sites, together with the do-nothing alternative of keeping the plant at its present location. Their findings are summarized as follows:

| Plant Location | First Cost ($000s) | Uniform Annual Benefit ($000s) |
|---|---|---|
| Denver | $300 | $ 52 |
| Dallas | 550 | 137 |
| San Antonio | 450 | 117 |
| Los Angeles | 750 | 167 |
| Cleveland | 150 | 18 |
| Atlanta | 200 | 49 |
| Chicago | 0 | 0 |

The annual benefits are expected to be constant over the 8-year analysis period.

(a) Construct a choice table for interest rates from 0% to 100%.

(b) If the firm uses 10% annual interest in its economic analysis, where should the manufacturing plant be located?  (*Answer:* Dallas)

**8-35** An oil company plans to purchase a piece of vacant land on the corner of two busy streets for $70,000. On properties of this type, the company installs businesses of four different types.

| Plan | Cost* | Type of Business |
|---|---|---|
| A | $ 75,000 | Conventional gas station with service facilities for lubrication, oil changes, etc. |
| B | 230,000 | Automatic carwash facility with gasoline pump island in front |
| C | 30,000 | Discount gas station (no service bays) |
| D | 130,000 | Gas station with low-cost, quick-carwash facility |

*Improvements cost does not include $70,000 for the land.

In each case, the estimated useful life of the improvements is 15 years. The salvage value for each is estimated to be the $70,000 cost of the land. The net annual income, after paying all operating expenses, is projected as follows:

| Plan | Net Annual Income |
|---|---|
| A | $23,300 |
| B | 44,300 |
| C | 10,000 |
| D | 27,500 |

(a) Construct a choice table for interest rates from 0% to 100%.

(b) If the oil company expects a 10% rate of return on its investments, which plan (if any) should be selected?

**8-36** *The Financial Advisor* is a weekly column in the local newspaper. Assume you must answer the following question. "I recently retired at age 65, and I have a tax-free retirement annuity coming due soon. I have three options. I can receive (A) $30,976 now, (B) $359.60 per month for the rest of my life (assume 20 years), or (C) $513.80 per month for the next 10 years. What should I do?" Ignore the timing of the monthly cash flows and assume that the payments are received at the end of year.
*Contributed by D. P. Loucks, Cornell University*

(a) Develop a choice table for interest rates from 0% to 50%. (You do not know what the reader's interest rate is.)

(b) If $i = 9\%$, use an incremental rate of return analysis to recommend which option should be chosen.

**8-37** A firm must decide which of three alternatives to adopt to expand its capacity. The firm wishes a minimum annual profit of 20% of the initial cost of each separable increment of investment. Any money not invested in capacity expansion can be invested elsewhere for an annual yield of 20% of initial cost.

| Alt. | Initial Cost | Annual Profit | Profit Rate |
|------|-------------|---------------|-------------|
| A | $100,000 | $30,000 | 30% |
| B | 300,000 | 66,000 | 22% |
| C | 500,000 | 80,000 | 16% |

Which alternative should be selected? Use a challenger–defender rate of return analysis.

**8-38** The New England Soap Company is considering adding some processing equipment to the plant to aid in the removal of impurities from some raw materials. By adding the processing equipment, the firm can purchase lower-grade raw material at reduced cost and upgrade it for use in its products.

Four different pieces of processing equipment are being considered:

| | A | B | C | D |
|---|---|---|---|---|
| Initial investment | $10,000 | $18,000 | 25,000 | $30,000 |
| Annual saving in materials costs | 4,000 | 6,000 | 7,500 | 9,000 |
| Annual operating cost | 2,000 | 3,000 | 3,000 | 4,000 |

The company can obtain a 15% annual return on its investment in other projects and is willing to invest money on the processing equipment only as long as it can obtain 15% annual return on each increment of money invested. Which one, if any, of the alternatives should be selected? Use a challenger–defender rate of return analysis.

## Cash vs. Loan vs. Lease

**8-39** Frequently we read in the newspaper that one should lease a car rather than buying it. For a typical 24-month lease on a car costing $9400, the monthly lease charge is about $267. At the end of the 24 months, the car is returned to the lease company (which owns the car). As an alternative, the same car could be bought with no down payment and 24 equal monthly payments, with interest at a 12% nominal annual percentage rate. At the end of 24 months the car is fully paid for. The car would then be worth about half its original cost.

(a) Over what range of nominal before-tax interest rates is leasing the preferred alternative?

(b) What are some of the reasons that would make leasing more desirable than is indicated in (a)?

**8-40** *The Financial Advisor* is a weekly column in the local newspaper. Assume you must answer the following question. "I need a new car that I will keep for 5 years. I have three options. I can (A) pay $19,999 now, (B) make monthly payments for a 6% 5-year loan with 0% down, or (C) make lease payments of $299.00 per month for the next 5 years. The lease option also requires an up-front payment of $1000. What should I do?"

Assume that the number of miles driven matches the assumptions for the lease, and the vehicle's value after 5 years is $6500. Remember that lease payments are made at the beginning of the month, and the salvage value is received only if you own the vehicle.

(a) Develop a choice table for nominal interest rates from 0% to 50%. (You do not know what the reader's interest rate is.)

(b) If $i = 9\%$, use an incremental rate of return analysis to recommend which option should be chosen.

(*Answer:* For nominal interest rates of 0 to 6% pay cash, of 6 to 30.0% use the loan, and of over 30.0% use the lease.)

8-41  *The Financial Advisor* is a weekly column in the local newspaper. Assume you must answer the following question. "I need a new car that I will keep for 5 years. I have three options. I can (*A*) pay $15,999 now, (*B*) make monthly payments for a 9% 5-year loan with 0% down, or (*C*) make lease payments of $269.00 per month for the next 5 years.

The lease option also requires an up-front payment of $500. What should I do?"

Assume that the number of miles driven matches the assumptions for the lease, and the vehicle's value after 5 years is $4500. Remember that lease payments are made at the beginning of the month, and the salvage value is received only if you own the vehicle.

(*a*)  Develop a choice table for nominal interest rates from 0% to 50%. (You do not know what the reader's interest rate is.)

(*b*)  If *i* = 9%, use an incremental rate of return analysis to recommend which option should be chosen.

8-42  Assume that the vehicle described in Problem 8-40 will be kept for 10 rather than 5 years. If leased, the vehicle can be purchased for its value at the end of 5 years. At the end of 10 years, the vehicle will be worth $2000.

(*a*)  Develop a choice table for nominal interest rates from 0% to 50%. (You do not know what the reader's interest rate is.)

(*b*)  If *i* = 9%, use an incremental rate of return analysis to recommend which option should be chosen.

(*c*)  Are your answers different than in Problem 8-40? Why or why not?

8-43  Contact a car dealer and choose a car to evaluate a buy-versus-lease decision (keep it reasonable—no Lamborghinis). Tell the people at the dealership that you are a student working on an assignment. Be truthful and don't argue; if they don't want to help you, leave and find a friendlier dealer. For both buying and leasing, show all assumptions, costs, and calculations. Do not include the cost of maintenance, gasoline, oil, water, fluids, and other routine expenses in your calculations.

Determine the car's sales price (no need to negotiate) and the costs for sales tax, license, and fees. Estimate the "Blue Book value" in 5 years. Determine the monthly payment based on a 5-year loan at 9% interest. Assume that your down payment is large enough to cover only the sales tax, license, and fees. Calculate the equivalent uniform monthly cost of owning the car.

Identify the costs to lease the car (if available assume a 5-year lease). This includes the monthly lease payment, required down payments, and any return fees that are required. Calculate the equivalent uniform monthly cost of leasing the car.

The salesperson probably does not have the answers to many of these questions. Write a one- to two-page memo detailing the costs. Make a recommendation: Should you own or lease your car? Include nonfinancial items and potential financial items in your conclusions, such as driving habits and whether you are likely to drive more than the allowed number of miles dictated in the lease.

*Contributed by Neal Lewis, University of Bridgeport*

**Minicases**

8-44  Develop the costs and benefits to compare owning a car versus depending on public transit, friends, and/or a bicycle. Place a monetary value on each advantage or disadvantage. Develop a choice table for interest rates between 0% and 25%.

8-45  Develop the costs and benefits to compare owning a new car with one that is 2 years old. Place a monetary value on each advantage or disadvantage. Develop a choice table for interest rates between 0% and 25%.

**8-46**   For a vehicle that you or a friend owns, determine the number of miles driven per year. Find three alternative sets of 4 tires that differ in their tread warranty. Assume that the life of the tires equals the tread warranty divided by the number of miles driven per year. Compare the EUACs of the tires.

(*a*) For what interest rates is each choice the best?

(*b*) Develop a graph equivalent to Figure 8-4 to illustrate the results.

(*c*) For the interest rate that is in the "middle" of your range, how low and how high can the number of miles each year be without changing the best choice?

# CASES

**The following cases from *Cases in Engineering Economy* $2^{nd}$ are suggested as matched with this chapter.**

CASE 6   **Lease a Lot**

Compares leasing and ownership. Results show importance of separating financing and investing decisions.

CASE 10   **The Cutting Edge**

Make vs. buy and machine selection.

CASE 11   **Harbor Delivery Service**

Annual comparison of diesel versus gasoline engines.

# OTHER ANALYSIS TECHNIQUES

## Clean and Green

The U.S. Green Building Council (USGBC) reported in 2010 that in the U.S. buildings account for 72% of electricity consumption, 39% of energy use, 38% of $CO_2$ emissions, 40% of raw materials use, 30% of waste output, and 14% of potable water consumption. To address this huge impact on the environment, designers and engineers have made great progress in developing environment-friendly construction materials and building techniques. In addition, "green" buildings offer numerous advantages. They can improve worker productivity, reduce health and safety costs, improve the indoor environmental quality, and reduce energy and maintenance costs.

Some still believe that green structures cost more to build, especially with respect to first costs. However, growing evidence is dispelling this myth, and what has been learned is that location and final finishing play a bigger role in cost differentiation of buildings. In fact, in many areas of the country the engineering, procurement, and construction groups have embraced sustainable design and building practices and no longer see "green" as an added cost to be priced in their bids. To many, "green design" is simply "good design." What remains a concern is the high cost of documentation, in particular in the pursuit of LEED (Leadership in Energy and Environmental Design) certification. Although LEED is considered to be the nationally accepted benchmark for design, construction, and operations of high-performance green buildings, many managers believe that the application process is too complicated and requires too much paperwork. Although construction costs have risen between 25% and 30% in the past three years, LEED certification is still being pursued, and the USGBC reports that over 4.2 billion square feet of commercial building space is involved with the LEED green building certification system.

McGraw-Hill Construction (2009) reports that the green market was 2% of nonresidential construction starts in 2005 and 10–12% in 2008 and is expected to reach 20–25% in 2013. Expansion is seen in many sectors, including education, government, industrial, office, health care, hospitality, and retail.

Despite this mounting evidence in the support of building green, some commercial developers believe that environment-friendly features add to the expense of construction. And even with small additional costs in the range of 1–2% they can make it harder for the builder to recoup its investment and break even on the project in the short term. Furthermore, there is a general reluctance by many in this industry to be innovative and embrace new ideas or new technologies.

Renters may like the idea of being cleaner and greener—but they may be unwilling to pay extra for it unless they can see a tangible impact to their bottom line. ▪ ▪ ▪
*Contributed by Kim LaScola Needy, University of Arkanas*

## QUESTIONS TO CONSIDER

1. Office leases frequently require building owners, rather than tenants, to pay heating and cooling costs. What effect might this have on the decision making of potential tenants who are considering renting space in green buildings?

2. The green building movement has had more success among developers who hold onto their buildings for years and rent them out, rather than selling them as soon as they are constructed. What factors might influence their views?

3. Many environment-friendly buildings are architecturally distinctive and feature better-quality materials and workmanship than traditional commercial structures. Advocates for the environment hope these characteristics will help green buildings attract a rent "premium." How might these features make the buildings more attractive to tenants? Is it the green features or the higher-quality materials and workmanship that add significantly to the perceived higher first costs of green buildings?

4. How can the costs and benefits of green buildings be economically validated by an independent party so that designers and engineers can make a fair assessment?

5. Does the economic attractiveness of green buildings depend on which measure is used for evaluation?

6. What ethical questions arise from state or municipal regulations intended to promote or require green building practices?

### After Completing This Chapter...

*You should be able to:*

- Use future worth, benefit–cost ratio, payback period, and sensitivity analysis methods to solve engineering economy problems.

- Link the use of *future worth* analysis to the present worth and annual worth methods developed earlier.
- Mathematically develop the *benefit–cost ratio,* and use this model to select alternatives and make economic choices.
- Understand the concept of the *payback period* of an investment, and be able to calculate this quantity for prospective projects.
- Demonstrate a basic understanding of *sensitivity* and *breakeven analyses* and the use of these tools in an engineering economic analysis.
- Use a spreadsheet to perform *sensitivity* and *breakeven analyses.*

Chapter 9 examines four topics:

- Future worth analysis
- Benefit–cost ratio or present worth index analysis
- Payback period
- Sensitivity, breakeven, and what-if analysis

Future worth analysis is very much like present worth analysis, dealing with *then* (future worth) rather than with *now* (present worth) situations.

Previously, we have written economic analysis relationships based on either:

$$\text{PW of cost} = \text{PW of benefit} \quad \text{or} \quad \text{EUAC} = \text{EUAB}$$

Instead of writing it in this form, we could define these relationships as

$$\frac{\text{PW of benefit}}{\text{PW of cost}} = 1 \quad \text{or} \quad \frac{\text{EUAB}}{\text{EUAC}} = 1$$

When economic analysis is based on these ratios, the calculations are called benefit–cost ratio analysis. The PW ratio is also known as a present worth index.

Payback period is an approximate analysis technique, generally defined as the time required for cumulative benefits to equal cumulative costs.

Sensitivity describes how much a problem element must change to reverse a particular decision. Closely related is breakeven analysis, which determines the conditions under which two alternatives are equivalent. What-if analysis changes one or all variables to see how the economic value and recommended decision change. Thus, breakeven and what-if analysis are forms of sensitivity analysis.

## FUTURE WORTH ANALYSIS

In present worth analysis, alternatives are compared in terms of their present consequences. In annual cash flow analysis, the comparison was in terms of equivalent uniform annual costs (or benefits). But the concept of resolving alternatives into comparable units is not restricted to a present or annual comparison. The comparison may be made at any point in time. In many situations we would like to know what the *future* situation will be, if we take some particular course of action *now*. This is called **future worth analysis.**

### EXAMPLE 9–1

Ron Jamison, a 20-year-old college student, smokes about a carton of cigarettes a week. He wonders how much money he could accumulate by age 65 if he quit smoking now and put his cigarette money into a savings account. Cigarettes cost $35 per carton. Ron expects that a savings account would earn 5% interest, compounded semiannually. Compute the future worth of Ron's savings at age 65.

**SOLUTION**

$$\text{Semiannual saving} = (\$35/\text{carton})(26 \text{ weeks}) = \$910$$
$$\text{Future worth (FW)} = A(F/A, 2^1/2\%, 90) = 910(329.2) = \$299,572$$

## EXAMPLE 9–2

An East Coast firm has decided to establish a second plant in Kansas City. There is a factory for sale for $850,000 that could be remodeled and used. As an alternative, the firm could buy vacant land for $85,000 and have a new plant constructed there. Either way, it will be 3 years before the firm will be able to get a plant into production. The timing of costs for the factory are:

| Year | Construct New Plant | | Remodel Available Factory | |
|------|--------------------|----|--------------------------|----|
| 0 | Buy land | $ 85,000 | Purchase factory | $ 850,000 |
| 1 | Design | 200,000 | Design | 250,000 |
| 2 | Construction | 1,200,000 | Remodeling | 250,000 |
| 3 | Production equipment | 200,000 | Production equipment | 250,000 |

If interest is 8%, which alternative has the lower equivalent cost when the firm begins production at the end of Year 3?

**SOLUTION**

**New Plant**

$$\text{FW of cost} = 85,000(F/P, 8\%, 3) + 200,000(F/A, 8\%, 3)$$
$$+ 1,000,000(F/P, 8\%, 1) = \$1,836,000$$

**Remodel Available Factory**

$$\text{FW of cost} = 850,000(F/P, 8\%, 3) + 250,000(F/A, 8\%, 3)$$
$$= \$1,882,000$$

The total cost of remodeling the available factory ($1,600,000) is smaller than the total cost of a new plant ($1,685,000). However, the timing of the expenditures is better with the new plant. The new plant is projected to have the smaller future worth of cost and thus is the preferred alternative.

## BENEFIT–COST RATIO ANALYSIS

At a given minimum attractive rate of return (MARR), we would consider an alternative acceptable, provided that

$$\text{PW of benefits} - \text{PW of costs} \geq 0 \qquad \text{or} \qquad \text{EUAB} - \text{EUAC} \geq 0$$

These could also be stated as a ratio of benefits to costs, or

$$\text{Benefit–cost ratio } \frac{B}{C} = \frac{\text{PW of benefit}}{\text{PW of costs}} = \frac{\text{EUAB}}{\text{EUAC}} \geq 1$$

Rather than using present worth or annual cash flow analysis to solve problems, we can base the calculations on the benefit–cost ratio, B/C. The criteria are presented in Table 9-1. In Table 9-1 the two special cases where maximizing the B/C ratio is correct are listed below the more common situation where incremental analysis is required. In Chapter 16 we will detail how this measure is applied in the public sector. Its use there is so pervasive, that the term *present worth index* is sometimes used to distinguish private-sector applications.

**TABLE 9–1  Benefit–Cost Ratio Analysis**

| | Situation | Criterion |
|---|---|---|
| Neither input nor output fixed | Neither amount of money or other inputs nor amount of benefits or other outputs are fixed | *Two alternatives:* Compute incremental benefit–cost ratio ($\Delta B/\Delta C$) on the increment of investment between the alternatives. If $\Delta B/\Delta C \geq 1$, choose higher-cost alternative; otherwise, choose lower-cost alternative. *Three or more alternatives:* Solve by benefit–cost ratio incremental analysis |
| Fixed input | Amount of money or other input resources are fixed | Maximize B/C |
| Fixed output | Fixed task, benefit, or other output to be accomplished | Maximize B/C |

### EXAMPLE 9–3

A firm is trying to decide which of two devices to install to reduce costs. Both devices have useful lives of 5 years and no salvage value. Device $A$ costs $1000 and can be expected to result in $300 savings annually. Device $B$ costs $1350 and will provide cost savings of $300 the first year, but savings will increase by $50 annually, making the second-year savings $350, the third-year savings $400, and so forth. With interest at 7%, which device should the firm purchase?

### SOLUTION

We have used three types of analysis thus far to solve this problem: present worth in Example 5-1, annual cash flow in Example 6-5, and rate of return in Example 7-14. First we correctly analyze this incrementally, then we look at each device's benefit–cost ratio.

**Incremental $B-A$**

$$\text{PW of cost} = \$350$$

$$\text{PW of benefits} = 50(P/G, 7\%, 5)$$

$$= 50(7.647) = \$382$$

$$\frac{B}{C} = \frac{\text{PW of benefit}}{\text{PW of costs}} = \frac{382}{350} = 1.09$$

The increment is justified at the MARR of 7%. Device $B$ should be purchased.

**Device $A$**

$$\text{PW of cost} = \$1000$$

$$\text{PW of benefits} = 300(P/A, 7\%, 5)$$

$$= 300(4.100) = \$1230$$

$$\frac{B}{C} = \frac{\text{PW of benefit}}{\text{PW of costs}} = \frac{1230}{1000} = 1.23$$

**Device $B$**

$$\text{PW of cost} = \$1350$$

$$\text{PW of benefit} = 300(P/A, 7\%, 5) + 50(P/G, 7\%, 5)$$

$$= 300(4.100) + 50(7.647) = 1230 + 382 = 1612$$

$$\frac{B}{C} = \frac{\text{PW of benefit}}{\text{PW of costs}} = \frac{1612}{1350} = 1.19$$

Maximizing the benefit–cost ratio indicates the wrong choice, Device $A$. Incremental analysis must be used.

### EXAMPLE 9–4

In Example 7-15 we analyzed two machines that were being considered for purchase. Assuming 10% interest, which machine should be bought?

|  | Machine X | Machine Y |
|---|---|---|
| Initial cost | $200 | $700 |
| Uniform, annual benefit | 95 | 120 |
| End-of-useful-life salvage value | 50 | 150 |
| Useful life, in years | 6 | 12 |

### SOLUTION

Assuming a 12-year analysis period, the cash flow table is:

| Year | Machine X | Machine Y |
|---|---|---|
| 0 | −$200 | −$700 |
| 1–5 | 95 | 120 |
| 6 | { 95<br>−200<br>50 | 120 |
| 7–11 | 95 | 120 |
| 12 | { 95<br>50 | 120<br>150 |

We will solve the problem using

$$\frac{B}{C} = \frac{EUAB}{EUAC}$$

and considering the salvage value of the machines to be reductions in cost, rather than increases in benefits. This choice affects the ratio value, but not the decision.

**Machine X**

$$EUAC = 200(A/P, 10\%, 6) - 50(A/F, 10\%, 6)$$
$$= 200(0.2296) - 50(0.1296) = 46 - 6 = \$40$$
$$EUAB = \$95$$

Note that this assumes the replacement for the last 6 years has identical costs. Under these circumstances, the EUAC for the first 6 years equals the EUAC for all 12 years.

**Machine Y**

$$EUAC = 700(A/P, 10\%, 12) - 150(A/F, 10\%, 12)$$
$$= 700(0.1468) - 150(0.0468) = 103 - 7 = \$96$$
$$EUAB = \$120$$

**Machine Y − Machine X**

$$\frac{\Delta B}{\Delta C} = \frac{120 - 95}{96 - 40} = \frac{25}{56} = 0.45$$

The incremental benefit–cost ratio of less than 1 represents an undesirable increment of investment. We therefore choose the lower-cost alternative—Machine $X$. If we had computed benefit–cost ratios for each machine, they would have been:

| Machine $X$ | Machine $Y$ |
|---|---|
| $\dfrac{B}{C} = \dfrac{95}{40} = 2.38$ | $\dfrac{B}{C} = \dfrac{120}{96} = 1.25$ |

Although B/C $= 1.25$ for Machine $Y$ (the higher-cost alternative), we must not use this fact as the basis for selecting the more expensive alternative. It only indicates that $Y$ would be acceptable if $X$ were unavailable. The incremental benefit–cost ratio, $\Delta$B/$\Delta$C, clearly shows that $Y$ is a less desirable alternative than $X$. Also, we must not jump to the conclusion that the best alternative is always the one with the largest B/C ratio. This, too, may lead to incorrect decisions—as we saw in Example 9-3, and we shall see again when we examine problems with three or more alternatives.

## EXAMPLE 9–5

Consider the five mutually exclusive alternatives from Examples 8-6 and 8-7 plus an additional alternative, $F$. They have 20-year useful lives and no salvage value. If the minimum attractive rate of return is 6%, which alternative should be selected?

|  | A | B | C | D | E | F |
|---|---|---|---|---|---|---|
| Cost | $4000 | $2000 | $6000 | $1000 | $9000 | $10,000 |
| PW of benefit | 7330 | 4700 | 8730 | 1340 | 9000 | 9,500 |
| $\dfrac{B}{C} = \dfrac{\text{PW of benefits}}{\text{PW of cost}}$ | 1.83 | 2.35 | 1.46 | 1.34 | 1.00 | 0.95 |

### SOLUTION

Incremental analysis is needed to solve the problem. The steps in the solution are the same as the ones presented in Example 8-7 for incremental rate of return, except here the criterion is $\Delta$B/$\Delta$C, and the cutoff is 1, rather than $\Delta$IRR with a cutoff of MARR.

1. Be sure all the alternatives are identified.
2. (Optional) Compute the B/C ratio for each alternative. Since there are alternatives for which B/C $\geq 1$, we will discard any with B/C $< 1$. Discard Alt. $F$.
3. Arrange the remaining alternatives in ascending order of investment.

|  | D | B | A | C | E |
|---|---|---|---|---|---|
| Cost ($=$ PW of cost) | $1000 | $2000 | $4000 | $6000 | $9000 |
| PW of benefits | 1340 | 4700 | 7330 | 8730 | 9000 |
| B/C | 1.34 | 2.35 | 1.83 | 1.46 | 1.00 |

|  | B − D Increment | A − B Increment | C − A Increment |
|---|---|---|---|
| $\Delta$Cost | $1000 | $2000 | $2000 |
| $\Delta$ Benefits | 3360 | 2630 | 1400 |
| $\Delta$B/C | 3.36 | 1.32 | 0.70 |

4. For each increment of investment, if $\Delta$B/$\Delta$C $\geq 1$ the increment is attractive. If $\Delta$B/$\Delta$C $< 1$ the increment of investment is not desirable. The increment $B - D$ is desirable, so $B$ is preferred to $D$. The increment $A - B$ is desirable. Thus, Alt. $A$ is preferred. Increment $C - A$ is not attractive since $\Delta$B/$\Delta$C $= 0.70$. Now we compare $A$ and $E$:

|  | E − A<br>Increment |
|---|---|
| $\Delta$Cost | $5000 |
| $\Delta$Benefit | 1670 |
| $\Delta$B/$\Delta$C | 0.33 |

The increment is undesirable. We choose Alt. *A* as the best of the six alternatives. [*Note:* The best alternative does not have the highest B/C ratio, nor is it the largest project with a B/C ratio > 1. Alternative *A* does have the largest difference between the PW of its benefits and costs (= $3330).]

Benefit–cost ratio analysis may be graphically represented. Figure 9-1 is a graph of Example 9-5. We see that *F* has a B/C < 1 and can be discarded. Alternative *D* is the starting point for examining the separable increments of investment. The slope of line *B–D* indicates a $\Delta$B/$\Delta$C ratio of >1. This is also true for line *A–B*. Increment *C–A* has a slope much flatter than B/C = 1, indicating an undesirable increment of investment. Alternative *C* is therefore discarded and *A* retained. Increment *E–A* is similarly unattractive. Alternative *A* is therefore the best of the six alternatives.

Note two additional things about Figure 9-1: first, even if alternatives with B/C ratio < 1 had not been initially excluded, they would have been systematically eliminated in the incremental analysis. Second, Alt. *B* had the highest B/C ratio (B/C = 2.35), but it is not the best of the six alternatives. We saw the same situation in rate of return analysis of three or more alternatives. The reason is the same in both analysis situations. We seek to maximize the *total* profit, not the profit rate.

FIGURE 9-1 Benefit–cost ratio graph of Example 9-5.

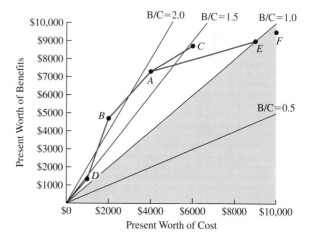

## Variations on the Theme of the Benefit–Cost Ratio

Two of the most common applications of the benefit–cost ratio typically use modified versions of the ratio. The basic ratio has been defined as placing *all* benefits in the numerator of the ratio and *all* costs in the denominator. Example 9-4 modified this slightly by considering the salvage values as reducing the costs rather than as increasing the benefits.

In the public sector, it is common to define the benefit–cost ratio so that the numerator includes all consequences to the users or the public and the denominator includes all consequences to the sponsor or government. For example, the numerator might include the positive benefits of improved highway traffic flow and the disbenefits of congestion during construction, since both accrue to the public or users. The denominator in this case includes consequences to the government, such as the costs of construction and the reduced maintenance cost for the new highway.

Example 9-6 illustrates this for a public project. Then in Example 9-7 exactly the same numbers are put in a private context. Here the benefit–cost ratio is typically called a present worth index. The calculation is modified so that the denominator is the project's first cost, and all other consequences are placed in the numerator. This formulation of the benefit–cost ratio emphasizes the "bang for buck" of how much return is gained for each dollar of investment.

We will examine the public-sector application of the benefit–cost ratio in more detail in Chapter 16, and the present worth index will be used in Chapter 15. Here and in those chapters, the same standard applies for all versions of the ratio. Is the ratio $\geq 1$? More importantly, if one version of the ratio is $\geq 1$, then all versions are $\geq 1$. As shown in Examples 9-6 and 9-7, the values of the ratios may differ, but whether they are above or below 1 and the recommended decisions do not change.

## EXAMPLE 9–6

Traffic congestion on Riverview Boulevard has reached a point where something must be done. Two suggested plans have a life of 15 years, because that is the scheduled time for completion of the new Skyway Highway. After that time traffic will fall well below current levels.

Adding right-turn lanes at key intersections will cost $8.9M (million) with annual maintenance costs for signals and lane painting of $150,000. Added congestion during construction is a disbenefit of $900,000, but the reduced congestion after construction is an annual benefit of $1.6M. This benefit actually starts lower and increases over time, but for a simple initial analysis we are assuming a uniform annual benefit.

Adding a second left-turn lane at a few key intersections will cost an additional $3M with an added annual maintenance cost of $75,000. This construction is more disruptive, and the total disbenefit for congestion during construction is $2.1M. Upon completion, the total benefit for reduced congestion will be $2.2M annually.

Which alternatives is preferred if the interest rate is 10%? Analyze using a government B/C ratio (public in numerator and government in denominator).

### SOLUTION

Since something *must* be done and we have only two identified alternatives, we could simply analyze the difference between the alternatives to see which is better. But we are going to start by analyzing the less expensive "right-turns" alternative to check that this is a reasonable choice for what *must* be done.

The user consequences include an annual benefit for reduced congestion and a first "cost" that is the disbenefit of increased congestion during construction.

$$\text{PW}_{\text{right turns}} = -900,000 + 1,600,000(P/A, 10\%, 15)$$
$$= -900,000 + 1,600,000(7.606) = \$11.27\text{M}$$

The government costs include a first cost for construction and annual maintenance costs. Note that these are calculated as present *costs*.

$$\text{PC}_{\text{right turns}} = 8,900,000 + 150,000(P/A, 10\%, 15)$$
$$= 8,900,000 + 150,000(7.606) = \$10.04\text{M}$$

The benefit–cost ratio for public divided by government consequences is

$$\text{B/C ratio} = \$11.27\text{M}/\$10.04\text{M} = 1.122$$

Thus, the right-turns-only alternative is better than doing nothing.

Now we evaluate the incremental investment for also doing the left-turn improvements. Because we are using a benefit–cost ratio, this evaluation must be done incrementally. The user consequences include an incremental annual benefit for reduced congestion and an incremental first "cost" that is the disbenefit of increased congestion during construction.

$$\text{PW}_{\text{left turns} - \text{right turns}} = -1,200,000 + 600,000(P/A, 10\%, 15)$$
$$= -1,200,000 + 600,000(7.606) = \$3.364\text{M}$$

The government costs include a first cost for construction and annual maintenance costs.

$$\text{PC}_{\text{left turns} - \text{right turns}} = 3,000,000 + 75,000(P/A, 10\%, 15)$$
$$= 3,000,000 + 75,000(7.606) = \$3.570\text{M}$$

The benefit–cost ratio for public divided by government consequences is

$$\text{B/C ratio} = \$3.364\text{M}/\$3.570\text{M} = 0.942$$

Thus, the right-turns-only alternative is better than adding the left-turn increment.

## EXAMPLE 9–7

The industrial engineering department of Amalgamated Widgets is considering two alternatives for improving material flow in its factory. Both plans have a life of 15 years, because that is the estimated remaining life for the factory.

A minimal reconfiguration will cost $8.9M (million) with annual maintenance costs of $150,000. During construction there is a cost of $900,000 for extra material movements and overtime, but more efficient movement of materials will save $1.6M annually. The cost savings actually start lower and increase over time, but for a simple initial analysis we are assuming a uniform annual cost savings.

Reconfiguring a second part of the plant will cost an additional $3M with an added annual maintenance cost of $75,000. This construction is more disruptive, and the total cost for material movement and overtime congestion during construction is $2.1M. Once complete, the total cost savings for more efficient movement of materials is $2.2M annually.

Which alternative is preferred if the interest rate is 10%? Analyze using a present worth index (all consequences in Years 1 to $n$ in numerator and all first costs in denominator).

### SOLUTION

Since something *must* be done and we have only two identified alternatives, we could simply analyze the difference between the alternatives to see which is better. But we are going to start by analyzing the less expensive minimal reconfiguration alternative to check that this is a reasonable choice for what *must* be done.

The consequences in Years 1 to $n$ include an annual cost savings for more efficient flow and annual maintenance costs.

$$PW_{\text{Years 1 to } n} = (1,600,000 - 150,000)(P/A, 10\%, 15)$$
$$= (1,600,000 - 150,000)(7.606) = \$11.03M$$

The first costs include a first cost for construction and the cost for disruption during construction.

$$PC = 8,900,000 + 900,000 = \$9.8M$$

The present worth index is

$$PW \text{ index} = \$11.03M/\$9.8M = 1.125$$

Thus, the minimal reconfiguration is better than doing nothing.

Now we evaluate the incremental investment for also reconfiguring the second part of the plant. Because we are using a present worth index, this evaluation must be done incrementally. The annual consequences include an incremental annual cost savings and incremental maintenance costs.

$$PW_{\text{Years 1 to } n} = (600,000 - 75,000)(P/A, 10\%, 15)$$
$$= 525,000(7.606) = \$3.993M$$

There is a first cost for construction and for the associated disruption.

$$PC = 3,000,000 + 1,200,000 = \$4.2M$$

The present worth index is

$$PW \text{ index} = \$3.993M/\$4.2M = 0.951$$

Thus, the minimal reconfiguration is better than reconfiguring the second part of the plant.

In Examples 9-6 and 9-7, the numbers that appeared in the numerator and denominator were changed, and the exact values of the B/C ratio and present worth index also changed. However, the conclusions did not. The ratios were above 1.0 for the minimal investment choice. The ratios were below 1.0 for the incremental investment. It was always best to make the minimal investment.

These examples demonstrate that present worth analysis and incremental benefit–cost ratio analysis lead to the same optimal decision. We saw in Chapter 8 that rate of return and present worth analysis led to identical decisions. Any of the exact analysis methods—present worth, annual cash flow, rate of return, or benefit–cost ratio—will lead to the same decision. Benefit–cost ratio analysis is extensively used in economic analysis at all levels of government.

## PAYBACK PERIOD

Payback period is the period of time required for the profit or other benefits from an investment to equal the cost of the investment. This is the general definition for payback period. Other definitions consider depreciation of the investment, interest, and income taxes; they, too, are simply called "payback period." We will limit our discussion to the simplest form.

> **Payback period** is the period of time required for the project's profit or other benefits to equal the project's cost.

The criterion in all situations is to minimize the payback period. The computation of payback period is illustrated in Examples 9-8 and 9-9.

### EXAMPLE 9–8

The cash flows for two alternatives are as follows:

| Year | A | B |
|------|--------|--------|
| 0 | −$1000 | −$2783 |
| 1 | 200 | 1200 |
| 2 | 200 | 1200 |
| 3 | 1200 | 1200 |
| 4 | 1200 | 1200 |
| 5 | 1200 | 1200 |

You may assume the benefits occur throughout the year rather than just at the end of the year. Based on payback period, which alternative should be selected?

### SOLUTION

**Alternative A**

Payback period is how long it takes for the profit or other benefits to equal the cost of the investment. In the first 2 years, only $400 of the $1000 cost is recovered. The remaining $600 cost is recovered in the first half of Year 3. Thus the payback period for Alt. A is 2.5 years.

**Alternative B**

Since the annual benefits are uniform, the payback period is simply

$$\$2783/\$1200 \text{ per year} = 2.3 \text{ years}$$

To minimize the payback period, choose Alt. B.

**EXAMPLE 9–9**    **(Example 5-4 revisited)**

A firm is trying to decide which of two weighing scales it should install to check a package-filling operation in the plant. If both scales have a 6-year life, which one should be selected? Assume an 8% interest rate.

| Alternative | Cost | Uniform Annual Benefit | End-of-Useful-Life Salvage Value |
|---|---|---|---|
| Atlas scale | $2000 | $450 | $100 |
| Tom Thumb scale | 3000 | 600 | 700 |

**SOLUTION**

**Atlas Scale**

$$\text{Payback period} = \frac{\text{Cost}}{\text{Uniform annual benefit}}$$

$$= \frac{2000}{450} = 4.4 \text{ years}$$

**Tom Thumb Scale**

$$\text{Payback period} = \frac{\text{Cost}}{\text{Uniform annual benefit}}$$

$$= \frac{3000}{600} = 5 \text{ years}$$

FIGURE 9–2  Payback period plots for Example 9-9.

Figure 9-2 illustrates the situation. To minimize payback period, select the Atlas scale.

There are four important points to be understood about payback period calculations:

1. This is an approximate, rather than an exact, economic analysis calculation.
2. All costs and all profits, or savings of the investment before payback are included *without* considering differences in their timing.
3. All the economic consequences beyond the payback period are completely ignored.
4. Being an approximate calculation, payback period may or may not select the correct alternative.

This last point—that payback period may select the *wrong* alternative—was illustrated by Example 9-9. When payback period is used, the Atlas scale appears to be the more attractive alternative. Yet, when the same problem was solved earlier by the present worth method (Example 5-4), the Tom Thumb scale was the chosen alternative. The reason for the different conclusions is the $700 salvage value at the end of 6 years. The salvage value occurs after the payback period; so it was ignored in the payback calculation. It *was* considered in the present worth analysis, which correctly showed that the Tom Thumb scale was in fact more desirable.

But if payback period calculations are approximate and may lead to selecting the wrong alternative, why are they used? First, the calculations can be readily made by people unfamiliar with economic analysis. Second, payback period is easily understood. Earlier we pointed out that this is also an advantage to rate of return.

Moreover, payback period *does* measure how long it will take for the cost of the investment to be recovered from its benefits. Firms are often very interested in this time period: a rapid return of invested capital means that the funds can be reused sooner for other purposes. But one must not confuse the *speed* of the return of the investment, as measured by the payback period, with economic *efficiency*. They are two distinctly separate concepts. The former emphasizes the quickness with which invested funds return to a firm; the latter considers the overall profitability of the investment.

Example 9-10 illustrates how using the payback period criterion may result in an unwise decision.

## EXAMPLE 9–10

A firm is buying production equipment for a new plant. Two alternative machines are being considered for a particular operation.

|  | Tempo Machine | Dura Machine |
|---|---|---|
| Installed cost | $30,000 | $35,000 |
| Net annual benefit after all annual expenses have been deducted | 12,000 the first year, *declining* $3000 per year thereafter | 1000 the first year, increasing $3000 per year thereafter |
| Useful life, in years | 4 | 8 |

Neither machine has any salvage value. Compute the payback period for each machine.

## SOLUTION BASED ON PAYBACK PERIOD

FIGURE 9–3 Payback period plots for Example 9-10.

**Tempo machine**

**Dura machine**

The Tempo machine has a declining annual benefit, while the Dura has an increasing annual benefit. Figure 9-3 shows the Tempo has a 4-year payback period and the Dura has a 5-year payback period. To minimize the payback period, the Tempo is selected.

Now, as a check on the payback period analysis, compute the rate of return for each alternative. Assume the minimum attractive rate of return is 10%.

## SOLUTION BASED ON RATE OF RETURN

The cash flows for the two alternatives are as follows:

| Year | Tempo Machine | Dura Machine |
|------|---------------|--------------|
| 0 | −$30,000 | −$35,000 |
| 1 | 12,000 | 1,000 |
| 2 | 9,000 | 4,000 |
| 3 | 6,000 | 7,000 |
| 4 | 3,000 | 10,000 |
| 5 | 0 | 13,000 |
| 6 | 0 | 16,000 |
| 7 | 0 | 19,000 |
| 8 | 0 | 22,000 |
| | 0 | 57,000 |

**Tempo Machine**

Since the sum of the cash flows for the Tempo machine is zero, we see immediately that the $30,000 investment just equals the subsequent benefits. The resulting rate of return is 0%.

**Dura Machine**

$$\text{PW}(i) = -35,000 + 1000(P/A, i, 8) + 3000(P/G, i, 8)$$

$$\text{PW}(20\%) = -35,000 + 1000(3.837) + 3000(9.883) = -\$1514$$

The 20% interest rate has discounted future benefits too much; it is too high. Try $i = 15\%$:

$$PW(15\%) = -35{,}000 + 1000(4.487) + 3000(12.481) = \$6930$$

This time, the interest rate is too low. Linear interpolation would show that the rate of return is approximately 19%; and a spreadsheet gives the exact answer.

Using an exact calculation—rate of return—it is clear that the Dura machine is far superior to the Tempo. On the other hand, the shorter payback period for the Tempo does measure the speed of the return of the investment. The conclusion to be drawn is that **liquidity** and **profitability** may be two quite different criteria.

From the discussion and the examples, we see that payback period can measure the speed of the return of the investment. This might be quite important, for example, for a company that is short of working capital or for a firm in an industry experiencing rapid changes in technology. Calculation of payback period alone, however, must not be confused with a careful economic analysis. Ignoring all cash flows after the payback period is seldom wise. We have shown that a short payback period does not always mean that the associated investment is desirable. Thus, payback period is not a suitable replacement for accurate economic analysis calculations.

## SENSITIVITY AND BREAKEVEN ANALYSIS

Since many data gathered in solving a problem represent *projections* of future consequences, there may be considerable uncertainty regarding the data's accuracy. Since the goal is to make good decisions, an appropriate question is: To what extent do variations in the data affect my decision? When small variations in a particular estimate would change which alternative is selected, the decision is said to be **sensitive to the estimate.** To better evaluate the impact of any particular estimate, we compute "how much a particular estimate would need to change in order to change a particular decision." This is called **sensitivity analysis.**

An analysis of the sensitivity of a problem's decision to its various parameters highlights the important aspects of that problem. For example, estimated annual maintenance and salvage values may vary substantially. Sensitivity analysis might indicate that a certain decision is insensitive to the salvage-value estimate over the full range of possible values. But, at the same time, we might find that the decision is sensitive to changes in the annual maintenance estimate. Under these circumstances, one should place greater emphasis on improving the annual maintenance estimate and less on the salvage-value estimate.

As indicated at the beginning of this chapter, breakeven analysis is a form of sensitivity analysis that is often presented as a **breakeven chart**. Another nomenclature that is sometimes used for the breakeven point is *point of indifference*. One application of these tools is **staged construction.** Should a facility be constructed now to meet its future full-scale requirement? Or should it be constructed in stages as the need for the increased capacity arises? What is the breakeven point on how soon the capacity is needed for this decision? Three examples are:

- Should we install a cable with 400 circuits now or a 200-circuit cable now and another 200-circuit cable later?

- A 10-cm water main is needed to serve a new area of homes. Should it be installed now, or should a 15-cm main be installed to ensure an adequate water supply to adjoining areas later, when other homes have been built?
- An industrial firm needs a new warehouse now and estimates that it will need to double its size in 4 years. The firm could have a warehouse built now and later enlarged, or have the warehouse with capacity for expanded operations built right away.

Examples 9-11 through 9-13 illustrate sensitivity and breakeven analysis.

## EXAMPLE 9-11

Consider a project that may be constructed to full capacity now or may be constructed in two stages.

| Construction Alternative | Costs |
|---|---|
| Two-stage construction | |
| Construct first stage now | $100,000 |
| Construct second stage | 120,000 |
| *n* years from now | |
| Full-capacity construction | |
| Construct full capacity now | 140,000 |

**Other Factors**

1. All facilities will last for 40 years regardless of when they are installed; after 40 years, they will have zero salvage value.
2. The annual cost of operation and maintenance is the same for both two-stage construction and full-capacity construction.
3. Assume an 8% interest rate.

Plot "age when second stage is constructed" versus "costs for both alternatives." Mark the breakeven point on your graph. What is the sensitivity of the decision to second-stage construction 16 or more years in the future?

### SOLUTION

Since we are dealing with a common analysis period, the calculations may be either annual cost or present worth. Present worth calculations appear simpler and are used here.

**Construct Full Capacity Now**

$$\text{PW of cost} = \$140,000$$

**Two-Stage Construction**

If the first stage is to be constructed now and the second stage *n* years hence, compute the PW of cost for several values of *n* (years).

$$\text{PW of cost} = 100{,}000 + 120{,}000(P/F, 8\%, n)$$

$$n = 5 \qquad \text{PW} = 100{,}000 + 120{,}000(0.6806) = \$181{,}700$$

$$n = 10 \qquad \text{PW} = 100{,}000 + 120{,}000(0.4632) = \phantom{0}155{,}600$$

$$n = 20 \qquad \text{PW} = 100{,}000 + 120{,}000(0.2145) = \phantom{0}125{,}700$$

$$n = 30 \qquad \text{PW} = 100{,}000 + 120{,}000(0.0994) = \phantom{0}111{,}900$$

These data are plotted in the form of a breakeven chart in Figure 9-4.

**FIGURE 9-4** Breakeven chart for Example 9-11.

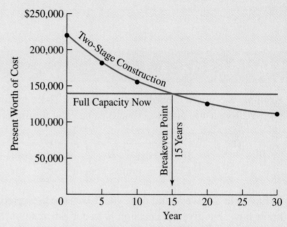

In Figure 9-4 we see that the PW of cost for two-stage construction naturally decreases as the second stage is deferred. The one-stage construction (full capacity now) is unaffected by the $x$-axis variable and, hence, is a horizontal line.

The breakeven point on the graph is where both alternatives have the same PW. We see that if the second stage is deferred for 15 years, then the PW of cost of two-stage construction is equal to one-stage construction; Year 15 is the breakeven point. The graph also shows that if the second stage were to be needed prior to Year 15, then one-stage construction, with its smaller PW of cost, would be preferred. On the other hand, if the second stage would not be required until after 15 years, two-stage construction is preferred.

This breakeven point can also be calculated by setting the two alternatives equal to each other.

$$\text{PW} = 140{,}000 = 100{,}000 + 120{,}000(P/F, 8\%, n)$$

$$(P/F, 8\%, n) = \frac{40{,}000}{120{,}000} = 0.3333$$

From the tables

$$n = 14 + (15 - 14)(0.3405 - 0.3333)/(0.3405 - 0.3152)$$

$$n = 14.3 \text{ years}$$

From Excel's NPER function or GOAL SEEK, or using a TVM calculator,

$$n = n(i, A, P, F) = n(8\%, 0, -40000, 120000) = 14.27 \text{ years}$$

The decision on how to construct the project is sensitive to the age at which the second stage is needed *only* if the range of estimates includes 15 years. For example, if one estimated that the second-stage capacity would be needed between 5 and 10 years hence, the decision is insensitive to that estimate. For any value within that range, the decision does not change. But, if the second-stage capacity were to be needed sometime between, say, 12 and 18 years, the decision would be sensitive to the estimate of when the full capacity would be needed.

One question posed by this example is *how* sensitive the decision is to the need for the second stage at 16 years or beyond. The graph shows that the decision is insensitive. In all cases for construction on or after 16 years, two-stage construction has a lower PW of cost.

---

## EXAMPLE 9–12

Example 8-3 posed the following situation. Three mutually exclusive alternatives are given, each with a 20-year life and no salvage value. The minimum attractive rate of return is 6%.

|                        | *A*    | *B*    | *C*    |
|------------------------|--------|--------|--------|
| Initial cost           | $2000  | $4000  | $5000  |
| Uniform annual benefit | 410    | 639    | 700    |

In Example 8-3 we found that Alt. *B* was the preferred alternative at 6%. Here we would like to know how sensitive the decision is to our estimate of the initial cost of *B*. If *B* is preferred at an initial cost of $4000, it will continue to be preferred at any smaller initial cost. But *how much* higher than $4000 can the initial cost be and still have *B* the preferred alternative? With neither input nor output fixed, maximizing net present worth is a suitable criterion.

**Alternative *A***

$$\text{NPW}_A = 410(P/A, 6\%, 20) - 2000$$
$$= 410(11.470) - 2000 = \$2703$$

**Alternative *B***

Let $x =$ initial cost of *B*.

$$\text{NPW}_B = 639(P/A, 6\%, 20) - x$$
$$= 639(11.470) - x$$
$$= 7329 - x$$

**Alternative *C***

$$\text{NPW}_C = 700(P/A, 6\%, 20) - 5000$$
$$= 700(11.470) - 5000 = \$3029$$

For the three alternatives, we see that *B* will only maximize NPW as long as its NPW is greater than 3029.

$$3029 = 7329 - x$$
$$x = 7329 - 3029 = \$4300$$

Therefore, *B* is the preferred alternative if its initial cost does not exceed $4300.

Figure 9-5 is a breakeven chart for the three alternatives. Here the criterion is to maximize NPW; as a result, the graph shows that $B$ is preferred if its initial cost is less than $4300. At an initial cost above $4300, $C$ is preferred. We have a breakeven point at $4300. When $B$ has an initial cost of $4300, $B$ and $C$ are equally desirable.

**FIGURE 9-5** Breakeven chart for Example 9-12.

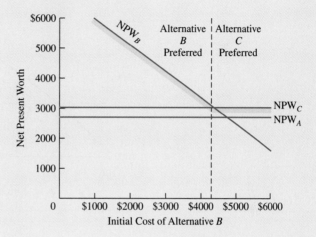

**EXAMPLE 9–13**

In both Examples 9-6 (traffic congestion on Riverview Boulevard) and 9-7 (reconfiguring the plant of Amalgamated Widgets), the life of 15 years is clearly subject to some uncertainty. While holding the other data constant, analyze the sensitivity of the recommended decisions to the project life. Use the present worth measure, since it is the same for both examples.

**SOLUTION**

The two alternatives have the following present worth values.

$$PW_{\text{right turns or minimal}} = -900,000 - 8,900,000 + (1,600,000 - 150,000)(P/A, 10\%, n)$$
$$= -9,800,000 + 1,450,000(P/A, 10\%, n)$$
$$PW_{\text{left turns or 2nd part of plant}} = -2,100,000 - 8,900,000 - 3,000,000$$
$$+ (2,200,000 - 225,000)(P/A, 10\%, n)$$
$$= -14,000,000 + 1,975,000(P/A, 10\%, n)$$

These could be analyzed for breakeven values of $n$. However, it is easier to use the graphing technique for multiple alternatives that was presented in Chapter 8. Instead of using the interest rate for the $x$ axis, use $n$.

As shown in Figure 9-6, the right-turn or minimal alternative is the best one for lives of 12 to 16 years. The left-turn increment or 2nd part of the plant, is the best choice for lives of 17 or more years. If the life is 11 years or less, doing nothing is better. To keep the graph readable, Figure 9-6 includes only Years 10 through 20.

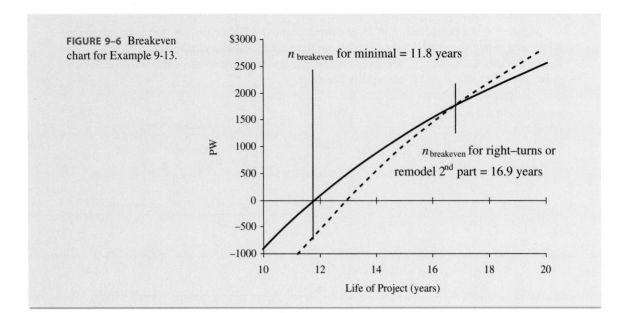

**FIGURE 9–6** Breakeven chart for Example 9-13.

Breakeven points can be estimated from graphs or calculated with formulas or GOAL SEEK. Sensitivity analysis and breakeven point calculations can be very useful in identifying how different estimates affect the calculations. It must be recognized that these calculations assume that all parameters except one are held constant, and the sensitivity of the decision to that one variable is evaluated. The next section presents ways to modify your Excel chart to make it more effective.

## GRAPHING WITH SPREADSHEETS FOR SENSITIVITY AND BREAKEVEN ANALYSIS

Chapter 4 introduced drawing $xy$ plots with spreadsheets, and Chapter 7 reviewed this procedure for plotting present worth versus $i$. The Chapter 7 plot (Figure 7-6) is an example of breakeven analysis because it is used to determine at what interest rate does the project break even or have a present worth of 0. This section will present some of the spreadsheet tools and options that can make the $xy$ plots more effective and attractive.

The spreadsheet tools and options can be used to:

- Modify the $x$ or $y$ axes

    Specify the minimum or maximum value

    Specify at what value the other axis intersects (default is 0)

- Match line types to data

    Use line types to distinguish one curve from another

    Use markers to show real data

    Use lines without markers to plot curves (straight segments or smooth curves)

- Match chart colors to how displayed

    Color defaults are fine for color computer screen

    Color defaults are OK for color printers

    Black-and-white printing is better with editing (use line types not colors)

- Annotate the graph

    Add text, arrows, and lines to graphs

    Add data labels

In most cases the menus of Excel are self-explanatory, so the main step is deciding what you want to achieve. Then you just look for the way to do it. Left clicks are used to select the item to modify, and right clicks are used to bring up the options for that item. Example 9-14 illustrates this process.

## EXAMPLE 9–14

The staged construction choice described in Example 9-11 used a broad range of $x$ values for the $x$ axis. Create a graph that focuses on the 10- to 20-year period and is designed for printing in a report. The costs are:

| Year | Full Capacity | Two Stages |
|------|---------------|------------|
| 0    | $140,000      | $100,000   |
| $n$  | 0             | 120,000    |

## SOLUTION

The first step is to create a table of values that shows the present worth of the costs for different values of $n \equiv$ the length of time until the second stage or full capacity is needed. Notice that the full capacity is calculated at $n = 0$. The only reason to calculate the corresponding value for staged construction is to see if the formula is properly entered, since building both stages at the same time will not really cost $220,000. The values for staged construction at 5, 10, 20, and 30 years check with the values in Example 9-11.

The next step is to select cells A8:C13, which includes the $x$ values and two series of $y$ values. Then the ChartWizard tool is selected. In the first step, the $xy$ (scatter) plot is selected with the option of smoothed lines without markers. In Step 2 no action is required, since the cells A8:C13 were selected first. In Step 3 labels are added for the $x$ and $y$ axes. In Step 4 the chart is moved around on the worksheet page, so that it does not overlap with the data. The result is shown in Figure 9-7.

Our first step in cleaning up the graph is to delete the formula in cell C9, since two-stage construction will not be done at Time 0. We also delete the label in the adjacent cell, which explains the formula. Then we create a new label for cell C10. As shown in Appendix A, the easy way to create that label is to insert an apostrophe, or space, as the first entry in cell C10. This converts the formula to a label which we can copy to D10. Then we delete the apostrophe in cell C10.

**FIGURE 9-7**
Automatic graph from spreadsheet.

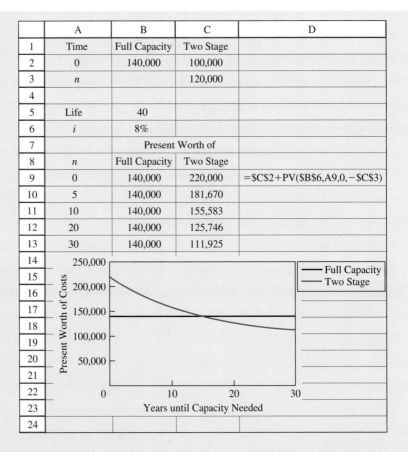

| | A | B | C | D |
|---|---|---|---|---|
| 1 | Time | Full Capacity | Two Stage | |
| 2 | 0 | 140,000 | 100,000 | |
| 3 | $n$ | | 120,000 | |
| 4 | | | | |
| 5 | Life | 40 | | |
| 6 | $i$ | 8% | | |
| 7 | | Present Worth of | | |
| 8 | $n$ | Full Capacity | Two Stage | |
| 9 | 0 | 140,000 | 220,000 | =$C$2+PV($B$6,A9,0,−$C$3) |
| 10 | 5 | 140,000 | 181,670 | |
| 11 | 10 | 140,000 | 155,583 | |
| 12 | 20 | 140,000 | 125,746 | |
| 13 | 30 | 140,000 | 111,925 | |

The axis scales must be modified to focus on the area of concern. Select the $x$ axis and change the minimum from automatic to 10 and the maximum to 20. Select the $y$ axis and change the minimum to 125,000 and the maximum to 160,000.

Left-click on the plot area to select it. Then right-click to bring up the options. Select Format Plot Area and change the area pattern to "none." This will eliminate the gray fill that made Figure 9-7 difficult to read.

Left-click on the two-stage curve to select it. Then right-click for the options. Format the data series using the Patterns tab. Change the line style from solid to dashed, the line color from automatic to black, and increase the line weight. Similarly, increase the line weight for the full-capacity line. Finally, select a grid line and change the line style to dotted. The result is far easier to read in black and white.

To further improve the graph, we can replace the legend with annotations on the graph. Left-click somewhere in the white area around the graph to select "chart area." Right-click and then choose the chart options on the menu. The legends tab will let us delete the legend by turning "show legend" off. Similarly, we can turn the $x$-axis gridlines on. The line style for these gridlines should be changed to match the $y$-axis gridlines. This allows us to see that the breakeven time is between 14 and 15 years.

To make the graph less busy, change the scale on the $x$ axis so that the interval is 5 years rather than automatic. Also eliminate the gridlines for the $y$ axis (by selecting the Chart Area, Chart

**FIGURE 9–8**
Spreadsheet of
Figure 9-7
with improved
graph.

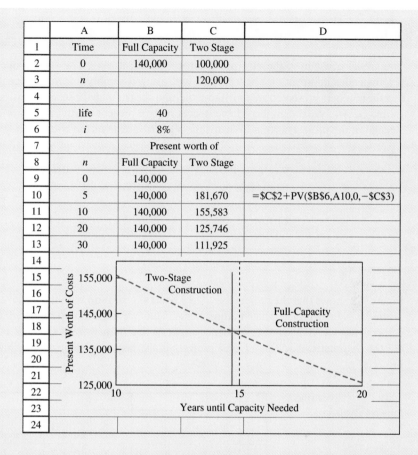

|   | A | B | C | D |
|---|---|---|---|---|
| 1 | Time | Full Capacity | Two Stage | |
| 2 | 0 | 140,000 | 100,000 | |
| 3 | n | | 120,000 | |
| 4 | | | | |
| 5 | life | 40 | | |
| 6 | i | 8% | | |
| 7 | | Present worth of | | |
| 8 | n | Full Capacity | Two Stage | |
| 9 | 0 | 140,000 | | |
| 10 | 5 | 140,000 | 181,670 | =$C$2+PV($B$6,A10,0,−$C$3) |
| 11 | 10 | 140,000 | 155,583 | |
| 12 | 20 | 140,000 | 125,746 | |
| 13 | 30 | 140,000 | 111,925 | |

Options, and Gridlines tabs). The graph size can be increased for easier reading, as well. This may require specifying an interval of 10,000 for the scale of the *y* axis.

Finally, to add the labels for the full-capacity curve and the two-stage curve, find the toolbar for graphics, which is open when the chart is selected (probably along the bottom of the spreadsheet). Select the text box icon, and click on a location close to the two-stage chart. Type in the label for two-stage construction. Notice how including a return and a few spaces can shape the label to fit the slanted line. Add the label for full construction. Figure 9-8 is the result.

## DOING WHAT- IF ANALYSIS WITH SPREADSHEETS

Breakeven charts change one variable at a time, while what-if analysis may change many of the variables in a problem. However, spreadsheets remain a very powerful tool for this form of sensitivity analysis. In Example 9-15 a project appears to be very promising. However, what-if analysis indicates that a believable scenario raises some questions about whether the project should be done.

## EXAMPLE 9–15

You are an assistant to the vice president for manufacturing. The staff at one of the plants has recommended approval for a new product with a new assembly line to produce it. The VP believes that the numbers presented are too optimistic, and she has added a set of adjustments to the original estimates. Analyze the project's benefit–cost ratio or present worth index as originally submitted. Reanalyze the project, asking "What if the VP's adjustments are correct?"

|  | Initial Estimate | Adjustment |
|---|---|---|
| First cost | $70,000 | +10% |
| Units/year | 1,200 | −20% |
| Net unit revenue | $   25 | −15% |
| Life, in years | 8 | −3 |
| Interest rate | 12% | None |

### SOLUTION

Figure 9-9 shows that the project has a 2.13 benefit–cost ratio with the initial estimates, but only a value of 0.96 with the what-if adjustments. Thus we need to determine which set of numbers is more realistic. Real-world experience suggests that in many organizations the initial estimates are too optimistic. Auditing of past projects is the best way to develop adjustments for future projects.

**FIGURE 9–9** Spreadsheet for what-if analysis.

|  | A | B | C | D |
|---|---|---|---|---|
| 1 |  | Initial Estimate | Adjust-ment | Adjusted Values |
| 2 | First cost | $70, 000 | 10% | $77, 000 |
| 3 | Units/year | 1,200 | −20% | 960 |
| 4 | Net unit revenue | $25 | −15% | $21 |
| 5 | Life (years) | 8 | −3 | 5 |
| 6 | Interest rate | 12% | none | 12% |
| 7 |  |  |  |  |
| 8 | Benefits | 149,029 |  | 73,537 |
| 9 | Cost | 70,000 |  | 77,000 |
| 10 | B/C Ratio | 2.13 |  | 0.96 |
| 11 |  |  |  |  |
| 12 |  | =PV(B6,B5-B3*B4) |  |  |

## SUMMARY

In this chapter, we have looked at four new analysis techniques.

*Future worth:* When the comparison between alternatives will be made in the future, the calculation is called future worth. This is very similar to present worth, which is based on the present, rather than a future point in time.

*Benefit–cost ratio analysis:* This technique is based on the ratio of benefits to costs using either present worth or annual cash flow calculations. The method is graphically similar to present worth analysis. When neither input nor output is fixed, incremental benefit–cost ratios ($\Delta B/\Delta C$) are required. The method is similar in this respect to rate of return analysis. Benefit–cost ratio analysis is often used at the various levels of government.

*Payback period:* Here we define payback as the period of time required for the profit or other benefits of an investment to equal the cost of the investment. Although simple to use and understand, payback is a poor analysis technique for ranking alternatives. While it provides a measure of the speed of the return of the investment, it is not an accurate measure of the profitability of an investment.

*Sensitivity, breakeven, and what-if analysis:* These techniques are used to see how sensitive a decision is to estimates for the various parameters. Breakeven analysis is done to locate conditions under which the alternatives are equivalent. This is often presented in the form of breakeven charts. Sensitivity analysis examines a range of values of some parameters to determine the effect on a particular decision. What-if analysis changes one or many estimates to see what results.

## PROBLEMS

### Future Worth

**9-1** Compute the future worth for the following cash flows.

$i = 12\%$

(*Answer: F = $1199*)

**9-2** For the following cash flows, compute the future worth.

$i = 15\%$

**9-3** For the following cash flows, compute the future worth.

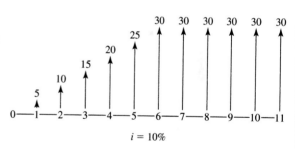

$i = 10\%$

**9-4** For a 12% interest rate, compute the value of $F$ so the following cash flows have a future worth of 0.

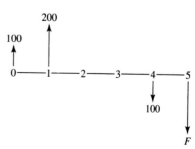

**9-5**   Compute $F$ so the following cash flows have a future worth of 0.

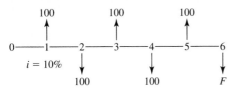

**9-6**   Calculate the present worth and the future worth of a series of 10 annual cash flows with the first cash flow equal to $15,000 and each successive cash flow increasing by $1200. The interest rate is 12%.

**9-7**   Sally deposited $100 a month in her savings account for 24 months. For the next 5 years she made no deposits. What is the future worth in Sally's savings account at the end of the 7 years, if the account earned 6% annual interest, compounded monthly? (*Answer:* $3430.78)

**9-8**   A 20-year-old student decided to set aside $100 on his 21$^{st}$ birthday for investment. Each subsequent year through his 55$^{th}$ birthday, he plans to increase the investment on a $100 arithmetic gradient. He will not set aside additional money after his 55$^{th}$ birthday. If the student can achieve a 12% rate of return, what is the future worth of the investments on his 65$^{th}$ birthday?   (*Answer:* $1,160,700)

**9-9**   You can buy a piece of vacant land for $30,000 cash. You plan to hold it for 15 years and then sell it at a profit. During this period, you would pay annual property taxes of $600. You would have no income from the property. Assuming that you want a 10% rate of return, at what net price would you have to sell the land 15 years hence?   (*Answer:* $144,373)

**9-10**   An individual who makes $32,000 per year anticipates retiring in 30 years. If his salary is increased by $600 each year and he deposits 10% of his yearly salary into a fund that earns 7% interest, what is the future worth at retirement?

**9-11**   Stamp collecting has become an increasingly popular—and expensive—hobby. One favorite method is to save plate blocks (usually four stamps with the printing plate number in the margin) of each new stamp as it is issued. With rising postage rates and increased numbers of new stamps being issued, this collecting plan costs more each year.

Stamps have been a good place to invest money over the last 10 years, as the demand for stamps previously issued has caused resale prices to increase 18% each year. Suppose a collector purchased

$100 worth of stamps 10 years ago, and increased his purchases by $50 per year in each subsequent year. After 10 years of stamp collecting, what is the future worth of the stamp collection?

**9-12**   The interest rate is 16% per year and there are 48 compounding periods per year. The principal is $50,000. What is the future worth in 5 years?

**9-13**   In the early 1980s, planners were examining alternate sites for a new London airport. The economic analysis included the value of structures that would need to be removed from various airport sites. At one potential site, the twelfth-century Norman church of St. Michaels, in the village of Stewkley, would have had to be demolished. The planners used the value of the fire insurance policy on the church—a few thousand pounds sterling—as the church's value.

An outraged antiquarian wrote to the London *Times* that an equally plausible computation would be to assume that the original cost of the church (estimated at 100 pounds sterling) be increased at the rate of 10% per year for 800 years. Based on his proposal, what would be the future worth of St. Michaels? (*Note:* There was great public objection to tearing down the church, and it was spared.)

**9-14**   Bill made a budget and planned to deposit $150 a month in a savings account, beginning September 1. He did this, but on the following January 1, he reduced the monthly deposits to $100. He made 18 deposits, four at $150 and 14 at $100. If the savings account paid 6% interest, compounded monthly, what was the future worth of his savings immediately after the last deposit?   (*Answer:* $2094.42)

**9-15**   A company deposits $1000 in a bank at the beginning of each year for 6 years. The account earns 8% interest, compounded every 6 months. What is in the account at the end of 6 years?

**9-16**   Don Ball is a 55-year-old engineer. According to mortality tables, a male at age 55 has an average life expectancy of 21 more years. Don has accumulated $48,500 toward his retirement. He is now adding $5000 per year to his retirement fund. The fund earns 12% interest. Don will retire when he can obtain an annual income from his retirement fund of $20,000, assuming he lives to age 76. He will make no provision for a retirement income after age 76. What is the youngest age at which Don can retire?

**9-17**   Jean invests $100 in Year 1 and doubles the amount each year after that (so the investment is $100, 200,

400, 800, ... ). If she does this for 10 years, and the investment pays 10% annual interest, what is the future worth of her investment?

9-18 If you invested $2500 in a 24-month bank certificate of deposit (CD) paying 8.65%, compounded monthly, what is the future worth of the CD when it matures?

9-19 After receiving an inheritance of $25,000 on her 21$^{st}$ birthday, Ayn deposited the inheritance in a savings account with an effective annual interest rate of 6%. She decided to make regular deposits, beginning with $1000 on her 22$^{nd}$ birthday and increasing by $200 each year (i.e., $1200 on her 23$^{rd}$ birthday, $1400 on her 24$^{th}$ birthday, etc.). What was the future worth of Ayn's deposits on her 56$^{th}$ birthday?

9-20 The Association of General Contractors (AGC) is endowing a fund of $1 million for the Construction Engineering Technology Program at Grambling State University. The AGC established an escrow account in which 10 equal end-of-year deposits that earn 7% compound interest were to be made. After seven deposits, the Louisiana legislature revised laws relating to the licensing fees AGC can charge its members, and there was no deposit at the end of Year 8. What must the amount of the remaining equal end-of-year deposits be, to ensure that the $1 million is available on schedule for the Construction Engineering Technology Program?

9-21 A new engineer is considering investing in an individual retirement account (IRA) with a mutual fund that has an average annual return of 10%. What is the future worth of her IRA at age 65 if she makes annual investments of $2000 into the fund beginning on her 25$^{th}$ birthday? Assume that the fund continues to earn an annual return of 10%.

9-22 IPS Corp. will upgrade its package-labeling machinery. It costs $150,000 to buy the machinery and have it installed. Operation and maintenance costs, which are $1500 per year for the first 3 years, increase by 500 per year for the machine's 10-year life. The machinery has a salvage value of 5% of its initial cost. Interest is 10%. What is the future worth of cost of the machinery?

9-23 A company is considering buying a new bottle-capping machine. The initial cost of the machine is $325,000 and it has a 10-year life. Monthly maintenance costs are expected to be $1200 per month for the first 7 years and $2000 per month for the remaining years. The machine

requires a major overhaul costing $55,000 at the end of the fifth year of service. Assume that all these costs occur at the end of the appropriate period. What is the future value of all the costs of owning and operating this machine if the nominal interest rate is 7.2%?

9-24 A family starts an education fund for their son Patrick when he is 8 years old, investing $150 on his eighth birthday, and increasing the yearly investment by $150 per year until Patrick is 18 years old. The fund pays 9% annual interest. What is the fund's future worth when Patrick is 18?

9-25 A bank account pays 19.2% interest with monthly compounding. A series of deposits started with a deposit of $5000 on January 1, 2007. Deposits in the series were to occur each 6 months. Each deposit in the series is for $150 less than the one before it. The last deposit in the series will be due on January 1, 2022. What is the future worth of the account on July 1, 2024, if the balance was zero before the first deposit and no withdrawals are made?

9-26 A recent college graduate got a good job and began a savings account. He authorized the bank to automatically transfer $75 each month from his checking account to the savings account. The bank made the first withdrawal on July 1, 2012 and is instructed to make the last withdrawal on January 1, 2030. The bank pays a nominal interest rate of 4.5% and compounds twice a month. What is the future worth of the account on January 1, 2030?

9-27 Bob, an engineer, decided to start a college fund for his son. Bob will deposit a series of equal, semi-annual cash flows with each deposit equal to $1500. Bob made the first deposit on July 1, 2011 and will make the last deposit on July 1, 2031. Joe, a friend of Bob's, received an inheritance on April 1, 2016, and has decided to begin a college fund for his daughter. Joe wants to send his daughter to the same college as Bob's son. Therefore, Joe needs to accumulate the same amount of money on July 1, 2031, as Bob will have accumulated from his semiannual deposits. Joe never took engineering economics and had no idea how to determine the amount that should be deposited. He decided to deposit $40,000 on July 1, 2016. Will Joe's deposit be sufficient? If not, how much should he have put in? Use a nominal interest of 7% with semiannual compounding on all accounts.

9-28 A business executive is offered a management job at Generous Electric Company, which offers him a 5-year contract that calls for a salary of $62,000 per year, plus 600 shares of GE stock at the end of the 5 years. This executive is currently employed by Fearless Bus Company, which also has offered him a 5-year contract. It calls for a salary of $65,000, plus 100 shares of Fearless stock each year. The Fearless stock is currently worth $60 per share and pays an annual dividend of $2 per share. Assume end-of-year payments of salary and stock. Stock dividends begin one year after the stock is received. The executive believes that the value of the stock and the dividend will remain constant. If the executive considers 9% a suitable rate of return in this situation, what must the Generous Electric stock be worth per share to make the two offers equally attractive? Use the future worth analysis method in your comparison.   (*Answer:* $83.76)

9-29   Pick a discretionary expense that you incur on a regular basis, such as buying cigarettes weekly, buying fashion items monthly, buying sports tickets monthly, or going to movies weekly. Assume that you instead place the money in an investment account that earns 9% annually. After 40 years, how much is in the account?

## Benefit–Cost Ratio

9-30   Cornell has two options for upgrading their athletic facilities. The off-campus option costs only $20 million, but it will require frequent bus service to those facilities at an annual cost that starts at $300,000 and increases by 4% per year. (buses, drivers and mechanics salaries, maintenance, road wear, etc.). Improving the on-campus facilities will cost $50 million, but no extra transportation costs are required. Both options involve an estimated annual maintenance cost of $1 million for about 40 years before new facilities will again be needed. Using benefit–cost ratio analysis, determine which option is more economically efficient. Use an interest rate of 8% per year.
*Contributed by D. P. Loucks, Cornell University*

9-31   Each of the three alternatives shown has a 5-year useful life. If the MARR is 10%, which alternative should be selected? Solve the problem by benefit–cost ratio analysis.

|  | A | B | C |
|---|---|---|---|
| Cost | $600.0 | $500.0 | $200.0 |
| Uniform annual benefit | 158.3 | 138.7 | 58.3 |

(*Answer: B*)

9-32   Consider three alternatives, each with a 10-year useful life. If the MARR is 10%, which alternative should be selected? Solve the problem by benefit–cost ratio analysis.

|  | A | B | C |
|---|---|---|---|
| Cost | $800 | $300 | $150 |
| Uniform annual benefit | 142 | 60 | 33.5 |

9-33   An investor is considering buying some land for $100,000 and constructing an office building on it. Three different buildings are being analyzed.

|  | Building Height | | |
|---|---|---|---|
|  | 2 Stories | 5 Stories | 10 Stories |
| Cost of building (excluding cost of land) | $400,000 | $800,000 | $2,100,000 |
| Resale value* of land + building after 20-year horizon | 200,000 | 300,000 | 400,000 |
| Annual net rental income | 70,000 | 105,000 | 256,000 |

*Resale value considered a reduction in cost—not a benefit.

Using benefit–cost ratio analysis and an 8% MARR, determine which alternative, if any, should be selected.

9-34   Using benefit–cost ratio analysis, determine which one of the three mutually exclusive alternatives should be selected. Each alternative has a 6-year useful life. Assume a 10% MARR.

|  | A | B | C |
|---|---|---|---|
| First cost | $560 | $340 | $120 |
| Uniform annual benefit | 140 | 100 | 40 |
| Salvage value | 40 | 0 | 0 |

9-35   Consider four alternatives, each of which has an 8-year useful life:

|  | A | B | C | D |
|---|---|---|---|---|
| Cost | $100.0 | $80.0 | $60.0 | $50.0 |
| Uniform annual benefit | 12.2 | 12.0 | 9.7 | 12.2 |
| Salvage value | 75.0 | 50.0 | 50.0 | 0 |

If the MARR is 8%, which alternative should be selected? Solve the problem by benefit–cost ratio analysis.

**9-36** A government agency is planning a new office building close to its current headquarters. Four proposed sites are to be evaluated. Any of these sites will save the agency $700,000 per year, since two of its current satellite offices will no longer need to be rented. The agency uses a 6% interest rate and assumes that the building and its benefits will last for 40 years. Based on a benefit–cost analysis what should the agency do?

| | Site | | | |
|---|---|---|---|---|
| | A | B | C | D |
| Initial cost | $8.6M | $8.1M | $7.5M | $6.8M |
| Annual operating cost | 0.12M | 0.155M | 0.2M | 0.3M |

*Contributed by Hamed Kashani, Saeid Sadri, and Baabak Ashuri, Georgia Institute of Technology*

**9-37** Using benefit–cost ratio analysis, a 5-year useful life, and a 15% MARR, determine which of the following alternatives should be selected.

| | A | B | C | D | E |
|---|---|---|---|---|---|
| Cost | $100 | $200 | $300 | $400 | $500 |
| Uniform annual benefit | 37 | 69 | 83 | 126 | 150 |

**9-38** Five mutually exclusive investment alternatives have been proposed. Based on benefit–cost ratio analysis, and a MARR of 15%, which alternative should be selected?

| Year | A | B | C | D | E |
|---|---|---|---|---|---|
| 0 | −$200 | −$100 | −$125 | −$150 | −$225 |
| 1–5 | 68 | 25 | 42 | 52 | 68 |

**9-39** A project will cost $50,000. The benefits at the end of the first year are estimated to be $10,000, increasing at a 10% uniform rate in subsequent years. Using an 8-year analysis period and a 10% interest rate, compute the benefit–cost ratio.

**9-40** A do-nothing and two mutually exclusive alternatives are being considered for reducing traffic congestion. User benefits come from reduced congestion once the project is complete, while user disbenefits are due to increased congestion during construction. The interest rate is 9%, and the life of each alternative is 15 years. Which alternative should be chosen?

| | A | B |
|---|---|---|
| User benefits ($M/yr) | 2.1 | 2.6 |
| User disbenefits ($M) | 1.2 | 2.1 |
| First cost ($M) | 6.9 | 9.9 |
| Operations and maintenance ($M/yr) | 0.75 | 0.825 |

(*a*) Use the benefit–cost ratio.

(*b*) Use the modified benefit–cost ratio.

(*c*) Use the public/government version of the B/C ratio.

(*d*) Assume these numbers apply to a private firm and use a present worth index.

(*e*) Are your recommendations for (*a*) through (*d*) consistent? Which measure gives the largest value? Why?

**9-41** A school is overcrowded and there are three options. The do-nothing alternative corresponds to continuing to use modular classrooms. The school can be expanded, or a new school can be built to "split the load" between the schools. User benefits come from improvements in school performance for the expanded or new schools. If a new school is built, there are more benefits because more students will be able to walk to school, the average distance for those who ride the school buses will be shorter, and the schools will be smaller and more "student friendly." The disbenefits for the expanded school are due to the impact of the construction process during the school year. The interest rate is 8%, and the life of each alternative is 20 years. Which alternative should be chosen? What is the incremental ratio for the preferred alternative?

| | A | B |
|---|---|---|
| User benefits ($M/yr) | 2.1 | 3.1 |
| User disbenefits ($M) | 0.8 | 0 |
| First cost ($M) | 8.8 | 10.4 |
| Operations and maintenance ($M/yr) | 0.95 | 1.7 |

(*a*) Use the benefit–cost ratio.

(*b*) Use the modified benefit–cost ratio.

(*c*) Use the public/government version of the B/C ratio.

(*d*) Assume these numbers apply to a private firm and use a present worth index.

(*e*) Are your recommendations for (*a*) through (*d*) consistent? Which measure gives the largest value? Why?

## Payback Period and Exact Methods

**9-42**  Able Plastics, an injection-molding firm, has negotiated a contract with a national chain of department stores. Plastic pencil boxes are to be produced for a 2-year period. Able Plastics has never produced the item before and requires all new dies. If the firm invests $67,000 for special removal equipment to unload the completed pencil boxes from the molding machine, one machine operator can be eliminated. This would save $26,000 per year. The removal equipment has no salvage value and is not expected to be used after the 2-year production contract is completed. The equipment, although useless to Able, would be serviceable for about 15 years. What is the payback period? Should Able Plastics buy the removal equipment?

**9-43**  A cannery is considering installing an automatic case-sealing machine to replace current hand methods. If they purchase the machine for $3800 in June, at the beginning of the canning season, they will save $400 per month for the 4 months each year that the plant is in operation. Maintenance costs of the case-sealing machine are expected to be negligible. The case-sealing machine is expected to be useful for five annual canning seasons and then will have no salvage value. What is the payback period? What is the nominal annual rate of return?

**9-44**  A project has the following costs and benefits. What is the payback period?

| Year | Costs | Benefits |
|------|-------|----------|
| 0 | $1400 | |
| 1 | 500 | |
| 2 | 300 | $400 |
| 3–10 | | $300 in each year |

**9-45**  A car dealer leases a small computer with software for $5000 per year. As an alternative he could buy the computer for $7000 and lease the software for $3500 per year. Any time he would decide to switch to some other computer system he could cancel the software lease and sell the computer for $500. If he buys the computer and leases the software,

(*a*)  What is the payback period?

(*b*)  If he kept the computer and software for 6 years, what would be the benefit–cost ratio, based on a 10% interest rate?

**9-46**  A large project requires an investment of $200 million. The construction will take 3 years: $30 million will be spent during the first year, $100 million

during the second year, and $70 million during the third year of construction. Two project operation periods are being considered: 10 years with the expected net profit of $40 million per year and 20 years with the expected net profit of $32.5 million per year. For simplicity of calculations it is assumed that all cash flows occur at end of year. The company minimum required return on investment is 10%.

Calculate for each alternative:

(*a*)  The payback period

(*b*)  The total equivalent investment cost at the end of the construction period

(*c*)  The equivalent uniform annual worth of the project (use the operation period of each alternative)

Which operation period should be chosen?

**9-47**  Two alternatives with identical benefits are being considered:

| | A | B |
|---|---|---|
| Initial cost | $500 | $800 |
| Uniform annual cost | 200 | 150 |
| Useful life, in years | 8 | 8 |

(*a*)  Compute the payback period if Alt. *B* is purchased rather than Alt. *A*.

(*b*)  Use a MARR of 12% and benefit–cost ratio analysis to identify the alternative that should be selected.

**9-48**  Tom Sewel has gathered data on the relative costs of a solar water heater system and a conventional electric water heater. The data are based on statistics for a mid-American city and assume that during cloudy days an electric heating element in the solar heating system will provide the necessary heat.

The installed cost of a conventional electric water tank and heater is $200. A family of four uses an average of 300 liters of hot water a day, which takes $230 of electricity per year. The glass-lined tank has a 20-year guarantee. This is probably a reasonable estimate of its actual useful life.

The installed cost of two solar panels, a small electric pump, and a storage tank with auxiliary electric heating element is $1400. It will cost $60 a year for electricity to run the pump and heat water on cloudy days. The solar system will require $180 of maintenance work every 4 years. Neither the conventional electric water heater nor the solar water heater will have any salvage value at the end of its useful life.

Using Tom's data, what is the payback period if the solar water heater system is installed, rather than the conventional electric water heater?

**9-49** Consider four mutually exclusive alternatives:

|  | A | B | C | D |
|---|---|---|---|---|
| Cost | $75.0 | $50.0 | $15.0 | $90.0 |
| Uniform annual benefit | 18.8 | 13.9 | 4.5 | 23.8 |

Each alternative has a 5-year useful life and no salvage value. The MARR is 10%. Which alternative should be selected, based on

(a) Future worth analysis
(b) Benefit–cost ratio analysis
(c) The payback period

**9-50** Consider three alternatives:

|  | A | B | C |
|---|---|---|---|
| First cost | $50 | $150 | $110 |
| Uniform annual benefit | 28.8 | 39.6 | 39.6 |
| Useful life, in years* | 2 | 6 | 4 |
| Rate of return | 10% | 15% | 16.4% |

*At the end of its useful life, an identical alternative (with the same cost, benefits, and useful life) may be installed.

All the alternatives have no salvage value. If the MARR is 12%, which alternative should be selected?

(a) Solve the problem by future worth analysis.
(b) Solve the problem by benefit–cost ratio analysis.
(c) Solve the problem by payback period.
(d) If the answers in parts (a), (b), and (c) differ, explain why this is the case.

**9-51** Consider three mutually exclusive alternatives. The MARR is 10%.

| Year | X | Y | Z |
|---|---|---|---|
| 0 | −$100 | −$50 | −$50 |
| 1 | 25 | 16 | 21 |
| 2 | 25 | 16 | 21 |
| 3 | 25 | 16 | 21 |
| 4 | 25 | 16 | 21 |

(a) For Alt. X, compute the benefit–cost ratio.
(b) Based on the payback period, which alternative should be selected?

(c) Determine the preferred alternative based on an exact economic analysis method.

**9-52** The cash flows for three alternatives are as follows:

| Year | A | B | C |
|---|---|---|---|
| 0 | −$500 | −$600 | −$900 |
| 1 | −400 | −300 | 0 |
| 2 | 200 | 350 | 200 |
| 3 | 250 | 300 | 200 |
| 4 | 300 | 250 | 200 |
| 5 | 350 | 200 | 200 |
| 6 | 400 | 150 | 200 |

(a) Based on payback period, which alternative should be selected?
(b) Using future worth analysis, and a 12% interest rate, determine which alternative should be selected.

**9-53** Three mutually exclusive alternatives are being considered:

|  | A | B | C |
|---|---|---|---|
| Initial cost | $500 | $400 | $300 |
| Benefit at end of the first year | 200 | 200 | 200 |
| Uniform benefit at end of subsequent years | 100 | 125 | 100 |
| Useful life, in years | 6 | 5 | 4 |

At the end of its useful life, an alternative is *not* replaced. If the MARR is 10%, which alternative should be selected

(a) Based on the payback period?
(b) Based on benefit–cost ratio analysis?

**9-54**

| Year | E | F | G | H |
|---|---|---|---|---|
| 0 | −$90 | −$110 | −$100 | −$120 |
| 1 | 20 | 35 | 0 | 0 |
| 2 | 20 | 35 | 10 | 0 |
| 3 | 20 | 35 | 20 | 0 |
| 4 | 20 | 35 | 30 | 0 |
| 5 | 20 | 0 | 40 | 0 |
| 6 | 20 | 0 | 50 | 180 |

(a) Based on future worth analysis, which of the four alternatives is preferred at 6% interest?
(b) Based on future worth analysis, which alternative is preferred at 15% interest?
(c) Based on the payback period, which alternative is preferred?

(*d*) At 7% interest, what is the benefit–cost ratio for Alt. *G*?

## Sensitivity

9-55 Tom Jackson is buying a new car. Alternative *A* is an American-built compact. It has an initial cost of $8900 and operating costs of 9¢/km, excluding depreciation. From resale statistics, Tom estimates the American car can be resold at the end of 3 years for $1700. Alternative *B* is a foreign-built Fiasco. Its initial cost is $8000, the operating cost, also excluding depreciation, is 8¢/km. How low could the resale value of the Fiasco be to provide equally economical transportation? Assume Tom will drive 12,000 km/year and considers 8% as an appropriate interest rate. (*Answer:* $175)

9-56 Victoria is choosing between a standard Honda Civic for $17,350 or a hybrid Civic for $20,875. She calculates her annual cost of ownership including payments but not including gasoline to be $5000 for the standard and $5800 for the hybrid. The standard Civic will cost Victoria 12¢/mile for gasoline, while the hybrid will cost her only 7¢/mile. How many miles must Victoria drive in a year before the hybrid vehicle becomes more cost efficient to her?
*Contributed by Paul R. McCright, University of South Florida*

9-57 A road can be paved with either asphalt or concrete. Concrete costs $15,000/km and lasts 20 years. Assume the annual maintenance costs are $500 for concrete and $800 for asphalt per kilometer per year. Use an interest rate of 8% per year.
*Contributed by D. P. Loucks, Cornell University*

(*a*) What is the maximum that should be spent for asphalt if it lasts only 10 years?

(*b*) Assume the asphalt road costs $7000 per kilometer. How long must it last to be the preferred alternative?

9-58 Ithaca is considering a new $50,000 snowplow that will save the city $600 per day of use compared to the existing one. It should last 10 years and have a resale value of $2000. To obtain a 12% rate of return what is the minimum number of days per year on average it will have to be used.
*Contributed by D. P. Loucks, Cornell University*

9-59 A car company has decided to spend $150,000,000 on a museum for exhibiting its classic cars. Land can be purchased for $500,000. The museum building will require 30,000 square feet of general space, while each car displayed will require an additional 1000 square feet. The design and planning process will cost $100,000, which should be paid immediately. The construction of the building will cost $600 per square foot, and the building will be completed within the next 2 years, while the cost of construction will be distributed evenly between the 2 years of construction. All cars will be purchased during the second year of construction at an average cost of $120,000 per car. The annual operation of the museum will cost $2,000,000 plus $30,000 per car. If the funds are invested at 9% per year and the museum is to exist forever, how many cars should the trustees plan to purchase?
*Contributed by Hamed Kashani, Saeid Sadri, and Baabak Ashuri, Georgia Institute of Technology*

9-60 Southwest Airlines flies nonstop between Tampa and Albuquerque with 137-passenger planes. Considering all the costs of owning each plane plus the salaries for their crews and the fuel costs and landing fees, SWA engineers have determined that the fixed cost for a single flight is $10,400. If the costs associated with each passenger (reservations cost, check-in cost, baggage handling cost, snack cost, etc.) total $48 per passenger and the average ticket price is $157 (before the various taxes are added), what percentage of seats must be filled for the flight to break even?
*Contributed by Paul R. McCright, University of South Florida*

9-61 Southwest Airlines (SWA) is planning to expand its fleet of jets to replace some old planes and to expand its routes. It has received a proposal from Boeing to purchase 112 737's over the next 4 years. What annual net revenue must each jet produce to break even on its operating cost? The analysis should be done by finding the EUAC for the 10-year planned ownership period. SWA has a MARR of 12%, purchases the jet for $22 million, has operating and maintenance costs of $3.2 million the first year, increasing 8% per year, and performs a major maintenance upgrade costing $4.5M at end of Year 5. Assume the plane has a salvage value at end of Year 10 of $13 million.
*Contributed by Paul R. McCright, University of South Florida*

9-62 A newspaper is considering buying locked vending machines to replace open newspaper racks in the downtown area. The vending machines cost $45 each. It is expected that the annual revenue from selling the same quantity of newspapers will increase $12 per vending machine. The useful life of the vending machine is unknown.

(a) To determine the sensitivity of rate of return to useful life, prepare a graph for rate of return versus useful life for lives up to 8 years.

(b) If the newspaper requires a 12% rate of return, what minimum useful life must it obtain from the vending machines?

(c) What would be the rate of return if the vending machines were to last indefinitely?

**9-63**  If the MARR is 12%, compute the value of $X$ that makes the two alternatives equally desirable.

|  | A | B |
|---|---|---|
| Cost | $800 | $1000 |
| Uniform annual benefit | 230 | 230 |
| Useful life, in years | 5 | X |

**9-64**  If the MARR is 12%, compute the value of $X$ that makes the two alternatives equally desirable.

|  | A | B |
|---|---|---|
| Cost | $150 | $ X |
| Uniform annual benefit | 40 | 65 |
| Salvage value | 100 | 200 |
| Useful life, in years | 6 | 6 |

**9-65**  Consider two alternatives:

|  | A | B |
|---|---|---|
| Cost | $500 | $300 |
| Uniform annual benefit | 75 | 75 |
| Useful life, in years | Infinity | X |

Assume that Alt. *B* is not replaced at the end of its useful life. If the MARR is 10%, what must be the useful life of *B* to make Alternatives *A* and *B* equally desirable?

**9-66**  Chris Cook studied the situation described in Problem 9-48 and decided that the solar system will *not* require the $180 of maintenance every 4 years. Chris believes future replacements of either the conventional electric water heater, or the solar water heater system can be made at the same costs and useful lives as the initial installation. Based on a 10% interest rate, what must be the useful life of the solar system to make it no more expensive than the electric water heater system?

**9-67**  Jane Chang is making plans for a summer vacation. She will take $1000 with her in the form of traveler's checks. From the newspaper, she finds that if she purchases the checks by May 31, she will not have to pay a service charge. That is, she will obtain $1000 worth of traveler's checks for $1000. But if she waits to buy the checks until just before starting her summer trip, she must pay a 1% service charge. (It will cost her $1010 for $1000 of traveler's checks.)

Jane can obtain a 13% interest rate, compounded weekly, on her money. How many weeks after May 31 can she begin her trip and still justify buying the traveler's checks on May 31?

**9-68**  Fence posts for a particular job cost $10.50 each to install, including the labor cost. They will last 10 years. If the posts are treated with a wood preservative, they can be expected to have a 15-year life. Assuming a 10% interest rate, how much could one afford to pay for the wood preservative treatment?

**9-69**  A piece of property is purchased for $10,000 and yields a $1000 yearly net profit. The property is sold after 5 years. What is its minimum price to breakeven with interest at 10%?

**9-70**  Rental equipment is for sale for $110,000. A prospective buyer estimates he would keep the equipment for 12 years and spend $6000 a year on maintaining it. Estimated annual net receipts from equipment rentals would be $14,400. It is estimated the rental equipment could be sold for $80,000 at the end of 12 years. If the buyer wants a 7% rate of return on his investment, what is the maximum price he should pay for the equipment?

**9-71**  *The Financial Advisor* is a weekly column in the local newspaper. Assume you must answer the following question. "I recently retired at age 65, and I have a tax-free retirement annuity coming due soon. I have three options. I can receive (*A*) $30,976 now, (*B*) $359.60 per month for the rest of my life, or (*C*) $513.80 per month for the next 10 years. What should I do?" Ignore the timing of the monthly cash flows and assume that the payments are received at the end of year. Assume the 10-year annuity will continue to be paid to loved heirs if the person dies before the 10-year period is over.
*Contributed by D. P. Loucks, Cornell University*

(a) If $i = 6\%$, develop a choice table for lives from 5 to 30 years. (You do not know how long this person or other readers may live.)

(b) If $i = 10\%$, develop a choice table for lives from 5 to 30 years. (You do not know how long this person or other readers may live.)

(c) How does increasing the interest rate change your recommendations?

**9-72** A motor with a 200-horsepower output is needed in the factory for intermittent use. A Graybar motor costs $7000 and has an electrical efficiency of 89%. A Blueball motor costs $6000 and has an 85% efficiency. Neither motor would have any salvage value, since the cost to remove it would equal its scrap value. The annual maintenance cost for either motor is estimated at $300 per year. Electric power costs $0.072/kWh (1 hp $=0.746$ kW). If a 10% interest rate is used in the calculations, what is the minimum number of hours the higher initial cost Graybar motor must be used each year to justify its purchase?

**9-73** Plan $A$ requires a $100,000 investment now. Plan $B$ requires an $80,000 investment now and an additional $40,000 investment at a later time. At 8% interest, compute the breakeven point for the timing of the $40,000 investment.

**9-74** A low-carbon-steel machine part, operating in a corrosive atmosphere, lasts 6 years, and costs $350 installed. If the part is treated for corrosion resistance, it will cost $500 installed. How long must the treated part last to be the preferred alternative, assuming 10% interest?

**9-75** Neither of the following machines has any net salvage value.

|                      | A        | B        |
| -------------------- | -------- | -------- |
| Original cost        | $55,000  | $75,000  |
| Annual expenses      |          |          |
|   Operation | 9,500    | 7,200    |
|   Maintenance | 5,000  | 3,000    |
|   Taxes and insurance | 1,700 | 2,250 |

At what useful life are the machines equivalent if

(a) 10% interest is used in the computations?

(b) 0% interest is used in the computations?

**9-76** A machine costs $5240 and produces benefits of $1000 at the end of each year for 8 years. Assume an annual interest rate of 10%.

(a) What is the payback period (in years)?

(b) What is the breakeven point (in years)?

(c) Since the answers in (a) and (b) are different, which one is "correct"?

**9-77** Analyze Problem 9-69 again with the following changes:

(a) What if the property is purchased for $12,000?

(b) What if the yearly net profit is $925?

(c) What if it is sold after 7 years?

(d) What if (a), (b), and (c) happen simultaneously?

**9-78** Analyze Problem 9-55 again with the following changes:

(a) What if the Fiasco is more reliable than expected, so that its operating cost is $0.075/km?

(b) What if Tom drives only 9000 km/year?

(c) What if Tom's interest rate is 6% annually?

(d) What if (a), (b), and (c) happen simultaneously?

**9-79** Analyze Problem 9-63 again with the following changes:

(a) What if $B$'s first cost is $1200?

(b) What if $B$'s annual benefit is $280?

(c) What if the MARR is 10% annually?

(d) What if (a), (b), and (c) happen simultaneously?

**9-80** Analyze Problem 9-74 again with the following changes:

(a) What if the installed cost of the corrosion-treated part is $600?

(b) What if the untreated part will last only 4 years?

(c) What if the MARR is 12% annually?

(d) What if (a), (b), and (c) happen simultaneously?

## Minicases

**9-81** A proposed steel mill may include a co-generation electrical plant. This plant will add $2.3M in first cost with net annual savings of $0.27M considering operating costs and electrical bills. The plant will have a $0.4M salvage value after 25 years. The firm uses an interest rate of 12% and present worth index (PWI) in its decision making.

The public utility offers a subsidy for co-generation facilities because it will not have to invest as much in new capacity. This subsidy is calculated as 20% of the co-generation facility's first cost, but it is paid annually. The utility calculates the subsidy using a benefit–cost ratio at 8% and a life of 20 years.

(a) Is the plant economically justifiable to the firm without the subsidy? What is the PWI?

(b) What is the annual subsidy?

(c) Is the plant economically justifiable to the firm with the subsidy? Now what is the PWI?

(*d*) How important is the difference in interest rates, and how does it affect these results?

(*e*) How important is the difference in horizons, and how does it affect these results?

**9-82** Assume a cost improvement project has only a first cost of $100,000 and a monthly net savings, *M*. There is no salvage value. Graph the project's IRR for payback periods from 6 months to the project's life of *N* years. The firm accepts projects with a 2-year payback period or a 20% IRR. When are these standards consistent and when are they not?

(*a*) Assume that $N = 3$ years.

(*b*) Assume that $N = 5$ years.

(*c*) Assume that $N = 10$ years.

(*d*) What recommendation do you have for the firm about its project acceptance criteria?

## CASES

**The following cases from *Cases in Engineering Economy 2^{nd}* are suggested as matched with this chapter.**

**CASE 16**   **Great White Hall**

Proposal comparison using B/C analysis for RFP with unclear specifications.

**CASE 17**   **A Free Lunch?**

Is proposal too good to be true from two perspectives? Realistic (unordered) statement of facts.

# UNCERTAINTY IN FUTURE EVENTS

## Video Game Development and Uncertainty

Today, most computing devices, from laptops to music players to cell phones, are being equipped with technologies that allow users to run a wide range of products. Consumers can download and install applications directly from the Internet, and firms rely on this capability to reach as wide an audience as possible. Unfortunately, developing products that run on multiple platforms adds a great deal of uncertainty to the product-development process.

The army has licensed several virtual software environment products to use for training, education, and outreach that are required to operate on many product architectures. The product developers must build the software environments so clients can use them across a wide range of different PC configurations and Internet speeds. In a closed environment, where every PC has the exact same configuration, testing is easy. However, software that is intended to be distributed to the public must be able to run on all common PC configurations.

One of the most successful projects has been America's Army. The series of video games was developed to help the U.S. Army with both public relations and with recruitment. Since its first release in 2002, America's Army has had over two dozen versions released. Like most software development projects, each version of America's Army has placed time and budget constraints on the project manager. A typical software development cycle allows for minor tweaks and bug fixes prior to product release. However, there are times when a test reveals a critical issue that requires more resources to fix. Most of the time, the source of the critical issues, like an operating system patch that creates a major change in how applications access the system, lies outside the development team's control.

When patches and upgrades are announced and released, it is uncertain whether there will be problems and how significant they may be.

Complexity of the software can also affect uncertainty in the project. Some small projects, such as improving how a vehicle performs in the game, may take only a few weeks to complete. Other projects, such as adding a new training scenario, could take several months and require a dozen or more programmers. A large new training scenario might also require programmers to develop multiple terrains, different vehicles, and a range of new equipment types for the soldiers; all of these programming efforts require extensive testing and add to the uncertainty in budgeting and scheduling the project. ■ ■ ■

*Contributed by Paul Componation, University of Alabama in Huntsville, and Marsha Berry, U.S. Army Research, Development & Engineering Command (AMRDEC), Software Engineering Directorate.*

## QUESTIONS TO CONSIDER

1. As a software developer how would you quantify the uncertainty in a specific software development effort?

2. How would you estimate the impact on your project's schedule and costs when you have to deal with uncertainty in your software testing program?

3. The number of new applications for software has increased dramatically over the past decade. If you are developing a new software product, how will you prepare it not only to meet all the applications today but also to deal with future unforeseen applications?

4. How do you economically justify projects whose payoff is in improved recruiting and training rather than revenue from product sales?

### After Completing This Chapter...

*You should be able to:*

- Use a range of estimated variables to evaluate a project.
- Describe possible outcomes with probability distributions.
- Combine probability distributions for individual variables into joint probability distributions.
- Use expected values for economic decision making.
- Use economic decision trees to describe and solve more complex problems.
- Measure and consider risk when making economic decisions.
- Understand how simulation can be used to evaluate economic decisions.

An assembly line is built after the engineering economic analysis has shown that the anticipated product demand will generate profits. A new motor, heat exchanger, or filtration unit is installed after analysis has shown that future cost savings will economically justify current costs. A new road, school, or other public facility is built after analysis has shown that the future demand and benefits justify the present cost to build. However, future performance of the assembly line, motor, and so on is uncertain, and demand for the product or public facility is more uncertain.

Engineering economic analysis is used to evaluate projects with long-term consequences when the time value of money matters. Thus, it must concern itself with future consequences; but describing the future accurately is not easy. In this chapter we consider the problem of evaluating the future. The easiest way to begin is to make a careful estimate. Then we examine the possibility of predicting a range of possible outcomes. Finally, we consider what happens when the probabilities of the various outcomes are known or may be estimated. We will show that the tools of probability are quite useful for economic decision making.

## ESTIMATES AND THEIR USE IN ECONOMIC ANALYSIS

Economic analysis requires evaluating the future consequences of an alternative. In practically every chapter of this book, there are cash flow tables and diagrams that describe precisely the costs and benefits for future years. We don't really believe that we can exactly foretell a future cost or benefit. Instead, our goal is to select a single value representing the *best* estimate that can be made.

We recognize that estimated future consequences are not precise and that the actual values will be somewhat different from our estimates. Even so, it is likely we have made the tacit assumption that these estimates *are* correct. We know estimates will not always turn out to be correct; yet we treat them like facts once they are *in* the economic analysis. We do the analysis as though the values were exact. This can lead to trouble. If actual costs and benefits are different from the estimates, an undesirable alternative may be selected. This is because the variability of future consequences is concealed by assuming that the estimates will actually occur. The problem is illustrated by Example 10-1.

### EXAMPLE 10–1

Two alternatives are being considered. The best estimates for the various consequences are as follows:

|  | A | B |
|---|---|---|
| Cost | $1000 | $2000 |
| Net annual benefit | 150 | 250 |
| End-of-useful-life salvage value | 100 | 400 |
| Useful life, in years | 10 | 10 |

If interest is $3\frac{1}{2}\%$, which alternative has the better net present worth (NPW)?

## SOLUTION

**Alternative A**

$$NPW = -1000 + 150(P/A, 3^{1}/2\%, 10) + 100(P/F, 3^{1}/2\%, 10) \quad \text{or} \quad = -1000 - PV(i, n, A, F)^{*}$$

$$= -1000 + 150(8.317) + 100(0.7089) \qquad\qquad = -1000 - PV(3.5\%, 10, 150, 100)$$

$$= -1000 + 1248 + 71 \qquad\qquad\qquad\qquad = -1000 - (-318.4)$$

$$= +\$319 \qquad\qquad\qquad\qquad\qquad\qquad = \$318.4$$

*Multiply PV by −1 to change sign.

**Alternative B**

$$NPW = -2000 + 250(P/A, 3^{1}/2\%, 10) + 400(P/F, 3^{1}/2\%, 10) \quad \text{or} \quad = -2000 - PV(i, n, A, F)$$

$$= -2000 + 250(8.317) + 400(0.7089) \qquad\qquad = -2000 - PV(3.5\%, 10, 250, 400)$$

$$= -2000 + 2079 + 284 \qquad\qquad\qquad\qquad = 2000 - (-2362.7)$$

$$= +\$363 \qquad\qquad\qquad\qquad\qquad\qquad = \$362.7$$

*Alternative B,* with its larger NPW, would be selected.

### Alternate Formation of Example 10-1

Suppose that at the end of 10 years, the actual salvage value for *B* were $300 instead of the $400, the best estimate. If all the other estimates are correct, is *B* still the preferred alternative?

## SOLUTION

**Revised B**

$$NPW = -2000 + 250(P/A, 3^{1}/2\%, 10) + 300(P/F, 3^{1}/2\%, 10) \quad \text{or} = -2000 - PV(i, n, A, F)$$

$$= -2000 + 250(8.317) + 300(0.7089) \qquad\qquad = -2000 - PV(3.5\%, 250, 300)$$

$$= -2000 + 2079 + 213 \qquad\qquad\qquad\qquad = -2000 - (-2291.8)$$

$$= +\$292 \qquad\qquad\qquad\qquad\qquad\qquad = \$291.8$$

$$\rightarrow A \text{ is now the preferred alternative.}$$

Example 10-1 shows that the change in the salvage value of Alternative *B* actually results in a change of preferred alternative. Thus, a more thorough analysis of Example 10-1 would consider (1) which values are uncertain, (2) whether the uncertainty is ±5% or −50 to +80%, and (3) which uncertain values lead to different decisions. A more thorough analysis, which is done with the tools of this chapter, determines which decision is better over the range of possibilities. Explicitly considering uncertainty lets us make better decisions. The tool of breakeven analysis is illustrated in Example 10-2.

## EXAMPLE 10–2

Use the data of Example 10-1 to compute the sensitivity of the decision to the Alt. $B$ salvage value by computing the breakeven value. For Alt. $A$, NPW $= +319$. For breakeven between the alternatives,

$$NPW_A = NPW_B$$

$$+319 = -2000 + 250(P/A, 3^1/2\%, 10) + \text{Salvage value}_B (P/F, 3^1/2\%, 10)$$

$$= -2000 + 250(8.317) + \text{Salvage value}_B (0.7089) \quad \text{or} \quad = -2319 + 250 + \text{Salvage value}$$

$$\boxed{PV} \quad \boxed{PMT} \qquad \boxed{FV}$$

$$= FV(i, n, A, P)$$

$$= FV(3.5\%, 10, 250, -2319)$$

$$= \$338.3$$

At the breakeven point

$$\text{Salvage value}_B = \frac{319 + 2000 - 2079}{0.7089} = \frac{240}{0.7089} = \$339$$

When Alt. $B$ salvage value $> \$339$, $B$ is preferred; when $< \$339$, $A$ is preferred.

Breakeven analysis, as shown in Example 10-2, is one means of examining the impact of the variability of some estimate on the outcome. It helps by answering the question, How much variability can a parameter have before the decision will be affected? While the preferred decision depends on whether the salvage value is above or below the breakeven value, the economic difference between the alternatives is small when the salvage value is "close" to breakeven. Breakeven analysis does not solve the basic problem of how to take the inherent variability of parameters into account in an economic analysis. This will be considered next.

## A RANGE OF ESTIMATES

It is usually more realistic to describe parameters with a range of possible values, rather than a single value. A range could include an **optimistic** estimate, the **most likely** estimate, and a **pessimistic** estimate. Then, the economic analysis can determine whether the decision is sensitive to the range of projected values.

## EXAMPLE 10–3

A firm is considering an investment. The most likely data values were found during the feasibility study. Analyzing past data of similar projects shows that optimistic values for the first cost and the annual benefit are 5% better than most likely values. Pessimistic values are 15% worse.

The firm's most experienced project analyst has estimated the values for the useful life and salvage value.

|  | Optimistic | Most Likely | Pessimistic |
|---|---|---|---|
| Cost | $950 | $1000 | $1150 |
| Net annual benefit | 210 | 200 | 170 |
| Salvage value | 100 | 0 | 0 |
| Useful life, in years | 12 | 10 | 8 |

Compute the rate of return for each estimate. If a 10% before-tax minimum attractive rate of return is required, is the investment justified under all three estimates? If it is justified only under some estimates, how can these results be used?

**SOLUTION**

**Optimistic Estimate**

$$PW = 0 = -\$950 + 210(P/A, \text{IRR}_{opt}, 12) + 100(P/F, \text{IRR}_{opt}, 12)$$

$$\text{IRR}_{opt} = i(n, A, P, F) = i(12, 210, -950, 100)$$

$$= 19.8\%$$

**Most Likely Estimate**

$$PW = 0 = -\$1000 + 200(P/A, \text{IRR}_{most\,likely}, 10)$$

$$(P/A, \text{IRR}_{most\,likely}, 10) = 1000/200 = 5 \rightarrow \text{IRR}_{most\,likely} = 15.1\%$$

or

$$\text{IRR}_{most\,likely} = i(n. A, P, F) = i(10, 200, -1000, 0) = 15.1\%$$

**Pessimistic Estimate**

$$PW = 0 = -\$1150 + 170(P/A, \text{IRR}_{pess}, 8)$$

$$(P/A, \text{IRR}_{pess}, 8) = 1150/170 = 6.76 \rightarrow \text{IRR}_{pess} = 3.9\%$$

or

$$\text{IRR}_{pess} = i(n, A, P, F) = i(8, 170, -1150, 0) = 3.9\%$$

From the calculations we conclude that the rate of return for this investment is most likely to be 15.1%, but might range from 3.9% to 19.8%. The investment meets the 10% MARR criterion for two of the estimates. These estimates can be considered to be scenarios of what may happen with this project. Since one scenario indicates that the project is not attractive, we need to have a method of weighting the scenarios or considering how likely each is.

Example 10-3 made separate calculations for the sets of optimistic, most likely, and pessimistic values. The range of scenarios is useful. However, if there are more than a few uncertain variables, it is unlikely that all will prove to be optimistic (best case) or most likely or pessimistic (worst case). It is more likely that many parameters are the most likely values, while some are optimistic and some are pessimistic.

This can be addressed by using Equation 10-1 to calculate average or mean values for each parameter. Equation 10-1 puts four times the weight on the most likely value than on the other two. This equation has a long history of use in project management to estimate activity completion times. It is an approximation with the beta distribution.

$$\text{Mean value} = \frac{\text{Optimistic value} + 4(\text{Most likely value}) + \text{Pessimistic value}}{6} \tag{10-1}$$

This approach is illustrated in Example 10-4.

## EXAMPLE 10–4

Solve Example 10-3 by using Equation 10-1. Compute the resulting mean rate of return.

### SOLUTION

Compute the mean for each parameter:

$$\text{Mean cost} = [950 + 4(1000) + 1150]/6 = \$1016.7$$

$$\text{Mean net annual benefit} = [210 + 4(200) + 170]/6 = \$196.7$$

$$\text{Mean useful life} = [12 + 4(10) + 8]/6 = 10.0$$

$$\text{Mean salvage life} = 100/6 = \$16.7$$

Compute the mean rate of return:

$$\text{PW} = 0 = -\$1016.7 + 196.7(P/A, \text{IRR}_{\text{beta}}, 10) + 16.7(P/F, \text{IRR}_{\text{beta}}, 10)$$

$$\text{IRR}_{\text{beta}} = i(n, A, P, F) = i(10, 196.7, -1016.7, 16.7)$$

$$= 14.3\%$$

Example 10-3 gave a most likely rate of return (15.1%) that differed from the mean rate of return (14.3%) computed in Example 10-4. These values are different because the former is based exclusively on the most likely values and the latter takes into account the variability of the parameters.

In examining the data, we see that the pessimistic values are further away from the most likely values than are the optimistic values. This is a common occurrence. For example, a savings of 10–20% may be the maximum possible, but a cost overrun can be 50%, 100%, or even more. This causes the resulting weighted mean values to be less favorable than the most likely values. As a result, the mean rate of return, in this example, is less than the rate of return based on the most likely values.

## PROBABILITY

We all have used probabilities. For example, what is the probability of getting a "head" when flipping a coin? Using a model that assumes that the coin is fair, both the head and tail outcomes occur with a probability of 50%, or $1/2$. This probability is the likelihood of an event in a single trial. It also describes the long-run relative frequency of getting heads in many trials (out of 50 coin flips, we expect to average 25 heads).

Probabilities can also be based on data, expert judgment, or a combination of both. Past data on weather and climate, on project completion times and costs, and on highway traffic are combined with expert judgment to forecast future events. These examples can be important in engineering economy.

Another example based on long-run relative frequency is the PW of a flood-protection dam that depends on the probabilities of different-sized floods over many years. This would be based on data from past floods and would include many years of observation. An example of a single event that may be estimated by expert judgment is the probability of a successful outcome for a research and development project, which will determine its PW.

All the data in an engineering economy problem may have some level of uncertainty. However, small uncertainties may be ignored, so that more analysis can be done with the large uncertainties. For example, the price of an off-the-shelf piece of equipment may vary by only ±5%. The price could be treated as a known or deterministic value. On the other hand, demand over the next 20 years will have more uncertainty. Demand should be analyzed as a random or stochastic variable. We should establish probabilities for different values of demand.

There are also logical or mathematical rules for probabilities. If an outcome can never happen, then the probability is 0. If an outcome will certainly happen, then the probability is 1, or 100%. This means that probabilities cannot be negative or greater than 1; in other words, they must be within the interval [0, 1], as indicated shortly in Equation 10-2.

Probabilities are defined so that the sum of probabilities for all possible outcomes is 1 or 100% (Equation 10-3). Summing the probability of 0.5 for a head and 0.5 for a tail leads to a total of 1 for the possible outcomes from the coin flip. An exploration well drilled in a potential oil field will have three outcomes (dry hole, noncommercial quantities, or commercial quantities) whose probabilities will sum to one.

Equations 10-2 and 10-3 can be used to check that probabilities are valid. If the probabilities for all but one outcome are known, the equations can be used to find the unknown probability for that outcome (see Example 10-5).

$$0 \leq \text{Probability} \leq 1 \tag{10-2}$$

$$\sum_{j=1 \text{ to } K} P(\text{outcome}_j) = 1, \quad \text{where there are } K \text{ outcomes} \tag{10-3}$$

In a probability course many probability distributions, such as the normal, uniform, and beta are presented. These continuous distributions describe a large population of data. However, for engineering economy it is more common to use 2 to 5 outcomes with discrete probabilities—even though the 2 to 5 outcomes only represent or approximate the range of possibilities.

This is done for two reasons. First, the data often are estimated by expert judgment, so that using 7 to 10 outcomes would be false accuracy. Second, each outcome requires more analysis. In most cases the 2 to 5 outcomes represents the best trade-off between representing the range of possibilities and the amount of calculation required. Example 10-5 illustrates these calculations.

## EXAMPLE 10–5

What are the probability distributions for the annual benefit and life for the following project?

The annual benefit's most likely value is $8000 with a probability of 60%. There is a 30% probability that it will be $5000, and the highest value that is likely $10,000. A life of 6 years is twice as likely as a life of 9 years.

### SOLUTION

For the annual benefit, probabilities are given for only two of the possible outcomes. The third value is found from the fact that the probabilities for the three outcomes must sum to 1 (Equation 10-3).

$$1 = P(\text{Benefit is } \$5000) + P(\text{Benefit is } \$8000) + P(\text{Benefit is } \$10,000)$$
$$P(\text{Benefit is } \$10,000) = 1 - 0.6 - 0.3 = 0.1$$

The probability distribution can then be summarized in a table. The histogram or relative frequency diagram is Figure 10-1.

| Annual benefit | $5000 | $8000 | $10,000 |
|---|---|---|---|
| Probability | 0.3 | 0.6 | 0.1 |

FIGURE 10–1 Probability distribution for annual benefit.

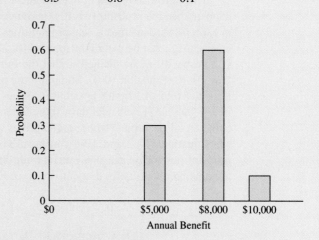

For the life's probability distribution, the problem statement tells us

$$P(\text{life is 6 years}) = 2P(\text{life is 9 years})$$

Equation 10-3 can be applied to write a second equation for the two unknown probabilities:

$$P(6) + P(9) = 1$$

Combining these, we write

$$2P(9) + P(9) = 1$$
$$P(9) = 1/3$$
$$P(6) = 2/3$$

The probability distribution for the life is $P(6) = 66.7\%$ and $P(9) = 33.3\%$.

# JOINT PROBABILITY DISTRIBUTIONS

Example 10-5 constructed the probability distributions for a project's annual benefit and life. These examples show how likely each value is for the input data. We would like to construct a similar probability distribution for the project's present worth. This is the distribution that we can use to evaluate the project. That present worth depends on both input probability distributions, so we need to construct the *joint* probability distribution for the different combinations of their values.

For this introductory text, we assume that two random variables such as the annual benefit and life are unrelated or statistically independent. This means that the *joint* probability of a combined event (Event A defined on the first variable and Event B on the second variable) is the product of the probabilities for the two events. This is Equation 10-4:

$$\text{If A and B are independent, then } P(A \text{ and } B) = P(A) \times P(B) \qquad (10\text{-}4)$$

For example, flipping a coin and rolling a die are statistically independent. Thus, the probability of {flipping a head and rolling a 4} equals the probability of a {heads} $= 1/2$ times the probability of a $\{4\} = 1/6$, for a joint probability $= 1/12$.

The number of outcomes in the joint distribution is the product of the number of outcomes in each variable's distribution. Thus, for the coin and the die, there are 2 times 6, or 12 combinations. Each of the 2 outcomes for the coin is combined with each of the 6 outcomes for the die.

Some variables are not statistically independent, and the calculation of their joint probability distribution is more complex. For example, a project with low revenues may be terminated early and one with high revenues may be kept operating as long as possible. In these cases annual cash flow and project life are not independent. While this type of relationship can sometimes be modeled with economic decision trees (covered later in this chapter), we will limit our coverage in this text to the simpler case of independent variables.

Example 10-6 uses the three values and probabilities for the annual benefit and the two values and probabilities for the life to construct the six possible combinations. Then the values and probabilities are constructed for the project's PW.

## EXAMPLE 10–6

The project described in Example 10-5 has a first cost of $25,000. The firm uses an interest rate of 10%. Assume that the probability distributions for annual benefit and life are unrelated or statistically independent. Calculate the probability distribution for the PW.

### SOLUTION

Since there are three outcomes for the annual benefit and two outcomes for the life, there are six combinations. The first four columns of the following table show the six combinations of life and annual benefit. The probabilities in columns 2 and 4 are multiplied to calculate the joint probabilities in column 5. For example, the probability of a low annual benefit and a short life is $0.3 \times 2/3$, which equals 0.2 or 20%.

The PW values include the $25,000 first cost and the results of each pair of annual benefit and life. For example, the PW for the combination of high benefit and long life is

$$PW_{\$10,000,9} = -25,000 + 10,000(P/A, 10\%, 9) = -25,000 + 10,000(5.759) = \$32,590$$

| Annual Benefit | Probability | Life (years) | Probability | Joint Probability | PW |
|---|---|---|---|---|---|
| $ 5,000 | 30% | 6 | 66.7% | 20.0% | −$ 3,224 |
| 8,000 | 60 | 6 | 66.7 | 40.0 | +9,842 |
| 10,000 | 10 | 6 | 66.7 | 6.7 | 18,553 |
| 5,000 | 30 | 9 | 33.3 | 10.0 | 3,795 |
| 8,000 | 60 | 9 | 33.3 | 20.0 | 21,072 |
| 10,000 | 10 | 9 | 33.3 | 3.3 | 32,590 |
| | | | | 100.0% | |

Figure 10-2 shows the probabilities for the PW in the form of a histogram for relative frequency distribution, or probability distribution function.

**FIGURE 10–2** Probability distribution function for PW.

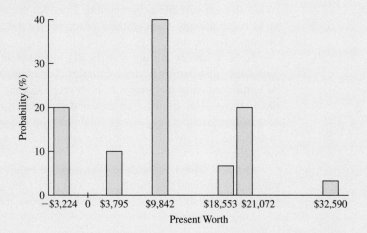

This probability distribution function shows that there is a 20% chance of having a negative PW. It also shows that there is a small (3.3%) chance of the PW being $32,590. The three values used to describe possible annual benefits for the project and the two values for life have been combined to describe the uncertainty in the project's PW.

Creating a distribution, as in Example 10-6, gives us a much better understanding of the possible PW values along with their probabilities. The three possibilities for the annual benefit and the two for the life are representative of the much broader set of possibilities that really exist. Optimistic, most likely, and pessimistic values are a good way to represent the uncertainty about a variable.

Similarly the six values for the PW represent the much broader set of possibilities. The 20% probability of a negative PW is one measure of risk that we will talk about later in the chapter.

Some problems, such as Examples 10-3 and 10-4, have so many variables or different outcomes that constructing the joint probability distribution is arithmetically burdensome. If the values in Equation 10-1 are treated as a discrete probability distribution function, the probabilities are $1/6, 2/3, 1/6$. With an optimistic, most likely, and pessimistic outcome for each of 4 variables, there are $3^4 = 81$ combinations. In Examples 10-3 and 10-4, the salvage value has only two distinct values, so there are still $3 \times 3 \times 3 \times 2 = 54$ combinations.

When the problem is important enough, the effort to construct the joint probability distribution is worthwhile. It gives the analyst and the decision maker a better understanding of what may happen. It is also needed to calculate measures of a project's risk. While spreadsheets can automate the arithmetic, simulation (described at the end of the chapter) can be a better choice when there are a large number of variables and combinations.

## EXPECTED VALUE

For any probability distribution we can compute the **expected value (EV)** or weighted arithmetic average (mean). To calculate the EV, each outcome is weighted by its probability, and the results are summed. This is NOT the simple average or unweighted mean. When the class average on a test is computed, this is an unweighted mean. Each student's test has the same weight. This simple "average" is the one that is shown by the button $\bar{x}$ on many calculators.

The expected value is a weighted average, like a student's grade point average (GPA). To calculate a GPA, the grade in each class is weighted by the number of credits. For the expected value of a probability distribution, the weights are the probabilities.

This is described in Equation 10-5. We saw in Example 10-4, that these expected values can be used to compute a rate of return. They can also be used to calculate a present worth as in Example 10-7.

$$\text{Expected value} = \text{Outcome}_A \times P(A) + \text{Outcome}_B \times P(B) + \cdots \qquad (10\text{-}5)$$

### EXAMPLE 10-7

The first cost of the project in Example 10-5 is $25,000. Use the expected values for annual benefits and life to estimate the present worth. Use an interest rate of 10%.

### SOLUTION

$$\text{EV}_{\text{benefit}} = 5000(0.3) + 8000(0.6) + 10{,}000(0.1) = \$7300$$
$$\text{EV}_{\text{life}} = 6(2/3) + 9(1/3) = 7 \text{ years}$$

The PW using these values is

$$\text{PW(EV)} = -25{,}000 + 7300(P/A, 10\%, 7) = -25{,}000 + 6500(4.868) = \$10{,}536$$

[*Note:* This is the present worth of the expected values, PW(EV), not the expected value of the present worth, EV(PW). It is an easy value to calculate that approximates the EV(PW), which will be computed from the joint probability distribution found in Example 10-6.]

Example 10-7 is a simple way to approximate the project's expected PW. But the true expected value of the PW is somewhat different. To find it, we must use the joint probability distribution for benefit and life, and the resulting probability distribution function for PW that was derived in Example 10-6. Example 10-8 shows the expected value of the PW or the EV(PW).

## EXAMPLE 10–8

Use the probability distribution function of the PW that was derived in Example 10–6 to calculate the EV(PW). Does this indicate an attractive project?

### SOLUTION

The table from Example 10-6 can be reused with one additional column for the weighted values of the PW (= PW × probability). Then, the expected value of the PW is calculated by summing the column of present worth values that have been weighted by their probabilities.

| Annual Benefit | Probability | Life (years) | Probability | Joint Probability | PW | PW×Joint Probability |
|---|---|---|---|---|---|---|
| $ 5,000 | 30% | 6 | 66.7% | 20.0% | −$ 3,224 | −$ 645 |
| 8,000 | 60 | 6 | 66.7 | 40.0 | 9,842 | 3,937 |
| 10,000 | 10 | 6 | 66.7 | 6.7 | 18,553 | 1,237 |
| 5,000 | 30 | 9 | 33.3 | 10.0 | 3,795 | 380 |
| 8,000 | 60 | 9 | 33.3 | 20.0 | 21,072 | 4,214 |
| 10,000 | 10 | 9 | 33.3 | 3.3 | 32,590 | 1,086 |
| | | | | 100.0% | | EV(PW) = $10,209 |

With an expected PW of $10,209, this is an attractive project. While there is a 20% chance of a negative PW, the possible positive outcomes are larger and more likely. Having analyzed the project under uncertainty, we are much more knowledgeable about the potential result of the decision to proceed.

The $10,209 value is more accurate than the approximate value calculated in Example 10-7. The values differ because PW is a nonlinear function of the life. The more accurate value of $10,209 is lower because the annual benefit values for the longer life are discounted by $1/(1+i)$ for more years.

In Examples 10-7 and 10-8, the question was whether the project had a positive PW. With two or more alternatives, the criterion would have been to maximize the PW. With equivalent uniform annual costs (EUACs) the goal is to minimize the EUAC. Example 10-9 uses the criterion of minimizing the EV of the EUAC to choose the best height for a dam.

## EXAMPLE 10–9

A dam is being considered to reduce river flooding. But if a dam is built, what height should it be? Increasing the dam's height will (1) reduce a flood's probability, (2) reduce the damage when floods occur, and (3) cost more. Which dam height minimizes the expected total annual cost? The state uses an interest rate of 5% for flood protection projects, and all the dams should last 50 years.

| Dam Height (ft) | First Cost |
|---|---|
| No dam | $ 0 |
| 20 | 700,000 |
| 30 | 800,000 |
| 40 | 900,000 |

### SOLUTION

The easiest way to solve this problem is to choose the dam height with the lowest equivalent uniform annual cost (EUAC). Calculating the EUAC of the first cost requires multiplying the first cost by $(A/P, 5\%, 50)$. For example, for the dam 20 ft high, this is $700,000(A/P, 5\%, 50) =$ $38,344.

Calculating the annual expected flood damage cost for each alternative is simplified because the term for the $P$(no flood) is zero, because the damages for no flood are $0. Thus we need to calculate only the term for flooding. This is done by multiplying the $P$(flood) times the damages if a flood happens. For example, the expected annual flood damage cost with no levee is $0.25 \times \$800,000$, or $200,000.

Then the EUAC of the first cost and the expected annual flood damage are added together to find the total EUAC for each height. The 30 ft dam is somewhat cheaper than the 40 ft dam.

| Dam Height (ft) | EUAC of First Cost | Annual P (flood) > Height | | Damages if Flood Occurs | | Expected Annual Flood Damages | Total Expected EUAC |
|---|---|---|---|---|---|---|---|
| No dam | $ 0 | 0.25 | × | $800,000 | = | $200,000 | $200,000 |
| 20 | 38,344 | 0.05 | × | 500,000 | = | 25,000 | 63,344 |
| 30 | 43,821 | 0.01 | × | 300,000 | = | 3000 | 46,821 |
| 40 | 49,299 | 0.002 | × | 200,000 | = | 400 | 49,699 |

## ECONOMIC DECISION TREES

Some engineering projects are more complex, and evaluating them properly is correspondingly more complex. For example, consider a new product with potential sales volumes ranging from low to high. If the sales volume is low, then the product may be discontinued early in its potential life. On the other hand, if sales volume is high, additional capacity may be added to the assembly line and new product variations may be added. This can be modeled with a decision tree.

The following symbols are used to model decisions with decision trees:

*Decision node* ▷——→$D_2$: Decision maker chooses 1 of the available paths.

*Chance node* ◯——→$C_2$: Represents a probabilistic (chance) event. Each pos-
sible outcome $(C_1, C_2, \ldots, C_Y)$ has a probability $(p_1, p_2, \ldots, p_y)$ associated with it.

*Outcome node* —→▢: Shows result for a particular path through the decision tree.

*Pruned branch* —‖→: The double hash mark indicates that a branch has been pruned because another branch has been chosen. This can happen only at decision nodes, not at chance nodes. The term "pruned" is chosen to correspond to the gardener's practice of trimming or pruning off branches to make a tree or bush healthier.

Figure 10-3 illustrates how decision nodes ▷, chance nodes ◯, and outcome nodes ▢ can be used to describe a problem's structure. Details such as the probabilities and costs can be added on the branches that link the nodes. With the branches from decision and chance nodes, the model becomes a decision tree.

Figure 10-3 illustrates that decision trees describe the problem by starting at the decision that must be made and then adding chance and decision nodes in the proper logical sequence. Thus describing the problem starts at the first step and goes forward in time with sequences of decision and chance nodes.

To make the decision, calculations begin with the final nodes in the tree. Since they are the final nodes, enough information is available to evaluate them. At decision nodes the criterion is either to maximize PW or to minimize EUAC. At chance nodes an expected value for PW or EUAC is calculated.

Once all nodes that branch from a node have been evaluated, the originating node can be evaluated. If the originating node is a decision node, choose the branch with the best PW or EUAC and place that value in the node. If the originating node is a chance node, calculate the expected value and place that value in the node. This process "rolls back" values from the terminal nodes in the tree to the initial decision. Example 10-10 illustrates this process.

## EXAMPLE 10–10

What decision should be made on the new product summarized in Figure 10-3? What is the expected value of the product's PW? The firm uses an interest rate of 10% to evaluate projects. If the product is terminated after one year, the capital equipment has a salvage value of $550,000 for use with other new products. If the equipment is used for 8 years, the salvage value is $0.

**FIGURE 10–3** Economic decision tree for new product.

## SOLUTION

Evaluating decision trees is done by starting with the end outcome nodes and the decisions that lead to them. In this case the decisions are whether to terminate after 1 year if sales volume is low and whether to expand after 1 year if sales volume is high.

The decision to terminate the product depends on which is more valuable, the equipment's salvage value of $550,000 or the revenue of $100,000 per year for 7 more years. The worth ($PW_1$) of the salvage value is $550,000. The worth ($PW_1$) of the revenue stream at the end of Year 1 shown in node 8 is

$$PW_1 \text{ for node } 8 = 100,000(P/A, 10\%, 7)$$

$$= 100,000(4.868) = \$486,800$$

Thus, terminating the product and using the equipment for other products is better. We enter the two "present worth" values at the end of Year 1 in nodes 7 and 8. We make the *arc to node 7 bold* to indicate that it is our preferred choice at node 4. We use a *double hash mark* to show that we're *pruning the arc to node 8* to indicate that it has been rejected as an inferior choice at node 4.

The decision to expand at node 6 could be based on whether the $800,000 first cost for expansion can be justified based on increasing annual revenues for 7 years by $200,000 per year. However, this is difficult to show on the tree. It is easier to calculate the "present worth" values at the end of Year 1 for each of the two choices. The worth ($PW_1$) of node 9 (expand) is

$$PW_1 \text{ for node } 9 = -800,000 + 600,000(P/A, 10\%, 7)$$

$$= -800,000 + 600,000(4.868)$$

$$= \$2,120,800$$

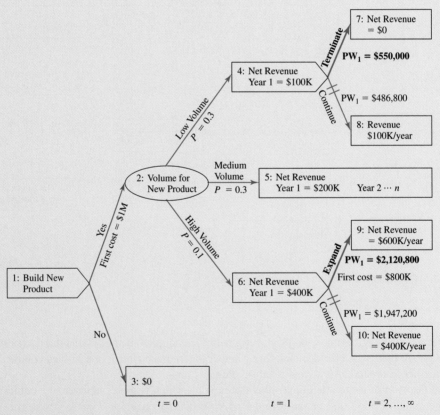

**FIGURE 10–4**  Partially solved decision tree for new product.

The value of node 10 (continue without expanding) is

$$PW_1 \text{ for node } 10 = 400,000(P/A, 10\%, 7)$$
$$= 400,000(4.868)$$
$$= \$1,947,200$$

This is $173,600 less than the expansion node, so the expansion should happen if volume is high. Figure 10-4 summarizes what we know at this stage of the process.

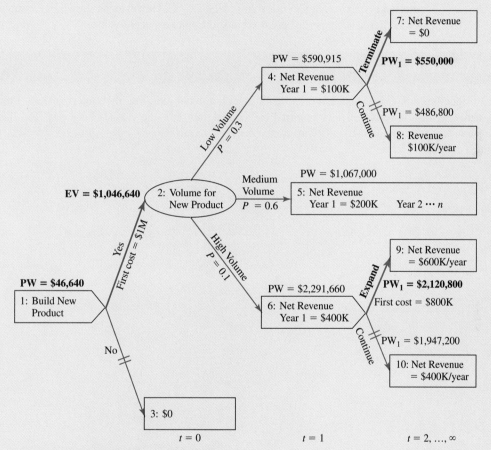

**FIGURE 10-5** Solved decision tree for new product.

The next step is to calculate the PW (Time 0) at nodes 4, 5, and 6.

PW at node $4 = (100,000 + 550,000)(P/F, 10\%, 1) = 650,000(0.9091) = \$590,915$

PW at node $5 = (200,000)(P/A, 10\%, 8) = 200,000(5.335) = \$1,067,000$

PW at node $6 = [400,000 - 800,000 + 600,000(P/A, 10\%, 7)] \, (P/F, 10\%, 1)$
$$= [-400,000 + 600,000(4.868)](0.9091) = \$2,291,660$$

Now the expected value at node 2 can be calculated:

$$\text{EV at node 2} = 0.3(590,915) + 0.6(1,067,000) + 0.1(2,291,660) = \$1,046,640$$

The cost of selecting node 2 is $1,000,000, so proceeding with the product has an expected PW of $46,640. This is greater than the $0 for not building the project. So the decision is to build. Figure 10-5 is the decision tree at the final stage.

Example 10-10 is representative of many problems in engineering economy. The main criterion is maximizing PW or minimizing EUAC. However, as shown in Example 10-11, other criteria, such as risk, are used in addition to expected value.

## EXAMPLE 10–11

Consider the economic evaluation of collision and comprehensive (fire, theft, etc.) insurance for a car. This insurance is typically required by lenders, but once the car has been paid for, this insurance is not required. (Liability insurance *is* a legal requirement.)

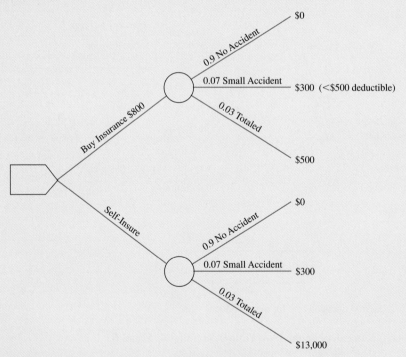

**FIGURE 10-6** Decision tree for buying auto collision insurance.

Figure 10-6 begins with a decision node with two alternatives for the next year. Insurance will cost $800 per year with a $500 deductible if a loss occurs. The other option is to self-insure, which means to go without buying collision and comprehensive insurance. Then if a loss occurs, the owner must replace the vehicle with money from savings or a loan, or do without a vehicle until the owner can afford to replace it.

Three accident severities are used to represent the range of possibilities: a 90% chance of no accident, a 7% chance of a small accident (at a cost of $300, which is less than the deductible), and a 3% chance of totaling the $13,000 vehicle. Since our driving habits are likely to be the same with and without insurance, the accident probabilities are the same for both chance nodes.

Even though this is a text on engineering economy, we have simplified the problem and ignored the difference in timing of the cash flows. Insurance payments are made at the beginning of the covered period, and accident costs occur during the covered period. Since car insurance is usually paid semiannually, the results of the economic analysis are not changed significantly by the simplification. We focus on the new concepts of expected value, economic decision trees, and risk.

What are the expected values for each alternative, and what decision is recommended?

### SOLUTION

The expected values are computed by using Equation 10-5. If insured, the maximum cost equals the deductible of $500. If self-insured, the cost is the cost of the accident.

$$EV_{accident \ w/ins.} = (0.9)(0) + (0.07)(300) + (0.03)(500) = \$36$$

$$EV_{accident \ w/o \ ins.} = (0.9)(0) + (0.07)(300) + (0.03)(13,000) = \$411$$

Thus, buying insurance lowers the expected cost of an accident by $375. To evaluate whether we should buy insurance, we must also account for the cost of the insurance. Thus, these expected costs are combined with the $0 for self-insuring (total $411) and the $800 for insuring (total $836). Thus self-insuring has an expected value cost that is $425 less per year (= $836 − $411). This is not surprising, since the premiums collected must cover both the costs of operating the insurance company and the expected value of the payouts.

This is also an example of *expected values alone not determining the decision*. Buying insurance has an expected cost that is $425 per year higher, but that insurance limits the maximum loss to $500 rather than $13,000. The $425 may be worth spending to avoid that risk.

## RISK

Risk can be thought of as the chance of getting an outcome other than the expected value—with an emphasis on something negative. One common measure of risk is the probability of a loss (see Example 10-6). The other common measure is the **standard deviation** ($\sigma$), which measures the dispersion of outcomes about the expected value. For example, many students have used the normal distribution in other classes. The normal distribution has 68% of its probable outcomes within ±1 standard deviation of the mean and 95% within ±2 standard deviations of the mean.

Mathematically, the standard deviation is defined as the square root of the variance. This term is defined as the weighted average of the squared difference between the outcomes of the random variable $X$ and its mean. Thus the larger the difference between the mean and the values, the larger are the standard deviation and the variance. This is Equation 10-6:

$$Standard \ deviation \ (\sigma) = \sqrt{[EV(X - mean)^2]} \qquad (10\text{-}6)$$

Squaring the differences between individual outcomes and the EV ensures that positive and negative deviations receive positive weights. Consequently, negative values for the standard deviation are impossible, and they instantly indicate arithmetic mistakes. The standard deviation equals 0 if only one outcome is possible. Otherwise, the standard deviation is positive.

This is not the standard deviation formula built into most calculators, just as the weighted average is not the simple average built into most calculators. The calculator formulas are for $N$ equally likely data points from a randomly drawn sample, so that each probability is $1/N$. In economic analysis we will use a weighted average for the squared deviations since the outcomes may not be equally likely.

The second difference is that for calculations (by hand or the calculator), it is easier to use Equation 10-7, which is shown to be equivalent to Equation 10-6 in introductory probability and statistics texts.

$$\text{Standard deviation } (\sigma) = \sqrt{\{EV(X^2) - [EV(X)]^2\}} \qquad (10\text{-}7)$$

$$= \sqrt{\{\text{Outcome}_A^2 \times P(A) + \text{Outcome}_B^2 \times P(B) + \cdots - \text{expected value}^2\}} \qquad (10\text{-}7')$$

This equation is the square root of the difference between the average of the squares and the square of the average. The standard deviation is used instead of the *variance* because the standard deviation is measured in the same units as the expected value. The variance is measured in "squared dollars"—whatever they are.

The calculation of a standard deviation by itself is only a descriptive statistic of limited value. However, as shown in the next section on risk/return trade-offs, it is useful when the standard deviation of each alternative is calculated and these results are compared. But first, some examples of calculating the standard deviation.

## EXAMPLE 10–12  (Example 10-11 continued)

Consider the economic evaluation of collision and comprehensive (fire, theft, etc.) insurance for an auto. One example was described in Figure 10-6. The probabilities and outcomes are summarized in the calculation of the expected values, which was done using Equation 10–5.

$$EV_{\text{accident w/ins.}} = (0.9)(0) + (0.07)(300) + (0.03)(500) = \$36$$

$$EV_{\text{accident w/o ins.}} = (0.9)(0) + (0.07)(300) + (0.03)(13{,}000) = \$411$$

Calculate the standard deviations for insuring and not insuring.

### SOLUTION

The first step is to calculate the EV(outcome$^2$) for each.

$$EV^2_{\text{accident w/ins.}} = (0.9)(0^2) + (0.07)(300^2) + (0.03)(500^2) = 13{,}800$$

$$EV^2_{\text{accident w/o ins.}} = (0.9)(0^2) + (0.07)(300^2) + (0.03)(13{,}000^2) = 5{,}076{,}300$$

Then the standard deviations can be calculated.

$$\sigma_{w/ins.} = \sqrt{EV^2_{w/ins.} - (EV_{w/ins.})^2}$$

$$= \sqrt{(13,800 - 36^2)} = \sqrt{12,504} = \$112$$

$$\sigma_{w/o\ ins.} = \sqrt{EV^2_{w/o\ ins.} - (EV_{w/o\ ins.})^2}$$

$$= \sqrt{(5,076,300 - 411^2)} = \sqrt{4,907,379} = \$2215$$

As described in Example 10-11, the expected value cost of insuring is $836 ($= \$36 + \$800$) and the expected value cost of self-insuring is $411. Thus the expected cost of not insuring is about half the cost of insuring. But the standard deviation of self-insuring is 20 times larger. It is clearly riskier.

Which choice is preferred depends on how much risk one is comfortable with.

As stated before, this is an example of *expected values alone not determining the decision.* Buying insurance has an expected cost that is $425 per year higher, but that insurance limits the maximum loss to $500 rather than $13,000. The $425 may be worth spending to avoid that risk.

## EXAMPLE 10–13  (Example 10-6 continued)

Using the probability distribution for the PW from Example 10–6, calculate the PW's standard deviation.

### SOLUTION

The following table adds a column for (PW$^2$) (probability) to calculate the EV(PW$^2$).

| Annual Benefit | Probability | Life (years) | Probability | Joint Probability | PW | PW× Probability | PW$^2$× Probability |
|---|---|---|---|---|---|---|---|
| $ 5,000 | 30% | 6 | 66.7% | 20.0% | −$ 3,224 | −$ 645 | 2,079,480 |
| 8,000 | 60 | 6 | 66.7 | 40.0 | 9,842 | 3937 | 38,747,954 |
| 10,000 | 10 | 6 | 66.7 | 6.7 | 18,553 | 1237 | 22,950,061 |
| 5,000 | 30 | 9 | 33.3 | 10.0 | 3,795 | 380 | 1,442,100 |
| 8,000 | 60 | 9 | 33.3 | 20.0 | 21,072 | 4214 | 88,797,408 |
| 10,000 | 10 | 9 | 33.3 | 3.3 | 32,590 | 1086 | 35,392,740 |
| | | | | | | EV = $10,209 | 189,409,745 |

$$\text{Standard deviation} = \sqrt{\{EV(X^2) - [EV(X)]^2\}}$$

$$\sigma = \sqrt{\{189,405,745 - [10,209]^2\}} = \sqrt{85,182,064} = \$9229$$

For those with stronger backgrounds in probability than this chapter assumes, let us consider how the standard deviation in Example 10-13 depends on the assumption of independence between the variables. While exceptions exist, a positive statistical dependence between variables often increases the PW's standard deviation. Similarly, a negative statistical dependence between variables often decreases the PW's standard deviation.

## RISK VERSUS RETURN

A graph of risk versus return is one way to consider these items together. Figure 10-7 in Example 10-14 illustrates the most common format. Risk measured by standard deviation is placed on the $x$ axis, and return measured by expected value is placed on the $y$ axis. This is usually done with internal rates of return of alternatives or projects.

### EXAMPLE 10–14

A large firm is discontinuing an older product; thus some facilities are becoming available for other uses. The following table summarizes eight new projects that would use the facilities. Considering expected return and risk, which projects are good candidates? The firm believes it can earn 4% on a risk-free investment in government securities (labeled as Project $F$).

| Project | IRR | Standard Deviation |
|---------|------|--------------------|
| 1 | 13.1% | 6.5% |
| 2 | 12.0 | 3.9 |
| 3 | 7.5 | 1.5 |
| 4 | 6.5 | 3.5 |
| 5 | 9.4 | 8.0 |
| 6 | 16.3 | 10.0 |
| 7 | 15.1 | 7.0 |
| 8 | 15.3 | 9.4 |
| $F$ | 4.0 | 0.0 |

### SOLUTION

Answering the question is far easier if we use Figure 10-7. Since a larger expected return is better, we want to select projects that are as "high up" as possible. Since a lower risk is better, we want to select projects that are as "far left" as possible. The graph lets us examine the trade-off of accepting more risk for a higher return.

**FIGURE 10–7** Risk-versus-return graph.

First, we can eliminate Projects 4 and 5. They are **dominated projects.** Dominated alternatives are no better than another alternative on all measures and inferior on at least one measure. Project 4 is dominated by Project 3, which has a higher expected return and a lower risk. Project 5 is dominated by Projects 1, 2, and 7. All three have a higher expected return and a lower risk.

Second, we look at the **efficient frontier.** This is the blue line in Figure 10-7 that connects Projects $F$, 3, 2, 7, and 6. Depending on the trade-off that we want to make between risk and return, any of these could be the best choice.

Project 1 appears to be inferior to Projects 2 and 7. Project 8 appears to be inferior to Projects 7 and 6. Projects 1 and 8 are inside and not on the efficient frontier.

There are models of risk and return that can allow us to choose between Projects $F$, 3, 2, 7, and 6; but those models are beyond what is covered here.

There is a simple rule of thumb for comparing a project's risk and return for which we would like to thank Joe Hartman. If the expected present worth is at least double the standard deviation of the present worth, then the project is relatively *safe*. For comparison, remember that a normal distribution has about 2.5% of its values less than 2 standard deviations below the mean.

## SIMULATION

Simulation is a more advanced approach to considering risk in engineering economy problems. As such, the following discussion focuses on what it is. As the examples show, spreadsheet functions and add-in packages make simulation easier to use for economic analysis.

Economic **simulation** uses random sampling from the probability distributions of one or more variables to analyze an economic model for many iterations. For each iteration, all variables with a probability distribution are randomly sampled. These values are used to calculate the PW, IRR, or EUAC. Then the results of all iterations are combined to create a probability distribution for the PW, IRR, or EUAC.

Simulation can be done by hand, using a table of random numbers—if there are only a few random variables and iterations. However, results are more reliable as the number of iterations increases, so in practice this is usually computerized. This can be done in Excel using the RAND() function to generate random numbers, as shown in Example 10-15.

Because we were analyzing each possible outcome, the probability distributions earlier in this chapter (and in the end-of-chapter problems) used two or three discrete outcomes. This limited the number of combinations that we needed to consider. Simulation makes it easy to use continuous probability distributions like the uniform, normal, exponential, log normal, binomial, and triangular. Examples 10-15 and 10-16 use the normal and the discrete uniform distributions.

### EXAMPLE 10–15

ShipM4U is considering installing a new, more accurate scale, which will reduce the error in computing postage charges and save $250 a year. The scale's useful life is believed to be uniformly distributed over 12, 13, 14, 15, and 16 years. The initial cost of the scale is estimated to be normally distributed with a mean of $1500 and a standard deviation of $150.

Use Excel to simulate 25 random samples of the problem and compute the rate of return for each sample. Construct a graph of rate of return versus frequency of occurrence.

**SOLUTION**

FIGURE 10-8 Excel spreadsheet for simulation ($N = 25$).

| | A | B | C | D |
|---|---|---|---|---|
| 1 | 250 | Annual Savings | | |
| 2 | | Life | First Cost | |
| 3 | Min | 12 | 1500 | Mean |
| 4 | Max | 16 | 150 | Std dev |
| 5 | | | | |
| 6 | Iteration | | | IRR |
| 7 | 1 | 12 | 1277 | 16.4% |
| 8 | 2 | 15 | 1546 | 13.9% |
| 9 | 3 | 12 | 1523 | 12.4% |
| 10 | 4 | 16 | 1628 | 13.3% |
| 11 | 5 | 14 | 1401 | 15.5% |
| 12 | 6 | 12 | 1341 | 15.2% |
| 13 | 7 | 12 | 1683 | 10.2% |
| 14 | 8 | 14 | 1193 | 19.2% |
| 15 | 9 | 15 | 1728 | 11.7% |
| 16 | 10 | 12 | 1500 | 12.7% |
| 17 | 11 | 16 | 1415 | 16.0% |
| 18 | 12 | 12 | 1610 | 11.2% |
| 19 | 13 | 15 | 1434 | 15.4% |
| 20 | 14 | 12 | 1335 | 15.4% |
| 21 | 15 | 14 | 1468 | 14.5% |
| 22 | 16 | 13 | 1469 | 13.9% |
| 23 | 17 | 14 | 1409 | 15.3% |
| 24 | 18 | 15 | 1484 | 14.7% |
| 25 | 19 | 14 | 1594 | 12.8% |
| 26 | 20 | 15 | 1342 | 16.8% |
| 27 | 21 | 14 | 1309 | 17.0% |
| 28 | 22 | 12 | 1541 | 12.1% |
| 29 | 23 | 16 | 1564 | 14.0% |
| 30 | 24 | 13 | 1590 | 12.2% |
| 31 | 25 | 16 | 1311 | 17.7% |
| 32 | | | | |
| 33 | Mean | 14 | 1468 | 14.4% |
| 34 | Std dev | 2 | 135 | 2.2% |

This problem is simple enough to allow us to construct a table with each iteration's values of the life and the first cost. From these values and the annual savings of $250, the IRR for each iteration can be calculated using the RATE function. These are shown in Figure 10-8. The IRR values are summarized in a relative frequency diagram in Figure 10-9.

*Note:* Each time Excel recalculates the spreadsheet, different values for all the random numbers are generated. Thus the results depend on the set of random numbers, and your results will be different if you create this spreadsheet.

*Note for students who have had a course in probability and statistics:* Creating the random values for life and first cost is done as follows. Select a random number in [0, 1] using Excel's RAND function. This is the value of the cumulative distribution function for the variable. Convert this to the variable's value by using an inverse function from Excel, or build the inverse function. For the discrete uniform life, the function is = min life + INT(range∗ RAND()). For the normally distributed first cost, the functions is = NORMINV(RAND(), mean, standard deviation).

**FIGURE 10-9** Graph of IRR values.

Stand-alone simulation programs and commercial spreadsheet add-in packages such as @Risk and Crystal Ball provide probability distribution functions to use for each input variable. In Example 10-16 the functions RiskUniform and RiskNormal are used. The packages also collect values for the output variables, such as the IRR for Example 10-16. In other problems the PW or EUAC could be collected. These values form a probability distribution for the PW, IRR, or EUAC. From this distribution the simulation package can calculate the expected return, $P$(loss), and the standard deviation of the return.

Example 10-16 uses @Risk to simulate 1000 iterations of PW for the data in Example 10-15. A simulation package makes it easy to do more iterations. More important still, since it is much easier to use different probability distributions and parameters, more accurate models can be built. Because the models are easier to build, they are less likely to contain errors.

## EXAMPLE 10–16

Consider the scale described in Example 10-15. Generate 1000 iterations and construct a frequency distribution for the scale's rate of return.

### SOLUTION

The first IRR (cell A8) of 14.01% that is computed in Figure 10-10 is based on the average life and the average first cost. The second IRR (cell A11) of 14.01% is computed by @Risk using the average of each distribution. The cell *content* is the RATE formula with its RiskUniform and RiskNormal function, however spreadsheets with @Risk functions *display* by default the results of using average values.

The RATE function contains two @Risk functions: RiskUniform and RiskNormal. The uniform distribution has the minimum and maximum values as parameters. The normal distribution has the average and standard deviation as parameters.

The third IRR (cell A13) is the average for 1000 iterations. It will change each time the simulation is done. The graph in Figure 10-10 with 1000 iterations is much smoother than Figure 10-9, the graph from Example 10-15, where 25 iterations were done.

**FIGURE 10–10** Simulation spreadsheet for Examples 10-15 and 10-16.

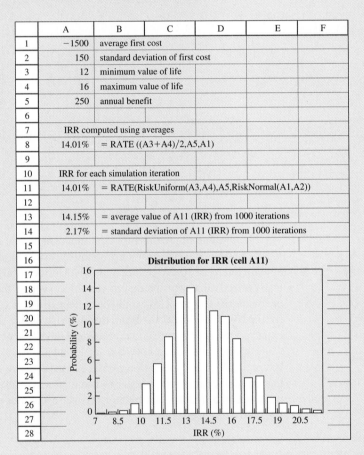

| | A | B | C | D | E | F |
|---|---|---|---|---|---|---|
| 1 | −1500 | average first cost | | | | |
| 2 | 150 | standard deviation of first cost | | | | |
| 3 | 12 | minimum value of life | | | | |
| 4 | 16 | maximum value of life | | | | |
| 5 | 250 | annual benefit | | | | |
| 6 | | | | | | |
| 7 | | IRR computed using averages | | | | |
| 8 | 14.01% | = RATE ((A3+A4)/2,A5,A1) | | | | |
| 9 | | | | | | |
| 10 | | IRR for each simulation iteration | | | | |
| 11 | 14.01% | = RATE(RiskUniform(A3,A4),A5,RiskNormal(A1,A2)) | | | | |
| 12 | | | | | | |
| 13 | 14.15% | = average value of A11 (IRR) from 1000 iterations | | | | |
| 14 | 2.17% | = standard deviation of A11 (IRR) from 1000 iterations | | | | |
| 15 | | | | | | |

# REAL OPTIONS

Real-options analysis is another approach for evaluating projects with significant future uncertainties. It is derived from the theory underlying financial options in the stock market. Rather than buying or selling a share of stock, the option gives you the right to buy or sell the stock at a given price for a given period. Since the option costs less than the stock, a $1000 investment in options has more potential gain than the same investment in the stock.

It also has more risk. For engineering projects, a real option is an alternative that keeps open the opportunity by delaying the decision.

For an intuitive understanding of why real options make sense, consider the example of a lease for gas and oil development rights. Many such projects are never developed because they are not "economic." Nevertheless, it may make sense to renew the lease for a noneconomic project because of the possibility that the project will *become* economically profitable. For example, prices may rise, or pipelines and facilities may be built to support a nearby tract.

The models and assumptions of real options are beyond the scope of this text. Interested readers are referred to the third edition of the advanced text *Economic Analysis of Industrial Projects* by Eschenbach, Lewis, Hartman, and Bussey.

These options may include delaying, abandoning, expanding, shrinking, changing, and replicating a project—with different models for different actions. However, some suggested guidelines may help you decide when you should consider applying real options to your project analysis.

- The value of exercising an option cannot be negative, since you can choose not to exercise it.
- An option to delay a project is valuable when the project is not currently economic but may become economic. Thus, if a project's present worth is positive or very negative, a delay option is unlikely to be worthwhile.
- Unlike projects, options become more valuable as risk (described as volatility) increases. For projects, we want more return but less risk. However, options are more likely to be worthwhile and to result in action if the risk increases.
- Volatility is the key parameter in real-option valuation, but it is difficult to calculate. In addition, much of the risk associated with a project may not be *actionable* volatility, and only actionable volatility can create value for a real option [Lewis, Eschenbach, and Hartman, 2008].
- Many published examples of real-option analysis ignore the cost of waiting, but this is rarely appropriate for engineering projects [Eschenbach, Lewis, and Hartman, 2009]. For example, if an R&D project, a new product, or a new building is delayed, then it will later face more competition and during the period of delay there are no revenues or cost savings since the project is not yet operating.
- In many cases decision trees and simulation are required to properly develop and describe the real option, and by then the real-option analysis may add little value to the decision-making process.

## SUMMARY

Estimating the future is required for economic analysis, and there are several ways to do this. Precise estimates will not ordinarily be exactly correct, but they are considered to be the best single values to represent what we think will happen.

A simple way to represent uncertainty is through a range of estimates for each variable, such as optimistic, most likely, and pessimistic. The full range of prospective results may be examined by using the optimistic values to solve the problem and then using the pessimistic values. Solving the problem with the most likely values is a good single value estimate.

However, the extremes with all optimistic values or all pessimistic values are less likely—it is more likely that a mix of optimistic, most likely, and pessimistic values will occur.

One approach taken from project management uses the following weighted values instead of a range of estimates.

| **Estimate** | Optimistic | Most Likely | Pessimistic |
|---|---|---|---|
| **Relative weight** | 1 | 4 | 1 |

The most commonly used approach for decision making relies on **expected values.** Here, known or estimated probabilities for future events are used as weights for the corresponding outcomes.

$$\text{Expected value} = \text{Outcome}_A \times \text{Probability}_A + \text{Outcome}_B \times \text{Probability}_B + \cdots$$

Expected value is the most useful and the most frequently used technique for estimating a project's attractiveness.

However, risk as measured by standard deviation and the probability of a loss is also important in evaluating projects. Since projects with higher expected returns also frequently have higher risk, evaluating the trade-offs between risk and return is useful in decision making.

More complicated problems can be summarized and analyzed by using decision trees, which allow logical evaluation of problems with sequential chance, decision, and outcome nodes.

Where the elements of an economic analysis are stated in terms of probability distributions, a repetitive analysis of a random sample is often done. This simulation-based approach relies on the premise that a random sampling of increasing size becomes a better and better estimate of the possible outcomes. The large number of computations means that simulation is usually computerized.

## PROBLEMS

### Range of Estimates

10-1  Telephone poles exemplify items that have varying useful lives. Telephone poles, once installed in a location, remain in useful service until one of a variety of events occur.

(a) Name three reasons why a telephone pole might be removed from useful service at a particular location.

(b) You are to estimate the total useful life of telephone poles. If the pole is removed from an original location while it is still serviceable, it will be installed elsewhere. Estimate the optimistic life, most likely life, and pessimistic life for telephone poles. What percentage of all telephone poles would you expect to have a total useful life greater than your estimated optimistic life?

10-2  The purchase of a used pickup for $9000 is being considered. Records for other vehicles show that costs for oil, tires, and repairs about equal the cost for fuel.

Fuel costs are $990 per year if the truck is driven 10,000 miles. The salvage value after 5 years of use drops about 8% per mile. Find the equivalent uniform annual cost if the interest rate is 8%. How much does this change if the annual mileage is 15,000? 5000?

10-3  A heat exchanger is being installed as part of a plant modernization program. It costs $80,000, including installation, and is expected to reduce the overall plant fuel cost by $20,000 per year. Estimates of the useful life of the heat exchanger range from an optimistic 12 years to a pessimistic 4 years. The most likely value is 5 years. Assume the heat exchanger has no salvage value at the end of its useful life.

(a) Determine the pessimistic, most likely, and optimistic rates of return.

(b) Use the range of estimates to compute the mean life and determine the estimated before-tax rate of return.

**10-4** For the data in Problem 10-2 assume that the 5000, 10,000, and 15,000 mileage values are, respectively, pessimistic, most likely, and optimistic estimates. Use a weighted estimate to calculate the equivalent annual cost.

**10-5** A new engineer is evaluating whether to use a higher-voltage transmission line. It will cost $250,000 more initially, but it will reduce transmission losses. The optimistic, most likely, and pessimistic projections for annual savings are $20,000, $15,000, and $8,000. The interest rate is 6%, and the transmission line should have a life of 30 years.

(*a*) What is the present worth for each estimated value?

(*b*) Use the range of estimates to compute the mean annual savings, and then determine the present worth.

(*c*) Does the answer to (*b*) match the present worth for the most likely value? Why or why not?

**10-6** A new 2-lane road is needed in a part of town that is growing. At some point the road will need 4-lanes to handle the anticipated traffic. If the city's optimistic estimate of growth is used, the expansion will be needed in 4 years. For the most likely and pessimistic estimates, the expansion will be needed in 8 and 15 years, respectively. The expansion will cost $4.2 million. Use an interest rate of 8%.

(*a*) What is the PW for each scenario, and what is the range of values?

(*b*) Use Equation 10-1 to find the mean value of the expansion's PW.

## Probabilities

**10-7** When a pair of dice are tossed, the results may be any whole number from 2 through 12. In the game of craps one can win by tossing either a 7 or an 11 on the first roll. What is the probability of doing this? (*Hint:* There are 36 ways that a pair of six-sided dice can be tossed. What portion of them result in either a 7 or an 11?) (*Answer:* $8/36$)

**10-8** Over the last 10 years, the hurdle or discount rate for projects from the firm's research and development division has been 10% twice, 15% three times, and 20% the rest of the time. There is no recognizable pattern. Calculate the probability distribution for next year's discount rate.

**10-9** The construction time for a bridge depends on the weather. The project is expected to take 250 days if the weather is dry and hot. If the weather is damp and cool, the project is expected to take 350 days. Otherwise, it is expected to take 300 days. Historical data suggest that the probability of cool, damp weather is 30% and that of dry, hot weather is 20%. Find the project's probability distribution.

**10-10** A new product's sales and profits are uncertain. The marketing department has predicted that sales might be as high as 10,000 units per year with a probability of 10%. The most likely value is 7000 units annually. The pessimistic value is estimated to be 5000 units annually with a probability of 20%. Manufacturing and marketing together have estimated the most likely unit profit to be $32. The pessimistic value of $24 has a probability of 0.3, and the optimistic value of $38 has a probability of 0.2. Construct the probability distributions for sales and unit profits.

**10-11** A road between Fairbanks and Nome, Alaska, will have a most likely construction cost of $4 million per mile. Doubling this cost is considered to have a probability of 30%, and cutting it by 25% is considered to have a probability of 10%. The state's interest rate is 8%, and the road should last 40 years. What is the probability distribution of the equivalent annual construction cost per mile?

**10-12** You recently had an auto accident that was your fault. If you have another accident or receive a another moving violation within the next 3 years, you will become part of the "assigned risk" pool, and you will pay an extra $600 per year for insurance. If the probability of an accident or moving violation is 20% per year, what is the probability distribution of your "extra" insurance payments over the next 4 years? Assume that insurance is purchased annually and that violations register at the end of the year—just in time to affect next year's insurance premium.

**10-13** Al took a midterm examination in physics and received a score of 65. The mean was 60 and the standard deviation was 20. Bill received a score of 14 in mathematics, where the exam mean was 12 and the standard deviation was 4. Which student ranked higher in his class? Explain.

## Joint Probabilities

**10-14** For the data in Problem 10-10, construct the probability distribution for the annual profit. Assume that the sales and unit profits are statistically independent.

**10-15** A project has a life of 10 years, and no salvage value. The firm uses an interest rate of 12% to evaluate engineering projects. The project has an uncertain first cost and net revenue.

| First Cost | P | Net Revenue | P |
|---|---|---|---|
| $300,000 | 0.2 | $ 70,000 | 0.3 |
| 400,000 | 0.5 | 90,000 | 0.5 |
| 600,000 | 0.3 | 100,000 | 0.2 |

(a) What is the joint probability distribution for first cost and net revenue?

(b) Define optimistic, most likely, and pessimistic scenarios by using both optimistic, both most likely, and both pessimistic estimates. What is the present worth for each scenario?

**10-16** A robot has just been installed at a cost of $81,000. It will have no salvage value at the end of its useful life.

| Savings per Year | Probability | Useful Life (years) | Probability |
|---|---|---|---|
| $18,000 | 0.2 | 12 | 1/6 |
| 20,000 | 0.7 | 5 | 2/3 |
| 22,000 | 0.1 | 4 | 1/6 |

(a) What is the joint probability distribution for savings per year and useful life?

(b) Define optimistic, most likely, and pessimistic scenarios by using both optimistic, both most likely, and both pessimistic estimates. What is the rate of return for each scenario?

**10-17** Modifying an assembly line has a first cost of $80,000, and its salvage value is $0. The firm's interest rate is 9%. The savings shown in the table depend on whether the assembly line runs one, two, or three shifts, and on whether the product is made for 3 or 5 years.

| Shifts/ day | Savings/ year | Probability | Useful Life (years) | Probability |
|---|---|---|---|---|
| 1 | $15,000 | 0.3 | 3 | 0.6 |
| 2 | 30,000 | 0.5 | 5 | 0.4 |
| 3 | 45,000 | 0.2 | | |

(a) Give the joint probability distribution for savings per year and the useful life.

(b) Define optimistic, most likely, and pessimistic scenarios by using both optimistic, both most likely, and both pessimistic estimates. Use a life of 4 years as the most likely value. What is the present worth for each scenario?

## Expected Value

**10-18** Annual savings due to an energy efficiency project have a most likely value of $30,000. The high estimate of $40,000 has a probability of 0.2, and the low estimate of $20,000 has a probability of 0.30. What is the expected value for the annual savings? (*Answer:* $29,000)

**10-19** Two instructors announced that they "grade on the curve," that is, give a fixed percentage of each of the various letter grades to each of their classes. Their curves are as follows:

| Grade | Instructor A | Instructor B |
|---|---|---|
| A | 10% | 15% |
| B | 15 | 15 |
| C | 45 | 30 |
| D | 15 | 20 |
| F | 15 | 20 |

If a random student came to you and said that his object was to enroll in the class in which he could expect the higher grade point average, which instructor would you recommend? (*Answer:* $GPA_B = 1.95$, Instructor A)

**10-20** For the data in Problem 10-8, compute the expected value for the next year's discount rate.

**10-21** For the data in Problem 10-9, compute the project's expected completion time.

**10-22** For the data in Problem 10-11 calculate the expected value of the equivalent annual construction cost per mile.

**10-23** A man wants to decide whether to invest $1000 in a friend's speculative venture. He will do so if he thinks he can get his money back in one year. He believes the probabilities of the various outcomes at the end of one year are as follows:

| Result | Probability |
|---|---|
| $2000 (double his money) | 0.3 |
| 1500 | 0.1 |
| 1000 | 0.2 |
| 500 | 0.3 |
| 0 (lose everything) | 0.1 |

What would be his expected outcome if he invests the $1000?

**10-24** The MSU football team has 10 games scheduled for next season. The business manager wishes to estimate how much money the team can be expected to have left over after paying the season's expenses, including any postseason "bowl game" expenses. From records for the past season and estimates by informed people, the business manager has assembled the following data:

| Situation | Prob- ability | Situation | Net Income |
|---|---|---|---|
| Regular season | | Regular season | |
| Win 3 games | 0.10 | Win 5 or | $250,000 |
| Win 4 games | 0.15 | fewer | |
| Win 5 games | 0.20 | games | |
| Win 6 games | 0.15 | Win 6 to 8 | 400,000 |
| Win 7 games | 0.15 | games | |
| Win 8 games | 0.10 | | |
| Win 9 games | 0.07 | Win 9 or 10 | 600,000 |
| Win 10 games | 0.03 | games | |
| Postseason | | Postseason | Additional |
| Bowl game | 0.10 | Bowl game | income of $100,000 |

What is the expected net income for the team next season? (*Answer:* $355,000)

**10-25** In the New Jersey and Nevada gaming casinos, craps is a popular gambling game. One of the many bets available is the "Hard-way 8." A $1 bet in this fashion will win the player $4 if in the game the pair of dice come up 4 and 4 before one of the other ways of totaling 8. For a $1 bet, what is the expected result? (*Answer:* 80¢)

**10-26** A man went to Atlantic City with $500 and placed 100 bets of $5 each, one after another, on the same number on the roulette wheel. There are 38 numbers on the wheel, and the gaming casino pays 35 times the amount bet if the ball drops into the bettor's numbered slot in the roulette wheel. In addition, the bettor receives back the original $5 bet. Estimate

how much money the man is expected to win or lose in Atlantic City.

**10-27** For the data in Problems 10-2 and 10-4, assume that the optimistic probability is 20%, the most likely is 50%, and the pessimistic is 30%.

(*a*) What is the expected value of the equivalent uniform annual cost?

(*b*) Compute the expected value for the number of miles, and the corresponding equivalent uniform annual cost.

(*c*) Do the answers to (*a*) and (*b*) match? Why or why not?

**10-28** For the data in Problem 10-3, assume that the optimistic probability is 15%, the most likely is 80%, and the pessimistic is 5%.

(*a*) What is the expected value of the rate of return?

(*b*) Compute the expected value for the life, and the corresponding rate of return.

(*c*) Do the answers to (*a*) and (*b*) match? Why or why not?

**10-29** For the data in Problem 10-5, assume that the optimistic probability is 20%, the most likely is 50%, and the pessimistic is 30%.

(*a*) What is the expected value of the present worth?

(*b*) Compute the expected value for annual savings, and the corresponding present worth.

(*c*) Do the answers to (*a*) and (*b*) match? Why or why not?

**10-30** For the data in Problem 10-10, calculate the expected value of sales and unit profits. For the data in Problem 10-14, calculate the expected value of annual profit. Are these results consistent?

**10-31** Assume that the pessimistic and optimistic estimates in Problem 10-6 have 40% and 20% probabilities, respectively.

(*a*) What is the expected PW the expansion costs?

(*b*) What is the expected number of years until the expansion?

(*c*) What is PW of the expansion cost using the expected number of years until the expansion?

(*d*) Do your answers to (*a*) and (*c*) match? If not, why not?

**10-32** The energy efficiency project described in Problem 10-18 has a first cost of $150,000, a life of 10 years, and no salvage value. Assume that the interest rate is 8%.

(a) What is the equivalent uniform annual worth for the expected annual savings?

(b) Compute the equivalent uniform annual worth for the pessimistic, most likely, and optimistic estimates of the annual savings. What is the expected value of the equivalent uniform annual worth?

(c) Do the answers to (a) and (b) match? Why or why not?

**10-33** An industrial park is being planned for a tract of land near the river. To prevent flood damage to the industrial buildings that will be built on this low-lying land, an earthen embankment can be constructed. The height of the embankment will be determined by an economic analysis of the costs and benefits. The following data have been gathered.

| Embankment Height Above Roadway (m) | Initial Cost |
|---|---|
| 2.0 | $100,000 |
| 2.5 | 165,000 |
| 3.0 | 300,000 |
| 3.5 | 400,000 |
| 4.0 | 550,000 |

| Flood Level Above Roadway (m) | Average Frequency That Flood Level Will Exceed Height in Col. 1 |
|---|---|
| 2.0 | Once in 3 years |
| 2.5 | Once in 8 years |
| 3.0 | Once in 25 years |
| 3.5 | Once in 50 years |
| 4.0 | Once in 100 years |

The embankment can be expected to last 50 years and will require no maintenance. Whenever the flood water flows over the embankment, $300,000 of damage occurs. Should the embankment be built? If so, to which of the five heights above the roadway? A 12% rate of return is required.

**10-34** If your interest rate is 8%, what is the expected value of the present worth of the "extra" insurance payments in Problem 10-12? *(Answer: $528.70)*

**10-35** Should the project in Problem 10-15 be undertaken if the firm uses an expected value of present worth to evaluate engineering projects?

(a) Compute the PW for each combination of first cost and revenue and the corresponding expected worth.

(b) What are the expected first cost, expected net revenue, and corresponding present worth of the expected values?

(c) Do the answers for (a) and (b) match? Why or why not?

*(Answer: (b) $45,900, do project)*

**10-36** A new engineer is evaluating whether to use a larger-diameter pipe for a water line. It will cost $350,000 more initially, but it will reduce pumping costs. The optimistic, most likely, and pessimistic projections for annual savings are $30,000, $20,000, and $5000, with respective probabilities of 20%, 50%, and 30%. The interest rate is 6–8%, and the water line should have a life of 40 years.

(a) What is the PW for each estimated value? What is the expected PW?

(b) Compute the expected annual savings and expected PW.

(c) Do the answers for the expected PW match? Why or why not?

**10-37** An energy efficiency project has a first cost of $300,000, a life of 10 years, and no salvage value. Assume that the interest rate is 9%. The most likely value for annual savings is $60,000. The optimistic value for annual savings is $80,000 with a probability of 0.2. The pessimistic value is $40,000 with a probability of 0.3.

(a) What is the expected annual savings and the expected PW?

(b) Compute the PW for the pessimistic, most likely, and optimistic estimates of the annual savings. What is the expected PW?

(c) Do the answers for the expected PW match? Why or why not?

**10-38** For the data in Problem 10-16:

(a) What are the expected savings per year, life, and corresponding rate of return for the expected values?

(b) Compute the rate of return for each combination of savings per year and life. What is the expected rate of return?

(c) Do the answers for (a) and (b) match? Why or why not?

**10-39** For the data in Problem 10-17:

(a) What are the expected savings per year, life, and corresponding present worth for the expected values?

(b) Compute the present worth for each combination of savings per year and life. What is the expected present worth?

(c) Do the answers for (a) and (b) match? Why or why not?

## Decision Trees

**10-40** The tree in Figure P10-40 has probabilities after each chance node and PW values for each terminal node. What decision should be made? What is the expected value?

**FIGURE P10-40**

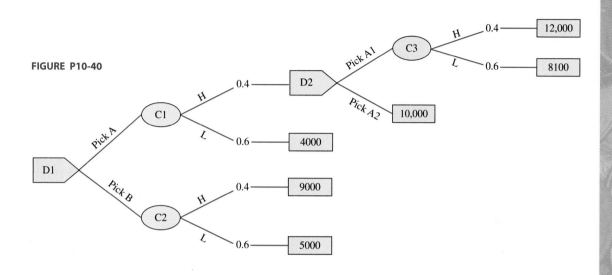

**10-41** The tree in Figure P10-41 has probabilities after each chance node and PW values for each terminal node. What decision should be made? What is the expected value?

**FIGURE P10-41**

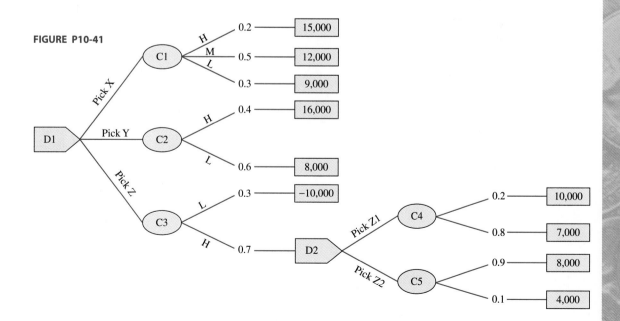

**10-42** A decision has been made to perform certain repairs on the outlet works of a small dam. For a particular 36-inch gate valve, there are three available alternatives:

A. Leave the valve as it is.

B. Repair the valve.

C. Replace the valve.

If the valve is left as it is, the probability of a failure of the valve seats, over the life of the project, is 60%; the probability of failure of the valve stem is 50%; and of failure of the valve body is 40%.

If the valve is repaired, the probability of a failure of the seats, over the life of the project, is 40%; of failure of the stem is 30%; and of failure of the body is 20%. If the valve is replaced, the probability of a failure of the seats, over the life of the project, is 30%; of failure of the stem is 20%; and of failure of the body is 10%.

The present worth of cost of future repairs and service disruption of a failure of the seats is $10,000; the present worth of cost of a failure of the stem is $20,000; the present worth of cost of a failure of the body is $30,000. The cost of repairing the valve now is $10,000; and of replacing it is $20,000. If the criterion is to minimize expected costs, which alternative is best?

**10-43** A factory building is located in an area subject to occasional flooding by a nearby river. You have been brought in as a consultant to determine whether flood-proofing of the building is economically justified. The alternatives are as follows:

A. Do nothing. Damage in a moderate flood is $10,000 and in a severe flood, $25,000.

B. Alter the factory building at a cost of $15,000 to withstand moderate flooding without damage and to withstand severe flooding with $10,000 damages.

C. Alter the factory building at a cost of $20,000 to withstand a severe flood without damage.

In any year the probability of flooding is as follows: 0.70, no flooding of the river; 0.20, moderate flooding; and 0.10, severe flooding. If interest is 15% and a 15-year analysis period is used, what do you recommend?

**10-44** Five years ago a dam was constructed to impound irrigation water and to provide flood protection for the area below the dam. Last winter a 100-year flood caused extensive damage both to the dam and to the surrounding area. This was not surprising, since the dam was designed for a 50-year flood.

The cost to repair the dam now will be $250,000. Damage in the valley below amounts to $750,000. If the spillway is redesigned at a cost of $250,000 and the dam is repaired for another $250,000, the dam may be expected to withstand a 100-year flood without sustaining damage. However, the storage capacity of the dam will not be increased and the probability of damage to the surrounding area below the dam will be unchanged. A second dam can be constructed up the river from the existing dam for $1 million. The capacity of the second dam would be more than adequate to provide the desired flood protection. If the second dam is built, redesign of the existing dam spillway will not be necessary, but the $250,000 of repairs must be done.

The development in the area below the dam is expected to be complete in 10 years. A new 100-year flood in the meantime would cause a $1 million loss. After 10 years the loss would be $2 million. In addition, there would be $250,000 of spillway damage if the spillway is not redesigned. A 50-year flood is also likely to cause about $200,000 of damage, but the spillway would be adequate. Similarly, a 25-year flood would cause about $50,000 of damage.

There are three alternatives: (1) repair the existing dam for $250,000 but make no other alterations, (2) repair the existing dam ($250,000) and redesign the spillway to take a 100-year flood ($250,000), and (3) repair the existing dam ($250,000) and build the second dam ($1 million). Based on an expected annual cash flow analysis, and a 7% interest rate, which alternative should be selected? Draw a decision tree to clearly describe the problem.

**10-45** In Problems 10-17 and 10-39, how much is it worth to the firm to be able to extend the product's life by 3 years, at a cost of $50,000, at the end of the product's initial useful life?

## Risk

**10-46** An engineer decided to make a careful analysis of the cost of fire insurance for his $200,000 home.

From a fire rating bureau he found the following risk of fire loss in any year.

| Outcome | Probability |
|---|---|
| No fire loss | 0.986 |
| $ 10,000 fire loss | 0.010 |
| 40,000 fire loss | 0.003 |
| 200,000 fire loss | 0.001 |

(a) Compute his expected fire loss in any year.

(b) He finds that the expected fire loss in any year is less than the $550 annual cost of fire insurance. In fact, an insurance agent explains that this is always true. Nevertheless, the engineer buys fire insurance. Explain why this is or is not a logical decision.

**10-47** For the data in Problems 10-8 and 10-20, compute the standard deviation of the interest rate.

**10-48** For the data in Problems 10-11 and 10-22, compute the standard deviation of the equivalent annual cost per mile.

**10-49** For the data in Problem 10-36, compute the standard deviation of the present worth.

**10-50** For the data in Problem 10-37, compute the standard deviation of the present worth.

**10-51** The Graham Telephone Company may invest in new switching equipment. There are three possible outcomes, having net present worth of $6570, $8590, and $9730. The outcomes have probabilities of 0.3, 0.5, and 0.2, respectively. Calculate the expected return and risk measured by the standard deviation associated with this proposal. (*Answer:* $E_{PW} = \$8212, \sigma_{PW} = \$1158$)

**10-52** A new machine will cost $25,000. The machine is expected to last 4 years and have no salvage value. If the interest rate is 12%, determine the return and the risk associated with the purchase.

| P | 0.3 | 0.4 | 0.3 |
|---|---|---|---|
| Annual savings | $7000 | $8500 | $9500 |

**10-53** A new product's chief uncertainty is its annual net revenue. So far, $35,000 has been spent on development, but an additional $30,000 is required to finish development. The firm's interest rate is 10%.

(a) What is the expected PW for deciding whether to proceed?

(b) Find the $P$(loss) and the standard deviation for proceeding.

| | State | | |
|---|---|---|---|
| | **Bad** | **OK** | **Great** |
| Probability | 0.3 | 0.5 | 0.2 |
| Net revenue | −$15,000 | $15,000 | $20,000 |
| Life, in years | 5 | 5 | 10 |

**10-54** (a) In Problem 10-53, how much is it worth to the firm to terminate the product after 1 year if the net revenues are negative?

(b) How much does the ability to terminate early change the $P$(loss) and the standard deviation?

**10-55** What is your risk associated with Problem 10-34?

**10-56** Measure the risk for Problems 10-15 and 10-35 using the $P(\text{loss})$, range of PW values, and standard deviation of the PWs. (*Answer:* $\sigma_{PW} = \$127,900$)

**10-57** (*a*) In Problems 10-17 and 10-39, describe the risk using the $P(\text{loss})$ and standard deviation of the PWs.

(*b*) How much do the answers change if the possible life extension in Problem 10-44 is allowed?

## Risk Versus Return

**10-58** A firm wants to select one new research and development project. The following table summarizes six possibilities. Considering expected return and risk, which projects are good candidates? The firm believes it can earn 5% on a risk-free investment in government securities (labeled as Project $F$).

| Project | IRR | Standard Deviation |
|---------|------|--------------------|
| 1 | 15.8% | 6.5% |
| 2 | 12.0 | 4.1 |
| 3 | 10.4 | 6.3 |
| 4 | 12.1 | 5.1 |
| 5 | 14.2 | 8.0 |
| 6 | 18.5 | 10.0 |
| F | 5.0 | 0.0 |

**10-59** A firm is choosing a new product. The following table summarizes six new potential products. Considering expected return and risk, which products are good candidates? The firm believes it can earn 4% on a risk-free investment in government securities (labeled as Product $F$).

| Product | IRR | Standard Deviation |
|---------|------|--------------------|
| 1 | 10.4% | 3.2% |
| 2 | 9.8 | 2.3 |
| 3 | 6.0 | 1.6 |
| 4 | 12.1 | 3.6 |
| 5 | 12.2 | 8.0 |
| 6 | 13.8 | 6.5 |
| F | 4.0 | 0.0 |

## Simulation

**10-60** A project's first cost is $25,000, and it has no salvage value. The interest rate for evaluation is 7%. The project's life is from a discrete uniform distribution that takes on the values 7, 8, 9, and 10. The annual benefit is normally distributed with a mean of $4400 and a standard deviation of $1000. Using Excel's RAND function, simulate 25 iterations. What are the expected value and standard deviation of the present worth?

**10-61** A factory's power bill is $55,000 a year. The first cost of a small geothermal power plant is normally distributed with a mean of $150,000 and a standard deviation of $50,000. The power plant has no salvage value. The interest rate for evaluation is 8%. The project's life is from a discrete uniform distribution that takes on the values 3, 4, 5, 6, and 7. (The life is relatively short due to corrosion.) The annual operating cost is expected to be about $10,000 per year. Using Excel's RAND function, simulate 25 iterations. What are the expected value and standard deviation of the present worth?

# CASES

**The following cases from *Cases in Engineering Economy* $2^{nd}$ are suggested as matched with this chapter.**

CASE 13   **Guaranteed Return**
Risk versus expected return and choice of interest rate.

CASE 18   **Gravity-Free High**
New product with development stages and probability of failure.

CASE 19   **Crummy Castings**
Decision tree problem with modest ambiguity. Options create more challenging problems.

CASE 21   **Glowing in the Dark**
Expected return and variance for facility of minimal size, of minimal size plus preparation for later expansion, and of the full size now.

## CHAPTER 11

# DEPRECIATION

| Form **4562** | **Depreciation and Amortization** (Including Information on Listed Property) | OMB No. 1545-0172 20**10** |
|---|---|---|
| Department of the Treasury Internal Revenue Service | ▶ See separate instructions. ▶ Attach to your tax return. | Attachment Sequence No. **67** |
| Name(s) shown on return | Business or activity to which this form relates | Identifying number |

**Part I**   **Election To Expense Certain Property Under Section 179**
Note: *If you have any listed property, complete Part V before you complete Part I.*

| | | | | | |
|---|---|---|---|---|---|
| 1 | Maximum amount. See the instructions for a higher limit for certain businesses . . . . . . . | | 1 | $108,000 |
| 2 | Total cost of section 179 property placed in service (see instructions) . . . . . . . . | | 2 | |
| 3 | Threshold cost of section 179 property before reduction in limitation . . . . . . . . | | 3 | $430,000 |
| 4 | Reduction in limitation. Subtract line 3 from line 2. If zero or less, enter -0- . . . . . . | | 4 | |
| 5 | Dollar limitation for tax year. Subtract line 4 from line 1. If zero or less, enter -0-. If married filing separately, see instructions | | 5 | |

| (a) Description of property | (b) Cost (business use only) | (c) Elected cost |
|---|---|---|
| 6 | | |

| | | | | |
|---|---|---|---|---|
| 7 | Listed property. Enter the amount from line 29 . . . . . . . . | 7 | | |
| 8 | Total elected cost of section 179 property. Add amounts in column (c), lines 6 and 7 . . . | 8 | |
| 9 | Tentative deduction. Enter the **smaller** of line 5 or line 8 . . . . . . . . . | 9 | |
| 10 | Carryover of disallowed deduction from line 13 of your 2005 Form 4562 . . . . . . | 10 | |
| 11 | Business income limitation. Enter the smaller of business income (not less than zero) or line 5 (see instructions) | 11 | |
| 12 | Section 179 expense deduction. Add lines 9 and 10, but do not enter more than line 11 . | 12 | |
| 13 | Carryover of disallowed deduction to 2007. Add lines 9 and 10, less line 12 ▶ | 13 | |

Note: *Do not use Part II or Part III below for listed property. Instead, use Part V.*

## Depreciation and Competitiveness

Depreciation sounds like a boring accounting term, but it is part of determining how much every firm pays in taxes. Reducing taxes is a common goal. Depreciation apportions an asset's capital cost into annual deductions from taxable income. Thus, factories, bulldozers, computers, sports stadiums, and even professional baseball players are depreciated. In each case what can be depreciated, and how fast, is linked to a firm's after-tax profits and competitiveness. More and faster depreciation improves competitiveness.

Why is depreciation important to governments? The answer is that depreciation rules and regulations can be crafted to promote specific outcomes such as total revenue received, increased capital investment, and improved employment rates. In 1981 and 1986, the federal government responded to concerns about U.S. firms building plants overseas by changing the tax code to allow depreciation over shorter recovery periods. The tax code was also changed to give capital investments full depreciation, using a salvage value of $0. In 2002 the tax code was again changed by the Job Creation and Worker Assistance Act to allow an additional first year's depreciation of 30% for a 3-year period. This was later amended to increase the added depreciation in the first year to 50% for tax year 2003. More recently the 2009 American Recovery and Reinvestment Act provided incentives to stimulate the economy. These acts are intended to increase the level of investment by firms in capital assets, and thus to create new jobs in the U.S. ▪▪▪

*Courtesy of the Internal Revenue Service.*

1. How can the importance of depreciation be measured?
2. What are some of the similarities and differences between depreciating a bulldozer and a ballplayer?
3. Is the added first-year depreciation still part of the tax code? How has it changed since it was legislated in 2002? What details can you find on the current version of the Section 179 deduction for *small* businesses (see p. 412)?
4. At times legislative bodies have been criticized for using potential changes in the depreciation and tax codes to stimulate lobbying and campaign donations. Is this truly a practical or ethical concern?

## After Completing This Chapter...

### The student should be able to:

- Describe depreciation, deterioration, and obsolescence.
- Distinguish various types of depreciable property and differentiate between depreciation expenses and other business expenses.
- Use *historical* depreciation methods to calculate the *annual depreciation charge* and *book value* over the asset's life.
- Explain the differences between the historical depreciation methods and the modified accelerated cost recovery system (MACRS).
- Use MACRS to calculate allowable *annual depreciation charge* and *book value* over the asset's life for various cost bases, property classes, and recovery periods.
- Fully account for *capital gains/losses, ordinary losses,* and *depreciation recapture* due to the disposal of a depreciated business asset.
- Use the *units of production* and *depletion* depreciation methods as needed in engineering economic analysis problems.
- Use spreadsheets to calculate depreciation.

We have so far dealt with a variety of economic analysis problems and many techniques for their solution. In the process we have avoided income taxes, which are an important element in many private sector economic analyses. Now, we move to more realistic—and more complex—situations.

Our government taxes individuals and businesses to support its processes—lawmaking, domestic and foreign economic policy making, even the making and issuing of money itself. The omnipresence of taxes requires that they be included in economic analyses, which means that we must understand the *way* taxes are imposed. For capital equipment, knowledge about depreciation is required to compute income taxes. Chapter 11 examines depreciation, and Chapter 12 illustrates how depreciation and other effects are incorporated in income tax computations. The goal is to support decision making on engineering projects, not to support final tax calculations.

## BASIC ASPECTS OF DEPRECIATION

The word **depreciation** is defined as a "decrease in value." This is somewhat ambiguous because *value* has several meanings. In economic analysis, value may refer to either *market value* or *value to the owner*. For example, an assembly line is far more valuable to the manufacturing firm that it was designed for, than it is to a used equipment market. Thus, we now have two definitions of depreciation: a decrease in value to the market or a decrease to the owner.

### Deterioration and Obsolescence

A machine may depreciate because it is **deteriorating**, or wearing out and no longer performing its function as well as when it was new. Many kinds of machinery require increased maintenance as they age, reflecting a slow but continuing failure of individual parts. In other types of equipment, the quality of output may decline due to wear on components and resulting poorer mating of parts. Anyone who has worked to maintain a car has observed deterioration due to failure of individual parts (such as fan belts, mufflers, and batteries) and the wear on components (such as bearings, piston rings, and alternator brushes).

Depreciation is also caused by **obsolescence.** A machine that is in excellent working condition, and serving a needed purpose, may still be obsolete. In the 1970s, mechanical business calculators with hundreds of gears and levers became obsolete. The advance of integrated circuits resulted in a completely different and far superior approach to calculator design. Thus, mechanical calculators rapidly declined or depreciated in value. Generations of computers have followed this pattern. The continuing stream of newer models makes older ones obsolete.

The accounting profession defines depreciation in yet another way, as allocating an asset's cost over its **useful** or **depreciable life.** Thus, we now have *three distinct definitions of depreciation:*

1. Decline in market value of an asset.
2. Decline in value of an asset to its owner.
3. Systematic allocation of an asset's cost over its depreciable life.

## Depreciation and Expenses

It is the third (accountant's) definition that is used to compute depreciation for business assets. Business costs are generally either **expensed** or **depreciated. Expensed** items, such as labor, utilities, materials, and insurance, are part of regular business operations and are "consumed" over short periods of time (sometimes recurring). These costs do not lose value gradually over time. For tax purposes they are subtracted from business revenues when they occur. Expensed costs reduce income taxes because businesses are able to *write off* their full amount when they occur. Section 179 of tax code allows profitable *small* businesses to expense some capital asset purchases.

In contrast, business costs due to capital assets (buildings, forklifts, chemical plants, etc.) are not fully written off when they occur. Capital assets lose value gradually and must be written off or **depreciated** over an extended period. For instance, consider an injection-molding machine used to produce the plastic beverage cups found at sporting events. The plastic pellets melted into the cup shape lose their value as raw material directly after manufacturing. The raw material cost for production material (plastic pellets) is expensed immediately. On the other hand, the injection-molding machine itself will lose value over time, and thus its costs (purchase price and installation expenses) are written off (or depreciated) over its **depreciable life** or **recovery period.** This is often different from the asset's useful or most economic life. Depreciable life is determined by the depreciation method used to spread out the cost—depreciated assets of many types operate well beyond their depreciable life.

Depreciation is a **noncash** cost that requires no exchange of dollars. Companies do not write a check to someone to **pay** their depreciation expenses. Rather, these are business expenses that are allowed by the government to offset the loss in value of business assets. Remember, the company has paid for assets up front; depreciation is simply a way to claim these "business expenses" over time. Depreciation deductions reduce the taxable income of businesses and thus reduce the amount of taxes paid. Since taxes are cash flows, depreciation must be considered in after-tax economic analyses.

In general, business assets can be depreciated only if they meet the following basic requirements:

1. The property must be used for business purposes to produce income.
2. The property must have a useful life that can be determined, and this life must be longer than one year.
3. The property must be an asset that decays, gets used up, wears out, becomes obsolete, or loses value to the owner from natural causes.

### EXAMPLE 11–1

Consider the costs that are incurred by a local pizza business. Identify each cost as either *expensed* or *depreciated* and describe why that term applies.

- Cost for pizza dough and toppings
- Cost to pay wages for janitor
- Cost of a new baking oven
- Cost of new delivery van
- Cost of furnishings in dining room
- Utility costs for soda refrigerator

| Cost Item | Type of Cost | Why |
|---|---|---|
| Pizza dough and toppings | Expensed | Life < 1 year; lose value immediately |
| New delivery van | Depreciated | Meets 3 requirements for depreciation |
| Wages for janitor | Expensed | Life < 1 year; lose value immediately |
| Furnishings in dining room | Depreciated | Meet 3 requirements for depreciation |
| New baking oven | Depreciated | Meets 3 requirements for depreciation |
| Utilities for soda refrigerator | Expensed | Life < 1 year; lose value immediately |

### Types of Property

The rules for depreciation are linked to the classification of business property as either tangible or intangible. Tangible property is further classified as either real or personal.

**Tangible property** can be seen, touched, and felt.

**Real property** includes land, buildings, and all things growing on, built upon, constructed on, or attached to the land.

**Personal property** includes equipment, furnishings, vehicles, office machinery, and anything that is tangible excluding assets defined as *real property*.

**Intangible property** is all property that has value to the owner but cannot be directly seen or touched. Examples include patents, copyrights, trademarks, trade names, and franchises.

Many different types of property that wear out, decay, or lose value can be depreciated as business assets. This wide range includes copy machines, helicopters, buildings, interior furnishings, production equipment, and computer networks. Almost all tangible property can be depreciated.

One important and notable exception is land, which is *never* depreciated. Land does not wear out, lose value, or have a determinable useful life and thus does not qualify as a depreciable property. Rather than decreasing in value, most land becomes more valuable as time passes. In addition to the land itself, expenses for clearing, grading, preparing, planting, and landscaping are not generally depreciated because they have no fixed useful life. Other tangible property that *cannot* be depreciated includes factory inventory, containers considered as inventory, and leased property. The leased property exception highlights the fact that only the owner of property may claim depreciation expenses.

Tangible properties used in *both* business and personal activities, such as a vehicle used in a consulting engineering firm that is also used to take one's kids to school, can be depreciated, but only in proportion to the use for business purposes.

### Depreciation Calculation Fundamentals

To understand the complexities of depreciation, the first step is to examine the fundamentals of depreciation calculations. Figure 11-1 illustrates the general depreciation problem of allocating the total depreciation charges over the asset's depreciable life. The vertical axis is labeled **book value.** At time zero the curve of book value starts at the cost basis (= the

**FIGURE 11–1** General depreciation.

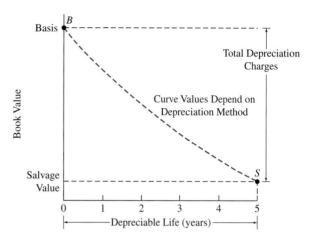

first cost plus installation cost). Over time, the book value declines to the salvage value. Thus, at any point in time:

**Book value = Asset cost − Depreciation charges made to date**

Looked at another way, book value is the asset's remaining unallocated cost.

In Figure 11-1, *book value* goes from a value of $B$ at time zero in the recovery period to a value of $S$ at the end of Year 5. Thus, book value is a *dynamic* variable that changes over an asset's recovery period. The equation used to calculate an asset's book value over time is

$$BV_t = \text{Cost basis} - \sum_{j=1}^{t} d_j \qquad (11\text{-}1)$$

where   $BV_t$ = book value of the depreciated asset at the end of time $t$

Cost basis = $B$ = dollar amount that is being depreciated; this includes the asset's purchase price as well as any other costs necessary to make the asset "ready for use"

$\sum_{j=1}^{t} d_j$ = sum of depreciation deductions taken from time 0 to time $t$, where $d_j$ is the depreciation deduction in Year $j$

Equation 11-1 shows that year-to-year depreciation charges reduce an asset's book value over its life. The following section describes methods that are or have been allowed under federal tax law for quantifying these yearly depreciation deductions.

## HISTORICAL DEPRECIATION METHODS

Allowing businesses to depreciate capital assets has long been part of the tax code. However, over time various depreciation methods have been used to calculate these deductions. In general, accounting depreciation methods can be categorized as follows.

*Pre-1981 historical methods:* These methods include the *straight-line, sum-of-the-years'-digits,* and *declining balance* methods. Each method required estimates of an asset's useful life and salvage value. Firms could elect to use any of these methods for assets, and thus there was little uniformity in how depreciation expenses were reported.

*1981–1986 method:* With the Economic Recovery Tax Act (ERTA) of 1981, Congress created the accelerated cost recovery system (ACRS). The ACRS method had three key features: (1) property class lives were created, and all depreciated assets were assigned to one particular category; (2) the need to estimate salvage values was eliminated because all assets were *fully* depreciated over their recovery period; and (3) shorter recovery periods were used to calculate annual depreciation, which *accelerated* the write-off of capital costs more quickly than did the historical methods—thus the name.

*1986 to present:* The modified accelerated cost recovery system (MACRS) has been in effect since the Tax Reform Act of 1986 (TRA-86). The MACRS method is similar to the ACRS system except that (1) the number of property classes was expanded and (2) the annual depreciation percentages were modified to include a half-year convention for the first and final years.

In this chapter, our primary focus is to describe the MACRS depreciation method. However, it is useful to first describe three historical depreciation methods. These methods are used in some countries, and MACRS is based on two of them.

### Straight-Line Depreciation

The simplest and best known depreciation method is **straight-line depreciation.** To calculate the constant **annual depreciation charge,** the total amount to be depreciated, $B - S$, is divided by the depreciable life, in years, $N$:[1]

$$\textbf{Annual depreciation charge} = d_t = \frac{B - S}{N} \qquad (11\text{-}2)$$

---

### EXAMPLE 11–2

Consider the following (in $1000):

| | |
|---|---|
| Cost of the asset, $B$ | $900 |
| Depreciable life, in years, $N$ | 5 |
| Salvage value, $S$ | $70 |

Compute the straight-line depreciation schedule.

#### SOLUTION

$$\textbf{Annual depreciation charge} = d_t = \frac{B - S}{N} = \frac{900 - 70}{5} = \$166$$

---

[1] $N$ is used for the depreciation period because it may be shorter than $n$, the horizon (or project life).

**FIGURE 11-2** Straight-line depreciation.

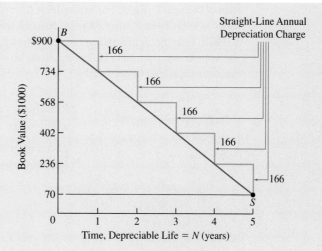

| | Depreciation for Year $t$ ($1000) | Sum of Depreciation Charges Up to Year $t$ ($1000) | Book Value at the End of Year $t$ ($1000) |
|---|---|---|---|
| Year | | | |
| $t$ | $d_t$ | $\sum_{j=1}^{t} d_j$ | $BV_t = B - \sum_{j=1}^{t} d_j$ |
| 1 | \$166 | \$166 | $900 - 166 = 734$ |
| 2 | 166 | 332 | $900 - 332 = 568$ |
| 3 | 166 | 498 | $900 - 498 = 402$ |
| 4 | 166 | 664 | $900 - 664 = 236$ |
| 5 | 166 | 830 | $900 - 830 = \ 70 = S$ |

This situation is illustrated in Figure 11-2. Notice the constant \$166,000 $d_t$ each year for 5 years, and that the asset has been depreciated down to a book value of \$70,000, which was the estimated salvage value.

The straight-line (SL) method is often used for intangible property. For example, Veronica's firm bought a patent in April that was not acquired as part of acquiring a business. She paid \$6800 for this patent and must use the straight-line method to depreciate it over 17 years with no salvage value. The annual depreciation is \$400 (= \$6800/17). Since the patent was purchased in April, the first year's deduction must be prorated over the 9 months of ownership. This year the deduction is \$300 (= \$400 × 9/12), and then next year she can begin taking the full \$400 per year.

## Sum-of-Years'-Digits Depreciation

Another method for allocating an asset's cost *minus* salvage value *over* its depreciable life is called **sum-of-years'-digits (SOYD) depreciation.** This method results in larger-than-straight-line depreciation charges during an asset's early years and smaller charges as the asset nears the end of its depreciable life. Each year, the depreciation charge equals a fraction of the total amount to be depreciated $(B - S)$. The denominator of the fraction is the sum of the years' digits. For example, if the depreciable life is 5 years, $1 + 2 + 3 + 4 + 5 = 15 = \text{SOYD}$. Then $5/15, 4/15, 3/15, 2/15$, and $1/15$ are the fractions from Year 1 to Year 5. Each year the depreciation charge shrinks by $1/15$ of $B - S$. Because this change is the same every year, SOYD depreciation can be modeled as an arithmetic gradient, $G$. The equations can also be written as

$$\begin{pmatrix} \text{Sum-of-years'-digits} \\ \text{depreciation charge for} \\ \text{any year} \end{pmatrix} = \frac{\begin{pmatrix} \text{Remaining depreciable life} \\ \text{at beginning of year} \end{pmatrix}}{\begin{pmatrix} \text{Sum of years' digits} \\ \text{for total depreciable life} \end{pmatrix}} (\text{Total amount depreciated})$$

$$d_t = \frac{N - t + 1}{\text{SOYD}} (B - S) \tag{11-3}$$

where $d_t$ = depreciation charge in any year $t$

$\qquad N$ = number of years in depreciable life

$\text{SOYD}$ = sum of years' digits, calculated as $N(N + 1)/2 = \text{SOYD}$

$\qquad B$ = cost of the asset made ready for use

$\qquad S$ = estimated salvage value after depreciable life

## EXAMPLE 11–3

Compute the SOYD depreciation schedule for the situation in Example 11.2 (in $1000):

| | |
|---|---|
| Cost of the asset, $B$ | $900 |
| Depreciable life, in years, $N$ | 5 |
| Salvage value, $S$ | $70 |

### SOLUTION

$$\text{SOYD} = \frac{5 \times 6}{2} = 15$$

Thus,

$$d_1 = \frac{5 - 1 + 1}{15}(900 - 70) = 277$$

$$d_2 = \frac{5 - 2 + 1}{15}(900 - 70) = 221$$

$$d_3 = \frac{5 - 3 + 1}{15}(900 - 70) = 166$$

$$d_4 = \frac{5 - 4 + 1}{15}(900 - 70) = 111$$

$$d_5 = \frac{5 - 5 + 1}{15}(900 - 70) = 55$$

**FIGURE 11–3** Sum-of-years'-digits depreciation.

| Year | Depreciation for Year $t$ ($1000) | Sum of Depreciation Charges Up to Year $t$ ($1000) | Book Value at End of Year $t$ ($1000) |
|------|------|------|------|
| $t$ | $d_t$ | $\sum_{j=1}^{t} d_j$ | $BV_t = B - \sum_{j=1}^{t} d_j$ |
| 1 | $277 | $277 | $900 - 277 = 623$ |
| 2 | 221 | 498 | $900 - 498 = 402$ |
| 3 | 166 | 664 | $900 - 664 = 236$ |
| 4 | 111 | 775 | $900 - 775 = 125$ |
| 5 | 55 | 830 | $900 - 830 = \ 70 = S$ |

These data are plotted in Figure 11-3.

## Declining Balance Depreciation

**Declining balance depreciation** applies a *constant depreciation rate* to the property's declining book value. Two rates were commonly used before the 1981 and 1986 tax revisions, and they are used to compute MACRS depreciation percentages. These are 150 and 200% of the straight-line rate. Since 200% is twice the straight-line rate, it is called **double declining balance,** or DDB; the general equation is

$$\text{Double declining balance} \quad d_t = \frac{2}{N}(\text{Book value}_{t-1}) \qquad (11\text{-}4a)$$

Since book value equals cost *minus* depreciation charges to date,

$$\text{DDB} \quad d_t = \frac{2}{N}(\text{Cost} - \text{Depreciation charges to date})$$

or

$$d_t = \frac{2}{N}\left(B - \sum_{j=1}^{t-1} d_j\right) \qquad (11\text{-}4b)$$

**EXAMPLE 11–4**

Compute the DDB depreciation schedule for the situations in Examples 11-2 and 11-3 ($1000):

| | |
|---|---|
| Cost of the asset, $B$ | $900 |
| Depreciable life, in years, $N$ | 5 |
| Salvage value, $S$ | $70 |

**SOLUTION**

| | Depreciation for Year $t$ Using Equation 11-4a ($1000) | Sum of Depreciation Charges Up to Year $t$ ($1000) | Book Value at End of Year $t$ ($1000) |
|---|---|---|---|
| Year | | | |
| $t$ | $d_t$ | $\displaystyle\sum_{j=1}^{t} d_j$ | $BV_t = B - \displaystyle\sum_{j=1}^{t} d_j$ |
| 1 | $(2/5)900 = 360$ | $360 | $900 - 360 = 540$ |
| 2 | $(2/5)540 = 216$ | 576 | $900 - 576 = 324$ |
| 3 | $(2/5)324 = 130$ | 706 | $900 - 706 = 194$ |
| 4 | $(2/5)194 = 78$ | 784 | $900 - 784 = 116$ |
| 5 | $(2/5)116 = 46$ | 830 | $900 - 830 = 70 = S$ |

Figure 11-4 illustrates the situation.

**FIGURE 11–4** Declining balance depreciation.

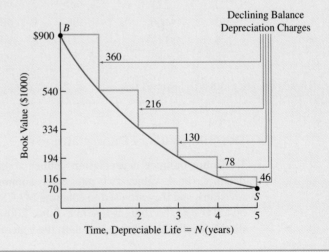

The final salvage value of $70,000 for Examples 11-2, 11-3, and 11-4 was chosen to match the ending value for the double declining balance method. This does not normally happen. If the final salvage value of Example 11-4 had not been $70,000, the double declining balance method would have had to be modified. One modification stops further depreciation once the book value has come to equal the salvage value—this prevents taking too much depreciation. The other modification would switch from declining balance depreciation to straight line—this ensures taking enough depreciation.

These modifications are not detailed here because (1) MACRS has been the legally appropriate system since 1986 and (2) MACRS incorporates the shift from declining balance to straight-line depreciation.

## MODIFIED ACCELERATED COST RECOVERY SYSTEM (MACRS)

The modified accelerated cost recovery system (MACRS) depreciation method, introduced by the Tax Reform Act of 1986, was continued with the Taxpayer Relief Act of 1997. Three major advantages of MACRS are that (1) the "property class lives" are less than the "actual useful lives," (2) salvage values are assumed to be zero, and (3) tables of annual percentages simplify computations.

The definition of the MACRS classes of depreciable property is based on work by the U.S. Treasury Department. In 1971 the Treasury published guidelines for about 100 broad asset classes. For each class there was a lower limit, a midpoint, and an upper limit of useful life, called the **asset depreciation range (ADR).** The ADR midpoint lives were somewhat shorter than the actual average useful lives. These guidelines have been incorporated into MACRS so that the property class lives are again shorter than the ADR midpoint lives.

Use of MACRS focuses on the general depreciation system (GDS), which is based on declining balance with switch to straight-line depreciation. The alternative depreciation system (ADS) provides for a longer period of recovery and uses straight-line depreciation. Thus it is much less economically attractive. Under law, ADS must be used for (1) any tangible property used primarily outside the U.S., (2) any property that is tax exempt or financed by tax-exempt bonds, and (3) farming property placed in service when uniform capitalization rules are not applied. The ADS may also be *elected* for property that can be depreciated using the GDS system. However, once ADS has been elected for an asset, it is not possible to switch back to the GDS system. Because the ADS makes the depreciation deductions less valuable, unless ADS is specifically mentioned, subsequent discussion assumes the GDS system when reference is made to MACRS.

Once a property has been determined to be eligible for depreciation, the next step is to calculate its depreciation deductions over its life. The following information is required to calculate these deductions:

- The asset's cost basis.
- The asset's *property class* and *recovery period.*
- The asset's placed-in-service date.

### Cost Basis and Placed-in-Service Date

The cost basis, $B$, is the cost to obtain and place the asset in service fit for use. However, for real property the basis may also include certain fees and charges that the buyer pays as part of the purchase. Examples of such fees include legal and recording fees, abstract fees, survey charges, transfer taxes, title insurance, and amounts that the seller owes that you pay (back taxes, interest, sales commissions, etc.).

Depreciation for a business asset begins when the asset is *placed in service* for a business purpose. If an asset is purchased and used in a personal context, depreciation may not be taken. If that asset is later used in business for income-producing activity, depreciation may begin with the change in usage.

## Property Class and Recovery Period

Each depreciated asset is placed in a *MACRS property class,* which defines the **recovery period** and the depreciation percentage for each year. Historically the IRS assigned each type of depreciable asset a *class life* or an *asset depreciation range.* With MACRS, asset class lives have been pooled together in the *property classes.* Table 11-1 lists the class lives and GDS and ADS property classes for several example depreciable assets. Table 11-2 lists the MACRS GDS property classes.

The MACRS GDS property classes are described in more detail in Table 11-2. The proper MACRS property class can be found several different ways. Of the five approaches listed, the first one that works should be used.

1. Property class given in problem.
2. Asset is named in Table 11-2.
3. IRS tables or Table 11-1.
4. Class life.
5. Seven-year property for "all other property not assigned to another class."

**TABLE 11–1  Example Class Lives and MACRS Property Classes***

| IRS Asset Class | Asset Description | Class Life (years) ADR | MACRS Property Class (years) | |
|---|---|---|---|---|
| | | | GDS | ADS |
| 00.11 | Office furniture, fixtures, and equipment | 10 | 7 | 10 |
| 00.12 | Information systems: computers/peripheral | 6 | 5 | 6 |
| 00.22 | Automobiles, taxis | 3 | 5 | 5 |
| 00.241 | Light general-purpose trucks | 4 | 5 | 5 |
| 00.25 | Railroad cars and locomotives | 15 | 7 | 15 |
| 00.40 | Industrial steam and electric distribution | 22 | 15 | 22 |
| 01.11 | Cotton gin assets | 12 | 7 | 12 |
| 01.21 | Cattle, breeding or dairy | 7 | 5 | 7 |
| 13.00 | Offshore drilling assets | 7.5 | 5 | 7.5 |
| 13.30 | Petroleum refining assets | 16 | 10 | 16 |
| 15.00 | Construction assets | 6 | 5 | 6 |
| 20.10 | Manufacture of grain and grain mill products | 17 | 10 | 17 |
| 22.2 | Manufacture of yarn, thread, and woven fabric | 11 | 7 | 11 |
| 24.10 | Cutting of timber | 6 | 5 | 6 |
| 32.20 | Manufacture of cement | 20 | 15 | 20 |
| 37.11 | Manufacture of motor vehicles | 12 | 7 | 12 |
| 48.11 | Telephone communications assets and buildings | 24 | 15 | 24 |
| 48.2 | Radio and television broadcasting equipment | 6 | 5 | 6 |
| 49.12 | Electric utility nuclear production plant | 20 | 15 | 20 |
| 49.13 | Electric utility steam production plant | 28 | 20 | 28 |
| 49.23 | Natural gas production plant | 14 | 7 | 14 |
| 50.00 | Municipal wastewater treatment plant | 24 | 15 | 24 |
| 80.00 | Theme and amusement park assets | 12.5 | 7 | 12.5 |

*See Table B.1, *Table of Class Lives and Recovery Periods* in IRS Publication 946, *How to Depreciate Property* (www.irs.gov) for a complete list of depreciable properties under MACRS.

### TABLE 11–2 MACRS GDS Property Classes

| Property Class | Personal Property (all property except real estate) |
|---|---|
| 3-year property | Special handling devices for food and beverage manufacture<br>Special tools for the manufacture of finished plastic products, fabricated metal products, and motor vehicles<br>Property with ADR class life of 4 years or less |
| 5-year property | Automobiles* and trucks<br>Aircraft (of non-air-transport companies)<br>Equipment used in research and experimentation<br>Computers<br>Petroleum drilling equipment<br>Property with ADR class life of more than 4 years and less than 10 years |
| 7-year property | All other property not assigned to another class<br>Office furniture, fixtures, and equipment<br>Property with ADR class life of 10 years or more and less than 16 years |
| 10-year property | Assets used in petroleum refining and certain food products<br>Vessels and water transportation equipment<br>Property with ADR class life of 16 years or more and less than 20 years |
| 15-year property | Telephone distribution plants<br>Municipal sewage treatment plants<br>Property with ADR class life of 20 years or more and less than 25 years |
| 20-year property | Municipal sewers<br>Property with ADR class life of 25 years or more |

| Property Class | Real Property (real estate) |
|---|---|
| 27.5 years | Residential rental property (does not include hotels and motels) |
| 39 years | Nonresidential real property |

*The depreciation deduction for passenger automobiles, trucks, and vans is limited and determined by year placed in service. For 2009, limits on passenger automobiles were $2960 for the first tax year (if special depreciation allowance is not elected), $4800 for Year 2, $2851 for Year 3, and $1775 in later years. For more detail, see details in IRS Publication 946, *How to Depreciate Property* (www.irs.gov).

Once the MACRS property class is known, as well as the placed-in-service date and the cost basis, the year-to-year depreciation deductions can be calculated for GDS assets over their depreciable life using

$$d_t = B \times r_t \qquad (11\text{-}5)$$

where $d_t$ = depreciation deduction in year $t$
   $B$ = cost basis being depreciated
   $r_t$ = appropriate MACRS percentage rate

## Percentage Tables

The IRS has prepared tables to assist in calculating depreciation charges when MACRS GDS depreciation is used. Table 11-3 gives the yearly depreciation percentages ($r_t$) that are used for the six personal property classes (3-, 5-, 7-, 10-, 15-, and 20-year property classes),

**TABLE 11–3  MACRS Depreciation for Personal Property: Half-Year Convention**

| Recovery Year | Applicable Percentage for Property Class | | | | | |
|---|---|---|---|---|---|---|
| | 3-Year Property | 5-Year Property | 7-Year Property | 10-Year Property | 15-Year Property | 20-Year Property |
| 1 | 33.33 | 20.00 | 14.29 | 10.00 | 5.00 | 3.750 |
| 2 | 44.45 | 32.00 | 24.49 | 18.00 | 9.50 | 7.219 |
| 3 | 14.81* | 19.20 | 17.49 | 14.40 | 8.55 | 6.677 |
| 4 | 7.41 | 11.52* | 12.49 | 11.52 | 7.70 | 6.177 |
| 5 | | 11.52 | 8.93* | 9.22 | 6.93 | 5.713 |
| 6 | | 5.76 | 8.92 | 7.37 | 6.23 | 5.285 |
| 7 | | | 8.93 | 6.55* | 5.90* | 4.888 |
| 8 | | | 4.46 | 6.55 | 5.90 | 4.522 |
| 9 | | | | 6.56 | 5.91 | 4.462* |
| 10 | | | | 6.55 | 5.90 | 4.461 |
| 11 | | | | 3.28 | 5.91 | 4.462 |
| 12 | | | | | 5.90 | 4.461 |
| 13 | | | | | 5.91 | 4.462 |
| 14 | | | | | 5.90 | 4.461 |
| 15 | | | | | 5.91 | 4.462 |
| 16 | | | | | 2.95 | 4.461 |
| 17 | | | | | | 4.462 |
| 18 | | | | | | 4.461 |
| 19 | | | | | | 4.462 |
| 20 | | | | | | 4.461 |
| 21 | | | | | | 2.231 |

**Computation method**

- The 3-, 5-, 7-, and 10-year classes use 200% and the 15- and 20-year classes use 150% declining balance depreciation.
- All classes convert to straight-line depreciation in the optimal year, shown with asterisk (*).
- A half-year of depreciation is allowed in the first and last recovery years.
- If more than 40% of the year's MACRS property is placed in service in the last 3 months, then a midquarter convention must be used with depreciation tables that are not shown here.

and Table 11-4 gives the percentages for nonresidential real property. Notice that the values are given in *percentages*—thus, for example, the value of 33.33% (given in Table 11-3 for Year 1 for a 3-year MACRS GDS property) is 0.3333.

Notice in Table 11-3 that the depreciation percentages continue for *one year beyond* the property class life. For example, a MACRS 10-year property has an $r_t$ value of 3.28% in Year 11. This is due to the *half-year convention* that also halves the percentage for the first year. The half-year convention assumes that all assets are placed in service at the midpoint of the first year.

**TABLE 11–4  MACRS Depreciation for Real Property (real estate)***

| Recovery Year | Recovery Percentages for Nonresidential Real Property (month placed in service) | | | | | | | | | | | |
|---|---|---|---|---|---|---|---|---|---|---|---|---|
| | **1** | **2** | **3** | **4** | **5** | **6** | **7** | **8** | **9** | **10** | **11** | **12** |
| 1 | 2.461 | 2.247 | 2.033 | 1.819 | 1.605 | 1.391 | 1.177 | 0.963 | 0.749 | 0.535 | 0.321 | 0.107 |
| 2–39 | 2.564 | 2.564 | 2.564 | 2.564 | 2.564 | 2.564 | 2.564 | 2.564 | 2.564 | 2.564 | 2.564 | 2.564 |
| 40 | 0.107 | 0.321 | 0.535 | 0.749 | 0.963 | 1.177 | 1.391 | 1.605 | 1.819 | 2.033 | 2.247 | 2.461 |

*The useful life is 39 years for nonresidential real property. Depreciation is straight line using the midmonth convention. Thus a property placed in service in January would be allowed $11\frac{1}{2}$ months depreciation for recovery Year 1.

Another characteristic of the MACRS percentage tables is that the $r_t$ values in any column sum to 100%. This means that assets depreciated using MACRS are *fully depreciated* at the end of the recovery period. This assumes a salvage value of zero. This is a departure from the pre-1981 historical methods, where an estimated salvage value was considered.

## Where MACRS Percentage Rates ($r_t$) Come From

This section describes the connection between historical depreciation methods and the MACRS percentages that are shown in Table 11-3. Before ACRS and MACRS, the most common depreciation method was declining balance with a switch to straight line. That combined method is used for MACRS with three further assumptions.

1. Salvage values are assumed to be zero for all assets.
2. The first and last years of the recovery period are each assumed to be *half-year*.
3. The declining balance rate is 200% for 3-, 5-, 7-, and 10-year property and 150% for 15- and 20-year property.

As shown in Example 11-5, the MACRS percentage rates can be derived from these rules and the declining balance and straight-line methods. However, it is obviously much easier to simply use the $r_t$ values from Tables 11-3 and 11-4.

## EXAMPLE 11–5

Consider a 5-year MACRS property asset with an installed and "made ready for use" cost basis of $100. (*Note*: The $100 value used here is for illustration purposes in developing the rates. One would not normally depreciate an asset with a cost basis of only $100.) Develop the MACRS percentage rates ($r_t$) for the asset based on the underlying depreciation methods.

### SOLUTION

To develop the 5-year MACRS property percentage rates, we use the 200% declining balance method, switching over to straight line at the optimal point. Since the assumed salvage value is zero, the entire cost basis of $100 is depreciated. Also the $100 basis mimics the 100% that is used in Table 11-3.

Let's explain the accompanying table year by year. In Year 1 the basis is $100 − 0$, and the $d_t$ values are halved for the initial half-year assumption. Double declining balance has a rate of 40% for 5 years ($= 2/5$). This is larger than straight line for Year 1. So one-half of the 40% is used for Year 1. The rest of the declining balance computations are simply $40\% \times$ (basis minus the cumulative depreciation).

In Year 2 there are 4.5 years remaining for straight line, so 4.5 is the denominator for dividing the remaining $80 in book value. Similarly in Year 3 there are 3.5 years remaining. In Year 4 the DDB and SL calculations happen to be identical, so the switch from DDB to SL can be done in either Year 4 or Year 5. Once we know that the SL depreciation is 11.52 at the switch point, then the only further calculation is to halve that for the last year.

Notice that the DDB calculations get smaller every year, so that at some point the straight-line calculations lead to faster depreciation. This point is the optimal switch point, and it is built into Table 11-3 for MACRs.

| Year | DDB Calculation | SL Calculation | MACRS $r_t$ (%) Rates | Cumulative Depreciation (%) |
|---|---|---|---|---|
| 1 | $1/(2/5)(100 - 0) = \mathbf{20.00}$ | $^1/_2(100 - 0)/5 = 10.00$ | 20.00 (DDB) | 20.00 |
| 2 | $(2/5)(100 - 20.00) = \mathbf{32.00}$ | $(100 - 20)/4.5 = 17.78$ | 32.00 (DDB) | 52.00 |
| 3 | $(2/5)(100 - 52.00) = \mathbf{19.20}$ | $(100 - 52)/3.5 = 13.71$ | 19.20 (DDB) | 71.20 |
| 4 | $(2/5)(100 - 71.20) = \mathbf{11.52}$ | $(100 - 71.20)/2.5 = \mathbf{11.52}$ | 11.52 (either) | 82.72 |
| 5 | | | $\mathbf{11.52}$  11.52 (SL) | 94.24 |
| 6 | | $(^1/_2)(11.52) = \mathbf{5.76}$ | 5.76 (SL) | 100.00 |

The values given in this example match the $r_t$ percentage rates given in Table 11-3 for a 5-year MACRS property.

## MACRS Method Examples

Remember the key points in using MACRS: (1) what type of asset you have, and whether it qualifies as depreciable property, (2) the amount you are depreciating [cost basis], and (3) when you are placing the asset in service. Let's look at several examples of using MACRS to calculate both depreciation deductions and book values.

### EXAMPLE 11–6

Use the MACRS GDS method to calculate the yearly depreciation allowances and book values for a firm that has purchased $150,000 worth of office equipment that qualifies as depreciable property. The equipment is estimated to have a salvage (market) value of $30,000 (20% of the original cost) after the end of its depreciable life.

### SOLUTION

1. The assets qualify as depreciable property.
2. The cost basis is given as $150,000.
3. The assets are being placed in service in Year 1 of our analysis.
4. MACRS GDS applies.
5. The salvage value is not used with MACRS to calculate depreciation or book value.

Office equipment is listed in Table 11-2 as a 7-year property. We now use the MACRS GDS 7-year property percentages from Table 11-3 and Equation 11-5 to calculate the year-to-year depreciation allowances. We use Equation 11-1 to calculate the book value of the asset.

| Year, $t$ | MACRS, $r_t$ | Cost Basis | $d_t$ | Cumulative $d_t$ | $BV_t = B - Cum.d_t$ |
|---|---|---|---|---|---|
| 1 | 14.29% × | $150,000 | $ 21,435 | $ 21,435 | $128,565 |
| 2 | 24.49 | 150,000 | 36,735 | 58,170 | 91,830 |
| 3 | 17.49 | 150,000 | 26,235 | 84,405 | 65,595 |
| 4 | 12.49 | 150,000 | 18,735 | 103,140 | 46,860 |
| 5 | 8.93 | 150,000 | 13,395 | 116,535 | 33,465 |
| 6 | 8.92 | 150,000 | 13,380 | 129,915 | 20,085 |
| 7 | 8.93 | 150,000 | 13,395 | 143,310 | 6,690 |
| 8 | 4.46 | 150,000 | 6,690 | 150,000 | 0 |
| | 100.00% | | $150,000 | | |

Notice in this example several aspects of the MACRS depreciation method: (1) the sum of the $r_t$ values is 100.00%, (2) this 7-year MACRS GDS property is depreciated over 8 years (=property class life $+$ 1), and (3) the book value after 8 years is $0.

## EXAMPLE 11–7

Investors in the JMJ Group purchased a hotel resort in April. The group paid $20 million for the hotel resort and $5 million for the grounds surrounding the resort. The group sold the resort 5 years later in August. Calculate the depreciation deductions for Years 1 through 6. What was the book value at the time the resort was sold?

### SOLUTION

Hotels are nonresidential real property and are depreciated over a 39-year life. Table 11-4 lists the percentages for each year. In this case the cost basis is $20 million, and the $5 million paid for the land is not depreciated. JMJ's depreciation is calculated as follows:

| | |
|---|---|
| Year 1 (obtained in April) | $d_1 = 20,000,000(1.819\%) = \$363,800$ |
| Year 2 | $d_2 = 20,000,000(2.564\%) = 512,800$ |
| Year 3 | $d_3 = 20,000,000(2.564\%) = 512,800$ |
| Year 4 | $d_4 = 20,000,000(2.564\%) = 512,800$ |
| Year 5 | $d_5 = 20,000,000(2.564\%) = 512,800$ |
| Year 6 (disposed of in August) | $d_6 = 20,000,000(1.605\%) = 320,100$ |

Thus the hotel's book value when it was sold was

$$BV_6 = B - (d_1 + d_2 + d_3 + d_4 + d_5 + d_6)$$
$$= 20,000,000 - (2,736,000) = \$17,264,000$$

The value of the land has not changed in terms of book value.

## Comparing MACRS and Historical Methods

In Examples 11-2 through 11-4 we used the *straight-line, sum-of-the-years'-digits,* and *declining balance* depreciation methods to illustrate how the book value of an asset that cost $900,000 and had a salvage value of $70,000 changed over its 5-year depreciation life. Figures 11-2 through 11-4 provided a graphical view of book value over the 5-year depreciation period using these methods. Example 11-8 compares the MACRS GDS depreciation method directly against the historical methods.

## EXAMPLE 11-8

Consider the equipment that was purchased in Example 11-6. Calculate the asset's depreciation deductions and book values over its depreciable life for MACRS and the historical methods.

### SOLUTION

Table 11-5 and Figure 11-5 compare MACRS and historical depreciation methods. MACRS depreciation is the most *accelerated* or fastest depreciation method—remember its name is the modified *accelerated* cost recovery system. The book value drops fastest and furthest with MACRS, thus the present worth is the largest for the MACRS depreciation deductions.

Depreciation deductions *benefit* a firm after taxes because they reduce taxable income and taxes. The time value of money ensures that it is better to take these deductions as soon as possible. In general, MACRS, which allocates larger deductions earlier in the depreciable life, provides more economic benefits than historical methods.

**TABLE 11-5   Comparison of MACRS and Historical Methods for Asset in Example 11-6**

| Year, $t$ | MACRS $d_t$ | MACRS $BV_t$ | Straight Line $d_t$ | Straight Line $BV_t$ | Double Declining $d_t$ | Double Declining $BV_t$ | Sum-of-Years' Digits $d_t$ | Sum-of-Years' Digits $BV_t$ |
|---|---|---|---|---|---|---|---|---|
| 1 | 21,435 | 128,565 | 12,000 | 138,000 | 30,000 | 120,000 | 21,818 | 128,182 |
| 2 | 36,735 | 91,830 | 12,000 | 126,000 | 24,000 | 96,000 | 19,636 | 108,545 |
| 3 | 26,235 | 65,595 | 12,000 | 114,000 | 19,200 | 76,800 | 17,455 | 91,091 |
| 4 | 18,735 | 46,860 | 12,000 | 102,000 | 15,360 | 61,440 | 15,273 | 75,818 |
| 5 | 13,395 | 33,465 | 12,000 | 90,000 | 12,288 | 49,152 | 13,091 | 62,727 |
| 6 | 13,380 | 20,085 | 12,000 | 78,000 | 9,830 | 39,322 | 10,909 | 51,818 |
| 7 | 13,395 | 6,690 | 12,000 | 66,000 | 7,864 | 31,457 | 8,727 | 43,091 |
| 8 | 6,690 | 0 | 12,000 | 54,000 | 1,457 | 30,000 | 6,545 | 36,545 |
| 9 | 0 | 0 | 12,000 | 42,000 | 0 | 30,000 | 4,364 | 32,182 |
| 10 | 0 | 0 | 12,000 | 30,000 | 0 | 30,000 | 2,182 | 30,000 |
| PW (10%) | $108,217 | | $73,734 | | $89,918 | | $84,118 | |

## DEPRECIATION AND ASSET DISPOSAL

When a depreciated asset is disposed of, the key question is, Which is larger, the asset's *book value, BV,* or the asset's *market value, MV*? If the book value is lower than the market

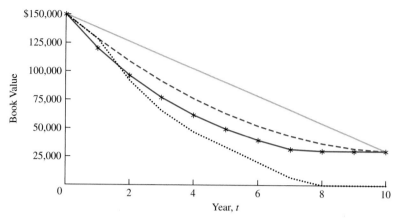

**FIGURE 11–5** Comparing MACRS (··············) and historical depreciation methods: double declining balance (——✱——), sum of the years' digits (– – – – –), and straight line (————).

value, then excess depreciation will be recaptured and taxed. On the other hand, if the book value is higher than the market value, there is a *loss* on the disposal. In either case, the level of taxes owed changes.

> *Depreciation recapture (ordinary gains):* Depreciation recapture, also called ordinary gains, is necessary when an asset is sold for more than an asset's current book value. If more than the original cost basis is received, only the amount up to the original cost basis is recaptured depreciation. Since MACRS assumes $S = 0$ for its annual calculations, MACRS often has recaptured depreciation at disposal.

> *Losses:* A *loss* occurs when less than book value is received for a depreciated asset. In the accounting records we've disposed of an asset for a dollar amount less than its book value, which is a loss.

> *Capital gains:* Capital gains occur when more than the asset's original cost basis is received for it. The excess over the original cost basis is the *capital gain*. As described in Chapter 12, the tax rate on such gains is sometimes lower than the rate on ordinary income, but this depends on how long the investment has been held ("short," $\leq 1$ year; "long," $\geq 1$ year). In most engineering economic analyses capital gains are very uncommon because business and production equipment and facilities almost always *lose* value over time. Capital gains are much more likely to occur for nondepreciated assets like stocks, bonds, real estate, jewelry, art, and collectibles.

The relationship between depreciation recapture, loss, and capital gain is illustrated in Figure 11-6. Each case given is at a point in time in the life of the depreciated asset, where the original cost basis is $10,000 and the book value is $5000. Case (a) represents depreciation recapture (ordinary gain), Case (b) represents a loss, and in Case (c) both recaptured depreciation and a capital gain are present.

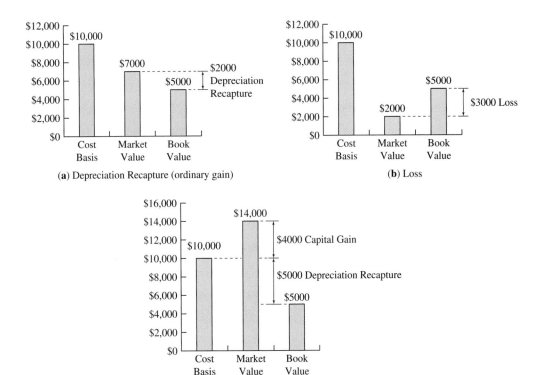

(a) Depreciation Recapture (ordinary gain)

(b) Loss

(c) Capital Gain and Depreciation Recapture

**FIGURE 11–6** Recaptured depreciation, loss on sale, and capital gain.

## EXAMPLE 11–9

Consider an asset with a cost basis of $10,000 that has been depreciated using the MACRS method. This asset is a 3-year MACRS property. What is the gain or loss if the asset is disposed of after 5 years of operation for (**a**) $7000, (**b**) $0, and (**c**) a cost of $2000?

### SOLUTION

To find *gain* or *loss* at disposal we compare *market* and *book value*. Since MACRS depreciates to a salvage value of 0, and 5 years is greater than the recovery period, the book value equals $0.

   (**a**)  Recaptured depreciation = $7000, since the book value is $7000 higher.

   (**b**)  Since market value equals salvage value, there is no recaptured depreciation or loss.

   (**c**)  Since the money is paid for disposal, this is less than the book value, and there is a loss.

This general method for calculating recaptured depreciation or loss applies to all of the depreciation methods described in this chapter.

If the asset is in the middle of its depreciable life, then recaptured depreciation and losses are calculated in a similar manner—compare the *market* and *book values* at the time of disposal. However, in computing the book value with MACRS depreciation, a special rule must be applied for assets disposed of before the end of the recovery period. The rule is to *take one half of the allowable depreciation deduction for that year.* This rule assumes that disposals take place on average halfway through the year. Thus for a 5-year asset disposed of in the middle of Year 4, the rate allowed for MACRS depreciation is half of 11.52% or 5.76%. If the asset is disposed of in Year 6, it is already past the recovery period, and a half-year assumption has already been built into the MACRS schedule. Thus, the full $r_6$ is taken. Example 11-10 illustrates several cases of disposal before the asset is fully depreciated.

## EXAMPLE 11–10

Consider the $10,000 asset in Example 11-9. Do the following:

1. Calculate the effect of disposal if this asset is sold during Year 2 for $2500.
2. Calculate the effect of disposal if the asset is sold during Year 3 for $2500.
3. Calculate the effect of disposal if the asset is sold after Year 3 for $4000 if straight-line depreciation is used over the asset's 5-year life, and a salvage value of $5000 is assumed.

## SOLUTION

1. Market value$_2$ = $2500

$$\text{Book value}_2 = 10{,}000 - 10{,}000[r_1 + r_2/2] = 10{,}000 - 10{,}000[0.3333 + 0.4445/2]$$
$$= \$4444.5$$
$$\text{Loss} = \$1944.5 \,(= 4444.5 - 2500) \qquad \text{(Loss, since BV greater than MV)}$$

*Note:* If the full rather than half-deduction were taken in Year 2, then the book value would be $2222.50 less and the loss would become a gain of $278. Since depreciation would increase by that $2222.50, the total of (depreciation) + (loss or gain) would be $4167 in either calculation.

2. Market value$_3$ = $2500

$$\text{Book value}_3 = 10{,}000 - 10{,}000[r_1 + r_2 + r_3/2]$$
$$= 10{,}000 - 10{,}000[0.3333 + 0.4445 + 0.1481/2] = \$1481.50$$

Recaptured depreciation = $2500 - 1481.50 = \$1018.50$   (Recaptured, since MV > BV)

3.
$$\text{Straight-line rate} = (B - S)/N = (10{,}000 - 5000)/5 = \$1000$$
$$\text{Market value}_4 = \$4000$$
$$\text{Book value}_4 = 10{,}000 - (4)(1000) = \$6000$$
$$\text{Loss} = 6000 - 4000 = \$2000$$

If the asset were disposed of *during the year rather than at year's end,* then the straight-line depreciation deduction would have to be prorated for the number of months during the year that the asset was in service. There is no half-year convention with the historical methods. For example, if disposal occurred on September 30 of Year 4, $d_4 = (9/12)(1000) = \$750$.

*Note:* In Case 1 it was shown that the required half-year convention did not affect the total deductions from taxable income. This is true for the other cases as well, since both recaptured depreciation and losses are treated as ordinary income.

## UNIT-OF-PRODUCTION DEPRECIATION

At times the recovery of depreciation on a particular asset is more closely related to use than to time. In these few situations (and they are rare), the **unit-of-production (UOP) depreciation** in any year is

$$\text{UOP depreciation in any year} = \frac{\text{Production for year}}{\text{Total lifetime production for asset}}(B - S) \quad \text{(11-6)}$$

This method might be useful for machinery that processes natural resources if the resources will be exhausted before the machinery wears out. Historically, this method was sometimes used for construction equipment that had very heavy use in some years and very light use in others. It is not considered an acceptable method for general use in depreciating industrial equipment.

### EXAMPLE 11–11

For numerical similarity with previous examples, assume that equipment costing $900,000 has been purchased for use in a sand and gravel pit. The pit will operate for 5 years, while a nearby airport is being reconstructed and paved. Then the pit will be shut down, and the equipment removed and sold for $70,000. Compute the unit-of-production (UOP) depreciation schedule if the airport reconstruction schedule calls for 40,000 m³ of sand and gravel as follows:

| Year | Required Sand and Gravel ($m^3$) |
|------|---------------------------------|
| 1 | 4,000 |
| 2 | 8,000 |
| 3 | 16,000 |
| 4 | 8,000 |
| 5 | 4,000 |

### SOLUTION

The cost basis, $B$, is $900,000. The salvage value, $S$, is $70,000. The total lifetime production for the asset is 40,000 m³ of sand and gravel. From the airport reconstruction schedule, the first-year UOP depreciation would be

$$\text{First-year UOP depreciation} = \frac{4000 \text{ m}^3}{40,000 \text{ m}^3}(900,000 - 70,000) = \$83,000$$

Similar calculations for the subsequent 4 years give the complete depreciation schedule:

| Year | UOP Depreciation (in $1000) |
|------|-----------------------------|
| 1 | $ 83 |
| 2 | 166 |
| 3 | 332 |
| 4 | 166 |
| 5 | 83 |
|   | $830 |

It should be noted that the actual unit-of-production depreciation charge in any year is based on the actual production for the year rather than the scheduled production.

# DEPLETION

**Depletion** is the exhaustion of natural resources as a result of their removal. Since depletion covers such things as mineral properties, oil and gas wells, and standing timber, removal may take the form of digging up metallic or nonmetallic minerals, producing petroleum or natural gas from wells, or cutting down trees.

Depletion is recognized for income tax purposes for the same reason depreciation is—capital investment is being consumed or used up. Thus a portion of the gross income should be considered to be a return of the capital investment. The calculation of the depletion allowance is different from depreciation because there are two distinct methods of calculating depletion: *cost depletion* and *percentage depletion*. Except for standing timber and most oil and gas wells, depletion is calculated by both methods and the larger value is taken as depletion for the year. For standing timber and most oil and gas wells, only cost depletion is permissible.

## Cost Depletion

Depreciation relied on an asset's cost, depreciable life, and salvage value to apportion the cost *minus* salvage value *over* the depreciable life. In some cases, where the asset is used at fluctuating rates, we might use the unit-of-production (UOP) method of depreciation. For mines, oil wells, and standing timber, fluctuating production rates are the usual situation. Thus, *cost depletion* is computed like unit-of-production depreciation using:

1. Property cost, less cost for land.
2. Estimated number of recoverable units (tons of ore, cubic meters of gravel, barrels of oil, million cubic feet of natural gas, thousand board-feet of timber, etc.).
3. Salvage value, if any, of the property.

## EXAMPLE 11–12

A small lumber company bought a tract of timber for $35,000, of which $5000 was the land's value and $30,000 was the value of the estimated 1.5 million board-feet of standing timber. The first year, the company cut 100,000 board-feet of standing timber. What was the year's depletion allowance?

**SOLUTION**

$$\text{Depletion allowance per 1000 board-ft} = \frac{\$35,000 - \$5000}{1500 \times 1000 \text{ board-ft}}$$

$$= \$20 \text{ per 1000 board-ft}$$

The depletion allowance for the year would be

$$100,000 \text{ board-ft} \times \$20 \text{ per 1000 board-ft} = \$2000$$

## Percentage Depletion

Percentage depletion is an alternate method for mineral property. The allowance is a certain percentage of the property's gross income during the year. This is an entirely different concept from depreciation. Unlike depreciation, which allocates cost *over* useful life, the **percentage depletion** allowance (see Table 11-6) is based on the property's gross income.

Since percentage depletion is computed on the *income* rather than the property's cost, the total depletion *may exceed the cost of the property*. In computing the *allowable percentage depletion* on a property in any year, the *percentage depletion allowance* cannot exceed 50% of the property's taxable income computed without the depletion deduction. The percentage depletion calculations are illustrated by Example 11-13.

TABLE 11–6 **Percentage Depletion Allowances for Selected Deposits**

| Deposits | Rate |
|---|---|
| Sulfur, uranium, and, if from deposits in the U.S., asbestos, lead ore, zinc ore, nickel ore, and mica | 22% |
| Gold, silver, copper, iron ore, and certain oil shale, if from deposits in the U.S. | 15% |
| Borax, granite, limestone, marble, mollusk shells, potash, slate, soapstone, and carbon dioxide produced from a well | 14% |
| Coal, lignite, and sodium chloride | 10% |
| Clay and shale used or sold for use in making sewer pipe or bricks or used or sold for use as sintered or burned lightweight aggregates | $7\frac{1}{2}\%$ |
| Clay used or sold for use in making drainage and roofing tile, flower pots, and kindred products, and gravel, sand, and stone (other than stone used or sold for use by a mine owner or operator as dimension or ornamental stone) | 5% |

*Source*: Internal Revenue Service, Publication 535, Chapter 9. Section 613(b) of the Internal Revenue Code gives a complete list of minerals and their percentage depletion rates.

**EXAMPLE 11–13**

A coal mine has a gross income of $250,000. Mining expenses equal $210,000. Compute the allowable percentage depletion deduction.

From Table 11-6, coal has a 10% depletion allowance based on gross mining income. The allowable percentage depletion deduction is also limited to a maximum of 50% of taxable income.

*Computed Percentage Depletion*

| | |
|---|---|
| Gross income from mine | $250,000 |
| Depletion percentage | ×    10% |
| Computed percentage depletion | $ 25,000 |

*Taxable Income Limitation*

| | |
|---|---|
| Gross income from mine | $250,000 |
| *Less*: Expenses other than depletion | −210,000 |
| Taxable income from mine | 40,000 |
| Deduction limitation | ×    50% |
| Taxable income limitation | $20,000 |

Since the taxable income limitation ($20,000) is less than the computed percentage depletion ($25,000), the allowable percentage depletion deduction is $20,000.

As previously stated, on mineral property the depletion deduction can be based on either cost or percentage depletion. Each year, depletion is computed by both methods, and the allowable depletion deduction is the larger of the two amounts.

## SPREADSHEETS AND DEPRECIATION

The spreadsheet functions for straight-line, double declining balance, and sum-of-years'-digits depreciation are listed in Table 11-7. Because these techniques are simple and were replaced by MACRS in 1986, they are not covered in detail here. All three functions include parameters for *cost* (initial book value), *salvage* (final salvage value), and *life* (depreciation period). Both DDB and SYD change depreciation amounts every year, so they include a parameter to pick the *period* (year). Finally, DDB includes a *factor*. The default value is 2 for 200% or double declining balance, but another commonly used value is 1.5 for 150%.

**TABLE 11-7  Spreadsheet Functions for Depreciation**

| Depreciation Technique | Excel |
|---|---|
| Straight line | SLN(cost, salvage, life) |
| Double declining balance | DDB(cost, salvage, life, period, factor) |
| Sum of years' digits | SYD(cost, salvage, life, period) |
| MACRS | VDB(cost, salvage, life, start_period, end_period, factor, no_switch) |

### Using VDB for MACRS

The Excel function VDB is a flexible or variable declining balance method. It includes the ability to specify the starting and ending periods, rather than simply a year. It also includes an optional no_switch for problems where a switch from declining balance to straight-line depreciation is *not* desired.

To use VDB to calculate MACRS depreciation, the following are true.

1. Salvage = 0, since MACRS assumes no salvage value.
2. Life = recovery period of 3, 5, 7, 10, 15, or 20 years.
3. First period runs from 0 to 0.5, 2nd period from 0.5 to 1.5, 3rd from 1.5 to 2.5, $t^{\text{th}}$ from $t - 1.5$ to $t - 0.5$, and last period from $life - 0.5$ to $life$.
4. Factor = 2 for recovery periods of 3, 5, 7, or 10 years and = 1.5 for recovery periods of 15 or 20 years.
5. Since MACRS includes a switch to straight line, no_switch can be omitted.

The start_period and end_period arguments are from $t - 1.5$ to $t - 0.5$, because MACRS uses a half-year convention for the first year. Thus the first year has 0 to 0.5 year of depreciation, and the second year starts where the first year stops. When one is writing the Excel function, either the first and last periods must be edited individually, or start_period must be defined with a minimum of 0 and end_period with a maximum of life. This prevents the calculation of depreciation from $-0.5$ to 0 and from $life$ to $life + 0.5$.

The results of using the VDB function match Table 11-4, except that the VDB function has more significant digits rather than being rounded to 2 decimals. Example 11-14 illustrates the use of the VDB function.

## EXAMPLE 11–14

Return to the data of Example 11-5 which had $150,000 of office equipment, which is 7-year MACRS property. Use VDB to compute the depreciation amounts.

### SOLUTION

|   | A | B | C | D | E | F | G |
|---|---|---|---|---|---|---|---|
| 1 | 150,000 | First Cost | | | | | |
| 2 | 0 | Salvage | | | | | |
| 3 | 7 | Life = recovery period | | | | | |
| 4 | 200% | Factor | | | | | |
| 5 | | | | | | | |
| 6 | Period | Depreciation | | | | | |
| 7 | 1 | $21,428.57 | =VDB($A$1,0,$A$3,MAX(0,A7−1.5),MIN($A$3,A7−0.5),$A$4) | | | | |
| 8 | 2 | $36,734.69 | or (cost, salvage, life, max(0, $t$−1.5), min (life, $t$−.5), factor) | | | | |
| 9 | 3 | $26,239.07 | | | | | |
| 10 | 4 | $18,742.19 | | | | | |
| 11 | 5 | $13,387.28 | | | | | |
| 12 | 6 | $13,387.28 | | | | | |
| 13 | 7 | $13,387.28 | | | | | |
| 14 | 8 | $6,693.64 | | | | | |
| 15 | | $150,000 | = Sum | | | | |

**FIGURE 11–7** Using VDB to calculate MACRS depreciation.

The spreadsheet in Figure 11-7 defines the start_period with a minimum of 0 and the end_period with a maximum of life. Remember that this is the *depreciable* life or recovery period. Thus this formula could be used for any year of any recovery schedule. Notice that the VDB formula uses the value 0 for the salvage value, rather than referring to the data cell for the salvage value. MACRS assumes a salvage value of zero—no matter what the value truly is.

## SUMMARY

Depreciation is part of computing income taxes in economic analysis. There are three distinct definitions of depreciation:

1. Decline in asset's market value.
2. Decline in asset's value to its owner.
3. Allocating the asset's cost *less* its salvage value *over* its recovery period or depreciable life.

While the first two definitions are used in everyday discussions, it is the third, or accountant's, definition that is used in tax computations and in this chapter. Book value is the remaining unallocated cost of an asset, or

$$\text{Book value} = \text{Asset cost} - \text{Depreciation charges made to date}$$

This chapter describes how depreciable assets are *written off* (or claimed as a business expense) over a period of years instead of *expensed* in a single period (like wages, material costs, etc.). The depreciation methods described include the historical pre-1981 methods: *straight line, sum of the years' digits,* and *declining balance.* These methods required estimating the asset's salvage value and depreciable life.

The current tax law specifies use of the modified accelerated capital recovery system (MACRS). This chapter has focused on the general depreciation system (GDS) with limited discussion of the less attractive alternative depreciation system (ADS). MACRS (GDS) specifies faster *recovery periods* and a salvage value of zero, so it is generally economically more attractive than the historical methods.

The MACRS system is the current tax law, and it assumes a salvage value of zero. This is in contrast with historical methods, which ensured the final book value would equal the predicted salvage value. Thus, when one is using MACRS it is often necessary to consider recaptured depreciation. This is the excess of salvage value over book value, and it is taxed as ordinary income. Similarly, losses on sale or disposal are taxed as ordinary income.

Unit-of-production (UOP) depreciation relies on usage to quantify the loss in value. UOP is appropriate for assets that lose value based on the number of units produced, the tons of gravel moved, and so on (vs. number of years in service). However, this method is not considered to be acceptable for most business assets.

Depletion is the exhaustion of natural resources like minerals, oil and gas, and standing timber. The owners of the natural resources are consuming their investments as the natural resources are removed and sold. Cost depletion is computed based on the fraction of the resource that is removed or sold. For minerals and some oil and gas wells, an alternate calculation called percentage depletion is allowed. Percentage depletion is based on income, so the total allowable depletion deductions may *exceed* the invested cost.

Integrating depreciation schedules with cash flows often involves a lot of arithmetic. Thus, the tool of spreadsheets can be quite helpful. The functions for the historical methods, straight line, sum of the years' digits, and declining balance, are straightforward. Rather than individually entering MACRS percentages into the spreadsheet, the function VDB can be used to calculate MACRS depreciation percentages.

# PROBLEMS

## Depreciation Schedules

**11-1** A depreciable asset costs $10,000 and has an estimated salvage value of $1600 at the end of its 6-year depreciable life. Compute the depreciation schedule for this asset by both SOYD depreciation and DDB depreciation.

**11-2** A million-dollar oil drilling rig has a 6-year depreciable life and a $75,000 salvage value at the end of that time. Determine which one of the following methods provides the preferred depreciation schedule: DDB or SOYD. Show the depreciation schedule for the preferred method.

**11-3** A new machine tool is being purchased for $16,000 and is expected to have a zero salvage value at the end of its 5-year useful life. Compute its DDB depreciation schedule. Assume any remaining depreciation is claimed in the last year.

**11-4** Some special handling devices can be obtained for $12,000. At the end of 4 years, they can be sold for $600. Compute the depreciation schedule for the devices using the following methods:

(a) Straight-line depreciation
(b) Sum-of-years'-digits depreciation
(c) Double declining balance depreciation
(d) MACRS depreciation

**11-5** The company treasurer must determine the best depreciation method for office furniture that costs $50,000 and has a zero salvage value at the end of a 10-year depreciable life. Compute the depreciation schedule using:

(a) Straight line
(b) Double declining balance
(c) Sum-of-years'-digits
(d) Modified accelerated cost recovery system

**11-6** The RX Drug Company has just purchased a capsulating machine for $76,000. The plant engineer estimates the machine has a useful life of 5 years and no salvage value. Compute the depreciation schedule using:

(a) Straight-line depreciation
(b) Sum-of-years'-digits depreciation
(c) Double declining balance depreciation (assume any remaining depreciation is claimed in the last year)

**11-7** The Acme Chemical Company paid $45,000 for research equipment, which it believes will have zero salvage value at the end of its 5-year life. Compute the depreciation schedule using:

(a) Straight line
(b) Sum-of-years'-digits
(c) Double declining balance
(d) Modified accelerated cost recovery system

**11-8** Consider a $6500 piece of machinery, with a 5-year depreciable life and an estimated $1200 salvage value. The projected utilization of the machinery when it was purchased, and its actual production to date, are as follows:

| Year | Projected Production (tons) | Actual Production (tons) |
|------|------|------|
| 1 | 3500 | 3000 |
| 2 | 4000 | 5000 |
| 3 | 4500 | [Not |
| 4 | 5000 | yet |
| 5 | 5500 | known] |

Compute the depreciation schedule using:

(a) Straight line
(b) Sum-of-years'-digits
(c) Double declining balance
(d) Unit of production (for first 2 years only)
(e) Modified accelerated cost recovery system

**11-9** A large profitable corporation bought a small jet plane for use by the firm's executives in January. The plane cost $1.5 million and, for depreciation purposes, is assumed to have a zero salvage value at the end of 5 years. Compute the MACRS depreciation schedule.

**11-10** For an asset that fits into the MACRS "all property not assigned to another class" designation, show in a table the depreciation and book value over the asset's 10-year life of use. The cost basis of the asset is $10,000.

**11-11** A company that manufactures food and beverages in the vending industry has purchased some handling equipment that cost $75,000 and will be depreciated using MACRS GDS; the class life of the asset is 4 years. Show in a table the yearly depreciation amount and book value of the asset over its depreciation life.

**11-12** Consider five depreciation schedules:

| Year | A | B | C | D | E |
|------|------|------|------|------|------|
| 1 | $45.00 | $35.00 | $29.00 | $58.00 | $43.50 |
| 2 | 36.00 | 20.00 | 46.40 | 34.80 | 30.45 |
| 3 | 27.00 | 30.00 | 27.84 | 20.88 | 21.32 |
| 4 | 18.00 | 30.00 | 16.70 | 12.53 | 14.92 |
| 5 | 9.00 | 20.00 | 16.70 | 7.52 | 10.44 |
| 6 | | | 8.36 | | |

They are based on the same initial cost, useful life, and salvage value. Identify each schedule as one of the following:

- Straight-line depreciation
- Sum-of-years'-digits depreciation
- 150% declining balance depreciation
- Double declining balance depreciation
- Unit-of-production depreciation
- Modified accelerated cost recovery system

**11-13** The depreciation schedule for an asset, with a salvage value of $90 at the end of the recovery period, has been computed by several methods. Identify the depreciation method used for each schedule.

| Year | A | B | C | D | E |
|------|------|------|------|------|------|
| 1 | $323.3 | $212.0 | $424.0 | $194.0 | $107.0 |
| 2 | 258.7 | 339.2 | 254.4 | 194.0 | 216.0 |
| 3 | 194.0 | 203.5 | 152.6 | 194.0 | 324.0 |
| 4 | 129.3 | 122.1 | 91.6 | 194.0 | 216.0 |
| 5 | 64.7 | 122.1 | 47.4 | 194.0 | 107.0 |
| 6 | | 61.1 | | | |
| | 970.0 | 1060.0 | 970.0 | 970.0 | 970.0 |

**11-14** The depreciation schedule for a machine has been arrived at by several methods. The estimated salvage value of the equipment at the end of its 6-year useful life is $600. Identify the resulting depreciation schedules.

| Year | A | B | C | D |
|------|------|------|------|------|
| 1 | $2114 | $2000 | $1600 | $1233 |
| 2 | 1762 | 1500 | 2560 | 1233 |
| 3 | 1410 | 1125 | 1536 | 1233 |
| 4 | 1057 | 844 | 922 | 1233 |
| 5 | 705 | 633 | 922 | 1233 |
| 6 | 352 | 475 | 460 | 1233 |

**11-15** A heavy construction firm has been awarded a contract to build a large concrete dam. It is expected that a total of 8 years will be required to complete the work. The firm will buy $600,000 worth of special equipment for the job. During the prepara-tion of the job cost estimate, the following uti-lization schedule was computed for the special equipment:

| Year | Utilization (hr/yr) | Year | Utilization (hr/yr) |
|------|------|------|------|
| 1 | 6000 | 5 | 800 |
| 2 | 4000 | 6 | 800 |
| 3 | 4000 | 7 | 2200 |
| 4 | 1600 | 8 | 2200 |

At the end of the job, it is estimated that the equip-ment can be sold at auction for $60,000.

(a) Compute the sum-of-years-digits' depreciation schedule.

(b) Compute the unit-of-production depreciation schedule.

**11-16** Hillsborough Architecture and Engineering, Inc. has purchased a blueprint printing machine for $18,800. This printer falls under the MACRS cat-egory "Office Equipment" and thus has a 7-year MACRS class life. Prepare a depreciation table for this printer.
*Contributed by Paul R. McCright, University of South Florida*

**11-17** If Hillsborough Architecture and Engineering (in Problem 11-16) purchases a new office building in May for $5.7M, determine the allowable depre-ciation for each year. (*Note*: This asset is non-residential real property.)
*Contributed by Paul R. McCright, University of South Florida*

**11-18** Al Jafar Jewel Co. purchased a crystal extraction machine for $50,000 that has an estimated salvage value of $10,000 at the end of its 8-year useful life. Compute the depreciation schedule using:

(a) MACRS depreciation
(b) Straight-line depreciation
(c) Sum-of-the-years' digits (SOYD) depreciation
(d) Double declining balance depreciation

**11-19** Gamma Cruise, Inc. purchased a new tender (a small motorboat) for $35,000. Its salvage value is $7500 after its useful life of 5 years. Calculate the depreciation schedule using (*a*) MACRS and (*b*) SOYD methods.

**11-20** A profitable company making earthmoving equip-ment is considering an investment of $100,000 on equipment that will have a 5-year useful life and a $20,000 salvage value. Use a spreadsheet func-tion to compute the MACRS depreciation schedule. Show the total depreciation taken (=sum( )) as well as the PW of the depreciation charges discounted at the MARR%.

**11-21** Office equipment whose initial cost is $100,000 has an estimated actual life of 6 years, with an estimated salvage value of $10,000. Prepare tables listing the annual costs of depreciation and the book value at the end of each 6 years, based on straight-line, sum-of-years'-digits, and MACRS depreciation. Use spreadsheet functions for the depreciation methods.

**11-22** You are equipping an office. The total office equipment will have a first cost of $1,750,000 and a salvage value of $200,000. You expect the equipment will last 10 years. Use a spreadsheet function to compute the MACRS depreciation schedule.

**11-23** Units-of-production depreciation is being used for a machine that, based on usage, has an allowable depreciation charge of $6500 the first year and increasing by $1000 each year until complete depreciation. If the machine's cost basis is $110,000, set up a depreciation schedule that shows depreciation charge and book value over the machine's 10-year useful life.

**11-24** A custom-built production machine is being depreciated using the units-of-production method. The machine costs $65,000 and is expected to produce 1.5 million units, after which it will have a $5000 salvage value. In the first 2 years of operation the machine was used to produce 140,000 units each year. In the $3^{rd}$ and $4^{th}$ years, production went up to 400,000 units. After that time annual production returned to 135,000 units. Use a spreadsheet to develop a depreciation schedule showing the machine's depreciation allowance and book value over its depreciable life.

## Comparing Depreciation Methods

**11-25** TELCO Corp. has leased some industrial land near its plant. It is building a small warehouse on the site at a cost of $250,000. The building will be ready for use January 1. The lease will expire 15 years after the building is occupied. The warehouse will belong at that time to the landowner, with the result that there will be no salvage value to TELCO. The warehouse is to be depreciated either by MACRS or SOYD depreciation. If 10% interest is appropriate, which depreciation method should be selected?

**11-26** For the data in Problem 11-20, if money is worth 10%, which one of the following three methods of depreciation would be preferable?

(*a*) Straight-line method

(*b*) Double declining balance method

(*c*) MACRS method

**11-27** The White Swan Talc Company paid $120,000 for mining equipment for a small talc mine. The mining engineer's report indicates the mine contains 40,000 cubic meters of commercial-quality talc. The company plans to mine all the talc in the next 5 years as follows:

| Year | Talc Production (m$^3$) |
|------|-------------------------|
| 1 | 15,000 |
| 2 | 11,000 |
| 3 | 4,000 |
| 4 | 6,000 |
| 5 | 4,000 |

At the end of 5 years, the mine will be exhausted and the mining equipment will be worthless. The company accountant must now decide whether to use sum-of-years'-digits depreciation or unit-of-production depreciation. The company considers 8% to be an appropriate time value of money. Compute the depreciation schedule for each method. Which would you recommend? Show the computations to justify your decision.

**11-28** A small used delivery van can be purchased for $18,000. At the end of its useful life (8 years), the van can be sold for $2500. Determine the PW of the depreciation schedule based on 10% interest using:

(*a*) Straight-line depreciation

(*b*) Sum-of-the-years'-digits depreciation

(*c*) MACRS depreciation

(*d*) Double declining balance depreciation

**11-29** The XYZ Block Company purchased a new office computer and other depreciable computer hardware for $4800. During the third year, the computer is declared obsolete and is donated to the local community college. Using an interest rate of 15%, calculate the PW of the depreciation deductions. Assume that no salvage value was initially declared and that the machine was expected to last 5 years.

(*a*) Straight-line depreciation

(*b*) Sum-of-the-years'-digits depreciation

(*c*) MACRS depreciation

(*d*) Double declining balance depreciation

**11-30** Loretta Livermore Labs purchased R&D equipment costing $200,000. The interest rate is 5%, salvage value is $20,000, and expected life is 10 years. Compute the PW of the depreciation deductions assuming:

(a) Straight-line depreciation

(b) Sum-of-the-years'-digits depreciation

(c) MACRS depreciation

(d) Double declining balance depreciation

**11-31** Some equipment that costs $1000 has a 5-year depreciable life and an estimated $50 salvage value at the end of that time. You have been assigned to determine whether to use straight-line or SOYD depreciation. If a 10% interest rate is appropriate, which is the preferred depreciation method for this profitable corporation? Use a spreadsheet to show your computations of the difference in present worths.

**11-32** The FOURX Corp. has purchased $12,000 of experimental equipment. The anticipated salvage value is $400 at the end of its 5-year depreciable life. This profitable corporation is considering two methods of depreciation: sum of years' digits and double declining balance. If it uses 7% interest in its comparison, which method do you recommend? Show computations to support your recommendation. Use a spreadsheet to develop your solution.

**Depreciation and Book Value**

**11-33** For its fabricated metal products, the Able Corp. is paying $10,000 for special tools that have a 4-year useful life and no salvage value. Compute the depreciation charge for the *second* year by each of the following methods:

(a) DDB

(b) Sum-of-years'-digits

(c) Modified accelerated cost recovery system

**11-34** The MACRS depreciation percentages for 7-year personal property are given in Table 11-3. Make the necessary computations to determine if the percentages shown are correct.

**11-35** The MACRS depreciation percentages for 10-year personal property are given in Table 11-3. Make the necessary computations to determine if the percentages shown are correct.

**11-36** Use MACRS GDS depreciation for each of the assets, 1–3, to calculate the following items, (a)–(c).

1. A light general-purpose truck used by a delivery business, cost = $17,000

2. Production equipment used by a Detroit automaker to produce vehicles, cost = $30,000

3. Cement production facilities used by a construction firm, cost = $130,000

(a) The MACRS GDS property class

(b) The depreciation deduction for Year 3

(c) The book value of the asset after 6 years

**11-37** On July 1, Nancy paid $600,000 for a commercial building and an additional $150,000 for the land on which it stands. Four years later, also on July 1, she sold the property for $850,000. Compute the modified accelerated cost recovery system depreciation for each of the *five* calendar years during which she had the property.

**11-38** A group of investors has formed Trump Corporation to purchase a small hotel. The asking price is $150,000 for the land and $850,000 for the hotel building. If the purchase takes place in June, compute the MACRS depreciation for the first three calendar years. Then assume the hotel is sold in June of the fourth year, and compute the MACRS depreciation in that year also.

**11-39** Mr. Donald Spade bought a computer in January to keep records on all the property he owns. The computer cost $70,000 and is to be depreciated using MACRS. Donald's accountant pointed out that under a special tax rule (the rule applies when the value of property placed in service in the last 3 months of the tax year exceeds 40% of all the property placed in service during the tax year), the computer and all property that year would be subject to the midquarter convention. The midquarter convention assumes that all property placed in service in any quarter-year is placed in service at the midpoint of the quarter. Use the midquarter convention to compute Donald's MACRS depreciation for the first year.

**11-40** A company is considering buying a new piece of machinery. A 10% interest rate will be used in the computations. Two models of the machine are available.

| | Machine I | Machine II |
|---|---|---|
| Initial cost | $80,000 | $100,000 |
| End-of-useful-life salvage value, *S* | 20,000 | 25,000 |
| Annual operating cost | 18,000 | 15,000 first 10 years |
| | | 20,000 thereafter |
| Useful life, in years | 20 | 25 |

(a) Determine which machine should be purchased, based on equivalent uniform annual cost.

(b) What is the capitalized cost of Machine I?

(c) Machine I is purchased and a fund is set up to replace Machine I at the end of 20 years. Compute the required uniform annual deposit.

(d) Machine I will produce an annual saving of material of $28,000. What is the rate of return if Machine I is installed?

(e) What will be the book value of Machine I after 2 years, based on sum-of-years'-digits depreciation?

(f) What will be the book value of Machine II after 3 years, based on double declining balance depreciation?

(g) Assuming that Machine II is in the 7-year property class, what would be the MACRS depreciation in the third year?

**11-41** Explain in your own words the difference between capital gains and ordinary gains. In addition, explain why it is important to our analysis as engineering economists. Do we see capital gains much in industry-based economic analyses or in our personal lives?

**11-42** Sarah Jarala recently purchased an asset that she intends to use for business purposes in her small Iceland Tourism business. The asset has MACRS class life of 5 years. Sarah purchased the asset for $85,000 and uses a *salvage value for tax purposes of $15,000* (when applicable). Also, the ADR life of the asset is 8 years. Please answer the following questions for Sarah.

(a) Using MACRS depreciation, what is the book value after 4 years ?

(b) Using MACRS depreciation, what is the depreciation for the 6th year ?

(c) Using MACRS depreciation, what is the book value after 8 years ?

(d) Using MACRS depreciation, what is the book value after 2 years ?

(e) Using MACRS depreciation, what is the sum of the depreciation charges through the 5th year ?

(f) Using straight-line depreciation (with no half-year convention), what is the book value after the 3rd year?

(g) Using straight-line depreciation (with no half-year convention), what is the book value after the 8th year?

**11-43** Mary, Medhi, Marcos, and Marguerite have purchased a small warehouse in St. Pete Beach. If they paid $745,000 for the unit in September, how much will their depreciation be in the 15th year? (*Note*: This asset is nonresidential real property.)
*Contributed by Paul R. McCright, University of South Florida*

**11-44** Metal Stampings, Inc., can purchase a new forging machine for $100,000. After 20 years of use the forge should have a salvage value of $15,000. What depreciation is allowed for this asset in Year 3 for

(a) MACRS depreciation?

(b) straight-line depreciation?

(c) double declining balance depreciation?

**11-45** Muddy Meadows Earthmoving can purchase a bulldozer for $80,000. After 7 years of use, the bulldozer should have a salvage value of $15000. What depreciation is allowed for this asset in Year 5 for

(a) MACRS depreciation?

(b) straight-line depreciation?

(c) sum-of-the-years'-digits (SOYD) depreciation?

(d) 150% declining balance depreciation?

**11-46** An asset costs $150,000 and has a salvage value of $15,000 after 10 years. What is the depreciation charge for the 4th year, and what is the book value at the end of the 8th year with

(a) MACRS depreciation?

(b) straight-line depreciation?

(c) sum-of-the-year' digits (SOYD) depreciation?

(d) double declining balance depreciation?

**11-47** A five-axis milling machine costs $180,000, and it will be scrapped after 10 years. Compute the book value and depreciation for the first 2 years using

(a) MACRS depreciation

(b) straight-line depreciation

(c) sum-of-the-years'-digits (SOYD) depreciation

(d) double declining balance depreciation

(e) 150% declining balance depreciation

**11-48** To meet increased delivery demands, Mary Moo Dairy just purchased 15 new delivery trucks. Each truck cost $30,000 and has an expected life of 4 years. The trucks can each be sold for $8000 after 4 years. Using MACRS depreciation (a) determine the depreciation allowance for Year 2 for the fleet, and (b) determine the book value of the fleet at the end of Year 3.

**11-49** A minicomputer purchased in 2010 costing $12,000 has no salvage value after 4 years. What is the depreciation allowance for Year 2 and book value at the end of that year?

**11-50** Blank Lobes, Inc. just purchased a new psychograph machine for $100,000. The expected resale value after 4 years is $15,000. Determine the book value after 2 years using

(a) MACRS depreciation.

(b) Straight-line depreciation.

(c) Sum-of-the-years' digits (SOYD) depreciation.

(d) 150% declining balance depreciation.

**11-51** A used drill press costs $55,000, and delivery and installation charges add $5000. The salvage value after 8 years is $15,000. Compute the accumulated depreciation through Year 4 using

(a) 7-year MACRS depreciation

(b) Straight-line depreciation

(c) Sum-of-the-years' digits (SOYD) depreciation

(d) Double declining balance depreciation

**11-52** A pump in an ethylene production plant costs $15,000. After 9 years, the salvage value is declared at $0.

(a) Determine depreciation charge and book value for Year 9 using straight-line, sum-of-the-years'-digits, and 7-year MACRS depreciation.

(b) Find the PW of each depreciation schedule if the interest rate is 5%.

**Gain/Loss on Disposal**

**11-53** Equipment costing $20,000 that is a MACRS 3-year property is disposed of during the second year for $14,000. Calculate any depreciation recapture, ordinary losses, or capital gains associated with disposal of the equipment.

**11-54** An asset with an 8-year ADR class life costs $50,000 and was purchased on January 1, 2001. Calculate any depreciation recapture, ordinary losses, or capital gains associated with selling the equipment on December 31, 2003, for $15,000, $25,000, and $60,000. Consider two cases of depreciation for the problem: if MACRS GDS is used, and if straight-line depreciation over the ADR class life is used with a $10,000 salvage value.

**11-55** A $150,000 asset has been depreciated with the straight-line method over an 8-year life. The estimated salvage value was $30,000. At the end of the 5$^{\text{th}}$ year the asset was sold for $90,000. From a tax perspective, what is happening at the time of disposal, and what is the dollar amount?

**11-56** O'Leary Engineering Corp. has been depreciating a $50,000 machine for the last 3 years. The asset was just sold for 60% of its first cost. What is the size of the recaptured depreciation or loss at disposal using the following depreciation methods?

(a) Sum-of-years'-digits with $N = 8$ and $S = 2000$

(b) Straight-line depreciation with $N = 8$ and $S = 2000$

(c) MACRS GDS depreciation, classified as a 7-year property

**11-57** A purchased machine cost $245,000, with delivery and installation charges amounting to $15,000. The declared salvage value was $20,000. Early in Year 4, the company changed its product mix and found that it no longer needed the machine. One of its competitors agreed to buy the machine for $110,000. Determine the loss, gain, or recapture of MACRS depreciation on the sale. The ADR is 12 years for this machine.

**11-58** A numerically controlled milling machine was purchased for $95,000 four years ago. The estimated salvage value was $15,000 after 15 years. What is the machine's book value after 5 years of depreciation? If the machine is sold for $20,000 early in Year 7, how much gain on sale or recaptured depreciation is there? Assume

(a) 7-year MACRS depreciation

(b) Straight-line depreciation

(c) Sum-of-the-years'-digits (SOYD) depreciation

(d) 150% declining balance depreciation

**11-59** A computer costs $9500 and its salvage value in 5 years is negligible. What is the book value after 3 years? If the machine is sold for $1500 in Year 5, how much gain or recaptured depreciation is there? Assume

(a) MACRS depreciation

(b) Straight-line depreciation

(c) Sum-of-the-years'-digits (SOYD) depreciation

(d) Double declining balance depreciation

**11-60** A belt-conveyor purchased for $140,000 has shipping and installation costs of $20,000. It was expected to last 6 years, when it would be sold for $25,000 after paying $5000 for dismantling. Instead, it lasted 4 years, and several workers were permitted to take it apart on their own time for reassembly at a private technical school. How much gain, loss, or recaptured depreciation is there? Assume

(a) 7-year MACRS depreciation

(b) Straight-line depreciation

(c) Sum-of-the-years'-digits (SOYD) depreciation

(d) 150% declining balance depreciation

## Depletion

11-61 When a major highway was to be constructed nearby, a farmer realized that a dry streambed running through his property might be a valuable source of sand and gravel. He shipped samples to a testing laboratory and learned that the material met the requirements for certain low-grade fill material. The farmer contacted the highway construction contractor, who offered 65¢ per cubic meter for 45,000 cubic meters of sand and gravel. The contractor would build a haul road and would use his own equipment. All activity would take place during a single summer.

The farmer hired an engineering student for $2500 to count the truckloads of material hauled away. The farmer estimated that 2 acres of streambed had been stripped of the sand and gravel. The 640-acre farm had cost him $300 per acre, and the farmer felt the property had not changed in value. He knew that there had been no use for the sand and gravel prior to the construction of the highway, and he could foresee no future use for any of the remaining 50,000 cubic meters of sand and gravel.

Determine the farmer's depletion allowance.

(*Answer:* $1462.50)

11-62 Mr. H. Salt purchased an $1/8$ interest in a producing oil well for $45,000. Recoverable oil reserves for the well were estimated at that time at 15,000 barrels, $1/8$ of which represented Mr. Salt's share of the reserves. During the subsequent year, Mr. Salt received $12,000 as his $1/8$ share of the gross income from the sale of 1000 barrels of oil. From this amount, he had to pay $3000 as his share of the expense of producing the oil. Compute Mr. Salt's depletion allowance for the year.     (*Answer:* $3000)

11-63 American Pulp Corp. (APC) has entered into a contract to harvest timber for $450,000. The total estimated available harvest is 150 million board-feet.

(a) What is the depletion allowance for Years 1 to 3, if 42, 45, and 35 million board-feet are harvested by APC in those years?

(b) After 3 years, the total available harvest for the original tract was reestimated at 180 million board-feet. Compute the depletion allowances for Years 4 and beyond.

11-64 A piece of machinery has a cost basis of $45,000. Its salvage value will be $5000 after 10,000 hours of operation. With units-of-production depreciation, what is the allowable depreciation rate per hour? What is the book value after 4000 hours of operation?

11-65 Mining recently began on a new deposit of 10 million metric tons of ore (2% nickel and 4% copper). Annual production of 350,000 metric tons begins this year. The market price of nickel is $3.75 per pound and $0.65 for copper. Mining operation costs are expected to be $0.50 per pound. XYZ Mining Company paid $600 million for the deposits. What is the maximum depletion allowance each year for the mine?

11-66 During the construction of a highway bypass, earth-moving equipment costing $40,000 was purchased for use in transporting fill from the borrow pit. At the end of the 4-year project, the equipment will be sold for $20,000. The schedule for moving fill calls for a total of 100,000 cubic feet during the project. In the first year, 40% of the total fill is required; in the second year, 30%; in the third year, 25%; and in the final year, the remaining 5%. Determine the units-of-production depreciation schedule for the equipment.

11-67 Western Carolina Coal Co. expects to produce 125,000 tons of coal annually for 15 years. The deposit cost $3M to acquire; the annual gross revenues are expected to be $9.50 per ton, and the net revenues are expected to be $4.25 per ton.

(a) Compute the annual depletion on a cost basis.

(a) Compute the annual depletion on a percentage basis.

11-68 Eastern Gravel expects to produce 60,000 tons of gravel annually for 5 years. The deposit cost $150K to acquire; the annual gross revenues are expected to be $9 per ton, and the net revenues are expected to be $4 per ton.

(a) Compute the annual depletion on a cost basis.

(a) Compute the annual depletion on a percentage basis.

11-69 A 2500-acre tract of timber is purchased by the Houser Paper Company for $1,200,000. The acquisitions department at Houser estimates the land will be worth $275 per acre once the timber is cleared. The materials department estimates that a total of 5 million board-feet of timber is available from the tract. The harvest schedule calls for equal amounts of the timber to be harvested each year for 5 years. Determine the depletion allowance for each year.

11-70 The Piney Copper Company purchased an ore-bearing tract of land for $7,500,000. The geologist for Piney estimated the recoverable copper reserves to be 450,000 tons. During the first year, 50,000 tons was mined and 40,000 tons was sold for $4,000,000. Expenses (not including depletion allowances) were $2,500,000. What are the percentage depletion and the cost depletion allowances?

11-71 The Red River oil field will become less productive each year. Rojas Brothers is a small company that owns Red River, which is eligible for percentage depletion. Red River costs $2.5M to acquire, and it will be produced over 15 years. Initial production costs are $4 per barrel, and the wellhead value is $10 per barrel. The first year's production is 90,000 barrels, which will decrease by 6000 barrels per year.

(a) Compute the annual depletion (each year may be cost-based or percentage-based).

(b) What is the PW at $i = 12\%$ of the depletion schedule?

11-72 An automated assembly line is purchased for $1,250,000. The company has decided to use units-of-production depreciation. At the end of 8 years, the line will be scrapped for an estimated $250,000. Using the following information, determine the depreciation schedule for the assembly line.

| Year | Production Level |
|------|------------------|
| 1 | 5,000 units |
| 2 | 10,000 units |
| 3 | 15,000 units |
| 4 | 15,000 units |
| 5 | 20,000 units |
| 6 | 20,000 units |
| 7 | 10,000 units |
| 8 | 5,000 units |

# CASE

**The following case from *Cases in Engineering Economy* 2$^{nd}$ is suggested as matched with this chapter.**

CASE 26 **Molehill & Mountain Movers**
  Compare depreciation methods with option for inflation.

# INCOME TAXES

## On with the Wind

For over three decades, environmental activists and community leaders have bemoaned American dependence on foreign oil. The war in Iraq again highlighted the precariousness of relying on an unstable region of the world for a major part of our energy requirements.

One solution to this dilemma is to rely more on renewable sources of energy, such as solar power and wind. The technology for such alternative energy sources has been around for many years, and most voters seem to favor more reliance on renewables. If good intentions were all it took, we'd be getting much of our electricity from windmills.

But, of course, making the transition to greater use of wind power requires a significant investment in infrastructure, especially costly wind turbines. And few investors are willing to plunk down their money unless they have a solid expectation of earning a competitive return.

As recently as the late 1980s, wind-generated power cost roughly twice as much to produce as energy from conventional sources. But by 2009, Texas had 9405 megawatts (MW) of capacity and Iowa's 3670 MW of capacity produced 14.2% of its power. The American Wind Energy Association (AWEA) reports that many modern, state-of-the-art wind plants can now produce power for less than 5¢ per kilowatt-hour, making them competitive with conventional sources. Unsurprisingly, investment in wind power production has also increased substantially.

How did this happen? In part, the trend was driven by advances in wind turbine technology. But it was helped significantly by a provision of the federal tax law contained in the Energy Policy Act of 1992. This statute allowed utilities and other electricity suppliers a "production tax credit" of 1.5 ¢ per kilowatt-hour (by 2009 adjusted to 2.1¢ to account for the effects of inflation).

This was a key incentive to the wind power industry which, like all energy producers, must expend large sums on capital assets. The tax credit had a dramatic effect on the wind energy market. During 2009, for instance, energy producers added almost 10,000 megawatts of wind-generating capacity—enough to power nearly 2.4 million homes.

But what the government giveth, the government can take away. When the production tax credit (PTC) briefly expired at the end of 2001, an estimated $3 billion worth of wind projects were suspended, and hundreds of workers were laid off. Fortunately for the industry, the credit, was subsequently extended, most recently through 2011, as part of the 2009 American Recovery and Reinvestment Act. ■ ■ ■

## QUESTIONS TO CONSIDER

1. The wind energy tax credit has both advocates and opponents. One camp argues that the government should not support wind energy through tax credits—instead, let the market determine the future of alternative energy sources. Others argue that government has historically supported the oil and coal industries through subsidies, and in addition we as a society "pay" for the negative effects that these forms of energy have on health and the environment. Which side has the stronger argument in your view? Why?

2. Clearly wind energy has both costs and benefits as an alternate energy technology, and government may have some role in enabling its development and promoting investments in it. Develop a table of the costs and benefits of wind energy technology. In doing so, think about the producers and consumers of the technology as well as society in general. Think about who the winners and losers would be if we as a nation could dramatically increase the fraction of our energy from wind sources. Add a column that identifies the ethical issues that arise? Can we solve any of the ethical issues that you've identified?

3. Developing a wind power project takes many years and requires the commitment of large sums of investment capital before the project begins to return a profit. What is the effect on investment when the wind power production tax credit is allowed to expire, or is extended for periods of only a few years?

4. The federal income tax was introduced in the U.S. in 1913. Can you determine how tax rates changed from 1900 to 2010? How has this affected the value of tax credits to industry?

## After Completing This Chapter...

*The student should be able to:*

- Calculate *taxes due* or *taxes owed* for both individuals and corporations.
- Understand the incremental nature of the individual and corporate tax rates used for calculating taxes on income.
- Calculate a combined income tax rate for state and federal income taxes and select an appropriate tax rate for engineering economic analyses.
- Utilize an *after-tax tax table* to find the after-tax cash flows for a prospective investment project.
- Calculate after-tax measures of merit, such as present worth, annual worth, payback period, internal rate of return, and benefit–cost ratio, from developed after-tax cash flows.
- Evaluate investment alternatives on an after-tax basis including asset disposal.
- Use spreadsheets for solving after-tax economic analysis problems.

As Benjamin Franklin said, two things are inevitable: death and taxes. In this chapter we will examine the structure of taxes in the U.S. These include sales taxes, gasoline taxes, property taxes, and state and federal income taxes. Here we will concentrate our attention on federal income taxes. Income taxes are part of most real problems and often have a substantial impact that must be considered.

First, we must understand the way in which taxes are imposed. Chapter 11 concerning depreciation is an integral part of this analysis, so the principles covered there must be well understood. Then, having understood the mechanism, we will see how federal income taxes affect our economic analysis.

## A PARTNER IN THE BUSINESS

Probably the most straightforward way to understand the role of federal income taxes is to consider the U.S. government as a partner in every business activity. As a partner, the government shares in the profits from every successful venture. In a somewhat more complex way, the government shares in the losses of unprofitable ventures too. The tax laws are complex, and it is not our purpose to fully explain them.[1] Instead, we will examine the fundamental concepts of the federal income tax laws—and we emphasize at the start that there are exceptions and variations to almost every statement we shall make!

## CALCULATION OF TAXABLE INCOME

At the mention of income taxes, one can visualize dozens of elaborate and complex calculations. There is some truth to that vision, for there can be complexities in computing income taxes. Yet incomes taxes are just another type of disbursement. Our economic analysis calculations in prior chapters have dealt with all sorts of disbursements: operating costs, maintenance, labor and materials, and so forth. Now we simply add one more prospective disbursement to the list—income taxes.

### Taxable Income of Individuals

The amount of federal income taxes to be paid depends on taxable income and the income tax rates. Therefore, our first concern is the definition of **taxable income.** To begin, one must compute his or her **gross income:**

Gross income = Wages, salary, etc. + Interest income + Dividends + Capital gains
+ Unemployment compensation + Other income

From gross income, we subtract any allowable retirement plan contributions and other **adjustments.** The result is **adjusted gross income (AGI).** From adjusted gross income, individuals may deduct the following items:[2]

---

[1]Many government and private sources exist that describe detailed taxation information. These include the Internal Revenue Service (www.irs.gov), U.S. Treasury (www.treasury.gov), Commerce Clearing House (www.cch.com), and Thomson Reuters (ria.thomsonreuters.com).
[2]The 2009 returns itemized deductions are limited if adjusted gross income is more than $166,800 ($83,400 if married filing separately).

1. **Personal Exemptions.** One exemption ($3650 for 2009 returns) is provided for each person who depends on the gross income for his or her living.[3]

2. **Itemized Deductions.** Some of these are:

   (a) Excessive medical and dental expenses (exceeding $7\frac{1}{2}\%$ of AGI)

   (b) State and local income, real estate and personal property tax

   (c) Home mortgage interest

   (d) Charitable contributions (limited to 50, 30, or 20% of AGI based on assets contributed.)

   (e) Casualty and theft losses (exceeding $500 + 10% of AGI)

   (f) Job expenses and certain miscellaneous deductions (some categories must exceed 2% of AGI)

3. **Standard Deduction**. Each taxpayer may either itemize his or her deductions, or instead take a standard deduction as follows:

   (a) Single taxpayers, $5700 (for 2009 returns)

   (b) Married taxpayers filing a joint return, $11,400 (for 2009 returns)

The result is **taxable income.**

For individuals, taxable income is computed as follows:

$$\text{Adjusted gross income} = \text{Gross income} - \text{Adjustments}$$

$$\begin{aligned}\textbf{Taxable income} = {} & \text{Adjusted gross income} \\ & - \text{Personal exemption(s)} \\ & - \text{Itemized deductions or standard deduction} \qquad (12\text{-}1)\end{aligned}$$

## Classification of Business Expenditures

When an individual or a firm operates a business, there are three distinct types of business expenditure:

1. For depreciable assets.
2. For nondepreciable assets.
3. All other business expenditures.

*Expenditures for Depreciable Assets:* When facilities or productive equipment with useful lives in excess of one year are acquired, the taxpayer will recover the investment through depreciation charges.[4] Chapter 11 detailed how to allocate an asset's cost over its useful life.

---

[3]For 2009 returns if adjusted gross income is over $125,100 and depending upon filing status, the personal exemption deduction may be reduced by a maximum of 1/3 of the allowed amount.

[4]There is an exception. Businesses may immediately deduct (*expense*) $250,000 of equipment costs (via the Section 179 deduction) in the year of purchase. In addition, 50% of purchases up to a maximum of $800,000 may be expensed. Above $800,000 the deduction is reduced dollar for dollar. In 2009 there is no Section 179 deduction for assets costing more than $1,050,000. Unless otherwise corrected, for the 2010 tax year the maximum amount of equipment placed in service that businesses can expense was scheduled to drop to $135,000.

*Expenditures for Nondepreciable Assets:* Land is considered to be a nondepreciable asset, for there is no finite life associated with it. Other nondepreciable assets are properties *not* used either in a trade, in a business, or for the production of income. An individual's home and car are generally nondepreciable assets. The final category of nondepreciable assets comprises those subject to *depletion,* rather than *depreciation.* Since business firms generally acquire assets for use in the business, their only nondepreciable assets normally are land and assets subject to depletion.

*All Other Business Expenditures:* This category is probably the largest of all, for it includes all the ordinary and necessary expenditures of operating a business. Labor costs, materials, all direct and indirect costs, and facilities and productive equipment with a useful life of one year or less are part of routine expenditures. They are charged as a business expense—*expensed*—when they occur.

Business expenditures in the first two categories—that is, for either depreciable or nondepreciable assets—are called **capital expenditures.** In the accounting records of the firm, they are **capitalized**; all ordinary and necessary expenditures in the third category are **expensed.**

## Taxable Income of Business Firms

The starting point in computing a firm's taxable income is *gross income.* All ordinary and necessary expenses to conduct the business—*except* capital expenditures—are deducted from gross income. Capital expenditures may *not* be deducted from gross income. Except for land, business capital expenditures are allowed on a period-by-period basis through depreciation or depletion charges.

For business firms, taxable income is computed as follows:

$$\textbf{Taxable income} = \text{Gross income}$$
$$- \text{ All expenditures except capital expenditures}$$
$$- \text{ Depreciation and depletion charges} \qquad (12\text{-}2)$$

Because of the treatment of capital expenditures for tax purposes, the taxable income of a firm may be quite different from the actual cash flows.

---

### EXAMPLE 12–1

During a 3-year period, a firm had the following cash flows (in millions of dollars):

|                                               | Year 1 | Year 2 | Year 3 |
|-----------------------------------------------|--------|--------|--------|
| Gross income from sales                       | $200   | $200   | $200   |
| Purchase of special tooling (useful life: 3 years) | −60    | 0      | 0      |
| All other expenditures                        | −140   | −140   | −140   |
| Cash flows for the year                       | $ 0    | $ 60   | $ 60   |

Compute the taxable income for each of the 3 years.

---

**SOLUTION**

The cash flows for each year would suggest that Year 1 was a poor one, while Years 2 and 3 were very profitable. A closer look reveals that the firm's cash flows were adversely affected in Year 1 by the purchase of special tooling. Since the special tooling has a 3-year useful life, it is a capital expenditure with its cost allocated over the useful life. If we assume that straight-line depreciation applies with no salvage value, we use Equation 11-2 to find the annual charge:

$$\text{Annual depreciation charge} = \frac{B - S}{N} = \frac{60 - 0}{3} = \$20 \text{ million}$$

Applying Equation 12-2, we write

$$\text{Taxable income} = 200 - 140 - 20 = \$40 \text{ million}$$

In each of the 3 years, the taxable income is $40 million.

---

An examination of the cash flows and the taxable income in Example 12-1 indicates that taxable income is a better indicator of the firm's annual performance.

## INCOME TAX RATES

Income tax rates for individuals changed many times between 1960 and 1995, as illustrated in Figure 12-1. Recent movements have been less dramatic. From 1995 to 2000 the maximum rate was 39.6%; it fell to 39.1% in 2001 and to 38.6% in 2002. From 2003 through 2010 it was 35%.

### Individual Tax Rates

There are four schedules of federal income tax rates for individuals. Single taxpayers use the Table 12-1a schedule. Married taxpayers filing a joint return use the Table 12-1b schedule.

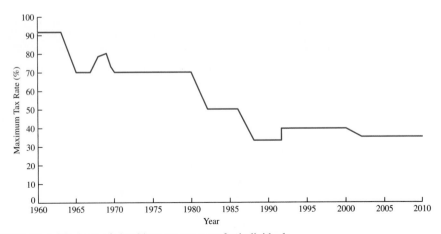

**FIGURE 12-1** Maximum federal income tax rates for individuals.

### TABLE 12–1a  2009 Tax Rate Schedules Schedule X–a If Filing Status Is Single

| If your taxable income is | | The tax is: | |
| --- | --- | --- | --- |
| Over– | But not over– | | Of the amount over– |
| $     0 | $   8,350 | 10% | $      0 |
| 8,350 | 33,950 | $     835.00 + 15% | 8,350 |
| 33,950 | 82,250 | 4,675.00 + 25% | 33,950 |
| 82,250 | 171,550 | 16,750.00 + 28% | 82,250 |
| 171,550 | 372,950 | 41,754.00 + 33% | 171,550 |
| 372,950 | | 108,216.00 + 35% | 372,950 |

### TABLE 12–1b  2009 Tax Rates Schedule Y-1—If Your Filing Status Is Married Filing Jointly or Qualifying Window(er)

| If your taxable income is | | The tax is: | |
| --- | --- | --- | --- |
| Over– | But not over– | | Of the amount over– |
| $     0 | $  16,700 | 10% | $      0 |
| 16,700 | 67,900 | $   1,670.00 + 15% | 16,700 |
| 67,900 | 137,050 | 9,350.00 + 25% | 67,900 |
| 137,050 | 208,850 | 26,637.50 + 28% | 137,050 |
| 208,850 | 372,950 | 46,741.50 + 33% | 208,850 |
| 372,950 | | 100,894.50 + 35% | 372,950 |

Two other schedules (not shown here) are applicable to unmarried individuals with dependent relatives ("head of household"), and married taxpayers filing separately.

## EXAMPLE 12-2

An unmarried student earned $10,000 in the summer plus another $6000 during the rest of the year. When he files an income tax return, he will be allowed one exemption (for himself). He estimates that he spent $1000 on allowable itemized deductions. How much income tax will he pay?

$$\textbf{Adjusted gross income} = \$10{,}000 + \$6000 = \$16{,}000$$

$$\textbf{Taxable income} = \text{Adjusted gross income}$$
$$- \text{ Deduction for one exemption (\$3400)}$$
$$- \text{ Standard deduction (\$5350)}$$
$$= 16{,}000 - 3400 - 5350 = \$7250$$
$$\textbf{Federal income tax} = 10\% \,(7250) = \$725$$

## Corporate Tax Rates

Income tax for corporations is computed in a manner similar to that for individuals. The data in Figure 12-2 can be presented in schedule form (Table 12-2).

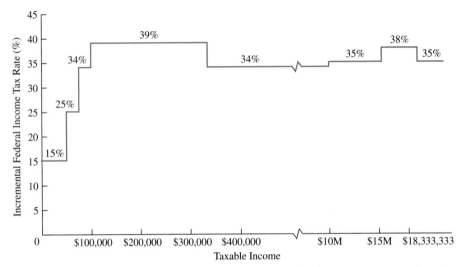

**FIGURE 12-2**  Corporation federal income tax rates (2009). Federal corporate tax rates have been constant over the past several years.

### TABLE 12-2  Corporate Income Tax Rates

| Taxable Income | Tax Rate | Corporate Income Tax |
|---|---|---|
| Not over $50,000 | 15% | 15% over $0 |
| $50,000–75,000 | 25% | $7,500 + 25% over $50,000 |
| $75,000–100,000 | 34% | $13,750 + 34% over $75,000 |
| $100,000–335,000 | 39%* | $22,250 + 39% over $100,000 |
| $335,000–10 million | 34% | $113,900 + 34% over $335,000 |
| $10 million–15 million | 35% | $3,400,000 + 35% over $10 million |
| $15 million–18,333,333 | 38% | $5,150,000 + 38% over $15 million |
| ≥$18,333,333 | 35% | Flat rate at 35% |

*The extra 5% from $100,000 to $335,000 was chosen so that firms in the $335,000 to $10 million bracket pay a flat 34% tax rate. [(0.39−0.34)](335,000− 100,000) = (0.34−0.15)(50,000)+(0.34−0.25)(75,000− 50,000)] so tax = 0.34 (tax income) in $335,000 to $10 million bracket.

### EXAMPLE 12-3

The French Chemical Corporation was formed to produce household bleach. The firm bought land for $220,000, had a $900,000 factory building erected, and installed $650,000 worth of chemical and packaging equipment. The plant was completed and operations begun on April 1. The gross income for the calendar year was $450,000. Supplies and all operating expenses, excluding the capital expenditures, were $100,000. The firm will use modified accelerated cost recovery system (MACRS) depreciation.

(a) What is the first-year depreciation charge?
(b) What is the first-year taxable income?
(c) How much will the corporation pay in federal income taxes for the year?

## SOLUTION TO PART a

*MACRS depreciation*: Chemical equipment is personal property. From Table 11-2, it is probably in the "7-year, all other property" class.

$$\text{First-year depreciation(equipment)} = \$650,000 \times 14.29\% = \$92,885$$

The building is in the 39-year real property class. Since it was placed in service on April 1, the first-year depreciation is

$$\text{First-year depreciation(building)} = \$900,000 \times 1.819\% = \$16,371 \quad \text{(see Table 11-4)}$$

The land is a nondepreciable asset. Thus we have

$$\text{Total first-year MACRS depreciation} = \$92,885 + \$16,371 = \$109,256$$

## SOLUTION TO PART b

Taxable income = Gross income
  − All expenditures except capital expenditures
  − Depreciation and depletion charges
  = $450,000 − 100,000 − 109,256 = $240,744

## SOLUTION TO PART c

Federal income tax = $22,250 + 39%(240,744 − 100,000)
             = $77,140

## Combined Federal and State Income Taxes

In addition to federal income taxes, most individuals and corporations pay state income taxes. It would be convenient if we could derive a single tax rate to represent both the state and federal incremental tax rates. In the computation of taxable income for federal taxes, the amount of state taxes paid is one of the allowable itemized deductions. Federal income taxes are not, however, generally deductible in the computation of state taxable income. Therefore, the state income tax is applied to a *larger* taxable income than is the federal income tax rate. As a result, the combined incremental tax rate will not be the sum of two tax rates.

For an increment of income ($\Delta$Income) and tax rate on incremental income ($\Delta$Tax rate):

$$\text{State income taxes} = (\Delta\text{State tax rate})(\Delta\text{Income})$$
$$\text{Federal taxable income} = (\Delta\text{Income})(1 - \Delta\text{State tax rate})$$
$$\text{Federal income taxes} = (\Delta\text{Federal tax rate})(\Delta\text{Income}) \times (1 - \Delta\text{State tax rate})$$

The total of state and federal income taxes is

$$[\Delta\text{State tax rate} + (\Delta\text{Federal tax rate})(1 - \Delta\text{State tax rate})](\Delta\text{Income})$$

The term in the brackets gives the combined incremental tax rate.

**Combined incremental tax rate**

$$= \Delta\text{State tax rate} + (\Delta\text{Federal tax rate})(1 - \Delta\text{State tax rate}) \quad (12\text{-}3)$$

## EXAMPLE 12–4

An engineer has an income that puts her in the 25% federal income tax bracket and at the 10% state incremental tax rate. She has an opportunity to earn an extra $500 by doing a small consulting job. What will be her combined state and federal income tax rate on the additional income?

### SOLUTION

Use Equation 12-3 to find the combined incremental tax rate: $0.10 + 0.25(1 - 0.10) = 32.5\%$.

### Selecting an Income Tax Rate for Economy Studies

Since income tax rates vary with the level of taxable income for both individuals and corporations, one must decide which tax rate to use in a particular situation. The simple answer is that the tax rate to use is the incremental tax rate that applies to the change in taxable income projected in the economic analysis. If a married couple filing jointly has a taxable income of $40,000 and can increase their income by $2000, what tax rate should be used for the $2000 of incremental income? From Table 12-1b, we see the $2000 falls within the 10% tax bracket.

Now suppose this couple could increase their $40,000 income by $20,000. In this situation, Table 12-1b shows that the 10% incremental tax rate should be applied to the first $16,700 and a 15% incremental tax rate to the last $3300 of extra income. The appropriate incremental tax rate for corporations is equally easy to determine. For larger corporations, the federal incremental tax rate is 35%. In addition, there may be up to a 12 to 15% state tax.

## ECONOMIC ANALYSIS TAKING INCOME TAXES INTO ACCOUNT

An important step in economic analysis has been to resolve the consequences of alternatives into a cash flow. Because income taxes have been ignored, the result has been a *before-tax cash flow*. This before-tax cash flow is an essential component in economic analysis that also considers the consequences of income tax. The principal elements in an *after-tax analysis* are as follows:

- Before-tax cash flow
- Depreciation
- Taxable income (Before-tax cash flow − Depreciation)
- Income taxes (Taxable income × Incremental tax rate)
- After-tax cash flow (Before-tax cash flow − Income taxes)

These elements are usually arranged to form an after-tax *cash flow table*. This is illustrated by Example 12–5.

## EXAMPLE 12–5

A medium-sized profitable corporation may buy a $15,000 used pickup truck for use by the shipping and receiving department. During the truck's 5-year useful life, it is estimated the firm will save $4000 per year after all the costs of owning and operating the truck have been paid. Truck salvage value is estimated at $4500.

(a) What is the before-tax rate of return?

(b) What is the after-tax rate of return on this capital expenditure? Assume straight-line depreciation.

### SOLUTION TO PART a

For a before-tax rate of return, we must first compute the before-tax cash flow.

| Year | Before-Tax Cash Flow |
|------|----------------------|
| 0 | −$15,000 |
| 1 | 4000 |
| 2 | 4000 |
| 3 | 4000 |
| 4 | 4000 |
| 5 | 4000 4500 |

Solve for the before-tax rate of return, $IRR_{BT}$:

$$15,000 = 4000(P/A, i, 5) + 4500(P/F, i, 5)$$

Try $i = 15\%$:

$$15,000 \neq 4000(3.352) + 4500(0.4972) \neq 13,408 + 2237 = 15,645$$

Since $i$ is slightly low, try $i = 18\%$:

$$15,000 \neq 4000(3.127) + 4500(0.4371) \neq 12,508 + 1967 = 14,475$$

$$IRR_{BT} = 15\% + 3\% \left( \frac{15,645 - 15,000}{15,645 - 14,475} \right) = 15\% + 3\%(0.55) = 16.7\%$$

## SOLUTION TO PART b

For an after-tax rate of return, we must set up an after-tax cash flow table (Table 12-3). The starting point is the before-tax cash flow. Then we will need the depreciation schedule for the truck:

$$\text{Straight-line depreciation} = \frac{B - S}{N} = \frac{15,000 - 4500}{5} = \$2100 \text{ per year}$$

**TABLE 12-3   After-Tax Cash Flow Table for Example 12-5**[*]

| Year | (a) Before-Tax Cash Flow | (b) Straight-Line Depreciation | (c) $\Delta$(Taxable Income) (a)−(b) | (d) 34% Income Taxes −0.34 (c) | (e) After-Tax Cash Flow (a) + (d)[†] |
|---|---|---|---|---|---|
| 0 | −$15,000 | | | | −$15,000 |
| 1 | 4000 | $2100 | $1900 | −$646 | 3354 |
| 2 | 4000 | 2100 | 1900 | −646 | 3354 |
| 3 | 4000 | 2100 | 1900 | −646 | 3354 |
| 4 | 4000 | 2100 | 1900 | −646 | 3354 |
| 5 | $\left\{\begin{array}{l} 4000 \\ 4500 \end{array}\right.$ | 2100 | 1900 | −646 | $\left\{\begin{array}{l} 3354 \\ 4500 \end{array}\right.$ |

[*]Sign convention for income taxes: a minus (−) represents a disbursement of money to pay income taxes; a plus (+) represents the receipt of money by a decrease in the tax liability.
[†]The after-tax cash flow is the before-tax cash flow minus income taxes. Based on the income tax sign convention, this is accomplished by *adding* columns (**a**) and (**d**).

Taxable income is the before-tax cash flow *minus* depreciation. For this medium-sized profitable corporation, let's assume an incremental federal income tax rate of 34%. Therefore income taxes are 34% of taxable income. Finally, the after-tax cash flow equals the before-tax cash flow *minus* income taxes. These data are used to compute Table 12-3.

The after-tax cash flow may be solved to find the after-tax rate of return, $\text{IRR}_{\text{AT}}$. Try $i = 10\%$:

$$15,000 \overset{?}{=} 3354(P/A, 10\%, 5) + 4500(P/F, 10\%, 5)$$

$$\overset{?}{=} 3354(3.791) + 4500(0.6209) = 15,509$$

Since $i$ is slightly low, try $i = 12\%$:

$$15,000 \overset{?}{=} 3354(3.605) + 4500(0.5674) = 14,644$$

$$\text{IRR}_{\text{AT}} = 10\% + 2\% \left( \frac{15,509 - 1500}{15,509 - 14,644} \right) = 11.2\%$$

Using a TVM calculator

$$\text{IRR}_{\text{AT}} = i(n, P, P, F) = i(5, 3354, -15000, 4500) = 11.15\%$$

The calculations required to compute the after-tax rate of return in Example 12-5 were certainly more elaborate than those for the before-tax rate of return. It must be emphasized, however, that often the after-tax rate of return is the key value, since income taxes are a major disbursement that cannot be ignored.

## EXAMPLE 12–6

An analysis of a firm's sales activities indicates that a number of profitable sales are lost each year because the firm cannot deliver some of its products quickly enough. By investing an additional $20,000 in inventory, it is believed that the firm will realize $1000 more in before-tax profits in the first year. In the second year, before-tax extra profit will be $1500. Profits for subsequent years are expected to continue to increase on a $500-per-year gradient. The investment in the additional inventory may be recovered at the end of a 4-year analysis period simply by selling and not replenishing the inventory. Compute:

(a) The before-tax rate of return.

(b) The after-tax rate of return assuming an incremental tax rate of 39%.

## SOLUTION

Inventory is not considered to be a depreciable asset. Therefore, the investment in additional inventory is not depreciated. The after-tax cash flow table for the problem is presented in Table 12-4. The sign convention described in the first footnote to Table 12-3 applies.

TABLE 12–4  **After-Tax Cash Flow Table for Example 12–6**

| Year | (a) Before-Tax Cash Flow | (b) Depreciation | (c) Δ(Taxable Income) (a)−(b) | (d) 39% Income Taxes (c) x −0.39 | (e) After-Tax Cash Flow (a) + (d) |
|---|---|---|---|---|---|
| 0 | −$20,000 | | | | −$20,000 |
| 1 | 1000 | — | $1000 | −$390 | 610 |
| 2 | 1500 | — | 1500 | −585 | 915 |
| 3 | 2000 | — | 2000 | −780 | 1220 |
| 4 | { 2500 <br> 20,000 | — | 2500 | −975 | { 1525 <br> 20,000 |

## SOLUTION TO PART a

Use the following equation to calculate the before-tax rate of return:

$$20,000 = 1000(P/A, i, 4) + 500(P/G, i, 4) + 20,000(P/F, i, 4)$$

Try $i = 8\%$:

$$20,000 \neq 1000(3.312) + 500(4.650) + 20,000(0.7350)$$

$$\neq 3312 + 2325 + 14,700 = 20,337$$

Since $i$ is too low, try $i = 10\%$:

$$20,000 \neq 1000(3.170) + 500(4.378) + 20,000(0.6830)$$

$$\neq 3170 + 2189 + 13,660 = 19,019$$

$$\text{Before-tax rate of return} = 8\% + 2\%\left(\frac{20,337 - 20,000}{20,337 - 19,019}\right) = 8.5\%$$

**SOLUTION TO PART b**

For a before-tax cash flow gradient of $500, the resulting after-tax cash flow gradient is $(1 - 0.39)(500) = \$305$.

$$20,000 = 610(P/A, i, 4) + 305(P/G, i, 4) + 20,000(P/F, i, 4)$$

Try $i = 5\%$:

$$20,000 \neq 610(3.546) + 305(5.103) + 20,000(0.8227) \neq 20,173$$

Since $i$ is too low, try $i = 6\%$:

$$20,000 \neq 610(3.465) + 300(4.4945) + 20,000(0.7921) \neq 19,304$$

$$\text{After-tax rate of return} = 5\% + 1\%\left(\frac{20,173 - 20,000}{20,173 - 19,304}\right) = 5.2\%$$

## CAPITAL GAINS AND LOSSES FOR NONDEPRECIATED ASSETS

When a nondepreciated capital asset is sold or exchanged, appropriate entries are made in the firm's accounting records. If the selling price of the capital asset exceeds the original cost basis, the excess is called a **capital gain.** If the selling price is less than the original cost basis, the difference is a **capital loss.** Examples of nondepreciated assets include stocks, land, art, and collectibles.

$$\text{Capital}\begin{Bmatrix} \text{Gain} \\ \text{Loss} \end{Bmatrix} = \text{Selling price} - \text{Original cost basis}$$

Current tax law sets the highest net capital gains tax at 15% for most assets held by individuals for more than one year. This is in contrast to recaptured depreciation, which is taxed at the same rate as other (ordinary) income. The tax treatment of capital gains and losses for nondepreciated assets is shown in Table 12-5.

### Investment Tax Credit

When the economy slows down and unemployment rises, the U.S. government frequently alters its tax laws to promote greater industrial activity. One technique used to stimulate capital investments has been the **investment tax credit (ITC).** Businesses were able to deduct a percentage of their new business equipment purchases as a *tax credit*. This meant that the firm's net cost of the equipment was reduced by the amount of the investment tax credit. Depending on the specific investment tax credit provisions, the credit might be subtracted from the basis for depreciation, or the basis for computing depreciation might

### TABLE 12–5  Tax Treatment of Capital Gains and Losses

**For Individuals**

| | |
|---|---|
| Capital gain | For most assets held for less than 1 year, taxed as ordinary income. |
| | For most assets held for more than 1 year, taxed at 15% tax rate.* |
| Capital loss | Subtract capital losses from any capital gains; balance may be deducted from ordinary income, but not more than $3000 per year ($1500 if married filing separately). Excess capital losses may be carried into future taxable years indefinitely. |

**For Corporations**

| | |
|---|---|
| Capital gain | Taxed as ordinary income. |
| Capital loss | Corporations may deduct capital losses only to the extent of capital gains. Any capital loss in the current year that exceeds capital gains can be carried back 2 years, and, if not completely absorbed, is then carried forward for up to 20 years. |

*Depending on tax bracket, type of asset, and duration of ownership, the capital gain tax rate can range from 0% to 28%. In addition, a special 0% capital gain tax rate applies to homeowners who treat their home as a primary residence for at least 2 years within a 5-year period before selling. In such cases, single taxpayers are allowed a maximum $250,000 in tax-free gain ($500,000 for married couples). Exclusions for nonqualified use apply.

remain the full cost of the equipment. The Tax Reform Act of 1986 eliminated the investment tax credit for most assets, although credits are allowed in some specialized cases such as historic building preservation and in the development of alternate energy sources. It is likely, however, that the general ITC will reappear at some future time.

## ESTIMATING THE AFTER-TAX RATE OF RETURN

There is no shortcut method for computing the after-tax rate of return from the before-tax rate of return. One possible exception to this statement is in the situation of nondepreciable assets. In this special case, we have

**After-tax rate of return** $= (1 -$ Incremental tax rate$) \times ($Before-tax rate of return$)$

For Example 12-6, we could estimate the after-tax rate of return from the before-tax rate of return as follows:

$$\text{After-tax rate of return} = (1 - 0.39)(8.5\%) = 5.2\%$$

This value agrees with the value computed in Example 12-6(b).

This relationship may be helpful for selecting a trial after-tax rate of return when the before-tax rate of return is known. It must be emphasized, however, this relationship is only a rough approximation in almost all situations.

## AFTER-TAX CASH FLOWS AND SPREADSHEETS

Realistic after-tax analysis requires spreadsheets. Even if costs and revenues are the same every year, MACRS depreciation percentages are not. The steps for calculating an after-tax internal rate of return are illustrated in Example 12-7. Because some cash flows are taxed and some are not, the spreadsheet is easier to build if these two types are separated. Spreadsheet construction is easier, as well, if recaptured depreciation or other gain/loss on disposal or sale are tabulated separately.

Taxes are considered even if only the costs of a project are known. The firm that does an engineering project must generate profits—or go out of business. Even if a firm has an unprofitable year, the tax law includes carry-forward and -backward provisions to transfer deductions to profitable years. The depreciation and revenues in Example 12-7 result in a negative taxable income for Year 2. Thus the *positive cash flow* due to taxes that is shown in Year 2 of Example 12-7 really represents tax savings for the firm.

## EXAMPLE 12–7

Return to the data of Example 12-5, where the used truck had a first cost of $15,000, a salvage value after 5 years of $4500, and savings of $4000 per year. Use MACRS depreciation and calculate the after-tax rate of return.

### SOLUTION

Under MACRS, vehicles have a 5-year recovery period. Thus the MACRS depreciation can be calculated by using a VDB function (see "Spreadsheets and Depreciation" at the end of Chapter 11) or by lookup in Table 11-3. The depreciation in Year 5 has been halved, since it is the year of disposal.

|  | A | B | C | D | E | F | G | H |
|---|---|---|---|---|---|---|---|---|
| 1 | 15,000 | First Cost |  |  |  |  |  |  |
| 2 | 4000 | Annual Benefit |  |  |  |  |  |  |
| 3 | 5 | Recovery Period |  |  |  |  |  |  |
| 4 | 4500 | Salvage Value |  |  |  |  |  |  |
| 5 | 0.34 | Tax Rate |  |  |  |  |  |  |
| 6 |  |  |  |  |  |  |  |  |
| 7 | Year | Untaxed BTCF | Taxed BTCF | MACRS | Recaptured Depreciation | Tax Income | Tax | ATCF |
| 8 | 0 | −15,000 |  |  |  |  |  | −15,000.0 |
| 9 | 1 |  | 4000 | 3000 |  | 1000 | 340.0 | 3660.0 |
| 10 | 2 |  | 4000 | 4800 |  | −800 | −272.0 | 4272.0 |
| 11 | 3 |  | 4000 | 2880 |  | 1120 | 380.8 | 3619.2 |
| 12 | 4 |  | 4000 | 1728 |  | 2272 | 772.5 | 3227.5 |
| 13 | 5 | 4500 | 4000 | 864 | 2772 | 5908 | 2008.7 | 6491.3 |
| 14 |  | Cum. Depr.= |  | 13,272 |  |  | IRR= | 11.96% |
| 15 |  |  |  |  |  |  |  |  |
| 16 |  |  |  |  |  | =Taxed BTCF − MACRS + Recapt. |  |  |
| 17 |  |  |  |  | =Salvage − BookValue |  |  |  |

FIGURE 12–3  Spreadsheet for after-tax IRR calculation.

Note from Figure 12-3 that using MACRS rather than straight-line depreciation increases the after-tax IRR from 11.15% to 11.96%. This is due solely to the faster write-off that is allowed under MACRS.

## SUMMARY

Since income taxes are part of most problems, no realistic economic analysis can ignore their consequences. Income taxes make the U.S. government a partner in every business venture. Thus the government benefits from all profitable ventures and shares in the losses of unprofitable ventures.

The first step in computing individual income taxes is to tabulate gross income. Any adjustments—for example, allowable taxpayer contributions to a retirement fund—are subtracted to yield adjusted gross income. Personal exemptions and either itemized deductions or the standard deduction are subtracted to find taxable income. This is used, together with a tax rate table, to compute the income tax liability for the year.

For corporations, taxable income equals gross income *minus* all ordinary and necessary expenditures (except capital expenditures) and depreciation and depletion charges. The income tax computation (whether for an individual or a corporation) is relatively simple, with rates ranging from 10 to 39%. The proper rate to use in an economic analysis is the incremental tax rate applicable to the increment of taxable income being considered.

Most individuals and corporations pay state income taxes in addition to federal income taxes. Since state income taxes are an allowable deduction in computing federal taxable income, it follows that the taxable income for the federal computation is lower than the state taxable income.

Combined state and federal incremental tax rate

$$= \Delta \text{State tax rate} + (\Delta \text{Federal tax rate})(1 - \Delta \text{State tax rate})$$

To introduce the effect of income taxes into an economic analysis, the starting point is a before-tax cash flow. Then the depreciation schedule is deducted from appropriate parts of the before-tax cash flow to obtain taxable income. Income taxes are obtained by multiplying taxable income by the proper tax rate. Before-tax cash flow less income taxes equals the after-tax cash flow. This data is all captured in an after-tax cash flow table.

Current tax law has decreased long-term capital gains on most nondepreciated assets for individuals to 15% when held for more than 1 year and provided an exclusion on the gain of the principal residence held for more than 2 years.

When dealing with nondepreciable assets, there is a nominal relationship between before-tax and after-tax rate of return. It is

$$\text{After-tax rate of return} = (1 - \Delta \text{Tax rate})(\text{Before-tax rate of return})$$

There is no simple relationship between before-tax and after-tax rate of return in the more usual case of investments involving depreciable assets.

## PROBLEMS

These problems can be solved by hand, but most will be solved much more easily with a spreadsheet.

### Individual/Joint Taxes

**12-1** An unmarried taxpayer with no dependents expects an adjusted gross income of $70,000 in a given year. His nonbusiness deductions are expected to be $6000.

(*a*) What will his federal income tax be?

(*b*) He is considering an additional activity expected to increase his adjusted gross income. If this increase should be $16,000 and there should be no change in nonbusiness deductions or exemptions, what will be the increase in his federal income tax?

**12-2** John Adams has a $65,000 adjusted gross income from Apple Corp. and allowable itemized deductions of $7200. Mary Eve has a $75,000 adjusted gross income and $2000 of allowable itemized deductions. Compute the total tax they would pay as unmarried individuals. Then compute their tax as a married couple filing a joint return. (*Answers:* 9725 + 12,687.50 = 22,412.50; 22,700)

**12-3** Bill Jackson had a total taxable income of $1800. Bill's employer wants him to work another month during the summer, but Bill had planned to spend the month hiking. If an additional month's work would increase Bill's taxable income by $1600, how much more money would he have after paying the income tax? (*Answer:* $1440)

**12-4** A married couple filing jointly have a combined total adjusted gross income of $75,000. They have computed that their allowable itemized deductions are $4000. Compute their federal income tax. (*Answer:* $7610)

**12-5** Jane Shay operates a management consulting business. The business has been successful and now produces a taxable income of $100,000 per year after all "ordinary and necessary" expenses and depreciation have been deducted. At present the business is operated as a proprietorship; that is, Jane pays personal federal income tax on the entire $100,000. For tax purposes, it is as if she had a job that pays her a $100,000 salary per year.

As an alternative, Jane is considering incorporating the business. If she does, she will pay herself a salary of $40,000 a year from the corporation. The corporation will then pay taxes on the remaining $60,000 and retain the balance of the money as a corporate asset. Thus Jane's two alternatives are to operate the business as a proprietorship or as a corporation. Jane is single and has $3500 of itemized personal deductions. Which alternative will result in a smaller total payment of taxes to the government? (*Answer:* Incorporation, $14,180 versus $19,102)

**12-6** Bill Alexander and his wife, Valerie, are both employed. Bill will have an adjusted gross income this year of $70,000. Valerie has an adjusted gross income of $2000 a month. Bill and Valerie have agreed that Valerie should continue working only until the federal income tax on their joint income tax return becomes $11,500. On what date should Valerie quit her job?

**12-7** An unmarried individual in California with a taxable income of about $80,000 has a federal incremental tax rate of 28% and a state incremental tax rate of 9.3%. What is his combined incremental tax rate?

**12-8** A $10,000 commercial bond that has a 10% bond rate and matures in 5 years can be purchased for $8000. Interest is paid at the end of each year for the next 5 years. Find the annual after-tax rate of return of this investment. Assume a 35% tax rate applies.
*Contributed by D. P. Loucks, Cornell University*

**12-9** Given the following data, compute your combined income tax rate (CTR) assuming you deduct allowable expenses on your income tax forms: a before-tax MARR of 5%, an inflation rate of 3%, a federal income tax rate of 28%, a state income tax rate of 6%, a local city income tax of 3%, and a capital gains tax rate of 15%, as applicable.
*Contributed by D. P. Loucks, Cornell University*

**12-10** Using the answer from Problem 12-9, set up the equation to compute the after-tax rate of return for each scenario:

(*a*) A $5000 corporate bond costs $4600, paying $500 interest at the end of each year for the next 5 years. At the end of 5 years you get $5000.

(*b*) A federal-income-tax-free (on interest only) municipal bond costing $3000, paying $250 at the end of each year for 5 years, and $2500 at the end of the 5 years. State and local income taxes must be paid.

(*c*) A tax-deferred savings plan in which you invest $4000 for 5 years. The plan earns 8% annual interest and you get back your $4000 plus interest at the end of 5 years.
*Contributed by D. P. Loucks, Cornell University*

## Corporate Taxes

**12-11** A company wants to set up a new office in a country where the corporate tax rate is as follows: 15% of first $50,000 profits, 25% of next $25,000, 34% of next $25,000, and 39% of everything over $100,000. Executives estimate that they will have gross revenues of $500,000, total costs of $300,000, $30,000 in allowable tax deductions, and a one-time business start-up credit of $8000. What is taxable income for the first year, and how much should the company expect to pay in taxes?

**12-12** ARKO oil company purchased two large compressors for $125,000 each. One compressor was installed in the firm's Texas refinery and is being depreciated by MACRS depreciation. The other compressor was placed in the Oklahoma refinery, where it is being depreciated by sum-of-years'-digits depreciation with zero salvage value. Assume the company pays federal income taxes each year and the tax rate is constant. The corporate accounting department noted that the two compressors are being depreciated differently and wonders whether the corporation will wind up paying more income taxes over the life of the equipment as a result of this. What do you tell them?

**12-13** Sole Brother Inc. is a shoe outlet to a major shoe manufacturing industry located in Chicago. Sole Brother uses accounts payable as one of its financing sources. Shoes are delivered to Sole Brother with a 3% discount if payment on the invoice is received within 10 days of delivery. By paying after the 10-day period, Sole is borrowing money and paying (giving up) the 3% discount. Although Sole Brother is not required to pay interest on delayed payments, the shoe manufacturers require that payments not be delayed beyond 45 days after the invoice date. To be sure of paying within 10 days, Sole Brothers decides to pay on the fifth day. Sole has a marginal corporate income tax of 40% (combined state and federal). By paying within the 10-day period, Sole is avoiding paying a fairly high price to retain the money owed shoe manufacturers. What would have been the effective annual after-tax interest rate?

**12-14** A major industrialized state has a state corporate tax rate of 9.6% of taxable income. If a corporation has a state taxable income of $150,000, what is the total state and federal income tax it must pay? Also, compute its combined incremental state and federal income tax rate.    (*Answers:* $50,534; 44.86%)

**12-15** To increase its market share, Sole Brother Inc. decided to borrow $50,000 from its banker for the purchase of newspaper advertising for its shoe retail line. The loan is to be paid in four equal annual payments with 15% interest. The loan is discounted 12 points. The first 6 "points" are an additional interest charge of 6% of the loan, deducted immediately. This additional interest $6\%(50,000) = \$3000$ means the actual amount received from the $50,000 loan is $47,000. Another 6 points or $3000 of additional interest is deducted as four $750 additional annual interest payments. What is the after-tax interest rate on this loan?

**12-16** The Lynch Bull investment company suggests that Steven Comstock, a wealthy New York City investor (his incremental income tax rate is 35%), consider the following investment.

Buy corporate bonds on the New York Stock Exchange with a face value (par value) of $100,000 and a 5% interest rate paid annually. These bonds can be purchased at their present market value of $75,000. Each year Steve will receive the $5000 interest, and after 5 years, when the bonds mature, he will receive $100,000 plus the last $5000 of interest.

Steve will pay for the bonds by borrowing $50,000 at 10% interest for 5 years. The $5000 interest paid on the loan each year will equal the $5000 of interest income from the bonds. As a result Steve will have no net taxable income during the five years due to this bond purchase and borrowing money scheme.

At the end of 5 years, Steve will receive $100,000 plus $5000 interest from the bonds and will repay the $50,000 loan and pay the last $5000 interest. The net result is that he will have a $25,000 capital gain; that is, he will receive $100,000 from a $75,000 investment. (*Note:* This situation represents an actual recommendation of a brokerage firm.)

(*a*) Compute Steve's after-tax rate of return on this dual bond-plus-loan investment package.

(*b*) What would be Steve's after-tax rate of return if he purchased the bonds for $75,000 cash and *did not* borrow the $50,000?

### Historical Depreciation

**12-17** Albert Chan decided to buy an old duplex as an investment. After looking for several months, he found a desirable duplex that could be bought for $300,000 cash. He decided that he would rent both sides of the duplex, and determined that the total expected income would be $1500 per month. The total annual expenses for property taxes, repairs, gardening, and so forth are estimated at $750 per year. For tax purposes, Al plans to depreciate the building by the sum-of-years'-digits method, assuming that the building has a 20-year remaining life and no salvage value. Of the total $300,000 cost of the property, $250,000 represents the value of the building and $50,000 is the value of the lot. Assume that Al is in the 38% incremental income tax bracket (combined state and federal taxes) throughout the 20 years.

In this analysis Al estimates that the income and expenses will remain constant at their present levels. If he buys and holds the property for 20 years, what after-tax rate of return can he expect to receive on his investment, using the following assumptions?

A. Al believes the building and the lot can be sold at the end of 20 years for the $50,000 estimated value of the lot.

B. A more optimistic estimate of the future value of the building and the lot is that the property can be sold for $380,000 at the end of 20 years.

**12-18** Zeon, a large, profitable corporation, is considering adding some automatic equipment to its production facilities. An investment of $120,000 will produce an initial annual benefit of $29,000, but the benefits are expected to decline $3000 per year, making second-year benefits $26,000, third-year benefits $23,000, and so forth. If the firm uses sum-of-years'-digits depreciation, an 8-year useful life, and $12,000 salvage value, will it obtain the desired 6% after-tax rate of return? Assume that the equipment can be sold for its $12,000 salvage value at the end of the 8 years. Also assume a 46% income tax rate for state and federal taxes combined.

**12-19** A group of businessmen formed a corporation to lease for 5 years a piece of land at the intersection of two busy streets. The corporation has invested $50,000 in car-washing equipment. They will depreciate the equipment by sum-of-years'-digits depreciation, assuming a $5000 salvage value at the end of the 5-year useful life. The corporation is expected to have a before-tax cash flow, after meeting all expenses of operation (except depreciation), of $20,000 the first year, declining $3000 per year in future years (second year = $17,000, third year = $14,000, etc.). The corporation has other income, so it is taxed at a combined corporate tax rate of 20%. If the projected income is correct, and the equipment can be sold for $5000 at the end of 5 years, what after-tax rate of return would the corporation receive from this venture? (*Answer: 14%*)

**12-20** The effective combined tax rate in an owner-managed corporation is 40%. An outlay of $2 million for certain new assets is under consideration. It is estimated that for the next 8 years, these assets will be responsible for annual receipts of $600,000 and annual disbursements (other than for income taxes) of $250,000. After this time, they will be used only for stand-by purposes, and no future excess of receipts over disbursements is estimated.

(a) What is the prospective rate of return before income taxes?

(b) What is the prospective rate of return after taxes if straight-line depreciation can be used to write off these assets for tax purposes in 8 years?

(c) What is the prospective rate of return after taxes if it is assumed that these assets must be written off for tax purposes over the next 20 years, using straight-line depreciation?

**12-21** A firm is considering the following investment project:

| Year | Before-Tax Cash Flow (thousands) |
|------|------|
| 0 | −$1000 |
| 1 | 500 |
| 2 | 340 |
| 3 | 244 |
| 4 | 100 |
| 5 | $\begin{cases} 100 \\ 125 \text{ Salvage value} \end{cases}$ |

The project has a 5-year useful life with a $125,000 salvage value, as shown. Double declining balance depreciation will be used, assuming the $125,000 salvage value. The combined income tax rate is 34%. If the firm requires a 10% after-tax rate of return, should the project be undertaken?

**12-22** The Shellout Corp. owns a piece of petroleum drilling equipment that costs $100,000 and will be depreciated in 10 years by double declining balance depreciation, with conversion to straight-line depreciation at the optimal point. Assume no salvage value in the depreciation computation and a combined 34% tax rate. Shellout will lease the equipment to others and each year receive $30,000 in rent. At the end of 5 years, the firm will sell the equipment for $35,000. (Note that this is different from the zero-salvage-value assumption used in computing the depreciation.) What is the after-tax rate of return Shellout will receive from this equipment investment?

**12-23** A mining corporation purchased $120,000 of production machinery and depreciated it using SOYD depreciation, a 5-year depreciable life, and zero salvage value. The corporation is a profitable one that has a 34% combined incremental tax rate.

At the end of 5 years the mining company changed its method of operation and sold the production machinery for $40,000. During the 5 years the machinery was used, it reduced mine operating costs by $32,000 a year, before taxes. If the company MARR is 12% after taxes, was the investment in the machinery a satisfactory one?

**12-24** An automaker is buying some special tools for $100,000. The tools are being depreciated by double declining balance depreciation using a 4-year depreciable life and a $6250 salvage value. It is expected the tools will actually be kept in service for 6 years and then sold for $6250. The before-tax benefit of owning the tools is as follows:

| Year | Before-Tax Cash Flow |
|------|---------------------|
| 1 | $30,000 |
| 2 | 30,000 |
| 3 | 35,000 |
| 4 | 40,000 |
| 5 | 10,000 |
| 6 | 10,000 |
| | 6,250 Selling price |

Compute the after-tax rate of return for this investment situation, assuming a 46% incremental tax rate. (*Answer:* 11.6%)

**12-25** This is the continuation of Problem 12-24. Instead of paying $100,000 cash for the tools, the corporation will pay $20,000 now and borrow the remaining $80,000. The depreciation schedule will remain unchanged. The loan will be repaid by 4 equal end-of-year payments of $25,240.

Prepare an expanded cash flow table that takes into account both the special tools and the loan.

(*a*) Compute the after-tax rate of return for the tools, taking into account the $80,000 loan.

(*b*) Explain why the rate of return obtained in part (*a*) is different from the rate of return obtained in Problem 12-24.

*Hints:* **1.** Interest on the loan is 10%, $25,240 = 80,000 $(A/P, 10\%, 4)$. Each payment is made up of part interest and part principal. Interest portion for any year is 10% of balance due at the beginning of the year.

**2.** Interest payments are tax deductible (i.e., they reduce taxable income and thus taxes paid). Principal payments are not. Separate each $25,240 payment into interest and principal portions.

**3.** The Year-0 cash flow is −$20,000.

**4.** After-tax cash flow will be before-tax cash flow − interest payment − principal payment − taxes.

**12-26** A project will require the investment of $108,000 in equipment (sum-of-years'-digits depreciation with a depreciable life of 8 years and zero salvage value) and $25,000 in raw materials (not depreciable). The annual project income after all expenses except depreciation have been paid is projected to be $24,000. At the end of 8 years the project will be discontinued and the $25,000 investment in raw materials will be recovered.

Assume a 34% combined income tax rate for this corporation. The corporation wants a 15% after-tax rate of return on its investments. Determine by present worth analysis whether this project should be undertaken.

**12-27** A profitable incorporated business is considering an investment in equipment having the following before-tax cash flow. The equipment will be depreciated by double declining balance depreciation with conversion, if appropriate, to straight-line depreciation at the preferred time. For depreciation purposes a $700,000 salvage value at the end of 6 years is assumed. But the actual value is thought to be $1,000,000, and it is this sum that is shown in the before-tax cash flow.

| Year | Before-Tax Cash Flow (in $1000) |
|------|--------------------------------|
| 0 | $12,000 |
| 1 | 1,727 |
| 2 | 2,414 |
| 3 | 2,872 |
| 4 | 3,177 |
| 5 | 3,358 |
| 6 | 1,997 |
| | 1,000 Salvage value |

If the firm wants a 9% after-tax rate of return and its combined incremental income tax rate is 34%, determine by annual cash flow analysis whether the investment is desirable.

**12-28** A salad oil bottling plant can either buy caps for the glass bottles at 5¢ each or install $500,000 worth of plastic molding equipment and manufacture the caps at the plant. The manufacturing engineer estimates the material, labor, and other costs would be 3¢ per cap.

(*a*) If 12 million caps per year are needed and the molding equipment is installed, what is the payback period?

(*b*) The plastic molding equipment would be depreciated by straight-line depreciation using a 5-year useful life and no salvage value. Assuming a combined 40% income tax rate, what is the after-tax payback period, and what is the after-tax rate of return?

**12-29** A firm has invested $14,000 in machinery with a 7-year useful life. The machinery will have no salvage value, as the cost to remove it will equal its scrap value. The uniform annual benefits from the machinery are $3600. For a combined 47% income tax rate, and sum-of-years'-digits depreciation, compute the after-tax rate of return.

**12-30** A firm manufactures padded shipping bags. A cardboard carton should contain 100 bags, but machine operators fill the cardboard cartons by eye, so a carton may contain anywhere from 98 to 123 bags (average = 105.5 bags).

Management realizes that they are giving away $5^{1}/2\%$ of their output by overfilling the cartons. The solution would be to weigh each filled shipping carton. Underweight cartons would have additional shipping bags added, and overweight cartons would have some shipping bags removed. If the weighing is done, it is believed that the average quantity of bags per carton could be reduced to 102, with almost no cartons containing fewer than 100 bags.

The weighing equipment would cost $18,600. The equipment would be depreciated by straight-line depreciation using a 10-year depreciable life and a $3600 salvage value at the end of 10 years. Assume the $18,600 worth of equipment qualifies for a 10% investment tax credit. One person, hired at a cost of $16,000 per year, would be required to operate the weighing equipment and to add or remove padded bags from the cardboard cartons. 200,000 cartons will be checked on the weighing equipment each year, with an average removal of 3.5 padded bags per carton with a manufacturing cost of 3¢ per bag. This large profitable corporation has a 50% combined federal-plus-state incremental tax rate. Assume a 10-year study period for the analysis and an after-tax MARR of 20%.

Compute:

(*a*) The after-tax present worth

(*b*) The after-tax internal rate of return

(*c*) The after-tax simple payback period

**12-31** ACDC Company is considering the installation of a new machine that costs $150,000. The machine is expected to lead to net income of $44,000 per year for the next 5 years. Using straight-line depreciation, $0 salvage value, and an effective income tax rate of 50%, determine the after-tax rate of return for this investment. If the company's after-tax MARR rate is 12%, would this be a good investment or not? (*Answer:* 7.42%, no)

*Contributed by Mukasa Ssemakula, Wayne State University*

**12-32** Florida Construction Equipment Rentals (FCER) purchases a new 10,000-pound-rated crane for rental to its customers. This crane costs $1,125,000 and is expected to last for 25 years, at which time it will have an expected salvage value of $147,000. FCER earns $195,000 before-tax cash flow each year in rental income from this crane, and its total taxable income each year is between $10M and $15M. If FCER uses straight-line depreciation and a MARR of 15%, what is the present worth of the after-tax cash flow for this equipment? Should the company invest in this crane? (*Answers:* $212,700, no)

*Contributed by Paul R. McCright, University of South Florida*

**12-33** A computer-controlled milling machine will cost Ajax Manufacturing $65,000 to purchase plus $4700 to install.

(*a*) If the machine would have a salvage value of $6600 at EOY 20, how much could Ajax charge annually to depreciation of this equipment? Ajax uses straight-line depreciation.

(*b*) What is the book value of the machine at EOY 3?

(*c*) Ajax Manufacturing earns a net profit before tax (also called a taxable income) of $28,800,000. How much tax would Ajax owe for this year?

*Contributed by Paul R. McCright, University of South Florida*

**12-34** A farmer bought a new harvester for $120,000. The harvester's operating expenses averaged $10,000 per year but the harvester saved $40,000 per year in labor costs. It was depreciated over a life of 5 years using the SOYD method, assuming a salvage value of $30,000. The farmer sold the harvester for only $10,000 at the end of the fifth year. Given an income tax rate of 30% and a MARR rate of 5% per year, determine the after-tax net present worth of this investment. (*Answer:* $5244)

*Contributed by Mukasa Ssemakula, Wayne State University*

## MACRS Depreciation

**12-35** Mr. Sam K. Jones, a successful businessman, is considering erecting a small building on a commercial lot. A local furniture company is willing to lease the building for $9000 per year, paid at the end of each year. It is a net lease, which means the furniture company must also pay the property taxes, fire insurance, and all other annual costs. The furniture company will require a 5-year lease with an option to buy the building and land on which it stands for $125,000 after 5 years. Mr. Jones could have the building constructed for $82,000. He could sell the commercial lot now for $30,000, the same price he paid for it. Mr. Jones files a joint return and has an annual taxable income from other sources of $63,900. He would depreciate the commercial building by modified accelerated cost recovery system (MACRS) depreciation. Mr. Jones believes that at the end of the 5-year lease he could easily sell the property for $125,000. What is the after-tax present worth of this 5-year venture if Mr. Jones uses a 10% after-tax MARR?

**12-36** One January Gerald Adair bought a small house and lot for $99,700. He estimated that $9700 of this amount represented the land's value. He rented the house for $6500 a year during the 4 years he owned the house. Expenses for property taxes, maintenance, and so forth were $500 per year. For tax purposes the house was depreciated by MACRS depreciation (27.5-year straight-line depreciation with a midmonth convention is used for rental property). At the end of 4 years the property was sold for $105,000. Gerald is married and works as an engineer. He estimates that his incremental state and federal combined tax rate is 24%. What after-tax rate of return did Gerald obtain on his investment?

**12-37** A corporation with a 34% combined income tax rate is considering the following investment in research equipment and has projected the benefits as follows:

| Year | Before-Tax Cash Flow |
|---|---|
| 0 | −$5,000,000 |
| 1 | 200,000 |
| 2 | 800,000 |
| 3 | 1,760,000 |
| 4 | 1,376,000 |
| 5 | 576,000 |
| 6 | 288,000 |

Prepare an after-tax cash flow table assuming MACRS depreciation.

(a) What is the after-tax rate of return?

(b) What is the before-tax rate of return?

**12-38** An engineer is working on the layout of a new research and experimentation facility. Two plant operators will be required. If, however, an additional $100,000 of instrumentation and remote controls were added, the plant could be run by a single operator. The total before-tax cost of each plant operator is projected to be $35,000 per year. The instrumentation and controls will be depreciated by means of the modified accelerated cost recovery system (MACRS).

If this corporation (34% combined corporate tax rate) invests in the additional instrumentation and controls, how long will it take for the after-tax benefits to equal the $100,000 cost? In other words, what is the after-tax payback period? (*Answer:* 3.24 years).

**12-39** A special power tool for plastic products costs $400,000 has a 4-year useful life, no salvage value, and a 2-year before-tax payback period. Assume uniform annual end-of-year benefits.

(a) Compute the before-tax rate of return.

(b) Compute the after-tax rate of return, based on MACRS depreciation and a 34% combined corporate income tax rate.

**12-40** The Ogi Corporation, a construction company, purchased used a pickup truck for $14,000 and used MACRS depreciation in the income tax return. During the time the company had the truck, they estimated that it saved $5000 a year. At the end of 4 years, Ogi sold the truck for $3000. The combined federal and state income tax rate for Ogi is 45%. Compute the after-tax rate of return for the truck. (*Answer:* 12.5%)

**12-41** A profitable wood products corporation is considering buying a parcel of land for $50,000, building a small factory building at a cost of $200,000, and equipping it with $150,000 of MACRS 5-year-class machinery.

If the project is undertaken, MACRS depreciation will be used. Assume the plant is put in service October 1. The before-tax net annual benefit from the project is estimated at $70,000 per year. The analysis period is to be 5 years, and planners assume the sale of the total property (land, building, and machinery) at the end of 5 years, also on October 1, for $328,000. Compute the after-tax cash flow based on a 34% combined income tax rate. If the corporation's criterion is a 15% after-tax rate of return, should it proceed with the project?

**12-42** A chemical company bought a small vessel for $55,000; it is to be depreciated by MACRS depreciation. When requirements changed suddenly, the chemical company leased the vessel to an oil company for 6 years at $10,000 per year. The lease also provided that the oil company could buy the vessel at the end of 6 years for $35,000. At the end of the 6 years, the oil company exercised its option and bought the vessel. The chemical company has a 34% combined incremental tax rate. Compute its after-tax rate of return on the vessel. (*Answer:* 9.86%)

**12-43** Xon, a small oil company, purchased a new petroleum drilling rig for $1,800,000. Xon will depreciate the drilling rig using MACRS depreciation. The drilling rig has been leased to a drilling company, which will pay Xon $450,000 per year for 8 years. At the end of 8 years the drilling rig will belong to the drilling company. If Xon has a 34% combined incremental tax rate and a 10% after-tax MARR, does the investment appear to be satisfactory?

**12-44** The profitable Palmer Golf Cart Corp. is considering investing $300,000 in special tools for some of the plastic golf cart components. Executives of the company believe the present golf cart model will continue to be manufactured and sold for 5 years, after which a new cart design will be needed, together with a different set of special tools.

The saving in manufacturing costs, owing to the special tools, is estimated to be $150,000 per year for 5 years. Assume MACRS depreciation for the special tools and a 39% combined income tax rate.

(*a*) What is the after-tax payback period for this investment?

(*b*) If the company wants a 12% after-tax rate of return, is this a desirable investment?

**12-45** Uncle Elmo is contemplating a $10,000 investment in a methane gas generator. He estimates his gross income would be $2000 the first year and increase by $200 each year over the next 10 years. His expenses of $200 the first year would increase by $200 each year over the next 10 years. He would depreciate the generator by MACRS depreciation, assuming a 7-year property class. A 10-year-old methane generator has no market value. The income

tax rate is 40%. (Remember that recaptured depreciation is taxed at the same 40% rate.)

(*a*) Construct the after-tax cash flow for the 10-year project life.

(*b*) Determine the after-tax rate of return on this investment. Uncle Elmo thinks it should be at least 8%.

(*c*) If Uncle Elmo could sell the generator for $7000 at the end of the fifth year, would his rate of return be better than if he kept the generator for 10 years? You don't have to actually find the rate of return, just do enough calculations to see whether it is higher than that of part (*b*).

**12-46** Granny's Butter and Egg Business is such that she pays an effective tax rate of 40%. Granny is considering the purchase of a new Turbo Churn for $25,000. This churn is a special handling device for food manufacture and has an estimated life of 4 years and a salvage value of $5000. The new churn is expected to increase net income by $8000 per year for each of the 4 years of use. If Granny works with an after-tax MARR of 10% and uses MACRS depreciation, should she buy the churn?

**12-47** Eric has a house and lot for sale for $210,000. It is estimated that $40,000 is the value of the land and $170,000 is the value of the house. On January 1 Bonnie is purchasing the house to rent and plans to own it for 5 years. After 5 years, it is expected that the house and land can be sold on December 31 for $225,000. Total annual expenses (maintenance, property taxes, insurance, etc.) are expected to be $5000 per year. The house would be depreciated by MACRS depreciation using a 27.5-year straight-line rate with midmonth convention for rental property. For depreciation, a salvage value of zero was used. Bonnie wants a 15% after-tax rate of return on her investment. You may assume that Bonnie has an incremental income tax rate of 28% in each of the 5 years. Capital gains are taxed at 15%. Determine the following:

(*a*) The annual depreciation

(*b*) The capital gain (loss) resulting from the sale of the house

(*c*) The annual rent Bonnie must charge to produce an after-tax rate of return of 15%. (*Hint:* Write an algebraic equation to solve for rent.)

**12-48** Bill owns a data processing company. He plans to buy an additional computer for $20,000, use the computer for 3 years, and sell it for $10,000. He expects that use of the computer will produce a net

TABLE P12-48  **Worksheet for Problem 12-48**

| Year | Before-Tax Cash Flow | MACRS Depreciation | Taxable Income | Income Tax (45%) | After-Tax Cash Flow | Present Worth (12%) |
|------|------|------|------|------|------|------|
| 0 | −$20,000 | | | | | |
| 1 | 8,000 | | | | | |
| 2 | 8,000 | | | | | |
| 3 | 8,000 | | | | | |
| | 10,000 | | | | | |
| | | | | Net Present Worth = | | |

income of $8000 per year. The combined federal and state incremental tax rate is 45%. Using MACRS depreciation, complete Table P12-48 to determine the net present worth of the after-tax cash flow using an interest rate of 12%.

**12-49** Refer to Problem 12-40. To help pay for the pickup truck, the Ogi Corp. obtained a $10,000 loan from the truck dealer, payable in four end-of-year payments of $2500 plus 10% interest on the loan balance each year.

(a) Compute the after-tax rate of return for the truck together with the loan. Note that the interest on the loan is tax deductible, but the $2500 principal payments are not.

(b) Why is the after-tax rate of return computed in part (a) so much different from the 12.5% obtained in Problem 12-40?

**12-50** Specialty Machining, Inc. bought a new multi turret turning center for $250,000. The machine generated new revenue of $80,000 per year. Operating costs for the machine averaged $10,000 per year. Following IRS regulations, the machine was depreciated using the MACRS method, with a recovery period of 7 years. The center was sold for $75,000 after 5 years of service. The company uses an after-tax MARR rate of 12% and is in the 35% tax bracket.

Determine the after-tax net present worth of this asset over the 5-year service period.    (*Answer:* $3275)
*Contributed by Mukasa Ssemakula, Wayne State University*

**12-51** Fleet Fleet rental car company purchased 10 new cars for a total cost of $180,000. The cars generated income of $150,000 per year and incurred operating expenses of $60,000 per year. The company uses MACRS depreciation and its marginal tax rate is 40% (*Note:* Per IRS regulations, cars have a class life of 5 years). The 10 cars were sold at the end of the third year for a total of $75,000. Assuming a MARR of 10% and using NPW, determine if this was a good investment on an after-tax basis. (*Answer:* NPW = $46,197)
*Contributed by Mukasa Ssemakula, Wayne State University*

**12-52** An investor bought a racehorse for $1 million. The horse's average winnings were $700,000 per year and expenses averaged $200,000 per year. The horse was retired after 3 years, at which time it was sold to a breeder for $175,000. Assuming MACRS depreciation and an income tax rate of 50%, determine the investor's after-tax rate of return on this investment.
*Contributed by Mukasa Ssemakula, Wayne State University*

**12-53** Vanguard Solar Systems is building a new manufacturing facility to be used for the production of solar panels. Vanguard uses a MARR of 15%, and MACRS depreciation on its assets and is in the 35% tax bracket. The building will cost $2.75 million, and the equipment (3-year property) will cost $1.55 million plus installation costs of $135,000. O&M costs are expected to be $1.3 million the first year, increasing 6% annually. If the facility opens in the month of March and sales are as shown, determine the present worth of the after-tax cash flows for the first 5 years of operation.
*Contributed by Paul R. McCright, University of South Florida*

| Year | Sales |
|------|-----------|
| 1 | $2,100,000 |
| 2 | 3,200,000 |
| 3 | 3,800,000 |
| 4 | 4,500,000 |
| 5 | 5,300,000 |

**12-54** Tampa Electric Company (TECO) is planning a major upgrade in its computerized demand management system. In order to accommodate this upgrade, a building will be constructed on land already owned by the company. The building is estimated to cost $1.8M and will be opened in August 2010. The computer equipment for the building will cost $2.75M, and all office equipment will cost $225,000. Annual expenses for operating this facility (labor, materials, insurance, energy, etc.) are expected to be $325,000 during 2010. Use of the new demand management system is expected to decrease fuel and other costs for the company by $1.8M in the first year (2010). If the company expects to earn 9% on its investments, is in the 35% tax bracket, and uses a 20-year planning horizon, determine the estimated after-tax cash flow from this project in 2010.    (*Answer:* $569,083)
*Contributed by Paul R. McCright, University of South Florida*

**12-55** Christopher wants to add a solar photovoltaic system to his home. He plans to install a 2-kW system and has received a quote from an installer who will install this unit for $19,750. The federal government will give him a tax credit of 30% of the cost, and the state will give a 10% tax credit. State law requires the utility company to buy back all excess power generated by the system. Christopher's annual power bill is estimated to be $2000, and this will be eliminated by the solar system. Christopher expects to receive a check from the power company for $600 each year for his excess production. If the tax credits are received at EOY 1 and Christopher receives a $2000 savings plus a $600 income at the end of each year, use present worth to determine if the system pays for itself in 8 years? Assume that Christopher earns 3% on all investments.
*Contributed by Paul R. McCright, University of South Florida*

**12-56** Mid-America Shipping is considering purchasing a new barge for use on its Ohio River routes. The new barge will cost $13.2 million and is expected to generate an income of $7.5 million the first year (growing $1M each year), with additional expenses of $2.6 million the first year (growing $400,000 per year). If Mid-America uses MACRS, is in the 38% tax bracket, and has a MARR of 12%, what is the present worth of the first 4 years of after-tax cash flows from this barge? Would you recommend that Mid-America purchase this barge? (*Hints:* Remember that all expenses are deducted from income prior to determining the taxes. Round to the nearest tenth of a million dollars.)
*Contributed by Paul R. McCright, University of South Florida*

**Solving for Unknowns**

**12-57** A store owner, Joe Lang, believes his business has suffered from the lack of adequate customer parking space. Thus, when he was offered an opportunity to buy an old building and lot next to his store, he was interested. He would demolish the old building and make off-street parking for 20 customers' cars. Joe estimates that the new parking would increase his business and produce an additional before-income-tax profit of $7000 per year. It would cost $2500 to demolish the old building.

Mr. Lang's accountant advised that both costs (the property and demolishing the old building) would be considered to comprise the total value of the land for tax purposes, and it would not be depreciable.

Mr. Lang would spend an additional $3000 right away to put a light gravel surface on the lot. This expenditure, he believes, may be charged as an operating expense immediately and need not be capitalized. To compute the tax consequences of adding the parking lot, Joe estimates that his combined state and federal incremental income tax rate will average 40%. If Joe wants a 15% after-tax rate of return from this project, how much could he pay to purchase the adjoining land with the old building?

Assume that the analysis period is 10 years and that the parking lot could always be sold to recover the costs of buying the property and demolishing the old building. (*Answer:* $23,100)

**12-58** The management of a private hospital is considering the installation of an automatic telephone switchboard, which would replace a manual switchboard and eliminate the attendant operator's position. The class of service provided by the new equipment is estimated to be at least equal to the present method of operation. To provide telephone service, five operators currently work three shifts per day, 365 days per year. Each operator earns $25,000 per year. Company-paid benefits and overhead are 25% of wages. Money costs 8% after income taxes. Combined federal and state income taxes are 40%. Annual property taxes and maintenance are $2\frac{1}{2}$ and 4% of investment, respectively. Depreciation is 15-year straight line. Disregarding inflation, how large an investment in the new equipment can be economically justified by savings obtained by eliminating the present equipment and labor costs? The existing equipment has zero salvage value.

**12-59** A contractor has to choose one of the following alternatives in performing earthmoving contracts:

*A.* Buy a heavy-duty truck for $35,000. Salvage value is expected to be $8000 at the end of the vehicle's 7-year depreciable life. Maintenance is $2500 per year. Daily operating expenses are $200.

*B.* Hire a similar unit for $550 per day.

Based on a 10% after-tax rate of return, how many days per year must the truck be used to justify its purchase? Base your calculations on straight-line depreciation and a 50% income tax rate.

**12-60** The Able Corporation is considering the installation of a small electronic testing device for use in conjunction with a government contract the firm has just won. The testing device will cost $20,000 and will have an estimated salvage value of $5000 in 5 years when the government contract is finished. The firm will depreciate the instrument by the sum-of-years'-digits method, using 5 years as the useful life and a $5000 salvage value. Assume that Able pays 50% federal and state corporate income taxes and uses 8% *after-tax* in economic analysis. What minimum equal annual benefit must Able obtain *before taxes* in each of the 5 years to justify purchasing the electronic testing device? (*Answer:* $5150)

**12-61** A house and lot are for sale for $155,000. It is estimated that $45,000 is the land's value and $110,000 is the value of the house. The net rental income would be $12,000 per year after taking all expenses, except depreciation, into account. The house would be depreciated by straight-line depreciation using a 27.5-year depreciable life and zero salvage value.

Mary Silva, the prospective purchaser, wants a 10% after-tax rate of return on her investment after considering both annual income taxes and a capital gain when she sells the house and lot. At what price would she have to sell the house at the end of 10 years to achieve her objective? You may assume that Mary has an incremental income tax rate of 28% in each of the 10 years.

**12-62** A corporation is considering buying a medium-sized computer that will eliminate a task that must be performed three shifts per day, 7 days per week, except for one 8-hour shift per week when the operation is shut down for maintenance. At present four people are needed to perform the day and night tasks. Thus the computer will replace four employees. Each employee costs the company $32,000 per year ($24,000 in direct wages plus $8000 per year in other company employee costs). It will cost $18,000 per year to maintain and operate the computer.

The computer will be depreciated by sum-of-years'-digits depreciation using a 6-year depreciable life, at which time it will be assumed to have zero salvage value.

The corporation has a combined federal and state incremental tax rate of 50%. If the firm wants a 15% rate of return, after considering both state and federal income taxes, how much can it afford to pay for the computer?

**12-63** A sales engineer has the following alternatives to consider in touring his sales territory.

A. Buy a 2-year old used car for $14,500. Salvage value is expected to be about $5000 after 3 more years. Maintenance and insurance cost is $1000 in the first year and increases at the rate of $500/year in subsequent years. Daily operating expenses are $50/day.

B. Rent a similar car for $80/day.

Based on a 12% after-tax rate of return, how many days per year must he use the car to justify its purchase? You may assume that this sales engineer is in the 28% incremental tax bracket. Use MACRS depreciation.    (*Answer:* 218 days)

**12-64** A large profitable company, in the 40% combined federal/state tax bracket, is considering the purchase of a new piece of equipment that will yield benefits of $10,000 in Year 1, $15,000 in Year 2, $20,000 in Year 3, and $20,000 in Year 4. The equipment is to be depreciated using 5-year MACRS depreciation starting in the year of purchase (Year 0). It is expected that the equipment will be sold at the end of Year 4 at 20% of its purchase price. What is the maximum equipment purchase price the company can pay if its after-tax MARR is 10%?

**12-65** A prosperous businessman is considering two alternative investments in bonds. In both cases the first interest payment would be received at the end of the first year. If his personal taxable income is fixed at $40,000 and he is single, which investment produces the greater after-tax rate of return? Compute the after-tax rate of return for each bond to within $1/4$ of 1 percent.

*Ann Arbor Municipal Bonds:* A bond with a face value of $1000 pays $60 per annum. At the end of 15 years, the bond becomes due ("matures"), at which time the owner of the bond will receive $1000 plus the final $60 annual payment. The bond may be purchased for $800. Since it is

a municipal bond, the annual interest is *not* subject to federal income tax. The difference between what the businessman would pay for the bond ($800) and the $1000 face value he would receive at the end of 15 years must be included in taxable income when the $1000 is received.

*Southern Coal Corporation Bonds:* A thousand-dollar bond yields $100 per year in annual interest payments. When the bonds mature at the end of 20 years, the bondholder will receive $1000 plus the final $100 interest. The bonds may be purchased now for $1000. The income from corporation bonds must be included in federal taxable income.

**12-66** You recently bought a mini-supercomputer for $10,000 to allow for tracking and analysis of real-time changes in stock and bond prices. Assume you plan on spending half your time tending to the stock market with this computer and the other half as personal use. Also assume you can depreciate your computer by 20% per year over 5 years (straight line rate). How much tax savings will you have in each of those 5 years, if any? Use a tax rate of 28%. (*Answer:* $280)

*Contributed by D. P. Loucks, Cornell University*

## Multiple Alternatives

**12-67** Use the after-tax IRR method to evaluate the following three alternatives for MACRS 3-year property, and offer a recommendation. The after-tax MARR is 25%, the project life is 5 years, and the firm has a combined incremental tax rate of 45%.

| Alt. | First Cost | Annual Costs | Salvage Value |
|---|---|---|---|
| A | $14,000 | $2500 | $ 5,000 |
| B | 18,000 | 1000 | 10,000 |
| C | 10,000 | 5000 | 0 |

**12-68** A small-business corporation is considering whether to replace some equipment in the plant. An analysis indicates there are five alternatives in addition to the do-nothing option, Alt. *A*. The alternatives have a 5-year useful life with no salvage value. Straight-line depreciation would be used.

| Alternatives | Cost (thousands) | Before-Tax Uniform Annual Benefits (thousands) |
|---|---|---|
| *A* | $ 0 | $ 0 |
| *B* | 25 | 7.5 |
| *C* | 10 | 3 |
| *D* | 5 | 1.7 |
| *E* | 15 | 5 |
| *F* | 30 | 8.7 |

The corporation has a combined federal and state income tax rate of 20%. If the corporation expects a 10% after-tax rate of return for any new investments, which alternative should be selected?

**12-69** A corporation with $7 million in annual taxable income is considering two alternatives:

| Year | Before-Tax Cash Flow ($1000) Alt. 1 | Alt. 2 |
|---|---|---|
| 0 | −$10,000 | −$20,000 |
| 1–10 | 4,500 | 4,500 |
| 11–20 | 0 | 4,500 |

Both alternatives will be depreciated by straight-line depreciation assuming a 10-year depreciable life and no salvage value. Neither alternative is to be replaced at the end of its useful life. If the corporation has a minimum attractive rate of return of 10% *after taxes,* which alternative should it choose? Solve the problem by:

(*a*) Present worth analysis

(*b*) Annual cash flow analysis

(*c*) Rate of return analysis

(*d*) Future worth analysis

(*e*) Benefit–cost ratio analysis

**12-70** Two mutually exclusive alternatives are being considered by a profitable corporation with an annual taxable income between $5 million and $10 million.

| Year | Before-Tax Cash Flow ($1,000) Alt. A | Alt. B |
|---|---|---|
| 0 | −$3000 | −$5000 |
| 1 | 1000 | 1000 |
| 2 | 1000 | 1200 |
| 3 | 1000 | 1400 |
| 4 | 1000 | 2600 |
| 5 | 1000 | 2800 |

Both alternatives have a 5-year useful and depreciable life and no salvage value. Alternative *A* would be depreciated by sum-of-years'-digits depreciation, and Alt. *B* by straight-line depreciation. If the MARR is 10% after taxes, which alternative should be selected?

(*Answer:* Alt. *B*)

**12-71** A large profitable corporation is considering two mutually exclusive capital investments:

| | Alt. A | Alt. B |
|---|---|---|
| Initial cost | $11,000 | $33,000 |
| Uniform annual benefit | 3,000 | 9,000 |
| End-of-depreciable-life salvage value | 2,000 | 3,000 |
| Depreciation method | SL | SOYD |
| End-of-useful-life salvage value obtained | 2,000 | 5,000 |
| Depreciable life, in years | 3 | 4 |
| Useful life, in years | 5 | 5 |

If the firm's after-tax minimum attractive rate of return is 12% and its combined incremental income tax rate is 34%, which project should be selected?

**12-72** A plant can be purchased for $1,000,000 or it can be leased for $200,000 per year. The annual income is expected to be $800,000 with the annual operating cost of $200,000. The resale value of the plant is estimated to be $400,000 at the end of its 10-year life. The company's combined federal and state income tax rate is 40%. A straight-line depreciation can be used over the 10 years with the full first-year depreciation rate.

(*a*) If the company uses the after-tax minimum attractive rate of return of 10%, should it lease or purchase the plant?

(*b*) What is the breakeven rate of return of purchase versus lease?

**12-73** VML Industries has need of specialized yarn manufacturing equipment for operations over the next 3 years. The firm could buy the machinery for $95,000 and depreciate it using MACRS. Annual maintenance would be $7500, and it would have a salvage value of $25,000 after 3 years. Another alternative would be to lease the same machine for $45,000 per year on an "all costs" inclusive lease (maintenance costs included in lease payment). These lease payments are due at the beginning of each year. VML Industries uses an after-tax MARR of 18% and a combined tax rate of 40%. Do an after-tax present worth analysis to determine which option is preferred.

**12-74** Padre Pio owns a small business and has taxable income of $150,000. He is considering four mutually exclusive alternative models of machinery. Which machine should be selected on an after-tax basis? The after-tax MARR is 15%. Assume that each machine is MACRS 5-year property and can be sold for a market value that is 25% of the purchase cost, and the project life is 10 years.

| Model | I | II | III | IV |
|---|---|---|---|---|
| First cost | $9000 | $8000 | $7500 | $6200 |
| Annual costs | 25 | 200 | 300 | 600 |

(*Answer:* Alt. III)

**12-75** LoTech Welding can purchase a machine for $175,000 and depreciate it as 5-year MACRS property. Annual maintenance would be $9800, and its salvage value after 8 years is $15,000. The machine can also be leased for $35,000 per year on an "all costs" inclusive lease (maintenance costs included). Lease payments are due at the beginning of each year, and they are tax-deductible. The firm's combined tax rate for state and federal income taxes is 40%. If the firm's after-tax interest rate is 9%, which alternative has the lower EAC and by how much?

**12-76** For Problem 12-75, assume that Section 179 depreciation must be considered as this is LoTech's only major equipment investment this year. Also assume that the firm is profitable. How much lower is the annual cost of purchasing the machine?

**12-77** For Problem 12-75, assume that the machine will be financed with a 4-year loan whose interest rate is 12%.

(*a*) Graph the EAC for the purchase for financing fractions ranging from 0% to 100%.

(*b*) Assume that 50% bonus depreciation is also available. Graph the EAC for the purchase for financing fractions ranging from 0% to 100%.

(*c*) Assume that Section 179 depreciation must be considered, as this is LoTech's only major equipment investment this year. Also assume that the firm is profitable. Graph the EAC for the purchase for financing fractions ranging from 0% to 100%.

# CASES

**The following cases from *Cases in Engineering Economy 2 <sup>nd</sup>* are suggested as matched with this chapter.**

CASE 26 **Molehill & Mountain Movers**
Compare depreciation methods with option for inflation.

CASE 28 **Olives in Your Backyard**
Emphasizes taxes and sensitivity analysis.

CASE 29 **New Fangled Manufacturing**
Emphasizes taxes and sensitivity analysis.

CASE 36 **Brown's Nursery (Part A)**
After-tax analysis of expansion opportunity.

CASE 38 **West Muskegon Machining and Manufacturing**
More complex inflation and tax problem with sunk cost and leverage.

CASE 53 **Problems in Pasta Land**
Long case statement. Includes taxes and limited uncertainty.

# REPLACEMENT ANALYSIS

## The $7 Billion Upgrade

I n 2009 Intel Corporation announced it was spending $7 billion in 2009 and 2010 to modernize and update its silicon wafer manufacturing plants in the U.S. The upgrade will allow Intel to manufacture 32-nm chips in fabs originally built for 45-nm and 65-nm chips. Upgrade work will include remodeling the plant's interior and purchasing new wafer fabrication tools.

Despite the price tag for the project, Intel planned on saving money overall by upgrading the existing plants instead of designing and building a new one. The upgraded plants use newer processes that allow chips to hold smaller and faster transistors, and the new larger wafer accommodates more chips. All these changes lead to lower production costs.

The company also noted that by deciding to remain in its current locations, it was able to retain its highly skilled workforce. ■ ■ ■

1. Intel made the decision to upgrade existing plant facilities versus building new ones. In their analysis they claimed to have saved money by doing so. Use the Internet and search for articles that discuss what new manufacturing plants cost in industries of different types. Does Intel's $7 billion upgrade decision seem in line? What issues are you not considering if you think the cost seems very high?

2. Companies that upgrade existing production assets may or may not scrap their current assets. Give a realistic scenario for each case.

3. The upgrade vs. replace decision often includes economic as well as noneconomic factors. List three economic and three noneconomic factors that may have been at the top of Intel's list in making the decision.

4. Companies that build new manufacturing plants often site them outside the U.S. Discuss the ethical impacts that are or should be part of a firm's decision process.

## After Completing This Chapter...

### The student should be able to:

- Recast an equipment reinvestment decision as a *challenger vs. defender* analysis.
- Use the *replacement analysis decision map* to select the correct economic analysis technique to apply.
- Calculate the *minimum cost life* of economic challengers.
- Incorporate concepts such as *repeatability assumption for replacement analysis* and *marginal cost data for the defender* to select the correct economic analysis techniques.
- Perform replacement problems on an after-tax basis, utilizing the *defender sign change procedure* when appropriate.
- Use spreadsheets for solving before-tax and after-tax replacement analysis problems.

Up to this point in our economic analysis we have considered the evaluation and selection of *new* alternatives. Which new car or production machine should we buy? What new material handling system or ceramic grinder should we install? More frequently, however, economic analysis weighs *existing* versus *new* facilities. For most engineers, the problem is less likely to be one of building a new plant; rather, the goal is more often keeping a present plant operating economically. We are not choosing between new ways to perform the desired task. Instead, we have equipment performing the task, and the question is: Should the existing equipment be retained or replaced? This adversarial situation has given rise to the terms **defender** and **challenger.** The defender is the existing equipment; the challenger is the best available replacement equipment. Economically evaluating the existing defender and its challengers is the domain of **replacement analysis.**

## THE REPLACEMENT PROBLEM

Replacement of an existing asset may be appropriate due to obsolescence, depletion, and deterioration due to aging. In each of these cases, the ability of a previously implemented business asset to produce a desired output is challenged. For cases of obsolescence, depletion, and aging, it may be economical to replace the existing asset. We define each of these situations.

> *Obsolescence:* occurs when an asset's technology is surpassed by newer and/or different technologies. As an example, today's personal computers (PCs) with more RAM, faster clock speeds, larger hard drives, and more powerful central processors have made older, less powerful PCs obsolete. Thus, obsolete assets may need to be replaced with newer, more technologically advanced ones.

> *Depletion:* the gradual loss of market value of an asset as it is being consumed or exhausted. Oil wells and timber stands are examples of such assets. In most cases the asset will be used until it is depleted, at which time a replacement asset will be obtained. Depletion was treated in Chapter 11.

> *Deterioration due to aging:* the general condition of loss in value of some asset due to the aging process. Production machinery and other business assets that were once new eventually become aged. To compensate for a loss in functionality due to the aging process, additional operating and maintenance expenses are usually incurred to maintain the asset at its operating efficiency.

Aging equipment often has a greater risk of breakdowns. Planned replacements can be scheduled to minimize the time and cost of disruptions. Unplanned replacements can be very costly or even, as with an airplane engine, potentially catastrophic.

There are variations of the replacement problem: An existing asset may be abandoned or retired, augmented by a new asset but kept in service, or overhauled to reduce its operating and maintenance costs. These variations are most easily considered as potential new challengers.

Replacement problems are normally analyzed by looking only at the *costs* of the existing and replacement assets. Since the assets typically perform the same function, the value of using the vehicle, machine, or other equipment can be ignored. If the new asset has new features or better performance, these can be included as a cost savings.

Alternatives in a replacement problem almost always have *different lives*, and the problem includes picking the best one. This is because an existing asset will often be kept for at most a few years longer, while the potential replacements may have lives of any length. Thus replacement problems are focused on annual marginal costs and on EUAC values. We can calculate present costs, but only as a step in calculating EUAC values.

In industry, as in government, expenditures are normally monitored by means of *annual budgets.* One important facet of a budget is the allocation of money for new capital expenditures, either new facilities or replacement and upgrading of existing facilities.

Replacement analysis may recommend that certain equipment be replaced, with the cost included in the capital expenditures budget. Even if no recommendation to replace is made, such a recommendation may be made the following year or subsequently. At *some* point, existing equipment will be replaced, either when it is no longer necessary or when better equipment is available. Thus, the question is not *if* the defender will be replaced, but *when* it will be replaced. This leads us to the first question in the defender–challenger comparison:

*Shall we replace the defender now, or shall we keep it for one or more additional years?*

If we do decide to keep the asset for another year, we will often reanalyze the problem next year. The operating environment and costs may change, or new challengers with lower costs or better performance may emerge.

## REPLACEMENT ANALYSIS DECISION MAP

Figure 13-1 is a basic decision map for conducting a replacement analysis.

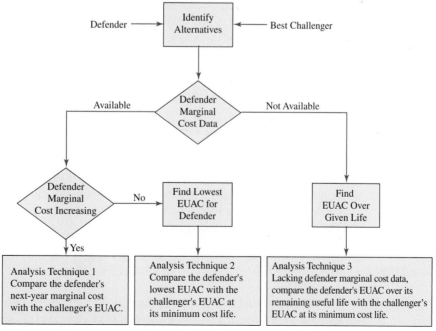

**FIGURE 13–1** Replacement analysis decision map.

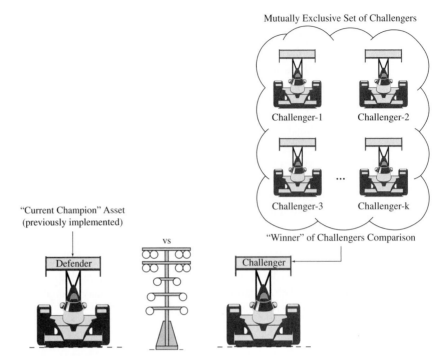

**FIGURE 13-2** Defender–challenger comparison.

Looking at the map, we can see there are three *replacement analysis techniques* that can be used under different circumstances. The correct replacement analysis technique depends on the data available for the alternatives and how the data behave over time.

By looking at the replacement analysis map, we see that the first step is to identify the alternatives. Again, in replacement analysis we are interested in comparing the previously implemented asset (the *defender*) against the best current available *challenger.*

> If the defender proves more economical, it will be retained. If the challenger proves more economical, it will be installed.

In this comparison the defender is being evaluated against a challenger that has been selected from a set of mutually exclusive competing challengers. Figure 13-2 illustrates this concept as a drag race between the defender and a challenger. The challenger that is competing against the defender has emerged from an earlier competition among a set of potential challengers. Any of the methods previously discussed in this text for evaluating sets of mutually exclusive alternatives could be used to identify the "best" challenger to race against the defender. However, it is important to note that the comparison of these potential challenger alternatives should be made at each alternative's respective *minimum cost life.* This concept is discussed next.

## MINIMUM COST LIFE OF A NEW ASSET—THE CHALLENGER

The **minimum cost life** of any new asset is the number of years at which the equivalent uniform annual cost (EUAC) of ownership is minimized. This minimum cost life is often shorter than either the asset's physical or useful life, because of increasing operating and maintenance costs in the later years of asset ownership. The challenger asset selected to "race" against the defender (in Figure 13-2) is the one having the lowest minimum cost of all the competing mutually exclusive challengers.

To calculate the minimum cost life of an asset, we determine the EUAC for each possible life less than or equal to the useful life. As illustrated in Example 13-1, the EUAC tends to be high if the asset is kept only a few years; then it decreases to some minimum value, and then increases again as the asset ages. By identifying the number of years at which the EUAC is a minimum and then keeping the asset for that number of years, we are minimizing the yearly cost of ownership.

## EXAMPLE 13–1

A piece of machinery costs $7500 and has no salvage value after it is installed. The manufacturer's warranty will pay the first year's maintenance and repair costs. In the second year, maintenance costs will be $900, and they will increase on a $900 arithmetic gradient in subsequent years. Also, operating expenses for the machinery will be $500 the first year and will increase on a $400 arithmetic gradient in the following years. If interest is 8%, compute the machinery's economic life that minimizes the EUAC. That is, find its minimum cost life.

### SOLUTION

| | If Retired at the End of Year $n$ | | | |
|---|---|---|---|---|
| Year, $n$ | EUAC of Capital Recovery Costs: $7500(A/P, 8\%, n)$ | EUAC of Maintenance and Repair Costs: $900(A/G, 8\%, n)$ | EUAC of Operating Costs: $500 + $400(A/G, 8\%, n)$ | EUAC Total |
| 1 | $8100 | $ 0 | $ 500 | $8600 |
| 2 | 4206 | 433 | 692 | 5331 |
| 3 | 2910 | 854 | 880 | 4644 |
| 4 | 2264 | 1264 | 1062 | 4589 ← |
| 5 | 1878 | 1661 | 1238 | 4779 |
| 6 | 1622 | 2048 | 1410 | 5081 |
| 7 | 1440 | 2425 | 1578 | 5443 |
| 8 | 1305 | 2789 | 1740 | 5834 |
| 9 | 1200 | 3142 | 1896 | 6239 |
| 10 | 1117 | 3484 | 2048 | 6650 |
| 11 | 1050 | 3816 | 2196 | 7063 |
| 12 | 995 | 4136 | 2338 | 7470 |
| 13 | 948 | 4446 | 2476 | 7871 |
| 14 | 909 | 4746 | 2609 | 8265 |
| 15 | 876 | 5035 | 2738 | 8648 |

The total EUAC data are plotted in Figure 13-3. From either the tabulation or the figure, we see that the machinery's minimum cost life is 4 years, with a minimum EUAC of $4589 for each of those 4 years.

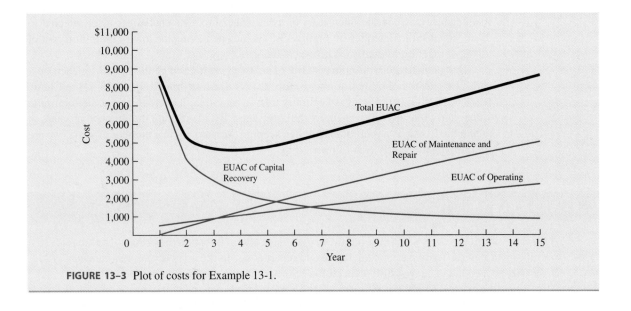

**FIGURE 13-3** Plot of costs for Example 13-1.

Looking at Figure 13-3 a bit more closely, we see the effects of each of the individual cost components on total EUAC (capital recovery, maintenance/repair, and operating expense EUACs) and how they behave over time. The total EUAC curve of most assets tends to follow this concave shape—high at the beginning due to capital recovery costs and high at the end due to increased maintenance/repair and operating expenses. The minimum EUAC occurs somewhere between these high points.

Like many pieces of installed equipment, the item considered in Example 13-1 had no salvage value. However, some assets, like your car, can easily be sold for a value that depends on the car's age and condition. Another possible complication is that repair costs may be reduced in early years by a warranty. The resulting cost curves will look like Figure 13-3, but the calculations are more work and a spreadsheet is very helpful. Example 13-11 in the last section of this chapter illustrates finding the minimum cost life for a new vehicle using spreadsheets.

## DEFENDER'S MARGINAL COST DATA

Once the basic participants in the defender–challenger comparison have been identified (see Figure 13-1), two specific questions regarding marginal costs must be answered: *Do we have marginal cost data for the defender?* and *Are the defender's marginal costs increasing on a year-to-year basis?* Let us first define marginal cost and then discuss why it is important to answer these two questions.

**Marginal costs,** as opposed to an EUAC, are the year-by-year costs of keeping an asset. Therefore, the "period" of any yearly marginal cost of ownership is always *1 year.* In our analysis, marginal cost is compared with EUAC, which is an end-of-year cash flow. Therefore, the marginal cost is also calculated as end-of-year cash flow.

On the other hand, an EUAC can apply to any number of consecutive years. Thus, the marginal cost of ownership for any year in an asset's life is the cost for *that year only.* In replacement problems, the total marginal cost for any year can include the capital recovery cost (loss in market value and lost interest for the year), yearly operating and maintenance costs, yearly taxes and insurance, and any other expense that occurs during that year. To calculate an asset's yearly marginal cost of ownership, it is necessary to have estimates of an asset's market value on a year-to-year basis over its useful life, as well as ordinary yearly expenses. Example 13-2 illustrates how total marginal cost can be calculated for an asset.

## EXAMPLE 13–2

A new piece of production machinery has the following costs.

$$\text{Investment cost} = \$25,000$$

Annual operating and maintenance cost $= \$2000$ in Year 1 and then increasing by $\$500$ per year

Annual cost for risk of breakdown $= \$5000$ per year for 3 years, then increasing by $\$1500$ per year

$$\text{Useful life} = 7 \text{ years}$$

$$\text{MARR} = 15\%$$

Calculate the marginal cost of keeping this asset over its useful life.

### SOLUTION

From the problem data we can easily find the marginal costs for O&M and risk of breakdowns. However, to calculate the marginal capital recovery cost, we need estimates of each year's market value.

| Year | Market Value |
|------|--------------|
| 1 | $18,000 |
| 2 | 13,000 |
| 3 | 9,000 |
| 4 | 6,000 |
| 5 | 4,000 |
| 6 | 3,000 |
| 7 | 2,500 |

We can now calculate the machinery's *marginal cost* (year-to-year cost of ownership) over its 7-year useful life.

| Year, $n$ | Loss in Market Value in Year $n$ | Interest in Year $n$ | O&M Cost in Year $n$ | Cost of Breakdown Risk in Year $n$ | Total Marginal Cost in Year $n$ |
|-----------|----------------------------------|----------------------|----------------------|-------------------------------------|----------------------------------|
| 1 | $25,000 − 18,000 = $7000 | $25,000(0.15) = $3750 | $2000 | $ 5,000 | $17,750 |
| 2 | 18,000 − 13,000 = 5000 | 18,000(0.15) = 2700 | 2500 | 5,000 | 15,200 |
| 3 | 13,000 − 9,000 = 4000 | 13,000(0.15) = 1950 | 3000 | 5,000 | 13,950 |
| 4 | 9,000 − 6,000 = 3000 | 9,000(0.15) = 1350 | 3500 | 6,500 | 14,350 |
| 5 | 6,000 − 4,000 = 2000 | 6,000(0.15) = 900 | 4000 | 8,000 | 14,900 |
| 6 | 4,000 − 3,000 = 1000 | 4,000(0.15) = 600 | 4500 | 9,500 | 15,600 |
| 7 | 3,000 − 2,500 = 500 | 3,000(0.15) = 450 | 5000 | 11,000 | 16,950 |

Notice that each year's total marginal cost includes loss in market value, interest, O&M cost, and cost for risk of breakdowns. For example, the Year-5 marginal cost of $14,900 is calculated as $2000 + 900 + 4000 + 8000$.

### Do We Have Marginal Cost Data for the Defender?

Our decision map indicates that it is necessary to know whether marginal cost data are available for the defender asset to determine the appropriate replacement technique to use. Usually in engineering economic problems, annual savings and expenses are given for all alternatives. However, as in Example 13-2, it is also necessary to have year-to-year salvage value estimates to calculate total marginal costs. If the total marginal costs for the defender can be calculated, and if the data are increasing from year to year, then *replacement analysis technique 1* should be used to compare the defender to the challenger.

### Are These Marginal Costs Increasing?

We have seen that it is important to know whether the marginal cost for the defender is increasing from year to year. This is determined by inspecting the total marginal cost of ownership of the defender over its remaining life. In most replacement analyses the defender is nearing the end of its economic life. The question usually is, Should we replace it now, next year, or perhaps the year after? In the early years of an asset's life we rarely analyze whether it is time for replacement. Thus, the defender's marginal costs are usually increasing, as shown in Example 13-3.

## EXAMPLE 13–3

An asset purchased 5 years ago for $75,000 can be sold today for $15,000. Operating expenses will be $10,000 this year, but these will increase by $1500 per year. It is estimated that the asset's market value will decrease by $1000 per year over the next 5 years. If the MARR used by the company is 15%, calculate the total marginal cost of ownership of this old asset (that is, the defender) for each of the next 5 years.

## SOLUTION

We calculate the total marginal cost of maintaining the old asset for the next 5-year period as follows:

| Year, $n$ | Loss in Market Value in Year $n$ | Interest in Year $n$ | Operating Cost in Year $n$ | Marginal Cost in Year $n$ |
|---|---|---|---|---|
| 1 | $15,000 - 14,000 = $1000 | $15,000(0.15) = $2250 | $10,000 | $13,250 |
| 2 | 14,000 - 13,000 = 1000 | 14,000(0.15) = 2100 | 11,500 | 14,600 |
| 3 | 13,000 - 12,000 = 1000 | 13,000(0.15) = 1950 | 13,000 | 15,950 |
| 4 | 12,000 - 11,000 = 1000 | 12,000(0.15) = 1800 | 14,500 | 17,300 |
| 5 | 11,000 - 10,000 = 1000 | 11,000(0.15) = 1650 | 16,000 | 18,650 |

We can see that marginal costs increase in each subsequent year of ownership. When the condition of increasing marginal costs for the defender has been met, then the defender–challenger comparison should be made with *replacement analysis technique 1*.

## REPLACEMENT ANALYSIS TECHNIQUE 1: DEFENDER MARGINAL COSTS CAN BE COMPUTED AND ARE INCREASING

When our first method of analyzing the defender asset against the best available challenger is used, the basic comparison involves *the defender's marginal cost data and the challenger's minimum cost life data.*

When the defender's marginal cost is increasing from year to year, we will maintain that defender as long as the marginal cost of keeping it one more year is less than the challenger's minimum EUAC. Thus our decision rule is as follows:

*Maintain the defender as long as the marginal cost of ownership for one more year is less than the challenger's minimum EUAC. When the defender's marginal cost becomes greater than the challenger's minimum EUAC, then replace the defender with the challenger.*

One can see that this technique assumes that the current best challenger, with its minimum EUAC, will be available and unchanged in the future. However, it is easy to update a replacement analysis when marginal costs for the defender change or when there is a change in the cost and/or performance of available challengers. Example 13-4 illustrates the use of this technique for comparing defender and challenger assets.

### EXAMPLE 13–4

Taking the machinery in Example 13-2 as the *challenger* and the machinery in Example 13-3 as the *defender,* use *replacement analysis technique 1* to determine when, if at all, a replacement decision should be made.

### SOLUTION

*Replacement analysis technique 1* should be used only in the condition of increasing marginal costs for the defender. Since these marginal costs are increasing for the defender (from Example 13-3), we can proceed by comparing defender marginal costs against the minimum EUAC of the challenger asset. In Example 13-2 we calculated only the marginal costs of the challenger; thus it is necessary to calculate the challenger's minimum EUAC. The EUAC of keeping this asset for each year of its useful life is worked out as follows.

| Year, $n$ | Challenger Total Marginal Cost in Year $n$ | Present Cost if Kept Through Year $n$ ($PC_n$) | | EUAC if Kept Through Year $n$ |
|---|---|---|---|---|
| 1 | $17,750 | $[17,750(P/F,15\%,1)]$ | $\times$ | $(A/P,15\%,1) = \$17,750$ |
| 2 | 15,200 | $PC_1 + 15,200(P/F,15\%,2)$ | $\times$ | $(A/P,15\%,2) = 16,560$ |
| 3 | 13,950 | $PC_2 + 13,950(P/F,15\%,3)$ | $\times$ | $(A/P,15\%,3) = 15,810$ |
| 4 | 14,350 | $PC_3 + 14,350(P/F,15\%,4)$ | $\times$ | $(A/P,15\%,4) = 15,520$ |
| 5 | 14,900 | $PC_4 + 14,900(P/F,15\%,5)$ | $\times$ | $(A/P,15\%,5) = 15,430$ |
| 6 | 15,600 | $PC_5 + 15,600(P/F,15\%,6)$ | $\times$ | $(A/P,15\%,6) = 15,450$ |
| 7 | 16,950 | $PC_6 + 16,950(P/F,15\%,7)$ | $\times$ | $(A/P,15\%,7) = 15,580$ |

A minimum EUAC of $15,430 is attained for the challenger at Year 5, which is the challenger's *minimum cost life*. We proceed by comparing this value against the *marginal* costs of the defender from Example 13-3:

| Year, *n* | Defender Total Marginal Cost in Year *n* | Challenger Minimum EUAC | Comparison Result and Recommendation |
|---|---|---|---|
| 1 | $13,250 | $15,430 | Since $13,250 is *less than* $15,430, keep defender. |
| 2 | 14,600 | 15,430 | Since $14,600 is *less than* $15,430, keep defender. |
| 3 | 15,950 | 15,430 | Since $15,950 is *greater than* $15,430, replace defender. |
| 4 | 17,300 | | |
| 5 | 18,650 | | |

Based on the data given for the challenger and for the defender, we would keep the defender for 2 more years and then replace it with the challenger because at that point the defender's marginal cost of another year of ownership would be greater than the challenger's minimum EUAC.

## REPLACEMENT REPEATABILITY ASSUMPTIONS

The decision to use the challenger's minimum EUAC reflects two assumptions: the best challenger will be available "with the same minimum EUAC" in the future; and the period of needed service is indefinitely long. In other words, we assume that once the decision has been made to replace, there will be an indefinite cycle of replacement with the current best challenger asset. These assumptions must be satisfied for our calculations to be correct into an indefinite future. However, because the near future is economically more important than the distant future, and because our analysis is done with the *best data currently available*, our results and recommendations are robust or stable for reasonable changes in the estimated data.

The repeatability assumptions together are much like the repeatability assumptions that allowed us to use the annual cost method to compare competing alternatives with different useful lives. Taken together, we call these the **replacement repeatability assumptions.** They allow us to greatly simplify comparing the defender and the challenger.

Stated formally, these two assumptions are:

1. The currently available best challenger will continue to be available in subsequent years and will be unchanged in its economic costs. When the defender is ultimately replaced, it will be replaced with this challenger. Any challengers put into service will also be replaced with the same currently available challenger.
2. The period of needed service of the asset is indefinitely long. Thus the challenger asset, once put into service, will continuously replace itself in repeating cycles.

If these two assumptions are satisfied completely, then our calculations are exact. Often, however, future challengers represent further improvements so that Assumption 1 is not satisfied. While the calculations are no longer exact, the repeatability assumptions allow us to make the best decision we can with the data we have.

If the defender's marginal cost is increasing, once it rises above the challenger's minimum EUAC, it will continue to be greater. Under the repeatability assumptions, we would never want to incur a defender's marginal cost that was greater than the challenger's minimum EUAC. Thus, we use replacement analysis technique 1 when the defender's marginal costs are increasing.

## REPLACEMENT ANALYSIS TECHNIQUE 2: DEFENDER MARGINAL COSTS CAN BE COMPUTED AND ARE NOT INCREASING

If the defender's marginal costs do not increase, we have no guarantee that *replacement analysis technique 1* will produce the alternative that is of the greatest economic advantage. Consider the new asset in Example 13-2, which has marginal costs that begin at a high of $17,750, then *decrease* over the next years to a low of $13,950, and then *increase* thereafter to $16,950 in Year 7. If evaluated *one year after implementation,* the asset would not have increasing marginal costs. Defenders in their early stages typically do not fit the requirements of *replacement analysis technique 1.* In the situation graphed in Figure 13-3, such defender assets would be in the downward slope of a concave marginal cost curve.

Example 13-5 details why *replacement analysis technique 1* cannot be applied when defenders do not have consistently increasing marginal cost curves. Instead we apply *replacement analysis technique 2.* That is, we calculate the defender's minimum EUAC to see whether the replacement should occur immediately. If not, as shown in Example 13-5, the replacement occurs after the defender's minimum cost life when the marginal costs are increasing. Then *replacement analysis technique 1* applies again.

## EXAMPLE 13–5

Let us look again at the defender and challenger assets in Example 13-4. This time let us arbitrarily change the defender's marginal costs for its 5-year useful life. Now when, if at all, should the defender be replaced with the challenger?

| Year, $n$ | Defender Total Marginal Cost in Year $n$ |
|-----------|------------------------------------------|
| 1 | $16,000 |
| 2 | 14,000 |
| 3 | 13,500 |
| 4 | 15,300 |
| 5 | 17,500 |

### SOLUTION

In this case the defender's total marginal costs are *not* consistently increasing from year to year. However, if we ignore this fact and apply *replacement analysis technique 1,* the recommendation would be to replace the defender now, because the defender's marginal cost for the first year

($16,000) is greater than the minimum EUAC of the challenger ($15,430). This would be the wrong choice.

Since the defender's marginal costs are below the challenger's minimum EUAC in the second through fourth years, we must calculate the EUAC of keeping the defender asset in each of its remaining 5 years, at $i = 15\%$.

| Year, $n$ | Present Cost if Kept $n$ Years ($PC_n$) | | EUAC if Kept $n$ Years |
|---|---|---|---|
| 1 | $16,000(P/F, 15\%, 1)$ | $\times$ | $(A/P, 15\%, 1) = \$16,000$ |
| 2 | $PC_1 + 14,000(P/F, 15\%, 2)$ | $\times$ | $(A/P, 15\%, 2) = 15,070$ |
| 3 | $PC_2 + 13,500(P/F, 15\%, 3)$ | $\times$ | $(A/P, 15\%, 3) = 14,618$ |
| 4 | $PC_3 + 15,300(P/F, 15\%, 4)$ | $\times$ | $(A/P, 15\%, 4) = 14,754$ |
| 5 | $PC_4 + 17,500(P/F, 15\%, 5)$ | $\times$ | $(A/P, 15\%, 5) = 15,162$ |

The minimum EUAC of the defender for 3 years is $14,618, which is less than that of the challenger's minimum EUAC of $15,430. Thus, under the replacement repeatability assumptions, we will keep the defender for at least 3 years. We must still decide how much longer.

The defender's EUAC begins to rise in Year 4, because the marginal costs are increasing, and because they are above the defender's minimum EUAC. Thus, we can use *replacement analysis technique 1* for Year 4 and later. The defender's marginal cost in Year 4 is $15,300, which is $130 below the challenger's minimum EUAC of $15,430. Since the defender's marginal cost of $17,500 is higher in Year 5, we replace it with the new challenger at the end of Year 4. Notice that we did *not* keep the defender for its minimum cost life of 3 years, we kept it for 4 years.

If the challenger's minimum EUAC were less than the defender's minimum EUAC of $14,618, then the defender would be immediately replaced.

Example 13-5 illustrates several potentially confusing points about replacement analysis.

- If the defender's marginal cost data is not increasing, the defender's minimum EUAC must be calculated.
- If the defender's minimum EUAC exceeds the challenger's minimum EUAC, then replace immediately. If the defender's minimum EUAC is lower than the challenger's minimum EUAC, then under the replacement repeatability assumptions the defender will be kept *at least* the number of years for its minimum EUAC.
- After this number of years, then replace when the defender's increasing marginal cost exceeds the challenger's minimum EUAC.

The problem statement for Example 13-6 illustrates a second approach to calculating the defender's marginal costs for its capital costs. The value at the year's beginning is multiplied by $(1 + i)$ and the salvage value at the year's end is subtracted. Each year's total marginal cost also includes the operations and maintenance costs.

Then the solution to Example 13-6 details the calculation of the minimum EUAC when the defender's data is presented as costs and salvage values in each year rather than as marginal costs. Notice that this is calculated the same way as the minimum cost life was calculated for new assets—the challengers.

## EXAMPLE 13–6

A 5-year-old machine, whose current market value is $5000, is being analyzed to determine its minimum EUAC at a 10% interest rate. Salvage value and maintenance estimates and the corresponding marginal costs are given in the following table.

| | Data | | Calculating Marginal Costs | | |
|---|---|---|---|---|---|
| Year | Salvage Value | O&M Cost | $S_{t-1}(1 + i)$ | $-S_t$ | Marginal Cost |
| 0 | $5000 | | | | |
| 1 | 4000 | $ 0 | $5500 | $–4000 | $1500 |
| 2 | 3500 | 100 | 4400 | –3500 | 1000 |
| 3 | 3000 | 200 | 3850 | –3000 | 1050 |
| 4 | 2500 | 300 | 3300 | –2500 | 1100 |
| 5 | 2000 | 400 | 2750 | –2000 | 1150 |
| 6 | 2000 | 500 | 2200 | –2000 | 700 |
| 7 | 2000 | 600 | 2200 | –2000 | 800 |
| 8 | 2000 | 700 | 2200 | –2000 | 900 |
| 9 | 2000 | 800 | 2200 | –2000 | 1000 |
| 10 | 2000 | 900 | 2200 | –2000 | 1100 |
| 11 | 2000 | 1000 | 2200 | –2000 | 1200 |

## SOLUTION

Because the marginal costs have a complex, nonincreasing pattern, we must calculate the defender's minimum EUAC.

| | | | If Retired at End of Year $n$ | | |
|---|---|---|---|---|---|
| Years Kept, $n$ | Salvage Value ($S$) at End of Year $n$ | Maintenance Cost for Year | EUAC of Capital Recovery $(P—S) \times$ $(A/P, 10\%, n) + Si$ | EUAC of Maintenance $100(A/G, 10\%, n)$ | Total EUAC |
| 0 | $P = $5000 | | | | |
| 1 | 4000 | $ 0 | $1100 + 400 | $ 0 | $1500 |
| 2 | 3500 | 100 | 864 + 350 | 48 | 1262 |
| 3 | 3000 | 200 | 804 + 300 | 94 | 1198 |
| 4 | 2500 | 300 | 789 + 250 | 138 | 1177 |
| 5 | 2000 | 400 | 791 + 200 | 181 | 1172 |
| 6 | 2000 | 500 | 689 + 200 | 222 | 1111 |
| 7 | 2000 | 600 | 616 + 200 | 262 | 1078 |
| 8 | 2000 | 700 | 562 + 200 | 300 | 1062 |
| 9 | 2000 | 800 | 521 + 200 | 337 | 1058← |
| 10 | 2000 | 900 | 488 + 200 | 372 | 1060 |
| 11 | 2000 | 1000 | 462 + 200 | 406 | 1068 |

A minimum EUAC of $1058 is computed at Year 9 for the existing machine. Notice that the EUAC begins to increase with $n$ when the marginal cost in Year 10 exceeds the EUAC for 9 years.

Now to apply *replacement analysis technique 2* to Example 13-6, we ask: Is the challenger's minimum EUAC higher or lower than the defender's minimum EUAC of $1058? If the challenger's minimum EUAC is lower, then we replace the defender now.

Under the repeatability assumptions, if the challenger's minimum EUAC is higher, we would keep the defender at least 9 years. Replacement would occur in Year 10 or later when the defender's marginal costs exceed the challenger's minimum EUAC. Relaxing the repeatability assumptions to allow for better challengers, we may replace the defender whenever a new challenger has an EUAC that is lower than $1058.

Example 13-7 illustrates the common situation of a current defender that may be kept if overhauled. This can also be analyzed as a potential new challenger.

## EXAMPLE 13–7

We must decide whether existing (defender) equipment in an industrial plant should be replaced. A $4000 overhaul must be done now if the equipment is to be retained in service. Maintenance is $1800 in each of the next 2 years, after which it increases by $1000 each year. The defender has no present or future salvage value. The equipment described in Example 13-1 is the challenger (EUAC = $4589). Should the defender be kept or replaced if the interest rate is 8%?

**FIGURE 13–4** Overhaul and maintenance costs for the defender in Example 13-7.

## SOLUTION

The first step is to determine the defender's lowest EUAC. The pattern of overhaul and maintenance costs (Figure 13-4) suggests that if the overhaul is done, the equipment should be kept for several years. The computation is as follows:

| | If Retired at End of Year $n$ | | |
| Year, $n$ | EUAC of Overhaul $4000(A/P, 8\%, n)$ | EUAC of Maintenance $1800 + 1000$ Gradient from Year 3 on | Total EUAC |
| --- | --- | --- | --- |
| 1 | $4320 | $1800 | $6120 |
| 2 | 2243 | 1800 | 4043 |
| 3 | 1552 | $1800 + 308$* | 3660 ← |
| 4 | 1208 | $1800 + 683^{\dagger}$ | 3691 |
| 5 | 1002 | $1800 + 1079$ | 3881 |

*For the first 3 years, the maintenance is $1800, $1800, and $2800. Thus, $\text{EUAC} = 1800 + 1000(A/F, 8\%, 3) = 1800 + 308$.

$^{\dagger}\text{EUAC} = 1800 + 1000(P/G, 8\%, 3)(P/F, 8\%, 1)(A/P, 8\%, 4) = 1800 + 683$.

The lowest EUAC of the overhauled defender is $3660. In Example 13-1, the challenger's minimum cost life was 4 years with an EUAC of $4589. If we assume the equipment is needed for at least 4 years, the overhauled defender's EUAC ($3660) is less than the challenger's EUAC ($4589). Overhaul the defender.

If the defender's and challenger's cost data do not change, we can use *replacement analysis technique 1* to determine when the overhauled defender should be replaced. We know from the minimum EUAC calculation that the defender should be kept at least 3 years. Is this the best life? The following table computes the marginal cost to answer this question.

| Year, $n$ | Overhaul Cost | Maintenance Cost | Marginal Cost to Extend Service |
|---|---|---|---|
| 0 | $4000 | $ 0 | |
| 1 | 0 | 1800 | $6120 = 4000(1.08) + 1800$ |
| 2 | 0 | 1800 | 1800 |
| 3 | 0 | 2800 | 2800 |
| 4 | 0 | 3800 | 3800 |
| 5 | 0 | 4800 | 4800 |

Year 5 is the first year after Year 3, which has the overhauled defender's lowest EUAC (in which the $4800 marginal cost exceeds the challenger's $4589 minimum EUAC). Thus, the overhauled defender should be kept 4 more years if costs do not change. (Note that if the defender can be overhauled again after 3 or 4 years, that might be an even better choice.)

## REPLACEMENT ANALYSIS TECHNIQUE 3: WHEN DEFENDER MARGINAL COST DATA ARE NOT AVAILABLE

In our third case, we simply compare the defender's *EUAC over its stated useful life, and the challenger's minimum EUAC.* Pick the *EUAC* that is lower.

If the defender's marginal cost data is not known and cannot be estimated, it is impossible to apply *replacement analysis techniques 1 or 2* to decide *when* the defender should be replaced. Instead we must assume that the defender's stated useful life is the only one to consider. From a student problem-solving perspective, the defender in Example 13-7 might be described as follows.

The defender can be overhauled for $4800 to extend its life for 5 years. Maintenance costs will average $3000 per year, and there will be no salvage value. In this case the only possibility is to compare the defender's EUAC for a 5-year life with the best challenger.

In the real world, the most likely scenario for this approach involves a facility-wide overhaul every 3, 5, 10, etc. years. Pipelines and many process plants, such as refineries, chemical plants, and steel mills, must shut down to do major maintenance. All equipment is overhauled or replaced with a new challenger as needed, and the facility is expected to operate until the next maintenance shutdown.

The defender's EUAC over its remaining useful life is compared with the challenger's EUAC at its minimum cost life, and the lower cost is chosen. However, in making this basic comparison an often complicating factor is deciding what first cost to assign to the challenger and the defender.

## COMPLICATIONS IN REPLACEMENT ANALYSIS

### Defining Defender and Challenger First Costs

Because the defender is already in service, analysts often misunderstand what first cost to assign it. Example 13-8 demonstrates this problem.

---

### EXAMPLE 13–8

A model SK-30 was purchased 2 years ago for $1600; it has been depreciated by straight-line depreciation using a 4-year life and zero salvage value. Because of recent innovations, the current price of the SK-30 is $995. An equipment firm has offered a trade-in allowance of $350 for the SK-30 on a new $1200 model EL-40. Some discussion revealed that without a trade-in, the EL-40 can be purchased for $1050. Thus, the originally quoted price of the EL-40 was overstated to allow a larger trade-in allowance. The true current market value of the SK-30 is probably only $200. In a replacement analysis, what value should be assigned to the SK-30?

### SOLUTION

In the example, five different dollar amounts relating to the SK-30 have been outlined:

1. *Original cost:*  It cost $1600 2 years ago.
2. *Present cost:*  It now sells for $995.
3. *Book value:*  The original cost less 2 years of depreciation is $1600 - \frac{2}{4}(1600 - 0) = \$800$.
4. *Trade-in value:*  The offer was $350.
5. *Market value:*  The estimate was $200.

We know that an economic analysis is based on the current situation, not on the past. We refer to past costs as *sunk* costs to emphasize this. These costs cannot be altered, and they are not relevant. (There is one exception: past costs may affect present or future income taxes.)

We want to use actual cash flows for each alternative. Here the question is, What value should be used in an economic analysis for the SK-30? The relevant cost is the equipment's present market value of $200. Neither the original cost, the present cost, the book value, nor the trade-in value is relevant.

---

At first glance, an asset's trade-in value would appear to be a suitable present value for the equipment. Often the trade-in price is inflated *along with* the price for the new item. (This practice is so common in new-car showrooms that the term *overtrade* is used to describe the excessive portion of the trade-in allowance. The buyer is also quoted a higher price for the new car.) Distorting the defender's present value, or the challenger's price can be serious because these distortions do not cancel out in an economic analysis.

Example 13-8 illustrated that of the several different values that can be assigned to the defender, the correct one is the present market value. If a trade-in value is obtained, care should be taken to ensure that it actually represents a fair market value.

Determining the value for the challenger's installed cost should be less difficult. In such cases the first cost is usually made up of purchase price, sales tax, installation costs, and other items that occur initially on a one-time basis if the challenger is selected. These values are usually rather straightforward to obtain. One aspect to consider in assigning a first cost to the challenger is the defender's potential market or salvage value. One must not arbitrarily subtract the defender's disposition value from the challenger's first cost, for this practice can lead to an incorrect analysis.

As described in Example 13-8, the correct first cost to assign to the defender SK-30 is its $200 current market value. This value represents the present economic benefit that we would be *forgoing* to keep the defender. This can be called our *opportunity first cost.* If, instead of assuming that this is the defender's *opportunity cost*, we assume it is a *cash benefit* to the challenger, a potential error arises.

The error lies in incorrect use of a *cash flow* perspective when the lives of the challenger and the defender are not equal, which is usually the case. The approach of subtracting the defender's salvage value from the challenger's first cost is called the *cash flow* perspective. From this perspective, keeping the defender in place causes $0 in cash to flow, but selecting the challenger causes the cash flow now to equal the challenger's first cost minus the defender's salvage value.

If the lives of the defender and the challenger are the same, then the cash flow perspective will lead to the correct answer. However, it is normally the case that the defender is an aging asset with a relatively short horizon of possible lives and the challenger is a new asset with a longer life. The *opportunity cost* perspective will always lead to the correct answer, so it is the one that should be used.

For example, consider the SK-30 and EL-40 from Example 13-8. It is reasonable to assume that the 2-year-old SK-30 has 3 years of life left and that the new EL-40 would have a 5-year life. Assume that neither will have any salvage value at the end of its life. Compare the difference in their annual capital costs with the correct *opportunity cost* perspective and the incorrect *cash flow* perspective.

| SK-30 | | EL-40 | |
|---|---|---|---|
| Market value | $200 | First cost | $1050 |
| Remaining life | 3 years | Useful life | 5 years |

Looking at this from an *opportunity cost* perspective, the annual cost comparison of the first costs is

$$\text{Annualized first cost}_{\text{SK-30}} = \$200(A/P, 10\%, 3) = \$80$$

$$\text{Annualized first cost}_{\text{EL-40}} = \$1050(A/P, 10\%, 5) = \$277$$

The *difference* in annualized first cost between the SK-30 and EL-40 is

$$\text{AFC}_{\text{EL-40}} - \text{AFC}_{\text{SK-30}} = \$277 - \$80 = \$197$$

Now using an incorrect *cash flow* perspective to look at the first costs, we can calculate the *difference* due to first cost between the SK-30 and EL-40.

$$\text{Annualized first cost}_{\text{SK-30}} = \$0(A/P, 10\%, 3) = \$0$$

$$\text{Annualized first cost}_{\text{EL-40}} = (\$1050 - 200)(A/P, 10\%, 5) = \$224$$

$$\text{AFC}_{\text{EL-40}} - \text{AFC}_{\text{SK-30}} = \$224 - \$0 = \$224$$

When the defender's remaining life (3 years) differs from the challenger's useful life (5 years), the two perspectives yield different results. The correct difference of $197 is shown by using the *opportunity cost* approach, and an inaccurate difference of $224 is obtained if the *cash flow* perspective is used. From the opportunity cost perspective, the $200 is spread out over 3 years as a cost to the defender, and in the cash flow perspective, the opportunity cost is spread out over 5 years as a benefit to the challenger. Spreading the $200 over 3 years in one case and 5 years in the other case does not produce equivalent annualized amounts. Because of the difference in the lives of the assets, the annualized $200 opportunity cost for the defender cannot be called an equivalent benefit to the challenger.

In the case of unequal lives, the correct method is to assign the defender's current market value as its Time-0 opportunity costs, rather than subtracting this amount from the challenger's first cost. Because the cash flow approach yields an incorrect value when challenger and defender have unequal lives, the *opportunity cost* approach for assigning a first cost to the challenger and defender assets should *always* be used.

## REPEATABILITY ASSUMPTIONS NOT ACCEPTABLE

Under certain circumstances, the repeatability assumptions described earlier may not apply. Then replacement analysis techniques 1, 2, and 3 may not be valid. For instance, there may be a specific study period instead of an indefinite need for the asset. For example, consider the case of phasing out production after a certain number of years—perhaps a person who is about to retire is closing down a business and selling all the assets. Another example is production equipment such as molds and dies that are no longer needed when a new model with new shapes is introduced. Yet another is a construction camp that may be needed for only a year or two or three.

This specific study period could potentially be any number of years relative to the lives of the defender and the challenger, such as equal to the defender's life, equal to the challenger's life, less than the defender's life, greater than the challenger's life, or somewhere between the lives of the defender and challenger. The analyst must be explicit about the challenger's and defender's economic costs and benefits, as well as residual or salvage values at the end of the specific study period. In this case the repeatability replacement assumptions do not apply; costs must be analyzed over the study period. The analysis techniques in the decision map also may not apply when future challengers are not assumed to be identical to the current best challenger. This concept is discussed in the next section.

### A Closer Look at Future Challengers

We defined the challenger as the best available alternative to replace the defender. But over time, the best available alternative can indeed change. And given the trend in our technological society, it seems likely that future challengers will be better than the present challenger. If so, the prospect of improved future challengers may affect the present decision between the defender and the challenger.

Figure 13-5 illustrates two possible estimates of future challengers. In many technological areas it seems likely that the equivalent uniform annual costs associated with future challengers will decrease by a constant amount each year. In other fields, however, a rapidly changing technology will produce a sudden and substantially improved challenger—with decreased costs or increased benefits. The uniform decline curve of Figure 13-5 reflects the assumption that each future challenger has a minimum EUAC that is a fixed amount less than the previous year's challenger. This assumption, of course, is only one of many possible assumptions that could be made regarding future challengers.

**FIGURE 13-5** Two possible ways the EUAC of future challengers may decline.

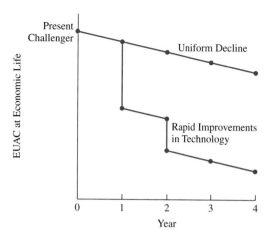

If future challengers will be better than the present challenger, what impact will this have on an analysis now? The prospect of better future challengers may make it more desirable to retain the defender and to reject the present challenger. By keeping the defender for now, we may be able to replace it later by a better future challenger. Or, to state it another way, the present challenger may be made less desirable by the prospect of improved future challengers.

As engineering economic analysts, we must familiarize ourselves with potential technological advances in assets targeted for replacement. This part of the decision process is much like the search for all available alternatives, from which we select the best. Upon finding out more about what alternatives and technologies are emerging, we will be better able to understand the repercussions of investing in the current best available challenger. Selecting the current best challenger asset can be particularly risky when (1) the costs are very high and/or (2) the challenger's useful minimum cost life is relatively long (5–10 years or more). When one or both of these conditions exist, it may be better to keep or even augment the defender asset until better future challengers emerge.

There are, of course, many assumptions that *could* be made regarding future challengers. However, if the replacement repeatability assumptions do not hold, the analysis needed becomes more complicated.

## AFTER-TAX REPLACEMENT ANALYSIS

As described in Chapter 12, an after-tax analysis provides greater realism and insight. Tax effects can alter before-tax recommendations. After-tax effects may influence calculations in the defender–challenger comparisons discussed earlier. Consequently, one should always perform or check these analyses on an after-tax basis.

### Marginal Costs on an After-Tax Basis

Marginal costs on an after-tax basis represent the cost that would be incurred through ownership of the defender *in each year.* On an after-tax basis we must consider the effects of ordinary taxes as well as gains and losses due to asset disposal. Consider Example 13-9.

## EXAMPLE 13-9

Refer to Example 13-2, where we calculated the before-tax marginal costs for a new piece of production machinery. Calculate the asset's after-tax marginal costs considering this additional information.

- Depreciation is by the straight-line method, with $S = \$0$ and $n = 5$ years, so $d_t = (\$25,000 - \$0)/5 = \$5000$.
- Ordinary income, recaptured depreciation, and losses on sales are taxed at a rate of 40%.
- The after-tax MARR is 10%.

Some classes skip or have not yet covered Chapter 10's explanation of expected value. Thus, the expected cost for risk of breakdowns is described here as an insurance cost.

## SOLUTION

The after-tax marginal cost of ownership will involve the following elements: incurred or forgone loss or recaptured depreciation, interest on invested capital, tax savings due to depreciation, and annual after-tax operating/maintenance and insurance. Figure 13-6 shows example cash flows for the marginal cost detailed in Table 13-1.

As a refresher of the recaptured depreciation calculations in Chapter 12:

The market value in Year $0 = 25,000$.

The market value decreases to $18,000 at Year 1.

The book value at Year $1 = 25,000 - 5000 = \$20,000$.

So loss on depreciation $= 20,000 - 18,000 = \$2000$.

This results in a tax savings of $(2000)(0.4) = \$800$.

$$\text{Marginal cost} = F_{\text{end-of-year}}$$
$$= 25,000\,(1 + 0.1) + 4200 - 800 - 18,000 - 2000$$
$$= \$10,900$$

$$\text{Marginal cost} = F_{\text{end-of-year}}$$
$$= 18,800\,(1.1) + 4500 - 800 - 13,000 - 2000 = \$9380$$

**FIGURE 13-6** Cash flow diagrams and calculations for marginal cost.

TABLE 13–1 **Marginal Costs of Ownership**

| Year | Market Value | Book Value | Recaptured Depr. or Loss | Taxes or Tax Savings | After-Tax Market Value |
|------|------|------|------|------|------|
| 0 | $25,000 | $25,000 | | | $25,000 |
| 1 | 18,000 | 20,000 | –$2000 | –$800 | 18,800 |
| 2 | 13,000 | 15,000 | –2000 | –800 | 13,800 |
| 3 | 9,000 | 10,000 | –1000 | –400 | 9,400 |
| 4 | 6,000 | 5,000 | 1000 | 400 | 5,600 |
| 5 | 4,000 | 0 | 4000 | 1600 | 2,400 |
| 6 | 3,000 | | 3000 | 1200 | 1,800 |
| 7 | 2,500 | | 2500 | 1000 | 1,500 |

| Year | Col. B After-Tax Market Value | Col. C Beg. Yr Value $\times (1+i)$ | Col. D Tax Savings from Depr. Deduct. | O&M and Insurance Cost | Col. F After-Tax Annual Expense | $= C + D + F - B$ Marginal Cost |
|------|------|------|------|------|------|------|
| 0 | $25,000 | | | | | |
| 1 | 18,800 | $27,500 | –$2000 | $ 7,000 | $4200 | $10,900 |
| 2 | 13,800 | 20,680 | –2000 | 7,500 | 4500 | 9,380 |
| 3 | 9,400 | 15,180 | –2000 | 8,000 | 4800 | 8,580 |
| 4 | 5,600 | 10,340 | –2000 | 10,000 | 6000 | 8,740 |
| 5 | 2,400 | 6,160 | –2000 | 12,000 | 7200 | 8,960 |
| 6 | 1,800 | 2,640 | 0 | 14,000 | 8400 | 9,240 |
| 7 | 1,500 | 1,980 | 0 | 16,000 | 9600 | 10,080 |

The marginal cost in each year is much lower after taxes than the pretax numbers shown in Example 13-2. This is true because depreciation and expenses can be subtracted from taxable income. However, the pattern of declining and then increasing marginal costs is the same, and Year 3 is still the year of lowest marginal costs.

## Minimum Cost Life Problems

Here we illustrate the effect of taxes on the calculation of the minimum cost life of the defender and the challenger. The after-tax minimum EUAC depends on both the depreciation method used and changes in the asset's market value over time. Using an accelerated depreciation method (like MACRS) tends to reduce the after-tax costs early in the asset's life. This alters the shape of the total EUAC curve—the concave shape can be shifted and the minimum EUAC changed. Example 13-10 illustrates the effect that taxes can have.

### EXAMPLE 13–10

Some new production machinery has a first cost of $100,000 and a useful life of 10 years. Its estimated operating and maintenance (O&M) costs are $10,000 the first year, which will increase annually by $4000. The asset's before-tax market value will be $50,000 at the end of the first year and then will decrease by $5000 annually. This property is a 7-year MACRS property. The company uses a 6% after-tax MARR and is subject to a combined federal/state tax rate of 40%.

Calculate the after-tax cash flows.

**SOLUTION**

To find this new production machinery's minimum cost life, we first find the after-tax cash flow (ATCF) effect of the O&M costs and depreciation (Table 13-2). Then, we find the ATCFs of disposal if the equipment is sold in each of the 10 years (Table 13-3). Finally in the closing section on spreadsheets, we combine these two ATCFs (in Figure 13-7) and choose the minimum cost life.

In Table 13-2, the O&M expense starts at $10,000 and increases at $4000 per year. The depreciation entries equal the 7-year $r_t$ MACRS depreciation values given in Table 11-3 multiplied by the $100,000 first cost. The taxable income, which is simply the O&M costs minus the depreciation values, is then multiplied by minus the tax rate to determine the tax savings. The O&M expense plus the tax savings is the Table 13-2 portion of the total ATCF.

Regarding the market value data in this problem, it should be pointed out that the initial decrease of $50,000 in Year 1 is not uncommon. This is especially true for custom-built equipment for a particular and unique application at a specific plant. Such equipment would not be valuable to others in the marketplace. Also, the $100,000 first cost (cost basis) could have included costs due to installation, facility modifications, or removal of old equipment. The $50,000 is realistic for the market value of one-year-old equipment.

The next step is to determine the ATCFs that would occur in each possible year of disposal. (The ATCF for Year 0 is easy; it is −$100,000.) For example, as shown in Table 13-3, in Year 1 there is a $35,710 loss as the book value exceeds the market value. The tax savings from this loss are added to the salvage (market) value to determine the ATCF (*if the asset is disposed of during Year 1*).

**TABLE 13–2  ATCF for O&M and Depreciation for Example 13-10**

| Year, t | O&M Expense | MACRS Depreciation, $d_t$ | Taxable Income | Tax Savings (at 40%) | O&M Depreciation ATCF |
|---|---|---|---|---|---|
| 1 | −$10,000 | $14,290 | −$24,290 | $ 9,716 | −$284 |
| 2 | −14,000 | 24,490 | −38,490 | 15,396 | 1,396 |
| 3 | −18,000 | 17,490 | −35,490 | 14,196 | −3,804 |
| 4 | −22,000 | 12,490 | −34,490 | 13,796 | −8,204 |
| 5 | −26,000 | 8,930 | −34,930 | 13,972 | −12,028 |
| 6 | −30,000 | 8,920 | −38,920 | 15,568 | −14,432 |
| 7 | −34,000 | 8,930 | −42,930 | 17,172 | −16,828 |
| 8 | −38,000 | 4,460 | −42,460 | 16,984 | −21,016 |
| 9 | −42,000 | 0 | −42,000 | 16,800 | −25,200 |
| 10 | −46,000 | 0 | −46,000 | 18,400 | −27,600 |

**TABLE 13–3  ATCF in Year of Disposal for Example 13-10**

| Year, t | Market Value | Book Value | Gain or Loss | Gain/Loss Tax (at 40%) | ATCF if Disposed of |
|---|---|---|---|---|---|
| 1 | $50,000 | $85,710 | −$35,710 | $14,284 | $64,284 |
| 2 | 45,000 | 61,220 | −16,220 | 6,488 | 51,488 |
| 3 | 40,000 | 43,730 | −3,730 | 1,492 | 41,492 |
| 4 | 35,000 | 31,240 | 3,760 | −1,504 | 33,496 |
| 5 | 30,000 | 22,310 | 7,690 | −3,076 | 26,924 |

TABLE 13-3 **Continued**

| Year, t | Market Value | Book Value | Gain or Loss | Gain/Loss Tax (at 40%) | ATCF if Disposed of |
|---|---|---|---|---|---|
| 6 | 25,000 | 13,390 | 11,610 | −4644 | 20,356 |
| 7 | 20,000 | 4,460 | 15,540 | −6216 | 13,784 |
| 8 | 15,000 | 0 | 15,000 | −6000 | 9,000 |
| 9 | 10,000 | 0 | 10,000 | −4000 | 6,000 |
| 10 | 5,000 | 0 | 5,000 | −2000 | 3,000 |

These tables assume that depreciation is taken during the year of disposal and then calculates the recaptured depreciation (gain) or loss on the year-end book value.

## SPREADSHEETS AND REPLACEMENT ANALYSIS

Spreadsheets are very useful in nearly all after-tax calculations. However, they are absolutely required for optimal life calculations in after-tax situations. Because MACRS is the tax law, the after-tax cash flows are different in every year. Thus, the NPV function and the PMT function are both needed to find the minimum EUAC after taxes. Figure 13-7 illustrates the calculation of the minimum cost life for Example 13-10.

In Figure 13-7, the NPV finds the present worth of the irregular cash flows from Period 1 through Period $t$ for $t = 1$ to life. The PV function is used to find the PW of the salvage value. Then PMT can be used to find the EUAC over each potential life. Before-tax replacement analysis can also be done this way. The spreadsheet block function NPV is used to find the PW of cash flows from Period 1 to Period $t$. Note that the cell for Period 1 is an

| | A | B | C | D | E | F |
|---|---|---|---|---|---|---|
| 1 | | **Table 13-2** | **Table 13-3** | 6% | Interest Rate | |
| 2 | | O&M & Depr. | if disposed of | | | |
| 3 | Year | ATCF | ATCF | PW | EUAC | |
| 4 | 0 | | | −$100,000 | | |
| 5 | 1 | −$284 | $64,284 | −39,623 | $42,000 | =PMT($D$1,A5,D5) |
| 6 | 2 | 1,396 | 51,488 | −53,201 | 29,018 | |
| 7 | 3 | −3,804 | 41,492 | −67,382 | 25,208 | |
| 8 | 4 | −8,204 | 33,496 | −82,186 | 23,718 | |
| 9 | 5 | −12,028 | 26,924 | −97,587 | 23,167 | |
| 10 | 6 | −14,432 | 20,356 | −113,530 | 23,088 | optimal life |
| 11 | 7 | −16,828 | 13,784 | −129,904 | 23,270 | |
| 12 | 8 | −21,016 | 9,000 | −146,610 | 23,610 | |
| 13 | 9 | −25,200 | 6,000 | −163,621 | 24,056 | |
| 14 | 10 | −27,600 | 3,000 | −180,909 | 24,580 | =PMT($D$1,A14,D14) |
| 15 | | | | | | |
| 16 | | | =NPV($D$1,$B$5:B14)+$D$4+PV($D$1,A14,0,−C14) | | | |
| 17 | | | =NPV($i$, B column) + year 0 + present value of a future salvage | | | |

**FIGURE 13-7** Spreadsheet for life with minimum after-tax cost.

absolute address and the cell for period $t$ is a relative address. This allows the formula to be copied.

Example 13-11 illustrates the before-tax optimal economic life of a new car.

## EXAMPLE 13–11

A new vehicle costs $19,999 plus $400 in fees. Its value drops 30% the first year, 20% per year for Years 2 through 4, and 15% each additional year. When the car is sold, detailing and advertising will cost $250. Repairs on similar vehicles have averaged $50 annually in lost time (driving to/from the dealer's shop) during the 3-year warranty period. After the warranty period, the cost of repairs and the associated inconvenience climbs at $400 annually. If the MARR is 8%, what is the optimal economic life for the vehicle?

### SOLUTION

Figure 13-8 is a spreadsheet that uses the same functions used in Figure 13-7. The present worth for each life is found by using the NPV function for the irregular costs from Years 1 to $n$, and the PV function for salvage value minus the cost to sell, and adding in the Time 0 costs to buy the vehicle. Then a PMT function is used to find the annual cost.

The optimal economic life is 10 years, but it is not much more expensive for lives that are 3 years shorter or 5 years longer. However, keeping a vehicle for less than 5 years is significantly more expensive.

| | A | B | C | D | E | F |
|---|---|---|---|---|---|---|
| 1 | $19,999 | first cost | | | | |
| 2 | $400 | cost to buy | | | | |
| 3 | $250 | cost to sell | | | | |
| 4 | 30% | salvage value drop yr 1 | | | | |
| 5 | 20% | salvage value drop yr 2–4 | | | | |
| 6 | 15% | salvage value drop yr 5+ | | | | |
| 7 | $50 | repair during 3 yr warranty | | | | |
| 8 | $400 | gradient for repair after 3 yr warranty | | | | |
| 9 | 8% | interest rate | | | | |
| 10 | year | cost | salvage value | PW | EUAC | |
| 11 | 0 | $20,399 | | | | |
| 12 | 1 | 50 | $13,999 | −$7,714 | | =PMT($A$9,A12,D12 |
| 13 | 2 | 50 | 11,199 | −11,101 | 6,225 | |
| 14 | 3 | 50 | 8,960 | −13,614 | 5,283 | |
| 15 | 4 | 450 | 7,168 | −15,774 | 4,762 | |
| 16 | 5 | 850 | 6,092 | −17,461 | 4,373 | |
| 17 | 6 | 1250 | 5,179 | −19,119 | 4,136 | |
| 18 | 7 | 1650 | 4,402 | −20,765 | 3,988 | |
| 19 | 8 | 2050 | 3,742 | −22,409 | 3,899 | |
| 20 | 9 | 2450 | 3,180 | −24,055 | 3,851 | |
| 21 | 10 | 2850 | 2,703 | −25,705 | 3,831 | economic life & min cost |
| 22 | 11 | 3250 | 2,298 | −27,356 | 3,832 | |
| 23 | 12 | 3650 | 1,953 | −29,008 | 3,849 | |
| 24 | 13 | 4050 | 1,660 | −30,655 | 3,879 | |
| 25 | 14 | 4450 | 1,411 | −32,293 | 3,917 | |
| 26 | 15 | 4850 | 1,199 | −33,918 | 3,963 | |
| 27 | | | | | | |
| 28 | =−$B$11−NPV($A$9,$B$12:B26)+PV($A$9,A26,,−C26+$A$3) | | | | | |
| 29 | = year 0 + NPV($i$, B column) + PV(salvage value − sale cost) | | | | | |

**FIGURE 13-8**  Spreadsheet for vehicle's optimal economic life.

Example 13-11 shows that even with complicated cost structures, it is still relatively easy to find the minimum cost EUAC with the power of spreadsheets.

## SUMMARY

The question in selecting new equipment is, Which machine will be more economical? But when there is an existing machine (called the **defender**), the question is, Shall we replace it now, or shall we keep it for one or more years? When replacement is indicated, the best available replacement equipment (called the **challenger**), will be acquired. When we already have equipment, there is a tendency to use past or sunk costs in the replacement analysis. But only present and future costs are relevant.

This chapter has presented three distinctly different **replacement analysis techniques.** All use the simplifying **replacement repeatability assumptions.** These state that the defender will ultimately be replaced by the current best challenger (as will any challengers implemented in the future), and that we have an indefinite need for the asset's service.

In the usual case, marginal cost data are both available and increasing on a year-to-year basis, and **replacement analysis technique 1** compares *the defender's marginal cost data against the challenger's minimum EUAC.* We keep the defender as long as its marginal cost is less than the challenger's minimum EUAC.

When marginal cost data are available for the defender but are not increasing from year to year, **replacement analysis technique 2** compares *the defender's lowest EUAC against the challenger's minimum EUAC.* If the challenger's EUAC is less, select this asset in place of the defender today. If the defender's lowest EUAC is smaller, we do not replace it yet. If the cost data for the challenger and the defender do not change, replace the defender after the life that minimizes its EUAC when its marginal cost data exceed the challenger's minimum EUAC.

In the case of no marginal cost data being available for the defender, **replacement analysis technique 3** compares *the defender's EUAC over its stated life, against the challenger's minimum EUAC.* As in the case of replacement analysis technique 2, select the alternative that has the smallest EUAC. An important concept when calculating the EUAC of both defender and challenger is the first cost to be assigned to each alternative for calculation purposes. When the lives of the two alternatives match, either an **opportunity cost** or a **cash flow approach** may be used. However, in the more common case of different useful lives, only the opportunity cost approach accurately assigns first costs to the defender and challenger.

It is important when performing engineering economic analyses to include the effects of taxes. This is much easier to accomplish with spreadsheets. Spreadsheets also make it easy to compute the optimal economic life of vehicles and equipment—even when this includes complex patterns of declining salvage values, warranty periods, and increasing repair costs.

Replacement analyses are vastly important, yet often ignored by companies as they invest in equipment and facilities. Investments in business and personal assets should not be forgotten once an initial economic evaluation has produced a "buy" recommendation. It is important to continue to evaluate assets over their respective life cycles to ensure that invested monies continue to yield the greatest benefit. Replacement analyses help us to ensure this.

# PROBLEMS

## Replacement Problems

**13-1** Typically there are two alternatives in a replacement analysis. One alternative is to replace the defender now. The other alternative is which one of the following?

A. Keep the defender for its remaining useful life.

B. Keep the defender for another year and then reexamine the situation.

C. Keep the defender until there is an improved challenger that is better than the present challenger.

D. The answer to this question depends on the data available for the defender and challenger as well as the assumptions made regarding the period of needed service and future challengers.

**13-2** The defender's economic life can be found if certain estimates about the defender can be made. Assuming those estimates prove to be exactly correct, one can accurately predict the year when the defender should be replaced, even if nothing is known about the challenger. True or false? Explain.

**13-3** A proposal has been made to replace a large heat exchanger (3 years ago, the initial cost was $85,000) with a new, more efficient unit at a cost of $120,000. The existing heat exchanger is being depreciated by the MACRS method. Its present book value is $20,400, but its scrap value just equals the cost to remove it from the plant. In preparing the before-tax economic analysis, should the $20,400 book value of the old heat exchanger be

A. *added* to the cost of the new exchanger?

B. *subtracted* from the cost of the new exchanger?

C. *ignored* in this before-tax economic analysis?

**13-4** Which one of the following is the proper dollar value of defender equipment to use in replacement analysis?

A. Original cost

B. Present market value

C. Present trade-in value

D. Present book value

E. Present replacement cost, if different from original cost

**13-5** Consider the following data for a defender asset. What is the correct replacement analysis technique to compare this asset against a competing challenger? How is this method used? That is, what comparison is made, and how do we choose?

| Year, n | BTCF in Year n (marginal costs) |
|---------|---------------------------------|
| 1 | −$2000 |
| 2 | −1750 |
| 3 | −1500 |
| 4 | −1250 |
| 5 | −1000 |
| 6 | −1000 |
| 7 | −1000 |
| 8 | −1500 |
| 9 | −2000 |
| 10 | −3000 |

**13-6** Describe an example in a replacement analysis scenario where the replacement is being considered due to

(a) reduced performance of the existing equipment.

(b) altered requirements.

(c) obsolescence of the existing equipment.

(d) risk of catastrophic failure or unplanned replacement of the existing equipment.

**13-7** A pulpwood-forming machine was purchased and installed 8 years ago for $45,000. The declared salvage value was $5000, with a useful life of 10 years. The machine can be replaced with a more efficient model that costs $75,000, including installation. The present machine can be sold on the open market for $14,000. The cost to remove the old machine is $2000. Which are the relevant costs for the old machine?

**13-8** A machine that has been used for one year has a salvage value of $10,000 now, which will drop by $2000 per year. The maintenance costs for the next 4 years are $1250, $1450, $1750, and $2250. Determine the marginal cost to extend service for each of the next 4 years if the MARR is 8%.

**13-9** A drill press was purchased 2 years ago for $40,000. The press can be sold for $15,000 today, or for $12,000, $10,000, $8000, $6000, $4000, or $2000 at the ends of each of the next 6 years. The annual operating and maintenance cost for the next 6 years will be $2700, $2900, $3300, $3700, $4200, and $4700. Determine the marginal cost to extend service for each of the next 6 years if the MARR is 12%. If a new drill press has an EAC of $7000, when should the drill press be replaced?

## Challenger Min EUAC

**13-10** A machine has a first cost of $50,000. Its market value declines by 20% annually. The operating and maintenance costs start at $3500 per year and climb by $2000 per year. The firm's MARR is 9%. Find the minimum EUAC for this machine and its economic life. (*Answer:* $17,240, 4 years)

**13-11** A machine has a first cost of $10,000. Its market value declines by 20% annually. The repair costs are covered by the warranty in Year 1, and then they increase $600 per year. The firm's MARR is 15%. Find the minimum EUAC for this machine and its economic life.

**13-12** A vehicle has a first cost of $20,000. Its market value declines by 15% annually. It is used by a firm that estimates the effect of older vehicles on the firm's image. A new car has no "image cost." But the image cost of older vehicles climbs by $700 per year. The firm's MARR is 10%. Find the minimum EUAC for this vehicle and its economic life.

**13-13** The Clap Chemical Company needs a large insulated stainless steel tank to expand its plant. Clap has located such a tank at a recently closed brewery. The brewery has offered to sell the tank for $15,000 delivered. The price is so low that Clap believes it can sell the tank at any future time and recover its $15,000 investment.

The outside of the tank is covered with heavy insulation that requires considerable maintenance.

| Year | Insulation Maintenance Cost |
|------|------------------------------|
| 0 | $2000 |
| 1 | 500 |
| 2 | 1000 |
| 3 | 1500 |
| 4 | 2000 |
| 5 | 2500 |

(*a*) Based on a 15% before-tax MARR, what life of the insulated tank has the lowest EUAC?

(*b*) Is it likely that the insulated tank will be replaced by another tank at the end of the period with the lowest EUAC? Explain.

**13-14** The plant manager may purchase a piece of unusual machinery for $10,000. Its resale value after 1 year is estimated to be $3000. Because the device is sought by antique collectors, resale value is rising $500 per year.

The maintenance cost is $300 per year for each of the first 3 years, and then it is expected to double each year. Thus the fourth-year maintenance will be $600; the fifth-year maintenance, $1200, and so on. Based on a 15% before-tax MARR, what life of this machinery has the lowest EUAC?

**13-15** Demonstrate how one would calculate the economic life of a truck costing $30,000 initially, and at the end of this and each following year ($y$) costing $OMR_y$ in operation, maintenance and repair costs. The truck is depreciated using the straight-line method over 5 years (i.e., $30,000/5 = D_y$). Its salvage value each year equals its book value. Develop an expression to show how to determine the truck's economic life—that is, the year when the truck's uniform equivalent annual cost is a minimum.
*Contributed by D. P. Loucks, Cornell University*

**13-16** Your firm is in need of a machine that costs $5000. During the first year the maintenance costs are estimated to be $400. The maintenance costs are expected to increase by $150 each year up to a total of $1750 at the end of Year 10. The machine can be depreciated over 5 years using the straight-line method. Assume each year's depreciation is a known amount $D_y$. There is no salvage value. Develop an expression to show how you would find the most economical useful life of this machine on a before-tax basis.
*Contributed by D. P. Loucks, Cornell University*

**13-17** A $40,000 machine will be purchased by a company whose interest rate is 12%. The installation cost is $5K, and removal costs are insignificant. What is its economic life if its salvage values and O&M costs are as follows?

| Year | 1 | 2 | 3 | 4 | 5 |
|------|------|------|------|------|------|
| S | $35K | $30K | $25K | $20K | $15K |
| O&M | $8K | $14K | $20K | $26K | $32K |

**13-18** Mytown's street department repaves a street every 8 years. Potholes cost $12,000 per mile beginning at the end of Year 3 after construction or repaving. The cost to fix potholes generally increases by $12,000 each year. Repaving costs are $180,000 per mile. Mytown uses an interest rate of 6%. What is the EAC for Mytown's policy? What is the EAC for the optimal policy? What is the optimal policy?

**13-19** A 2000-pound, counterbalanced, propane forklift can be purchased for $30,000. Due to the intended service use, the forklift's market value drops 20% of its prior year's value in Years 1 and 2 and then declines by 15% until Year 10 when it will have a scrap/market value of $1000. Maintenance of the forklift is $400 per year during Years 1 and 2 while the warranty is in place. In Year 3 it jumps to $750 and increases $200 per year thereafter. What is the optimal life of the forklift?

13-20 A 2000-pound, counterbalanced, electric forklift can be purchased for $25,000 plus $3000 for the charger and $3000 for a battery. The forklift's market value is 10% less for each of its first 6 years of service. After this period the market value declines at the rate of 7.5% for the next 6 years.

The battery has a life of 4 years and a salvage value of $300. The charger has a 12-year life and a $100 salvage value. The charger's market value declines 20% per year of use. The battery's market value declines by 30% of its purchase price each year. Maintenance of the charger and battery are minimal. The battery will most likely not work with a replacement forklift.

Maintenance of the forklift is $200 per year during Years 1 and 2 while the warranty is in place. In Year 3 it jumps to $600 and increases $50 per year thereafter. What is the optimal ownership policy?

13-21 J&E Fine Wines recently purchased a new grape press for $150,000. The annual operating and maintenance costs for the press are estimated to be $7500 the first year. These costs are expected to increase by $2200 each year after the first. The salvage value is expected to decrease by $25,000 each year to a value of zero. Using an interest rate of 8%, determine the economic life of the press.

13-22 A chemical process in your plant leaves scale deposits on the inside of pipes. The scale cannot be removed, but increasing the pumping pressure maintains flow through the narrower diameter. The pipe costs $25 per foot to install, and it has no salvage value when it is removed. The pumping costs are $8.50 per foot of pipe initially, and they increase annually by $6 per year starting in Year 2. What is the economic life of the pipe if the interest rate is 15%?

13-23 An electric oil pump's first cost is $45,000, and the interest rate is 10%. The pump's end-of-year salvage values over the next 5 years are $42K, $40K, $38K, $32K, and $26K. Determine the pump's economic life. (*Answer:* 3 years)

13-24 A $20,000 machine will be purchased by a company whose interest rate is 10%. It will cost $5000 to install, but its removal costs are insignificant. What is its economic life if its salvage values and O&M costs are as follows?

| Year | 1 | 2 | 3 | 4 | 5 |
|------|------|------|------|------|------|
| S | $16K | $13K | $11K | $10K | $9.5K |
| O&M | $5K | $8K | $11K | $14K | $17 |

(*Answer:* 3 years)

13-25 A $50,000 machine will be purchased by a company whose interest rate is 15%. The installation cost is $8K, and removal costs are insignificant. What is its economic life if its salvage values and O&M costs are as follows?

| Year | 1 | 2 | 3 | 4 | 5 |
|------|------|------|------|------|------|
| S | $35K | $30K | $25K | $20K | $15K |
| O&M | $8K | $14K | $20K | $26K | $32K |

**Replacement Technique 1**

13-26 A machine tool, which has been used in a plant for 10 years, is being considered for replacement. It cost $9500 and was depreciated by MACRS depreciation using a 5-year recovery period. An equipment dealer indicates that the machine has no resale value. Maintenance on the machine tool has been a problem, with an $800 cost this year. Future annual maintenance costs are expected to be higher. What is the economic life of this machine tool if it is kept in service?

13-27 In a replacement analysis problem, the following facts are known:

| Initial cost | $12,000 |
|---|---|
| Annual maintenance | Year |
| None | 1–3 |
| $2000 | 4–5 |
| $4500 | 6 |
| $7000+ | 7 |
| $2500/yr | |

Salvage value in any year is zero. Assume a 10% interest rate and ignore income taxes. Compute the life for this challenger having the lowest EUAC. (*Answer:* 5 years)

13-28 An injection-molding machine has a first cost of $1,050,000 and a salvage value of $225,000 in any year. The maintenance and operating cost is $235,000 with an annual gradient of $75,000. The MARR is 10%. What is the most economic life?

13-29 Five years ago, Thomas Martin installed production machinery that had a first cost of $25,000. At that time initial yearly costs were estimated at $1250, increasing by $500 each year. The market value of this machinery each year would be 90% of the previous year's value. There is a new machine available now that has a first cost of $27,900 and no yearly costs over its 5-year minimum cost life. If Thomas Martin uses an 8% before-tax MARR, when, if at all, should he replace the existing machinery with the new unit?

**13-30** Consider Problem 13-29 involving Thomas Martin. When, if at all, should the old machinery be replaced with the new, given the following changes in the data. The old machine retains only 70% of its value in the market from year to year. The yearly costs of the old machine were $3000 in Year 1 and increase at 10% thereafter.

**13-31** Mary O'Leary's company ships fine wool garments from County Cork, Ireland. Five years ago she purchased some new automated packing equipment having a first cost of $125,000 and a MACRS class life of 7 years. The annual costs for operating, maintenance, and insurance, as well as market value data for each year of the equipment's 10-year useful life are as follows.

| Year | Annual Costs in Year $n$ for | | | Market Value |
| $n$ | Operating | Maintenance | Insurance | in Year $n$ |
|---|---|---|---|---|
| 1 | $16,000 | $ 5,000 | $17,000 | $80,000 |
| 2 | 20,000 | 10,000 | 16,000 | 78,000 |
| 3 | 24,000 | 15,000 | 15,000 | 76,000 |
| 4 | 28,000 | 20,000 | 14,000 | 74,000 |
| 5 | 32,000 | 25,000 | 12,000 | 72,000 |
| 6 | 36,000 | 30,000 | 11,000 | 70,000 |
| 7 | 40,000 | 35,000 | 10,000 | 68,000 |
| 8 | 44,000 | 40,000 | 10,000 | 66,000 |
| 9 | 48,000 | 45,000 | 10,000 | 64,000 |
| 10 | 52,000 | 50,000 | 10,000 | 62,000 |

Now Mary is looking at the remaining 5 years of her investment in this equipment, which she had initially evaluated on the basis of an after-tax MARR of 25% and a tax rate of 35%. Assume that the replacement repeatability assumptions are valid.

(a) What is the before-tax marginal cost for the remaining 5 years?

(b) When, if at all, should Mary replace this packing equipment if a new challenger, with a minimum EUAC of $110,000, has been identified?

**13-32** SHOJ Enterprises has asked you to look at the following data. The interest rate is 10%.

| Year, $n$ | Marginal Cost Data Defender | EUAC if Kept $n$ Years Challenger |
|---|---|---|
| 1 | $3000 | $4500 |
| 2 | 3150 | 4000 |
| 3 | 3400 | 3300 |
| 4 | 3800 | 4100 |
| 5 | 4250 | 4400 |
| 6 | 4950 | 6000 |

(a) What is the *defender's* lowest EUAC?

(b) What is the *challenger's* economic life?

(c) When, if at all, should we replace the *defender* with the *challenger*?

**13-33** Bill's father read that each year a car's value declines by 25%. After a car is 3 years old, the rate of decline falls to 15%. Maintenance and operating costs increase as the car ages. Because of the manufacturer's warranty, first-year maintenance is very low.

| Age of Car (years) | Maintenance Expense |
|---|---|
| 1 | $ 50 |
| 2 | 150 |
| 3 | 180 |
| 4 | 200 |
| 5 | 300 |
| 6 | 390 |
| 7 | 500 |

Bill's dad wants to keep his annual cost of car ownership low. The car Bill's dad prefers costs $11,200 new. Should he buy a new or a used car and, if used, when would you suggest he buy it, and how long should it be kept? Give a practical, rather than a theoretical, solution. (*Answer:* Buy a 3-year-old car and keep it 3 years.)

**13-34** Should NewTech's computer system be replaced this year? The system has a salvage value now of $5000, which will fall to $4000 by the end of the year. The cost of lower productivity linked to the older computer is $3000 this year. NewTech uses an interest rate of 15%. What is the cost advantage of the best system? A potential new system costs $12,000 and has the following salvage values and lost productivity for each year.

| Year | S | Lost Productivity |
|------|--------|-------------------|
| 0 | $12,000 | |
| 1 | 9,000 | $ 0 |
| 2 | 7,000 | 1000 |
| 3 | 5,000 | 2000 |
| 4 | 3,000 | 3000 |

**13-35** In evaluating projects, LeadTech's engineers use a rate of 15%. One year ago a robotic transfer machine was installed at a cost of $38,000. At the time, a 10-year life was estimated, but the machine has had a downtime rate of 28% which is unacceptably high. A $12,000 upgrade should fix the problem, or a labor-intensive process costing $3500 in direct labor per year can be substituted. The plant estimates indirect plant expenses at 60% of direct labor, and it allocates front office overhead at 40% of plant expenses (direct and indirect). The robot has a value in other uses of $15,000. What is the difference between the EACs for upgrading and switching to the labor-intensive process?

## Replacement Technique 2

**13-36** A new $40,000 bottling machine has just been installed in a plant. It will have no salvage value when it is removed. The plant manager has asked you to estimate the machine's economic service life, ignoring income taxes. He estimates that the annual maintenance cost will be constant at $2500 per year. What service life will result in the lowest equivalent uniform annual cost?

**13-37** Big-J Construction Company, Inc. (Big-J CC) is conducting a routine periodic review of existing field equipment. They use a MAAR of 20%. This includes a replacement evaluation of a paving machine now in use. The machine was purchased 3 years ago for $200,000, The paver's current market value is $120,000, and yearly operating and maintenance costs are as follows.

| Year, n | Operating Cost in Year n | Maintenance Cost in Year n | Market Value if Sold in Year n |
|---------|--------------------------|----------------------------|--------------------------------|
| 1 | $15,000 | $ 9,000 | $85,000 |
| 2 | 15,000 | 10,000 | 65,000 |
| 3 | 17,000 | 12,000 | 50,000 |
| 4 | 20,000 | 18,000 | 40,000 |
| 5 | 25,000 | 20,000 | 35,000 |
| 6 | 30,000 | 25,000 | 30,000 |
| 7 | 35,000 | 30,000 | 25,000 |

Data for a new paving machine have been analyzed. Its most economic life is at 8 years, with a minimum EUAC of $62,000. When should the existing paving machine be replaced?

**13-38** VMIC Corp. has asked you to look at the following data. The interest rate is 10%.

| Year, n | Marginal Cost Data Defender | EUAC If Kept n Years Challenger |
|---------|------------------------------|----------------------------------|
| 1 | $2500 | $4500 |
| 2 | 2400 | 3600 |
| 3 | 2300 | 3000 |
| 4 | 2550 | 2600 |
| 5 | 2900 | 2700 |
| 6 | 3400 | 3500 |
| 7 | 4000 | 4000 |

(a) What is the *defender's* lowest EUAC?

(b) What is the *challenger's* minimum cost life?

(c) When, if at all, should we replace the *defender* with the *challenger*?

**13-39** Eight years ago, the Blank Block Building Company installed an automated conveyor system for $38,000. When the conveyor is replaced, the net cost of removal will be $2500. The minimum EAC of a new conveyor is $5500. When should the conveyor be replaced if BBB's MARR is 12%? The O&M costs for the next 5 years are $5K, $6K, $7K, $8K, and $9K.

## Replacement Technique 3

**13-40** You are considering the purchase of a new high-efficiency machine to replace older machines now. The new machine can replace four of the older machines, each with a current market value of $600. The new machine will cost $5000 and will save the equivalent of 10,000 kWh of electricity per year. After a period of 10 years, neither option (new or old) will have any market value. If you use a before-tax MARR of 25% and pay $0.075 per kilowatt-hour, would you replace the old machines today with the new one?

**13-41** The Quick Manufacturing Company, a large profitable corporation, may replace a production machine tool. A new machine would cost $3700, have a 4-year useful and depreciable life, and have no salvage value. For tax purposes, sum-of-years'-digits depreciation would be used. The existing machine tool cost $4000 4 years ago and has been depreciated by straight-line depreciation assuming an 8-year life and no salvage value. The tool could be sold now to a used equipment dealer for $1000 or be kept in service for another 4 years. It would then have no salvage value. The new machine tool would save about $900 per year in operating costs compared to the existing machine.

Assume a 40% combined state and federal tax rate. Compute the before-tax rate of return on the replacement proposal of installing the new machine rather than keeping the existing machine.    (*Answer:* 12.6%)

**13-42** A professor of engineering economics owns a 1996 car. In the past 12 months, he has paid $2000 to replace the transmission, bought two new tires for $160, and installed a CD player for $110. He wants to keep the car for 2 more years because he invested money 3 years ago in a 5-year certificate of deposit, which is earmarked to pay for his dream machine, a red European sports car. Today the old car's engine failed. The professor has two alternatives. He can have the engine overhauled at a cost of $1800 and then most likely have to pay another $800 per year for the next 2 years for maintenance. The car will have no salvage value at that time. Alternatively, a colleague offered to make the professor a $5000 loan to buy another used car. He must pay the loan back in two equal installments of $2500 due at the end of Year 1 and Year 2, and at the end of the second year he must give the colleague the car. The "new" used car has an expected annual maintenance cost of $300. If the professor selects this alternative, he can sell his current vehicle to a junkyard for $1500. Interest is 5%. Using present worth analysis, which alternative should he select and why?

**13-43** The Plant Department of the local telephone company purchased four special pole hole diggers 8 years ago for $14,000 each. They have been in constant use to the present. Owing to an increased workload, additional machines will soon be required.

Recently an improved model of the digger was announced. The new machines have a higher production rate and lower maintenance expense than the old machines but will cost $32,000 each with a service life of 8 years and salvage value of $750 each. The four original diggers have an immediate salvage of $2000 each and an estimated salvage value of $500 each 8 years hence. The average annual maintenance expense of the old machines is

about $1500 each, compared with $600 each for the new machines.

A field study and trial show that the workload would require three additional new machines if the old machines continue in service. However, if the old machines were all retired from service, the workload could be carried by six new machines with an annual savings of $12,000 in operation costs. A training program to prepare employees to run the machines will be necessary at an estimated cost of $700 per new machine. If the MARR is 9% before taxes, what should the company do?

**13-44** JMJ Inc. bought a manufacturing line 5 years ago for $35,000,000. At that time it was estimated to have a service life of 10 years and salvage value at the end of its service life of $10,000,000. JMJ's CFO recently proposed to replace the old line with a modern line expected to last 15 years and cost $95,000,000. This line will provide $5,000,000 savings in annual operating and maintenance costs, increase revenues by $2,000,000, and have a $15,000,000 salvage value (after 15 years). The seller of the new line is willing to accept the old line as a trade-in for its current fair market value, which is $12,000,000. The CFO estimates that if the old line is kept for 5 more years, its salvage value will be $6,000,000. If the JMJ's MARR is 8% per year, should the company keep the old line or replace it with the new line?
*Contributed by Hamed Kashani, Saeid Sadri, and Baabak Ashuri, Georgia Institute of Technology*

**13-45** A used car can be kept for two more years and then sold for an estimated value of $3000, or it can be sold now for $7500. The average annual maintenance cost over the past 7 years has been $500 per year. However, if the car is kept for two more years, this cost is expected to be $1800 the first year and $2000 the second year. As an alternative, a new car can be purchased for $22,000 and be used for 4 years, after which it will be sold for $8,000. The new car will be under warranty the first 4 years, and no extra maintenance cost will be incurred during those years. If the MARR is 15% per year, what is the best option?
*Contributed by Hamed Kashani, Saeid Sadri, and Baabak Ashuri, Georgia Institute of Technology*

**13-46** A couple bought their house 10 years ago for $165,000. At the time of purchase, they made a $35,000 down payment, and the balance was

financed by a 30-year mortgage with monthly payments of $988.35. They expect to live in this house for 20 years, after which time they plan to sell the house and move to another state. Alternatively, they can sell the house now and live in a rental unit for the next 20 years. The house can be sold now for $210,000, from which an 8% real estate commission and $110,000 remaining loan balance and miscellaneous expenses will be deducted. If they stay in the house, the house can be sold after 20 years for $320,000 from which a 10% real estate commission and $10,000 miscellaneous expenses will be deducted. A comparable rental unit rents for $960 payable at the beginning of every month. No security deposit will be required of them to rent the unit, and the rent will not increase if they maintain a good payment record. They use an interest rate of 0.5% per month for analyzing this financial opportunity. Should they stay in the house or should they sell it and move into a rental unit?
*Contributed by Hamed Kashani, Saeid Sadri, and Baabak Ashuri, Georgia Institute of Technology*

**13-47** Sacramento Cab Company owns several taxis that were purchased for $25,000 each 4 years ago. The cabs' current market value is $12,000 each, and if they are kept for another 6 years they can be sold for $2000 per cab. The annual maintenance cost per cab is $1000 per year. Sacramento Cab has been approached about a leasing plan that would replace the cabs. The leasing plan calls for payments of $6000 per year. The annual maintenance cost for each leased cab is $750 per year. Should the cabs be replaced if the interest rate is 10%?

**After-Tax**

**13-48** The Ajax Corporation purchased a railroad tank car 8 years ago for $60,000. It is being depreciated by SOYD depreciation, assuming a 10-year depreciable life and a $7000 salvage value. The tank car needs to be reconditioned now at a cost of $35,000. If this is done, it is estimated the equipment will last for 10 more years and have a $10,000 salvage value at the end of the 10 years.

On the other hand, the existing tank car could be sold now for $10,000 and a new tank car purchased for $85,000. The new tank car would be depreciated by MACRS depreciation. Its estimated actual salvage value after 10 years would be

$15,000. In addition, the new tank car would save $7000 per year in maintenance costs, compared to the reconditioned tank car.

Based on a 15% before-tax rate of return, determine whether the existing tank car should be reconditioned or a new one purchased. *Note:* The problem statement provides more data than are needed, which is typical of real situations. (*Answer:* Recondition the old tank car.)

**13-49** State the advantages and disadvantages with respect to after-tax benefits of the following options for a major equipment unit:

A. Buy new.

B. Trade in and buy a similar, rebuilt equipment from the manufacturer.

C. Have the manufacturer rebuild your equipment with all new available options.

D. Have the manufacturer rebuild your equipment to the original specifications.

E. Buy used equipment.

**13-50** Fifteen years ago the Acme Manufacturing Company bought a propane-powered forklift truck for $4800. The company depreciated the forklift using straight-line depreciation, a 12-year life, and zero salvage value. Over the years, the forklift has been a good piece of equipment, but lately the maintenance cost has risen sharply. Estimated end-of-year maintenance costs for the next 10 years are as follows:

| Year | Maintenance Cost |
|------|------------------|
| 1 | $ 400 |
| 2 | 600 |
| 3 | 800 |
| 4 | 1000 |
| 5–10 | 1400/year |

The old forklift has no present or future net salvage value, since its scrap metal value just equals the cost to haul it away. A replacement is now being considered for the old forklift. A modern unit can be purchased for $6500. It has an economic life equal to its 10-year depreciable life. Straight-line depreciation will be employed, with zero salvage value at the end of the 10-year depreciable life. At any time the new forklift can be sold for its book value. Maintenance on the new forklift is estimated to be a constant $50 per year for the next 10 years, after which maintenance is expected to increase sharply. Should Acme Manufacturing keep its old forklift truck for the present, or replace it now with a new one? The firm expects an 8% after-tax rate of return on its investments. Assume a 40% combined state-and-federal tax rate. (*Answer:* Keep the old forklift truck.)

**13-51** Machine *A* has been completely overhauled for $9000 and is expected to last another 12 years. The $9000 was treated as an expense for tax purposes last year. Machine *A* can be sold now for $30,000 net after selling expenses, but will have no salvage value 12 years hence. It was bought new 9 years ago for $54,000 and has been depreciated since then by straight-line depreciation using a 12-year depreciable life.

Because less output is now required, Machine *A* can be replaced with a smaller machine: Machine *B* costs $42,000, has an anticipated life of 12 years, and would reduce operating costs $2500 per year. It would be depreciated by straight-line depreciation with a 12-year depreciable life and no salvage value.

The income tax rate is 40%. Compare the after-tax annual costs and decide whether Machine *A* should be retained or replaced by Machine *B*. Use a 10% after-tax rate of return.

**13-52** (*a*) A new employee at CLL Engineering Consulting Inc., you are asked to join a team performing an economic analysis for a client. Your task is to find the Time-0 ATCFs. CLL Inc. has a combined federal/state tax rate of 45% on ordinary income, depreciation recapture, and losses.

*Defender:* This asset was placed in service 7 years ago. At that time the $50,000 cost basis was set up on a straight-line depreciation schedule with an estimated salvage value of $15,000 over its 10-year ADR life. This asset has a present market value of $30,000.

*Challenger:* The new asset has a first cost of $85,000 and will be depreciated by MACRS depreciation over its 10-year class life. This asset qualifies for a 10% investment tax credit.

(*b*) How would your calculations change if the present market value of the *defender* is $25,500?

(*c*) How would your calculations change if the present market value of the *defender* is $18,000?

**13-53** Foghorn Leghorn may replace an old egg-sorting machine used by his business, Foggy's Farm Fresh Eggs. The old egg machine is not quite running eggs-actly the way it was originally designed and will require an additional investment now of $2500 (expensed at Time 0) to get it back in working shape. This old machine was purchased 6 years ago for $5000 and has been depreciated by the straight-line method at $500 per year. Six years ago the estimated salvage value for tax purposes was $1000. Operating expenses for the old machine are projected at $600 this next year and are increasing by $150 per year each year thereafter. Foggy projects that with refurbishing, the machine will last another 3 years. Foggy believes that he could sell the old machine as-is today for $1000 to his friend Fido to sort bones. He also believes he could sell it 3 years from now at the barnyard flea market for $500.

The new egg-sorting machine, a deluxe model, has a purchase price of $10,000 and will last 6 years, at which time it will have a salvage value of $1000. The new machine qualifies as a MACRS 7-year property and will have operating expenses of $100 the first year, increasing by $50 per year thereafter. Foghorn uses an after-tax MARR of 18% and a tax rate of 35% on original income.

(*a*) What was the depreciation life used with the defender asset (the old egg sorter)?

(*b*) Calculate the after-tax cash flows for both the defender and challenger assets.

(*c*) Use the annual cash flow method to offer a recommendation to Foggy. What assumptions did you make in this analysis?

**13-54** A firm is concerned about the condition of some of its plant machinery. Bill James, a newly hired engineer, reviewed the situation and identified five feasible, mutually exclusive alternatives.

*Alternative A:* Spend $44,000 now repairing various items. The $44,000 can be charged as a current operating expense (rather than capitalized) and deducted from other taxable income immediately. These repairs will keep the plant functioning for 7 years with current operating costs.

*Alternative B:* Spend $49,000 to buy general-purpose equipment. Depreciation would be straight line over the equipment's 7-year useful life. The equipment has no salvage value. The new equipment will reduce annual operating costs by $6000.

*Alternative C:* Spend $56,000 to buy new specialized equipment. This equipment would be depreciated by sum-of-years'-digits depreciation over its 7-year useful life. This equipment would reduce annual operating costs by $12,000. It will have no salvage value.

*Alternative D:* This alternative is the same as Alternative *B*, except it reduces annual operating costs by $7000.

*Alternative E:* This is the "do nothing" alternative, with annual operating costs $8000 above the present level.

This profitable firm pays 40% corporate income taxes and uses a 10% after-tax rate of return. Which alternative should the firm adopt?

13-55 Fred's Rodent Control Corporation has been using a low-frequency sonar device to locate subterranean pests. This device was purchased 5 years ago for $18,000. The device has been depreciated using SOYD depreciation with an 8-year depreciable life and a salvage value of $3600. Presently, it could be sold for $7000. If it is kept for the next 3 years, its market value is expected to drop to $1600.

A new lightweight subsurface heat-sensing searcher (SHSS) that is available for $10,000 would improve the annual net income by $500 for each of the next 3 years. The SHSS would be depreciated as a 5-year class property, using MACRS. At the end of 3 years, the SHSS should have a market value of $4000. Fred's Rodent Control is a profitable enterprise subject to a 40% tax rate.

(*a*) Construct the after-tax cash flow for the old sonar unit for the next 3 years.

(*b*) Construct the after-tax cash flow for the SHSS unit for the next 3 years.

(*c*) Construct the after-tax cash flow for the difference between the SHSS unit and the old sonar unit for the next 3 years.

(*d*) Should Fred buy the new SHSS unit if his MARR is 20%? You do not have to calculate the incremental rate of return; just show how you reach your decision.

13-56 Compute the after-tax rate of return on the replacement proposal for Problem 13-23.

13-57 BC Junction purchased some embroidering equipment for their Denver facility 3 years ago for $15,000. This equipment qualified as MACRS 5-year property. Maintenance costs are estimated to be $1000 this next year and will increase by $1000 per year thereafter. The market (salvage) value for the equipment is $10,000 at the end of this year and declines by $1000 per year in the future. If BC Junction has an after-tax MARR of 30%, a marginal tax rate of 45% on ordinary income, depreciation recapture, and losses, what after-tax life of this previously purchased equipment has the lowest EUAC? Use a spreadsheet to develop your solution.

13-58 Reconsider the acquisition of packing equipment for Mary O'Leary's business, as described in Problem 13-31. Given the data tabulated there, and again using an after-tax MARR of 25% and a tax rate of 35% on ordinary income to evaluate the investment, determine the after-tax lowest EUAC of the equipment. Use a spreadsheet to develop your solution.

# CASES

The following cases from *Cases in Engineering Economy 2^{nd}* are suggested as matched with this chapter.

CASE 27    **To Use or Not to Use?**

Focus is treatment of sunk costs. More complicated than most. Some discoveries in the data gathering process. Solution uses equation rather than cash flow table.

CASE 31    **Freeflight Superdiscs**

Inflation and sensitivity analysis for three alternatives. Includes taxes.

CASE 32    **Mr. Speedy**

Includes two memos using different inappropriate financial comparisons. Choose optimal life for replacement of vehicles.

**CASE 33   Piping Plus**

Data from case intro, three memos, income statement, and balance sheet. Computer improvement in a professional services firm. Assumptions will lead to an instructive variety of results.

**CASE 34   R&D Device at EBP**

Equipment replacement cost comparison with unequal lives. Continuing demand requires careful analysis statement by the student because of detailed cost data. Before or after taxes.

# INFLATION AND PRICE CHANGE

## The Athabasca Oil Sands

For centuries, people have known about the sticky bitumen that lines the banks of the Athabasca River in northern Alberta. Even before the coming of the European fur traders, Native people used it to seal canoes. Over the years, many people dreamed of producing usable oil from the bitumen, but the sand and oil are not easily separated, and the recovery was not viewed as economically viable. In the 1950s, the world price of oil was around $3 a barrel and the estimated cost of mining and separating oil from the sands was over $30 a barrel.

In 1964 the Sun Oil Company, with government support, formed the Great Canadian Oil Sands Company, and in 1967 it started to mine and process shallow deposits of oil sands. The target production was 31,000 barrels per day, and the initial production cost would be about $25 a barrel. The world price was then about $3.50, but the planners were predicting that production costs would decline and market prices would increase.

Forty years later, in 2004, two major oil firms together supply one-third of Canada's oil from these bitumen deposits. Other firms extract bitumen from deeper deposits by steam heating, and new oil sands projects worth $40 billion are on the books. The successor to Great Canadian Oil Sands, Suncor Energy, had 4500 employees and was producing 130,000 barrels of oil a day. The production costs were in the range of $12 a barrel, and the world price of oil was nearly $50.

In 2004 the price of oil rose above $40, then $50. A series of world events led the price to climb to the $70 region in 2006 and to break the $100 mark in early 2008. Innovative ways of extracting the oil from the sand, such as in-situ steam-assisted gravity drainage (SAGD), were developed, and by 2007 there were new oil sands projects worth $100 billion.

The research challenge now is to develop more environmentally friendly methods of extraction. In 2008 Syncrude announced that through technology it had developed, about

80% of the water it uses is recycled, and that the amount of water it uses per cubic meter of oil is less than half the industry average. ■ ■ ■

*Contributed by Peter Flynn and John Whittaker, University of Alberta*

## QUESTIONS TO CONSIDER

1. In the 1960s, market analysts estimated that at a world price of $25 per barrel, oil sands projects could provide about a 10% rate of return. How might inflation have affected these estimates?

2. The prices of some items—for example, gas at the pumps, houses—increase over time while others, such as calculators and computers, decline in price. Given these variations, how can we know if inflation is occurring, and how could we measure it?

3. In 1967 the Canadian consumer price index was 21.5; in 2004, the CPI was 123.2. What was the 2004 production cost in 1967 dollars?

4. Use the Internet and other resources to learn more about the mining of Athabasca oil sands. Are there any environmental and ethical dimensions to this story?

### After Completing This Chapter...

#### The student should be able to:

- Describe inflation, explain how it happens, and list its effects on purchasing power.
- Define real and actual dollars and interest rates.
- Conduct constant dollar and nominal dollar analyses.
- Define and use composite and commodity-specific price indexes.
- Develop and use cash flows that inflate at different interest rates and cash flows subject to different interest rates per period.
- Incorporate the effects of inflation in before-tax and after-tax calculations.
- Develop spreadsheets to incorporate the effects of inflation and price change.

Thus far we have assumed that the dollars in our analyses were unaffected by inflation or price change. However, this assumption is not always valid or realistic. If inflation occurs in the general economy, or if there are price changes in economic costs and benefits, the impact can be substantial on both before- and after-tax analyses. In this chapter we develop several key concepts and illustrate how inflation and price changes may be incorporated.

## MEANING AND EFFECT OF INFLATION

Inflation is an important concept because the purchasing power of money used in most world economies rarely stays constant. Rather, over time the amount of goods and services that can be bought with a fixed amount of money tends to change. Inflation causes money to lose **purchasing power.** That is, when prices inflate we can buy less with the same amount of money. **Inflation makes future dollars less valuable than present dollars.** Think about examples in your own life, or for an even starker comparison, ask your grandparents how much a loaf of bread or a new car cost 50 years ago. Then compare those prices with what you would pay today for the same items. This exercise will reveal the effect of inflation: as time passes, goods and services cost more, and more monetary units are needed to buy the same goods and services.

Because of inflation, dollars in one period of time are not equivalent to dollars in another. We know that engineering economic analysis requires that comparisons be made on an **equivalent** basis. So, it is important for us to be able to incorporate the effects of inflation.

When the purchasing power of a monetary unit *increases* rather than decreases as time passes, the result is **deflation.** Deflation, very rare in the modern world, nonetheless can exist. Deflation has the opposite effect of inflation—one can buy **more** with money in future years than can be bought today. Thus, deflation makes future dollars more valuable than current dollars.

### How Does Inflation Happen?

Economists generally believe that inflation depends on the following, either in isolation or in combination.

*Money supply:* The amount of money in our national economy has an effect on its purchasing power. If there is too much money in the system (the Federal Reserve controls the flow of money) versus goods and services to purchase with that money, the value of dollars tends to decrease. When there are fewer dollars in the system, they become more valuable. The Federal Reserve seeks to increase the volume of money in the system at the same rate that the economy is growing.

*Exchange rates:* The strength of the dollar in world markets affects the profitability of international companies. Prices may be adjusted to compensate for the dollar's relative strength or weakness in the world market. As corporations' profits are weakened or eliminated in some markets owing to fluctuations in exchange rates, prices may be raised in other markets to compensate.

*Cost–push:* This cause of inflation develops as producers of goods and services "push" their increasing operating costs along to the customer through higher prices. These operating costs include fabrication/manufacturing, marketing, and sales.

*Demand–pull:* This cause is realized when consumers spend money freely on goods and services. As more and more people demand certain goods and services, the prices of those goods and services will rise (demand exceeding supply).

A further consideration in analyzing how inflation works is the usually different rates at which prices and wages rise. Do workers benefit if, as their wages increase, the prices of goods and services increase? To determine the net effect of differing rates of inflation, we must be able to make comparisons and understand costs and benefits from an equivalent perspective. In this chapter we will learn how to make such comparisons.

## Definitions for Considering Inflation in Engineering Economy

The following definitions are used throughout this chapter to illustrate how inflation and price change affect two quantities: interest rates and cash flows.

*Inflation rate (f):* The inflation rate captures the effect of goods and services costing more—a decrease in the purchasing power of dollars. More money is required to buy a good or service whose price has inflated. The inflation rate is measured as the annual rate of increase in the number of dollars needed to pay for the same amount of goods and services.

*Real interest rate (i′):* This interest rate measures the "real" growth of money excluding the effect of inflation. Because it does not include inflation, it is sometimes called the *inflation-free interest rate.*

*Market interest rate (i):* This is the rate of interest that one obtains in the general marketplace. For instance, the interest rates on passbook savings, checking plus, and certificates of deposit quoted by banks are all market rates. The lending interest rate for autos and boats is also a market rate. This rate is sometimes called the *combined interest* rate because it incorporates the effect of both real money growth **and** inflation. We can view *i* as follows:

| Market interest rate | has in it | "Real" growth of money | and | Effect of inflation |
|---|---|---|---|---|

The mathematical relationship between the inflation, real and market interest rates is given as

$$i = i' + f + (i')(f) \tag{14-1}$$

This is the first point where we have defined a real interest rate and a market or combined interest rate. This naturally leads to the question of what meaning should be attached to the interest rate *i*, which is found throughout the text. In fact, both meanings have been used. In problems about savings accounts and loans, the interest rate is usually a market rate. In problems about engineering projects where costs and benefits are often estimated as $*x* per year, the interest rate is a real rate.

## EXAMPLE 14–1

Suppose a professional golfer wants to invest some recent golf winnings in her hometown bank for one year. Currently, the bank is paying a rate of 5.5% *compounded annually*. Assume inflation is expected to be 2% per year. Identify $i$, $f$, and $i'$. Repeat for inflation of 8% per year.

## SOLUTION

### If Inflation Is 2% per Year

The bank is paying a *market rate* ($i$). The *inflation rate* ($f$) is given. What then is the *real interest rate* ($i'$)?

$$i = 5.5\%, \qquad f = 2\%, \qquad i' = ?$$

Solving for $i'$ in Equation 14-1, we have

$$i = i' + f + (i')(f)$$
$$i - f = i'(1 + f)$$
$$i' = (i - f)/(1 + f)$$
$$= (0.055 - 0.02)/(1 + 0.02) = 0.034 \quad \text{or} \quad \textbf{3.4\% per year}$$

This means that the golfer will have 3.4% **more** purchasing power than she had a year ago. At the end of the year she can buy 3.4% more goods and services than she could have at the beginning of the year. For example, assume she was buying golf balls that cost $5 each and that she had invested $1000.

At the *beginning* of the year she could buy

$$\text{Number of balls purchased today} = \frac{\text{Dollars today available to buy balls}}{\text{Cost of balls today}}$$

$$= 1000/\$5 = 200 \text{ golf balls}$$

At the *end* of the year she could buy

$$\text{Number of balls bought at end of year} = \frac{\text{Dollars available for purchase at end of year}}{\text{Cost per ball at end of year}}$$

$$= \frac{(\$1000)(F/P, 5.5\%, 1)}{(\$5)(1 + 0.02)^1} = \frac{\$1055}{\$5.10} = 207 \text{ golf balls}$$

The golfer can, after one year, buy 3.4% more golf balls than she could before. With rounding, this is 207 balls.

**If Inflation Is 8%**

As with the lower inflation rate, we would solve for $i'$:

$$i' = (i - f)/(1 + f)$$

$$= (0.055 - 0.08)/(1 + 0.08)$$

$$= -0.023 \quad \text{or} \quad -2.3\% \text{ per year}$$

In this case the real growth in money has *decreased* by 2.3%, so that the golfer can now buy 2.3% fewer balls with the money she had invested. Even though she has more money year-end, it is worth less, so she can purchase less.

Regardless of how inflation behaves over the year, the bank will pay the golfer $1055 at the end of the year. However, as we have seen, inflation can greatly affect the "real" growth of dollars over time. In a presidential speech, inflation has been called "that thief" because it steals real purchasing power from our dollars.

Let us continue the discussion of inflation by focusing on cash flows. We define dollars of two types:

*Actual dollars (A\$):* This is the type of dollar that we ordinarily think of when we think of money. They circulate in our economy and are used for investments and payments. We can touch these dollars and often keep them in our purses and wallets—they are "actual" and exist physically. Sometimes they are called *inflated dollars* because they carry any inflation that has reduced their worth. These are also the dollars shown on paychecks, credit card receipts, and normal financial transactions.

*Real dollars (R\$):* This type of dollar is a bit harder to define. Real dollars are always expressed in terms of some constant purchasing power "base" year, for example, 2011-based dollars. Real dollars are sometimes called *constant dollars* or *constant purchasing power dollars,* and because they do not carry the effects of inflation, they are also known as *inflation-free dollars.*

Having defined *market, inflation,* and *real interest rates* as well as *actual* and *real dollars,* let us describe how these quantities relate. Figure 14-1 illustrates the relationship between these quantities.

Figure 14-1 illustrates the following principles:

When dealing with actual dollars ($A\$$), use a market interest rate ($i$), and when discounting $A\$$ over time, also use $i$.

When dealing with real dollars ($R\$$), use a real interest rate ($i'$), and when discounting $R\$$ over time, also use $i'$.

Figure 14-1 shows the relationships between $A\$$ and $R\$$ that occur **at the same period of time.** Actual and real dollars are related by the *inflation rate,* in this case, over the period of years defined by $n - t$. To translate between dollars of one type to dollars of the other

**FIGURE 14–1** Relationship between $i$, $f$, $i'$, A$, and R$.

(A$ to R$ or R$ to A$), use the inflation rate for the right number of periods. The following example illustrates many of these relationships.

## EXAMPLE 14–2

When the university's stadium was completed in 1960, the total cost was $1.2 million. At that time a wealthy alumnus gifted the university with $1.2 million to be used for a future replacement. University administrators are now considering building the new facility in the year 2015. Assume that:

- Inflation is 6.0% per year from 1960 to 2015.
- In 1960 the university invested the gift at a market interest rate of 8.0% per year.

(a) Define $i$, $i'$, $f$, A$, and R$.
(b) How many actual dollars in the year 2015 will the gift be worth?
(c) How much would the actual dollars in 2015 be in terms of 1960 *purchasing power?*
(d) How much better or worse should the new stadium be?

### SOLUTION TO PART a

Since 6.0% is the inflation rate ($f$) and 8.0% is the market interest rate ($i$), we can write

$$i' = (0.08 - 0.06)/(1 + 0.06) = 0.01887, \quad \text{or} \quad 1.887\%$$

The building's cost in 1960 was $1,200,000, which were the actual dollars (A$) spent in 1960.

### SOLUTION TO PART b

From Figure 14-1 we are going from *actual dollars at t, in 1960,* to *actual dollars at n, in 2015.* To do so, we use the *market interest rate* and compound this amount forward 55 years, as illustrated in Figure 14-2.

$$\text{Actual dollars in 2015} = \text{Actual dollars in 1960} \, (F/P, i, 55 \text{ years})$$
$$= \$1,200,000(F/P, 8\%, 55)$$
$$= \$82,701,600$$

**FIGURE 14–2** Compounding A$ in 1960 to A$ in 2015.

## SOLUTION TO PART c

Now we want to determine the amount of *real 1960 dollars that occur in the year 2015,* which are equivalent to the $82.7 million from the solution to part **b**. Let us solve this problem two ways.

1. Translate *actual dollars in the year 2015 to real 1960 dollars in the year 2015.* From Figure 14-1 we can use the inflation rate to **strip 55 years of inflation** from the actual dollars. We do this by using the $P/F$ factor for 55 years at the inflation rate. We are not physically moving the dollars in time; rather, we are simply removing inflation from these dollars one year at a time—the $P/F$ factor does that for us. This is illustrated in the following equation and Figure 14-3.

$$\text{Real 1960 dollars in 2015} = (\text{Actual dollars in 2015})(P/F, \ f, 55)$$
$$= (\$82,701,600)(P/F, 6\%, 55)$$
$$= \$3,357,000$$

**FIGURE 14–3** Translation of A$ in 2015 to R 1960-based dollars in 2015.

2. Translate real 1960 dollars in 1960 to real 1960 dollars in 2015. The $1.2 million can also be said to be *real 1960 dollars that circulated in 1960.* So, let us translate those real dollars from 1960 to the year 2015 (Figure 14-4). Since they are *real dollars,* we use the *real interest rate.*

$$\text{Real 1960 dollars in 2015} = (\text{Real 1960 dollars in 1960})(P/F, i', 55)$$
$$= (\$1,200,000)(F/P, 1.887\%, 55)$$
$$= \$3,355,000$$

**FIGURE 14–4** Translation of $R$ 1960 dollars in 1960 to $R$ 1960 dollars in 2015.

$3,355,000

$1,200,000

1.887%

1960 ———————————————————— 2015

Time

*Note:* The answers differ due to rounding the market interest rate to 1.887% versus carrying it out to more significant digits. The difference due to this rounding is less than 0.1%. If $i'$ and the factors have enough digits, the answers to the two parts would be identical.

## SOLUTION TO PART d

Assuming that construction costs increased at the rate of 6% per year, then the amount available for the project *in terms of 1960 dollars* is almost $3.4 million. This means that the new stadium will be about 3.4/1.2 or approximately 2.8 times "better" than the original one using *real dollars*.

## EXAMPLE 14–3

In 1924 Mr. O'Leary buried $1000 worth of quarters in his backyard. Over the years he had always thought that the money would be a nice nest egg to give to his first grandchild. His first granddaughter, Gabrielle, arrived in 1994. From 1924 to 1994, inflation averaged 4.5%, the stock market increased an average of 15% per year, and investments in guaranteed government bonds averaged 6.5% return per year. What was the relative purchasing power of the jar of quarters that Mr. O'Leary gave to his granddaughter Gabrielle at birth? What might have been a better choice for his "backyard investment"?

### SOLUTION

Mr. O'Leary's $1000 was *actual dollars* both in 1924 **and** in 1994.

To obtain the *real 1924 dollar equivalent* of the $1000 that Gabrielle received in 1994, we would **strip 70 years of inflation out of those dollars.** As it turned out, Gabrielle's grandfather gave her $45.90 worth of 1924 purchasing power. Because inflation has "stolen" purchasing power from his stash of quarters during the 70-year period, Mr. O'Leary gave his grandaughter much less than the amount he first spaded underground. This loss of purchasing power caused by inflation can be calculated as follows:

$$\text{Real 1924 dollars in 1994} = (\text{Actual dollars in 1994})(P/F, \; f, 1994 - 1924)$$

$$= \$1000(P/F, 4.5\%, 70) = \$45.90$$

On the other hand, if Mr. O'Leary had put his $1000 in the stock market in 1924, he would have made baby Gabrielle an instant multimillionaire by giving her $17,735,000. We calculate this as follows:

$$\text{Actual dollars in } 1994 = (\text{Actual dollars in } 1924)(F/P, i, 1994 - 1924)$$

$$= \$1000(F/P, 15\%, 70) = \$17,735,000$$

At the time of Gabrielle's birth that $17.7 million translates to $814,069 in 1924 purchasing power. This is quite a bit different from the $45.90 in 1924 purchasing power calculated for the unearthed jar of quarters.

$$\text{Real 1924 dollars in } 1994 = \$17,735,000(P/F, 4.5\%, 70) = \$814,069$$

Mr. O'Leary was never a risk taker, so it is doubtful he would have chosen the stock market for his future grandchild's nest egg. If he had chosen guaranteed government bonds instead of his backyard, by 1994 the investment would have grown to $59,076 (actual dollars)—the equivalent of $2712 in 1924 purchasing power.

$$\text{Actual dollars in } 1994 = (\text{Actual dollars in } 1924)(F/P, i, 1994 - 1924)$$

$$= \$1000(F/P, 6\%, 70) = \$59,076$$

$$\text{Real 1924 dollars in } 1994 = \$59,076(P/F, 4.5\%, 70) = \$2712$$

Obviously, either option would have been better than the choice Mr. O'Leary made. This example illustrates the effects of inflation and purchasing power, as well as the power of compound interest. However, in Mr. O'Leary's defense, if the country had experienced 70 years of *deflation* instead of *inflation,* he might have had the last laugh!

There are in general two ways to approach an economic analysis problem after the effects of inflation have been recognized. The first is to ignore these effects in conducting the analysis, as we've done so far in the text.

*Ignoring inflation in the analysis:* Use **real dollars** and a **real interest rate** that does not reflect inflation.

The second approach is to systematically include the effects of inflation, as studied in this chapter.

*Incorporating inflation in the analysis:* Use a **market interest rate** and **actual dollars** that include inflation.

Since inflation is so common, why do many economic analyses of engineering projects and most of this text choose *not* to explicitly address inflation? This question is best answered by referring to the many examples and problems that contain statements like "Operations and maintenance costs are expected to be $30,000 annually for the equipment's 20-year life."

Does such a statement mean that accounting records for the next 20 years will show constant costs? Obviously not. Instead, it means that in constant-dollar terms the O&M costs are not expected to increase. In real dollars, O&M costs are uniform. In actual or inflated dollars, we will pay more each year, but each of those dollars will be worth less.

Most costs and benefits in the real world and in this text have prices that increase at about the same rate of inflation as the economy as a whole. In most analyses these inflation increases are addressed by simply stating everything in real dollar terms and using a real interest rate.

There are specific items, such as computers and depreciation deductions, where inflation is clearly expected to differ from the general rate of inflation. It is for these cases, that this chapter is included in this text.

## ANALYSIS IN CONSTANT DOLLARS VERSUS THEN-CURRENT DOLLARS

Performing an analysis requires that we distinguish cash flows as being either constant dollars (real dollars, expressed in terms of some purchasing power base) or then-current dollars (actual dollars that are then-current when they occur). As previously stated, constant (real) dollars require the use of a *real interest rate* for discounting, then-current dollars require a *market (or combined) interest rate*. We must not mix these two dollar types when performing an analysis. If both types are stated in the problem, one type must be converted to the other, so that a consistent comparison can be made.

### EXAMPLE 14-4

The Waygate Corporation is interested in evaluating a major new video display technology (VDT). Two competing computer innovation companies have approached Waygate to develop the technology. Waygate believes that both companies will be able to deliver equivalent products at the end of a 5-year period. From the yearly development costs of the VDT for each firm, determine which one Waygate should choose if the corporate MARR (investment market rate) is 25% and price inflation is assumed to be 3.5% per year over the next 5 years.

*Company Alpha costs:*  Development costs will be $150,000 the first year and will increase at a rate of 5% over the 5-year period.

*Company Beta costs:*  Development costs will be a constant $150,000 per year in terms of today's dollars over the 5-year period.

### SOLUTION

The costs for each of the two alternatives are as follows:

| Year | Then-Current Costs Stated by Alpha | Constant Dollar Costs Stated by Beta |
|------|------------------------------------|--------------------------------------|
| 1 | $150,000 \times (1.05)^0 = \$150,000$ | $150,000 |
| 2 | $150,000 \times (1.05)^1 = \ \ 157,500$ | 150,000 |
| 3 | $150,000 \times (1.05)^2 = \ \ 165,375$ | 150,000 |
| 4 | $150,000 \times (1.05)^3 = \ \ 173,644$ | 150,000 |
| 5 | $150,000 \times (1.05)^4 = \ \ 182,326$ | 150,000 |

We inflate (or escalate) the stated yearly cost given by Company Alpha by 5% per year to obtain the then-current (actual) dollars each year. Company Beta's costs are given in terms of today-based constant dollars.

**Using a Constant Dollar Analysis**

Here we must convert the then-current costs given by Company Alpha to constant today-based dollars. We do this by stripping the number of years of general inflation from each year's cost using $(P/F, f, n)$ or $(1 + f)^{-n}$.

| Year | Constant Dollar Costs Stated by Alpha | Constant Dollar Costs Stated by Beta |
|------|--------------------------------------|--------------------------------------|
| 1 | $\$150{,}000 \times (1.035)^{-1} = \$144{,}928$ | $\$150{,}000$ |
| 2 | $157{,}500 \times (1.035)^{-2} = 147{,}028$ | $150{,}000$ |
| 3 | $165{,}375 \times (1.035)^{-3} = 149{,}159$ | $150{,}000$ |
| 4 | $173{,}644 \times (1.035)^{-4} = 151{,}321$ | $150{,}000$ |
| 5 | $182{,}326 \times (1.035)^{-5} = 153{,}514$ | $150{,}000$ |

We use the *real interest rate* $(i')$ calculated from Equation 14-1 to calculate the present worth of costs for each alternative:

$$i' = (i - f)/(1 + f) = (0.25 - 0.035)/(1 + 0.035) = 0.208$$

$$\begin{aligned}
\text{PW of cost (Alpha)} = \ & 144{,}928(P/F, 20.8\%, 1) + 147{,}028(P/F, 20.8\%, 2) \\
& + 149{,}159(P/F, 20.8\%, 3) + 151{,}321(P/F, 20.8\%, 4) \\
& + 153{,}514(P/F, 20.8\%, 5) = \$436{,}000
\end{aligned}$$

$$\text{PW of cost (Beta)} = \$150{,}000(P/A, 20.8\%, 5) = \$150{,}000(2.9387) = \$441{,}000$$

**Using a Then-Current Dollar Analysis**

Here we must convert the constant dollar costs of Company Beta to then-current dollars. We do this by using $(F/P, f, n)$ or $(1 + f)^n$ to "add in" the appropriate number of years of general inflation to each year's cost.

| Year | Then-Current Costs Stated by Alpha | Then-Current Costs Stated by Beta |
|------|-----------------------------------|-----------------------------------|
| 1 | $\$150{,}000 \times (1.05)^{0} = \$150{,}000$ | $\$150{,}000 \times (1.035)^{1} = \$155{,}250$ |
| 2 | $150{,}000 \times (1.05)^{1} = 157{,}500$ | $150{,}000 \times (1.035)^{2} = 160{,}684$ |
| 3 | $150{,}000 \times (1.05)^{2} = 165{,}375$ | $150{,}000 \times (1.035)^{3} = 166{,}308$ |
| 4 | $150{,}000 \times (1.05)^{3} = 173{,}644$ | $150{,}000 \times (1.035)^{4} = 172{,}128$ |
| 5 | $150{,}000 \times (1.05)^{4} = 182{,}326$ | $150{,}000 \times (1.035)^{5} = 178{,}153$ |

Calculate the present worth of costs for each alternative using the *market interest rate* $(i)$.

$$\begin{aligned}
\text{PW of cost (Alpha)} = \ & 150{,}000(P/F, 25\%, 1) + 157{,}500(P/F, 25\%, 2) \\
& + 165{,}375(P/F, 25\%, 3) + 173{,}644(P/F, 25\%, 4) \\
& + 182{,}326(P/F, 25\%, 5) = \$436{,}000
\end{aligned}$$

$$\text{PW of cost (Beta)} = 155{,}250(P/F, 25\%, 1) + 160{,}684(P/F, 25\%, 2)$$
$$+ \ 166{,}308(P/F, 25\%, 3) + 172{,}128(P/F, 25\%, 4)$$
$$+ \ 178{,}153(P/F, 25\%, 5) = \$441{,}000$$

Using either a constant dollar or then-current dollar analysis, Waygate should chose Company Alpha's offer, which has the lower present worth of costs. There may, of course, be intangible elements in the decision that are more important than a 1% difference in the costs.

## PRICE CHANGE WITH INDEXES

We have already described the effects that inflation can have on money over time. Also, several definitions and relationships regarding dollars and interest rates have been given. We have seen that it is not correct to compare the benefits of an investment in 2010-based dollars with costs in 2015-based dollars. This is like comparing apples and oranges. Such comparisons of benefits and costs can be meaningful only if a standard purchasing power base of money is used. Thus we ask, How do I know what inflation rate to use in my studies? and, How can we measure price changes over time?

### What Is a Price Index?

Price indexes (introduced in Chapter 2) describe the relative price fluctuation of goods and services. They provide a *historical* record of prices over time. Price indexes are tracked for **specific commodities** as well as for **bundles (composites) of commodities.** As such, price indexes can be used to measure historical price changes for individual cost items (like labor and material costs) as well as general costs (like consumer products). We use **past** price fluctuations to predict **future** prices.

Table 14-1 lists the historic prices of sending a first-class letter in the U.S. from 1970 to 2010. The cost is given both in terms of dollars (cents) and as measured by a fictitious price index that we could call the letter cost index (LCI).

Notice two important aspects of the LCI. First, as with all cost or price indexes, there is a **base year,** which is assigned a value of 100. Our LCI has a base year of 1970—thus for 1970, LCI = 100. Values for subsequent years are stated in relation to the 1970 value. Second, the LCI changes only when the cost of first-class postage changes. In years when this quantity does not change, the LCI is not affected. These general observations apply to all price indexes.

In general, engineering economists are the "users" of cost indexes such as our hypothetical LCI. That is, cost indexes are calculated or tabulated by some other party, and our interest is in assessing what the index tells us about the historical prices and how these may affect our estimate of future costs. However, we should understand how the LCI in Table 14-1 was calculated.

In Table 14-1, the LCI is assigned a value of 100 because 1970 serves as our base year. In the following years the LCI is calculated on a year-to-year basis based on the annual percentage increase in first-class mail. Equation 14-2 illustrates the arithmetic used.

$$\text{LCI year, } n = \frac{\text{cost } (n)}{\text{cost 1970}} \times 100 \tag{14-2}$$

**TABLE 14–1  Historic Prices of First-Class Mail, 1970–2010, and Letter Cost Index**

| Year, n | Cost of First-Class Mail | LCI | Annual Increase for n | Year, n | Cost of First-Class Mail | LCI | Annual Increase for n |
|---------|--------------------------|-----|------------------------|---------|--------------------------|-----|------------------------|
| 1970 | $0.06 | 100 |       | 1991 | 0.29 | 483 | 16.00 |
| 1971 | 0.08 | 133 | 33.33 | 1992 | 0.29 | 483 | 0.00 |
| 1972 | 0.08 | 133 | 0.00  | 1993 | 0.29 | 483 | 0.00 |
| 1973 | 0.08 | 133 | 0.00  | 1994 | 0.29 | 483 | 0.00 |
| 1974 | 0.10 | 166 | 25.00 | 1995 | 0.32 | 533 | 10.34 |
| 1975 | 0.13 | 216 | 30.00 | 1996 | 0.32 | 533 | 0.00 |
| 1976 | 0.13 | 216 | 0.00  | 1997 | 0.32 | 533 | 0.00 |
| 1977 | 0.13 | 216 | 0.00  | 1998 | 0.33 | 550 | 3.13 |
| 1978 | 0.15 | 250 | 15.74 | 1999 | 0.33 | 550 | 0.00 |
| 1979 | 0.15 | 250 | 0.00  | 2000 | 0.33 | 550 | 0.00 |
| 1980 | 0.15 | 250 | 0.00  | 2001 | 0.34 | 567 | 3.03 |
| 1981 | 0.20 | 333 | 33.33 | 2002 | 0.37 | 617 | 8.82 |
| 1982 | 0.20 | 333 | 0.00  | 2003 | 0.37 | 617 | 0.00 |
| 1983 | 0.20 | 333 | 0.00  | 2004 | 0.37 | 617 | 0.00 |
| 1984 | 0.20 | 333 | 0.00  | 2005 | 0.37 | 617 | 0.00 |
| 1985 | 0.22 | 367 | 10.00 | 2006 | 0.39 | 650 | 5.41 |
| 1986 | 0.22 | 367 | 0.00  | 2007 | 0.41 | 683 | 5.13 |
| 1987 | 0.22 | 367 | 0.00  | 2008 | 0.42 | 700 | 2.44 |
| 1988 | 0.25 | 417 | 13.64 | 2009 | 0.42 | 700 | 0.00 |
| 1989 | 0.25 | 417 | 0.00% | 2010 | 0.44 | 733 | 4.76 |
| 1990 | 0.25 | 417 | 0.00  |      |      |     |      |

For example, consider the LCI for the year 1980. We calculate the LCI as follows.

$$\text{LCI year } 1980 = \frac{0.15}{0.06} \times 100 = 250$$

As mentioned, engineering economists often use cost indexes to project future cash flows. As such, our first job is to use a cost index to **calculate** the *year-to-year* percentage increase (or *inflation*) of prices tracked by an index. We can use Equation 14-3:

$$\text{Annual percentage increase, } n = \frac{\text{Index}(n) - \text{Index}(n-1)}{\text{Index}(n-1)} \times 100\% \qquad (14\text{-}3)$$

To illustrate, let us look at the percent change from 1977 to 1978 for the LCI.

$$\text{Annual percentage increase (1978)} = \frac{250 - 216}{216} \times 100\% = 15.74\%$$

For 1978 the price of mailing a first-class letter increased by 15.74% over the previous year. This is tabulated in Table 14-1.

An engineering economist often wants to know how a particular cost quantity changes over time. Often we are interested in calculating the *average* rate of price increases over a period of time. For instance, we might want to know the average yearly increase in postal prices from 2000 to 2010. If we generalize Equation 14-3 to calculate the percent change from 2000 to 2010, we obtain

$$\% \text{ Increase (2000 to 2010)} = \frac{700 - 550}{550} \times 100\% = 273\%$$

How do we use this to obtain the **average** rate of increase over those 10 years? Should we divide 273% by 10 years ($273/10 = 27.3\%$)? Of course not! Inflation, like interest, compounds. Such a simple division treats inflation like simple interest—without compounding. So the question remains: How do we calculate an *equivalent average rate of increase* in postage rates over a period of time? If we think of the index numbers as cash flows, we have

$$P = 550, \qquad F = 1700, \qquad n = 10, \text{ years}, \qquad i = ?$$

Using $F = P(1 + i)^n$   $700 = 500(1 + i)^{10}$   $i = (700/550)^{1/10} - 1$   $i = 0.0244 = 2.44\%$

We can use a cost index to calculate the average rate of increase over any period of years, which should provide insight into how prices may behave in the future.

## Composite Versus Commodity Indexes

Cost indexes come in two types: commodity-specific indexes and composite indexes. Commodity-specific indexes measure the historical change in price for specific items—such as construction labor or iron ore. Commodity indexes, like our letter cost index, are useful when an economic analysis includes individual cost items that are tracked by such indexes. For example, if we need to estimate the direct-labor cost portion of a construction project, we could use an index that tracks the inflation, or escalation, of labor costs. The U.S. Departments of Commerce and Labor track many cost quantities through the Department of Economic Analysis and Bureau of Labor Statistics. Example 14-5 uses data from a California government website (www.resd.dgs.ca.gov/CaliforniaConstructionCostIndexPage.htm) to demonstrate using a commodity index. This data is compiled from the *Engineering News-Record*.

### EXAMPLE 14–5

From the website just referenced (CCCI), in January 2010 bids were opened for a new building in Los Angeles. The low bid and the final construction cost were $5.25 million. Another building of the same size, quality, and purpose is planned with a bid opening in January 2015. Estimate the new building's low bid and cost.

### SOLUTION

In January 2010 the California Construction Cost Index (CCCI) had a value of 5260 and in January 2000 the value was 3746. If we wanted a cost estimate for January 2010, we could simply use the ratio of these values and Equation 14-2. But we want a value for January 2015, which is outside our data set. (This is true for all future estimates.)

The solution is to estimate the average annual rate of increase, and then to apply this for the longer period.

$$F = 5260, P = 3746, n = 10, \text{ find } f$$

$$F = P(1 + f)^n$$

$$f = (5260/3746)^{1/10} - 1 = 3.45\% \text{ per year}$$

Now we can apply the inflation rate for $n = 15$ years to the building's cost in 2000.

$$F = 5.25 \text{ million} \times (1.0345)^{15} = \$8.73 \text{ million}$$

**Composite cost indexes** do not track historical prices for individual classes of items. Instead, they measure the historical prices of *bundles* or *market baskets* of assets. Examples of composite indexes include the *Consumer Price Index* (CPI) and the *Producer Price Index* (PPI). The CPI measures prices for consumers in the U.S. marketplace, and each PPI measures prices for categories of producers in the U.S. economy.

The CPI, an index calculated by the Bureau of Labor Statistics, tracks the cost of a standard *bundle of consumer goods* from year to year. This "consumer bundle" or "market basket" includes housing, clothing, food, transportation, and entertainment. The CPI enjoys popular identification as the "inflation" indicator. Table 14-2 gives yearly index values and annual percent increases in the CPI. Figure 14-5 charts the CPI inflation rate for the same period.

Composite indexes are used the same way as commodity-specific indexes. That is, we can pick a single value from the table if we are interested in measuring the historic price for a single year, or we can calculate an *average inflation rate* or *average rate of price increase* as measured by the index over a time period extending several years.

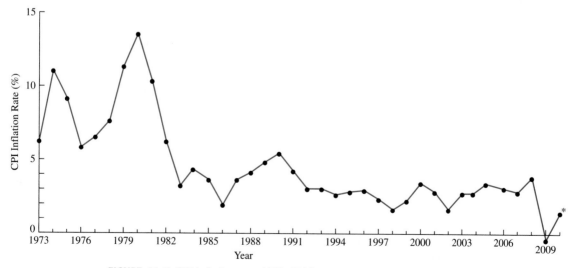

**FIGURE 14-5** CPI inflation rate: 1973–2010.          *March data only.

**TABLE 14–2 CPI Index Values and Yearly Percentage Increases, 1973–2010**

| Year | CPI Value* | CPI Increase | Year | CPI Value* | CPI Increase |
|------|-----------|--------------|------|-----------|--------------|
| 1973 | 44.4 | 6.2% | 1992 | 140.3 | 3.0% |
| 1974 | 49.3 | 11.0 | 1993 | 144.5 | 3.0 |
| 1975 | 53.8 | 9.1 | 1994 | 148.2 | 2.6 |
| 1976 | 56.9 | 5.8 | 1995 | 152.4 | 2.8 |
| 1977 | 60.6 | 6.5 | 1996 | 156.9 | 2.9 |
| 1978 | 65.2 | 7.6 | 1997 | 160.5 | 2.3 |
| 1979 | 72.6 | 11.3 | 1998 | 163.0 | 1.6 |
| 1980 | 82.4 | 13.5 | 1999 | 166.6 | 2.2 |
| 1981 | 90.9 | 10.3 | 2000 | 172.2 | 3.4 |
| 1982 | 96.5 | 6.2 | 2001 | 177.1 | 2.8 |
| 1983 | 99.6 | 3.2 | 2002 | 179.9 | 1.6 |
| 1984 | 103.9 | 4.3 | 2003 | 184.0 | 2.3 |
| 1985 | 107.6 | 3.6 | 2004 | 188.9 | 2.7 |
| 1986 | 109.6 | 1.9 | 2005 | 195.3 | 3.4 |
| 1987 | 113.6 | 3.6 | 2006 | 201.6 | 3.2 |
| 1988 | 118.3 | 4.1 | 2007 | 207.3 | 2.8 |
| 1989 | 124.0 | 4.8 | 2008 | 215.3 | 3.8 |
| 1990 | 130.7 | 5.4 | 2009 | 214.5 | −0.4 |
| 1991 | 136.2 | 4.2 | 2010† | 217.6 | 1.4 |

*Reference base: 1982–1984 = 100.
† Data as of March, not end of year 2010.

## How to Use Price Indexes in Engineering Economic Analysis

One may question the usefulness of *historical* data (as provided by price indexes) when engineering economic analysis deals with economic effects projected to occur in the *future*. However, historical index data are often better predictors of future prices than official government predictions, which may be influenced by political considerations. The engineering economist can use *average historical percentage increases (or decreases)* from commodity-specific and composite indexes, along with data from market analyses and other sources, to estimate future costs and benefits.

> When the estimated quantities are items that are tracked by commodity specific indexes, then those indexes should be used to calculate *average historical percentage increases (or decreases)*. If no commodity-specific indexes are kept, use an appropriate composite index to make this calculation.

For example, to estimate electric usage costs for a turret lathe over a 5-year period, you would first want to refer to a commodity-specific index for electric power in your area. If such an index does not exist, you might use a specific index for a very closely related commodity—perhaps an index of electric power costs nationally. In the absence of such substitute or related commodity indexes, you could use a composite index for national energy prices. The key point is that you should try to identify and use the price index that most closely relates to the quantity being estimated.

## CASH FLOWS THAT INFLATE AT DIFFERENT RATES

Engineering economic analysis requires the estimation of various parameters. Over time, it is not uncommon for these parameters to *inflate* or *increase* (or even decrease) at different rates. For instance, one parameter might *increase* 5% per year and another 15% per year, and a third might *decrease* 3.5% per year. Since we are looking at the behavior of cash flows over time, we must have a way of handling this effect.

In Example 14-6 several commodity prices change at different rates. By using the respective individual inflation rates, the **actual dollar** amounts for each commodity are obtained in each year. Then we use a market interest rate to discount these actual dollar amounts.

---

### EXAMPLE 14–6

On your first assignment as an engineer, your boss asks you to develop the utility cost estimate for a new manufacturing facility. After some research you define the problem as finding the present worth of utility costs given the following data:

- Your company uses a minimum attractive rate of return (MARR) = 35% as $i$ (not adjusted for inflation.)
- The project has a useful life of 25 years.
- Utilities to be estimated include electricity, water, and natural gas.
- The 35-year historical data reveal:

    Electricity costs increase at 8.5% per year

    Water costs increase at 5.5% per year

    Natural gas costs increase at 6.5% per year

- First-year estimates of the utility costs (in today's dollars) are as follows:

    Electricity will cost $55,000

    Water will cost $18,000

    Natural gas will cost $38,000

#### SOLUTION

For this problem we will take each of the utilities used in our manufacturing facility and inflate them independently at their respective historical annual rates. Once we have these actual dollar amounts ($A\$$), we can total them and then discount each year's total at 35% back to the present.

| Year | Electricity | | Water | | Natural Gas | | Total |
|---|---|---|---|---|---|---|---|
| 1 | $55,000(1.085)^0$ = | $55,000 | $18,000(1.055)^0$ | = $18,000 | $38,000(1.065)^0$ = | $38,000 | $111,000 |
| 2 | $55,000(1.085)^1$ = | 59,675 | $18,000(1.055)^1$ | = 18,990 | $38,000(1.065)^1$ = | 40,470 | 119,135 |
| 3 | $55,000(1.085)^2$ = | 64,747 | $18,000(1.055)^2$ | = 20,034 | $38,000(1.065)^2$ = | 43,101 | 127,882 |
| 4 | $55,000(1.085)^3$ = | 70,251 | $18,000(1.055)^3$ | = 21,136 | $38,000(1.065)^3$ = | 45,902 | 137,289 |
| 5 | $55,000(1.085)^4$ = | 76,222 | $18,000(1.055)^4$ | = 22,299 | $38,000(1.065)^4$ = | 48,886 | 147,407 |
| 6 | $55,000(1.085)^5$ = | 82,701 | $18,000(1.055)^5$ | = 23,525 | $38,000(1.065)^5$ = | 52,063 | 158,290 |
| 7 | $55,000(1.085)^6$ = | 89,731 | $18,000(1.055)^6$ | = 24,819 | $38,000(1.065)^6$ = | 55,447 | 169,997 |
| 8 | $55,000(1.085)^7$ = | 97,358 | $18,000(1.055)^7$ | = 26,184 | $38,000(1.065)^7$ = | 59,051 | 182,594 |
| . | . | | . | | . | | . |
| . | . | | . | | . | | . |
| . | . | | . | | . | | . |
| 24 | $55,000(1.085)^{23}$ = | 359,126 | $18,000(1.055)^{23}$ | = 61,671 | $38,000(1.065)^{23}$ = | 161,743 | 582,539 |
| 25 | $55,000(1.085)^{24}$ = | 389,652 | $18,000(1.055)^{24}$ | = 65,063 | $38,000(1.065)^{24}$ = | 172,256 | 626,970 |

The present worth of the total yearly utility costs is

$$PW = 111,000(P/F, 35\%, 1) + 119,135(P/F, 35\%, 2) + \cdots + 626,970(P/F, 35\%, 25)$$
$$= \$5,540,000$$

## DIFFERENT INFLATION RATES PER PERIOD

In this section we address the situation of inflation rates that are changing over the study period. Rather than different inflation rates for different cash flows, in Example 14-7 the *interest rate* for the same cash flow is changing over time. A method for handling this situation is much like that of the preceding section. We can simply apply the inflation rates in the years in which they are projected to occur. We would do this for each cash flow. Once we have all these actual dollar amounts, we can use the market interest rate to apply PW, EUAC, or other measures of merit.

### EXAMPLE 14–7

While working as a clerk at Piggly Wiggly, Elvis has learned much about the cost of different vegetables. The kitchen manager at Heartbreak Hotel called recently, requesting Elvis to estimate the raw material cost over the next 5 years to introduce succotash (lima beans and corn) to the buffet line. To develop his estimate, Elvis has used his advanced knowledge of soil growing conditions, world demand, and government subsidy programs for these two crops. He has estimated the following data:

- Costs for lima beans will inflate at 3% per year for the next 3 years and then at 4% for the following 2 years.
- Costs for corn will inflate at 8% per year for the next 2 years and then will decrease 2% in the following 3 years.

The kitchen manager wants to know the equivalent annual cost of providing succotash on the buffet line over the 5-year period. His before-tax MARR is 20%. An average of 50 pounds each of beans and corn will be needed each day. The hotel kitchen operates 6 days per week, 52 weeks per year. Current costs are $0.35/lb for lima beans and $0.80/lb for corn.

Today's cost for one year's supply of vegetables is

$$\text{Lima beans} \quad 0.35 \text{ \$/lb} \times 50 \text{ lb/day} \times 6 \text{ day/wk} \times 52 \text{ wk/yr} = \$ \ 5{,}460/\text{yr}$$
$$\text{Corn} \quad 0.80 \text{ \$/lb} \times 50 \text{ lb/day} \times 6 \text{ day/wk} \times 52 \text{ wk/yr} = \ 12{,}480/\text{yr}$$

| Year | Lima Beans | Corn | Total |
|------|-----------|------|-------|
| 0 | $5460 | $12,480 | |
| 1 | 5460(1.03) = 5624 | 12,480(1.08) = 13,478 | $19,102 |
| 2 | 5624(1.03) = 5793 | 13,478(1.08) = 14,556 | 20,349 |
| 3 | 5793(1.03) = 5967 | $14{,}556(1.02)^{-1} = 14{,}271$ | 20,238 |
| 4 | 5967(1.04) = 6206 | $14{,}271(1.02)^{-1} = 13{,}991$ | 20,197 |
| 5 | 6206(1.04) = 6454 | $13{,}991(1.02)^{-1} = 13{,}717$ | 20,171 |

$$\text{EUAC} = [19{,}102(P/F, 20\%, 1) + 20{,}349(P/F, 20\%, 2) + 20{,}238(P/F, 20\%, 3)$$
$$+ \ 20{,}197(P/F, 20\%, 4) + 20{,}171(P/F, 20\%, 5)](A/P, 20\%, 5)$$
$$= \$19{,}900 \text{ per year}$$

In Example 14-7, both today's cost of each vegetable and the respective inflation rates were used to calculate the yearly costs of purchasing the desired quantities over the 5-year period. As in Example 14-6, we obtained a total marginal cost (in terms of actual dollars) by combining the two individual yearly costs. We then calculated the equivalent uniform annual cost (EUAC) using the given market interest rate.

Example 14-8 provides another example of how the effect of changes in inflation rates over time can affect an analysis.

## EXAMPLE 14–8

If general price inflation is estimated to be 5% for the next 5 years, 7.5% for the 3 years after that, and 3% the following 5 years, at what market interest rate ($i$) would you have to invest your money to maintain a real purchasing power growth rate ($i'$) of 10% during those years?

In Years 1–5 you must invest at $0.10 + 0.050 + (0.10)(0.050) = 0.1150 = 11.50\%$ per year.

In Years 6–8 you must invest at $0.10 + 0.075 + (0.10)(0.075) = 0.1825 = 18.25\%$ per year.

In Years 9–13 you must invest at $0.10 + 0.030 + (0.10)(0.030) = 0.1330 = 13.30\%$ per year.

*Note:* Most interest-bearing investments have fixed, up-front rates that the investor well understands going in. On the other hand, inflation is not quantified, and its effect on real return is not measured until the end of the year. Therefore the real investment return ($i'$) may not turn out to be what was originally required.

## INFLATION EFFECT ON AFTER-TAX CALCULATIONS

Earlier we noted the impact of inflation on before-tax calculations. If the future benefits keep up with the rate of inflation, the rate of return will not be adversely affected by the inflation. Unfortunately, we are not so lucky when we consider a situation with income taxes, as illustrated by Example 14-9. The value of the depreciation deduction is diminished by inflation.

### EXAMPLE 14–9

A $12,000 investment with no salvage value will return annual benefits for 6 years. Assume straight-line depreciation and a 46% income tax rate. Solve for both before- and after-tax rates of return for two situations:

1. *No inflation:* the annual benefits are constant at $2918 per year.
2. *Inflation equal to 5%:* the benefits from the investment increase at this same rate, so that they continue to be the equivalent of $2918 in Year-0 dollars.

The benefit schedules are as follows:

| Year | Annual Benefit for Both Situations (Year-0 dollars) | No Inflation, Actual Dollars Received | 5% Inflation Factor* | 5% Inflation, Actual Dollars Received |
|------|------|------|------|------|
| 1 | $2918 | $2918 | $(1.05)^1$ | $3064 |
| 2 | 2918 | 2918 | $(1.05)^2$ | 3217 |
| 3 | 2918 | 2918 | $(1.05)^3$ | 3378 |
| 4 | 2918 | 2918 | $(1.05)^4$ | 3547 |
| 5 | 2918 | 2918 | $(1.05)^5$ | 3724 |
| 6 | 2918 | 2918 | $(1.05)^6$ | 3910 |

*May be read from the 5% compound interest table as $(F/P, 5\%, n)$.

### SOLUTIONS

**Before-Tax Rate of Return**

Since both situations (no inflation and 5% inflation) have an annual benefit, stated in Year-0 dollars of $2918, they have the same before-tax rate of return.

PW of cost = PW of benefit

$$12,000 = 2918(P/A, i, 6), \qquad (P/A, i, 6) = \frac{12,000}{2918} = 4.11$$

From compound interest tables: before-tax rate of return equals 12%.

### After-Tax Rate of Return, No Inflation

| Year | Before-Tax Cash Flow | Straight-Line Depreciation | Taxable Income | 46% Income Taxes | Actual Dollars, and Year-0 Dollars, After-Tax Cash Flow |
|------|------|------|------|------|------|
| 0 | −$12,000 | | | | −$12,000 |
| 1–6 | 2,918 | $2000 | $918 | −$422 | 2,496 |

PW of cost = PW of benefit

$$12,000 = 2496(P/A, i, 6), \qquad (P/A, i, 6) = \frac{12,000}{2496} = 4.81$$

From compound interest tables: after-tax rate of return equals 6.7%.

### After-Tax Rate of Return, 5% Inflation

| Year | Before-Tax Cash Flow | Straight-Line Depreciation | Taxable Income | 46% Income Taxes | Actual Dollars, After-Tax Cash Flow |
|------|------|------|------|------|------|
| 0 | −$12,000 | | | | −$12,000 |
| 1 | 3,064 | $2000 | $1064 | −$489 | 2,575 |
| 2 | 3,217 | 2000 | 1217 | −560 | 2,657 |
| 3 | 3,378 | 2000 | 1378 | −634 | 2,744 |
| 4 | 3,547 | 2000 | 1547 | −712 | 2,835 |
| 5 | 3,724 | 2000 | 1724 | −793 | 2,931 |
| 6 | 3,910 | 2000 | 1910 | −879 | 3,031 |

### Converting to Year-0 Dollars and Solving for Rate of Return

| Year | Actual Dollars, After-Tax Cash Flow | Conversion Factor | Year-0 Dollars, After-Tax Cash Flow | Present Worth at 4% | Present Worth at 5% |
|------|------|------|------|------|------|
| 0 | −$12,000 | | −$12,000 | −$12,000 | −$12,000 |
| 1 | 2,575 × $(1.05)^{-1}$ = | | 2,452 | 2,358 | 2,335 |
| 2 | 2,657 × $(1.05)^{-2}$ = | | 2,410 | 2,228 | 2,186 |
| 3 | 2,744 × $(1.05)^{-3}$ = | | 2,370 | 2,107 | 2,047 |
| 4 | 2,835 × $(1.05)^{-4}$ = | | 2,332 | 1,993 | 1,919 |
| 5 | 2,931 × $(1.05)^{-5}$ = | | 2,297 | 1,888 | 1,800 |
| 6 | 3,031 × $(1.05)^{-6}$ = | | 2,262 | 1,788 | 1,688 |
| | | | | +362 | −25 |

Linear interpolation between 4 and 5%:

After-tax rate of return = 4% + 1%[362/(362 + 25)] = 4.9%

From Example 14-9, we see that the before-tax rate of return for both situations (no inflation and 5% inflation) is the same. Equal before-tax rates of return are expected because the benefits in the inflation situation increased in proportion to the inflation. No special calculations are needed in before-tax calculations when future benefits are expected to respond to inflation or deflation rates.

The after-tax calculations illustrate that equal before-tax rates of return do not produce equal after-tax rates of return considering inflation.

| | Rate of Return (%) | |
|---|---|---|
| Situation | Before Taxes | After Taxes |
| No inflation | 12 | 6.7 |
| 5% inflation | 12 | 4.9 |

Inflation reduces the after-tax rate of return, even though the benefits increase at the same rate as the inflation. A review of the cash flow table reveals that while benefits increase, the depreciation schedule does not. Thus, the inflation results in increased taxable income and, hence, larger income tax payments.

The result is that while the after-tax cash flow in actual dollars increases, the augmented amount is not high enough to offset *both* inflation and increased income taxes. The Year-0-dollar after-tax cash flow is smaller with inflation than the Year-0-dollar after-tax cash flow without inflation.

## USING SPREADSHEETS FOR INFLATION CALCULATIONS

Spreadsheets are the perfect tool for incorporating consideration of inflation into analyses of economic problems. For example, next year's labor costs are likely to be estimated as equal to this year's costs times $(1 + f)$, where $f$ is the inflation rate. Thus each year's value is different, so we can't use factors for uniform flows, $A$. Also the formulas that link different years are easy to write. The result is problems that are very tedious to do by hand, but easy by spreadsheet.

Example 14-10 illustrates two different ways to write the equation for inflating costs. Example 14-11 illustrates that inflation reduces the after-tax rate of return because inflation makes the depreciation deduction less valuable.

### EXAMPLE 14–10

Two costs for construction of a small, remote mine are for labor and transportation. Labor costs are expected to be $350,000 the first year, with inflation of 6% annually. Unit transportation costs are expected to inflate at 5% annually, but the volume of material being moved changes each year. In Time-0 dollars, the transportation costs are estimated to be $40,000, $60,000, $50,000, and $30,000 in Years 1 through 4. The inflation rate for the value of the dollar is 3%. If the firm uses an $i'$ of 7%, what is the equivalent annual cost for this 4-year project?

## SOLUTION

The data for labor costs can be stated so that no inflation needs to be applied in Year 1: the cost is $350,000. In contrast, the transportation costs for Year 1 are determined by multiplying $40,000 by $1.05 \, (= 1 + f)$.

Also in later years the labor $\text{cost}_t = \text{labor cost}_{t-1}(1 + f)$, while each transportation cost must be computed as the Time-0 value times $(1 + f)^t$. In Figure 14-6, the numbers in the Year-0 (or real) dollar column equal the values in the actual dollars column divided by $(1.03)^t$.

| | A | B | C | D | E | F | G | H |
|---|---|---|---|---|---|---|---|---|
| 1 | | | | | | 7% | Inflation-Free Interest | |
| 2 | Inflation Rate | 6% | | 5% | | 3% | | |
| 3 | | | Transportation Costs | | Total | Total | | |
| 4 | Year | Labor Costs | Year 0 $s | Actual $s | Actual $s | Real $s | | |
| 5 | 1 | 120,000 | 40,000 | 42,000 | 162,000 | 157,282 | = E5/(1+$F$2)^A5 | |
| 6 | 2 | 127,200 | 60,000 | 66,150 | 193,350 | 182,251 | | |
| 7 | 3 | 134,832 | 50,000 | 57,881 | 192,713 | 176,360 | | |
| 8 | 4 | 142,922 | 30,000 | 36,465 | 179,387 | 159,383 | | |
| 9 | | | | | | $571,732 | = NPV(F1,F5:F8) | |
| 10 | | | | | =B8+D8 | $168,791 | = −PMT(F1,4,F9) | |
| 11 | =B7*(1+$B$2) | | =C8*(1+$D$2)^A8 | | | | | |

**FIGURE 14–6**  Spreadsheet for inflation.

The equivalent annual cost equals $168,791.

## EXAMPLE 14–11

For the data of Example 14-9, calculate the IRR with and without inflation with MACRS depreciation. How are the results affected by inflation by comparison with the earlier results.

## SOLUTION

Most of the formulas for this spreadsheet are given in rows 11 and 12 of Figure 14-6 for the data in Year 6. The benefits received are computed from the base value in cell B5. The depreciation is the MACRS percentage times the $12,000 spent in Year 0. This value is not influenced by inflation, so the depreciation deduction is less valuable as inflation increases. The tax paid equals the tax rate times the taxable income, which equals dollars received minus the depreciation charge. Then ATCF (after-tax cash flow) equals the before-tax cash flow minus the tax paid.

In Figure 14-7, notice that in Year 2 the depreciation charge is large enough to cause this project to pay "negative" tax. For a firm, this means that the deduction on this project will be used to offset income from other projects.

| | A | B | C | D | E | F | G | H |
|---|---|---|---|---|---|---|---|---|
| 1 | | 0% | = Inflation Rate | | 46% | = Tax Rate | | |
| 2 | | Actual $s | MACRS | Actual $s | Actual $s | Actual $s | Real $s | |
| 3 | Year | Received | Deprec. % | Deprec. | Tax | ATCF | ATCF | |
| 4 | 0 | −12000 | | | | −12000 | −12000 | |
| 5 | 1 | 2918 | 20.00% | 2400 | 238 | 2680 | 2680 | = F5/(1+$B$1)^A5 |
| 6 | 2 | 2918 | 32.00% | 3840 | −424 | 3342 | 3342 | |
| 7 | 3 | 2918 | 19.20% | 2304 | 282 | 2636 | 2636 | |
| 8 | 4 | 2918 | 11.52% | 1382 | 706 | 2212 | 2212 | |
| 9 | 5 | 2918 | 11.52% | 1382 | 706 | 2212 | 2212 | |
| 10 | 6 | 2918 | 5.76% | 691 | 1024 | 1894 | 1894 | |
| 11 | Formulas | | = −$B$4*C10 | | =B10−E10 | | | |
| 12 | for Yr 6 | =$B$5*(1+$B$1)^A10 | | =(B10−D10)*$E$1 | | | 7.29% | = IRR |
| 13 | | | | | | | | |
| 14 | | 5% | = Inflation Rate | | 46% | = Tax Rate | | |
| 15 | | Actual $s | MACRS | Actual $s | Actual $s | Actual $s | Real $s | |
| 16 | Year | Received | Deprec. % | Deprec. | Tax | ATCF | ATCF | |
| 17 | 0 | −12000 | | | | −12000 | −12000 | |
| 18 | 1 | 3064 | 20.00% | 2400 | 305 | 2759 | 2627 | |
| 19 | 2 | 3217 | 32.00% | 3840 | −287 | 3504 | 3178 | |
| 20 | 3 | 3378 | 19.20% | 2304 | 494 | 2884 | 2491 | |
| 21 | 4 | 3547 | 11.52% | 1382 | 996 | 2551 | 2099 | |
| 22 | 5 | 3724 | 11.52% | 1382 | 1077 | 2647 | 2074 | |
| 23 | 6 | 3910 | 5.76% | 691 | 1481 | 2430 | 1813 | |
| 24 | | | | | | | 5.68% | = IRR |

**FIGURE 14–7** After-tax IRRs with MACRS and inflation.

The IRRs are higher in this example (7.29% without inflation vs. 6.7% with straight-line depreciation in Example 14-9, and 5.68% with inflation vs. 4.9%) because MACRS supports faster depreciation, so the depreciation deductions are more valuable. Also because the depreciation is faster, the results are affected somewhat less by inflation. Specifically, with MACRS 5% inflation lowers the IRR by 1.6% and with straight-line depreciation, 5% inflation lowers the IRR by 1.8%.

# SUMMARY

Inflation is characterized by rising prices for goods and services, whereas deflation produces a fall in prices. An inflationary trend makes future dollars have less **purchasing power** than present dollars. Deflation has the opposite effect. If money is borrowed over a period of time in which deflation is occurring, then debt will be repaid with dollars that have **more** purchasing power than those originally borrowed. Inflation and deflation have opposite effects on the purchasing power of a monetary unit over time.

To distinguish and account for the effect of inflation in our engineering economic analysis, we define *inflation, real,* and *market* interest rates. These interest rates are related by the following expression:

$$i = i' + f + i'f$$

Each rate applies in a different circumstance, and it is important to apply the correct rate to the correct circumstance. Cash flows are expressed in terms of either *actual* or *real dollars*. The *market interest* rate should be used with *actual dollars* and the *real interest rate* should be used with *real dollars*.

The different cash flows in our analysis may inflate or change at different interest rates when we look over the life cycle of the investment. Also, a single cash flow may inflate or deflate at different rates over time. These two circumstances are handled easily by applying the correct inflation rates to each cash flow over the study period to obtain the actual dollar amounts occurring in each year. After the actual dollar quantities have been calculated, the analysis proceeds using the market interest rate to calculate the measure of merit of interest.

Historical price change for single commodities and bundles of commodities are tracked with price indexes. The Consumer Price Index (CPI) is an example of a composite index formed by a bundle of consumer goods. The CPI serves as a surrogate for general inflation in our economy. Indexes can be used to calculate the *average annual increase* (or decrease) of the costs and benefits in our analysis. The historical data provide valuable information about how economic quantities may behave in the future over the long run.

The effect of inflation on the computed rate of return for an investment depends on how future benefits respond to the inflation. Usually the costs and benefits increase at the same rate as inflation, so the before-tax rate of return will not be adversely affected by the inflation. This outcome is not found when an after-tax analysis is made because the allowable depreciation schedule does not increase. The result will be increased taxable income and income tax payments, which reduce the available after-tax benefits and, therefore, the after-tax rate of return. The important conclusion is that estimates of future inflation or deflation may be important in evaluating capital expenditure proposals.

# PROBLEMS

### Meaning and Effect

**14-1** Define inflation in terms of the purchasing power of dollars.

**14-2** Define and describe the relationships between the following: actual and real dollars, inflation, and real and market (combined) interest rates.

**14-3** How does inflation happen? Describe a few circumstances that cause prices in an economy to increase.

**14-4** Is it necessary to account for inflation in an engineering economy study? What are the two approaches for handling inflation in such analyses?

**14-5** In Chapters 5 (Present Worth Analysis) and 6 (Annual Cash Flow Analysis) it is assumed that prices are stable and a machine purchased today for $5000 can be replaced for the same amount many

years hence. In fact, prices have generally been rising, so the stable price assumption tends to be incorrect. Under what circumstances is it correct to use the "stable price" assumption when prices actually are changing?

**14-6** An economist has predicted 7% inflation during the next 10 years. How much will an item that presently sells for $10 bring a decade hence?
(*Answer:* $19.67)

**14-7** A man bought a 5% tax-free municipal bond. It cost $1000 and will pay $50 interest each year for 20 years. At maturity the bond returns the original $1000. If there is 2% annual inflation, what real rate of return will the investor receive?

**14-8** A man wishes to set aside some money for his daughter's college education. His goal is to have a bank savings account containing an amount equivalent to $20,000 in today's dollars at the girl's 18$^{th}$ birthday.

The estimated inflation rate is 8%. If the bank pays 5% compounded annually, what lump sum should he deposit on the child's 4th birthday?
(*Answer:* $29,670)

**14-9** A newspaper reports that in the last 5 years, prices have increased a total of 50%. This is equivalent to what annual inflation rate, compounded annually?
(*Answer:* 8.45%)

**14-10** An economist has predicted that for the next 5 years, the U.S. will have an 8% annual inflation rate, followed by 5 years at a 6% inflation rate. This is equivalent to what average price change per year for the entire 10-year period?

**14-11** An investor wants a real rate of return $i'$ of 10% per year. If the expected annual inflation rate for the next several years is 6%, what interest rate $i$ should be used in project analysis calculations?

**14-12** A South American country has had a high rate of inflation. Recently, its exchange rate was 15 cruzados per dollar; that is, one dollar will buy 15 cruzados. It is likely that the country will continue to experience a 25% inflation rate and that the U.S. will continue at a 7% inflation rate. Assume that the exchange rate will vary the same as the inflation. In this situation, one dollar will buy how many cruzados 5 years from now?
(*Answer:* 32.6)

**14-13** An automaker has a car that gets 10 kilometers per liter of gasoline. Gas prices will increase 12% per year, compounded annually, for the next 8 years. The manufacturer believes that the fuel consumption for its new cars should decline as fuel prices increase to keep the fuel costs constant. To achieve this, what must be the fuel rating, in kilometers per liter, of the cars 8 years hence?

**14-14** An economist has predicted that during the next 6 years, prices in the U.S. will increase 55%. He expects a further increase of 25% in the subsequent 4 years, so that prices at the end of 10 years will have increased to 180% of the present level. Compute the inflation rate, $f$, for the entire 10-year period.

**14-15** Sally Johnson loaned a friend $10,000 at 15% interest, compounded annually. The loan will be paid in five equal end-of-year payments. Sally expects the inflation rate to be 12%. After taking inflation into account, what rate of return is Sally receiving on the loan? Compute your answer to the nearest 0.1%.
(*Answer:* 2.7%)

**14-16** You may pay $15,000 for an annuity that pays $2500 per year for the next 10 years. You want a real rate of return of 5%, and you estimate inflation will average 6% per year. Should you buy the annuity?

**14-17** Inflation has been reality for the general economy of the U.S. in many years. Given this assumption, calculate the number of years it will take for the purchasing power of today's dollars to equal *one-fifth* of their present value. Assume that inflation will average 6% per year.

**14-18** A homebuilder's advertising had the caption, "Inflation to Continue for Many Years." The ad explained that if one buys a home now for $297,000, and inflation continues at 7%, the home will be worth $819,400 in 15 years. Thus, by buying a new home now, one can realize a profit of $522,400 in 15 years. Do you find this logic persuasive? Explain.

**14-19** You were recently looking at the historical prices paid for homes in a neighborhood that interests you. Calculate on a year-to-year basis how home prices in this neighborhood have inflated (*a–e* in the table).

| Year | Average Home Price | Inflation Rate for That Year |
|---|---|---|
| 5 years ago | $265,000 | (*a*) |
| 4 years ago | 267,000 | (*b*) |
| 3 years ago | 272,000 | (*c*) |
| 2 years ago | 280,000 | (*d*) |
| Last year | 283,000 | (*e*) |
| This year | 290,000 | (*f*, see below) |

(*f*) What is your estimate of the inflation rate for this year?

**14-20** The average cost of a certain model car was $18,000 ten years ago. This year the average cost is $30,000.

(*a*) Calculate the average monthly inflation rate ($f_m$) for this model.

(*b*) Given the monthly rate $f_m$, what is the effective annual rate, $f$, of inflation for this model.

(*c*) Estimate what these will sell for 10 years from now, expressed in today's dollars.

*Contributed by D. P. Loucks, Cornell University*

**14-21** Dale saw that the campus bookstore is having a special on pads of computation paper normally priced at $3 a pad, now on sale for $2.50 a pad. This sale is unusual and Dale assumes the paper will not be put on sale again. On the other hand, Dale expects that there will be no increase in the $3 regular price, even though the inflation rate is 2% every 3 months. Dale believes that competition in the paper industry will keep wholesale and retail prices constant. He uses a pad of computation paper every 3 months. Dale considers 19.25% a suitable minimum attractive rate of return. Dale will buy one pad of paper for his immediate needs. How many extra pads of computation paper should he buy?   (*Answer:* 4)

**14-22** If inflation averages 2.5% each year from 2000 to 2009, what is the purchasing power in 2000 dollars of $25,000 in 2009?
*Contributed by Paul R. McCright, University of South Florida*

**14-23** If inflation is currently 3.35% and a bank is lending money at 6.65% interest, what is the real interest rate the bank is earning on its loans?
*Contributed by Paul R. McCright, University of South Florida*

**14-24** Samantha receives a starting salary offer of $60,000 for Year 1. If inflation is 4% each year, what must her salary be to have the same purchasing power in Year 10? Year 20? Year 30? Year 40?

**14-25** Assume that Samantha (Problem 14-24) receives an annual 5% raise. How much more, in Year-1 dollars, is her salary in Year 10? Year 20? Year 30? Year 40?

**14-26** Inflation is 4%. If $1000 is invested in an account paying 6% compounded semiannually, what is the Year-0 dollar value of the account at the end of the 5 years?

**14-27** Assume your salary is $55,000 in 2012 and $160,000 in 2042. If inflation has averaged 2% per year, what is the real or differential inflation rate of salary increases?

**14-28** In the 1920 Sears Roebuck catalog, an oak chest of drawers costs $8 plus freight. In 1990 this same chest of drawers, in good condition, is $1200. If the average rate of inflation over that 70-year period was 3%, what was the average yearly rate of appreciation, adjusted for inflation?

**14-29** Assume inflation is 5% per year. What is the price tag in 6 years for an item that has an inflation rate of 4% that costs $400 today?   (*Answer:* $671)

**14-30** The expected rise in prices due to inflation over the next 5 years is expected to be 30%. Determine the average annual inflation rate over the 5-year period.

**14-31** Explain how high inflation in a booming real estate market could benefit an engineer who sells a home 5 years after she buys it.

**14-32** Felix Jones, a recent engineering graduate, expects a starting salary of $65,000 per year. His future employer has averaged 5% per year in salary increases for the last several years. If inflation is estimated to be 4% per year for the next 3 years, how much, in Year-1 dollars, will Felix be earning each year? What is the inflation rate in Felix's salary?

**14-33** The price of a HeeHaw Model BR549 computer is presently $2200. If deflation of 2% per quarter is expected on this computer, what will its price be in nominal dollars at the end of 1 year? If inflation is 4.5% per year, what will the price be in Year-0 dollars?   (*Answer:* $1942 Year-0 dollars)

**14-34** You place $4000 into an account paying 8% compounded annually. Inflation is 5% during each of the next 3 years. What is the account's value at the end of the 3 years in Year-0 dollars?

**14-35** The following series from the *Historical Statistics of the United States* can be combined with data in Table 14-2 to construct a long-term measure of inflation. You are asked to develop an average inflation rate from 1779 through 2010. Describe and defend your solution.

| 1910 = 100 (base) | | | | | | | |
|---|---|---|---|---|---|---|---|
| **Year** | 1779 | 1785 | 1800 | 1803 | 1830 | 1850 | 1864 | 1880 |
| **Index** | 226 | 92 | 129 | 118 | 91 | 84 | 193 | 100 |

| 1927 = 100 (base) | | | | | | | |
|---|---|---|---|---|---|---|---|
| **Year** | 1890 | 1910 | 1920 | 1921 | 1932 | 1940 | 1950 | 1955 |
| **Index** | 56 | 90 | 154 | 98 | 65 | 79 | 162 | 174 |

**14-36** If $10,000 is deposited in a 5% savings account and inflation is 3%, what is the value of the account at the end of Year 20 in Year-0 dollars? If the time value of money is 4%, what is the present worth?

**14-37** The cost of garbage pickup in Green Gulch is $4,500,000 for Year 1. Estimate the cost in Year 3 in Year-1 dollars and in nominal dollars. The population is increasing at 6%, the nominal cost per ton is increasing at 5%, and the inflation rate is estimated at 4%.   (*Answer:* $5.57M in nominal $s)

**14-38** Your beginning salary is $70,000. You deposit 10% each year in a savings account that earns 6% interest. Your salary increases by 5% per year and inflation is 3% per year. What value does your savings account show after 40 years? What is the value in Year-1 dollars?

**14-39** The market for widgets is increasing by 15% per year from current profits of $200,000. Investing in a design change will allow the profit per widget to stay steady; otherwise the price will drop by 3% per year. If inflation in the economy is 2%, what is the present worth in Year-1 dollars of the savings over the next 5 years? 10 years? The interest rate is 10%.

**14-40** Enrollment at City University is increasing 3% per year, its cost per credit hour is increasing 8% per year, and state funds are decreasing by 4% per year. State funds currently pay half of the costs for City U., while tuition pays the rest. What annual increase in tuition is required?

**14-41** A homeowner is considering an upgrade of a fuel-oil-based furnace to a natural gas unit. The investment in the fixed equipment, such as a new boiler, will be $2500 installed. The cost of the natural gas will average $60 per month over the year, instead of the $145 per month that the fuel oil costs. If funds cost 9% per year and cost inflation in fossil fuels will be 3% per year, how long will it take to recover the initial investment? Solve on a monthly basis.

**14-42** Joan earns a salary of $110,000 per year, and she expects to receive increases at a rate of 4% per year for the next 30 years. She is purchasing a home for $380,000 at 7% for 30 years (under a special veterans, preference loan with 0% down). She expects the home to appreciate at a rate of 3% per year. She will also save 10% of her gross salary in savings certificates that earn 5% per year. Assume that her payments and deposits are made annually. If inflation is assumed to have a constant 5% rate, what is the value (in Year-1 dollars) of each of Joan's two investments at the end of the 30-year period? Use a before-tax analysis.

**Before-Tax Cases**

**14-43** Using a market interest rate of 15% and an inflation rate of 8%, calculate the future equivalent in Year 15 of:

(*a*) Dollars having today's purchasing power.

(*b*) Then-current purchasing power dollars, of $10,000 today.

**14-44** The City of Columbia is trying to attract a new manufacturing business. It has offered to install and operate a water pumping plant to provide service to the proposed plant site. This would cost $50,000 now, plus $5000 per year in operating costs for the next 10 years, all measured in Year-0 dollars.

To reimburse the city, the new business must pay a fixed uniform annual fee, *A*, at the end of each year

for 10 years. In addition, it is to pay the city $50,000 at the end of 10 years. It has been agreed that the city should receive a 3% rate of return, after taking an inflation rate, *f*, of 7% into account.

Determine the amount of the uniform annual fee. (*Answer:* $12,100)

**14-45** A firm is having a large piece of equipment overhauled. It expects that the machine will be needed for the next 12 years. The firm has an 8% minimum attractive rate of return. The contractor has suggested three alternatives:

*A*. A complete overhaul for $6000 that should permit 12 years of operation.

*B*. A major overhaul for $4500 that can be expected to provide 8 years of service. At the end of 8 years, a minor overhaul would be needed.

*C*. A minor overhaul now. At the end of 4 and 8 years, additional minor overhauls would be needed.

If minor overhauls cost $2500, which alternative should the firm select? If minor overhauls, which now cost $2500, increase in cost at +5% per year, but other costs remain unchanged, which alternative should the firm select?   (*Answers:* Alt. *C*; Alt. *A*)

**14-46** A group of students decided to lease and run a gasoline service station. The lease is for 10 years. Almost immediately the students were confronted with the need to alter the gasoline pumps to read in liters. The Dayton Company has a conversion kit available for $900 that may be expected to last 10 years. The firm also sells a $500 conversion kit that has a 5-year useful life. The students believe that any money not invested in the conversion kits may be invested elsewhere at a 10% interest rate. Income tax consequences are to be ignored in this problem.

(*a*) Assuming that future replacement kits cost the same as today, which alternative should be selected?

(*b*) If one assumes a 7% inflation rate, which alternative should be selected?

**14-47** Pollution control equipment must be purchased to remove the suspended organic material from liquid being discharged from a vegetable packing plant. Two alternative pieces of equipment are available that would accomplish the task. A Filterco unit costs $7000 and has a 5-year useful life. A Duro unit, on the other hand, costs $10,000 but will have a 10-year useful life.

With inflation, equipment costs are rising at 8% per year, compounded annually, so when the Filterco unit needed to be replaced, the cost would be much more than $7000. Based on a 10-year analysis period, and a 20% minimum attractive rate of return, which pollution control equipment should be purchased?

**14-48** Sam bought a house for $150,000 with some creative financing. The bank, which agreed to lend Sam $120,000 for 6 years at 15% interest, took a first mortgage on the house. The Joneses, who sold Sam the house, agreed to lend Sam the remaining $30,000 for 6 years at 12% interest. They received a second mortgage on the house. Thus Sam became the owner without putting up any cash. Sam pays $1500 a month on the first mortgage and $300 a month on the second mortgage. In both cases these are "interest only" loans, and the principal is due at the end of the loan.

Sam rented the house to Justin and Shannon, but after paying the taxes, insurance, and so on, he had only $800 left, so he was forced to put up $1000 a month to make the monthly mortgage payments. At the end of 3 years, Sam sold the house for $205,000. After paying off the two loans and the real estate broker, he had $40,365 left. After taking an 8% inflation rate into account, what was his before-tax rate of return?

**14-49** Ima Luckygirl recently found out that her grandfather had passed away and left her his Rocky Mountain Gold savings account. The only deposit was 50 years ago, when Ima's grandfather deposited $2500. If the account has earned an average rate of 10% and inflation has been 4%, answer the following:

(*a*) How much money is now in the account in *actual dollars*?

(*b*) Express the answer to part (*a*) in terms of the purchasing power of dollars from 50 years ago.

**14-50** Auntie Frannie wants to provide tuition for her twin nephews to attend a private school. She intends to send a check for $5000 at the end of each of the next 8 years.

(*a*) If general price inflation, as well as tuition price inflation, is expected to average 5% per year for those 8 years, calculate the present worth of the gifts. Assume that the real interest rate will be 3% per year.

(*b*) If Auntie Frannie wants her gifts to keep pace with inflation, what would be the present worth of her gifts? Again assume inflation is 5% and the real interest rate is 3%.

**14-51** As a recent graduate, you are considering employment offers from three different companies. However, in an effort to confuse you and perhaps make their offers seem better, each company has used a different *purchasing power base* for expressing your annual salary over the next 5 years. If you expect inflation to be 6% for the next 5 years and your personal (real) MARR is 8%, which plan would you choose?

*Company A:* A constant $50,000 per year in terms of today's purchasing power.

*Company B:* $45,000 the first year, with increases of $2500 per year thereafter.

*Company C:* A constant $65,000 per year in terms of Year-5-based purchasing power.

**14-52** If McDonnell Manufacturing has a MARR of 20%, inflation is 2.75%, and the company uses present worth analysis with a planning horizon of 15 years in making economic decisions, which of the following alternatives would be preferred?

| | Alternative A | Alternative B |
|---|---|---|
| Initial costs | $236,000 | $345,000 |
| Annual operating costs | 64,000 | 38,000 |
| Annual maintenance costs | 4,000 | 5,000 |
| Salvage value (EOY 20) | 23,000 | 51,000 |

*Contributed by Paul R. McCright, University of South Florida*

**14-53** Due to competition from a new polycarbon, revenues for the mainstay product of Toys-R-Polycarbon are declining at 5% per year. Revenues will be $5M for this year. The product will be discontinued at the end of Year 6. If the firm's interest rate is 10%, calculate the PW of the revenue stream.

**14-54** The ABC Block Company anticipates receiving $4000 per year from its investments (inflation rate of 0%) over the next 5 years. If ABC's interest rate is 8% and the inflation rate is 3%, determine the present value of the cash flows.

**14-55** The Wildwood Widget Company needs a milling machine for its new assembly line. The machine presently costs $85,000, but has a cost inflation rate of 2%. Widget will not need to purchase the machine for 2 years. If general inflation is expected to be 4% per year during those 2 years, determine the price of the machine. What is the present worth of the machinery if the market rate of interest for Widget is 9%?

**14-56** Bob has lost his job and had to move back in with his mother. She agreed to let Bob have his old room back on the condition that he pay her $1000 rent per year, and an additional $1000 every other year to pay for her biannual jaunt to Florida. Since he is down on his luck, she will allow him to pay his rent at the end of the year. If inflation is 6% and Bob's interest rate is 15%, how much is the present cost (in Year-1 dollars) for a 5-year contract? (*Note*: Mom's trips are in Years 2 and 4).

**14-57** A 30-year mortgage for $100,000 has been issued. The interest rate is 10% and payments are made annually. If your time value of money is 12%, what is the PW of the payments in Year-1 dollars if inflation is 0%? 3%? 6%? 9%?    *Answer*: $PW_{3\%} = \$70,156$)

## Indexes

**14-58** What is the Consumer Price Index (CPI)? What is the difference between commodity-specific and composite price indexes? Can each be used in engineering economic analysis?

**14-59** A composite price index for the cost of vegetarian foods called *Eggs, Artichokes, and Tofu* (EAT) was 330 ten years ago and has averaged an annual increase of 12% since. Calculate the current value of the index.

**14-60** From the data in Table 14-1 in the text, calculate the *overall rate change* of first-class postage as measured by the LCI for the following decades:

(*a*) The 1970s (1970–1979)

(*b*) The 1980s (1980–1989)

(*c*) The 1990s (1990–1999)

(*d*) The 2000s (2000–2010)

**14-61** From the data in Table 14-1 in the text, calculate the *average annual inflation rate* of first class postage as measured by the LCI for the following years:

(*a*) End of 1970 to end of 1979

(*b*) End of 1980 to end of 1989

(*c*) End of 1990 to end of 1999

(*d*) End of 1999 to end of 2009

**14-62** From the data in Table 14-2 in the text, calculate the *average annual inflation rate* as measured by the CPI for the following years:

(*a*) End of 1973 to end of 1982

(*b*) End of 1980 to end of 1989

(*c*) End of 1985 to end of 2002

(*d*) End of 1994 to end of 2009

**14-63** (*a*) Compute the equivalent annual inflation rate, based on the Consumer Price Index, for the period from 1990 to 2010.

(*b*) Using the equivalent annual inflation rate computed in part (*a*), estimate the consumer price index for 2015.

**14-64** Redo Problem 14-17, but estimate the annual inflation rate using the period from 1996 to 2006 and the CPI values in Table 14-2.

**14-65** Here is some information about a professor salary index (PSI).

| Year | PSI | Change in PSI |
|------|-----|---------------|
| 2004 | 82  | 3.22%         |
| 2005 | 89  | 8.50          |
| 2006 | 100 | a             |
| 2007 | b   | 4.00          |
| 2008 | 107 | c             |
| 2009 | 116 | d             |
| 2010 | e   | 5.17          |
| 2011 | 132 | 7.58          |

(*a*) Calculate the unknown quantities *a, b, c, d, e* in the table. Review Equation 14-3.

(*b*) What is the *base year* of the PSI? How did you determine it?

(*c*) Given the data for the PSI, calculate the *average annual price increase* in salaries paid to professors for between 2004 and 2008 and between 2006 and 2011.

**14-66** Homeowner Henry is building a fireplace for his house. The fireplace will require 800 bricks.

(*a*) If the cost of a Chimney Brick in 1978 was $2.10, calculate the material cost of Henry's project in 1998. The Chimney Brick Index (CBI) was 442 in 1978 and is expected to be 618 in 1998.

(*b*) Estimate the material cost of a similar fireplace to be built in the year 2012. What assumption did you make?

## Different Rates

**14-67** General price inflation is estimated to be 3% for the next 5 years, 5% the 5 years after that, and 8% the following 5 years. If you invest $10,000 at 10% for those 15 years, what is the future worth of your investment in actual dollars at that time and in Year-0 dollars at that time?

**14-68** Due to cost structures, trade policies, and corporate changes, three big automakers have different inflation rates for the purchase prices of their vehicles. Which car should Mary Clare buy at graduation 3 years from now, assuming everything but purchase price is equivalent?

| Automaker | Current Price | Price Will Inflate x% per Year |
|-----------|---------------|-------------------------------|
| X | $27,500 | 4 % |
| Y | 30,000 | 1.5 |
| Z | 25,000 | 8 |

**14-69** Granny Viola has been saving money in the Bread & Butter mutual fund for 15 years. She has been a steady depositor over those years and has a pattern of putting $100 into the account every 3 months. If her original investment 15 years ago was $500 and interest in the account has varied as shown, what is the current value of her savings?

| Years | Interest Earned in the Account |
|-------|-------------------------------|
| 1–5 | 12% compounded quarterly |
| 6–10 | 16 compounded quarterly |
| 10–15 | 8 compounded quarterly |

**14-70** Andrew just bought a new boat for $15,000 to use on the river near his home. He has taken delivery of the boat and agreed to the terms of the following loan: all principal and interest is due in 3 years (balloon loan), first-year annual interest (on the purchase price) is set at 5%, this is to be adjusted up 1.5% per year for each of the following years of the loan.

(a) How much does Andrew owe to pay off the loan in 3 years?

(b) If inflation is 4%, what is the payment in Year-0 dollars?

**14-71** Given the following data, calculate the present worth of the investment.

First cost = $60,000    Project life = 10 years
Salvage value = $15,000    MARR = 25%

General price inflation = 4% per year
    Annual cost 1 = $4500 in Year 1 and
        inflating at 2.5% per year
    Annual cost 2 = $7000 in Year 1 and
        inflating at 10.0% per year
    Annual cost 3 = $10,000 in Year 1 and
        inflating at 6.5% per year
    Annual cost 4 = $8500 in Year 1 and
        inflating at −2.5% per year

**14-72** As the owner of Beanie Bob's Basement Brewery, you are interested in a construction project to increase production to offset competition from your crosstown rival, Bad Brad's Brewery and Poolhall. Construction cost percentage increases, as well as current cost estimates, are given for a 3-year period. Use a market interest rate of 25%, and assume that general price inflation is 5% over the 3-year period.

| Item | Cost if Incurred Today | Cost Percentage Increase | | |
|------|------------------------|------------|--------|--------|
| | | Year 1 | Year 2 | Year 3 |
| Structural metal/concrete | $120,000 | 4.3% | 3.2% | 6.6% |
| Roofing materials | 14,000 | 2.0 | 2.5 | 3.0 |
| Heating/plumbing equipment/fixtures | 35,000 | 1.6 | 2.1 | 3.6 |
| Insulation material | 9,000 | 5.8 | 6.0 | 7.5 |
| Labor | 85,000 | 5.0 | 4.5 | 4.5 |

(a) What would the costs be for labor in Years 1, 2, and 3?

(b) What is the *average percentage increase* of labor cost over the 3-year period?

(c) What is the present worth of the insulation cost of this project?

(d) Calculate the future worth of the labor and insulation material cost portion of the project.

(e) Calculate the present worth of the total construction project for Beanie Bob.

**14-73** Philippe Marie wants to race in the Tour de France 10 years from now. He wants to know what the cost of a custom-built racing bicycle will be 10 years from today.

| Item | Current Cost | Cost Will Inflate x% per Year |
|------|-------------|-------------------------------|
| Frame | $800 | 2 % |
| Wheels | 350 | 10 |
| Gearing system | 200 | 5 |
| Braking system | 150 | 3 |
| Saddle | 70 | 2.5 |
| Finishes | 125 | 8 |

### After-Tax Cases

**14-74** Sally Seashell bought a lot at the Salty Sea for $18,000 cash. She does not plan to build on the lot, but instead will hold it as an investment for 10 years. She wants a 10% after-tax rate of return after taking the 6% annual inflation rate into account. If income taxes amount to 15% of the capital gain, at what price must she sell the lot at the end of the 10 years? (*Answer:* $95,188)

**14-75** The U.S. tax laws provide for the depreciation of equipment based on original cost. Yet owing to substantial inflation, the replacement cost of equipment is often much greater than the original cost. What effect, if any, does this have on a firm's ability to buy new equipment to replace old equipment?

**14-76** Tom Ward put $10,000 in a 5-year certificate of deposit that pays 12% interest per year. At maturity he will receive his $10,000 back. Tom's marginal income tax rate is 42%. If the inflation rate is 7% per year, find his

(a) before-tax rate of return, ignoring inflation

(b) after-tax rate of return, ignoring inflation

(c) after-tax rate of return, after taking inflation into account

**14-77** Dick DeWolf and his wife have a total taxable income of $60,000 this year and file a joint federal income tax return. If inflation continues for the next 20 years at a 7% annual rate, Dick wonders what their taxable income must be to provide the same purchasing power after taxes. Assuming the federal income tax rate table is unchanged, what must their taxable income be 20 years from now?

**14-78** Assume that your private university's tuition is $28,000.

(a) If the inflation rate for tuition is 5% per year, calculate what the tuition will cost 20 years from now.

(b) If the general inflation rate for the economy is 3% per year, express that future tuition in today's dollars.

(c) Calculate the amount you would have to invest today to pay for tuition costs 20, 21, 22, and 23 years from now. Assume you can invest at 7% per year, your income tax rate is 40% per year, and the tuition has to be paid at the beginning of the year.

*Contributed by D. P. Loucks, Cornell University*

**14-79** You must decide when to go on a vacation. One option is right after graduation. The other option is to wait and go after spending 2 years with the Peace Corps. Assume the vacation costs $2500 now, and your annual income tax rate is 20% now, and is expected to continue to be 20% during the next 2 years. Also assume the annual inflation rate for a week's trip to Hawaii (hotel included) is 4%.

(a) Calculate the additional money you could spend on your vacation, after taxes, by putting your vacation money ($2500) into a taxable investment at 6% per year (before taxes) and waiting 2 years until after you come out of the Peace Corps compared to taking your well-deserved vacation now.

(b) If your tax rate drops to 0% while you're in the Peace Corps, how much additional money will you have for your vacation?

*Contributed by D. P. Loucks, Cornell University*

**14-80** Sam Johnson inherited $85,000 from his father. Sam is considering investing the money in a house, which he will then rent to tenants. The $85,000 cost of the property consists of $17,500 for the land, and $67,500 for the house. Sam believes he can rent the house and have $8000 a year net income left after paying the property taxes and other expenses. The house will be depreciated by straight-line depreciation using a 45-year depreciable life.

(a) If the property is sold at the end of 5 years for its book value at that time, what after-tax rate of return will Sam receive? Assume that his marginal personal income tax rate is 34% for federal and state taxes.

(b) Now assume there is 7% per year inflation, compounded annually. Sam will increase the rent 7% per year to match the inflation rate, so that after considering increased taxes and other expenses, the annual net income will go up 7% per year. Assume Sam's marginal income tax rate remains at 34% for all ordinary taxable income related to the property. The value of the property is now projected to increase from its present $85,000 at a rate of 10% per year, compounded annually.

If the property is sold after 5 years, compute the rate of return on the after-tax cash flow in actual dollars. Also compute the rate of return on the after-tax cash flow in Year-0 dollars.

**14-81** A small research device can be purchased for $10,000 and depreciated by MACRS depreciation. The net benefits from the device, before deducting depreciation, are $2000 at the end of the first year and increasing $1000 per year after that (second year equals $3000, third year equals $4000, etc.), until the device is hauled to the junkyard at the end of 7 years. During the 7-year period there will be an inflation rate $f$ of 7%.

This profitable corporation has a 40% combined federal and state income tax rate. If it requires a real 12% after-tax rate of return on its investment, should the device be purchased?

**14-82** A couple in Ruston, Louisiana, must decide whether it is more economical to buy a home or to continue to rent during an inflationary period. The couple rents a one-bedroom duplex for $450 a month plus $139 a month in basic utilities (heating and cooling). These costs have a projected inflation rate of 5%, so the couple's monthly costs per year over a 10-year planning horizon are:

| $n =$ | 1 | 2 | 3 | 4 | 5 | 6 | 7 | 8 | 9 | 10 |
|---|---|---|---|---|---|---|---|---|---|---|
| Rent | 450 | 473 | 496 | 521 | 547 | 574 | 603 | 633 | 665 | 698 |
| Utilities | 139 | 146 | 153 | 161 | 169 | 177 | 186 | 196 | 205 | 216 |

The couple would like to buy a home that costs $75,000. A local mortgage company will provide a loan that requires a down payment of 5% plus estimated closing costs of 1% cash. The couple prefers a 30-year fixed-rate mortgage with an 8% interest rate. The couple falls in the 30% marginal income tax rate (federal plus state), and as such, buying a home will provide them some tax write-off. It is estimated that the basic utilities for the home inflating at 5% will cost $160 per month; insurance and maintenance also inflating at 5% will cost $50 per month. The home will appreciate in value about 6% per year. Assuming a nominal interest rate of 15.5%, which alternative should the couple select? Use a present worth analysis. (*Note:* Realtor's sales commission here is 5%.)

**14-83** When there is little or no inflation, a homeowner can expect to rent an unfurnished home for 12% of its market value. About $1/8$ of the rental income is paid out for property taxes, insurance, and other operating expenses. Thus the net annual income to the owner is 10.5% of the market value. Since prices are relatively stable, the future selling price of the property often equals the original price paid by the owner.

For a $150,000 property (where the land is estimated at $46,500 of the $150,000), compute the after-tax rate of return, assuming the selling price 59 months later (in December) equals the original purchase price. Use modified accelerated cost recovery system depreciation beginning January 1. Also, assume a 35% income tax rate.   (*Answer:* 6.84%)

**14-84** (This continues Problem 14-83.) As inflation has increased throughout the world, the rental income of homes has decreased and a net annual rental income of 8% of the market value is common. On the other hand, the market value of homes tends to rise about 2% per year more than the inflation rate. As a result, both annual net rental income, and the resale value of the property rise faster than the inflation rate. Consider the following situation.

A $150,000 property (with the house valued at $103,500 and the land at $46,500) is purchased for cash in Year 0. The market value of the property increases at a 12% annual rate. The annual rental income is 8% of the beginning-of-year market value of the property. Thus the rental income also increases each year. The general inflation rate $f$ is 10%.

The individual who purchased the property has an average income tax rate of 35%.

(*a*) Use MACRS depreciation, beginning January 1, to compute the actual dollar after-tax rate of return for the owner, assuming he sells the property 59 months later (in December).

(*b*) Similarly, compute the after-tax rate of return for the owner, after taking the general inflation rate into account, assuming he sells the property 59 months later.

**14-85** Consider two mutually exclusive alternatives stated in Year-0 dollars. Both alternatives have a 3-year life with no salvage value. Assume the annual inflation rate is 5%, an income tax rate of 25%, and straight-line depreciation. The minimum attractive rate of return (MARR) is 7%. Use rate of return analysis to determine which alternative is preferable.

| Year | A | B |
|------|--------|--------|
| 0 | −$420 | −$300 |
| 1 | 200 | 150 |
| 2 | 200 | 150 |
| 3 | 200 | 150 |

**14-86** The purchase of a large-volume copier is being considered. Use MACRS and the current tax law. The corporation's taxable income exceeds $20M. The first cost is $18,000. Maintenance cost per year is $1300 in Year-1 dollars. The salvage value after 10 years is $2000 in Year-10 dollars. Assume the real MARR is 8%, and the inflation rate is 5%. Find the equivalent annual cost after taxes. Does inflation increase or decrease the annual cost?

**14-87** Ruralville is suffering a 1% annual loss of population and property values. Even so, Ruralville must maintain its tax collections at a constant value of $3.2 million. If the inflation rate is 4.5%, what inflation rate in taxes for the remaining taxpayers is required for Ruralville to attain its goal? (*Note:* Although Ruralville uses a rate of 6% for discounting purposes, that rate is irrelevant to this problem.)

# CASES

The following case from *Cases in Engineering Economy 2$^{nd}$* are suggested as matched with this chapter.

**CASE 4   Northern Windows**
Tax affected energy conservation project for a home with simplifying assumptions made clear. Includes inflation.

**CASE 37   Brown's Nursery (Part B)**
Adds inflation to Case 36.

**CASE 38   West Muskegon Machining and Manufacturing**
More complex inflation and tax problem with sunk cost and leverage.

**CASE 39   Uncertain Demand at WM$^3$**
Includes inflation, taxes, and uncertainty.

**CASE 41   Freeflight Superdiscs**
Inflation and sensitivity analysis for three alternatives. Includes taxes.

# SELECTION OF A MINIMUM ATTRACTIVE RATE OF RETURN

## What's the Rate of Return on a Dam?

The broadly defined "energy" industry has done well in recent years even when the U.S. and world economies have not. It's now quite evident that humans are becoming increasingly dependent on various forms of energy to power our world. Yet we clearly understand that the use of traditional means (oil, coal, etc.), over time do pose environmental damage threats to the world. Thus government and businesses alike are scrambling to come up with "green" solutions to the world's energy needs. *But is the economic investment in renewable forms of energy (solar, wind, water) worth the risks? Do private investors need to reconsider their rates of return when evaluating risky investments?*

Consider Tajikistan, a former Soviet republic with a population of about 7 million people. While the country is lacking in most natural resources and has few major industries (cotton and aluminum), it is the source of over 40% of Central Asia's water. Tajikistan's government is relying on hydroelectric power for the country's economic and political future. The Rogun Dam is an unfinished dam across the Vakhsh River in southern Tajikistan with a planned capacity of 3600 megawatts of power. Construction began in 1976 but was halted after the collapse of the Soviet Union. In 2010, hoping to revitalize this project and complete the dam, Tajikistan made a public offering of bonds worth $1.4 billion. Because this is an extremely expensive undertaking, the country must rely heavily on foreign investment if the project is to be completed, and the investors are likely to expect significant rates of returns.

Upstream and downstream countries have different perspectives on how completion of dam projects will affect their own economies and water supplies. The more hydroelectric projects

Tajikistan develops, the more control they will have over water in the region; thus political tensions are a significant issue. The Nurek Dam, downstream on the Vakshsh River, is one of the world's highest. Its associated 9-unit power plant provides about 3000 megawatts of power, and the water impounded is used to irrigate about 1.6 million acres of farmland.

In February 2010, Uzbekistan's prime minister sent a letter to his Tajikistan counterpart demanding an independent examination of the possible consequences of the Rogun Dam, including the impact on Uzbek irrigation systems. In addition, Tajikistan is in an earthquake zone, and any dam projects must be built to withstand significant seismic shocks, not to mention the effects that these projects will have on the plant and animal life in the region. This kind of investment is very risky for any business or government.

*Contributed by Karen M. Bursic, University of Pittsburgh; acknowledgement to NSF DUE 071780; for more see www.modelsandmodeling.net.*

## QUESTIONS TO CONSIDER

1. What are the advantages of relying on hydroelectric power in a country such as Tajikistan?

2. What are the disadvantages of relying on hydroelectric power in a country such as Tajikistan? What are the possible consequences of new dam construction?

3. How would you balance the benefits against the costs in order to determine whether a new hydroelectric generating facility should be built in a given location?

4. What are some of the ethical issues that arise when one is considering investment in foreign countries?

5. What risks are associated with the Rogun Dam for Tajikistan? For its neighboring countries downstream? For foreign investors?

6. How would a private company determine the MARR to be used in evaluating such a project?

## After Completing This Chapter...

### The student should be able to:

- Define various sources of capital and the costs of those funds to the firm.
- Discuss the impact of inflation and the cost of borrowed money.
- Select a firm's MARR based on the opportunity cost approach for analyzing investments.
- Adjust the firm's MARR to account for risk and uncertainty.
- Use spreadsheets to develop cumulative investments and the opportunity cost of capital.

The preceding chapters have said very little about what interest rate or minimum attractive rate of return is suitable for use in a particular situation. This problem is complex, and no single answer is always appropriate. A discussion of what interest rate to use must inevitably begin by examining the sources of capital, followed by looking at the prospective investment opportunities and risk. Only in this way can an interest rate or minimum attractive rate of return (MARR) be chosen intelligently.

## SOURCES OF CAPITAL

In broad terms there are three sources of capital available to a firm: money generated from the firm's operation, borrowed money, and money from selling stock.

### Money Generated from the Firm's Operations

A major source of capital investment money is retained profits from the firm's operation. Overall, industrial firms retain about half of their profits and pay out the rest to stockholders. In addition to profit, the firm generates money equal to the annual depreciation charges on existing capital assets. In other words, a profitable firm will generate money equal to its depreciation charges *plus* its retained profits. Even a firm that earns zero profit will still generate money from operations equal to its depreciation charges. (A firm with a loss, of course, will have still less funds.)

### External Sources of Money

When a firm requires money for a few weeks or months, it typically borrows from banks. Longer-term unsecured loans (of, say, 1–4 years) may also be arranged through banks. While banks undoubtedly finance a lot of capital expenditures, regular bank loans cannot be considered a source of permanent financing.

Longer-term financing is done by selling bonds to banks, insurance firms, pension funds, and the public. A wide variety of bonds exist, but most are interest-only loans, where interest is paid every 6 months or once a year and the principal is due at the bond's maturity. Common maturities are 10 to 30 years, although some extend to 100 years and a few even longer. Usually interest rates are explicitly stated, and Chapter 7 includes examples of how to calculate the interest rates.

A firm can also raise funds by issuing new stock (shares of ownership in the firm). Many firms have also repurchased their own stock in the past, which is called *treasury stock*. Another way firms can raise funds is to sell this treasury stock.

One of the finance questions each firm must address is maintaining an appropriate balance between debt (loans and bonds) and equity (stock and retained earnings). The debt has a maturity date, and there are legal obligations to repay it unless the firm declares bankruptcy. On the other hand, stockholders expect a higher rate of return to compensate them for the risks of ownership. Those who are interested in the models used to calculate the cost of equity capital are referred to *The Economic Analysis of Industrial Projects*, 3rd edition, by Eschenbach, Lewis Hartman, and Bussey, published by Oxford University Press.

### Choice of Sources of Funds

Choosing the source of funds for capital expenditures is a decision for the firm's top executives, and it may require approval of the board of directors. When internal operations generate adequate funds for the desired capital expenditures, external sources of money are not likely to be used. But when the internal sources are inadequate, external sources must be employed or the capital expenditures will have to be deferred or canceled.

# COST OF FUNDS

## Cost of Borrowed Money

A first step in deciding on a minimum attractive rate of return might be to determine the interest rate at which money can be borrowed. Longer-term loans or bonds may be obtained from banks, insurance companies, or the variety of places in which substantial amounts of money accumulates (for example, the sovereign wealth funds of the oil-producing nations).

A large, profitable corporation might be able to borrow money at the **prime rate,** that is, the interest rate that banks charge their best and most sought-after customers. All other firms are charged an interest rate that is higher by one-half to several percentage points. In addition to the firm's financial strength and ability to repay the debt, the interest rate will depend on the debt's duration and on whether the debt has collateral or is unsecured.

## Cost of Capital

Another relevant interest rate is the **cost of capital.** This is also called the weighted average cost of capital (WACC). This is the rate from *all* sources of funds in the firm's overall capitalization. The mechanics for computing the cost of capital or WACC are given in Example 15-1.

---

### EXAMPLE 15–1

For a particular firm, the purchasers of common stock require an 11% rate of return, bonds are sold at a 7% interest rate, and bank loans are available at 9%. Compute the cost of capital or WACC for the following capital structure:

|  |  | Rate of Return |
|---|---|---|
| $ 20 million | Bank loan | 9% |
| 20 | Bonds | 7 |
| 60 | Common stock and retained earnings | 11 |
| $100 million |  |  |

### SOLUTION

The weighted cost of capital "weights" the return on each source of capital by the fraction of the total capital it represents. In this case, 20% of the total capital is from the bank loan, 20% of the capital is from bonds, and 60% of the capital is from equity sources—that is, common stock and retained earnings.

$$\text{WACC}_{\text{before-taxes}} = (0.2)(9\%) + (0.2)(7\%) + (0.6)(11\%)$$
$$= 1.8\% + 1.4\% + 6.6\% = 9.8\%$$

Note that since this is an *average*, the result must be between the lowest rate of return (7%) and the largest (11%). Since it is a *weighted* average, the return with the largest weight (60%) has the most impact on the final average. We recommend some "mental" arithmetic to approximate the expected answer, which will help catch any errors with your calculator.

The cost of capital is also computed after considering that interest payments on debt, like bank loans and bonds, are tax-deductible business expenses. Thus,

$$\text{After-tax interest cost} = (\text{Before-tax interest cost}) \times (1 - \text{Tax rate})$$

If we assume that the firm pays 40% income taxes, the computations become

Bank loan     After-tax interest cost = $9\%(1 - 0.40) = 5.4\%$

Bonds          After-tax interest cost = $7\%(1 - 0.40) = 4.2\%$

Dividends paid on the ownership in the firm (common stock + retained earnings) are not tax deductible. The cost of capital can also be computed by dividing the total amount of interest by the total amount of capital. Combining the three components, the after-tax interest cost for the $100 million of capital is

$$\$20 \text{ million } (5.4\%) + \$20 \text{ million } (4.2\%) + \$60 \text{ million } (11\%) = \$8.52 \text{ million}$$

$$\text{WACC}_{\text{after-taxes}} = \frac{\$8.52 \text{ million}}{\$100 \text{ million}} = 8.52\%$$

In practical situations, the cost of capital is often difficult to compute. The fluctuation in the price of common stock, for example, makes it difficult to pick a cost, and because of the fluctuating prospects of the firm, it is even more difficult to estimate the future benefits that purchasers of the stock might expect to receive. Given the fluctuating costs and prospects of future benefits, what rate of return do stockholders require? There is no precise answer, but we can obtain an approximate answer. As described in the next section, inflation complicates the task of finding the *real* interest rate.

### Inflation and the Cost of Borrowed Money

As inflation varies, what is its effect on the cost of borrowed money? A widely held view has been that interest rates on long-term borrowing, like 20-year Treasury bonds, will be about 3% more than the inflation rate. For borrowers this is the real—that is, after-inflation—cost of money, and for lenders the real return on loans. If inflation rates were to increase, it would follow that borrowing rates would also increase. All this suggests a rational and orderly situation, about as we might expect.

Unfortunately, things have not worked out this way. Figure 15-1 shows that the real interest rate has not always been about 3% and, in fact, there have been long periods during

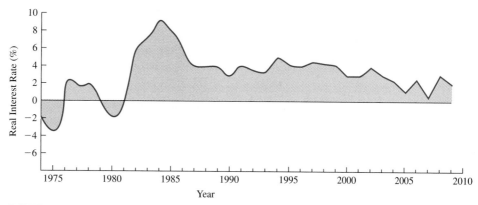

**FIGURE 15-1** The real interest rate. The interest rate on 20-year Treasury bonds *minus* the inflation rate, $f$, as measured by changes in the Consumer Price Index.

which the real interest rate was negative. Can this be possible? Would anyone invest money at an interest rate several percentage points below the inflation rate? Well, consider this: when the U.S. inflation rate was 12%, savings banks were paying 5½% on regular passbook deposits—and there was a lot of money in those accounts. While there must be a relationship between interest rates and inflation, Figure 15-1 suggests that it is complex.

The relationship between the inflation rate and the rate of return on investments is quite important, because inflation reduces the real rate of return (as shown in Chapter 14). In addition, many interest rates are reported without adjusting for inflation. For example, you know the interest rate on your car loan, student loan, house loan, savings account, and so on—but all those rates are stated without adjusting for inflation.

## INVESTMENT OPPORTUNITIES

An industrial firm can invest its money in many more places than are available to an individual. A firm has larger amounts of money, which allows investments that are unavailable to individuals, with their more limited investment funds.

More important, however, is the fact that a firm conducts a business, which itself offers many investment opportunities. While exceptions can be found, a good generalization is that the opportunities for investment of money within the firm are superior to the investment opportunities outside the firm. Consider the available investment opportunities for a particular firm as outlined in Table 15-1. The cumulative investment required for all projects at or above a given rate of return is given in Figure 15-2.

Figure 15-2 illustrates that a firm may have a broad range of investment opportunities available at varying rates of return and with varying lives and uncertainties. It may take some study and searching to identify the better investment projects available to a firm. Typically, the good projects will almost certainly require more money than the firm budgets for capital investment projects.

**TABLE 15–1   A Firm's Available Investment Opportunities**

| Project Number | Project | Cost $(\times 10^3)$ | Estimated Rate of Return |
|---|---|---|---|
| | **Investment Related to Current Operations** | | |
| 1 | New equipment to reduce labor costs | $150 | 30% |
| 2 | Other new equipment to reduce labor costs | 50 | 45 |
| 3 | Overhaul particular machine to reduce material costs | 50 | 38 |
| 4 | New test equipment to reduce defective products produced | 100 | 40 |
| | **New Operations** | | |
| 5 | Manufacture parts that previously had been purchased | 200 | 35 |
| 6 | Further processing of products previously sold in semifinished form | 100 | 28 |
| 7 | Further processing of other products | 200 | 18 |
| | **New Production Facilities** | | |
| 8 | Relocate production to new plant | 250 | 25 |
| | **External Investments** | | |
| 9 | Investment in a different industry | 300 | 20 |
| 10 | Other investment in a different industry | 300 | 10 |
| 11 | Overseas investment | 400 | 15 |
| 12 | Purchase of Treasury bills | Unlimited | 8 |

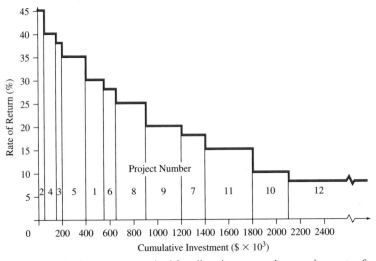

**FIGURE 15–2** Cumulative investment required for all projects at or above a given rate of return.

## Opportunity Cost

We see that there are two aspects of investing that are basically independent. One factor is the source and quantity of money available for capital investment projects. The other aspect is the firm's investment opportunities.

These two situations are typically out of balance, with investment opportunities exceeding the available money supply. Thus some investment opportunities can be selected and others must be rejected. Obviously, we want to ensure that *all the selected projects are better than the best rejected project.* To do this, we must know something about the rate of return on the best rejected project. The best rejected project is the best opportunity forgone, and this in turn is called the **opportunity cost.**

**Opportunity cost = Cost of the best opportunity forgone**

**= Rate of return on the best rejected project**

If one could predict the opportunity cost for some future period (like the next 12 months), this rate of return could be one way to judge whether to accept or reject any proposed capital expenditure. Examples 15-2 and 15-3 illustrate this.

### EXAMPLE 15–2

Consider the situation represented by Table 15-1 and Figure 15-2. For a capital expenditure budget of $1.2 million ($1.2 × $10^6$), what is the opportunity cost?

### SOLUTION

From Figure 15-2 we see that the eight projects with a rate of return of 20% or more require a cumulative investment of $1.2 (× $10^6$). We would take on these projects and reject the other four (7, 11, 10, and 12) with rates of return of 18% or less. The best rejected project is 7, and it has an 18% rate of return. Thus the opportunity cost is 18%.

### EXAMPLE 15–3

Nine independent projects are being considered. Figure 15-3 may be prepared from the following data.

| Project | Cost (thousands) | Uniform Annual Benefit (thousands) | Useful Life (years) | Salvage Value (thousands) | Computed Rate of Return |
|---|---|---|---|---|---|
| 1 | $100 | $23.85 | 10 | $0 | 20% |
| 2 | 200 | 39.85 | 10 | 0 | 15 |
| 3 | 50 | 34.72 | 2 | 0 | 25 |
| 4 | 100 | 20.00 | 6 | 100 | 20 |
| 5 | 100 | 20.00 | 10 | 100 | 20 |
| 6 | 100 | 18.00 | 10 | 100 | 18 |
| 7 | 300 | 94.64 | 4 | 0 | 10 |
| 8 | 300 | 47.40 | 10 | 100 | 12 |
| 9 | 50 | 7.00 | 10 | 50 | 14 |

If a capital budget of $650,000 is available, what is the opportunity cost of capital? With this model, which projects should be selected?

**FIGURE 15-3**  Cumulative cost of projects versus rate of return.

Looking at the nine projects, we see that some are expected to produce a larger rate of return than others. It is natural that if we are to select from among them, we will pick those with a higher rate of return. When the projects are arrayed by rate of return, as in Figure 15-3, Project 2 is the last one funded. Thus, the opportunity cost of capital is 14% from Project 9, the highest ranked unfunded project. This model implies funding Projects 3, 1, 4, 5, 6, and 2.

## SELECTING A MINIMUM ATTRACTIVE RATE OF RETURN

Focusing on the three concepts on the cost of money (the cost of borrowed money, the cost of capital, and opportunity cost), which, if any, of these values should be used as the minimum attractive rate of return (MARR) in economic analyses?

Fundamentally, we know that unless the benefits of a project exceed its cost, we cannot add to the profitability of the firm. A lower boundary for the minimum attractive rate of return must be the cost of the money invested in the project. It would be unwise, for example, to borrow money at 8% and invest it in a project yielding a 6% rate of return.

Further, we know that no firm has an unlimited ability to borrow money. Bankers—and others who evaluate the limits of a firm's ability to borrow money—look at both the profitability of the firm and the relationship between the components in the firm's capital structure. This means that continued borrowing of money will require that additional stock be sold to maintain an acceptable ratio between **ownership** (equity) and **debt.** In other words, borrowing for a particular investment project is only a block of money from the overall capital structure of the firm. This suggests that the MARR should not be less than the cost of capital. Finally, we know that the MARR should not be less than the rate of return on the best opportunity forgone. Stated simply,

Minimum attractive rate of return should be equal to the largest of: cost of borrowed money, cost of capital, or opportunity cost.

## ADJUSTING MARR TO ACCOUNT FOR RISK AND UNCERTAINTY

We know from our study of estimating the future that what actually occurs is often different from the estimate. When we are fortunate enough to be able to assign probabilities to a set of possible future outcomes, we call this a **risk** situation. We saw in Chapter 10 that techniques like expected value and simulation may be used when the probabilities are known.

**Uncertainty** is the term used to describe the condition when the probabilities are *not* known. Thus, if the probabilities of future outcomes are known, we have *risk*, and if the probabilities are unknown, we have *uncertainty*. With uncertainty, adjustments for risk are more subjective.

In projects accompanied by normal business risk and uncertainty, the MARR is used without adjustment. For projects with greater than average risk or uncertainty, most firms increase the MARR. As reported in Block (2005), the percentage of firms using risk-adjusted rates varied from 66% in retail to 82% in health care. Some of the percentages for other industries are 70% for manufacturing, 73% for energy, and 78% for technology firms. Table 15-2 lists an example of risk-adjusted MARRs in manufacturing.

**TABLE 15–2  Example Risk-Adjusted MARR Values in Manufacturing**

| Rate (%) | Applies to: |
|---|---|
| 6 | Equipment replacement |
| 8 | New equipment |
| 10 | New product in normal market |
| 12 | New product in related market |
| 16 | New product in new market |
| 20 | New product in foreign market |

Some firms use the same rates for all divisions and groups. Other firms vary the rates by division for strategic reasons. There are even cases when a project-specific rate based on that project's financing may be justified. For example, a firm or joint venture may be founded to develop a specific mine, pipeline, or other resource development project.

However, as shown in Example 15-4, risk-adjusted rates may not work well. A preferable way deals explicitly with the probabilities using the techniques from Chapter 10. When the interest rate (MARR) used in economic analysis calculations is raised to adjust for risk or uncertainty, greater emphasis is placed on immediate or short-term results and less emphasis on longer-term results.

## EXAMPLE 15–4

Consider the two following alternatives. The MARR of Alt. *B* has been raised from 10% to 15% to take into account the greater risk and uncertainty associated with that alternative. What is the impact of this change of MARR on the decision?

| Year | Alt. A | Alt. B |
|---|---|---|
| 0 | −$80 | −$80 |
| 1–10 | 10 | 13.86 |
| 11–20 | 20 | 10 |

### SOLUTION

|  |  | NPW | | |
|---|---|---|---|---|
| Year | Alt. A | At 14.05% | At 10% | At 15% |
| 0 | −$80 | −$80.00 | −$80.00 | −$80.00 |
| 1–10 | 10 | 52.05 | 61.45 | 50.19 |
| 11–20 | 20 | 27.95 | 47.38 | 24.81 |
|  |  | 0 | +28.83 | −5.00 |

|  |  | NPW | | |
|---|---|---|---|---|
| Year | Alt. B | At 15.48% | At 10% | At 15% |
| 0 | −$80 | −$80.00 | −$80.00 | −$80.00 |
| 1–10 | 13.86 | 68.31 | 85.14 | 69.56 |
| 11–20 | 10 | 11.99 | 23.69 | 12.41 |
|  |  | 0 | +28.83 | +1.97 |

**Computations at MARR of 10% Ignoring Risk and Uncertainty**

Both alternatives have the same positive NPW (+$28.83) at a MARR of 10%. Also, the differences in the benefits schedules $(A−B)$ produce a 10% incremental rate of return. (This calculation is not shown here.) This must be true if NPW for the two alternatives is to remain constant at a MARR of 10%.

**Considering Risk and Uncertainty with MARR of 10%**

At 10%, the alternatives are equally desirable. Since Alt. $B$ is believed to have greater risk and uncertainty, a logical conclusion is to select Alt. $A$ rather than $B$.

**Increase MARR to 15%**

At a MARR of 15%, Alt. $A$ has a negative NPW and Alt. $B$ has a positive NPW. Alternative $B$ is preferred under these circumstances.

**Conclusion**

Based on a business-risk MARR of 10%, the two alternatives are equivalent. Recognizing some greater risk of failure for Alt. $B$ makes $A$ the preferred alternative. If the MARR is increased to 15%, to add a margin of safety against risk and uncertainty, the computed decision is to select $B$. Since Alt. $B$ has been shown to be less desirable than $A$, the decision, based on a MARR of 15%, may be an unfortunate one. The difficulty is that the same risk adjustment (increase the MARR by 5%) is applied to both alternatives even though they have different amounts of risk.

The conclusion to be drawn from Example 15-4 is that increasing the MARR to compensate for risk and uncertainty is only an approximate technique and may not always achieve the desired result. Nevertheless, it is common practice in industry to adjust the MARR upward to compensate for increased risk and uncertainty.

## REPRESENTATIVE VALUES OF MARR USED IN INDUSTRY

We argued that the minimum attractive rate of return should be established at the highest one of the following: cost of borrowed money, cost of capital, or the opportunity cost.

The cost of borrowed money will vary from enterprise to enterprise, with the lowest rate being the prime interest rate. The prime rate may change several times in a year; it is widely reported in newspapers and business publications. As we pointed out, the interest rate for firms that do not qualify for the prime interest rate may be $1/2\%$ to several percentage points higher.

The cost of capital of a firm is an elusive value. There is no widely accepted way to compute it; we know that as a *composite value* for the capital structure of the firm, it conventionally is higher than the cost of borrowed money. The cost of capital must consider the market valuation of the shares (common stock, etc.) of the firm, which may fluctuate widely, depending on future earnings prospects of the firm. We cannot generalize on representative costs of capital.

Somewhat related to cost of capital is the computation of the return on total capital (long-term debt, capital stock, and retained earnings) actually achieved by firms. *Fortune* magazine, among others, does an annual analysis of the rate of return on total capital. The after-tax rate of return on total capital for individual firms ranges from 0% to about 40% and averages 8%. *Business Week* does a periodic survey of corporate performance. This magazine reports an after-tax rate of return on common stock and retained earnings. We would expect the values to be higher than the rate of return on total capital, and this is the case. The after-tax return on common stock and retained earnings ranges from 0% to about 65% with an average of 14%.

Higher values for the MARR are used by firms that are short of capital, such as high-technology start-ups. They are also used in industries, such as petroleum and mining, where volatile prices increase the risk of poor returns for projects. Rates of 25 to 30% are relatively common, and even higher rates are sometimes used. For companies with more normal levels of risk, rates of 12 to 15% are more typical.

Note that the values of MARR given earlier are approximations. But the values quoted appear to be opportunity costs, rather than cost of borrowed money or cost of capital. This indicates that firms cannot or do not obtain money to fund projects whose anticipated rates of return are nearer to the cost of borrowed money or cost of capital. One reason that firms operate as they do is that they can focus limited resources of people, management, and time on a smaller number of good projects.

One cannot leave this section without noting that the MARR used by enterprises is much higher than can be obtained by individuals. (Where can you get a 30% after-tax rate of return without excessive risk?) The reason appears to be that businesses are not obliged to compete with the thousands of individuals in any region seeking a place to invest $2000 with safety, whereas the number of people who could or would want to invest $500,000 in a business is far smaller. This diminished competition, combined with a higher risk, appears to explain at least some of the difference.

## CAPITAL BUDGETING OR SELECTING THE BEST PROJECTS

The opportunity cost of capital approach of ranking projects by their rate of return introduced a new type of problem. Up to that point we'd been analyzing mutually exclusive alternatives, where only one could be chosen. Engineering design problems are this type of problem, where younger engineers use engineering economy to choose the best alternative design.

At higher levels in the organization, engineering economy is applied to solve a different problem. For example, 30 projects have passed initial screening and are being proposed for funding. Every one of the 30 meets the MARR. The firm can afford to invest in only some of them. So, which ones should be chosen and how? This is called the *capital budgeting* problem.

Examples 15-2 and 15-3 applied the opportunity cost of capital approach to the capital budgeting problem. Firms often use this approach as a starting point to rank the projects from best to worst. In some cases the ranking by rate of return is used to make the decision.

More often, managers then meet and decide which projects will be funded by obtaining a consensus, or a decision by the highest-ranking manager, which will modify the rate of return ranking. At this meeting, business units argue for a larger share of the capital budget, as do plants in the same business, groups at the same plants, and individuals within the groups. Some considerations, such as strategy, necessity, and the availability and capability of particular resources and people are difficult to represent in the project's *numbers*, which are the subject of economic analysis.

Other firms rank projects using a benefit–cost ratio or present worth index. As shown in Example 15-5, the present worth index is the NPW divided by the cost's present value.

Anyone who has ever bought firecrackers probably used the practical ranking criterion of "biggest bang for the buck" in making a selection. This same criterion—stated more elegantly—is used by some firms to rank independent projects.

Rank independent projects according to their value of net present worth divided by the present worth of cost. The appropriate interest rate is MARR (as a reasonable estimate of the cutoff rate of return).

### EXAMPLE 15–5

Rank the following nine independent projects in their order of desirability, based on a 14.5% minimum attractive rate of return.

| Project | Cost (thousands) | Uniform Annual Benefit (thousands) | Useful Life (years) | Salvage Value (thousands) | Computed Rate of Return | Computed NPW at 14.5% (thousands) | Computed NPW/Cost (thousands) |
|---|---|---|---|---|---|---|---|
| 1 | $100 | $23.85 | 10 | $0 | 20% | $22.01 | 0.2201 |
| 2 | 200 | 39.85 | 10 | 0 | 15 | 3.87 | 0.0194 |
| 3 | 50 | 34.72 | 2 | 0 | 25 | 6.81 | 0.1362 |
| 4 | 100 | 20.00 | 6 | 100 | 20 | 21.10 | 0.2110 |
| 5 | 100 | 20.00 | 10 | 100 | 20 | 28.14 | 0.2814 |
| 6 | 100 | 18.00 | 10 | 100 | 18 | 17.91 | 0.1791 |
| 7 | 300 | 94.64 | 4 | 0 | 10 | −27.05 | −0.0902 |
| 8 | 300 | 47.40 | 10 | 100 | 12 | −31.69 | −0.1056 |
| 9 | 50 | 7.00 | 10 | 50 | 14 | −1.28 | −0.0256 |

Ranked NPW/PW of cost, the projects are listed as follows:

| Project | NPW/PW of Cost | Rate of Return |
|---------|----------------|----------------|
| 5 | 0.2814 | 20% |
| 1 | 0.2201 | 20 |
| 4 | 0.2110 | 20 |
| 6 | 0.1791 | 18 |
| 3 | 0.1362 | 25 |
| 2 | 0.0194 | 15 |
| 9 | −0.0256 | 14 |
| 7 | −0.0902 | 10 |
| 8 | −0.1056 | 12 |

With a 14.5% MARR, Projects 1 to 6 are recommended for funding and 7 to 9 are not. However, they are ranked in a different order by the present worth index and by the rate of return approaches. For example, Project 3 has the highest ranking for the rate of return and is fifth by the present worth index.

Some consider the present worth index to be a better measure, but this can be true only if PW is applied at the correct interest rate. It is more common for firms to simply rank on the rate of return. If independent projects can be ranked in their order of desirability, then the selection of projects to be included in a capital budget is a simple task. One may proceed down the list of ranked projects until the capital budget has been exhausted. The only difficulty with this scheme occurs, occasionally, when the capital budget is more than enough for $n$ projects but too little for $n + 1$ projects.

In Example 15-5, suppose the capital budget is $550,000. This is more than enough for the top five projects (sum = $450,000) but not enough for the top six projects (sum = $650,000). When we have this situation, it may not be possible to say with certainty that the best use of a capital budget of $550,000 is to fund the top five projects. There may be some other set of projects that makes better use of the available $550,000. While some trial-and-error computations may indicate the proper set of projects, more elaborate techniques are needed to prove optimality.

As a practical matter, a capital budget probably has some flexibility. If in Example 15-5 the tentative capital budget is $550,000, then a careful examination of Project 2 will dictate whether to expand the capital budget to $650,000 (to be able to include Project 2) or to drop back to $450,000 (and leave Project 2 out of the capital budget). Or perhaps Project 2 can be started in this budget year and finished next year.

## Spreadsheets, Cumulative Investments, and the Opportunity Cost of Capital

As shown in earlier chapters, spreadsheets make computing rates of return dramatically easier. In addition, spreadsheets can be used to sort the projects by rate of return and then calculate the cumulative first cost. This is accomplished through the following steps.

1. Enter or calculate each project's rate of return.
2. Select the data to be sorted. Do *not* include headings, but do include all information on the row that goes with each project.
3. Select the SORT tool (found in the menu under DATA), identify the rate of return column as the first key, and a sort order of descending. Also ensure that row sorting is selected. Sort.
4. Add a column for the cumulative first cost. This column is compared with the capital limit to identify the opportunity cost of capital and which projects should be funded.

Example 15-6 illustrates these steps.

### EXAMPLE 15–6

A firm has a budget of $800,000 for projects this year. Which of the following projects should be accepted? What is the opportunity cost of capital?

| Project | First Cost | Annual Benefit | Salvage Value | Life (years) |
|---------|-----------|----------------|---------------|--------------|
| A | $200,000 | $25,000 | $50,000 | 15 |
| B | 250,000 | 47,000 | −25,000 | 10 |
| C | 150,000 | 17,500 | 20,000 | 15 |
| D | 100,000 | 20,000 | 15,000 | 10 |
| E | 200,000 | 24,000 | 25,000 | 20 |
| F | 300,000 | 35,000 | 15,000 | 15 |
| G | 100,000 | 18,000 | 0 | 10 |
| H | 200,000 | 22,500 | 15,000 | 20 |
| I | 350,000 | 50,000 | 0 | 25 |

### SOLUTION

The first step is to use the RATE function to find the rate of return for each project. The results of this step are shown in the top portion of Figure 15-4. Next the projects are sorted in descending order by their rates of return. Finally, the cumulative first cost is computed. Projects *D*, *I*, *B*, and *G* should be funded. The opportunity cost of capital is 10.6% the rate for the first project rejected.

| | A | B | C | D | E | F | G | H |
|---|---|---|---|---|---|---|---|---|
| 1 | Project | First Cost | Annual Benefit | Salvage Value | Life | IRR | | |
| 2 | A | 200,000 | 25,000 | 50,000 | 15 | 10.2% | =RATE(E2,C2,−B2,D2) | |
| 3 | B | 250,000 | 47,000 | −25,000 | 10 | 12.8% | | |
| 4 | C | 150,000 | 17,000 | 20,000 | 15 | 8.6% | | |
| 5 | D | 100,000 | 20,000 | 15,000 | 10 | 16.0% | | |
| 6 | E | 200,000 | 24,000 | 25,000 | 20 | 10.6% | | |
| 7 | F | 300,000 | 35,000 | 15,000 | 15 | 8.2% | | |
| 8 | G | 100,000 | 18,000 | 0 | 10 | 12.4% | | |
| 9 | H | 200,000 | 22,500 | 15,000 | 20 | 9.6% | | |
| 10 | I | 350,000 | 50,000 | 0 | 25 | 13.7% | | |
| 11 | Projects Sorted by IRR | | | | | | Cumulative First Cost | |
| 12 | D | 100,000 | 20,000 | 15,000 | 10 | 16.0% | 100,000 | |
| 13 | I | 350,000 | 50,000 | 0 | 25 | 13.7% | 450,000 | |
| 14 | B | 250,000 | 47,000 | −25,000 | 10 | 12.8% | 700,000 | |
| 15 | G | 100,000 | 18,000 | 0 | 10 | 12.4% | 800,000 | |
| 16 | E | 200,000 | 24,000 | 25,000 | 20 | 10.6% | 1,000,000 | |
| 17 | A | 200,000 | 25,000 | 50,000 | 15 | 10.2% | 1,200,000 | |
| 18 | H | 200,000 | 22,500 | 15,000 | 20 | 9.6% | 1,400,000 | |
| 19 | C | 150,000 | 17,500 | 20,000 | 15 | 8.6% | 1,550,000 | |
| 20 | F | 300,000 | 35,000 | 15,000 | 15 | 8.2% | 1,850,000 | |

**FIGURE 15–4** Spreadsheet for finding opportunity cost of capital.

Example 15-7 outlines the more complicated case of independent projects with mutually exclusive alternatives.

## EXAMPLE 15–7

A company is preparing its capital budget for next year. The amount has been set at $250,000 by the board of directors. Rank the following project proposals for the board's consideration and recommend which should be funded.

| Project Proposals | Cost (thousands) | Uniform Annual Benefit (thousands) | Salvage Value (thousands) | Useful Life (years) |
|---|---|---|---|---|
| Proposal 1 | | | | |
| Alt. *A* | $100 | $23.85 | $0 | 10 |
| Alt. *B* | 150 | 32.20 | 0 | 10 |
| Alt. *C* | 200 | 39.85 | 0 | 10 |
| Proposal 2 | 50 | 14.92 | 0 | 5 |
| Proposal 3 | | | | |
| Alt. *A* | 100 | 18.69 | 25 | 10 |
| Alt. *B* | 150 | 19.42 | 125 | 10 |

**SOLUTION**

For project proposals with two or more alternatives, incremental rate of return analysis is required.

| | | Uniform | | | | Uniform | | |
| | | Annual | Salvage | | Incremental Analysis | | | |
| Combination | Cost | Benefit | Value | Rate of | Cost | Annual Benefit | Salvage Value | Rate of |
| of Alternatives | ($1000s) | ($1000s) | ($1000s) | Return | ($1000s) | ($1000s) | ($1000s) | Return |
|---|---|---|---|---|---|---|---|---|
| Proposal 1 | | | | | | | | |
| A | $100 | $23.85 | $0 | 20.0% | | | | |
| B − A | | | | | $50 | 8.35 | $0 | 10.6% |
| B | 150 | 32.20 | 0 | 17.0 | | | | |
| C − B | | | | | 50 | 7.65 | 0 | 8.6 |
| C − A | | | | | 100 | 16.00 | 0 | 9.6 |
| C | 200 | 39.85 | 0 | 15.0 | | | | |
| Proposal 2 | 50 | 14.92 | 0 | 15.0 | | | | |
| Proposal 3 | | | | | | | | |
| A | 100 | 18.69 | 25 | 15.0 | | | | |
| B − A | | | | | 50 | 0.73 | 100 | 8.3 |
| B | 150 | 19.42 | 125 | 12.0 | | | | |

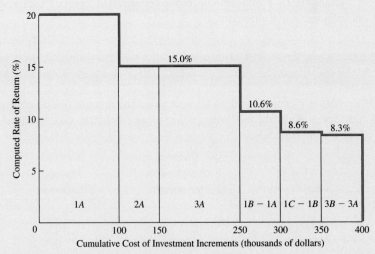

**FIGURE 15–5** Cumulative cost versus incremental rate of return.

The various separable increments of investment may be ranked by the rate of return. They are plotted in a graph of cumulative cost versus rate of return in Figure 15-5. The ranking of projects by rate of return gives the following:

| Project | Rate of Return |
|---|---|
| 1A | 20.0% |
| 2 | 15.0% |
| 3A | 15.0% |
| 1B in place of 1A | 10.6% |
| 1C in place of 1B | 8.6% |
| 3B in place of 3A | 8.3% |

For a budget of $250,000 the selected projects are 1A, 2, and 3A. Note that if a budget of $300,000 were available, 1B would replace 1A, making the proper set of projects 1B, 2, and 3A. At a budget of $400,000 1C would replace 1B; and 3B would replace 3A, making the selected projects 1C, 2, and 3B.

In Example 15-7, each of the more expensive mutually exclusive alternatives (1B, 1C, and 3B) had lower rates of return than the less expensive alternatives (1A and 3A). This is common because many projects exhibit decreasing returns to scale. As in Example 15-7, when it occurs, then it makes the ranking of increments easy to use. Sometimes more expensive mutually exclusive alternatives have higher rates of return, then several combinations or more advanced techniques must be tried.

## SUMMARY

There are three general sources of capital available to a firm. The most important one is money generated from the firm's operations. This has two components: there is the portion of profit that is retained in the business; in addition, funds equal to its depreciation charges are available for reinvestment.

The two other sources of capital are from outside the firm's operations.

*Debt:* borrowed as loans from banks, insurance companies, and so forth.

Longer-term borrowing: from selling bonds.

*Equity:* sale of equity securities like common or preferred stock.

Retained profits and cash equal to depreciation charges are the primary sources of investment capital for most firms, and the only sources for many enterprises.

In selecting a value of MARR, three values are frequently considered:

1. Cost of borrowed money.
2. Cost of capital. This is a composite cost of the components of the firm's overall capitalization.
3. Opportunity cost, which is the rate of return on the best investment project that is rejected.

The MARR should be equal to the highest one of these three values.

When there is a risk aspect to the problem (probabilities are known or reasonably estimated), this can be handled by techniques like expected value and simulation. Where there is uncertainty (probabilities of the various outcomes are not known), there are analytical techniques, but they are less satisfactory. A method commonly used to adjust for risk and uncertainty is to increase the MARR. This method can distort the time-value-of-money relationship. The effect is to discount longer-term consequences more heavily than short-term consequences, which may or may not be desirable. Prior to this chapter we had assumed that all worthwhile projects are approved and implemented. But industrial firms, like individuals and governments, are typically faced with more good projects than can be funded with the money available. The task is to select the best projects and reject, or at least delay, the rest.

Capital may be rationed among competing investment opportunities by either rate of return or present worth methods. The results may not always be the same for these two methods in many practical situations.

If projects are ranked by rate of return, a proper procedure is to go down the list until the capital budget has been exhausted. The rate of return at this point is the cutoff rate of return. This procedure gives the best group of projects, but does not necessarily have them in the proper priority order.

It has been shown in earlier chapters that the usual business objective is to maximize NPW, and this is not necessarily the same as maximizing rate of return. One suitable procedure is to use the ratio (NPW/PW of cost) to rank the projects, letting the MARR equal the cutoff rate of return (which is the opportunity cost of capital). This present worth ranking method will order the projects so that, for a limited capital budget, NPW will be maximized. The MARR must equal the cutoff rate of return for the rate of return and present worth methods to yield compatible results.

# PROBLEMS

## Cost of Funds

**15-1**  Determine the current interest rate on the following securities, and explain why the interest rates are different for these different bonds.

(a)  U.S. Treasury bond due in 5 years

(b)  General obligation bond of a municipal district, city, or a state due in 20 years

(c)  Corporate debenture bond of a U.S. industrial firm due in 20 years

**15-2**  A small engineering firm has borrowed $125,000 at 8%. The partners have invested another $75,000. If the partners require a 12% rate of return, what is the firm's cost of capital:

(a)  Before taxes?

(b)  After taxes with a tax rate of 30%?

   (*Answer:* 9.50%)

**15-3**  An engineering firm has borrowed $725,000 at 7%. The stockholders have invested another $600,000. The firm's retained earnings total $1.2M. The return on equity is estimated to be 11%. What is the firm's cost of capital:

(a)  Before taxes?

(b)  After taxes with a tax rate of 40%?

**15-4**  A firm's stockholders expect a 15% rate of return, and there is $12M in common stock and retained earnings. The firm has $5M in loans at an average rate of 7%. The firm has raised $8M by selling bonds at an average rate of 6%. What is the firm's cost of capital:

(a)  Before taxes?

(b)  After taxes with a tax rate of 35%?

**15-5**  A firm's stockholders expect an 18% rate of return, and there is $22M in common stock and retained earnings. The firm has $9M in loans at an average rate of 8%. The firm has raised $14M by selling bonds at an average rate of 4%. What is the firm's cost of capital:

(a)  Before taxes?

(b)  After taxes with a tax rate of 34%?

**15-6**  A firm has 40,000 shares whose current price is $80.75. Those stockholders expect a return of 15%. The firm has a 2-year loan of $900,000 at 6.4%. It has issued 12,500 bonds with a face value of 1000, 15 years left to maturity, semiannual compounding, a coupon interest rate of 6%, and a current price of $1090. Using market values for debt and equity, what is the firm's cost of capital:

(a)  Before taxes?

(b)  After taxes with a tax rate of 40%?

**15-7** A firm has 60,000 shares whose current price is $45.90. Those stockholders expect a return of 14%. The firm has a 3-year loan of $1,900,000 at 7.3%. It has issued 22,000 bonds with a face value of 1000, 20 years left to maturity, semiannual compounding, a coupon interest rate of 7%, and a current price of $925. Using market values for debt and equity, what is the firm's cost of capital:

  (*c*) Before taxes?

  (*d*) After taxes with a tax rate of 34%?

**15-8** A public university wants to apply the concept of the WACC to developing its interest rate for analyzing capital projects. It has an endowment of $68 million which is earning 6.3% interest. It is paying 4.5% interest on $29 million in bonds. It believes that $94 million in general funds from the taxpayers should be assigned an interest rate of 13%. What is the university's cost of capital? Note that only the interest on the endowment is available to fund capital projects.

**15-9** Assume you have $2000 available for investment for a 5-year period. You wish to *invest* the money—not just spend it on fun things. There are obviously many alternatives available. You should be willing to assume a modest amount of risk of loss of some or all of the money if this is necessary, but not a great amount of risk (no investments in poker games or at horse races). How would you invest the money? What is your minimum attractive rate of return? Explain.

**15-10** There are many venture capital syndicates that consist of a few (say, eight or ten) wealthy people who combine to make investments in small and (hopefully) growing businesses. Typically, the investors hire a young investment manager (often an engineer with an MBA) who seeks and analyzes investment opportunities for the group. Would you estimate that the MARR sought by this group is more or less than 12%? Explain.

**Inflation**

**15-11** What is the interest rate on a 2-year certificate of deposit at a bank or credit union in your area? What is the most recent value of the Consumer Price Index (CPI)? If inflation matches that rate, what is the real rate of return on the 2-year CD? Include references for the sources of your data.

**15-12** What is the interest rate on a 4- or 5-year new car loan at a bank or credit union in your area? What is

the most recent value of the Consumer Price Index (CPI)? If inflation matches that rate, what is the real interest rate you would pay on the car loan? Include references for the sources of your data.

**15-13** Over the last 10 years, what has the inflation rate been? Compare this with the rate of return on the "Dow" average over the same period. What has the real rate of return for investing in this mix of stocks been? Include references for the sources of your data.

**15-14** Over the last 10 years, what has the inflation rate been? Compare this with the rate of return on the NASDAQ average over the same period. What has the real rate of return for investing in this mix of stocks been? Include references for the sources of your data.

**Opportunity Cost of Capital**

**15-15** A factory has a $100,000 capital budget. Determine which project(s) should be funded and the opportunity cost of capital.

| Project | First Cost | Annual Benefits | Life (years) | Salvage Value |
|---------|-----------|-----------------|--------------|---------------|
| A | $50,000 | $13,500 | 5 | $5000 |
| B | 50,000 | 9,000 | 10 | 0 |
| C | 50,000 | 13,250 | 5 | 1000 |
| D | 50,000 | 9,575 | 8 | 6000 |

**15-16** Chips USA is considering the following projects to improve their production process. Chips have a short life, so a 3-year horizon is used in evaluation. Which projects should be done if the budget is $70,000? What is the opportunity cost of capital?

| Project | First Cost | Benefit |
|---------|-----------|---------|
| 1 | $20,000 | $11,000 |
| 2 | 30,000 | 14,000 |
| 3 | 10,000 | 6,000 |
| 4 | 5,000 | 2,400 |
| 5 | 25,000 | 13,000 |
| 6 | 15,000 | 7,000 |
| 7 | 40,000 | 21,000 |

**15-17** The National Motors Rock Creek plant is considering the following projects to improve the company's production process. Which projects should be done if the budget is $500,000? What is the opportunity cost of capital?

| Project | First Cost | Annual Benefit | Life (years) |
|---------|-----------|---------------|-------------|
| 1 | $200,000 | $50,000 | 15 |
| 2 | 300,000 | 70,000 | 10 |
| 3 | 100,000 | 40,000 | 5 |
| 4 | 50,000 | 12,500 | 10 |
| 5 | 250,000 | 75,000 | 5 |
| 6 | 150,000 | 32,000 | 20 |
| 7 | 400,000 | 125,000 | 5 |

**15-18** The WhatZit Company has decided to fund six of nine project proposals for the coming budget year. Determine the next capital budget for WhatZit. What is the MARR?

| Project | First Cost | Annual Benefits | Life (years) |
|---------|-----------|----------------|-------------|
| A | $15,000 | $ 4,429 | 4 |
| B | 20,000 | 6,173 | 4 |
| C | 30,000 | 9,878 | 4 |
| D | 25,000 | 6,261 | 5 |
| E | 40,000 | 11,933 | 5 |
| F | 50,000 | 11,550 | 5 |
| G | 35,000 | 6,794 | 8 |
| H | 60,000 | 12,692 | 8 |
| I | 75,000 | 14,058 | 8 |

**15-19** Which projects should be done if the budget is $100,000? What is the opportunity cost of capital?

| Project | Life (years) | First Cost | Annual Benefit | Salvage Value |
|---------|-------------|-----------|---------------|---------------|
| 1 | 20 | $20,000 | $4000 | |
| 2 | 20 | 20,000 | 3200 | $20,000 |
| 3 | 30 | 20,000 | 3300 | 10,000 |
| 4 | 15 | 20,000 | 4500 | |
| 5 | 25 | 20,000 | 4500 | −20,000 |
| 6 | 10 | 20,000 | 5800 | |
| 7 | 15 | 20,000 | 4000 | 10,000 |

**Risk-Adjusted MARR**

**15-20** Use the example risk-adjusted interest rates for manufacturing projects in Table 15-2. Assume Project B in Problem 15-15 is a new product in a new market. What is the interest rate for evaluating this project? Should it be done?

**15-21** Use the example risk-adjusted interest rates for manufacturing projects in Table 15-2. Assume Project E in Problem 15-18 is a new product in an existing market. What is the interest rate for evaluating this project? Should it be done?

**15-22** Use the example risk-adjusted interest rates for manufacturing projects in Table 15-2. Assume Project 1 in Problem 15-19 is a new product in a foreign market. What is the interest rate for evaluating this project? Should it be done?

**15-23** Use the example risk-adjusted interest rates for manufacturing projects in Table 15-2 and the project data in Problem 15-18. Assume that each project is to launch a new product. New products for the existing market will have a life of 8 years. New products for a related market will have a life of 5 years. New products for a new market will have a life of 4 years. Which new products are recommended for funding? What is the total cost of these projects?

**Capital Budgeting**

**15-24** Each of the following 10 independent projects has a 10-year life and no salvage value.

| Project | Cost (thousands) | Uniform Annual Benefits (thousands) | Rate of Return |
|---------|-----------------|------------------------------------|---------------|
| 1 | $ 5 | $1.03 | 16% |
| 2 | 15 | 3.22 | 17 |
| 3 | 10 | 1.77 | 12 |
| 4 | 30 | 4.88 | 10 |
| 5 | 5 | 1.19 | 20 |
| 6 | 20 | 3.83 | 14 |
| 7 | 5 | 1.00 | 15 |
| 8 | 20 | 3.69 | 13 |
| 9 | 5 | 1.15 | 19 |
| 10 | 10 | 2.23 | 18 |

The projects have been proposed by the staff of the Ace Card Company. The MARR of Ace has been 12% for several years.

(a) If there is ample money available, what projects should Ace approve?

(b) Rank-order all the acceptable projects in according to desirability.

(c) If only $55,000 is available, which projects should be approved?

**15-25** Ten capital spending proposals have been made to the budget committee as the members prepare the annual budget for their firm. Each independent project has a 5-year life and no salvage value.

| Project | Initial Cost (thousands) | Uniform Annual Benefit (thousands) | Rate of Return |
|---|---|---|---|
| A | $10 | $2.98 | 15% |
| B | 15 | 5.58 | 25 |
| C | 5 | 1.53 | 16 |
| D | 20 | 5.55 | 12 |
| E | 15 | 4.37 | 14 |
| F | 30 | 9.81 | 19 |
| G | 25 | 7.81 | 17 |
| H | 10 | 3.49 | 22 |
| I | 5 | 1.67 | 20 |
| J | 10 | 3.20 | 18 |

(a) Based on a MARR of 14%, which projects should be approved?

(b) Rank-order all the projects according to desirability.

(c) If only $85,000 is available, which projects should be approved?

**15-26** At Miami Products, four project proposals (three with mutually exclusive alternatives) are being considered. All the alternatives have a 10-year useful life and no salvage value.

| Project Proposal | Cost (thousands) | Uniform Annual Benefits (thousands) | Rate of Return |
|---|---|---|---|
| Project 1 | | | |
| Alt. A | $25 | $4.61 | 13% |
| Alt. B | 50 | 9.96 | 15 |
| Alt. C | 10 | 2.39 | 20 |
| Project 2 | | | |
| Alt. A | 20 | 4.14 | 16 |
| Alt. B | 35 | 6.71 | 14 |
| Project 3 | | | |
| Alt. A | 25 | 5.56 | 18 |
| Alt. B | 10 | 2.15 | 17 |
| Project 4 | 10 | 1.70 | 11 |

(a) Use rate of return methods to determine which set of projects should be undertaken if the capital budget is limited to about $100,000.

(b) For a budget of about $100,000, what interest rate should be used in rationing capital by present worth methods? (Limit your answer to a value for which there is a compound interest table available in Appendix C).

(c) Using the interest rate determined in part (b), rank-order the eight different investment opportunities by means of the present worth method.

(d) For a budget of about $100,000 and the ranking in part (c), which of the investment opportunities should be selected?

**15-27** Al Dale is planning his Christmas shopping for seven people. To quantify how much his various relatives would enjoy receiving items from a list of prospective gifts, Al has assigned appropriateness units (called "ohs") for each gift for each person (Table P15-27). A rating of 5 ohs represents a gift that the recipient would really like. A rating of 4 ohs indicates the recipient would like it four-fifths as much; 3 ohs, three-fifths as much, and so forth. A zero rating indicates an inappropriate gift that cannot be given to that person. Everyone must get a gift.

**TABLE P15-27** **Data**

| Prospective Gift | "Oh" Rating of Gift If Given to Various Family Members | | | | | | |
|---|---|---|---|---|---|---|---|
| | Father | Mother | Sister | Brother | Aunt | Uncle | Cousin |
| 1. $20 box of candy | 4 | 4 | 2 | 1 | 5 | 2 | 3 |
| 2. $12 box of cigars | 3 | 0 | 0 | 1 | 0 | 1 | 2 |
| 3. $16 necktie | 2 | 0 | 0 | 3 | 0 | 3 | 2 |
| 4. $20 shirt or blouse | 5 | 3 | 4 | 4 | 4 | 1 | 4 |
| 5. $24 sweater | 3 | 4 | 5 | 4 | 3 | 4 | 2 |
| 6. $30 camera | 1 | 5 | 2 | 5 | 1 | 2 | 0 |
| 7. $ 6 calendar | 0 | 0 | 1 | 0 | 1 | 0 | 1 |
| 8. $16 magazine subscription | 4 | 3 | 4 | 4 | 3 | 1 | 3 |
| 9. $18 book | 3 | 4 | 2 | 3 | 4 | 0 | 3 |
| 10. $16 game | 2 | 2 | 3 | 2 | 2 | 1 | 2 |

The objective is to maximize total ohs that can be obtained with the selected budget.

(a) How much will it cost to buy the seven gifts the people would like best, if there is ample money for Christmas shopping?

(b) If the Christmas shopping budget is set at $112, which gifts should be purchased, and what is their total appropriateness rating in ohs?

(c) If the Christmas shopping budget must be cut to $90, which gifts should be purchased, and what is their total appropriateness rating in ohs?
(*Answer:* (a) $168)

15-28 A financier has a staff of three people whose job it is to examine possible business ventures for him. Periodically they present their findings concerning business opportunities. On a particular occasion, they presented the following investment opportunities:

*Project A:* This is a project for the use of commercial land the financier already owns. There are three mutually exclusive alternatives.

  • *A1.* Sell the land for $500,000.

  • *A2.* Lease the property for a car-washing business. An annual income, after all costs (property taxes, etc.) of $98,700 would be received at the end of each year for 20 years. At the end of the 20 years, it is believed that the property could be sold for $750,000.

  • *A3.* Construct an office building on the land. The building will cost $4.5 million to construct and will not produce any net income for the first 2 years. The probabilities of various levels of rental income, after all expenses, for the subsequent 18 years are as follows:

| Annual Rental Income | Probability |
|---|---|
| $1,000,000 | 0.1 |
| 1,100,000 | 0.3 |
| 1,200,000 | 0.4 |
| 1,900,000 | 0.2 |

The property (building and land) probably can be sold for $3 million at the end of 20 years.

*Project B:* An insurance company is seeking to borrow money for 90 days at $13\frac{3}{4}\%$ per annum, compounded continuously.

*Project C:* A financier owns a manufacturing company. The firm desires additional working capital to allow it to increase its inventories of raw materials and finished products. An investment of $2 million will allow the company to obtain sales that in the past the company had to forgo. The additional capital will increase company profits by $500,000 a year. The financier can recover this additional investment by ordering the company to reduce its inventories and to return the $2 million. For planning purposes, assume the additional investment will be returned at the end of 10 years.

*Project D:* The owners of *Sunrise* magazine are seeking a loan of $500,000 for 10 years at a 16% interest rate.

*Project E:* The Galveston Bank has indicated a willingness to accept a deposit of any sum of money over $100,000, for any desired duration, at a 14.06% interest rate, compounded monthly. It seems likely that this interest rate will be available from Galveston, or some other bank, for the next several years.

*Project F:* A car rental firm is seeking a loan of $2 million to expand its fleet. The firm offers to repay the loan by paying $1 million at the end of Year 1 and $1,604,800 at the end of Year 2.

- If there is $4 million available for investment now (or $4.5 million if the Project *A* land is sold), which projects should be selected? What is the MARR in this situation?

- If there is $9 million available for investment now (or $9.5 million if the Project *A* land is sold), which projects should be selected?

15-29 The Raleigh Soap Company has been offered a 5-year contract to manufacture and package a leading brand of soap for Taker Bros. It is understood that the contract will not be extended past the 5 years because Taker Bros. plans to build its own plant nearby. The contract calls for 10,000 metric tons (one metric ton equals 1000 kg) of soap a year. Raleigh normally produces 12,000 metric tons of soap a year, so production for the 5-year period would be increased to 22,000 metric tons. Raleigh must decide what changes, if any, to make to accommodate this increased production. Five projects are under consideration.

*Project 1:* Increase liquid storage capacity. Raleigh has been forced to buy caustic soda in tank truck quantities owing to inadequate storage capacity. If another liquid caustic soda tank is installed to hold 1000 cubic meters, the caustic soda may be purchased in railroad tank car quantities at a more favorable price. The result would be a saving of 0.1¢ per kilogram of soap. The tank, which would cost $83,400, has no net salvage value.

*Project 2:* Acquire another sulfonation unit. The present capacity of the plant is limited by the sulfonation unit. The additional 12,000 metric tons of soap cannot be produced without an additional sulfonation unit. Another unit can be installed for $320,000.

*Project 3:* Expand the packaging department. With the new contract, the packaging department must either work two 8-hour shifts or have another packaging line installed. If the two-shift operation is used, a 20% wage premium must be paid for the second shift. This premium would amount to $35,000 a year. The second packaging line could be installed for $150,000. It would have a $42,000 salvage value at the end of 5 years.

*Project 4:* Build a new warehouse. The existing warehouse will be inadequate for the greater production. It is estimated that 400 square meters of additional warehouse is needed. A new warehouse can be built on a lot beside the existing warehouse for $225,000, including the land. The annual taxes, insurance, and other ownership costs would be $5000 a year. It is believed the warehouse could be sold at the end of 5 years for $200,000.

*Project 5:* Lease a warehouse. An alternative to building an additional warehouse would be to lease warehouse space. A suitable warehouse one mile away could be leased for $15,000 per year. The $15,000 includes taxes, insurance, and so forth. The annual cost of moving materials to this more remote warehouse would be $34,000 a year.

The contract offered by Taker Bros. is a favorable one, which Raleigh Soap plans to accept. Raleigh management has set a 15% before-tax minimum attractive rate of return as the criterion for any of the projects. Which projects should be undertaken?

15-30 Mike Moore's microbrewery is considering production of a new ale called Mike's Honey Harvest Brew. To produce this new offering, Mike is considering two independent projects. Each of these projects has two mutually exclusive alternatives, and each alternative has a useful life of 10 years and no salvage value. Mike's MARR is 8%. Information regarding the projects and alternatives are given in the following table

| Project/Alternative | Cost | Annual Benefit |
|---|---|---|
| Project 1. Purchase new fermenting tanks | | |
| Alt. *A*: 5000-gallon tank | $ 5,000 | $1192 |
| Alt. *B*: 15,000-gallon tank | 10,000 | 1992 |
| Project 2. Purchase bottle filler and capper | | |
| Alt. *A*: 2500-bottle/hour machine | 15,000 | 3337 |
| Alt. *B*: 5000-bottle/hour machine | 25,000 | 4425 |

Use incremental rate of return analysis to complete the following worksheet.

| Proj./Alt. | Cost, P | Annual Benefit, A | (A/P, i, 10) | IRR |
|---|---|---|---|---|
| 1*A* | $ 5,000 | $1192 | 0.2385 | 20% |
| 1*B*–1*A* | 5,000 | 800 | 0.1601 | |
| 2*A* | 15,000 | 3337 | | |
| 2*B*–2*A* | 10,000 | | | |

Use this information to determine:

(*a*) which projects should be funded if only $15,000 is available.

(*b*) the cutoff rate of return if only $15,000 is available.

(*c*) which projects should be funded if $25,000 is available.

# CASES

**The following cases from *Cases in Engineering Economy* 2*nd* are suggested as matched with this chapter.**

CASE 46  **Aero Tech**

Case budgeting with memos for three different approaches to ranking.

CASE 47  **Bigstate Highway Department**

Capital budgeting including mutually exclusive alternatives. Includes uncertain value of a life.

CASE 48  **Dot Puff Project Selection**

Public-sector capital budgeting with added constraint on man-years required.

CASE 49   **The Arbitrator**

Includes two memos of suggested solutions for four projects with nine alternatives. Data presented in text discussion rather than tabulated.

CASE 50   **Capital Planning Consultants**

Capital budgeting including mutually exclusive alternatives. Includes uncertainties in first cost, annual benefit, and lives.

CASE 51   **Refrigerator Magnet Company**

Capital budgeting including mutually exclusive alternatives.

## CHAPTER 16

# ECONOMIC ANALYSIS IN THE PUBLIC SECTOR

## From Waste to Power and Money

**M**any of the world's leading climate scientists have concluded that global warming is "unequivocal" and that human activity is the main driver. One such activity is the dumping into landfills of organic material that decomposes over time, producing a significant amount of landfill gas (LFG). About 50% of LFG is methane, which is a greenhouse gas that contributes to global warming. Methane is of particular concern because it traps heat in the atmosphere with an effect 20 times greater than that of carbon dioxide—also a large contributor to global climate change.

Landfills that do not control the escape of landfill gas are the largest single human source of methane emissions in the U.S. Of additional concern are emissions of volatile organic compounds that create smog in the form of ground-level ozone and produce respiratory problems. Landfills also generate hazardous air pollutants such as benzene and toluene, and odorous compounds such as hydrogen sulfide, that negatively impact the quality of life for individuals nearby.

But humans must produce waste to live—it is one of the basic laws of thermodynamics. Bad news all around—except that using engineering principles, bringing together various "constituencies," and evaluating alternatives with engineering economy—can bring benefits out of almost every one of these issues.

If extracted and properly utilized, landfill gas can be an asset. This medium-BTU gas has a heating value of between 350 and 600 BTU per cubic foot, about half the heating value of natural gas. As a result, more and more owners of municipal solid waste landfills extract this gas and use it to generate electricity. Such electricity is used directly for the landfills' power needs or is sold to the general power grid.

The environment benefits from the gas extraction and combustion because the release to the environment of methane, volatile organics, hazardous air pollutants, and odorous gases is greatly reduced. The generation of electrical power by utilizing a previously wasted source (i.e., methane from LFG) reduces the use of fossil fuel and the associated emissions. The income from selling power can assist in paying landfill operating expenses.

Extracting and using a landfill gas to generate electricity is a "*6-win*" situation.

1. It benefits the environment by reducing unwanted gas emission.
2. It adds electrical power to the grid.
3. It produces cash flow for the landfill owner.
4. It lowers the use of nonrenewable fossil fuels.
5. It reduces financial costs to the local population.
6. It reduces the hazardous, noxious, and odorous gases for those nearby and downwind.

*Contributed by William R. Truran, Stevens Institute of Technology, and Peter A. Cerenzio, Cerenzio & Panaro Consulting Engineers*

## QUESTIONS TO CONSIDER

1. How can a proposed landfill methane project best be summarized for a nontechnical public audience? This is an important "skill" of an engineer—conveying sometimes complex and confusing scientific issues to a less informed, less scientific, and less interested audience with a short attention span.

2. What engineering economic principles would you apply in analyzing this application? Which measures would be most effective for the public audience? The important measures, and the metrics used to describe them, may be considered "key performance indicators."

3. This waste-to-power application cannot be answered with a simple equation with a closed-ended answer that is "correct" or "wrong." This problem requires the cooperation of several (at least) agencies and communities of interest. Name some, and identify how their objectives are aligned or in conflict.

4. Like this vignette, the case of Enron, a "poster child" of what can go wrong in business, is related to electric power generation. Discuss how there may be ways for some of the stakeholders to be "shady" here. Discuss how you—as an engineer—can not only influence factual answers but project an image that is not reality. You as an engineer can be honored (or guilty) of your representations.

## After Completing This Chapter...

*The student should be able to:*

- Distinguish the unique objective and viewpoint of public decisions.
- Explain methods for determining the interest rates for evaluating public projects.
- Use the benefit–cost ratio to analysis projects.
- Distinguish between the conventional and modified versions of the benefit–cost ratio.
- Use an incremental benefit–cost ratio to evaluate a set of mutually exclusive projects.
- Discuss the impact of financing, duration, quantifying and valuing consequences, and politics in public investment analysis.

Public organizations, such as federal, state, and local governments, port authorities, school districts, and government agencies, no less than firms and individuals in the private sector, make investment decisions. For these decision-making bodies, economic analysis is complicated by several factors that do not affect companies in the private sector. These factors include the overall purpose of investment, the viewpoint for analysis, and how to select the interest rate. Other factors include project financing sources, expected project duration, quantifying and valuing benefits and disbenefits, effects of politics, beneficiaries of investment, and the multipurpose nature of investments. The overall mission in the public sector is the same as that in the private sector—to make prudent investment decisions that promote the organization's overall objectives.

The primary economic measure used in the public sector is the benefit–cost (B/C) ratio, which was introduced in Chapter 9. This measure is calculated as a ratio of the equivalent worth of the project's benefits to the equivalent worth of the project's costs. If the B/C ratio is *greater than 1.0*, the project under evaluation is accepted; if not, it is rejected. The B/C ratio is used to evaluate both single investments and sets of mutually exclusive projects (where the incremental B/C ratio is used). The uncertainties of quantifying cash flows, long project lives, and low interest rates all tend to lessen the reliability of public sector engineering economic analysis. There are two versions of the B/C ratio: *conventional* and *modified*. Both provide consistent recommendations to decision makers for single investment decisions and for decisions involving sets of mutually exclusive alternatives. The B/C ratio is a widely used and accepted measure in government economic analysis and decision making.

## INVESTMENT OBJECTIVE

Organizations exist to promote the overall goals of those they serve. In private sector firms, investment decisions are based on increasing the firm's wealth and economic stability. Beneficiaries of investments generally are clearly identified as the firm's owners and/or stockholders.

In the public sector, the purpose of investment decisions is sometimes ambiguous. For people in the United States of America, the Preamble to the Constitution establishes the overall theme or objective for why the public body exists:

> We the People of the United States, in order to form a more perfect Union, establish justice, insure domestic tranquillity, provide for the common defense, *promote the general welfare* [italics supplied], and secure the blessings of liberty to ourselves and our posterity, do ordain and establish this Constitution for the United States of America.

The catch phrase *promote the general welfare* serves as a guideline for public decision making. But what does this phrase mean? At best it is a general guideline; at worst it is a vague slogan that can be used to justify any action. Projects some citizens want may be opposed by other citizens. In government economic analysis, it is not always easy to distinguish which investments promote the "general welfare" and which do not.

Consider the case of a dam construction project to provide water, electricity, flood control, and recreational facilities. Such a project might seem to be advantageous for a region's entire population. But on closer inspection, decision makers must consider that the dam will require the loss of land upstream due to backed-up water. Farmers will lose pastures or cropland, and nature lovers will lose canyon lands. Or perhaps the land to be lost is a pivotal breeding ground for protected species, and environmentalists will oppose the project. The project may also have a negative impact on towns, cities, and states downstream. How will it affect their water supply? Thus, a project initially deemed to have many benefits, on closer inspection, reveals many conflicts. Projects' conflicting aspects are characteristic of public-sector investment and decision making.

Public investment decisions are more difficult than those in the private sector owing to the many people, organizations, and political units potentially affected. Opposition to a proposal is more likely in public investment decisions than in those made by private-sector companies because for every group that benefits from a particular project, there is usually an opposing group. Many conflicts in opinion arise when the project involves the use of public lands, including industrial parks, housing projects, business districts, roadways, sewage and power facilities, and landfills. Opposition may be based on the belief that development of *any* kind is bad or that the proposed development should not be near "our" homes, schools, or businesses.

Consider the decision that a small town might face when considering whether to establish a municipal rose garden, seemingly a beneficial public investment with no adverse consequences. However, an economic analysis of the project must consider *all* effects of the project, including potential unforeseen outcomes. Where will visitors park their vehicles? Will increased travel around the park necessitate new traffic lights and signage? Will traffic and visitors to the park increase noise levels for adjacent homes? Will special varieties of roses create a disease hazard for local gardens? Will the garden require high levels of fertilizers and insecticides, and where will these substances wind up after they have been applied? Clearly, many issues must be addressed. What appeared to be a simple proposal for a city rose garden, in fact, brings up many aspects to be considered.

Our simple rose garden illustrates how effects on *all parties involved* must be identified, even for projects that seem very useful. Public decision makers must reach a compromise between the positive effects enjoyed by some groups and the negative effects on other groups. The overall objective is to make prudent decisions that *promote the general welfare*, but in the public sector the decision process is not so straightforward as in the private sector.

The **Flood Control Act of 1936** specified that waterway improvements for flood control could be made as long as "the benefits *to whomsoever they accrue* [italics supplied] are in excess of the estimated costs." Perhaps the overall general objective of investment decision analysis in government should be a dual one: to promote the general welfare and to ensure that the value to those who can potentially benefit exceeds the overall costs to those who do not benefit.

## VIEWPOINT FOR ANALYSIS

When governmental bodies do economic analysis, an important concern is the proper viewpoint of the analysis. A look at industry will help to explain how the viewpoint from which an analysis is conducted influences the final recommendation. A firm pays its costs and counts *its* benefits. Thus, both the costs and benefits are measured from the firm's perspective.

Costs and benefits that occur outside the firm are referred to as external consequences (Figure 16-1). In years past, private-sector companies generally ignored the external consequences of their actions. Ask anyone who has lived near a paper plant, a slaughterhouse, or a steel mill about external consequences! More recently, governments

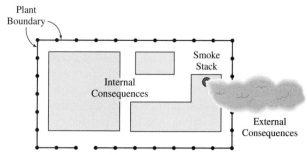

**FIGURE 16-1** Internal and external consequences of an industrial plant.

have forced industries to reduce pollution and other undesirable external consequences, with the result that today many companies are evaluating the consequences of their action from a broader, or community-oriented, viewpoint.

The council members of a small town that levies taxes can be expected to take the "viewpoint of the town" in making decisions: unless it can be shown that the money from taxes can be used *effectively*, the town council is unlikely to spend it. But what happens when the money is contributed to the town by the federal government, as in "revenue sharing" or by means of some other federal grant? Often the federal government pays a share of project costs varying from 10 to 90%. Example 16-1 illustrates the viewpoint problem that is created.

## EXAMPLE 16–1

A municipal project will cost $1 million. The federal government will pay 50% of the cost if the project is undertaken. Although the original economic analysis showed that the PW of benefits was $1.5 million, a subsequent detailed analysis by the city engineer indicates a more realistic estimate of the PW of benefits is $750,000. The city council must decide whether to proceed with the project. What would you advise?

### SOLUTION

From the viewpoint of the city, the project is still a good one. If the city puts up half the cost ($500,000) it will receive all the benefits ($750,000). On the other hand, from an *overall* viewpoint, the revised estimate of $750,000 of benefits does not justify the $1 million expenditure. This illustrates the dilemma caused by varying viewpoints. For economic efficiency, one does not want to encourage the expenditure of money, regardless of the source, unless the benefits at least equal the costs.

Possible viewpoints that may be taken include those of an individual, a business firm or corporation, a town or city district, a city, a state, a nation, or a group of nations. To avoid suboptimizing, the proper approach is to *take a viewpoint at least as broad as those who pay the costs and those who receive the benefits*. When the costs and benefits are totally confined to a town, for example, the town's viewpoint seems to be an appropriate basis for the analysis. But when the costs or the benefits are spread beyond the proposed viewpoint, then the viewpoint should be enlarged to this broader population.

Other than investments in defense and social programs, most of the benefits provided by government projects are realized at a regional or local level. Projects, such as dams for electricity, flood control, and recreation, as well as transportation facilities, such as roads, bridges, and harbors, all benefit most those in the region in which they are constructed. Even smaller-scale projects, such as the municipal rose garden, although funded by public monies at a local or state level, provide most benefit to those nearby. As in the case of private decision making, it is important to adopt an appropriate and *consistent* viewpoint and to designate all the costs and benefits that arise from the prospective investment from that perspective. To shift perspective when quantifying costs and benefits could greatly skew the results of the analysis and subsequent decision, and thus such changes in perspective are inappropriate.

## SELECTING AN INTEREST RATE

Several factors, not present for private-sector firms, influence selecting an interest rate for economic analysis in the government sector. Recall that for private-sector firms the overall objective is wealth maximization, and the interest rate is selected consistent with this goal. Most firms use *cost of capital* or *opportunity cost* concepts when setting an interest rate. The objective of public investment involves the use of public resources to *promote the general welfare* and to secure the benefits of a given project *to whomsoever they may accrue*, as long as those benefits outweigh the costs. How to set an interest rate is less clear-cut in this case. Possibilities include no interest rate, cost of capital concepts, and an opportunity cost concept.

### No Time-Value-of-Money Concept

In government, monies are obtained through taxation and spent about as quickly as they are obtained. Often, there is little time delay between collecting money from taxpayers and spending it. (Remember that the federal government and many states collect taxes every paycheck in the form of withholding tax.) The collection of taxes, like their disbursement, although based on an annual budget, is actually a continuous process. Using this line of reasoning, some would argue that there is little or no time lag between collecting and spending tax dollars. Thus, they would advocate the use of a 0% interest rate for economic analysis of public projects. Not surprisingly, this viewpoint is most often expressed by people who are *not* engineering economists and who are pushing a marginal project.

### Cost of Capital Concept

Another approach in determining interest rates in public investments is that most levels of government (federal, state, and local) borrow money for capital expenditures in addition to collecting taxes. Where money is borrowed for a specific project, one line of reasoning is to use an interest rate equal to the *cost of borrowed money*. This argument is less valid for state and local governments than for private firms because the federal government, through the income tax laws, subsidizes state and local bonded debt. If a state, county, city, or special assessment district raises money by selling bonds, the interest paid on these bonds is exempt from federal taxes. In this way the federal government is *subsidizing* the debt, thereby encouraging investors to purchase such bonds.

These bonds, called municipal bonds, can be either general obligation or revenue bonds. **General obligation municipal bonds** pay interest and are retired (paid off) through taxes raised by the issuing government unit. A school district may use property taxes it receives to finance bond debt for construction of new language labs. **Revenue bonds** are not supported by the taxing authority of the government unit; rather, they are supported by revenues earned by the project being funded. As an example, a city could use toll revenues from a new bridge to retire debt on revenue bonds sold for the bridge's construction.

For those who purchase municipal bonds, the instruments' tax-free status means that the expected return on this investment is somewhat less than that required of fully taxed bond investments (of similar risk). As a rough estimate, when fully taxed bonds yield an 8% interest rate, municipal bonds might make interest payments at a rate of 6%. The difference of 2% is due to the preferred treatment for federal taxation, and the federal subsidy on tax-free bonds. This can skew the *cost of capital* approach by lowering the apparent cost of long-term bonds. The true cost of the bonds must include the federal subsidy, which is paid by all taxpayers.

## Opportunity Cost Concept

Opportunity cost, which is related to the interest rate on the best opportunity forgone, may take two forms in governmental economic analysis: government opportunity cost and taxpayer opportunity cost. In public decision making, if the interest rate is based on the opportunity cost to a government agency or other governing body, this interest rate is known as **government opportunity cost.** In this case the interest rate is set at that of the best prospective project for which funding is not available. One disadvantage of the government opportunity cost concept is that different agencies and subdivisions of government will have different opportunities. Therefore, political units could use different interest rates for economic analysis, and a project that is rejected in one branch might be accepted in another. Differing interest rates lead to inconsistent evaluation and decision making across government.

Dollars used for public investments are generally gathered by taxing the citizenry. The concept of **taxpayer opportunity cost** suggests that a correct interest rate for evaluating public investments is that which the *taxpayer* could have received if the government had not collected those dollars through taxation. This philosophy holds that through taxation the government is taking away the taxpayers' opportunity to use the same dollars for investment. The interest rate that the government requires should not be less than what the taxpayer would have received.

The most widely followed standard is found in the Office of Management and Budget (OMB) A94 directive, which stipulates that a 7% interest rate be used in economic analysis for a wide range of federal projects. It is not economically desirable to take money from a taxpayer with a 12% opportunity cost, for example, and invest it in a government project yielding 4%.

## Recommended Concept

The general rule of thumb in setting an interest rate for government investments has been to select the *largest* of the cost of capital, the government opportunity cost, or the taxpayer opportunity cost interest rates. However, as is the case in the private sector, there is no hard and fast rule, universally applied in all decision circumstances. Setting an interest rate for use in economic analysis is at the discretion of the government entity performing the analysis.

## THE BENEFIT–COST RATIO

The benefit–cost ratio was described briefly in Chapter 9. This method is used almost exclusively in public investment analysis, and because of the magnitude of the amount of public dollars committed each year through such analysis, the benefit–cost ratio deserves our attention and understanding.

One of the primary reasons for using the benefit–cost ratio (B/C ratio) in public decision making is its simplicity. The ratio is formed by calculating the equivalent worth of the project's benefits divided by the equivalent worth of the project's costs. The benefit–cost ratio can be shown as follows:

$$\text{B/C ratio} = \frac{\text{Equivalent worth of net benefits}}{\text{Equivalent worth of costs}}$$

$$= \frac{\text{PW benefits}}{\text{PW costs}} = \frac{\text{FW benefits}}{\text{FW costs}} = \frac{\text{AW benefits}}{\text{AW costs}}$$

Notice that *any* of the equivalent worth methods (present, future, and annual) can be used to calculate this ratio. Each formulation of the ratio will produce an identical result, as illustrated in Example 16-2.

---

### EXAMPLE 16-2

Demonstrate that for this highway expansion project, the same B/C ratio is obtained using the present, future, and annual worth formulations.

| | |
|---|---|
| Initial costs of expansion | $1,500,000 |
| Annual costs for operating/maintenance | 65,000 |
| Annual savings and benefits to travelers | 225,000 |
| Scrap value after useful life | 300,000 |
| Useful life of investment | 30 years |
| Interest rate | 8% |

#### SOLUTION

**Using Present Worth**

$$\text{PW benefits} = 225{,}000(P/A, 8\%, 30) + 300{,}000(P/F, 8\%, 30) = \$2{,}563{,}000$$

$$\text{PW costs} = 1{,}500{,}000 + 65{,}000(P/A, 8\%, 30) = \$2{,}232{,}000$$

**or using a TVM calculator**

$$\text{PW benefits} = -PV(i, n, A, F) = -PV(8\%, 30, 225000, 300000) = \$2{,}562{,}814$$

$$\text{PW costs} = 1{,}500{,}000 - PV(8\%, 30, 65000, 0) = \$2{,}231{,}756$$

**Using Future Worth**

$$\text{FW benefits} = 225,000(F/A, 8\%, 30) + 300,000 = \$25,790,000$$

$$\text{FW costs} = 1,500,000(F/P, 8\%, 30) + 65,000(F/A, 8\%, 30) = \$22,460,000$$

**Using Annual Worth**

$$\text{AW benefits} = 225,000 + 300,000(A/F, 8\%, 30) = \$227,600$$

$$\text{AW costs} = 1,500,000(A/P, 8\%, 30) + 65,000 = \$198,200$$

$$\text{B/C ratio} = \frac{2,563,000}{2,232,000} = \frac{25,790,000}{22,460,000} = \frac{227,600}{198,200} = 1.15$$

One can see that the ratio provided by each of these methods produces the same result: 1.15.

An economic analysis is performed to assist in the objective of making a decision. When one is using the B/C ratio, the decision rule is in two parts:

If the B/C ratio is > 1.0, then the decision should be to invest.

If the B/C ratio is < 1.0, then the decision should be not to invest.

Cases of a B/C ratio *near* 1.0 are analogous to the case of a calculated net present worth of $0 or an IRR analysis that yields $i = \text{MARR}\%$. In other words, the decision measure is *near* the breakeven criteria. In such cases a detailed analysis of the input variables and their estimates is necessary, and one should consider the merits of other available opportunities for the targeted funds. But, if the B/C ratio is clearly greater than or less than 1.0, the recommendation is clear.

The B/C ratio is a numerator/denominator relationship between the equivalent worths (EW) of *benefits* and *costs:*

$$\text{B/C ratio} = \frac{\text{EW of net benefits to whomsoever they may accrue}}{\text{EW of costs to the sponsors of the project}}$$

The numerator and denominator aspects of the ratio are sometimes interpreted and used in different fashions. For instance, the **conventional B/C ratio** (see Example 16-2) can be restated as follows:

$$\text{Conventional B/C ratio} = \frac{\text{EW of net benefits}}{\text{EW of initial costs} + \text{EW of operating and maintenance costs}}$$

However, there is also the **modified B/C ratio.** To use a *modified* version, the *annual operating and maintenance costs* are subtracted in the numerator, rather than being added as a cost in the denominator. This *modified B/C ratio* is mathematically similar to the present worth index defined in Chapter 9. The ratio becomes

$$\text{Modified B/C ratio} = \frac{\text{EW of net benefits} - \text{EW of operating and maintenance costs}}{\text{EW of initial costs}}$$

For decision making, the two versions of the benefit–cost ratio will produce the same recommendation on whether to *invest* or *not invest* in the project being considered. The *numeric values for the B/C ratio* will not be the same, but the recommendation will be. This is illustrated in Example 16-3.

---

### EXAMPLE 16–3

Consider the highway expansion project from Example 16-2. Let us use the present worth formulation of *conventional* and *modified* versions to calculate the B/C ratio.

#### SOLUTION

**Using the Conventional B/C Ratio**

$$\text{B/C ratio} = \frac{225{,}000(P/A, 8\%, 30) + 300{,}000(P/F, 8\%, 30)}{1{,}500{,}000 + 65{,}000(P/A, 8\%, 30)} = 1.15$$

**Using the Modified B/C Ratio**

$$\text{B/C ratio} = \frac{225{,}000(P/A, 8\%, 30) + 300{,}000(P/F, 8\%, 30) - 65{,}000(P/A, 8\%, 30)}{1{,}500{,}000} = 1.22$$

Whether the conventional or the modified ratio is used, the recommendation is to invest in the highway expansion project. The ratios are not identical in magnitude (1.15 vs. 1.22), but the decision is the same.

---

It is important when one is using the conventional and modified B/C ratios not to directly compare the magnitudes of the two versions. Evaluating a project with one version may produce a higher ratio than is produced with the other version, but this does not imply that the project is somehow better.

The *net benefits to the users* of government projects are the difference between the expected *benefits* from investment minus the expected *disbenefits*. Disbenefits are the negative effects of government projects felt by some individuals or groups. For example, consider the U.S. National Park System. Development projects by the skiing or lumber industries might provide enormous benefits to the recreation or construction sectors while creating simultaneous disbenefits for environmental groups.

## INCREMENTAL BENEFIT–COST ANALYSIS

In Chapter 9 we discussed using the incremental benefit–cost ratio in economic decision analysis. As with the internal rate of return (IRR) measure, the incremental B/C ratio should be used in comparing *mutually exclusive alternatives*. Incremental B/C ratio analysis is consistent with maximizing the present worth of the alternatives. As with the incremental IRR method, it is *not* proper to simply calculate the B/C ratio for each alternative and choose the one with the highest value. Rather, an *incremental* approach is called for.

## Elements of the Incremental Benefit–Cost Ratio Method

**1.** *Identify all relevant alternatives.* Decision rules or models can recommend a *best* course of action *only* from the set of identified alternatives. If a better alternative exists but is not considered, then it will never be selected, and when available, the solution will be suboptimal. For benefit–cost ratio problems, the "do-nothing" option is always the "base case" from which the incremental methodology proceeds.

**2.** *(Optional) Calculate the B/C ratio of each alternative.* Once the individual B/C ratios have been calculated, the alternatives with a ratio *less than 1.0* are eliminated from further consideration. This step gets the "poor performers" out of the way before the incremental procedure is initiated. This step may be omitted, however, because the incremental analysis method will eliminate the subpar alternatives in due time.

*Note:* There is a case where this step *must* be skipped. If doing nothing is not an a alternative and if *all* alternatives have a B/C ratio less than 1.0, then the best of the unattractive alternatives will be selected through incremental analysis.

**3.** *Rank-order the projects.* The alternatives must be ordered from smallest to largest size of the *denominator of the B/C ratio.* (The rank order will be the same regardless of whether one uses the present worth, annual worth, or future worth of costs to calculate the *denominator.*) When available, the "do-nothing" alternative always becomes the first on the ordered list.

**4.** *Identify the increment under consideration.* The first increment considered is always that of going from the lowest cost alternative (when it is available, this is the do-nothing option) to the next higher cost alternative. As the analysis proceeds, any identified increment is always in reference to some previously justified alternative.

**5.** *Calculate the B/C ratio for the incremental cash flows.* First, calculate the *incremental benefits* and the *incremental costs.* This is done by finding the cash flows that represent the difference ($\Delta$) between the two alternatives under consideration. For two alternatives $X$ and $Y$, where $X$ is the defender, or base case, and $Y$ is the challenger, the increment can be written as ($X \rightarrow Y$) to signify *going from X to Y* or as ($Y - X$) to signify the *cash flows of Y **minus** cash flows of X.* Both modes identify the incremental costs ($\Delta C$) and benefits ($\Delta B$) of investing in Alternative $Y$, where $X$ is a previously justified (or base) alternative. The *incremental B/C ratio* equals $\Delta B / \Delta C$.

**6.** *Use the incremental B/C ratio to decide which alternative is better.* If the incremental B/C ratio ($\Delta B / \Delta C$) calculated in Step 5 is greater than 1.0, then the increment is desirable or justified; if the ratio is less than 1.0, it is not desirable, or is not justified. If an increment is accepted, the alternative (the challenger) associated with that increment of investment becomes the base from which the next increment is formed. When the increment is not justified, the alternative associated with the additional increment is rejected and the previously justified alternative (the defender) is maintained as the base for formation of the next increment.

**7.** *Iterate to Step 4 until all increments (projects) have been considered.* The incremental method requires that the entire list of ranked feasible alternatives be evaluated. All pairwise comparisons are made such that the additional increment being considered is examined with respect to a previously justified alternative. The incremental method continues until all alternatives have been evaluated.

**8.** *Select the best alternative from the set of mutually exclusive competing projects.* After all alternatives (and associated increments) have been considered, the incremental

B/C ratio method calls for selecting the alternative *associated with the last justified incre-ment.* This assures that a maximum investment is made such that each ratio of equivalent worth of incremental benefits to equivalent worth of incremental costs is greater than 1.0. (A common error in applying the incremental B/C method is selecting the alternative with the *largest* incremental B/C ratio, which is inconsistent with the objective of maximizing investment size with incremental B/C ratios above 1.0.)

Both the conventional and modified versions of the B/C ratio can be used with the incremental B/C ratio methodology just described, but the two versions should not be mixed in the same problem. Such an approach could cause confusion and errors. Instead, *one* of the two versions should be *consistently* used throughout the analysis.

In Examples 16-4 and 16-5 the incremental B/C ratio (conventional and modified version) is used to evaluate a set of mutually exclusive alternatives.

## EXAMPLE 16–4

A midwestern industrial state may construct and operate two coal-burning power plants and a distribution network to provide electricity to several state-owned properties. The following costs and benefits have been identified.

*Primary costs:* Construction of the power plant facilities; cost of installing the power distribution network; life-cycle maintenance and operating costs.

*Primary benefits:* Elimination of payments to the current electricity provider; creation of jobs for construction, operation, and maintenance of the facilities and distribution network; revenue from selling excess power to utility companies; increased employment at coal mines in the state.

There have been four competing designs identified for the power plants. Each has a life of 45 years. Use the *conventional* B/C ratio with an interest rate of 8% to recommend a course of action.

| | Values ($\times$ $10^4$) for Competing Design Alternatives | | | |
| --- | --- | --- | --- | --- |
| | I | II | III | IV |
| Project costs | | | | |
| Plant construction cost | $12,500 | $11,000 | $12,500 | $16,800 |
| Annual operating and maintenance cost | 120 | 480 | 450 | 145 |
| Project benefits | | | | |
| Annual savings from utility payments | 580 | 700 | 950 | 1,300 |
| Revenue from overcapacity | 700 | 550 | 200 | 250 |
| Annual effect of jobs created | 400 | 750 | 150 | 500 |

## SOLUTION

Alternatives I through IV and the do-nothing alternative are *mutually exclusive* choices because one and only one will be selected. Therefore, an incremental B/C ratio method is used to obtain the solution.

**Step 1** *Identify alternatives.* The alternatives are do nothing, and designs I, II, III, and IV.

**Step 2** *Calculate B/C ratio for each alternative.*

$$\text{B/C ratio (I)}    = (580 + 700 + 400)\,(P/A, 8\%, 45)/[12{,}500 + 120\,(P/A, 8\%, 45)] = 1.46$$
$$\text{B/C ratio (II)}   = (700 + 550 + 750)\,(P/A, 8\%, 45)/[11{,}000 + 480\,(P/A, 8\%, 45)] = 1.44$$
$$\text{B/C ratio (III)} = (200 + 950 + 150)\,(P/A, 8\%, 45)/[12{,}500 + 325\,(P/A, 8\%, 45)] = 0.96$$
$$\text{B/C ratio (IV)} = (1300 + 250 + 500)(P/A, 8\%, 45)/[16{,}800 + 145(P/A, 8\%, 45)] = 1.34$$

Alternatives I, II, and IV all have B/C ratios greater than 1.0 and thus merit further consideration. Alternative III does not meet the acceptability criterion and could be eliminated from further consideration. However, to illustrate that Step 2 is optional, all four design alternatives will be analyzed incrementally.

**Step 3** *Rank-order projects.* Here we calculate the PW of costs for each alternative. The denominator of the *conventional* B/C ratio includes first cost and annual O&M costs, so the PW of costs for the alternatives are:

$$\text{PW costs (I)}   = 12{,}500 + 120(P/A, 8\%, 45) = \$13{,}953$$
$$\text{PW costs (III)} = 12{,}500 + 325(P/A, 8\%, 45) = \$16{,}435$$
$$\text{PW costs (II)}  = 11{,}000 + 480(P/A, 8\%, 45) = \$16{,}812$$
$$\text{PW costs (IV)} = 16{,}800 + 145(P/A, 8\%, 45) = \$18{,}556$$

The rank order from low to high value of the B/C ratio *denominator* is as follows: do nothing, I, III, II, IV.

**Step 4** *Identify increment under consideration.*

**Step 5** *Calculate B/C ratio.*

**Step 6** *Which alternative is better?*

|  | Increment Under Consideration | 1st Iteration (Do Nothing → I) | 2nd Iteration (I → III) | 3rd Iteration (I → II) | 4th Iteration (II → IV) |
|---|---|---|---|---|---|
| **Step 4** | ΔPlant construction cost | $12,500 | $    0 | $−1500 | $5800 |
|  | ΔAnnual O&M cost | 120 | 205 | 360 | −335 |
|  | PW of ΔCosts | 13,953 | 2482 | 2859 | 1744 |
|  | ΔAnnual utility payment savings | 580 | 370 | 120 | 600 |
|  | ΔAnnual overcapacity revenue | 700 | 500 | −150 | −300 |
|  | ΔAnnual benefits of new jobs | 400 | −250 | 350 | −250 |
|  | PW of ΔBenefits | 20,342 | −4601 | 3875 | 605 |
| **Step 5** | ΔB/C ratio (PW ΔB)/(PW ΔC) | 1.46 | −1.15 | 1.36 | 0.35 |
| **Step 6** | Is increment justified? | Yes | No | Yes | No |

As an example of these calculations, consider the third increment (I → II).

| | |
|---|---|
| ΔPlant construction cost | $= 11{,}000 - 12{,}500 = -\$1500$ |
| ΔAnnual O&M cost | $= 480 - 120 = \$360$ |
| PW of ΔCosts | $= -1500 + 360(P/A, 8\%, 45) = \$2859$ |
| or | $= 16{,}812 - 13{,}953 = \$2859$ |
| ΔAnnual utility payment savings | $= 700 - 580 = \$120$ |
| ΔAnnual overcapacity revenue | $= 550 - 700 = -\$150$ |
| ΔAnnual benefits of new jobs | $= 750 - 400 = \$350$ |
| PW of ΔBenefits | $= (120 - 150 + 350)(P/A, 8\%, 45)$ |
|  | $= \$3875$ |
| ΔB/C ratio (PW ΔB)/(PW ΔC) | $= 3875/2859 = 1.36$ |

The analysis in the table proceeded as follows: do nothing to Alternative I was justified ($\Delta B/C$ ratio $= 1.46$), Alternative I became the new base; Alternative I to Alternative III was not justified ($\Delta$B/C ratio $= -1.15$), Alternative I remained base; Alternative I to Alternative II was justified ($\Delta$B/C ratio $= 1.36$), Alternative II became the base; Alternative II to Alternative IV was not justified ($\Delta$B/C ratio $= 0.35$).

**Step 7** *Select best alternative.* Alternative II became the recommended power plant design alternative because it is the one associated with the last justified increment. Notice that Alternative III, did not affect the recommendation and was eliminated through the incremental method. Notice also that the first increment considered (do nothing $\rightarrow$ I) was not selected even though it had the *largest* $\Delta$B/C ratio (1.45). The alternative associated with the *last justified increment* (in this case, Alt. II) should be selected.

## EXAMPLE 16–5

Let us reconsider Example 16-4, this time using the modified B/C ratio to analyze the alternatives. Again we will use the present worth method.

### SOLUTION

Here we use the modified B/C ratio.

**Step 1** *Identify alternatives.* The alternatives are still do nothing and designs I, II, III, and IV.

**Step 2** *Calculate modified B/C ratio for each alternative.*

$$\text{B/C ratio(I)} = (580 + 700 + 400 - 120)\,(P/A, 8\%, 45)/12{,}500 = 1.51$$
$$\text{B/C ratio(II)} = (700 + 550 + 750 - 480)\,(P/A, 8\%, 45)/11{,}000 = 1.67$$
$$\text{B/C ratio (III)} = (200 + 950 + 150 - 325)\,(P/A, 8\%, 45)/12{,}500 = 0.95$$
$$\text{B/C ratio (IV)} = (1300 + 250 + 500 - 145)\,(P/A, 8\%, 45)/16{,}800 = 1.37$$

Again Alternative III can be eliminated from further consideration because its B/C ratio is *less than 1.0*. In this case we will eliminate it. The remaining alternatives are do nothing and alternative designs I, II, and IV.

**Step 3** *Rank-order projects.* The PW of costs for each alternative:

$$\text{PW Costs (I)} = \$12{,}500$$
$$\text{PW Costs (II)} = \$11{,}000$$
$$\text{PW Costs (IV)} = \$16{,}800$$

The correct rank order is now do nothing, II, I IV. Notice that the *modified* B/C ratio produces an order of comparison different from that yielded by the *conventional* version in Example 16-4.

**Step 4** *Identify increment being considered.*

**Step 5** *Calculate B/C ratio.*

**Step 6**  *Which alternative is better?*

| Step 4  Incremental Effects | 1ˢᵗ Iteration (Do Nothing → II) | 2ⁿᵈ Iteration (II → I) | 3ʳᵈ Iteration (II → IV) |
|---|---|---|---|
| ΔPlant construction cost | $11,000 | $ 1500 | $ 5800 |
| PW of ΔCosts | 11,000 | 1500 | 5800 |
| ΔAnnual utility payment savings | 700 | −120 | 600 |
| ΔAnnual overcapacity revenue | 550 | 150 | −300 |
| ΔAnnual benefits of new jobs | 750 | −350 | −250 |
| ΔAnnual O&M disbenefit | 480 | −360 | −335 |
| PW of ΔBenefits | 18,405 | 484 | 4662 |
| **Step 5**  ΔB/C ratio (PW ΔB)/(PW ΔC) | 1.67 | 0.32 | 0.80 |
| **Step 6**  Is increment justified? | Yes | No | No |

As an example of the calculations in the foregoing table, consider the third increment (II → IV).

| | |
|---|---|
| ΔPlant construction cost | $= 16,800 - 11,000 = \$5800$ |
| PW of ΔCosts | $= \$5800$ |
| ΔAnnual utility payment savings | $= 1300 - 700 = \$600$ |
| ΔAnnual overcapacity revenue | $= 250 - 550 = -\$300$ |
| ΔAnnual benefits of new jobs | $= 500 - 750 = -\$250$ |
| ΔAnnual O&M disbenefit | $= 145 - 480 = -\$335$ |
| PW of ΔBenefits | $= (600 - 300 - 250 + 335)(P/A, 8\%, 45) = \$4662$ |
| ΔB/C ratio, (PW ΔB)/(PW ΔC) | $= 4662/5800 = 0.80$ |

When the modified version of the B/C ratio is used, Alt. II emerges as the recommended power plant design—just as it did when we used the conventional B/C ratio.

## OTHER EFFECTS OF PUBLIC PROJECTS

Four areas remain that merit discussion in describing the differences between government and nongovernment economic analysis: (1) financing government versus nongovernment projects, (2) the typical length of government versus nongovernment project lives, (3) quantifying and valuing benefits and disbenefits, and (4) the general effects of politics on economic analysis.

### Project Financing

Governmental organizations and market-driven firms differ in the way investments in equipment, facilities, and other projects are financed. In general, firms rely on monies from investors (through stock and bond issuance), private lenders, and retained earnings from operations. These sources serve as the pool from which investment dollars for projects come. Management's job is to match financial resources with projects in a way that keeps the firm growing, results in an efficient and productive environment, and continues to attract investors and future lenders of capital.

On the other hand, the government sector often uses taxation and municipal bond issuance as the source of investment capital. In government, taxation and revenue from operations are adequate to finance only modest projects. However, public projects tend to be large in scale (roadways, bridges, etc.), which means that for many public projects 100% of the investment costs must be borrowed. To prevent excessive public borrowing and to assure timely debt repayment, the U.S. government, through constitutional and legislative channels, has restricted government debt. These restrictions include:

1. Local government bodies are limited in their borrowing to a specified percentage of the assessed property value in their taxation district.

2. For new construction, borrowed funds attained through the sale of bonds require the approval of local voters (sometimes by a two-thirds majority). For example, a $2 million bond proposition for a new municipal jail might increase property taxes in the city's tax district by $1.50 for every $1000 of assessed property value. These added tax revenues would then be used to retire the debt on the bonds.

3. Repayment of public debt must be made following a specific plan over a pre-set period of time. For monies borrowed by issuing bonds, interest payments and maturity dates are set at the time of issuance.

Limitations on the use and sources of borrowed monies make funding public sector projects much different from in the private sector. Private sector firms are seldom able to borrow 100% of required funds for projects, as can be done in the public sector, but at the same time, private entities do not face restrictions on debt retirement or the uncertainty of voter approval. Passing the bond proposition is the public *go-ahead*.

## Project Duration

Government projects often have longer lives than those in the private sector. In the private sector, projects most often have a projected or intended life ranging between 5 and 15 years. Some markets and technologies change more rapidly and some more slowly, but a majority of projects fall in this interval. Complex advanced manufacturing technologies, like computer-aided manufacturing or flexible automated manufacturing cells, typically have project lives at the longer end of this range.

Government projects typically have lives in the range of 20 to 50 years (or longer). Typical projects include federal highways, city water/sewer infrastructure, county dumps, and state libraries and museums. These projects, by nature, have a longer useful life than a typical private-sector project. There are exceptions to this rule because private firms invest in facilities and other long-range projects, and government entities also invest in projects with shorter-term lives. But, in general, investment duration in the government sector is longer.

Government projects, because they tend to be long range and large scale, usually require substantial funding in the early stages. Highway, water/sewer, and library projects can require millions of dollars in design, surveying, and construction costs. Therefore, it is in the best interest of decision makers who are advocates of such projects to spread that first cost over as many years as possible to reduce the annual cost of capital recovery. Using longer project lives to downplay the effects of a large first cost increases the desirability of the project, as measured by the B/C ratio. Another aspect closely associated with managing the size of the capital recovery cost in a B/C ratio analysis is the interest rate used for discounting. Lower interest rates reduce the capital recovery cost of having money tied up in a project. Example 16-6 illustrates the effects that project life and interest rate can have on the analysis and acceptability of a project.

## EXAMPLE 16–6

Consider a project that has been approved by local voters to build a new junior high school, needed because of increased (and projected) population growth. Analyze the project with interest rates of 3, 10, and 15% and with horizons of 15, 30, and 60 years.

| | |
|---|---|
| Building first costs (design, planning, and construction) | $10,000,000 |
| Initial cost for roadway and parking facilities | 5,500,000 |
| First cost to equip and furnish facility | 500,000 |
| Annual operating and maintenance costs | 350,000 |
| Annual savings from rented space | 400,000 |
| Annual benefits to community | 1,600,000 |

### SOLUTION

With this project we examine the effect that varying project lives and interest rates have on the economic value of a public project. In each case, the formula is

$$\text{Conventional B/C ratio} = \frac{1{,}600{,}000 + 400{,}000}{(10{,}000 + 5{,}500{,}000 + 500{,}000)(A/P, i, n) + 350{,}000}$$

The B/C ratio for each combination of project life and interest rate is tabulated as follows:

**Conventional Benefit–Cost Ratio for Various
Combinations of Project Life and Interest Rate**

| Project Life (years) | Interest | | |
|---|---|---|---|
| | 3% | 10% | 15% |
| 15 | 1.24 | 0.86 | 0.69 |
| 30 | 1.79 | 1.03 | 0.76 |
| 60 | 2.24 | 1.08 | 0.77 |

From these numbers one can see the effect of project life and interest on the analysis and recommendation. At the lower interest rate, the project has B/C ratios above 1.0 in all cases of project life, while at the higher rate the ratios are all less than 1.0. At an interest rate of 10% the recommendation to invest changes from *no* at a life of 15 years to *yes* at 30 and 60 years. A higher interest rate discounts the benefits in later years more heavily, so that they may not matter. In this case, the benefits from Years 31 to 60 add only 0.01 to the B/C ratio at 15% and 0.05 at 10%. At 3% those benefits add 0.45 to the ratio. By manipulating these two parameters (project life and interest rate), it is possible to reach entirely different conclusions regarding the desirability of the project.

Example 16-6 demonstrates that we must ensure that a long life and a low interest rate for a public project are truly appropriate and not chosen solely to make a marginal project look better.

## Quantifying and Valuing Benefits and Disbenefits

The junior high school in Example 16-6 included annual benefits to the community of $1.6 million. If you were evaluating the school, how would you estimate this? Many public sector projects like the junior high school and the examples in Table 16-1 have consequences that are difficult to state in monetary terms.

First the number of people affected by the project have to be counted—now and through the project's horizon. Then a dollar value for each person is required. For the junior high school it may be easy to estimate the number of students. But how much better will the educational outcomes be with the new school, and how valuable is that improvement?

On the other hand, consider the levees around New Orleans that needed to be rebuilt in the aftermath of Katrina. An economic evaluation of the different alternatives required estimating the number of residents that would be protected by improving the levees in New Orleans, which is extremely difficult. This required estimating not only how the rebuilding of the city might progress but also storm surges from future hurricanes, whose frequency and intensity may be changing. Once the number of people and homes had been estimated over the next 30 to 100 years, it was necessary to put a value on property, on disrupted lives, and on human lives.

While many individuals find it difficult to put a dollar value on a human life, there are many public projects whose main intent is to reduce the number of deaths due to floods, cancer, auto accidents, and other causes. Those projects are often justified by the value

**TABLE 16–1  Example Benefits and Disbenefits for Public Investments**

| Public Project | Primary Benefits | Primary Disbenefits |
|---|---|---|
| New airport outside city | More flights, new businesses | Increased travel time to airport, more traffic on outer belt |
| Interstate bypass around town | Quicker commute times, reduced congestion on surface roads | Lost sales to businesses on surface roads, lost agricultural lands |
| New metro subway system | Faster commute times, less pollution | Lost jobs due to bus line closing, less access to service (fewer stops) |
| Creation of a city waste disposal facility versus sending waste out of state | Less costly; faster and more responsive to customers | Objectionable sight and smells, lost market value to homeowners, lost forestland |
| Construction of a nuclear power plant | Cheaper energy costs, new industry in area | Environmental risks |

of preventing deaths. Thus, valuing human lives is an inescapable part of public-sector engineering economy.

Because the benefits and disbenefits of public projects are often difficult to quantify and value, the estimated values will have more uncertainty than is typical for private-sector projects. Thus those who favor a project and those who oppose it will often push to have values used that support their position.

## Project Politics

To some degree political influences are felt in nearly every decision made in any organization. Predictably, some individual or group will support particular interests over competing views. This actuality exists in both firms and government organizations. In government the effects of politics are continuously felt at all levels because of the large-scale and multipurpose nature of projects, because government decision making involves the use of the citizenry's common pool of money, and because individuals and groups have different values and views. For example, how important is economic development relative to environmental protection?

The guideline for public decision making, as set forward in the Preamble to the Constitution, is to *promote the general welfare* of citizens. However, it is impossible to please everyone all the time. The term "general welfare" implies that the architects of this document understood that the political process would produce opposing viewpoints, but at the same time they empowered decision makers to act in a representative way.

Since government projects tend to be large in scale, the time required to plan, design, fund, and construct them is usually several years. However, the political process tends to produce government leaders who support short-term decision making (because many government terms of office, either elected or appointed, are relatively short). Therein lies another difference between firm and government decision making—short-term decision making, long-term projects.

Because government decision makers are in the public eye more than those in the private sector, governmental decisions are generally more affected by "politics." Thus, the decisions that public officials make may not always be the best from an *overall* perspective. If a particular situation exposes a public official to ridicule, he may choose an expedient action to eliminate negative exposure (whereas a more careful analysis might have been better). Or, such a decision maker may placate a small, but vocal, political group over the interest of the majority of citizens by committing funds to a favored project (at the expense of other better projects). Or, a public decision maker may avoid controversy by declining to make a decision on an important, but politically charged, issue (whereas it would be in the overall interest of the citizenry if action were taken). Indeed, the role of politics in government decision making is more complex and far ranging than in the private sector.

## EXAMPLE 16–7

Consider again Example 16–4, where we evaluated power plants designs. Remember that government projects are often opposed and supported by different groups in the populace. Thus, decision makers become very aware of potential political aspects when they are considering such projects.

For the electric power plant decision, several political considerations may affect any evaluation of funding this project.

- The governor has been a strong advocate of workers' rights and has received abundant campaign support from organized labor (which is especially important in an industrialized state). By championing this project, the governor should be seen as pro-labor, thereby benefiting his bid for reelection, even if the project is not funded.
- The regulated electric utility providers in the state are strongly against this project, claiming that it would directly compete with them and take away some of their biggest customers. The providers have a strong lobby and key contacts with the state's utilities commission. A senior state senator has already protested that this project is the first step toward "rampant socialism in this great state."
- Business leaders in the municipalities where the two facilities would be constructed favor the project because it would create more jobs and increase the tax base. These leaders promote the project as a win–win opportunity for government and industry, where the state can benefit by reducing costs, and the electric utilities can improve their service by focusing more effectively on residential customers and their needs.
- The lieutenant governor is promoting this project, proclaiming that it is an excellent example of "initiating proactive and creative solutions to the problems that this state faces."
- Federal and state regulatory agencies are closely watching this project with respect to the Clean Air Act. Speculation is that the state plans to use a high-sulfur grade of local coal exclusively. Thus "stack scrubbers" would be required, or the high-sulfur coal would have to be mixed with lower-sulfur coal imported from other states to bring the overall air emissions in line with federal standards. The governor is using this opportunity to make the point that "the people of this state don't need regulators to tell us if we can use our own coal!"
- The state's coal operators and mining unions strongly support this project. They see the increased demands for coal and the governor's pro-labor advocacy as very positive. They plan to lobby the legislature strongly in favor of the project.
- Land preservation and environmental groups are strongly opposing the proposed project. They have studied the potential negative impacts of this project on the land and on water and air quality, as well as on the ecosystem and wildlife, in the areas where the two facilities would be constructed. Environmentalists have started a public awareness campaign urging the governor to act as the "chief steward" of the state's natural beauty and resources.

Will the project be funded? We can only guess. Clearly, however, we can see the competing influences that can be, and often are, part of decision making in the public sector.

## SUMMARY

Economic analysis and decision making in government is notably different from these processes in the private sector because the basic objectives of the public and private sectors are fundamentally different. Government investments in projects seek to maximize benefits to the *greatest number of citizens,* while minimizing the *disbenefits to citizens* and *costs to the government.* Private firms, on the other hand, are focused primarily on maximizing stockholder wealth.

Several factors, not affecting private firms, enter into the decision-making process in government. The source of capital for public projects is limited primarily to taxes and bonds. Government bonds issued for project construction are subject to legislative restrictions on debt not required for private firms. Also, raising tax and bond monies involves sometimes long and politically charged processes not present in the private sector. In addition, government projects tend to be larger in scale than the projects of competitive firms and the government projects affect many more people and groups in the population. The benefits and disbenefits to the many people affected are difficult to quantify and value, which is unlike the private sector, where products and services are sold and the revenue to the firm is clearly defined. All these factors slow down the process and make investment decision analysis more difficult for government decision makers than for those in the private sector. Another difference between the public and private sectors lies in how the interest rate (MARR) is set for economic studies. In the private sector, considerations for setting the rate include the cost of capital and opportunity costs. In government, establishing the interest rate is complicated by uncertainty in specifying the cost of capital and the issue of assigning opportunity costs to taxpayers or to the government.

The benefit–cost ratio is widely used to evaluate and justify government-funded projects. This measure of merit is the ratio of the equivalent worth of benefits to the equivalent worth of costs. This ratio can be calculated by PW, AW, or FW methods. A B/C ratio *greater than 1.0* indicates that a project should be invested in if funding sources are available. For considering *mutually exclusive alternatives,* an incremental analysis is required. This method results in the recommendation of the project with the highest investment cost that can be incrementally justified. Two versions of the B/C ratio, the *conventional* and *modified* B/C ratios, produce identical recommendations. The conventional B/C ratio treats annual operating and maintenance costs as a cost in the denominator, while the modified B/C ratio subtracts those costs from the benefits in the numerator.

# PROBLEMS

### Objective and Viewpoint

**16-1**  Public-sector economic analysis and decision making is often called "a multi-actor or multi-stakeholder decision problem." Explain.

**16-2**  Compare the general underlying objective of public decision making versus private decision making.

**16-3**  The text recommends a viewpoint that is appropriate in public decision making. What is suggested? What example is given to highlight the dilemma of viewpoint in public decision making? Provide another example.

**16-4**  In government projects, what is meant by the phrase "most of the benefits are local"? What conflict does this create for the federal government in the funding projects from public monies.

**16-5**  List the potential costs, benefits, and disbenefits that should be considered in evaluating a potential nuclear power plant. What stakeholder viewpoints will need to be considered?

**16-6**  A municipal landfill and incineration facility is planned. Name at least three benefits, three disbenefits, and three costs. What stakeholder viewpoints will need to be considered?

**16-7**  An interstate bypass will completely circle a city. Name at least three benefits, three disbenefits, and three costs. What stakeholder viewpoints will need to be considered?

**16-8**  A light rail system will connect the airport, the city center, and a cluster of high-density housing on the other side of the river. Name at least three benefits, three disbenefits, and three costs. What stakeholder viewpoints will need to be considered?

**16-9**  Improvements at a congested intersection require that the government acquire the properties on the four corners: two gas stations, a church, and a bank. Construction will take a year, and the costs will be shared 70% by the state and 30% by the city. Traffic through the intersection is mainly commuters, local residents, and deliveries to and by local businesses. There is some through traffic from other parts of the metropolitan area.

Identify the benefits, disbenefits, and costs that must be considered in evaluating this project. From the city's viewpoint, which must be included? From the state's viewpoint? What viewpoint should be used to evaluate this project?

**16-10**  The state may eliminate a railroad grade crossing by building an overpass. The new structure, together with the needed land, would cost $1.8 million. The analysis period is assumed to be 30 years because either the railroad or the highway above it will be relocated by then. Salvage value of the bridge (actually, the net value of the land on either side of the railroad tracks) 30 years hence is estimated to be $100,000. A 6% interest rate is to be used.

About 1000 vehicles per day are delayed by trains at the grade crossing. Trucks represent 40%, and 60% are other vehicles. Time for truck drivers is valued at $18 per hour and for other drivers at $5 per hour. Average time saving per vehicle will be 2 minutes if the overpass is built. No time saving occurs for the railroad.

The railroad spends $48,000 annually for crossing guards. During the preceding 10-year period, the railroad has paid out $600,000 in settling lawsuits and accident cases related to the grade crossing. The proposed project will entirely eliminate both these expenses. The state estimates that the new overpass will save it about $6000 per year in expenses directly due to the accidents. The overpass, if built, will belong to the state.

Should the overpass be built? If the overpass is built, how much should the railroad be asked to contribute to the state as its share of the $1.8 million construction cost?

**Selecting Rate**

**16-11**  Discuss the alternative concepts that can be employed when setting the discounting rate for economic analysis in the public sector. What is the authors' final recommendation for setting this rate?

**16-12**  The city's landfill department has a capital budget of $600,000. What is the government's opportunity cost of capital if it has the following projects to consider?

What does this indicate about which projects should be done?

| Project | First Cost | Rate of Return (%) | B/C Ratio at 7% |
|---|---|---|---|
| A | $100,000 | 23 | 1.30 |
| B | 200,000 | 22 | 1.40 |
| C | 300,000 | 17 | 1.50 |
| D | 200,000 | 19 | 1.35 |
| E | 100,000 | 18 | 1.56 |

**16-13**  The state's fish and game department has a capital budget of $9 million. What is the government's opportunity cost of capital if it has the following projects to consider? What does this indicate about which projects should be done?

| Project | First Cost | Rate of Return (%) | B/C Ratio at 7% |
|---|---|---|---|
| A | $2,000,000 | 9 | 1.23 |
| B | 1,000,000 | 14 | 1.42 |
| C | 2,000,000 | 10 | 1.17 |
| D | 3,000,000 | 16 | 1.45 |
| E | 2,000,000 | 13 | 1.56 |
| F | 3,000,000 | 15 | 1.35 |
| G | 3,000,000 | 12 | 1.32 |
| H | 1,000,000 | 11 | 1.26 |

**6-14**  A proposed bridge would cost $4 million to build and $180,000 per year in maintenance. The bridge should last 40 years. Benefits to the driving public are estimated to be $900,000 per year. Damages (not paid) to adjacent property owners due to noise is estimated to be $250,000 per year. Because of the uncertainty on what interest rate should be used to evaluate this project, calculate the breakeven annual interest rate that results in a B/C ratio of 1.
*Contributed by D. P. Loucks, Cornell University*

**16-15**  Is the 7% interest rate specified in OMB A94 a real or a nominal interest rate? Should it be used with costs expressed as constant-value dollars or with costs inflated using carefully selected inflation rates?

**16-16**  A municipal bond has a face value of $10,000. Interest of $400 is paid every 6 months. The bond has a life of 20 years. What is the effective rate of interest on this bond? Is this rate adjusted for inflation? What is the municipal government's cost of capital for this bond? Estimate the rate, considering all levels of government.

## B/C and Modified Ratio

**16-17** Consider the following investment opportunity:

| | |
|---|---|
| Initial cost | $100,000 |
| Additional cost at end of Year 1 | 150,000 |
| Benefit at end of Year 1 | 0 |
| Annual benefit per year at end of Years 2–10 | 20,000 |

With interest at 7%, what is the benefit–cost ratio for this project?    (*Answer:* 0.51)

**16-18** Calculate the conventional and modified benefit–cost ratios for the following project.

| | |
|---|---|
| Required first costs | $1,200,000 |
| Annual benefits to users | $500,000 |
| Annual disbenefits to users | $25,000 |
| Annual cost to government | $125,000 |
| Project life | 35 years |
| Interest rate | 10% |

**16-19** For the data given in Problem 16-18, for handling benefits and costs, demonstrate that the calculated B/C ratio is the same using each of the following methods: present worth, annual worth, and future worth.

**16-20** What is the essential difference between the *conventional* and *modified* versions of the benefit–cost ratio? Is it possible for these two measures to provide conflicting recommendations regarding invest/do-not-invest decisions?

**16-21** A government agency has estimated that a flood control project has costs and benefits that are parabolic, according to the equation

$$(\text{Present worth of benefits})^2$$

$$- 22(\text{Present worth of cost}) + 44 = 0$$

where both benefits and costs are stated in millions of dollars. What is the present worth of cost for the optimal size project?

**16-22** Chungyang Dam is being constructed across the Hungshui River in southern China. The dam will produce electricity to serve over 500,000 people in the region. The initial cost is 3.7 billion yuan and annual operating costs are 39.2 million yuan. A major overhaul of the electric generation facilities estimated to cost 650 million yuan will occur at the end of Year 25. The dam and generating plant have no salvage value, but must be torn down and removed at the end of Year 50 for 175 million yuan. Ishan Electric has a MARR of 10%. What annual

benefit in yuan is needed for a B/C ratio of 1? Use the Internet and current exchange rates to find the annual benefit in U.S. dollars.
*Contributed by Paul R. McCright, University of South Florida*

## Incremental Analysis

**16-23** The Highridge Water District needs an additional supply of water from Steep Creek. The engineer has selected two plans for comparison:

> *Gravity plan:* Divert water at a point 10 miles up Steep Creek and carry it through a pipeline by gravity.
>
> *Pumping plan:* Divert water at a point near the district and pump it through 2 miles of pipeline. The pumping plant can be built in two stages: half now and half 10 years later.

| | Gravity | Pumping |
|---|---|---|
| Initial investment | $2,800,000 | $1,400,000 |
| Investment in 10th year | 0 | 200,000 |
| Operation, maintenance, replacements, per year | 10,000 | 25,000 |
| Average annual power cost | | |
| First 10 years | 0 | 50,000 |
| Next 30 years | 0 | 100,000 |

Use a 40-year analysis period and 8% interest. Salvage values can be ignored. Use the conventional benefit–cost ratio method to select the more economical plan.    (*Answer:* Pumping plan)

**16-24** Two different routes are being considered for a mountain highway. The **high road** would require building several bridges and would navigate around the highest mountain points, thus requiring more roadway. The **low road** would construct several tunnels for a more direct route through the mountains. Projected travel volume for this new section of road is 2500 cars per day. Use the *modified* B/C ratio to determine which alternative should be recommended. Assume that project life is 45 years and $i = 6\%$.

| | High Road | Low Road |
|---|---|---|
| Construction cost/mile | $200,000 | $450,000 |
| Number of miles required | 35 | 10 |
| Annual benefit/car-mile | $0.015 | $0.045 |
| Annual O&M costs/mile | $2,000 | $10,000 |

16-25 The city engineer has prepared two plans for roads in the city park. Both plans meet anticipated requirements for the next 40 years. The city's minimum attractive rate of return is 7%.

Plan *A* is a three-stage development program: $300,000 is to be spent now followed by $250,000 at the end of 15 years and $300,000 at the end of 30 years. Annual maintenance will be $75,000 for the first 15 years, $125,000 for the next 15 years, and $250,000 for the final 10 years.

Plan *B* is a two-stage program: $450,000 is required now, followed by $50,000 at the end of 15 years. Annual maintenance will be $100,000 for the first 15 years and $125,000 for the subsequent years. At the end of 40 years, this plan has a salvage value of $150,000.

Use a conventional benefit–cost ratio analysis to determine which plan should be chosen.

6-26 The Arkansas Department of Transportation may build a new highway between Texarkana and Fort Smith, currently a distance of 181 miles. Design 1 is a four-lane highway built entirely on the existing route. Design 2 includes a significant rerouting through a mountainous region that would reduce the mileage to 166 miles. Design 3 is a fully access-controlled interstate-quality highway with more rerouting, which would reduce the total mileage to 148 miles.

Benefits for this project depend on mileage saved times the number of vehicles, plus the estimated value for the larger number of trips that will occur with the shorter and faster routes. The estimated benefits and costs of the three potential designs are shown in the table. Doing nothing yields no costs and no benefits. Using incremental analysis for the B/C ratio, a planning horizon of 75 years, and a MARR of 9%, which design would you recommend?

| | Initial Cost | Annual Maintenance Cost | Annual Benefit |
|---|---|---|---|
| Design 1 | $ 456M | $17M | $107M |
| Design 2 | 810M | 28M | 198M |
| Design 3 | 1552M | 58M | 287M |

*Contributed by Paul R. McCright, University of South Florida*

16-27 A two-lane highway between two cities, 10 miles apart, is to be converted to a four-lane divided freeway. The average daily traffic (ADT) on the new freeway is forecast to average 20,000 vehicles per day over the next 20 years. Trucks represent 5% of the total traffic. Annual maintenance on the existing highway is $1500 per lane-mile. The existing accident rate is 4.58 per million vehicle-miles (MVM). Three alternate plans are under consideration.

*Plan A:* Improve along the existing development by adding two lanes adjacent to the existing lanes at a cost of $450,000 per mile. This will reduce auto travel time by 2 minutes and truck travel time by 1 minute. The estimated accident rate is 2.50 per MVM. Annual maintenance is $1250 per lane-mile.

*Plan B:* Improve the highway along the existing alignment with grade improvements at a cost of $650,000 per mile. Plan *B* adds two lanes and would reduce auto and truck travel time by 3 minutes each. The accident rate on this improved road is estimated to be 2.40 per MVM. Annual maintenance is $1000 per lane-mile.

*Plan C:* Construct a new freeway on a new alignment at a cost of $800,000 per mile. This plan would reduce auto travel time by 5 minutes and truck travel time by 4 minutes. Plan *C* is 0.3 mile longer than *A* or *B*. The estimated accident rate for *C* is 2.30 per MVM. Annual maintenance is $1000 per lane-mile. Plan *C* includes abandoning the existing highway with no salvage value.

**Incremental operating cost**

| | |
|---|---|
| Autos | 6¢ per mile |
| Trucks | 18¢ per mile |

**Time saving**

| | |
|---|---|
| Autos | 3¢ per minute |
| Trucks | 15¢ per minute |
| **Average accident cost** | $1200 |

If a 5% interest rate is used, which of the three proposed plans should be adopted?   (*Answer: Plan C*)

16-28 Evaluate these mutually exclusive alternatives with a horizon of 15 years and a MARR of 12%.

| | A | B | C |
|---|---|---|---|
| Initial investment | $9500 | $18,500 | $22,000 |
| Annual savings | 3200 | 5,000 | 9,800 |
| Annual costs | 1000 | 2,750 | 6,400 |
| Salvage value | 6000 | 4,200 | 14,000 |

Use each of these approaches:

(*a*) *Conventional* B/C ratio

(*b*) *Modified* B/C ratio

(*c*) Present worth analysis

(*d*) Internal rate of return analysis

(*e*) Payback period

**16-29** A 50-meter tunnel must be constructed for a new city aqueduct. One alternative is to build a full-capacity tunnel now for $500,000. The other alternative is to build a half-capacity tunnel now for $300,000 and then to build a second parallel half-capacity tunnel 20 years hence for $400,000. The cost to repair the tunnel lining every 10 years is $20,000 for the full-capacity tunnel and $16,000 for each half-capacity tunnel.

Determine whether the full-capacity tunnel or the half-capacity tunnel should be constructed now. Solve the problem by the conventional benefit–cost ratio analysis, using a 5% interest rate and a 50-year analysis period. There will be no tunnel lining repair at the end of the 50 years.

**16-30** The Fishery and Wildlife Agency of Ireland is considering four mutually exclusive design alternatives for a major salmon hatchery. This agency of the Irish government uses the following B/C ratio for decision making:

$$\text{B/C ratio} = \frac{\text{EW(Net benefits)}}{\text{EW(Capital recovery cost)} + \text{EW(O\&M cost)}}$$

Using an interest rate of 8% and a project life of 30 years, recommend which design is best.

**(Values in 1000s)**

| | Irish Fishery Design Alternatives | | | |
|---|---|---|---|---|
| | A | B | C | D |
| First cost | $9500 | $12,500 | $14,000 | $15,750 |
| Annual benefits | 2200 | 1,500 | 1,000 | 2,500 |
| Annual O&M costs | 550 | 175 | 325 | 145 |
| Annual disbenefits | 350 | 150 | 75 | 700 |
| Salvage value | 1000 | 6,000 | 3,500 | 7,500 |

**16-31** Six mutually exclusive investments have been identified for evaluation by means of the benefit–cost ratio method. Assume a MARR of 10% and an equal project life of 25 years for all alternatives.

| | Mutually Exclusive Alternatives | | | | | |
|---|---|---|---|---|---|---|
| Annualized | 1 | 2 | 3 | 4 | 5 | 6 |
| Net costs to sponsor ($M) | 15.5 | 13.7 | 16.8 | 10.2 | 17.0 | 23.3 |
| Net benefits to users ($M) | 20.0 | 16.0 | 15.0 | 13.7 | 22.0 | 25.0 |

(a) Use annual worth and the B/C ratio to identify the better alternative.

(b) If this were a set of *independent* alternatives, how would you conduct a comparison?

**16-32** A section of state highway needs repair. At present, the traffic volume is low and few motorists would benefit. However, traffic is expected to increase, with resulting increased motorist benefits. The repair work will produce benefits for 10 years after it is completed.

Should the road be repaired and, if so, when should the work be done? Use a 15% MARR.

**(Costs in 1000s)**

| Year | Repair Now | Repair in 2 Years | Repair in 4 Years | Repair in 5 Years |
|---|---|---|---|---|
| 0 | −$150 | | | |
| 1 | 5 | | | |
| 2 | 10 | −$150 | | |
| 3 | 20 | 20 | | |
| 4 | 30 | 30 | −$150 | |
| 5 | 40 | 40 | 40 | −$150 |
| 6 | 50 | 50 | 50 | 50 |
| 7 | 50 | 50 | 50 | 50 |
| 8 | 50 | 50 | 50 | 50 |
| 9 | 50 | 50 | 50 | 50 |
| 10 | 50 | 50 | 50 | 50 |
| 11 | 0 | 50 | 50 | 50 |
| 12 | 0 | 50 | 50 | 50 |
| 13 | 0 | 0 | 50 | 50 |
| 14 | 0 | 0 | 50 | 50 |
| 15 | 0 | 0 | 0 | 50 |

**16-33** The local highway department is analyzing reconstruction of a mountain road. The vehicle traffic increases each year, hence the benefits to the motoring public also increase. Based on a traffic count, the benefits are projected as follows:

| Year | End-of-Year Benefit |
|---|---|
| 2011 | $10,000 |
| 2012 | 12,000 |
| 2013 | 14,000 |
| 2014 | 16,000 |
| 2015 | 18,000 |
| 2016 | 20,000 |
| | and so on, increasing $2000 per year |

The reconstructed pavement will cost $275,000 when it is installed and will have a 15-year useful life. The construction period is short, hence a beginning-of-year reconstruction will result in the listed end-of-year benefits. Assume a 6% interest rate. The reconstruction, if done at all, must be done not later than 2016. Should it be done, and if so, in what year?

## Challenges for Public Sector

**16-34** Big City Carl, a local politician, is pushing a new dock and pier system at the river to attract commerce. A committee appointed by the mayor (an opponent of Carl's) has developed the following estimates.

| | |
|---|---|
| Cost to remove current facilities | $ 750,000 |
| Material, labor, and overhead for new construction | 2,750,000 |
| Annual operating & maintenance costs | 185,000 |
| Annual benefits from new commerce | 550,000 |
| Annual disbenefits to sportsmen | 35,000 |
| Project life | 20 years |
| Interest rate | 8% |

(*a*) Using the *conventional* B/C ratio, determine whether the project should be funded.

(*b*) After studying the numbers given by the committee, Big City Carl argued that the project life should be *at least* 25 years and more likely closer to 30 years. How did he arrive at this estimate, and why is he making this statement?

**16-35** The federal government proposes to construct a multipurpose water project to provide water for irrigation and municipal use. In addition, there are flood control and recreation benefits. The benefits are given below. The annual benefits are one-tenth of the decade benefits. The operation and maintenance cost is $15,000 per year. Assume a 50-year analysis period with no net project salvage value.

(*a*) If an interest rate of 5% is used, and a benefit–cost ratio of unity, what capital expenditure can be justified to build the water project now?

(*b*) If the interest rate is changed to 8%, how does this change the justified capital expenditure?

**(Benefits in 1000s)**

| | Decades | | | | |
|---|---|---|---|---|---|
| Purpose | First | Second | Third | Fourth | Fifth |
| Municipal | $ 40 | $ 50 | $ 60 | $ 70 | $110 |
| Irrigation | 350 | 370 | 370 | 360 | 350 |
| Flood control | 150 | 150 | 150 | 150 | 150 |
| Recreation | 60 | 70 | 80 | 80 | 90 |
| Totals | $600 | $640 | $660 | $660 | $700 |

**16-36** If you favored Plan *B* in Problem 16-25, what value of MARR would you use in the computations? Explain.

**16-37** Describe how a decision maker can use each of the following to "skew" the results of a B/C ratio analysis in favor of his or her own position on funding projects.

(*a*) Conventional versus modified ratios

(*b*) Interest rates

(*c*) Project duration

(*d*) Benefits, costs, and disbenefits

**16-38** Briefly describe your sources and methods for estimating the value of

(*a*) a saved hour of commuting time.

(*b*) converting 15 miles of unused railroad tracks near a community of 300,000 into a new bike path.

(*c*) reducing annual flood risks from the Mississippi River for St. Louis by 5%.

(*d*) a human life.

**16-39** Discuss potential data sources and methods for estimating each of the costs, benefits, and disbenefits identified in Problem 16-5 for the nuclear power plant.

**16-40** For the municipal landfill and incineration facility in Problem 16–6, discuss potential data sources and methods for estimating each of the benefits, disbenefits, and costs.

**16-41** For the interstate bypass in Problem 16-7, discuss potential data sources and methods for estimating each of the benefits, disbenefits, and costs.

**16-42** For the light rail system in Problem 16-8, discuss potential data sources and methods for estimating each of the benefits, disbenefits, and costs.

**16-43** Discuss potential data sources and methods for estimating each of the costs, benefits, and disbenefits identified in Problem 16-9 for the congested intersection.

**16-44** Think about a major government construction project under way in your state, city, or region. Are the decision makers who originally analyzed and initiated the project currently in office? How can politicians use "political posturing" with respect to government projects?

## Minicases

**16-45** Research and report on how an agency of your state evaluates public project proposals. What interest rate and what economic measures are used?

**16-46** Research and report on how an agency of your municipality evaluates public project proposals. What interest rate and what economic measures are used?

The data supplied adds detail to the chapter-opening vignette.

The three problems that follow the data were contributed by *William R. Truran, Stevens Institute of Technology, and Peter A. Cerenzio, Cerenzio & Panaro Consulting Engineers*

Daily and intermediate cover material is 20% of the landfill's usable space. Density of solid waste is 1500 pounds per cubic yard. LFG recovery rates:

> 3000 cubic feet per ton for municipal solid waste (MSW)
>
> 1500 cubic feet per ton for construction and demolition waste (C&D)

Assume waste composition is 80% MSW and 20% C&D.

Methane content in the LFG is 50% for MSW and 20% for C&D.

Heating value

> For methane: 1,030,000 BTU/1000 cubic feet
> For fuel oil: 138,800 BTU/gallon

Assumed furnace efficiency is 88% for methane and 82% for fuel oil.

Cost of fuel oil is $2.50/gallon.

Heating load for residential dwelling is

> 100 million BTU per year.
> $1.7 \times 10^4$ BTU per kilowatt-hours
> Value = $0.05/kWh

16-47 An economic analysis is needed for a new municipal solid waste landfill. This includes determining the potential economic benefit from the gas that is generated when solid waste decomposes. The landfill is proposed at 14 acres, with a design capacity of 1 million cubic yards of capacity. The final capping system will require a 3-foot layer over the 14 acres, and the waste flow rate is 120,000 tons per year. Calculate:

(*a*) The life of the landfill.

(*b*) The average annual methane production (assume all methane production has ceased by 15 years after landfill closure).

(*c*) The dollar value of annual methane converted to electricity (neglect collection and energy production costs).

16-48 A developer has proposed a 650-unit residential development adjacent to the landfill. The proposal includes using the gas generated from the landfill to heat the homes (and mitigate odors). To determine the economic feasibility of the proposal, the value of the gas for heating purposes must be determined. Determine whether the quantity is sufficient for heating the development. Is this economically more attractive than using fuel oil? Is it operationally feasible?

16-49 A 4.6-acre landfill must be evaluated for economic viability. As part of the cost–benefit analysis, the cost to extract and treat the landfill gas must be determined. Using the following, design a landfill gas extraction system and estimate the cost to implement such a system:

> Landfill dimensions are 1000 feet by 200 feet.
> Area of influence of a landfill gas extraction well is a 50-foot radius.
> Cost of well construction is $3000 each.
> Cost of wellhead (necessary for each well) is $2500 each.
> Cost of collection header piping 8″ diameter HDPE is $35/linear foot.
> Cost of condensate knockouts (located at low points) is $5000 each.
> Cost of blower/flare station is $500,000.

## CASES

**The following cases from *Cases in Engineering Economy* 2ⁿᵈ are suggested as matched with this chapter.**

CASE 16 **Great White Hall**
Proposal comparison using B/C analysis for RFP with unclear specifications.

CASE 41 **Metropolitan Highway**
Realistic variety of benefits and costs. Requires that assumptions be made.

CASE 40 **Olympic Bid Perspectives**
Public sector with data and questions from three perspectives.

CASE 42 **Protecting the Public**
$P$(injury) and fraction of contact wearers must be "guesstimated." Considers opportunity losses to other venues in visitor-days.

CASE 43 **Bigstate Bridging the Gap**
Difficulty ranges from comparing three alternative bridges with different lives to considering growth in traffic and uncertain construction costs.

CASE 44 **Sunnyside—Up or Not?**
Uncertain growth rates over 30 years and setting utility rates.

CASE 45 **Transmission Intertie**
Electric power project with primary and secondary benefits.

CASE 47 **Bigstate Highway Department**
Capital budgeting including mutually exclusive alternatives. Includes uncertain value of a life.

CASE 48 **Dot Puff Project Selection**
Public-sector capital budgeting with added constraint on man-years required.

# ACCOUNTING AND ENGINEERING ECONOMY

## The Two Faces of ABB

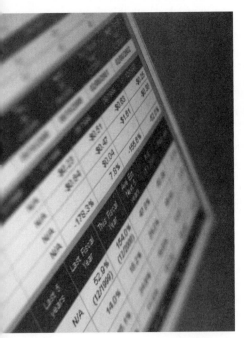

I n the late 1990s, ABB Ltd was flying high. The European conglomerate was an engineering giant with a global network of operations and forward-looking management. Unlike many of its competitors, ABB was also committed to modernizing its accounting system. ABB had followed the traditional practice of assigning overhead costs to its divisions on a roughly equal basis. But this practice tended to obscure the fact that some activities incurred far more costs than others.

So the company spent substantial time and resources switching over to an activity-based costing (ABC) system, which assigns costs to the activities that actually produce them. The results were quite positive, allowing ABB to zero in on areas where it could cut costs most effectively.

But in other respects, ABB's accounting practices were less effective. One of its most serious problems arose at Combustion Engineering, an American subsidiary that was exposed to numerous asbestos liability claims. For years, ABB had downplayed this potential liability, despite warnings from outside analysts. Finally, in late 2002, ABB admitted that its asbestos liability exceeded the subsidiary's total asset value.

But there was more bad news still. The company also issued a very poor third-quarter earnings report, despite having earlier assured investors that ABB was on target to improve its earnings and decrease its debt. When incredulous investors asked how this could have happened, ABB management blamed "poor internal reporting."

By this point, ABB's stock price had nose-dived, and credit rating agencies viewed the company's bonds as little better than junk. In 2003, Combustion Engineering went through bankruptcy and reached a settlement for the asbestos claims.

ABB recovered by 2007 when its stock price had climbed sevenfold over its low, but it is "down again in the global recession." ▪ ▪ ▪

1. ABB's adoption of activity-based costing received widespread publicity and boosted the company's reputation for innovative management. How might this have affected investors' assumptions about the company's other accounting practices?

2. Outside analysts estimated that ABB lost over $690 million in 2001. Yet many investors were still stunned by its poor showing in late 2002. What does this say about the relationship between financial accounting and investor confidence?

3. ABB was not alone in its financial misery, of course. How did ABB's accounting problems compare with those of well-known American companies such as Enron and WorldCom?

4. In each of these well-publicized situations, were there ethical lapses? What were the lapses, and who was responsible?

## After Completing This Chapter...

### The student should be able to:

- Describe the links between engineering economy and accounting.
- Describe the objectives of general accounting, explain what financial transactions are, and show how they are important.
- Use a firm's balance sheet and associated financial ratios to evaluate the firm's health.
- Use a firm's income statement and associated financial ratios to evaluate the firm's performance.
- Use traditional absorption costing to calculate product costs.
- Understand the greater accuracy in product costs available with activity-based costing (ABC).

Engineering economy focuses on the financial aspects of projects, while accounting focuses on the financial aspects of firms. Thus the application of engineering economy is much easier if one has some understanding of accounting principles. In fact, one important accounting topic, depreciation, was the subject of an earlier chapter.

## THE ROLE OF ACCOUNTING

Accounting data are used to value capital equipment, to decide whether to make or buy a part, to determine costs and set prices, to set indirect cost rates, and to make product mix decisions. Accounting is used in private-sector firms and public-sector agencies, but for simplicity this chapter uses "the firm" to designate both. Accountants track the costs of projects and products, which are the basis for estimating future costs and revenues.

The engineering economy, accounting, and managerial functions are interdependent. As shown in Figure 17-1, data and communications flow between them. Whether carried out by a single person in a small firm or by distinct divisions in a large firm, all are needed.

- Engineering economy analyzes the economic impact of design alternatives and projects over their life cycles.
- Accounting determines the dollar impact of past decisions, reports on the economic viability of a unit or firm, and evaluates potential funding sources.
- Management allocates available investment funds to projects, evaluates unit and firm performance, allocates resources, and selects and directs personnel.

### Accounting for Business Transactions

A business transaction involves two parties and the exchange of dollars (or the promise of dollars) for a product or service. Each day, millions of transactions occur between firms and their customers, suppliers, vendors, and employees. Transactions are the lifeblood of the business world and are most often stated in monetary terms. The accounting function records, analyzes, and reports these exchanges.

Transactions can be as simple as payment for a water bill, or as complex as the international transfer of millions of dollars of buildings, land, equipment, inventory, and

| Accounting | Management | Engineering Economy |
|---|---|---|
| About past | About past and future | About future |
| Analyzing | Capital budgeting | Feasibility of alternatives |
| Summarizing | Decision making | Collect & analyze data |
| Reporting | Setting goals | Estimate |
| Financial indicators | Assessing impacts | Evaluate projects |
| Economic trends | Analyzing risk | Recommend |
| Cost acquisitions | Planning | Audit |
| | Controlling | Identify needs |
| | Record keeping | Trade-offs & constraints |

Data and Communication ⟷    Data and Communication ⟷

Budgeting ⟷    Data and Communication ⟷    Estimating ⟷

**FIGURE 17–1**  The accounting, managerial, and engineering economy functions.

other assets. Also, with transactions, one business event may lead to another—all of which need to be accounted for. Consider, for example, the process of selling a robot or a bulldozer. This simple act involves several related transactions: (1) releasing equipment from inventory, (2) shipping equipment to the purchaser, (3) invoicing the purchaser, and finally (4) collecting from the purchaser.

Transaction accounting involves more than just reporting: it includes finding, synthesizing, summarizing, and analyzing data. For the engineering economist, historical data housed in the accounting function are the foundation for estimates of future costs and revenues.

Most accounting is done in nominal or *stable* dollars. Higher market values and costs due to inflation are less objective than cost data, and with a going concern, accountants have decided that objectivity should be maintained. Similarly, most assets are valued at their acquisition cost adjusted for depreciation and improvements. To be conservative, when market value is lower than this adjusted cost, the lower value is used. This restrains the interests of management in maximizing a firm's apparent value. If a firm must be liquidated, then current market value must be estimated.

The accounting function provides data for *general accounting* and *cost accounting*. This chapter's presentation begins with the balance sheet and income statement, which are the two key summaries of financial transactions for general accounting. This discussion includes some of the basic financial ratios used for short- and long-term evaluations. The chapter concludes with a key topic in cost accounting—allocating indirect expenses.

## THE BALANCE SHEET

The primary accounting statements are the **balance sheet** and the **income statement.** The **balance sheet** describes the firm's financial condition at a specific time, while the income statement describes the firm's performance over a period of time—usually a year.

The balance sheet lists the firm's assets, liabilities, and equity on a specified date. This is a picture of the organization's financial health or a snapshot in time. Usually, balance sheets are taken at the end of the quarter and fiscal year. The balance sheet is based on the **fundamental accounting equation:**

$$\text{Assets} = \text{Liabilities} + \text{Equity} \qquad (17\text{-}1)$$

Figure 17-2 illustrates the basic format of the balance sheet. Notice in the balance sheet, as in Equation 17-1, that **assets** are listed on the left-hand side and **liabilities** and **equity** are on the right-hand side. The fact that the firm's resources are *balanced* by the sources of funds is the basis for the name of the balance sheet.

### Assets

In Equation 17-1 and Figure 17-2, **assets** are owned by the firm and have monetary value. **Liabilities** are the dollar claims against the firm. **Equity** represents funding from the firm and its owners (the shareholders). In Equation 17-1, assets are always balanced by the sum of the liabilities and the equity. The value for retained earnings is set so that equity equals assets minus liabilities.

On a balance sheet, assets are listed in order of decreasing liquidity, that is, according to how quickly each one can be converted to cash. Thus, *current assets* are listed first, and within that category in order of decreasing liquidity are listed cash, receivables, securities,

## Balance Sheet for Engineered Industries, December 31, 2010 (all amounts in $1000s)

| Assets | | Liabilities | |
|---|---|---|---|
| Current assets | | Current liabilities | |
| Cash | $ 1940 | Accounts payable | $ 1150 |
| Accounts receivable | 950 | Notes payable | 80 |
| Securities | 4100 | | |
| Inventories | 1860 | Accrued expense | 950 |
| (*minus*) Bad debt provision | −80 | Total current liabilities | 2180 |
| Total current assets | 8770 | | |
| | | Long-term liabilities | 1200 |
| | | **Total liabilities** | **3380** |
| Fixed assets | | | |
| Land | 335 | | |
| Plant and equipment | 6500 | | |
| (*minus*) Accumulated depr. | −2350 | | |
| Total fixed assets | 4485 | **Equity** | |
| | | Preferred stock | 110 |
| Other assets | | Common stock | 650 |
| Prepays/deferred charges | 140 | Capital surplus | 930 |
| Intangibles | 420 | Retained earnings | 8,745 |
| Total other assets | 560 | Total equity | 10,435 |
| **Total assets** | **13,815** | **Total liabilities and equity** | **13,815** |

**FIGURE 17–2** Sample balance sheet.

and inventories. *Fixed assets,* or *property, plant, and equipment,* are used to produce and deliver goods and/or services, and they are not intended for sale. Items such as prepayments and intangibles such as patents are listed last.

The term "receivables" comes from the manner of handling billing and payment for most business sales. Rather than requesting immediate payment for every transaction by check or credit card, most businesses record each transaction and then once a month bill for all transactions. The total that has been billed less payments already received is called accounts receivable, or receivables.

## Liabilities

On the balance sheet, liabilities are divided into two major classifications—short term and long term. The *short-term* or *current liabilities* are expenses, notes, and other payable accounts that are due within one year from the balance sheet date. *Long-term liabilities* include mortgages, bonds, and loans with later due dates. For Engineered Industries in Figure 17-2, total current and long-term liabilities are $2,180,000 and $1,200,000, respectively. Often in performing engineering economic analyses, the **working capital** for a project must be estimated. The total amount of working capital available may be calculated with Equation 17-2 as the difference between current assets and current liabilities.

$$\text{Working capital} = \text{Current assets} - \text{Current liabilities} \qquad (17\text{-}2)$$

For Engineered Industries, there would be $8,770,000 − $2,180,000 = $6,590,000 available in working capital.

## Equity

Equity is also called *owner's equity* or *net worth*. It includes the par value of the owners' stockholdings and the capital surplus, which are the excess dollars brought in over par value when the stock was issued. The capital surplus can also be called *additional paid-in capital*, or APIC. Retained earnings are dollars a firm chooses to retain rather than paying out as dividends to stockholders.

Retained earnings within the equity component is the dollar quantity that always brings the balance sheet, and thus the fundamental accounting equation, into balance. For Engineered Industries, *total equity* value is listed at $10,435,000. From Equation 17-1 and the assets, liabilities, and equity values in Figure 17-2, we can write the balance as follows:

$$\text{Assets} = \text{Liabilities} + \text{Equity}$$

$$\text{Assets (current, fixed, other)} = \text{Liabilities (current and long-term)} + \text{Equity}$$

$$8,770,000 + 4,485,000 + 560,000 = 2,180,000 + 1,200,000 + 10,435,000$$

$$\$13,815,000 = \$13,815,000$$

An example of owner's equity is ownership of a home. Most homes are purchased by means of a mortgage loan that is paid off at a certain interest rate over 15 to 30 years. At any point in time, the difference between what is owed to the bank (the remaining balance on the mortgage) and what the house is worth (its appraised market value) is the *owner's equity*. In this case, the loan balance is the *liability*, and the home's value is the *asset*—with *equity* being the difference. Over time, as the house loan is paid off, the owner's equity increases.

The balance sheet is a very useful tool that shows one view of the firm's financial condition at a particular point in time.

## Financial Ratios Derived from Balance Sheet Data

One common way to evaluate the firm's health is through ratios of quantities on the balance sheet. Firms in a particular industry will typically have similar values, and exceptions will often indicate firms with better or worse performance. Two common ratios used to analyze the firm's current position are the current ratio and the acid-test ratio.

A firm's **current ratio** is the ratio of current assets to current liabilities, as in Equation 17-3.

$$\text{Current ratio} = \text{Current assets}/\text{Current liabilities} \qquad (17\text{-}3)$$

This ratio provides insight into the firm's solvency over the short term by indicating its ability to cover current liabilities. Historically, firms aim to be at or above a ratio of 2.0; however, this depends heavily on the industry as well as the individual firm's management practices and philosophies. The current ratio for Engineered Industries in Figure 17-2 is above 2 (8,770,000/2,180,000 = 4.02).

Both working capital and the current ratio indicate the firm's ability to meet currently maturing obligations. However, neither describes the type of assets owned. The **acid-test ratio** or **quick ratio** becomes important when one wishes to consider the firm's ability to pay debt "instantly." The acid-test ratio is computed by dividing a firm's **quick assets** (cash, receivables, and market securities) by total current liabilities, as in Equation 17-4.

$$\text{Acid-test ratio} = \text{Quick assets/Total current liabilities} \qquad (17\text{-}4)$$

Current inventories are excluded from quick assets because of the time required to sell these inventories, collect the receivables, and subsequently have the cash on hand to reduce debt. For Engineered Industries in Figure 17-2, the calculated acid-test ratio is well above 1 $[(1,940,000 + 950,000 + 4,100,000)/2,180,000 = 3.21]$.

Working capital, current ratio, and acid-test ratio are all indications of the firm's financial health (status). A thorough financial evaluation would consider all three, including comparisons with values from previous periods and with broad-based industry standards. When trends extend over multiple periods, the trends may be more important than the current values.

## THE INCOME STATEMENT

The **income statement** or **profit and loss statement** summarizes the firm's revenues and expenses over a month, quarter, or year. Rather than being a snapshot like the balance sheet, the income statement encompasses a *period* of business activity. The income statement is used to evaluate revenue and expenses that occur in the interval *between* consecutive balance sheet statements. The income statement reports the firm's *net income (profit)* or *loss* by subtracting expenses from revenues. If revenues minus expenses is positive in Equation 17-5, there has been a profit, if negative a loss has occurred.

$$\text{Revenues} - \text{Expenses} = \text{Net profit (Loss)} \qquad (17\text{-}5)$$

To aid in analyzing performance, the income statement in Figure 17-3 separates operating and nonoperating activities and shows revenues and expenses for each. Operating revenues are made up of sales revenues (minus returns and allowances), while nonoperating revenues come from rents and interest receipts.

Operating expenses produce the products and services that generate the firm's revenue stream of cash flows. Typical operating expenses include cost of goods sold, selling and promotion costs, depreciation, general and administrative costs, and lease payments. *Cost of goods sold (COGS)* includes the labor, materials, and indirect costs of production.

Engineers design production systems, specify materials, and analyze make/buy decisions. All these items affect a firm's cost of goods sold. Good engineering design focuses not only on technical functionality but also on cost-effectiveness as the design *integrates* the entire production system. Also of interest to the engineering economist is *depreciation* (see Chapter 11)—which is the systematic "writing off" of a capital expense over a period of years. This noncash expense is important because it represents a decrease in value of the firm's capital assets.

## Income Statement for Engineered Industries for End of Year 2011 (all amounts in $1000)

| | |
|---|---:|
| **Operating revenues and expenses** | |
| Operating revenues | |
| Sales | $ 18,900 |
| (*minus*) Returns and allowances | −870 |
| **Total operating revenues** | **18,030** |
| | |
| Operating expenses | |
| Cost of goods and services sold | |
| Labor | 6140 |
| Materials | 4640 |
| Indirect cost | 2,280 |
| Selling and promotion | 930 |
| Depreciation | 450 |
| General and administrative | 2,160 |
| Lease payments | 510 |
| **Total operating expense** | **17,110** |
| | |
| **Total operating income** | **920** |
| | |
| **Nonoperating revenues and expenses** | |
| Rents | 20 |
| Interest receipts | 300 |
| Interest payments | −120 |
| **Total nonoperating income** | **200** |
| | |
| **Net income before taxes** | **1120** |
| Income taxes | −390 |
| **Net profit (loss) for 2011** | **730** |

**FIGURE 17-3** Sample income statement.

The operating revenues and expenses are shown first, so that the firm's operating income from its products and services can be calculated. Also shown on the income statement are nonoperating expenses such as interest payments on debt in the form of loans or bonds.

From the data in Figure 17-3, Engineered Industries has total expenses (operating = $17,110,000 and nonoperating = $120,000) of $17,230,000. Total revenues are $18,350,000 (= $18,030,000 + $20,000 + $300,000). The net after-tax profit for year 2011 shown in Figure 17-3 as $730,000, but it can also be calculated using Equation 17-5 as

$$\text{Net profits (Loss)} = \text{Revenues} - \text{Expenses [before taxes]}$$

$$\$1,120,000 = 18,350,000 - 17,230,000 \text{ [before taxes] and with}$$

$$\$390,000 \text{ taxes paid}$$

thus

$$\$730,000 = 1,120,000 - 390,000 \text{ [after taxes]}$$

## Financial Ratios Derived from Income Statement Data

The **net profit ratio** (Equation 17-6) equals net profits divided by net sales revenue. Net sales revenue equals sales minus returns and allowances.

$$\text{Net profit ratio} = \text{Net profit}/\text{Net sales revenue} \qquad (17\text{-}6)$$

This ratio provides insight into the cost efficiency of operations as well as a firm's ability to convert sales into profits. For Engineered Industries in Figure 17-3, the net profit ratio is $730,000/18,030,000 = 0.040 = 4.0\%$. Like other financial measures, the net profit ratio is best evaluated by comparisons with other time periods and industry benchmarks, and trends may be more significant than individual values.

**Interest coverage,** as given in Equation 17-7, is calculated as the ratio of total income to interest payments—where *total income* is total revenues minus all expenses except interest payments.

$$\text{Interest coverage} = \text{Total income}/\text{Interest payments} \qquad (17\text{-}7)$$

The interest coverage ratio (which for industrial firms should be at least 3.0) indicates how much revenue must drop to affect the firm's ability to finance its debt. With an interest coverage ratio of 3.0, a firm's revenue would have to decrease by two-thirds (unlikely) before it became impossible to pay the interest on the debt. The larger the interest coverage ratio the better. Engineered Industries in Figure 17-3 has an interest coverage ratio of

$$(18,350,000 - 17,110,000)/120,000 = 10.3$$

## Linking the Balance Sheet, Income Statement, and Capital Transactions

The balance sheet and the income statement are separate but linked documents. Understanding how the two are linked together helps clarify each. Accounting describes such links as the *articulation* between these reports.

The balance sheet shows a firm's assets, liabilities, and equity at a particular point in time, whereas the income statement summarizes revenues and expenses over a time interval. These tabulations can be visualized as a snapshot at the period's beginning (a balance sheet), a video summary over the period (the income statement), and a snapshot at the period's end (another balance sheet). The income statement and changes in the balance sheets summarize the business transactions that have occurred during that period.

There are many links between these statements and the cash flows that make up business transactions, but for engineering economic analysis the following are the most important.

1. Overall profit or loss (income statement) and the starting and ending equity (balance sheets).
2. Acquisition of capital assets.
3. Depreciation of capital assets.

The overall profit or loss during the year (shown on the income statement) is reflected in the change in retained earnings between the balance sheets at the beginning and end of the year. To find the change in retained earnings (RE), one must also subtract any dividends distributed to the owners and add the value of any new capital stock sold:

$$RE_{beg} + \text{Net income}/\text{Loss} + \text{New stock} - \text{Dividends} = RE_{end}$$

When capital equipment is purchased, the balance sheet changes, but the income statement does not. If cash is paid, then the cash asset account decrease equals the increase in the capital equipment account—there is no change in total assets. If a loan is used, then the capital equipment account increases, and so does the liability item for loans. In both cases the equity accounts and the income statement are unchanged.

The depreciation of capital equipment is shown as a line on the income statement. The depreciation for that year equals the change in accumulated depreciation between the beginning and the end of the year—after subtraction of the accumulated depreciation for any asset that is sold or disposed of during that year.

Example 17-1 applies these relationships to the data in Figures 17-1 and 17-2.

## EXAMPLE 17–1

For simplicity, assume that Engineered Industries will not pay dividends in 2011 and did not sell any capital equipment. It did purchase $400,000 in capital equipment. What can be said about the values on the balance sheet at the end of 2011, using the linkages just described?

### SOLUTION

First, the net profit of $730,000 will be added to the retained earnings from the end of 2010 to find the new retained earnings at the end of 2011:

$$RE_{12/31/2011} = \$730,000 + \$8,745,000 = \$9,475,000$$

Second, the fixed assets shown at the end of 2011 would increase from $6,500,000 to $6,900,000.

Third, the accumulated depreciation would increase by the $450,000 in depreciation shown in the 2011 income statement from the $2,350,000 posted in 2010. The new accumulated depreciation on the 2011 balance sheet would be $2,800,000. Combined with the change in the amount of capital equipment, the new fixed asset total for 2011 would equal:

$$\$335,000 + \$6,900,000 - \$2,800,000 = \$4,435,000$$

## TRADITIONAL COST ACCOUNTING

A firm's *cost-accounting system* collects, analyzes, and reports operational performance data (costs, utilization rates, etc.). Cost-accounting data are used to develop product costs, to determine the mix of labor, materials, and other costs in a production setting, and to evaluate outsourcing and subcontracting possibilities.

### Direct and Indirect Costs

Costs incurred to produce a product or service are traditionally classified as either *direct* or *indirect (overhead)*. Direct costs come from activities directly associated with the final product or service produced. Examples include material costs and labor costs for engineering design, component assembly, painting, and drilling.

Some organizational activities are difficult to link to specific projects, products, or services. For example, the receiving and shipping areas of a manufacturing plant are used by all incoming materials and all outgoing products. Materials and products differ in their weight, size, fragility, value, number of units, packaging, and so on, and the receiving and shipping costs depend on all these factors. Also, different materials arrive together and different products are shipped together, so these costs are intermingled and often cannot be tied directly to each product or material.

Other costs, such as the organization's management, sales, and administrative expenses, are difficult to link directly to individual products or services. These indirect or overhead expenses also include machine depreciation, engineering and technical support, and customer warranties.

### Indirect Cost Allocation

To allocate indirect costs to different departments, products, and services, accountants use quantities such as direct-labor hours, direct-labor costs, material costs, and total direct cost. One of these is chosen to be the burden vehicle. The total of all indirect or overhead costs is divided by the total for the burden vehicle. For example, if direct-labor hours is the burden vehicle, then overhead will be allocated based on overhead dollar per direct-labor hour. Then each product, project, or department will *absorb* (or be allocated) overhead costs, based on the number of direct-labor hours each has.

This is the basis for calling traditional costing systems **absorption costing.** For decision making, the problem is that the absorbed costs represent average, not incremental, performance.

Four common ways of allocating overhead are direct-labor hours, direct-labor cost, direct-materials cost, and total direct cost. The first two differ significantly only if the cost per hour of labor differs for different products. Example 17-2 uses direct-labor and direct-materials cost to illustrate the different choices of burden vehicle.

## EXAMPLE 17–2

Industrial Robots does not manufacture its own motors or computer chips. Its premium product differs from its standard product in having heavier-duty motors and more computer chips for greater flexibility.

As a result, Industrial Robots manufactures a higher fraction of the standard product's value itself, and it purchases a higher fraction of the premium product's value. Use the following data to allocate $850,000 in overhead on the basis of labor cost and materials cost.

| | Standard | Premium |
|---|---|---|
| Number of units per year | 750 | 400 |
| Labor cost (each) | $400 | $500 |
| Materials cost (each) | 50 | 900 |

### SOLUTION

First, the labor and material costs for the standard product, the premium product, and in total are calculated.

|  | Standard | Premium | Total |
|---|---|---|---|
| Number of units per year | 750 | 400 | |
| Labor cost (each) | $ 400 | $ 500 | |
| Materials cost (each) | 550 | 900 | |
| Labor cost | 300,000 | 200,000 | $500,000 |
| Materials cost | 412,500 | 360,000 | 772,500 |

Then the allocated cost per labor dollar, $1.70, is found by dividing the $850,000 in overhead by the $500,000 in total labor cost. The allocated cost per material dollar, $1.100324, is found by dividing the $850,000 in overhead by the $772,500 in materials cost. Now, the $850,000 in allocated overhead is split between the two products using labor costs and material costs.

|  | Standard | Premium | Total |
|---|---|---|---|
| Labor cost | $300,000 | $200,000 | $500,000 |
| Overhead/labor | 1.70 | 1.70 | |
| Allocation by labor | 510,000 | 340,000 | 850,000 |
| | | | |
| Material cost | 412,500 | 360,000 | 0 |
| Overhead/material | 1.1003 | 1.1003 | |
| Allocation by material | 453,884 | 396,117 | 850,000 |

If labor cost is the burden vehicle, then 60% of the $850,000 in overhead is allocated to the standard product. If material cost is the burden vehicle, then 53.4% is allocated to the standard product. In both cases, the $850,000 has been split between the two products. Using total direct costs would produce another overhead allocation between these two values. However, for decision making about product mix and product prices, incremental overhead costs must be analyzed. All the allocation or burden vehicles are based on an average cost of overhead per unit of burden vehicle.

## Problems with Traditional Cost Accounting

Allocation of indirect costs can distort product costs and the decisions based on those costs. To be accurate, the analyst must determine which indirect or overhead expenses will be changed because of an engineering project. In other words, what are the incremental cash flows? For example, vacation and sick leave accrual may be part of overhead, but will they change if the labor content is changed? The changes in costs incurred must be estimated. Loadings, or allocations, of overhead expenses cannot be used.

This issue has become very important because in some firms, automation has reduced direct-labor content to less than 5% of the product's cost. Yet in some of these firms, the basis for allocating overhead is still direct-labor hours or cost.

Other firms are shifting to activity-based costing (ABC), where each activity is linked to specific cost drivers, and the number of dollars allocated as overhead is minimized (see Liggett, Trevino, and Lavelle, 1992). Figure 17-4 illustrates the difference between activity-based costing and traditional overhead allocations (see Tippet and Hoekstra, 1993).

**FIGURE 17–4** Activity-based costing versus traditional overhead allocation. (Based on an example by Kim LaScola Needy.)

## Other Problems to Watch For

Project managers have often accused centralized accounting systems of being too slow or being "untimely." Because engineering economy is not concerned with the problem of daily project control, this is a less critical issue. However, if an organization establishes multiple files and systems so that project managers (and others) have the timely data they need, then the level of accuracy in one or all systems may be low. As a result, analysts making cost estimates will have to consider other internal data sources.

There are several cases in which data on equipment or inventory values may be questionable. When inventory is valued on a "last in, first out" basis, the remaining inventory may be valued too low. Similarly, land valued at its acquisition cost is likely to be significantly undervalued. Finally, capital equipment may be valued at either a low or a high value, depending on allowable depreciation techniques and company policy.

## PROBLEMS

### Accounting

**17-1**  Why is it important for engineers and managers to understand accounting principles? Name a few ways that they can do so.

**17-2**  Explain the accounting function within a firm. What does this function do, and why is it important? What types of data does it provide?

**17-3**  Manipulation of financial data by the Enron Corporation was revealed in October 2001. Firm executives were sentenced to prison. Arthur Andersen, which had been one of the "Big 5" accounting firms and Enron's auditor, surrendered its licenses to practice as certified public accountants in August 2002. This and other scandals led to the passage of the federal Sarbanes–Oxley (SarbOx) legislation in July 2002. What are the key components of this law?

**17-4**  How would the information and activities in Figure 17-1 relate to potential contributions by external professional services that firms will often hire

(a)  consulting engineers?

(b)  management consulting firms?

(c)  accounting firms for auditing?

**17-5**  Consider Figure 17-1 and external professional services that firms will often hire. What ethical questions seem likely to arise when a firms hires

(a)  consulting engineers for design?

(b)  management consulting firms?

(c)  accounting firms for auditing?

**17-6**  Insolvency or cash flow problems in the U.S. banking industry started the financial crisis of 2007. Have significant changes in accounting standards and practices been forced through legislation? If so, what are these changes?

**17-7**  Accounting and finance are required topics for most management degrees, and many engineers do become managers during their careers. After graduation with a B.S. in engineering, what courses are available to you from your or another nearby university?

## Balance Sheet

**17-8** Develop short definitions for the following terms: balance sheet, income statement, and fundamental accounting equation.

**17-9** Explain the difference between short-term and long-term liabilities.

**17-10** Calculate the equity of the Gravel Construction Company if it has $1 million worth of assets. Gravel has $127,000 in current liabilities and $210,000 in long-term liabilities.

**17-11** Matbach Industries has $870,000 in current assets and $430,000 in fixed assets less $180,000 in accumulated depreciation. The firm's current liabilities total $330,000, and the long-term liabilities $115,000.

(*a*) What is the firm's equity?

(*b*) If the firm's stock and capital surplus total $305,000, what is the value for retained earnings?

**17-12** CalcTech has $930,000 in current assets and $320,000 in fixed assets less $108,000 in accumulated depreciation. The firm's current liabilities total $350,000, and the long-term liabilities $185,000.

(*a*) What is the firm's equity?

(*b*) If the firm's stock and capital surplus total $402,000, what is the value for retained earnings?

**17-13** Mama L's Baby Monitor Company has current assets of $5 million and current liabilities of $2 million. Give the company's working capital and current ratio. (*Answers:* $3 million; 2.50)

**17-14** For Gee-Whiz Devices, calculate the following: working capital, current ratio, and acid-test ratio.

### Gee-Whiz Devices Balance Sheet Data

| | |
|---|---|
| Cash | $100,000 |
| Market securities | 45,000 |
| Net accounts and notes receivable | 150,000 |
| Retailers' inventories | 200,000 |
| Prepaid expenses | 8,000 |
| Accounts and notes payable (short term) | 315,000 |
| Accrued expenses | 90,000 |

(*Answers*: $90,000; 1.22; 0.73)

**17-15** From the following data, taken from the balance sheet of Petey's Widget Factory, determine the working capital, current ratio, and acid-test ratio.

| | |
|---|---|
| Cash | $ 90,000 |
| Net accounts and notes receivable | 175,000 |
| Retailers' inventories | 210,000 |
| Prepaid expenses | 6,000 |
| Accounts and notes payable (short term) | 322,000 |
| Accrued expenses | 87,000 |

**17-16** (*a*) For Evergreen Environmental Engineering (EEE), determine the working capital, current ratio, and acid-test ratio. Evaluate the company's economic situation with respect to its ability to pay off debt.

### EEE Balance Sheet Data ($1000s)

| | |
|---|---|
| Cash | $110,000 |
| Securities | 40,000 |
| Accounts receivable | 160,000 |
| Inventories | 250,000 |
| Prepaid expenses | 3,000 |
| Accounts payable | 351,000 |
| Accrued expenses | 89,000 |

(*b*) The entries to complete EEE's balance sheet include:

### More EEE Balance Sheet Data ($1000s)

| | |
|---|---|
| Long-term liabilities | $ 220,000 |
| Land | 25,000 |
| Plant and equipment | 510,000 |
| Accumulated depreciation | 210,000 |
| Stock | 81,000 |
| Capital surplus | 15,000 |
| Retained earnings | Value not given |

Construct EEE's balance sheet.

(*c*) What are EEE's values for total assets, total liabilities, and retained earnings?

**17-17** Turbo Start has current assets totaling $1.5 million (this includes $500,000 in current inventory) and current liabilities totaling $50,000. Find the current ratio and acid-test ratio. Are the ratios at desirable levels? Explain.

**17-18** (*a*) For J&W Graphics Supply, compute the current ratio. Is this a financially healthy company? Explain.

### J&W Graphics Supply Balance Sheet Data ($1000s)

| Assets | |
|---|---|
| Cash | $1740 |
| Accounts receivable | 2500 |
| Inventories | 900 |
| Bad debt provision | −75 |
| Liabilities | |
| Accounts payable | 1050 |
| Notes payable | 500 |
| Accrued expenses | 125 |

(*b*) The entries of complete J&W's balance sheet include:

### More J&W Balance Sheet Data ($1000s)

| | |
|---|---|
| Long-term liabilities | $950 |
| Land | 475 |
| Plant and equipment | 3100 |
| Accumulated depreciation | 1060 |
| Stock | 680 |
| Capital surplus | 45 |
| Retained earnings | Value not given |

Construct J&W's balance sheet.

(*c*) What are J&W's values for total assets, total liabilities, and total earnings?

**17-19** For Sutton Manufacturing, determine the current ratio and the acid-test ratio. Are these values acceptable? Why or why not?

### Sutton Manufacturing Balance Sheet Data ($1000s)

| Assets | | | Liabilities | |
|---|---|---|---|---|
| Current assets | | | Current liabilities | |
| Cash | $ 870 | | Notes payable | $ 500 |
| Accounts receivable | 450 | | Accounts payable | 600 |
| Inventory | 1200 | | Accruals | 200 |
| Prepaid expenses | 60 | | Taxes payable | 30 |
| | | | Current portion long-term debt | 100 |
| Total current assets | 2670 | | | |
| Net fixed assets | | | Total current liabilities | 1430 |
| Land | 1200 | | | |
| Plant and equipment | 3800 | | Long-term debt | 2000 |
| (less accumulated depreciation) | −1000 | | Officer debt (subordinated) | 200 |
| | | | Total liabilities | 3630 |
| Other assets | | | **Equity** | |
| Notes receivable | 200 | | Common stock | 1670 |
| Intangibles | 120 | | Capital surplus | 400 |
| | | | Retained earnings | 1200 |
| | | | Total equity | 3270 |
| Total assets | 6900 | | Total liabilities and equity | 6900 |

**17-20** If a firm has a current ratio less than 2.0 and an acid-test ratio less than 1.0, will the company eventually go bankrupt and out of business? Explain your answer.

**17-21** What is the advantage of comparing financial statements across periods or against industry benchmarks over looking at statements associated with a single date or period?

### Income Statement

**17-22** List the two primary general accounting statements. What is each used for, and how do the two differ? Which is most important?

**17-23** Scarmack's Paint Company has annual sales of $500,000 per year. If there is a profit of $1000 per day, 6 days per week operation, what is the total yearly business expense? All calculations are on a before-tax basis.  (*Answer:* $188,000)

**17-24** Find the net income of Turbo Start (Problem 17-17) given the following data from the balance sheet and income statement.

### Turbo Start Data ($1000s)

| | |
|---|---|
| Accounts payable | $ 1,000 |
| Selling expense | 5,000 |
| Sales revenue | 50,000 |
| Owner's equity | 4,500 |
| Income taxes | 2,000 |
| Cost of goods sold | 30,000 |
| Accounts receivable | 15,000 |

(*Answer:* $13,000,000)

**17-25** Laila's Surveying Inc. had revenues of $100,000 in 2010. Expenses totaled $60,000. What was her net profit (or loss)?

**17-26** The general ledger of the Fly-Buy-Nite (FBN) Engineering Company contained the following account balances. Construct an income statement. What is the net income before taxes and the net profit (or loss) after taxes? FBN has a tax rate of 27%.

| | Amount<br>($1000s) |
|---|---|
| Administrative expenses | $ 2,750 |
| Subcontracted services | 18,000 |
| Development expenses | 900 |
| Interest expense | 200 |
| Sales revenue | 30,000 |
| Selling expenses | 4,500 |

**17-27** For Magdalen Industries, compute the net income before taxes and net profit (or loss). Taxes for the year were $1 million.

(*a*) Calculate net profit for the year.

(*b*) Construct the income statement.

(*c*) Calculate the interest coverage and net profit ratio. Is the interest coverage acceptable? Explain why or why not.

### Magdalen Industries Income Statement Data ($M)

| | |
|---|---|
| Revenues | |
| Total operating revenue | $81 |
| (including sales of $48 million) | |
| Total nonoperating revenue | 5 |
| Expenses | |
| Total operating expenses | 70 |
| Total nonoperating expenses | 7 |
| (interest payments) | |

**17-28** For Andrew's Electronic Instruments, calculate the interest coverage and net profit ratio. Is Andrew's business healthy?

### Income Statement for Andrew's Electronics for End of Year 2010 ($1000s)

| | |
|---|---|
| **Revenues** | |
| Operating revenues | |
| Sales | $395 |
| (*minus*) Returns | −15 |
| Total operating revenues | 380 |
| Nonoperating revenues | |
| Interest receipts | 50 |
| Stock revenues | 25 |
| Total nonoperating revenues | 75 |
| Total revenues, R | 455 |
| **Expenses** | |
| Operating expenses | |
| Cost of goods and services sold | |
| Labor | 200 |
| Materials | 34 |
| Indirect cost | 68 |
| Selling and promotion | 20 |
| Depreciation | 30 |
| General and administrative | 10 |
| Lease payments | 10 |
| Total operating expenses | 372 |
| Nonoperating expenses | |
| Interest payments | 22 |
| Total nonoperating expenses | 22 |
| Total expenses, E | 394 |
| Net income before taxes, R − E | 61 |
| Incomes taxes | 30 |
| Net profit (Loss) for the year 2010 | 31 |

### Linking Balance Sheet and Income Statement

**17-29** Sutton Manufacturing (balance sheet at the end of last year in Problem 17-19) had the following entries in this year's income statement.

| | |
|---|---|
| Depreciation | $420,000 |
| Profit | 480,000 |

In addition, we also know the firm purchased $800,000 of equipment with cash. The firm paid $200,000 in dividends this year.

What are the entries in the balance sheet at the end of this year for

(*a*) plant and equipment?

(*b*) accumulated depreciation?

(*c*) retained earnings?

**17-30** Magdalen Industries (Problem 17-27) had the following entries in its balance sheet at the end of last year.

| | |
|---|---|
| Plant and equipment | $15 million |
| (less accumulated depreciation) | 8 million |
| Retained earnings | 60 million |

In addition to the income statement data for this year in Problem 17-27, we also know that the firm purchased $3 million of equipment with cash and that depreciation expenses were $2 million of the $70 million in operating expenses listed in Problem 17-27. The firm paid no dividends this year.

What are the entries in the balance sheet at the end of this year for

(*a*) plant and equipment?

(*b*) accumulated depreciation?

(*c*) retained earnings?

### Allocating Costs

**17-31** Categorize each of the following costs as direct or indirect. Assume that a traditional costing system is in place.

Machine run costs
Machine depreciation
Material handling costs
Cost of materials
Overtime expenses
Machine operator wages
Utility costs
Support (administrative) staff salaries
Cost to market the product
Cost of storage
Insurance costs
Cost of product sales force
Engineering drawings
Machine labor
Cost of tooling and fixtures

**17-32** LeGaroutte Industries makes industrial pipe manufacturing equipment. Use direct-labor hours as the burden vehicle, and compute the total cost per unit for each model given in the table. Total manufacturing indirect costs are $15,892,000, and there are 100,000 units manufactured per year for Model *S*, 50,000 for Model *M*, and 82,250 for Model *G*.

| Item | Model *S* | Model *M* | Model *G* |
|---|---|---|---|
| Direct-material costs | $3,800,000 | $1,530,000 | $2,105,000 |
| Direct-labor costs | 600,000 | 380,000 | 420,000 |
| Direct-labor hours | 64,000 | 20,000 | 32,000 |

(*Answers:* $132; $93; $84)

**17-33** RLW-II Enterprises estimated that indirect manufacturing costs for the year would be $60 million and that 12,000 machine-hours would be used.

(*a*) Compute the predetermined indirect cost application rate using machine hours as the burden vehicle.

(*b*) Determine the total cost of production for a product with direct material costs of $1 million, direct-labor costs of $600,000, and 200 machine-hours.

(*Answers:* $5000; $2.6 million)

**17-34** Par Golf Equipment Company produces two types of golf bag: the standard and deluxe models. The total indirect cost to be allocated to the two bags is $35,000. Determine the net revenue that Par Golf can expect from the sale of each bag.

(*a*) Use direct-labor cost to allocate indirect costs.

(*b*) Use direct-materials cost to allocate indirect costs.

| Data Item | Standard | Deluxe |
|---|---|---|
| Direct-labor cost | $50,000 | $65,000 |
| Direct-material cost | 35,000 | 47,500 |
| Selling price | 60 | 95 |
| Units produced | 1800 | 1400 |

# CASES 💿

**The following cases from *Cases in Engineering Economy* $2^{nd}$ are suggested as matched with this chapter.**

CASE 4    **Balder-Dash Inc.**
Standard cost and allocated costs versus true marginal cost (IEs).

CASE 52    **Aunt Allee's Jams and Jellies**
Product costing needs activity based costs.

# INTRODUCTION TO SPREADSHEETS

C omputerized spreadsheets are available nearly everywhere, and they can be easily applied to economic analysis. In fact, spreadsheets were originally developed to analyze financial data, and they are often credited with initiating the explosive growth in demand for desktop computing.

A spreadsheet is a two-dimensional table whose cells can contain numerical values, labels, or formulas. The software automatically updates the table when an entry is changed, and there are powerful tools for copying formulas, creating graphs, and formatting results.

## THE ELEMENTS OF A SPREADSHEET

A spreadsheet is a two-dimensional table that labels the columns in alphabetical order A to Z, AA to AZ, BA to BZ, etc. The rows are numbered from 1 to 65,536 or higher. Thus a *cell* of the spreadsheet is specified by its column letter and row number. For example, A3 is the third row in column A, and AA6 is the sixth row in the twenty-seventh column. Each cell can contain a label, a numerical value, or a formula.

A *label* is any cell where the contents should be treated as text. Arithmetic cannot be performed on labels. Labels are used for variable names, row and column headings, and explanatory notes. In Excel any cell that contains more than a simple number, such as 3.14159, is treated as a label, unless it begins with an =, which is the signal for a formula. Thus 2*3 and B1+B2 are labels. Meaningful labels can be wider than a normal column. One solution is to allow those cells to "wrap" text, which is one of the "alignment" options. The table heading row (row 8) in Example A-1 has turned this on by selecting row 8, right-clicking on the row, and selecting wrap text under the alignment tab.

A *numerical value* is any number. Acceptable formats for entry or display include percentages, currency, accounting, scientific, fractions, date, and time. In addition, the number of decimal digits, the display of $ symbols, and commas for "thousands" separators can be adjusted. The format for cells can be changed by selecting a cell, a block of cells, a row, a column, or the entire spreadsheet. Then right-click on the selected area, and a menu that includes "format cells" will appear. Then number formats, alignment, borders, fonts, and patterns can be selected.

*Formulas* must begin with an =, such as = 3*4^2 or =B1+B2. They can include many functions—financial, statistical, trigonometric, etc. (and others can be defined by the user). The formula for the "current" cell is displayed in the formula bar at the top of the spreadsheet. The value resulting from the formula is displayed in the cell in the spreadsheet.

Often the printed-out spreadsheet will be part of a report or a homework assignment and the formulas must be explained. Here is an easy way to place a copy of the formula in an adjacent or nearby cell. (1) Convert the cell with the formula to a label by inserting a space before the = sign. (2) Copy that label to an adjacent cell by using cut and paste. Do not drag the cell to copy it, as any formula ending with a number (even an address like B4) will have the number automatically incremented. (3) Convert the original formula back into a formula by deleting the space.

## DEFINING VARIABLES IN A DATA BLOCK

The cell A1, top left corner, is the HOME cell for a spreadsheet. Thus, the top left area is where the data block should be placed. This data block should have every variable in the spreadsheet with an adjacent label for each. This data block supports a basic principle of good spreadsheet modeling, which is to use variables in your models.

The data block in Example A-1 contains *entered data*—the loan amount (A1), the number of payments (A2), and the interest rate (A3), and *computed data*—the payment (A4). Then instead of using the loan amount of $5000 in a formula, the cell reference A1 is used. Even if a value is referenced only once, it is better to include it in the data block. By using one location to define each variable, you can change any value at one place in the spreadsheet and have the entire spreadsheet instantly recomputed.

Even for simple homework problems you should use a data block.

1. You may be able to use it for another problem.
2. Solutions to simple problems may grow into solutions for complex problems.
3. Good habits, like using data blocks, are easy to maintain once they are established.
4. It makes the assumptions clear if you've estimated a value or for grading.

In the real world, data blocks are even more important. Most problems are solved more than once, as more and more accurate values are estimated. Often the spreadsheet is revised to add other variables, time periods, locations, etc. Without data blocks, it is hard to change a spreadsheet, and the likelihood of missing a required change skyrockets.

If you want your formulas to be easier to read, you can name your variables. *Note*: In Excel, the cell's location or name is displayed at the left of the formula bar. Variable names can be entered here. They will then automatically be applied if cell addresses are entered by the point-and-click method. If cell addresses are entered as A2, then A2 is what is displayed. To change a displayed A2 to the name of the cell (LoanAmount), the process is to click on Insert, click on Name, click on Apply, and then select the names to be applied.

## COPY COMMAND

The copy command and relative/absolute addressing make spreadsheet models easy to build. If the range of cells to be copied contains only labels, numbers, and functions, then the copy command is easy to use and understand. For example, the formula =EXP(1.9) would be copied unchanged to a new location. However, cell addresses are usually part of the range being copied, and their absolute and relative addresses are treated differently.

An *absolute address* is denoted by adding $ signs before the column and/or row. For example in Figure A-1a, $A$4 is the absolute address for the interest rate. When an absolute address is copied, the column and/or row that is fixed is copied unchanged. Thus $A$4 is completely fixed, $A4 fixes the column, and A$4 fixes the row. One common use for absolute addresses is any data block entry, such as the interest rate. When entering or editing a formula, changing between A4, $A$4, A$4, $A4, and A4 is most easily done using the F4 key, which scrolls an address through the choices.

In contrast, a *relative address* is best interpreted as directions from one cell to another. For example, in Figure A-1a, the balance due in year $t$ equals the balance due in year $t - 1$ minus the principal payment in Year $t$. Specifically for the balance due in Year 1, D10 contains = D9 − C10. From cell D10, cell D9 is one row up and C10 is one column to the left, so the formula is really (contents of 1 up) minus (contents of 1 to the left). When a cell containing a relative address is copied to a new location, it is these directions that are copied to determine any new relative addresses. So if cell D10 is copied to cell F14, the formula is = F13 − E14.

Thus to calculate a loan repayment schedule, as in Figure A-1, the row of formulas is created and then copied for the remaining years.

## EXAMPLE A–1

Four repayment schedules for a loan of $5000 to be repaid over 5 years at an interest rate of 8% were shown in Table 3-1. Use a spreadsheet to calculate the amortization schedule for the constant principal payment option.

### SOLUTION

The first step is to enter the loan amount, number of periods, and interest rate into a data block in the top left part of the spreadsheet. The next step is to calculate the constant principal payment amount, which was given as $1252.28 in Table 3-1. The factor approach to finding this value is given in Chapter 3 and the spreadsheet function is explained in Chapter 4.

The next step is to identify the columns for the amortization schedule. These are the year, interest owed, principal payment, and balance due. Because some of these labels are wider than a normal column, the cells are formatted so that the text wraps (row height increases automatically). The initial balance is shown in the Year-0 row.

Next, the formulas for the first year are written, as shown in Figure A-1a. The interest owed (cell B10) equals the interest rate ($A$4) times the balance due for Year 0 (D9). The principal payment (cell C10) equals the annual payment ($A$6) minus the interest owed and paid (B10). Finally, the balance due (cell D10) equals the balance due for the previous year (D9) minus the principal payment (C10). The results are shown in Figure A-1a.

Now cells A10 to D10 are selected for Year 1. By dragging down on the right corner of D10, the entire row can be copied for Years 2 through 5. Note that if you use cut and paste, then you must complete the year column separately (dragging increments the year, but cutting and pasting does not). The results are shown in Figure A-1b.

|    | A | B | C | D | E |
|----|---|---|---|---|---|
| 1  | Entered Data | | | | |
| 2  | 5000 | Loan Amount | | | |
| 3  | 5 | Number of Payments | | | |
| 4  | 8% | Interest Rate | | | |
| 5  | Computed Data | | | | |
| 6  | $1,252.28 | Loan Payment | | | |
| 7  | | | | | |
| 8  | Year | Interest Owed | Principal Payment | Balance Due | |
| 9  | 0 | | | 5000.00 | |
| 10 | 1 | 400.00 | 852.28 | 4147.72 | =D9−C10 |
| 11 | | | | | |
| 12 | | =$A$4*D9 | | =$A$6−B10 | |

(a)

|    | A | B | C | D | E |
|----|---|---|---|---|---|
| 8  | Year | Interest Owed | Principal Payment | Balance Due | |
| 9  | 0 | | | 5000.00 | |
| 10 | 1 | 400.00 | 852.28 | 4147.72 | |
| 11 | 2 | 331.82 | 920.46 | 3227.25 | |
| 12 | 3 | 258.18 | 994.10 | 2233.15 | |
| 13 | 4 | 178.65 | 1073.63 | 1159.52 | |
| 14 | 5 | 92.76 | 1159.52 | 0.00 | =D13−C14 |
| 15 | | | | | |
| 16 | | =$A$4*D13 | | =$A$6−B14 | |

(b)

**FIGURE A-1** **(a)** Year-1 amortization schedule. **(b)** Completed amortization schedule.

This appendix has introduced the basics of spreadsheets. Chapter 2 uses spreadsheets and simple bar charts to draw cash flow diagrams. Chapters 4 to 15 each have spreadsheet sections. These are designed to develop spreadsheet modeling skills and to reinforce your understanding of engineering economy. As spreadsheet packages are built around using the computer mouse to click on cells and items in charts, there is usually an intuitive connection between what you would like to do and how to do it. The best way to learn how to use the spreadsheet package is to simply play around with it. In addition, as you look at the menu choices, you will find new commands that you hadn't thought of but will find useful.

# TIME VALUE OF MONEY (TVM) CALCULATORS

## Coauthored with Neal A. Lewis, University of Bridgeport

## ADVANTAGES AND TYPES OF TVM CALCULATORS

There are calculators for solving problems with $i$, $n$, $A$, $P$, and $F$. They require the same conceptual understanding as the tables of engineering economy factors, but they are often faster and easier to use. Given the values for any four variables; the calculators will solve for the fifth variable. There are many advantages to using these calculators instead of or in addition to the tables.

- Time is saved by entering fewer numbers into the calculator and by the increased TVM capabilities of the calculator. Thus, using a TVM calculator, such as a properly programmed HP 33s or 35s, on the time-limited FE exam can leave more time for other questions.

- Because the tabulated factor values do not have to be entered into the calculator, there is a smaller chance of transposed digits or wrong row lookup or similar errors destroying the validity of the calculation.

- If all three of $P$, $A$, and $F$ are present, a single set of calculator values is needed rather than two tabulated factors. See Example B-2.

- There is no need to interpolate for interest rates or lives that are not tabulated, such as a life of 120 monthly periods, a mortgage at 5.85%, a car loan at 8.95%, or a credit card at 14.85%. See Examples B-1 and B-3.

- There is no need for iterative solutions to solve for $i$ or $n$. See Examples B-4 and B-5.

- For an *A* that represents beginning-of-period cash flows, such as a lease, a financial calculator can be set for "beginning" rather than "end," rather than adding a factor of $(F/P, i, 1)$. See Example B-6.

There is also one disadvantage of the TVM calculators.

- The tabulated factors include the two factors for the arithmetic gradient, $(P/G, i, n)$ and $(A/G, i, n)$, but there is no equivalent in the calculators.

There are three classes of TVM calculators. First, there are many *financial* or *business* calculators that have buttons labeled *i*, *N*, *PMT*, *PV*, and *FV* or equivalents. The button for the uniform series cash flow, *A*, is labeled *PMT*, because one of the most important uses of these calculators is for calculating loan payments (PMT). Examples include the Texas Instrument BAII Plus and BAII Plus Professional, the Sharp EL-738C, and the Hewlett Packard 10bII. Note that we no longer recommend the HP 12C for new purchases since when solving for *n*; the 12C rounds fractional values of *n* up to an integer value (see Example B-5). These financial calculators also typically have the capability to find the NPV of complex cash flow patterns, but we suggest using spreadsheets for more complex patterns.

There are also *graphing* calculators that have the TVM calculations in a menu. There are menu entries for *i*, *N*, *PMT*, *PV*, and *FV* or equivalents. We will refer to the calculator buttons or the menu entries as the TVM keys. Example calculators include the Texas Instrument 83 and 84 Plus.

Finally, there are programmable scientific calculators where the TVM equation (Equation B-1) can be entered to create the menu that is built into the graphing calculators. For example, Chapter 17 of the manuals for the Hewlett Packard 33s and 35s details this. These chapters use a nonstandard notation of *B* or balance for *PV* and *P* or payment for *PMT*. We suggest using standard notation and substituting *P* for *B* and *A* for *P*. Note that on the HP 35s this changes the checksum from 382E to 8DD6. These calculators are currently allowed for use on the Fundamentals of Engineering (FE) exam, while the financial and graphing calculators are not.

Financial calculators use Equation B-1 to solve for the unknown variable. Note that Equation B-1 is written so that *i* is entered as a percentage, so 8 is entered for 8%, which matches how the interest rate is entered for the financial calculators. Equation B-2 rewrites this in standard factor notation.

$$PMT \left[ \frac{1 - (1 + i/100)^{-n}}{i/100} \right] + FV(1 + i/100)^{-n} + PV = 0 \qquad \text{(B-1)}$$

$$A(P/A, i, n) + F(P/F, i, n) + P = 0 \qquad \text{(B-2)}$$

Since the factors are positive, one of *PMT, FV,* and *PV* (or *A, F,* and *P*) must be different in sign from the other two. If two positive cash flows are entered to solve for the third cash flow, then that third value must be negative to solve the equation. Equation B-1 is the foundation for the sign convention that is used in the TVM calculators *and* the spreadsheet annuity functions, PV, FV, and PMT. This sign change convention in the earliest financial calculators is why spreadsheet functions assume that if positive values are entered for the cash flows, then the solution is negative.

## NOTATION FOR TVM CALCULATORS

There are too many variations of labeling and physical layout of the keys and too many TVM calculators for a text to provide details. However, the website TVMcalcs.com provides good tutorials on the financial calculators, and calculator manuals are available for download from the manufacturers. The notation we will use is based on the most common choices for the calculators, on the standard notation for engineering economy, and on the notation for the very similar spreadsheet functions. The typical equivalents are:

- $i$, $I$, I/PY, RATE
- $n$, $N$, NPER
- $A$, PMT
- $P$, PV, PW
- $F$, FV, FW

For labeling of the variable that is being solved for, we will use the most common calculator key terms $\{i, N, PMT, PV, \text{and } FV\}$. For labeling of the data given in the examples, we will use the standard engineering economy notation of $\{i, n, A, P, \text{and } F\}$. Our reason for using both is to facilitate the automatic recognition of alternate terminology that the field treats as synonyms.

For the interest rate we will state it, for example, as 8%, not .08 or 8. For entry into the TVM calculator the integer value 8 is entered. Then the calculator through the equivalent of Equation B-1 assumes that the interest is the integer value without the % sign. Stating it as 8% will help us avoid the error of confusing it with the integer number of periods. Note that when we get to spreadsheets, we will find that the interest rate must be entered as 8% or .08 and that 8 = 800%.

Some calculators are shipped from the factory with a setting of 12 months per year, so that $n$ will be entered as years for loans with monthly payments. We recommend that this be changed to 1 payment per "year or period" and left on this setting. Then you know to enter $n$ as the number of periods or payments, and there is no confusion for problems with different length periods.

Finally, the order of the items in each function follows the order of the items in typical spreadsheet annuity functions $\{i, N, PMT, PV, \text{and } FV\}$. However, each function removes the variable that is being solved for from this list. So our notation is:

- $PV(i, n, A, F)$     or     $PV\ (i, n, A, \quad F)$
- $PMT(i, n, P, F)$     or     $PMT\ (i, n, \quad P, F)$
- $FV(i, n, A, P)$     or     $FV\ (i, n, A, P \quad)$
- $i(n, A, P, F)$     or     $i\ (\quad n, A, P, F)$
- $N(i, A, P, F)$     or     $N\ (i, \quad A, P, F)$

## EXAMPLES WITH TVM CALCULATORS AND TABULATED FACTORS

### EXAMPLE B–1

Kris is buying a used car and needs to find the monthly payment. The loan will be for $14,000, and the dealer is willing to offer financing for 5 years at 6.5% interest. What is the monthly payment?

**SOLUTION**

First, we need to convert the period and interest rate to months. There are 60 (= 5 × 12) monthly payments at a monthly interest rate of 0.5417% (= 6.5%/12). Using a TVM calculator, $PMT(i, n, P, F) = PMT(0.5417\%, 60, 14000, 0) = -\$273.93$. This is the payment that is due at the end of each month.

## EXAMPLE B–2

Automating a process will cost $8000 now, but it will save $2200 annually for 5 years. The machinery will have a salvage value of $1500. What is the present worth of the machinery, if the interest rate is 10%?

**SOLUTION**

Let us first solve this with the tabulated factors.

$$PW = -8000 + 2200(P/A, 10\%, 5) + 1500(P/F, 10\%, 5)$$
$$= -8000 + 2200(3.791) + 1500(0.6209)$$
$$= -8000 + 8340 + 931 = \$1271$$

Now using a TVM calculator, $PV(i, n, A, F) = PV(10, 5, 2200, 1500) = -9271$. The calculator find the combined present worth of the $2200 annual savings and the salvage value of $1500.

Notice that the $PV$ value is negative when the $A$ and $F$ values are entered as positive cash flows. This comes from solving Equation B-1 internally in the calculator. Since in this case we want the *positive* PW of the returns to the project, we simply change the sign of the displayed answer by pressing the +/− or CHS key. Now simply subtract the $8000 initial cash flow to find the PW of $1271.

## EXAMPLE B–3

Two engineering graduates of Middle State U. are buying a house. Their $143,200 mortgage is at 5.25% for 30 years. What is the monthly payment?

### SOLUTION

Because payments are monthly, the interest rate and the number of periods must be in terms of months.

$$\text{Monthly rate} = 5.25\%/12 = 0.4375\%/\text{month}$$

$$\text{Number of periods} = (12)(30) = 360 \text{ months}$$

Using a TVM calculator, $PMT(i, n, P, F) = PMT(0.4375\%, 360, 143200, 0) = -\$790.76$.

To solve this with the tabulated factors, interpolation between the results at $1/4$ and $1/2\%$ is needed, and the result is approximate, not exact.

$$\text{Payment} = 143,200(A/P, i_{\text{mon}}, 360)$$

$$\text{Payment}_{1/4} = 143,200(0.00422) = 604.30$$

$$\text{Payment}_{1/2} = 143,200(0.00600) = 859.20$$

$$\text{Payment}_{0.4375\%} = 604.30 + \frac{0.4375 - 0.25}{0.5 - 0.25}(859.20 - 604.30) = \$795.48$$

---

### EXAMPLE B–4

A facility renovation will cost \$75,000, but it will save \$15,000 annually in operating costs. The firm expects to sell the facility in 6 years to move into a larger one. The firm will recover \$20,000 of the renovation cost through a higher value when the facility is sold. What is the rate of return on the project?

### SOLUTION

Using a TVM calculator, $i(n, A, P, F) = i(6, 15000, -75000, 20000) = 10.65\%$.

Without knowledge that the rate is between 10 and 11%, using the tabulated factors would require trial solutions. If 5 and 10% are tried first, then probably 12% and 11% will be tried next, and then interpolation is required.

---

### EXAMPLE B–5

A firm may purchase a computerized quality control system. The proposed system will cost \$76,000 but is expected to save \$22,000 each year in reduced overtime. The firm requires that all cost reduction projects have a discounted payback of no more than 4 years with a 10% interest rate. Should the firm invest in the new system?

### SOLUTION

The time for the system to pay for itself can be found using

$$N(i, A, P, F) = N(10\%, 22000, -76000, 0) = 4.45 \text{ years}$$

The new system does not meet the firm's criteria. Note that the HP 12C calculator will report 5 years for increased savings up to $23,952 which has a payback period of 4.005 periods.

## EXAMPLE B-6

Some equipment is needed for a 4-year project. It can be leased for $15,000 annually, or it can be purchased for $60,000 at the beginning and sold for $35,000 at the end. What is the rate of return for owning the equipment rather than leasing it?

### SOLUTION

Because the lease payments occur at the beginning of each period, the calculator is shifted into *begin* rather than *end* mode. This affects only the $A$ and not the $F$. Using a financial calculator, $i(n, A, P, F) = i(4, 15000, -60000, 35000) = 23.49\%$.

Without knowledge that the rate is nearly 25%, using the tabulated factors would require multiple trial solutions to solve the following 3-factor equation.

$$0 = -60,000 + 15,000(F/P, i, 1)(P/A, i, 4) + 35,000(P/F, i, 4)$$

For any calculation written in factor notation (that does not have a $G$), it is possible to use a TVM calculator to solve the problem (see Example B-7).

## EXAMPLE B-7

What is the EUAC to keep a piece of new equipment operating? The warranty covers repair costs for the first 2 years, but expected repair costs are $1500 per year for the rest of the 7-year life. Normal maintenance is expected to cost $1800 annually. There is an overhaul costing $8500 at the end of Year 4. The firm's interest rate is 8%.

### SOLUTION

Whether solved with tables or a TVM calculator, the first step is to draw the cash flow diagram.

Let us first solve this with the tabulated factors.

$$\text{EUAC} = 1800 + 1500(F/A, 8\%, 5)(A/F, 8\%, 7) + 8500(P/F, 8\%, 4)(A/P, 8\%, 7)$$

$$= 1800 + 1500(5.867)(0.1121) + 8500(0.7350)(0.1921)$$

$$= 1800 + 986.5 + 1200 = \$3986.5$$

Now using a TVM calculator, it is easiest to separately calculate the EUAC values for the repair costs and the overhaul.

For the repair costs, $FV(i, n, A, P) = FV(8, 5, 1500, 0) = -8800$. Since the $i$ and $P$ values are unchanged, they do not need to be reentered. The only entry needed is $\{N = 7\}$, since the $F$ was just calculated. $PMT(i, n, P, F) = PMT(8\%, 7, 0, -8800) = \$986.23$.

For the overhaul costs, the $i$ value need not be reentered, $PV(i, n, A, F) = PV(8\%, 4, 0, 8500) = -6247.75$. Now $N$ is changed to 7 and $F$ to 0, then solving for $PMT(i, n, P, F) = PMT(8\%, 7, -6247.75, 0) = \$1200.02$.

Maintenance is $1800 per year.

Adding the three values is the same as before, and the EUAC is $3986.25.

Using TVM calculators requires the same understanding as using the tabulated factors, but far fewer keystrokes and no table lookups are required. For many problems, using the calculator reduces to entering four values and solving for the fifth value. If Equation B-1 is programmed into a calculator allowed for the FE exam, then the tables need be used only for problems involving the arithmetic gradient, $G$.

More complex problems may also be solved using TVM calculators, such as those involving combining factors (as in Example B-7). Many calculators can also handle problems that have multiple unequal cash flows; however, we recommend using a spreadsheet for these problems. Example B-8 is another example of the common conceptual understanding required for TVM calculators and factor solutions.

## EXAMPLE B–8

A $10,000 loan was repaid over 5 years at $2000/year except that $5000 was paid in Year 5. What interest rate was paid on the loan?

### SOLUTION

The payment in Year 5 may be separated into a $2000 portion that is included as an annual payment and a $3000 portion that is included as an $FV$. This allows for $A$ and $F$ to be entered into the calculator; $i(n, A, P, F) = i(5, 2000, -10000, 3000) = 8.09\%$.

To solve this with factors requires iterative solutions of either of these two equations (note that the first one is equivalent to the TVM calculator solution):

$$10,000 = -2000(P/A, i, 5) - 3000(P/F, i, 5)$$

or

$$10,000 = -2000(P/A, i, 4) - 5000(P/F, i, 5)$$

# PROBLEMS

## Present Worth

**B-1**  A process redesign will cost $70,000 now, but it will save $18,000 annually for 6 years. The new machinery will have a salvage value of $12,500. What is the present worth of the machinery, if the interest rate is 12%?     (*Answer*: $10,338)

**B-2**  Moving to a new facility will save $38,000 annually for 20 years. The cost of building the facility and moving is $240,000 now. The facility will have a salvage value of $75,000. What is the facility's PW, if the interest rate is 10%?

**B-3**  A lottery pays the winner $1 million in 20 equal annual payments of $50,000. The first payment will be made at the end of the second year. What is the present worth if the winner's annual interest rate is 5.25%?     (*Answer*: $579,680)

## Equivalent Uniform Annual Worth and Cost

**B-4**  You need to save $15,000 to buy a car in 3 years. At a 4% nominal annual interest rate, how much do you need to save each month?     (*Answer*: $396.20)

**B-5**  You can receive lottery winnings of either $800,000 now or $100,000 per year for the next 10 years. If your interest rate is 5% per year, which do you prefer?

**B-6**  A new road will cost $45 million to build, and $2 million annually to maintain and operate over its 50-year life. The roadbed and right-of-way are estimated to have a salvage value of $15 million. If the state highway department uses an interest rate of 5%, what is the EUAC for the road?

**B-7**  A new car will cost $22,000 to buy and $5500 annually to operate. If it is sold for $9300 after 6 years, what is the EUAC? Assume that the owner's interest rate is 8% for the time value of money.

## Price of a Bond

**B-8**  A $1000 bond has 15 more years to maturity, a coupon interest rate of 8%, and it pays interest semiannually. If the current market interest rate is 10%, what is the price of the bond?     (*Answer*: $846)

**B-9**  A $1000 bond has 12 more years to maturity, a coupon interest rate of 6%, and it pays interest semiannually. If the current market interest rate is 8%, what is the price of the bond?

## Loan Payment

**B-10**  What is the monthly payment for a 5-year car loan at a nominal interest rate of 8.5%? The loan's initial balance is $16,885.

## Number of Periods

**B-11**  A student owing $17,565 on a credit card has decided to use only a different debit card in the future. The nominal annual rate on the credit card is 13.8%. If the student makes monthly payments of $250, how long is it until the credit card is paid off?     (*Answer*: 13.5 years)

**B-12**  A new security system is expected to cost $85,000, but it should reduce employee and customer theft by about $12,000 annually. The salvage value will be only about 10% of the initial cost, whenever the system is replaced. If the firm's interest rate is 12%, how long must the system last to pay for itself?

## Rate of Return

**B-13**  An automated storage and retrieval system will cost $135,000, but it will save $33,000 annually in labor costs. The system's salvage value is expected to be $20,000, when it is renovated in 10 years. What is the rate of return on the project?

**B-14**  Moving to a new facility will save $38,000 annually for 20 years. The cost of building the facility and moving is $240,000 now. The facility will have a salvage value of $75,000. What is the rate of return on this project?

**B-15**  A new road will cost $45 million to build, and $2 million annually to maintain and operate over its 50-year life. The roadbed and right-of-way are estimated to have a salvage value of $15 million. If user benefits are estimated at $5.5 million annually, what is the rate of return on this road?

**B-16**  Some equipment is needed for a construction project. It can be leased for $150,000 annually, or it can be purchased for $900,000 at the beginning and sold for $225,000 at the end of 3 years. What is the rate of return for owning the equipment rather than leasing it?

# COMPOUND INTEREST TABLES

**Values of Interest Factors When *n* Equals Infinity**

Single Payment:

$(F/P, i, \infty) = \infty$

$(P/F, i, \infty) = 0$

Arithmetic Gradient Series:

$(A/G, i, \infty) = 1/i$

$(P/G, i, \infty) = 1/i^2$

Uniform Payment Series:

$(A/F, i, \infty) = 0$

$(A/P, i, \infty) = i$

$(F/A, i, \infty) = \infty$

$(P/A, i, \infty) = 1/i$

| | Single Payment | | Uniform Payment Series | | | | Arithmetic Gradient | | |
|---|---|---|---|---|---|---|---|---|---|
| | Compound Amount Factor Find *F* Given *P* | Present Worth Factor Find *P* Given *F* | Sinking Fund Factor Find *A* Given *F* | Capital Recovery Factor Find *A* Given *P* | Compound Amount Factor Find *F* Given *A* | Present Worth Factor Find *P* Given *A* | Gradient Uniform Series Find *A* Given *G* | Gradient Present Worth Find *P* Given *G* | |
| *n* | *F/P* | *P/F* | *A/F* | *A/P* | *F/A* | *P/A* | *A/G* | *P/G* | *n* |
| 1 | 1.003 | .9975 | 1.0000 | 1.0025 | 1.000 | 0.998 | 0.000 | 0.000 | 1 |
| 2 | 1.005 | .9950 | .4994 | .5019 | 2.003 | 1.993 | 0.499 | 0.995 | 2 |
| 3 | 1.008 | .9925 | .3325 | .3350 | 3.008 | 2.985 | 0.998 | 2.980 | 3 |
| 4 | 1.010 | .9901 | .2491 | .2516 | 4.015 | 3.975 | 1.497 | 5.950 | 4 |
| 5 | 1.013 | .9876 | .1990 | .2015 | 5.025 | 4.963 | 1.995 | 9.901 | 5 |
| 6 | 1.015 | .9851 | .1656 | .1681 | 6.038 | 5.948 | 2.493 | 14.826 | 6 |
| 7 | 1.018 | .9827 | .1418 | .1443 | 7.053 | 6.931 | 2.990 | 20.722 | 7 |
| 8 | 1.020 | .9802 | .1239 | .1264 | 8.070 | 7.911 | 3.487 | 27.584 | 8 |
| 9 | 1.023 | .9778 | .1100 | .1125 | 9.091 | 8.889 | 3.983 | 35.406 | 9 |
| 10 | 1.025 | .9753 | .0989 | .1014 | 10.113 | 9.864 | 4.479 | 44.184 | 10 |
| 11 | 1.028 | .9729 | .0898 | .0923 | 11.139 | 10.837 | 4.975 | 53.913 | 11 |
| 12 | 1.030 | .9705 | .0822 | .0847 | 12.167 | 11.807 | 5.470 | 64.589 | 12 |
| 13 | 1.033 | .9681 | .0758 | .0783 | 13.197 | 12.775 | 5.965 | 76.205 | 13 |
| 14 | 1.036 | .9656 | .0703 | .0728 | 14.230 | 13.741 | 6.459 | 88.759 | 14 |
| 15 | 1.038 | .9632 | .0655 | .0680 | 15.266 | 14.704 | 6.953 | 102.244 | 15 |
| 16 | 1.041 | .9608 | .0613 | .0638 | 16.304 | 15.665 | 7.447 | 116.657 | 16 |
| 17 | 1.043 | .9584 | .0577 | .0602 | 17.344 | 16.624 | 7.944 | 131.992 | 17 |
| 18 | 1.046 | .9561 | .0544 | .0569 | 18.388 | 17.580 | 8.433 | 148.245 | 18 |
| 19 | 1.049 | .9537 | .0515 | .0540 | 19.434 | 18.533 | 8.925 | 165.411 | 19 |
| 20 | 1.051 | .9513 | .0488 | .0513 | 20.482 | 19.485 | 9.417 | 183.485 | 20 |
| 21 | 1.054 | .9489 | .0464 | .0489 | 21.534 | 20.434 | 9.908 | 202.463 | 21 |
| 22 | 1.056 | .9465 | .0443 | .0468 | 22.587 | 21.380 | 10.400 | 222.341 | 22 |
| 23 | 1.059 | .9442 | .0423 | .0448 | 23.644 | 22.324 | 10.890 | 243.113 | 23 |
| 24 | 1.062 | .9418 | .0405 | .0430 | 24.703 | 23.266 | 11.380 | 264.775 | 24 |
| 25 | 1.064 | .9395 | .0388 | .0413 | 25.765 | 24.206 | 11.870 | 287.323 | 25 |
| 26 | 1.067 | .9371 | .0373 | .0398 | 26.829 | 25.143 | 12.360 | 310.752 | 26 |
| 27 | 1.070 | .9348 | .0358 | .0383 | 27.896 | 26.078 | 12.849 | 335.057 | 27 |
| 28 | 1.072 | .9325 | .0345 | .0370 | 28.966 | 27.010 | 13.337 | 360.233 | 28 |
| 29 | 1.075 | .9301 | .0333 | .0358 | 30.038 | 27.940 | 13.825 | 386.278 | 29 |
| 30 | 1.078 | .9278 | .0321 | .0346 | 31.114 | 28.868 | 14.313 | 413.185 | 30 |
| 36 | 1.094 | .9140 | .0266 | .0291 | 37.621 | 34.387 | 17.231 | 592.499 | 36 |
| 40 | 1.105 | .9049 | .0238 | .0263 | 42.014 | 38.020 | 19.167 | 728.740 | 40 |
| 48 | 1.127 | .8871 | .0196 | .0221 | 50.932 | 45.179 | 23.021 | 1 040.055 | 48 |
| 50 | 1.133 | .8826 | .0188 | .0213 | 53.189 | 46.947 | 23.980 | 1 125.777 | 50 |
| 52 | 1.139 | .8782 | .0180 | .0205 | 55.458 | 48.705 | 24.938 | 1 214.588 | 52 |
| 60 | 1.162 | .8609 | .0155 | .0180 | 64.647 | 55.653 | 28.751 | 1 600.085 | 60 |
| 70 | 1.191 | .8396 | .0131 | .0156 | 76.395 | 64.144 | 33.481 | 2 147.611 | 70 |
| 72 | 1.197 | .8355 | .0127 | .0152 | 78.780 | 65.817 | 34.422 | 2 265.557 | 72 |
| 80 | 1.221 | .8189 | .0113 | .0138 | 88.440 | 72.427 | 38.169 | 2 764.457 | 80 |
| 84 | 1.233 | .8108 | .0107 | .0132 | 93.343 | 75.682 | 40.033 | 3 029.759 | 84 |
| 90 | 1.252 | .7987 | .00992 | .0124 | 100.789 | 80.504 | 42.816 | 3 446.870 | 90 |
| 96 | 1.271 | .7869 | .00923 | .0117 | 108.349 | 85.255 | 45.584 | 3 886.283 | 96 |
| 100 | 1.284 | .7790 | .00881 | .0113 | 113.451 | 88.383 | 47.422 | 4 191.242 | 100 |
| 104 | 1.297 | .7713 | .00843 | .0109 | 118.605 | 91.480 | 49.252 | 4 505.557 | 104 |
| 120 | 1.349 | .7411 | .00716 | .00966 | 139.743 | 103.563 | 56.508 | 5 852.112 | 120 |
| 240 | 1.821 | .5492 | .00305 | .00555 | 328.306 | 180.312 | 107.586 | 19 398.985 | 240 |
| 360 | 2.457 | .4070 | .00172 | .00422 | 582.745 | 237.191 | 152.890 | 36 263.930 | 360 |
| 480 | 3.315 | .3016 | .00108 | .00358 | 926.074 | 279.343 | 192.670 | 53 820.752 | 480 |

¼%     Compound Interest Factors     ¼%

Compound Interest Factors

| | Single Payment | | Uniform Payment Series | | | | Arithmetic Gradient | | |
|---|---|---|---|---|---|---|---|---|---|
| | Compound Amount Factor Find F Given P F/P | Present Worth Factor Find P Given F P/F | Sinking Fund Factor Find A Given F A/F | Capital Recovery Factor Find A Given P A/P | Compound Amount Factor Find F Given A F/A | Present Worth Factor Find P Given A P/A | Gradient Uniform Series Find A Given G A/G | Gradient Present Worth Find P Given G P/G | |
| n | | | | | | | | | n |
| 1 | 1.005 | .9950 | 1.0000 | 1.0050 | 1.000 | 0.995 | 0 | 0 | 1 |
| 2 | 1.010 | .9901 | .4988 | .5038 | 2.005 | 1.985 | 0.499 | 0.991 | 2 |
| 3 | 1.015 | .9851 | .3317 | .3367 | 3.015 | 2.970 | 0.996 | 2.959 | 3 |
| 4 | 1.020 | .9802 | .2481 | .2531 | 4.030 | 3.951 | 1.494 | 5.903 | 4 |
| 5 | 1.025 | .9754 | .1980 | .2030 | 5.050 | 4.926 | 1.990 | 9.803 | 5 |
| 6 | 1.030 | .9705 | .1646 | .1696 | 6.076 | 5.896 | 2.486 | 14.660 | 6 |
| 7 | 1.036 | .9657 | .1407 | .1457 | 7.106 | 6.862 | 2.980 | 20.448 | 7 |
| 8 | 1.041 | .9609 | .1228 | .1278 | 8.141 | 7.823 | 3.474 | 27.178 | 8 |
| 9 | 1.046 | .9561 | .1089 | .1139 | 9.182 | 8.779 | 3.967 | 34.825 | 9 |
| 10 | 1.051 | .9513 | .0978 | .1028 | 10.228 | 9.730 | 4.459 | 43.389 | 10 |
| 11 | 1.056 | .9466 | .0887 | .0937 | 11.279 | 10.677 | 4.950 | 52.855 | 11 |
| 12 | 1.062 | .9419 | .0811 | .0861 | 12.336 | 11.619 | 5.441 | 63.218 | 12 |
| 13 | 1.067 | .9372 | .0746 | .0796 | 13.397 | 12.556 | 5.931 | 74.465 | 13 |
| 14 | 1.072 | .9326 | .0691 | .0741 | 14.464 | 13.489 | 6.419 | 86.590 | 14 |
| 15 | 1.078 | .9279 | .0644 | .0694 | 15.537 | 14.417 | 6.907 | 99.574 | 15 |
| 16 | 1.083 | .9233 | .0602 | .0652 | 16.614 | 15.340 | 7.394 | 113.427 | 16 |
| 17 | 1.088 | .9187 | .0565 | .0615 | 17.697 | 16.259 | 7.880 | 128.125 | 17 |
| 18 | 1.094 | .9141 | .0532 | .0582 | 18.786 | 17.173 | 8.366 | 143.668 | 18 |
| 19 | 1.099 | .9096 | .0503 | .0553 | 19.880 | 18.082 | 8.850 | 160.037 | 19 |
| 20 | 1.105 | .9051 | .0477 | .0527 | 20.979 | 18.987 | 9.334 | 177.237 | 20 |
| 21 | 1.110 | .9006 | .0453 | .0503 | 22.084 | 19.888 | 9.817 | 195.245 | 21 |
| 22 | 1.116 | .8961 | .0431 | .0481 | 23.194 | 20.784 | 10.300 | 214.070 | 22 |
| 23 | 1.122 | .8916 | .0411 | .0461 | 24.310 | 21.676 | 10.781 | 233.680 | 23 |
| 24 | 1.127 | .8872 | .0393 | .0443 | 25.432 | 22.563 | 11.261 | 254.088 | 24 |
| 25 | 1.133 | .8828 | .0377 | .0427 | 26.559 | 23.446 | 11.741 | 275.273 | 25 |
| 26 | 1.138 | .8784 | .0361 | .0411 | 27.692 | 24.324 | 12.220 | 297.233 | 26 |
| 27 | 1.144 | .8740 | .0347 | .0397 | 28.830 | 25.198 | 12.698 | 319.955 | 27 |
| 28 | 1.150 | .8697 | .0334 | .0384 | 29.975 | 26.068 | 13.175 | 343.439 | 28 |
| 29 | 1.156 | .8653 | .0321 | .0371 | 31.124 | 26.933 | 13.651 | 367.672 | 29 |
| 30 | 1.161 | .8610 | .0310 | .0360 | 32.280 | 27.794 | 14.127 | 392.640 | 30 |
| 36 | 1.197 | .8356 | .0254 | .0304 | 39.336 | 32.871 | 16.962 | 557.564 | 36 |
| 40 | 1.221 | .8191 | .0226 | .0276 | 44.159 | 36.172 | 18.836 | 681.341 | 40 |
| 48 | 1.270 | .7871 | .0185 | .0235 | 54.098 | 42.580 | 22.544 | 959.928 | 48 |
| 50 | 1.283 | .7793 | .0177 | .0227 | 56.645 | 44.143 | 23.463 | 1035.70 | 50 |
| 52 | 1.296 | .7716 | .0169 | .0219 | 59.218 | 45.690 | 24.378 | 1113.82 | 52 |
| 60 | 1.349 | .7414 | .0143 | .0193 | 69.770 | 51.726 | 28.007 | 1448.65 | 60 |
| 70 | 1.418 | .7053 | .0120 | .0170 | 83.566 | 58.939 | 32.468 | 1913.65 | 70 |
| 72 | 1.432 | .6983 | .0116 | .0166 | 86.409 | 60.340 | 33.351 | 2012.35 | 72 |
| 80 | 1.490 | .6710 | .0102 | .0152 | 98.068 | 65.802 | 36.848 | 2424.65 | 80 |
| 84 | 1.520 | .6577 | .00961 | .0146 | 104.074 | 68.453 | 38.576 | 2640.67 | 84 |
| 90 | 1.567 | .6383 | .00883 | .0138 | 113.311 | 72.331 | 41.145 | 2976.08 | 90 |
| 96 | 1.614 | .6195 | .00814 | .0131 | 122.829 | 76.095 | 43.685 | 3324.19 | 96 |
| 100 | 1.647 | .6073 | .00773 | .0127 | 129.334 | 78.543 | 45.361 | 3562.80 | 100 |
| 104 | 1.680 | .5953 | .00735 | .0124 | 135.970 | 80.942 | 47.025 | 3806.29 | 104 |
| 120 | 1.819 | .5496 | .00610 | .0111 | 163.880 | 90.074 | 53.551 | 4823.52 | 120 |
| 240 | 3.310 | .3021 | .00216 | .00716 | 462.041 | 139.581 | 96.113 | 13415.56 | 240 |
| 360 | 6.023 | .1660 | .00100 | .00600 | 1004.5 | 166.792 | 128.324 | 21403.32 | 360 |
| 480 | 10.957 | .0913 | .00050 | .00550 | 1991.5 | 181.748 | 151.795 | 27588.37 | 480 |

3/4%                                    **Compound Interest Factors**                                3/4%

| | Single Payment | | Uniform Payment Series | | | | Arithmetic Gradient | | |
| | Compound Amount Factor Find $F$ Given $P$ $F/P$ | Present Worth Factor Find $P$ Given $F$ $P/F$ | Sinking Fund Factor Find $A$ Given $F$ $A/F$ | Capital Recovery Factor Find $A$ Given $P$ $A/P$ | Compound Amount Factor Find $F$ Given $A$ $F/A$ | Present Worth Factor Find $P$ Given $A$ $P/A$ | Gradient Uniform Series Find $A$ Given $G$ $A/G$ | Gradient Present Worth Find $P$ Given $G$ $P/G$ | |
| $n$ | | | | | | | | | $n$ |
|---|---|---|---|---|---|---|---|---|---|
| 1 | 1.008 | .9926 | 1.0000 | 1.0075 | 1.000 | 0.993 | 0 | 0 | 1 |
| 2 | 1.015 | .9852 | .4981 | .5056 | 2.008 | 1.978 | 0.499 | 0.987 | 2 |
| 3 | 1.023 | .9778 | .3308 | .3383 | 3.023 | 2.956 | 0.996 | 2.943 | 3 |
| 4 | 1.030 | .9706 | .2472 | .2547 | 4.045 | 3.926 | 1.492 | 5.857 | 4 |
| 5 | 1.038 | .9633 | .1970 | .2045 | 5.076 | 4.889 | 1.986 | 9.712 | 5 |
| 6 | 1.046 | .9562 | .1636 | .1711 | 6.114 | 5.846 | 2.479 | 14.494 | 6 |
| 7 | 1.054 | .9490 | .1397 | .1472 | 7.160 | 6.795 | 2.971 | 20.187 | 7 |
| 8 | 1.062 | .9420 | .1218 | .1293 | 8.213 | 7.737 | 3.462 | 26.785 | 8 |
| 9 | 1.070 | .9350 | .1078 | .1153 | 9.275 | 8.672 | 3.951 | 34.265 | 9 |
| 10 | 1.078 | .9280 | .0967 | .1042 | 10.344 | 9.600 | 4.440 | 42.619 | 10 |
| 11 | 1.086 | .9211 | .0876 | .0951 | 11.422 | 10.521 | 4.927 | 51.831 | 11 |
| 12 | 1.094 | .9142 | .0800 | .0875 | 12.508 | 11.435 | 5.412 | 61.889 | 12 |
| 13 | 1.102 | .9074 | .0735 | .0810 | 13.602 | 12.342 | 5.897 | 72.779 | 13 |
| 14 | 1.110 | .9007 | .0680 | .0755 | 14.704 | 13.243 | 6.380 | 84.491 | 14 |
| 15 | 1.119 | .8940 | .0632 | .0707 | 15.814 | 14.137 | 6.862 | 97.005 | 15 |
| 16 | 1.127 | .8873 | .0591 | .0666 | 16.932 | 15.024 | 7.343 | 110.318 | 16 |
| 17 | 1.135 | .8807 | .0554 | .0629 | 18.059 | 15.905 | 7.822 | 124.410 | 17 |
| 18 | 1.144 | .8742 | .0521 | .0596 | 19.195 | 16.779 | 8.300 | 139.273 | 18 |
| 19 | 1.153 | .8676 | .0492 | .0567 | 20.339 | 17.647 | 8.777 | 154.891 | 19 |
| 20 | 1.161 | .8612 | .0465 | .0540 | 21.491 | 18.508 | 9.253 | 171.254 | 20 |
| 21 | 1.170 | .8548 | .0441 | .0516 | 22.653 | 19.363 | 9.727 | 188.352 | 21 |
| 22 | 1.179 | .8484 | .0420 | .0495 | 23.823 | 20.211 | 10.201 | 206.170 | 22 |
| 23 | 1.188 | .8421 | .0400 | .0475 | 25.001 | 21.053 | 10.673 | 224.695 | 23 |
| 24 | 1.196 | .8358 | .0382 | .0457 | 26.189 | 21.889 | 11.143 | 243.924 | 24 |
| 25 | 1.205 | .8296 | .0365 | .0440 | 27.385 | 22.719 | 11.613 | 263.834 | 25 |
| 26 | 1.214 | .8234 | .0350 | .0425 | 28.591 | 23.542 | 12.081 | 284.421 | 26 |
| 27 | 1.224 | .8173 | .0336 | .0411 | 29.805 | 24.360 | 12.548 | 305.672 | 27 |
| 28 | 1.233 | .8112 | .0322 | .0397 | 31.029 | 25.171 | 13.014 | 327.576 | 28 |
| 29 | 1.242 | .8052 | .0310 | .0385 | 32.261 | 25.976 | 13.479 | 350.122 | 29 |
| 30 | 1.251 | .7992 | .0298 | .0373 | 33.503 | 26.775 | 13.942 | 373.302 | 30 |
| 36 | 1.309 | .7641 | .0243 | .0318 | 41.153 | 31.447 | 16.696 | 525.038 | 36 |
| 40 | 1.348 | .7416 | .0215 | .0290 | 46.447 | 34.447 | 18.507 | 637.519 | 40 |
| 48 | 1.431 | .6986 | .0174 | .0249 | 57.521 | 40.185 | 22.070 | 886.899 | 48 |
| 50 | 1.453 | .6882 | .0166 | .0241 | 60.395 | 41.567 | 22.949 | 953.911 | 50 |
| 52 | 1.475 | .6780 | .0158 | .0233 | 63.312 | 42.928 | 23.822 | 1 022.64 | 52 |
| 60 | 1.566 | .6387 | .0133 | .0208 | 75.425 | 48.174 | 27.268 | 1 313.59 | 60 |
| 70 | 1.687 | .5927 | .0109 | .0184 | 91.621 | 54.305 | 31.465 | 1 708.68 | 70 |
| 72 | 1.713 | .5839 | .0105 | .0180 | 95.008 | 55.477 | 32.289 | 1 791.33 | 72 |
| 80 | 1.818 | .5500 | .00917 | .0167 | 109.074 | 59.995 | 35.540 | 2 132.23 | 80 |
| 84 | 1.873 | .5338 | .00859 | .0161 | 116.428 | 62.154 | 37.137 | 2 308.22 | 84 |
| 90 | 1.959 | .5104 | .00782 | .0153 | 127.881 | 65.275 | 39.496 | 2 578.09 | 90 |
| 96 | 2.049 | .4881 | .00715 | .0147 | 139.858 | 68.259 | 41.812 | 2 854.04 | 96 |
| 100 | 2.111 | .4737 | .00675 | .0143 | 148.147 | 70.175 | 43.332 | 3 040.85 | 100 |
| 104 | 2.175 | .4597 | .00638 | .0139 | 156.687 | 72.035 | 44.834 | 3 229.60 | 104 |
| 120 | 2.451 | .4079 | .00517 | .0127 | 193.517 | 78.942 | 50.653 | 3 998.68 | 120 |
| 240 | 6.009 | .1664 | .00150 | .00900 | 667.901 | 111.145 | 85.422 | 9 494.26 | 240 |
| 360 | 14.731 | .0679 | .00055 | .00805 | 1 830.8 | 124.282 | 107.115 | 13 312.50 | 360 |
| 480 | 36.111 | .0277 | .00021 | .00771 | 4 681.5 | 129.641 | 119.662 | 15 513.16 | 480 |

**1%** Compound Interest Factors **1%**

| | Single Payment | | Uniform Payment Series | | | | Arithmetic Gradient | | |
|---|---|---|---|---|---|---|---|---|---|
| | Compound Amount Factor Find F Given P F/P | Present Worth Factor Find P Given F P/F | Sinking Fund Factor Find A Given F A/F | Capital Recovery Factor Find A Given P A/P | Compound Amount Factor Find F Given A F/A | Present Worth Factor Find P Given A P/A | Gradient Uniform Series Find A Given G A/G | Gradient Present Worth Find P Given G P/G | n |
| n | | | | | | | | | |
| 1 | 1.010 | .9901 | 1.0000 | 1.0100 | 1.000 | 0.990 | 0 | 0 | 1 |
| 2 | 1.020 | .9803 | .4975 | .5075 | 2.010 | 1.970 | 0.498 | 0.980 | 2 |
| 3 | 1.030 | .9706 | .3300 | .3400 | 3.030 | 2.941 | 0.993 | 2.921 | 3 |
| 4 | 1.041 | .9610 | .2463 | .2563 | 4.060 | 3.902 | 1.488 | 5.804 | 4 |
| 5 | 1.051 | .9515 | .1960 | .2060 | 5.101 | 4.853 | 1.980 | 9.610 | 5 |
| 6 | 1.062 | .9420 | .1625 | .1725 | 6.152 | 5.795 | 2.471 | 14.320 | 6 |
| 7 | 1.072 | .9327 | .1386 | .1486 | 7.214 | 6.728 | 2.960 | 19.917 | 7 |
| 8 | 1.083 | .9235 | .1207 | .1307 | 8.286 | 7.652 | 3.448 | 26.381 | 8 |
| 9 | 1.094 | .9143 | .1067 | .1167 | 9.369 | 8.566 | 3.934 | 33.695 | 9 |
| 10 | 1.105 | .9053 | .0956 | .1056 | 10.462 | 9.471 | 4.418 | 41.843 | 10 |
| 11 | 1.116 | .8963 | .0865 | .0965 | 11.567 | 10.368 | 4.900 | 50.806 | 11 |
| 12 | 1.127 | .8874 | .0788 | .0888 | 12.682 | 11.255 | 5.381 | 60.568 | 12 |
| 13 | 1.138 | .8787 | .0724 | .0824 | 13.809 | 12.134 | 5.861 | 71.112 | 13 |
| 14 | 1.149 | .8700 | .0669 | .0769 | 14.947 | 13.004 | 6.338 | 82.422 | 14 |
| 15 | 1.161 | .8613 | .0621 | .0721 | 16.097 | 13.865 | 6.814 | 94.481 | 15 |
| 16 | 1.173 | .8528 | .0579 | .0679 | 17.258 | 14.718 | 7.289 | 107.273 | 16 |
| 17 | 1.184 | .8444 | .0543 | .0643 | 18.430 | 15.562 | 7.761 | 120.783 | 17 |
| 18 | 1.196 | .8360 | .0510 | .0610 | 19.615 | 16.398 | 8.232 | 134.995 | 18 |
| 19 | 1.208 | .8277 | .0481 | .0581 | 20.811 | 17.226 | 8.702 | 149.895 | 19 |
| 20 | 1.220 | .8195 | .0454 | .0554 | 22.019 | 18.046 | 9.169 | 165.465 | 20 |
| 21 | 1.232 | .8114 | .0430 | .0530 | 23.239 | 18.857 | 9.635 | 181.694 | 21 |
| 22 | 1.245 | .8034 | .0409 | .0509 | 24.472 | 19.660 | 10.100 | 198.565 | 22 |
| 23 | 1.257 | .7954 | .0389 | .0489 | 25.716 | 20.456 | 10.563 | 216.065 | 23 |
| 24 | 1.270 | .7876 | .0371 | .0471 | 26.973 | 21.243 | 11.024 | 234.179 | 24 |
| 25 | 1.282 | .7798 | .0354 | .0454 | 28.243 | 22.023 | 11.483 | 252.892 | 25 |
| 26 | 1.295 | .7720 | .0339 | .0439 | 29.526 | 22.795 | 11.941 | 272.195 | 26 |
| 27 | 1.308 | .7644 | .0324 | .0424 | 30.821 | 23.560 | 12.397 | 292.069 | 27 |
| 28 | 1.321 | .7568 | .0311 | .0411 | 32.129 | 24.316 | 12.852 | 312.504 | 28 |
| 29 | 1.335 | .7493 | .0299 | .0399 | 33.450 | 25.066 | 13.304 | 333.486 | 29 |
| 30 | 1.348 | .7419 | .0287 | .0387 | 34.785 | 25.808 | 13.756 | 355.001 | 30 |
| 36 | 1.431 | .6989 | .0232 | .0332 | 43.077 | 30.107 | 16.428 | 494.620 | 36 |
| 40 | 1.489 | .6717 | .0205 | .0305 | 48.886 | 32.835 | 18.178 | 596.854 | 40 |
| 48 | 1.612 | .6203 | .0163 | .0263 | 61.223 | 37.974 | 21.598 | 820.144 | 48 |
| 50 | 1.645 | .6080 | .0155 | .0255 | 64.463 | 39.196 | 22.436 | 879.417 | 50 |
| 52 | 1.678 | .5961 | .0148 | .0248 | 67.769 | 40.394 | 23.269 | 939.916 | 52 |
| 60 | 1.817 | .5504 | .0122 | .0222 | 81.670 | 44.955 | 26.533 | 1 192.80 | 60 |
| 70 | 2.007 | .4983 | .00993 | .0199 | 100.676 | 50.168 | 30.470 | 1 528.64 | 70 |
| 72 | 2.047 | .4885 | .00955 | .0196 | 104.710 | 51.150 | 31.239 | 1 597.86 | 72 |
| 80 | 2.217 | .4511 | .00822 | .0182 | 121.671 | 54.888 | 34.249 | 1 879.87 | 80 |
| 84 | 2.307 | .4335 | .00765 | .0177 | 130.672 | 56.648 | 35.717 | 2 023.31 | 84 |
| 90 | 2.449 | .4084 | .00690 | .0169 | 144.863 | 59.161 | 37.872 | 2 240.56 | 90 |
| 96 | 2.599 | .3847 | .00625 | .0163 | 159.927 | 61.528 | 39.973 | 2 459.42 | 96 |
| 100 | 2.705 | .3697 | .00587 | .0159 | 170.481 | 63.029 | 41.343 | 2 605.77 | 100 |
| 104 | 2.815 | .3553 | .00551 | .0155 | 181.464 | 64.471 | 42.688 | 2 752.17 | 104 |
| 120 | 3.300 | .3030 | .00435 | .0143 | 230.039 | 69.701 | 47.835 | 3 334.11 | 120 |
| 240 | 10.893 | .0918 | .00101 | .0110 | 989.254 | 90.819 | 75.739 | 6 878.59 | 240 |
| 360 | 35.950 | .0278 | .00029 | .0103 | 3 495.0 | 97.218 | 89.699 | 8 720.43 | 360 |
| 480 | 118.648 | .00843 | .00008 | .0101 | 11 764.8 | 99.157 | 95.920 | 9 511.15 | 480 |

| 1¹/₄% | | | Compound Interest Factors | | | | | | 1¹/₄% |

| | Single Payment | | Uniform Payment Series | | | | Arithmetic Gradient | | |
|---|---|---|---|---|---|---|---|---|---|
| | Compound Amount Factor Find F Given P F/P | Present Worth Factor Find P Given F P/F | Sinking Fund Factor Find A Given F A/F | Capital Recovery Factor Find A Given P A/P | Compound Amount Factor Find F Given A F/A | Present Worth Factor Find P Given A P/A | Gradient Uniform Series Find A Given G A/G | Gradient Present Worth Find P Given G P/G | |
| n | | | | | | | | | n |
| 1 | 1.013 | .9877 | 1.0000 | 1.0125 | 1.000 | 0.988 | 0 | 0 | 1 |
| 2 | 1.025 | .9755 | .4969 | .5094 | 2.013 | 1.963 | 0.497 | 0.976 | 2 |
| 3 | 1.038 | .9634 | .3292 | .3417 | 3.038 | 2.927 | 0.992 | 2.904 | 3 |
| 4 | 1.051 | .9515 | .2454 | .2579 | 4.076 | 3.878 | 1.485 | 5.759 | 4 |
| 5 | 1.064 | .9398 | .1951 | .2076 | 5.127 | 4.818 | 1.976 | 9.518 | 5 |
| 6 | 1.077 | .9282 | .1615 | .1740 | 6.191 | 5.746 | 2.464 | 14.160 | 6 |
| 7 | 1.091 | .9167 | .1376 | .1501 | 7.268 | 6.663 | 2.951 | 19.660 | 7 |
| 8 | 1.104 | .9054 | .1196 | .1321 | 8.359 | 7.568 | 3.435 | 25.998 | 8 |
| 9 | 1.118 | .8942 | .1057 | .1182 | 9.463 | 8.462 | 3.918 | 33.152 | 9 |
| 10 | 1.132 | .8832 | .0945 | .1070 | 10.582 | 9.346 | 4.398 | 41.101 | 10 |
| 11 | 1.146 | .8723 | .0854 | .0979 | 11.714 | 10.218 | 4.876 | 49.825 | 11 |
| 12 | 1.161 | .8615 | .0778 | .0903 | 12.860 | 11.079 | 5.352 | 59.302 | 12 |
| 13 | 1.175 | .8509 | .0713 | .0838 | 14.021 | 11.930 | 5.827 | 69.513 | 13 |
| 14 | 1.190 | .8404 | .0658 | .0783 | 15.196 | 12.771 | 6.299 | 80.438 | 14 |
| 15 | 1.205 | .8300 | .0610 | .0735 | 16.386 | 13.601 | 6.769 | 92.058 | 15 |
| 16 | 1.220 | .8197 | .0568 | .0693 | 17.591 | 14.420 | 7.237 | 104.355 | 16 |
| 17 | 1.235 | .8096 | .0532 | .0657 | 18.811 | 15.230 | 7.702 | 117.309 | 17 |
| 18 | 1.251 | .7996 | .0499 | .0624 | 20.046 | 16.030 | 8.166 | 130.903 | 18 |
| 19 | 1.266 | .7898 | .0470 | .0595 | 21.297 | 16.849 | 8.628 | 145.119 | 19 |
| 20 | 1.282 | .7800 | .0443 | .0568 | 22.563 | 17.599 | 9.088 | 159.940 | 20 |
| 21 | 1.298 | .7704 | .0419 | .0544 | 23.845 | 18.370 | 9.545 | 175.348 | 21 |
| 22 | 1.314 | .7609 | .0398 | .0523 | 25.143 | 19.131 | 10.001 | 191.327 | 22 |
| 23 | 1.331 | .7515 | .0378 | .0503 | 26.458 | 19.882 | 10.455 | 207.859 | 23 |
| 24 | 1.347 | .7422 | .0360 | .0485 | 27.788 | 20.624 | 10.906 | 224.930 | 24 |
| 25 | 1.364 | .7330 | .0343 | .0468 | 29.136 | 21.357 | 11.355 | 242.523 | 25 |
| 26 | 1.381 | .7240 | .0328 | .0453 | 30.500 | 22.081 | 11.803 | 260.623 | 26 |
| 27 | 1.399 | .7150 | .0314 | .0439 | 31.881 | 22.796 | 12.248 | 279.215 | 27 |
| 28 | 1.416 | .7062 | .0300 | .0425 | 32.280 | 23.503 | 12.691 | 298.284 | 28 |
| 29 | 1.434 | .6975 | .0288 | .0413 | 34.696 | 24.200 | 13.133 | 317.814 | 29 |
| 30 | 1.452 | .6889 | .0277 | .0402 | 36.129 | 24.889 | 13.572 | 337.792 | 30 |
| 36 | 1.564 | .6394 | .0222 | .0347 | 45.116 | 28.847 | 16.164 | 466.297 | 36 |
| 40 | 1.644 | .6084 | .0194 | .0319 | 51.490 | 31.327 | 17.852 | 559.247 | 40 |
| 48 | 1.845 | .5509 | .0153 | .0278 | 65.229 | 35.932 | 21.130 | 759.248 | 48 |
| 50 | 1.861 | .5373 | .0145 | .0270 | 68.882 | 37.013 | 21.930 | 811.692 | 50 |
| 52 | 1.908 | .5242 | .0138 | .0263 | 72.628 | 38.068 | 22.722 | 864.960 | 52 |
| 60 | 2.107 | .4746 | .0113 | .0238 | 88.575 | 42.035 | 25.809 | 1 084.86 | 60 |
| 70 | 2.386 | .4191 | .00902 | .0215 | 110.873 | 46.470 | 29.492 | 1 370.47 | 70 |
| 72 | 2.446 | .4088 | .00864 | .0211 | 115.675 | 47.293 | 30.205 | 1 428.48 | 72 |
| 80 | 2.701 | .3702 | .00735 | .0198 | 136.120 | 50.387 | 32.983 | 1 661.89 | 80 |
| 84 | 2.839 | .3522 | .00680 | .0193 | 147.130 | 51.822 | 34.326 | 1 778.86 | 84 |
| 90 | 3.059 | .3269 | .00607 | .0186 | 164.706 | 53.846 | 36.286 | 1 953.85 | 90 |
| 96 | 3.296 | .3034 | .00545 | .0179 | 183.643 | 55.725 | 38.180 | 2 127.55 | 96 |
| 100 | 3.463 | .2887 | .00507 | .0176 | 197.074 | 56.901 | 39.406 | 2 242.26 | 100 |
| 104 | 3.640 | .2747 | .00474 | .0172 | 211.190 | 58.021 | 40.604 | 2 355.90 | 104 |
| 120 | 4.440 | .2252 | .00363 | .0161 | 275.220 | 61.983 | 45.119 | 2 796.59 | 120 |
| 240 | 19.716 | .0507 | .00067 | .0132 | 1 497.3 | 75.942 | 67.177 | 5 101.55 | 240 |
| 360 | 87.543 | .0114 | .00014 | .0126 | 6 923.4 | 79.086 | 75.840 | 5 997.91 | 360 |
| 480 | 388.713 | .00257 | .00003 | .0125 | 31 017.1 | 79.794 | 78.762 | 6 284.74 | 480 |

| 1½% | | | | | Compound Interest Factors | | | | | 1½% |

| | Single Payment | | Uniform Payment Series | | | | Arithmetic Gradient | | | |
|---|---|---|---|---|---|---|---|---|---|---|
| | Compound Amount Factor Find F Given P | Present Worth Factor Find P Given F | Sinking Fund Factor Find A Given F | Capital Recovery Factor Find A Given P | Compound Amount Factor Find F Given A | Present Worth Factor Find P Given A | Gradient Uniform Series Find A Given G | Gradient Present Worth Find P Given G | |
| n | F/P | P/F | A/F | A/P | F/A | P/A | A/G | P/G | n |
| 1 | 1.015 | .9852 | 1.0000 | 1.0150 | 1.000 | 0.985 | 0 | 0 | 1 |
| 2 | 1.030 | .9707 | .4963 | .5113 | 2.015 | 1.956 | 0.496 | 0.970 | 2 |
| 3 | 1.046 | .9563 | .3284 | .3434 | 3.045 | 2.912 | 0.990 | 2.883 | 3 |
| 4 | 1.061 | .9422 | .2444 | .2594 | 4.091 | 3.854 | 1.481 | 5.709 | 4 |
| 5 | 1.077 | .9283 | .1941 | .2091 | 5.152 | 4.783 | 1.970 | 9.422 | 5 |
| 6 | 1.093 | .9145 | .1605 | .1755 | 6.230 | 5.697 | 2.456 | 13.994 | 6 |
| 7 | 1.110 | .9010 | .1366 | .1516 | 7.323 | 6.598 | 2.940 | 19.400 | 7 |
| 8 | 1.126 | .8877 | .1186 | .1336 | 8.433 | 7.486 | 3.422 | 25.614 | 8 |
| 9 | 1.143 | .8746 | .1046 | .1196 | 9.559 | 8.360 | 3.901 | 32.610 | 9 |
| 10 | 1.161 | .8617 | .0934 | .1084 | 10.703 | 9.222 | 4.377 | 40.365 | 10 |
| 11 | 1.178 | .8489 | .0843 | .0993 | 11.863 | 10.071 | 4.851 | 48.855 | 11 |
| 12 | 1.196 | .8364 | .0767 | .0917 | 13.041 | 10.907 | 5.322 | 58.054 | 12 |
| 13 | 1.214 | .8240 | .0702 | .0852 | 14.237 | 11.731 | 5.791 | 67.943 | 13 |
| 14 | 1.232 | .8118 | .0647 | .0797 | 15.450 | 12.543 | 6.258 | 78.496 | 14 |
| 15 | 1.250 | .7999 | .0599 | .0749 | 16.682 | 13.343 | 6.722 | 89.694 | 15 |
| 16 | 1.269 | .7880 | .0558 | .0708 | 17.932 | 14.131 | 7.184 | 101.514 | 16 |
| 17 | 1.288 | .7764 | .0521 | .0671 | 19.201 | 14.908 | 7.643 | 113.937 | 17 |
| 18 | 1.307 | .7649 | .0488 | .0638 | 20.489 | 15.673 | 8.100 | 126.940 | 18 |
| 19 | 1.327 | .7536 | .0459 | .0609 | 21.797 | 16.426 | 8.554 | 140.505 | 19 |
| 20 | 1.347 | .7425 | .0432 | .0582 | 23.124 | 17.169 | 9.005 | 154.611 | 20 |
| 21 | 1.367 | .7315 | .0409 | .0559 | 24.470 | 17.900 | 9.455 | 169.241 | 21 |
| 22 | 1.388 | .7207 | .0387 | .0537 | 25.837 | 18.621 | 9.902 | 184.375 | 22 |
| 23 | 1.408 | .7100 | .0367 | .0517 | 27.225 | 19.331 | 10.346 | 199.996 | 23 |
| 24 | 1.430 | .6995 | .0349 | .0499 | 28.633 | 20.030 | 10.788 | 216.085 | 24 |
| 25 | 1.451 | .6892 | .0333 | .0483 | 30.063 | 20.720 | 11.227 | 232.626 | 25 |
| 26 | 1.473 | .6790 | .0317 | .0467 | 31.514 | 21.399 | 11.664 | 249.601 | 26 |
| 27 | 1.495 | .6690 | .0303 | .0453 | 32.987 | 22.068 | 12.099 | 266.995 | 27 |
| 28 | 1.517 | .6591 | .0290 | .0440 | 34.481 | 22.727 | 12.531 | 284.790 | 28 |
| 29 | 1.540 | .6494 | .0278 | .0428 | 35.999 | 23.376 | 12.961 | 302.972 | 29 |
| 30 | 1.563 | .6398 | .0266 | .0416 | 37.539 | 24.016 | 13.388 | 321.525 | 30 |
| 36 | 1.709 | .5851 | .0212 | .0362 | 47.276 | 27.661 | 15.901 | 439.823 | 36 |
| 40 | 1.814 | .5513 | .0184 | .0334 | 54.268 | 29.916 | 17.528 | 524.349 | 40 |
| 48 | 2.043 | .4894 | .0144 | .0294 | 69.565 | 34.042 | 20.666 | 703.537 | 48 |
| 50 | 2.105 | .4750 | .0136 | .0286 | 73.682 | 35.000 | 21.428 | 749.955 | 50 |
| 52 | 2.169 | .4611 | .0128 | .0278 | 77.925 | 35.929 | 22.179 | 796.868 | 52 |
| 60 | 2.443 | .4093 | .0104 | .0254 | 96.214 | 39.380 | 25.093 | 988.157 | 60 |
| 70 | 2.835 | .3527 | .00817 | .0232 | 122.363 | 43.155 | 28.529 | 1 231.15 | 70 |
| 72 | 2.921 | .3423 | .00781 | .0228 | 128.076 | 43.845 | 29.189 | 1 279.78 | 72 |
| 80 | 3.291 | .3039 | .00655 | .0215 | 152.710 | 46.407 | 31.742 | 1 473.06 | 80 |
| 84 | 3.493 | .2863 | .00602 | .0210 | 166.172 | 47.579 | 32.967 | 1 568.50 | 84 |
| 90 | 3.819 | .2619 | .00532 | .0203 | 187.929 | 49.210 | 34.740 | 1 709.53 | 90 |
| 96 | 4.176 | .2395 | .00472 | .0197 | 211.719 | 50.702 | 36.438 | 1 847.46 | 96 |
| 100 | 4.432 | .2256 | .00437 | .0194 | 228.802 | 51.625 | 37.529 | 1 937.43 | 100 |
| 104 | 4.704 | .2126 | .00405 | .0190 | 246.932 | 52.494 | 38.589 | 2 025.69 | 104 |
| 120 | 5.969 | .1675 | .00302 | .0180 | 331.286 | 55.498 | 42.518 | 2 359.69 | 120 |
| 240 | 35.632 | .0281 | .00043 | .0154 | 2 308.8 | 64.796 | 59.737 | 3 870.68 | 240 |
| 360 | 212.700 | .00470 | .00007 | .0151 | 14 113.3 | 66.353 | 64.966 | 4 310.71 | 360 |
| 480 | 1 269.7 | .00079 | .00001 | .0150 | 84 577.8 | 66.614 | 66.288 | 4 415.74 | 480 |

| | Single Payment | | Uniform Payment Series | | | | Arithmetic Gradient | | |
|---|---|---|---|---|---|---|---|---|---|
| 1¾% | Compound Interest Factors | | | | | | | | 1¾% |

**1¾%**  Compound Interest Factors  **1¾%**

| | Single Payment | | Uniform Payment Series | | | | Arithmetic Gradient | | |
|---|---|---|---|---|---|---|---|---|---|
| n | Compound Amount Factor Find F Given P F/P | Present Worth Factor Find P Given F P/F | Sinking Fund Factor Find A Given F A/F | Capital Recovery Factor Find A Given P A/P | Compound Amount Factor Find F Given A F/A | Present Worth Factor Find P Given A P/A | Gradient Uniform Series Find A Given G A/G | Gradient Present Worth Find P Given G P/G | n |
| 1 | 1.018 | .9828 | 1.0000 | 1.0175 | 1.000 | 0.983 | 0 | 0 | 1 |
| 2 | 1.035 | .9659 | .4957 | .5132 | 2.018 | 1.949 | 0.496 | 0.966 | 2 |
| 3 | 1.053 | .9493 | .3276 | .3451 | 3.053 | 2.898 | 0.989 | 2.865 | 3 |
| 4 | 1.072 | .9330 | .2435 | .2610 | 4.106 | 3.831 | 1.478 | 5.664 | 4 |
| 5 | 1.091 | .9169 | .1931 | .2106 | 5.178 | 4.748 | 1.965 | 9.332 | 5 |
| 6 | 1.110 | .9011 | .1595 | .1770 | 6.269 | 5.649 | 2.450 | 13.837 | 6 |
| 7 | 1.129 | .8856 | .1355 | .1530 | 7.378 | 6.535 | 2.931 | 19.152 | 7 |
| 8 | 1.149 | .8704 | .1175 | .1350 | 8.508 | 7.405 | 3.409 | 25.245 | 8 |
| 9 | 1.169 | .8554 | .1036 | .1211 | 9.656 | 8.261 | 3.885 | 32.088 | 9 |
| 10 | 1.189 | .8407 | .0924 | .1099 | 10.825 | 9.101 | 4.357 | 39.655 | 10 |
| 11 | 1.210 | .8263 | .0832 | .1007 | 12.015 | 9.928 | 4.827 | 47.918 | 11 |
| 12 | 1.231 | .8121 | .0756 | .0931 | 13.225 | 10.740 | 5.294 | 56.851 | 12 |
| 13 | 1.253 | .7981 | .0692 | .0867 | 14.457 | 11.538 | 5.758 | 66.428 | 13 |
| 14 | 1.275 | .7844 | .0637 | .0812 | 15.710 | 12.322 | 6.219 | 76.625 | 14 |
| 15 | 1.297 | .7709 | .0589 | .0764 | 16.985 | 13.093 | 6.677 | 87.417 | 15 |
| 16 | 1.320 | .7576 | .0547 | .0722 | 18.282 | 13.851 | 7.132 | 98.782 | 16 |
| 17 | 1.343 | .7446 | .0510 | .0685 | 19.602 | 14.595 | 7.584 | 110.695 | 17 |
| 18 | 1.367 | .7318 | .0477 | .0652 | 20.945 | 15.327 | 8.034 | 123.136 | 18 |
| 19 | 1.390 | .7192 | .0448 | .0623 | 22.311 | 16.046 | 8.481 | 136.081 | 19 |
| 20 | 1.415 | .7068 | .0422 | .0597 | 23.702 | 16.753 | 8.924 | 149.511 | 20 |
| 21 | 1.440 | .6947 | .0398 | .0573 | 25.116 | 17.448 | 9.365 | 163.405 | 21 |
| 22 | 1.465 | .6827 | .0377 | .0552 | 26.556 | 18.130 | 9.804 | 177.742 | 22 |
| 23 | 1.490 | .6710 | .0357 | .0532 | 28.021 | 18.801 | 10.239 | 192.503 | 23 |
| 24 | 1.516 | .6594 | .0339 | .0514 | 29.511 | 19.461 | 10.671 | 207.671 | 24 |
| 25 | 1.543 | .6481 | .0322 | .0497 | 31.028 | 20.109 | 11.101 | 223.225 | 25 |
| 26 | 1.570 | .6369 | .0307 | .0482 | 32.571 | 20.746 | 11.528 | 239.149 | 26 |
| 27 | 1.597 | .6260 | .0293 | .0468 | 34.141 | 21.372 | 11.952 | 255.425 | 27 |
| 28 | 1.625 | .6152 | .0280 | .0455 | 35.738 | 21.987 | 12.373 | 272.036 | 28 |
| 29 | 1.654 | .6046 | .0268 | .0443 | 37.363 | 22.592 | 12.791 | 288.967 | 29 |
| 30 | 1.683 | .5942 | .0256 | .0431 | 39.017 | 23.186 | 13.206 | 306.200 | 30 |
| 36 | 1.867 | .5355 | .0202 | .0377 | 49.566 | 26.543 | 15.640 | 415.130 | 36 |
| 40 | 2.002 | .4996 | .0175 | .0350 | 57.234 | 28.594 | 17.207 | 492.017 | 40 |
| 48 | 2.300 | .4349 | .0135 | .0310 | 74.263 | 32.294 | 20.209 | 652.612 | 48 |
| 50 | 2.381 | .4200 | .0127 | .0302 | 78.903 | 33.141 | 20.932 | 693.708 | 50 |
| 52 | 2.465 | .4057 | .0119 | .0294 | 83.706 | 33.960 | 21.644 | 735.039 | 52 |
| 60 | 2.832 | .3531 | .00955 | .0271 | 104.676 | 36.964 | 24.389 | 901.503 | 60 |
| 70 | 3.368 | .2969 | .00739 | .0249 | 135.331 | 40.178 | 27.586 | 1 108.34 | 70 |
| 72 | 3.487 | .2868 | .00704 | .0245 | 142.127 | 40.757 | 28.195 | 1 149.12 | 72 |
| 80 | 4.006 | .2496 | .00582 | .0233 | 171.795 | 42.880 | 30.533 | 1 309.25 | 80 |
| 84 | 4.294 | .2329 | .00531 | .0228 | 188.246 | 43.836 | 31.644 | 1 387.16 | 84 |
| 90 | 4.765 | .2098 | .00465 | .0221 | 215.166 | 45.152 | 33.241 | 1 500.88 | 90 |
| 96 | 5.288 | .1891 | .00408 | .0216 | 245.039 | 46.337 | 34.756 | 1 610.48 | 96 |
| 100 | 5.668 | .1764 | .00375 | .0212 | 266.753 | 47.062 | 35.721 | 1 681.09 | 100 |
| 104 | 6.075 | .1646 | .00345 | .0209 | 290.028 | 47.737 | 36.652 | 1 749.68 | 104 |
| 120 | 8.019 | .1247 | .00249 | .0200 | 401.099 | 50.017 | 40.047 | 2 003.03 | 120 |
| 240 | 64.308 | .0156 | .00028 | .0178 | 3 617.6 | 56.254 | 53.352 | 3 001.27 | 240 |
| 360 | 515.702 | .00194 | .00003 | .0175 | 29 411.5 | 57.032 | 56.443 | 3 219.08 | 360 |
| 480 | 4 135.5 | .00024 | | .0175 | 236 259.0 | 57.129 | 57.027 | 3 257.88 | 480 |

| 2% | | | | Compound Interest Factors | | | | | 2% |
|---|---|---|---|---|---|---|---|---|---|

| | Single Payment | | Uniform Payment Series | | | | Arithmetic Gradient | | |
|---|---|---|---|---|---|---|---|---|---|
| | Compound Amount Factor Find F Given P F/P | Present Worth Factor Find P Given F P/F | Sinking Fund Factor Find A Given F A/F | Capital Recovery Factor Find A Given P A/P | Compound Amount Factor Find F Given A F/A | Present Worth Factor Find P Given A P/A | Gradient Uniform Series Find A Given G A/G | Gradient Present Worth Find P Given G P/G | |
| n | | | | | | | | | n |
| 1 | 1.020 | .9804 | 1.0000 | 1.0200 | 1.000 | 0.980 | 0 | 0 | 1 |
| 2 | 1.040 | .9612 | .4951 | .5151 | 2.020 | 1.942 | 0.495 | 0.961 | 2 |
| 3 | 1.061 | .9423 | .3268 | .3468 | 3.060 | 2.884 | 0.987 | 2.846 | 3 |
| 4 | 1.082 | .9238 | .2426 | .2626 | 4.122 | 3.808 | 1.475 | 5.617 | 4 |
| 5 | 1.104 | .9057 | .1922 | .2122 | 5.204 | 4.713 | 1.960 | 9.240 | 5 |
| 6 | 1.126 | .8880 | .1585 | .1785 | 6.308 | 5.601 | 2.442 | 13.679 | 6 |
| 7 | 1.149 | .8706 | .1345 | .1545 | 7.434 | 6.472 | 2.921 | 18.903 | 7 |
| 8 | 1.172 | .8535 | .1165 | .1365 | 8.583 | 7.325 | 3.396 | 24.877 | 8 |
| 9 | 1.195 | .8368 | .1025 | .1225 | 9.755 | 8.162 | 3.868 | 31.571 | 9 |
| 10 | 1.219 | .8203 | .0913 | .1113 | 10.950 | 8.983 | 4.337 | 38.954 | 10 |
| 11 | 1.243 | .8043 | .0822 | .1022 | 12.169 | 9.787 | 4.802 | 46.996 | 11 |
| 12 | 1.268 | .7885 | .0746 | .0946 | 13.412 | 10.575 | 5.264 | 55.669 | 12 |
| 13 | 1.294 | .7730 | .0681 | .0881 | 14.680 | 11.348 | 5.723 | 64.946 | 13 |
| 14 | 1.319 | .7579 | .0626 | .0826 | 15.974 | 12.106 | 6.178 | 74.798 | 14 |
| 15 | 1.346 | .7430 | .0578 | .0778 | 17.293 | 12.849 | 6.631 | 85.200 | 15 |
| 16 | 1.373 | .7284 | .0537 | .0737 | 18.639 | 13.578 | 7.080 | 96.127 | 16 |
| 17 | 1.400 | .7142 | .0500 | .0700 | 20.012 | 14.292 | 7.526 | 107.553 | 17 |
| 18 | 1.428 | .7002 | .0467 | .0667 | 21.412 | 14.992 | 7.968 | 119.456 | 18 |
| 19 | 1.457 | .6864 | .0438 | .0638 | 22.840 | 15.678 | 8.407 | 131.812 | 19 |
| 20 | 1.486 | .6730 | .0412 | .0612 | 24.297 | 16.351 | 8.843 | 144.598 | 20 |
| 21 | 1.516 | .6598 | .0388 | .0588 | 25.783 | 17.011 | 9.276 | 157.793 | 21 |
| 22 | 1.546 | .6468 | .0366 | .0566 | 27.299 | 17.658 | 9.705 | 171.377 | 22 |
| 23 | 1.577 | .6342 | .0347 | .0547 | 28.845 | 18.292 | 10.132 | 185.328 | 23 |
| 24 | 1.608 | .6217 | .0329 | .0529 | 30.422 | 18.914 | 10.555 | 199.628 | 24 |
| 25 | 1.641 | .6095 | .0312 | .0512 | 32.030 | 19.523 | 10.974 | 214.256 | 25 |
| 26 | 1.673 | .5976 | .0297 | .0497 | 33.671 | 20.121 | 11.391 | 229.196 | 26 |
| 27 | 1.707 | .5859 | .0283 | .0483 | 35.344 | 20.707 | 11.804 | 244.428 | 27 |
| 28 | 1.741 | .5744 | .0270 | .0470 | 37.051 | 21.281 | 12.214 | 259.936 | 28 |
| 29 | 1.776 | .5631 | .0258 | .0458 | 38.792 | 21.844 | 12.621 | 275.703 | 29 |
| 30 | 1.811 | .5521 | .0247 | .0447 | 40.568 | 22.396 | 13.025 | 291.713 | 30 |
| 36 | 2.040 | .4902 | .0192 | .0392 | 51.994 | 25.489 | 15.381 | 392.036 | 36 |
| 40 | 2.208 | .4529 | .0166 | .0366 | 60.402 | 27.355 | 16.888 | 461.989 | 40 |
| 48 | 2.587 | .3865 | .0126 | .0326 | 79.353 | 30.673 | 19.755 | 605.961 | 48 |
| 50 | 2.692 | .3715 | .0118 | .0318 | 84.579 | 31.424 | 20.442 | 642.355 | 50 |
| 52 | 2.800 | .3571 | .0111 | .0311 | 90.016 | 32.145 | 21.116 | 678.779 | 52 |
| 60 | 3.281 | .3048 | .00877 | .0288 | 114.051 | 34.761 | 23.696 | 823.692 | 60 |
| 70 | 4.000 | .2500 | .00667 | .0267 | 149.977 | 37.499 | 26.663 | 999.829 | 70 |
| 72 | 4.161 | .2403 | .00633 | .0263 | 158.056 | 37.984 | 27.223 | 1 034.050 | 72 |
| 80 | 4.875 | .2051 | .00516 | .0252 | 193.771 | 39.744 | 29.357 | 1 166.781 | 80 |
| 84 | 5.277 | .1895 | .00468 | .0247 | 213.865 | 40.525 | 30.361 | 1 230.413 | 84 |
| 90 | 5.943 | .1683 | .00405 | .0240 | 247.155 | 41.587 | 31.793 | 1 322.164 | 90 |
| 96 | 6.693 | .1494 | .00351 | .0235 | 284.645 | 42.529 | 33.137 | 1 409.291 | 96 |
| 100 | 7.245 | .1380 | .00320 | .0232 | 312.230 | 43.098 | 33.986 | 1 464.747 | 100 |
| 104 | 7.842 | .1275 | .00292 | .0229 | 342.090 | 43.624 | 34.799 | 1 518.082 | 104 |
| 120 | 10.765 | .0929 | .00205 | .0220 | 488.255 | 45.355 | 37.711 | 1 710.411 | 120 |
| 240 | 115.887 | .00863 | .00017 | .0202 | 5 744.4 | 49.569 | 47.911 | 2 374.878 | 240 |
| 360 | 1 247.5 | .00080 | .00002 | .0200 | 62 326.8 | 49.960 | 49.711 | 2 483.567 | 360 |
| 480 | 13 429.8 | .00007 | | .0200 | 671 442.0 | 49.996 | 49.964 | 2 498.027 | 480 |

2¹/₂ %                             **Compound Interest Factors**                            2¹/₂ %

| | Single Payment | | Uniform Payment Series | | | | Arithmetic Gradient | | |
| | Compound Amount Factor Find F Given P F/P | Present Worth Factor Find P Given F P/F | Sinking Fund Factor Find A Given F A/F | Capital Recovery Factor Find A Given P A/P | Compound Amount Factor Find F Given A F/A | Present Worth Factor Find P Given A P/A | Gradient Uniform Series Find A Given G A/G | Gradient Present Worth Find P Given G P/G | |
| n | | | | | | | | | n |
|---|---|---|---|---|---|---|---|---|---|
| 1 | 1.025 | .9756 | 1.0000 | 1.0250 | 1.000 | 0.976 | 0 | 0 | 1 |
| 2 | 1.051 | .9518 | .4938 | .5188 | 2.025 | 1.927 | 0.494 | 0.952 | 2 |
| 3 | 1.077 | .9286 | .3251 | .3501 | 3.076 | 2.856 | 0.984 | 2.809 | 3 |
| 4 | 1.104 | .9060 | .2408 | .2658 | 4.153 | 3.762 | 1.469 | 5.527 | 4 |
| 5 | 1.131 | .8839 | .1902 | .2152 | 5.256 | 4.646 | 1.951 | 9.062 | 5 |
| 6 | 1.160 | .8623 | .1566 | .1816 | 6.388 | 5.508 | 2.428 | 13.374 | 6 |
| 7 | 1.189 | .8413 | .1325 | .1575 | 7.547 | 6.349 | 2.901 | 18.421 | 7 |
| 8 | 1.218 | .8207 | .1145 | .1395 | 8.736 | 7.170 | 3.370 | 24.166 | 8 |
| 9 | 1.249 | .8007 | .1005 | .1255 | 9.955 | 7.971 | 3.835 | 30.572 | 9 |
| 10 | 1.280 | .7812 | .0893 | .1143 | 11.203 | 8.752 | 4.296 | 37.603 | 10 |
| 11 | 1.312 | .7621 | .0801 | .1051 | 12.483 | 9.514 | 4.753 | 45.224 | 11 |
| 12 | 1.345 | .7436 | .0725 | .0975 | 13.796 | 10.258 | 5.206 | 53.403 | 12 |
| 13 | 1.379 | .7254 | .0660 | .0910 | 15.140 | 10.983 | 5.655 | 62.108 | 13 |
| 14 | 1.413 | .7077 | .0605 | .0855 | 16.519 | 11.691 | 6.100 | 71.309 | 14 |
| 15 | 1.448 | .6905 | .0558 | .0808 | 17.932 | 12.381 | 6.540 | 80.975 | 15 |
| 16 | 1.485 | .6736 | .0516 | .0766 | 19.380 | 13.055 | 6.977 | 91.080 | 16 |
| 17 | 1.522 | .6572 | .0479 | .0729 | 20.865 | 13.712 | 7.409 | 101.595 | 17 |
| 18 | 1.560 | .6412 | .0447 | .0697 | 22.386 | 14.353 | 7.838 | 112.495 | 18 |
| 19 | 1.599 | .6255 | .0418 | .0668 | 23.946 | 14.979 | 8.262 | 123.754 | 19 |
| 20 | 1.639 | .6103 | .0391 | .0641 | 25.545 | 15.589 | 8.682 | 135.349 | 20 |
| 21 | 1.680 | .5954 | .0368 | .0618 | 27.183 | 16.185 | 9.099 | 147.257 | 21 |
| 22 | 1.722 | .5809 | .0346 | .0596 | 28.863 | 16.765 | 9.511 | 159.455 | 22 |
| 23 | 1.765 | .5667 | .0327 | .0577 | 30.584 | 17.332 | 9.919 | 171.922 | 23 |
| 24 | 1.809 | .5529 | .0309 | .0559 | 32.349 | 17.885 | 10.324 | 184.638 | 24 |
| 25 | 1.854 | .5394 | .0293 | .0543 | 34.158 | 18.424 | 10.724 | 197.584 | 25 |
| 26 | 1.900 | .5262 | .0278 | .0528 | 36.012 | 18.951 | 11.120 | 210.740 | 26 |
| 27 | 1.948 | .5134 | .0264 | .0514 | 37.912 | 19.464 | 11.513 | 224.088 | 27 |
| 28 | 1.996 | .5009 | .0251 | .0501 | 39.860 | 19.965 | 11.901 | 237.612 | 28 |
| 29 | 2.046 | .4887 | .0239 | .0489 | 41.856 | 20.454 | 12.286 | 251.294 | 29 |
| 30 | 2.098 | .4767 | .0228 | .0478 | 43.903 | 20.930 | 12.667 | 265.120 | 30 |
| 31 | 2.150 | .4651 | .0217 | .0467 | 46.000 | 21.395 | 13.044 | 279.073 | 31 |
| 32 | 2.204 | .4538 | .0208 | .0458 | 48.150 | 24.849 | 13.417 | 293.140 | 32 |
| 33 | 2.259 | .4427 | .0199 | .0449 | 50.354 | 22.292 | 13.786 | 307.306 | 33 |
| 34 | 2.315 | .4319 | .0190 | .0440 | 52.613 | 22.724 | 14.151 | 321.559 | 34 |
| 35 | 2.373 | .4214 | .0182 | .0432 | 54.928 | 23.145 | 14.512 | 335.886 | 35 |
| 40 | 2.685 | .3724 | .0148 | .0398 | 67.402 | 25.103 | 16.262 | 408.221 | 40 |
| 45 | 3.038 | .3292 | .0123 | .0373 | 81.516 | 26.833 | 17.918 | 480.806 | 45 |
| 50 | 3.437 | .2909 | .0103 | .0353 | 97.484 | 28.362 | 19.484 | 552.607 | 50 |
| 55 | 3.889 | .2572 | .00865 | .0337 | 115.551 | 29.714 | 20.961 | 622.827 | 55 |
| 60 | 4.400 | .2273 | .00735 | .0324 | 135.991 | 30.909 | 22.352 | 690.865 | 60 |
| 65 | 4.978 | .2009 | .00628 | .0313 | 159.118 | 31.965 | 23.660 | 756.280 | 65 |
| 70 | 5.632 | .1776 | .00540 | .0304 | 185.284 | 32.898 | 24.888 | 818.763 | 70 |
| 75 | 6.372 | .1569 | .00465 | .0297 | 214.888 | 33.723 | 26.039 | 878.114 | 75 |
| 80 | 7.210 | .1387 | .00403 | .0290 | 248.382 | 34.452 | 27.117 | 934.217 | 80 |
| 85 | 8.157 | .1226 | .00349 | .0285 | 286.278 | 35.096 | 28.123 | 987.026 | 85 |
| 90 | 9.229 | .1084 | .00304 | .0280 | 329.154 | 35.666 | 29.063 | 1036.54 | 90 |
| 95 | 10.442 | .0958 | .00265 | .0276 | 377.663 | 36.169 | 29.938 | 1082.83 | 95 |
| 100 | 11.814 | .0846 | .00231 | .0273 | 432.548 | 36.614 | 30.752 | 1125.97 | 100 |

**3%** — Compound Interest Factors — **3%**

| | Single Payment | | Uniform Payment Series | | | | Arithmetic Gradient | | |
|---|---|---|---|---|---|---|---|---|---|
| | Compound Amount Factor Find $F$ Given $P$ $F/P$ | Present Worth Factor Find $P$ Given $F$ $P/F$ | Sinking Fund Factor Find $A$ Given $F$ $A/F$ | Capital Recovery Factor Find $A$ Given $P$ $A/P$ | Compound Amount Factor Find $F$ Given $A$ $F/A$ | Present Worth Factor Find $P$ Given $A$ $P/A$ | Gradient Uniform Series Find $A$ Given $G$ $A/G$ | Gradient Present Worth Find $P$ Given $G$ $P/G$ | |
| $n$ | | | | | | | | | $n$ |
| 1 | 1.030 | .9709 | 1.0000 | 1.0300 | 1.000 | 0.971 | 0 | 0 | 1 |
| 2 | 1.061 | .9426 | .4926 | .5226 | 2.030 | 1.913 | 0.493 | 0.943 | 2 |
| 3 | 1.093 | .9151 | .3235 | .3535 | 3.091 | 2.829 | 0.980 | 2.773 | 3 |
| 4 | 1.126 | .8885 | .2390 | .2690 | 4.184 | 3.717 | 1.463 | 5.438 | 4 |
| 5 | 1.159 | .8626 | .1884 | .2184 | 5.309 | 4.580 | 1.941 | 8.889 | 5 |
| 6 | 1.194 | .8375 | .1546 | .1846 | 6.468 | 5.417 | 2.414 | 13.076 | 6 |
| 7 | 1.230 | .8131 | .1305 | .1605 | 7.662 | 6.230 | 2.882 | 17.955 | 7 |
| 8 | 1.267 | .7894 | .1125 | .1425 | 8.892 | 7.020 | 3.345 | 23.481 | 8 |
| 9 | 1.305 | .7664 | .0984 | .1284 | 10.159 | 7.786 | 3.803 | 29.612 | 9 |
| 10 | 1.344 | .7441 | .0872 | .1172 | 11.464 | 8.530 | 4.256 | 36.309 | 10 |
| 11 | 1.384 | .7224 | .0781 | .1081 | 12.808 | 9.253 | 4.705 | 43.533 | 11 |
| 12 | 1.426 | .7014 | .0705 | .1005 | 14.192 | 9.954 | 5.148 | 51.248 | 12 |
| 13 | 1.469 | .6810 | .0640 | .0940 | 15.618 | 10.635 | 5.587 | 59.419 | 13 |
| 14 | 1.513 | .6611 | .0585 | .0885 | 17.086 | 11.296 | 6.021 | 68.014 | 14 |
| 15 | 1.558 | .6419 | .0538 | .0838 | 18.599 | 11.938 | 6.450 | 77.000 | 15 |
| 16 | 1.605 | .6232 | .0496 | .0796 | 20.157 | 12.561 | 6.874 | 86.348 | 16 |
| 17 | 1.653 | .6050 | .0460 | .0760 | 21.762 | 13.166 | 7.294 | 96.028 | 17 |
| 18 | 1.702 | .5874 | .0427 | .0727 | 23.414 | 13.754 | 7.708 | 106.014 | 18 |
| 19 | 1.754 | .5703 | .0398 | .0698 | 25.117 | 14.324 | 8.118 | 116.279 | 19 |
| 20 | 1.806 | .5537 | .0372 | .0672 | 26.870 | 14.877 | 8.523 | 126.799 | 20 |
| 21 | 1.860 | .5375 | .0349 | .0649 | 28.676 | 15.415 | 8.923 | 137.549 | 21 |
| 22 | 1.916 | .5219 | .0327 | .0627 | 30.537 | 15.937 | 9.319 | 148.509 | 22 |
| 23 | 1.974 | .5067 | .0308 | .0608 | 32.453 | 16.444 | 9.709 | 159.656 | 23 |
| 24 | 2.033 | .4919 | .0290 | .0590 | 34.426 | 16.936 | 10.095 | 170.971 | 24 |
| 25 | 2.094 | .4776 | .0274 | .0574 | 36.459 | 17.413 | 10.477 | 182.433 | 25 |
| 26 | 2.157 | .4637 | .0259 | .0559 | 38.553 | 17.877 | 10.853 | 194.026 | 26 |
| 27 | 2.221 | .4502 | .0246 | .0546 | 40.710 | 18.327 | 11.226 | 205.731 | 27 |
| 28 | 2.288 | .4371 | .0233 | .0533 | 42.931 | 18.764 | 11.593 | 217.532 | 28 |
| 29 | 2.357 | .4243 | .0221 | .0521 | 45.219 | 19.188 | 11.956 | 229.413 | 29 |
| 30 | 2.427 | .4120 | .0210 | .0510 | 47.575 | 19.600 | 12.314 | 241.361 | 30 |
| 31 | 2.500 | .4000 | .0200 | .0500 | 50.003 | 20.000 | 12.668 | 253.361 | 31 |
| 32 | 2.575 | .3883 | .0190 | .0490 | 52.503 | 20.389 | 13.017 | 265.399 | 32 |
| 33 | 2.652 | .3770 | .0182 | .0482 | 55.078 | 20.766 | 13.362 | 277.464 | 33 |
| 34 | 2.732 | .3660 | .0173 | .0473 | 57.730 | 21.132 | 13.702 | 289.544 | 34 |
| 35 | 2.814 | .3554 | .0165 | .0465 | 60.462 | 21.487 | 14.037 | 301.627 | 35 |
| 40 | 3.262 | .3066 | .0133 | .0433 | 75.401 | 23.115 | 15.650 | 361.750 | 40 |
| 45 | 3.782 | .2644 | .0108 | .0408 | 92.720 | 24.519 | 17.156 | 420.632 | 45 |
| 50 | 4.384 | .2281 | .00887 | .0389 | 112.797 | 25.730 | 18.558 | 477.480 | 50 |
| 55 | 5.082 | .1968 | .00735 | .0373 | 136.072 | 26.774 | 19.860 | 531.741 | 55 |
| 60 | 5.892 | .1697 | .00613 | .0361 | 163.053 | 27.676 | 21.067 | 583.052 | 60 |
| 65 | 6.830 | .1464 | .00515 | .0351 | 194.333 | 28.453 | 22.184 | 631.201 | 65 |
| 70 | 7.918 | .1263 | .00434 | .0343 | 230.594 | 29.123 | 23.215 | 676.087 | 70 |
| 75 | 9.179 | .1089 | .00367 | .0337 | 272.631 | 29.702 | 24.163 | 717.698 | 75 |
| 80 | 10.641 | .0940 | .00311 | .0331 | 321.363 | 30.201 | 25.035 | 756.086 | 80 |
| 85 | 12.336 | .0811 | .00265 | .0326 | 377.857 | 30.631 | 25.835 | 791.353 | 85 |
| 90 | 14.300 | .0699 | .00226 | .0323 | 443.349 | 31.002 | 26.567 | 823.630 | 90 |
| 95 | 16.578 | .0603 | .00193 | .0319 | 519.272 | 31.323 | 27.235 | 853.074 | 95 |
| 100 | 19.219 | .0520 | .00165 | .0316 | 607.287 | 31.599 | 27.844 | 879.854 | 100 |

| 3½% | | | | Compound Interest Factors | | | | | 3½% |
|---|---|---|---|---|---|---|---|---|---|
| | Single Payment | | Uniform Payment Series | | | | Arithmetic Gradient | | |
| | Compound Amount Factor Find F Given P F/P | Present Worth Factor Find P Given F P/F | Sinking Fund Factor Find A Given F A/F | Capital Recovery Factor Find A Given P A/P | Compound Amount Factor Find F Given A F/A | Present Worth Factor Find P Given A P/A | Gradient Uniform Series Find A Given G A/G | Gradient Present Worth Find P Given G P/G | |
| n | | | | | | | | | n |
| 1 | 1.035 | .9662 | 1.0000 | 1.0350 | 1.000 | 0.966 | 0 | 0 | 1 |
| 2 | 1.071 | .9335 | .4914 | .5264 | 2.035 | 1.900 | 0.491 | 0.933 | 2 |
| 3 | 1.109 | .9019 | .3219 | .3569 | 3.106 | 2.802 | 0.977 | 2.737 | 3 |
| 4 | 1.148 | .8714 | .2373 | .2723 | 4.215 | 3.673 | 1.457 | 5.352 | 4 |
| 5 | 1.188 | .8420 | .1865 | .2215 | 5.362 | 4.515 | 1.931 | 8.719 | 5 |
| 6 | 1.229 | .8135 | .1527 | .1877 | 6.550 | 5.329 | 2.400 | 12.787 | 6 |
| 7 | 1.272 | .7860 | .1285 | .1635 | 7.779 | 6.115 | 2.862 | 17.503 | 7 |
| 8 | 1.317 | .7594 | .1105 | .1455 | 9.052 | 6.874 | 3.320 | 22.819 | 8 |
| 9 | 1.363 | .7337 | .0964 | .1314 | 10.368 | 7.608 | 3.771 | 28.688 | 9 |
| 10 | 1.411 | .7089 | .0852 | .1202 | 11.731 | 8.317 | 4.217 | 35.069 | 10 |
| 11 | 1.460 | .6849 | .0761 | .1111 | 13.142 | 9.002 | 4.657 | 41.918 | 11 |
| 12 | 1.511 | .6618 | .0685 | .1035 | 14.602 | 9.663 | 5.091 | 49.198 | 12 |
| 13 | 1.564 | .6394 | .0621 | .0971 | 16.113 | 10.303 | 5.520 | 56.871 | 13 |
| 14 | 1.619 | .6178 | .0566 | .0916 | 17.677 | 10.921 | 5.943 | 64.902 | 14 |
| 15 | 1.675 | .5969 | .0518 | .0868 | 19.296 | 11.517 | 6.361 | 73.258 | 15 |
| 16 | 1.734 | .5767 | .0477 | .0827 | 20.971 | 12.094 | 6.773 | 81.909 | 16 |
| 17 | 1.795 | .5572 | .0440 | .0790 | 22.705 | 12.651 | 7.179 | 90.824 | 17 |
| 18 | 1.857 | .5384 | .0408 | .0758 | 24.500 | 13.190 | 7.580 | 99.976 | 18 |
| 19 | 1.922 | .5202 | .0379 | .0729 | 26.357 | 13.710 | 7.975 | 109.339 | 19 |
| 20 | 1.990 | .5026 | .0354 | .0704 | 28.280 | 14.212 | 8.365 | 118.888 | 20 |
| 21 | 2.059 | .4856 | .0330 | .0680 | 30.269 | 14.698 | 8.749 | 128.599 | 21 |
| 22 | 2.132 | .4692 | .0309 | .0659 | 32.329 | 15.167 | 9.128 | 138.451 | 22 |
| 23 | 2.206 | .4533 | .0290 | .0640 | 34.460 | 15.620 | 9.502 | 148.423 | 23 |
| 24 | 2.283 | .4380 | .0273 | .0623 | 36.666 | 16.058 | 9.870 | 158.496 | 24 |
| 25 | 2.363 | .4231 | .0257 | .0607 | 38.950 | 16.482 | 10.233 | 168.652 | 25 |
| 26 | 2.446 | .4088 | .0242 | .0592 | 41.313 | 16.890 | 10.590 | 178.873 | 26 |
| 27 | 2.532 | .3950 | .0229 | .0579 | 43.759 | 17.285 | 10.942 | 189.143 | 27 |
| 28 | 2.620 | .3817 | .0216 | .0566 | 46.291 | 17.667 | 11.289 | 199.448 | 28 |
| 29 | 2.712 | .3687 | .0204 | .0554 | 48.911 | 18.036 | 11.631 | 209.773 | 29 |
| 30 | 2.807 | .3563 | .0194 | .0544 | 51.623 | 18.392 | 11.967 | 220.105 | 30 |
| 31 | 2.905 | .3442 | .0184 | .0534 | 54.429 | 18.736 | 12.299 | 230.432 | 31 |
| 32 | 3.007 | .3326 | .0174 | .0524 | 57.334 | 19.069 | 12.625 | 240.742 | 32 |
| 33 | 3.112 | .3213 | .0166 | .0516 | 60.341 | 19.390 | 12.946 | 251.025 | 33 |
| 34 | 3.221 | .3105 | .0158 | .0508 | 63.453 | 19.701 | 13.262 | 261.271 | 34 |
| 35 | 3.334 | .3000 | .0150 | .0500 | 66.674 | 20.001 | 13.573 | 271.470 | 35 |
| 40 | 3.959 | .2526 | .0118 | .0468 | 84.550 | 21.355 | 15.055 | 321.490 | 40 |
| 45 | 4.702 | .2127 | .00945 | .0445 | 105.781 | 22.495 | 16.417 | 369.307 | 45 |
| 50 | 5.585 | .1791 | .00763 | .0426 | 130.998 | 23.456 | 17.666 | 414.369 | 50 |
| 55 | 6.633 | .1508 | .00621 | .0412 | 160.946 | 24.264 | 18.808 | 456.352 | 55 |
| 60 | 7.878 | .1269 | .00509 | .0401 | 196.516 | 24.945 | 19.848 | 495.104 | 60 |
| 65 | 9.357 | .1069 | .00419 | .0392 | 238.762 | 25.518 | 20.793 | 530.598 | 65 |
| 70 | 11.113 | .0900 | .00346 | .0385 | 288.937 | 26.000 | 21.650 | 562.895 | 70 |
| 75 | 13.199 | .0758 | .00287 | .0379 | 348.529 | 26.407 | 22.423 | 592.121 | 75 |
| 80 | 15.676 | .0638 | .00238 | .0374 | 419.305 | 26.749 | 23.120 | 618.438 | 80 |
| 85 | 18.618 | .0537 | .00199 | .0370 | 503.365 | 27.037 | 23.747 | 642.036 | 85 |
| 90 | 22.112 | .0452 | .00166 | .0367 | 603.202 | 27.279 | 24.308 | 663.118 | 90 |
| 95 | 26.262 | .0381 | .00139 | .0364 | 721.778 | 27.483 | 24.811 | 681.890 | 95 |
| 100 | 31.191 | .0321 | .00116 | .0362 | 862.608 | 27.655 | 25.259 | 698.554 | 100 |

**4%**                                    **Compound Interest Factors**                               **4%**

| | Single Payment | | Uniform Payment Series | | | | Arithmetic Gradient | | |
| | Compound Amount Factor Find $F$ Given $P$ $F/P$ | Present Worth Factor Find $P$ Given $F$ $P/F$ | Sinking Fund Factor Find $A$ Given $F$ $A/F$ | Capital Recovery Factor Find $A$ Given $P$ $A/P$ | Compound Amount Factor Find $F$ Given $A$ $F/A$ | Present Worth Factor Find $P$ Given $A$ $P/A$ | Gradient Uniform Series Find $A$ Given $G$ $A/G$ | Gradient Present Worth Find $P$ Given $G$ $P/G$ | |
| $n$ | | | | | | | | | $n$ |
|---|---|---|---|---|---|---|---|---|---|
| 1 | 1.040 | .9615 | 1.0000 | 1.0400 | 1.000 | 0.962 | 0 | 0 | 1 |
| 2 | 1.082 | .9246 | .4902 | .5302 | 2.040 | 1.886 | 0.490 | 0.925 | 2 |
| 3 | 1.125 | .8890 | .3203 | .3603 | 3.122 | 2.775 | 0.974 | 2.702 | 3 |
| 4 | 1.170 | .8548 | .2355 | .2755 | 4.246 | 3.630 | 1.451 | 5.267 | 4 |
| 5 | 1.217 | .8219 | .1846 | .2246 | 5.416 | 4.452 | 1.922 | 8.555 | 5 |
| 6 | 1.265 | .7903 | .1508 | .1908 | 6.633 | 5.242 | 2.386 | 12.506 | 6 |
| 7 | 1.316 | .7599 | .1266 | .1666 | 7.898 | 6.002 | 2.843 | 17.066 | 7 |
| 8 | 1.369 | .7307 | .1085 | .1485 | 9.214 | 6.733 | 3.294 | 22.180 | 8 |
| 9 | 1.423 | .7026 | .0945 | .1345 | 10.583 | 7.435 | 3.739 | 27.801 | 9 |
| 10 | 1.480 | .6756 | .0833 | .1233 | 12.006 | 8.111 | 4.177 | 33.881 | 10 |
| 11 | 1.539 | .6496 | .0741 | .1141 | 13.486 | 8.760 | 4.609 | 40.377 | 11 |
| 12 | 1.601 | .6246 | .0666 | .1066 | 15.026 | 9.385 | 5.034 | 47.248 | 12 |
| 13 | 1.665 | .6006 | .0601 | .1001 | 16.627 | 9.986 | 5.453 | 54.454 | 13 |
| 14 | 1.732 | .5775 | .0547 | .0947 | 18.292 | 10.563 | 5.866 | 61.962 | 14 |
| 15 | 1.801 | .5553 | .0499 | .0899 | 20.024 | 11.118 | 6.272 | 69.735 | 15 |
| 16 | 1.873 | .5339 | .0458 | .0858 | 21.825 | 11.652 | 6.672 | 77.744 | 16 |
| 17 | 1.948 | .5134 | .0422 | .0822 | 23.697 | 12.166 | 7.066 | 85.958 | 17 |
| 18 | 2.026 | .4936 | .0390 | .0790 | 25.645 | 12.659 | 7.453 | 94.350 | 18 |
| 19 | 2.107 | .4746 | .0361 | .0761 | 27.671 | 13.134 | 7.834 | 102.893 | 19 |
| 20 | 2.191 | .4564 | .0336 | .0736 | 29.778 | 13.590 | 8.209 | 111.564 | 20 |
| 21 | 2.279 | .4388 | .0313 | .0713 | 31.969 | 14.029 | 8.578 | 120.341 | 21 |
| 22 | 2.370 | .4220 | .0292 | .0692 | 34.248 | 14.451 | 8.941 | 129.202 | 22 |
| 23 | 2.465 | .4057 | .0273 | .0673 | 36.618 | 14.857 | 9.297 | 138.128 | 23 |
| 24 | 2.563 | .3901 | .0256 | .0656 | 39.083 | 15.247 | 9.648 | 147.101 | 24 |
| 25 | 2.666 | .3751 | .0240 | .0640 | 41.646 | 15.622 | 9.993 | 156.104 | 25 |
| 26 | 2.772 | .3607 | .0226 | .0626 | 44.312 | 15.983 | 10.331 | 165.121 | 26 |
| 27 | 2.883 | .3468 | .0212 | .0612 | 47.084 | 16.330 | 10.664 | 174.138 | 27 |
| 28 | 2.999 | .3335 | .0200 | .0600 | 49.968 | 16.663 | 10.991 | 183.142 | 28 |
| 29 | 3.119 | .3207 | .0189 | .0589 | 52.966 | 16.984 | 11.312 | 192.120 | 29 |
| 30 | 3.243 | .3083 | .0178 | .0578 | 56.085 | 17.292 | 11.627 | 201.062 | 30 |
| 31 | 3.373 | .2965 | .0169 | .0569 | 59.328 | 17.588 | 11.937 | 209.955 | 31 |
| 32 | 3.508 | .2851 | .0159 | .0559 | 62.701 | 17.874 | 12.241 | 218.792 | 32 |
| 33 | 3.648 | .2741 | .0151 | .0551 | 66.209 | 18.148 | 12.540 | 227.563 | 33 |
| 34 | 3.794 | .2636 | .0143 | .0543 | 69.858 | 18.411 | 12.832 | 236.260 | 34 |
| 35 | 3.946 | .2534 | .0136 | .0536 | 73.652 | 18.665 | 13.120 | 244.876 | 35 |
| 40 | 4.801 | .2083 | .0105 | .0505 | 95.025 | 19.793 | 14.476 | 286.530 | 40 |
| 45 | 5.841 | .1712 | .00826 | .0483 | 121.029 | 20.720 | 15.705 | 325.402 | 45 |
| 50 | 7.107 | .1407 | .00655 | .0466 | 152.667 | 21.482 | 16.812 | 361.163 | 50 |
| 55 | 8.646 | .1157 | .00523 | .0452 | 191.159 | 22.109 | 17.807 | 393.689 | 55 |
| 60 | 10.520 | .0951 | .00420 | .0442 | 237.990 | 22.623 | 18.697 | 422.996 | 60 |
| 65 | 12.799 | .0781 | .00339 | .0434 | 294.968 | 23.047 | 19.491 | 449.201 | 65 |
| 70 | 15.572 | .0642 | .00275 | .0427 | 364.290 | 23.395 | 20.196 | 472.479 | 70 |
| 75 | 18.945 | .0528 | .00223 | .0422 | 448.630 | 23.680 | 20.821 | 493.041 | 75 |
| 80 | 23.050 | .0434 | .00181 | .0418 | 551.244 | 23.915 | 21.372 | 511.116 | 80 |
| 85 | 28.044 | .0357 | .00148 | .0415 | 676.089 | 24.109 | 21.857 | 526.938 | 85 |
| 90 | 34.119 | .0293 | .00121 | .0412 | 827.981 | 24.267 | 22.283 | 540.737 | 90 |
| 95 | 41.511 | .0241 | .00099 | .0410 | 1012.8 | 24.398 | 22.655 | 552.730 | 95 |
| 100 | 50.505 | .0198 | .00081 | .0408 | 1237.6 | 24.505 | 22.980 | 563.125 | 100 |

**4¹/₂%**  Compound Interest Factors  **4¹/₂%**

| | Single Payment | | Uniform Payment Series | | | | Arithmetic Gradient | | |
|---|---|---|---|---|---|---|---|---|---|
| | Compound Amount Factor Find F Given P F/P | Present Worth Factor Find P Given F P/F | Sinking Fund Factor Find A Given F A/F | Capital Recovery Factor Find A Given P A/P | Compound Amount Factor Find F Given A F/A | Present Worth Factor Find P Given A P/A | Gradient Uniform Series Find A Given G A/G | Gradient Present Worth Find P Given G P/G | |
| n | | | | | | | | | n |
| 1 | 1.045 | .9569 | 1.0000 | 1.0450 | 1.000 | 0.957 | 0 | 0 | 1 |
| 2 | 1.092 | .9157 | .4890 | .5340 | 2.045 | 1.873 | 0.489 | 0.916 | 2 |
| 3 | 1.141 | .8763 | .3188 | .3638 | 3.137 | 2.749 | 0.971 | 2.668 | 3 |
| 4 | 1.193 | .8386 | .2337 | .2787 | 4.278 | 3.588 | 1.445 | 5.184 | 4 |
| 5 | 1.246 | .8025 | .1828 | .2278 | 5.471 | 4.390 | 1.912 | 8.394 | 5 |
| 6 | 1.302 | .7679 | .1489 | .1939 | 6.717 | 5.158 | 2.372 | 12.233 | 6 |
| 7 | 1.361 | .7348 | .1247 | .1697 | 8.019 | 5.893 | 2.824 | 16.642 | 7 |
| 8 | 1.422 | .7032 | .1066 | .1516 | 9.380 | 6.596 | 3.269 | 21.564 | 8 |
| 9 | 1.486 | .6729 | .0926 | .1376 | 10.802 | 7.269 | 3.707 | 26.948 | 9 |
| 10 | 1.553 | .6439 | .0814 | .1264 | 12.288 | 7.913 | 4.138 | 32.743 | 10 |
| 11 | 1.623 | .6162 | .0722 | .1172 | 13.841 | 8.529 | 4.562 | 38.905 | 11 |
| 12 | 1.696 | .5897 | .0647 | .1097 | 15.464 | 9.119 | 4.978 | 45.391 | 12 |
| 13 | 1.772 | .5643 | .0583 | .1033 | 17.160 | 9.683 | 5.387 | 52.163 | 13 |
| 14 | 1.852 | .5400 | .0528 | .0978 | 18.932 | 10.223 | 5.789 | 59.182 | 14 |
| 15 | 1.935 | .5167 | .0481 | .0931 | 20.784 | 10.740 | 6.184 | 66.416 | 15 |
| 16 | 2.022 | .4945 | .0440 | .0890 | 22.719 | 11.234 | 6.572 | 73.833 | 16 |
| 17 | 2.113 | .4732 | .0404 | .0854 | 24.742 | 11.707 | 6.953 | 81.404 | 17 |
| 18 | 2.208 | .4528 | .0372 | .0822 | 26.855 | 12.160 | 7.327 | 89.102 | 18 |
| 19 | 2.308 | .4333 | .0344 | .0794 | 29.064 | 12.593 | 7.695 | 96.901 | 19 |
| 20 | 2.412 | .4146 | .0319 | .0769 | 31.371 | 13.008 | 8.055 | 104.779 | 20 |
| 21 | 2.520 | .3968 | .0296 | .0746 | 33.783 | 13.405 | 8.409 | 112.715 | 21 |
| 22 | 2.634 | .3797 | .0275 | .0725 | 36.303 | 13.784 | 8.755 | 120.689 | 22 |
| 23 | 2.752 | .3634 | .0257 | .0707 | 38.937 | 14.148 | 9.096 | 128.682 | 23 |
| 24 | 2.876 | .3477 | .0240 | .0690 | 41.689 | 14.495 | 9.429 | 136.680 | 24 |
| 25 | 3.005 | .3327 | .0224 | .0674 | 44.565 | 14.828 | 9.756 | 144.665 | 25 |
| 26 | 3.141 | .3184 | .0210 | .0660 | 47.571 | 15.147 | 10.077 | 152.625 | 26 |
| 27 | 3.282 | .3047 | .0197 | .0647 | 50.711 | 15.451 | 10.391 | 160.547 | 27 |
| 28 | 3.430 | .2916 | .0185 | .0635 | 53.993 | 15.743 | 10.698 | 168.420 | 28 |
| 29 | 3.584 | .2790 | .0174 | .0624 | 57.423 | 16.022 | 10.999 | 176.232 | 29 |
| 30 | 3.745 | .2670 | .0164 | .0614 | 61.007 | 16.289 | 11.295 | 183.975 | 30 |
| 31 | 3.914 | .2555 | .0154 | .0604 | 64.752 | 16.544 | 11.583 | 191.640 | 31 |
| 32 | 4.090 | .2445 | .0146 | .0596 | 68.666 | 16.789 | 11.866 | 199.220 | 32 |
| 33 | 4.274 | .2340 | .0137 | .0587 | 72.756 | 17.023 | 12.143 | 206.707 | 33 |
| 34 | 4.466 | .2239 | .0130 | .0580 | 77.030 | 17.247 | 12.414 | 214.095 | 34 |
| 35 | 4.667 | .2143 | .0123 | .0573 | 81.497 | 17.461 | 12.679 | 221.380 | 35 |
| 40 | 5.816 | .1719 | .00934 | .0543 | 107.030 | 18.402 | 13.917 | 256.098 | 40 |
| 45 | 7.248 | .1380 | .00720 | .0522 | 138.850 | 19.156 | 15.020 | 287.732 | 45 |
| 50 | 9.033 | .1107 | .00560 | .0506 | 178.503 | 19.762 | 15.998 | 316.145 | 50 |
| 55 | 11.256 | .0888 | .00439 | .0494 | 227.918 | 20.248 | 16.860 | 341.375 | 55 |
| 60 | 14.027 | .0713 | .00345 | .0485 | 289.497 | 20.638 | 17.617 | 363.571 | 60 |
| 65 | 17.481 | .0572 | .00273 | .0477 | 366.237 | 20.951 | 18.278 | 382.946 | 65 |
| 70 | 21.784 | .0459 | .00217 | .0472 | 461.869 | 21.202 | 18.854 | 399.750 | 70 |
| 75 | 27.147 | .0368 | .00172 | .0467 | 581.043 | 21.404 | 19.354 | 414.242 | 75 |
| 80 | 33.830 | .0296 | .00137 | .0464 | 729.556 | 21.565 | 19.785 | 426.680 | 80 |
| 85 | 42.158 | .0237 | .00109 | .0461 | 914.630 | 21.695 | 20.157 | 437.309 | 85 |
| 90 | 52.537 | .0190 | .00087 | .0459 | 1 145.3 | 21.799 | 20.476 | 446.359 | 90 |
| 95 | 65.471 | .0153 | .00070 | .0457 | 1 432.7 | 21.883 | 20.749 | 454.039 | 95 |
| 100 | 81.588 | .0123 | .00056 | .0456 | 1 790.9 | 21.950 | 20.981 | 460.537 | 100 |

| 5% | | | | Compound Interest Factors | | | | | | 5% |
|---|---|---|---|---|---|---|---|---|---|---|
| | Single Payment | | Uniform Payment Series | | | | Arithmetic Gradient | | | |
| | Compound Amount Factor Find F Given P F/P | Present Worth Factor Find P Given F P/F | Sinking Fund Factor Find A Given F A/F | Capital Recovery Factor Find A Given P A/P | Compound Amount Factor Find F Given A F/A | Present Worth Factor Find P Given A P/A | Gradient Uniform Series Find A Given G A/G | Gradient Present Worth Find P Given G P/G | | |
| n | F/P | P/F | A/F | A/P | F/A | P/A | A/G | P/G | n | |
| 1 | 1.050 | .9524 | 1.0000 | 1.0500 | 1.000 | 0.952 | 0 | 0 | 1 | |
| 2 | 1.102 | .9070 | .4878 | .5378 | 2.050 | 1.859 | 0.488 | 0.907 | 2 | |
| 3 | 1.158 | .8638 | .3172 | .3672 | 3.152 | 2.723 | 0.967 | 2.635 | 3 | |
| 4 | 1.216 | .8227 | .2320 | .2820 | 4.310 | 3.546 | 1.439 | 5.103 | 4 | |
| 5 | 1.276 | .7835 | .1810 | .2310 | 5.526 | 4.329 | 1.902 | 8.237 | 5 | |
| 6 | 1.340 | .7462 | .1470 | .1970 | 6.802 | 5.076 | 2.358 | 11.968 | 6 | |
| 7 | 1.407 | .7107 | .1228 | .1728 | 8.142 | 5.786 | 2.805 | 16.232 | 7 | |
| 8 | 1.477 | .6768 | .1047 | .1547 | 9.549 | 6.463 | 3.244 | 20.970 | 8 | |
| 9 | 1.551 | .6446 | .0907 | .1407 | 11.027 | 7.108 | 3.676 | 26.127 | 9 | |
| 10 | 1.629 | .6139 | .0795 | .1295 | 12.578 | 7.722 | 4.099 | 31.652 | 10 | |
| 11 | 1.710 | .5847 | .0704 | .1204 | 14.207 | 8.306 | 4.514 | 37.499 | 11 | |
| 12 | 1.796 | .5568 | .0628 | .1128 | 15.917 | 8.863 | 4.922 | 43.624 | 12 | |
| 13 | 1.886 | .5303 | .0565 | .1065 | 17.713 | 9.394 | 5.321 | 49.988 | 13 | |
| 14 | 1.980 | .5051 | .0510 | .1010 | 19.599 | 9.899 | 5.713 | 56.553 | 14 | |
| 15 | 2.079 | .4810 | .0463 | .0963 | 21.579 | 10.380 | 6.097 | 63.288 | 15 | |
| 16 | 2.183 | .4581 | .0423 | .0923 | 23.657 | 10.838 | 6.474 | 70.159 | 16 | |
| 17 | 2.292 | .4363 | .0387 | .0887 | 25.840 | 11.274 | 6.842 | 77.140 | 17 | |
| 18 | 2.407 | .4155 | .0355 | .0855 | 28.132 | 11.690 | 7.203 | 84.204 | 18 | |
| 19 | 2.527 | .3957 | .0327 | .0827 | 30.539 | 12.085 | 7.557 | 91.327 | 19 | |
| 20 | 2.653 | .3769 | .0302 | .0802 | 33.066 | 12.462 | 7.903 | 98.488 | 20 | |
| 21 | 2.786 | .3589 | .0280 | .0780 | 35.719 | 12.821 | 8.242 | 105.667 | 21 | |
| 22 | 2.925 | .3419 | .0260 | .0760 | 38.505 | 13.163 | 8.573 | 112.846 | 22 | |
| 23 | 3.072 | .3256 | .0241 | .0741 | 41.430 | 13.489 | 8.897 | 120.008 | 23 | |
| 24 | 3.225 | .3101 | .0225 | .0725 | 44.502 | 13.799 | 9.214 | 127.140 | 24 | |
| 25 | 3.386 | .2953 | .0210 | .0710 | 47.727 | 14.094 | 9.524 | 134.227 | 25 | |
| 26 | 3.556 | .2812 | .0196 | .0696 | 51.113 | 14.375 | 9.827 | 141.258 | 26 | |
| 27 | 3.733 | .2678 | .0183 | .0683 | 54.669 | 14.643 | 10.122 | 148.222 | 27 | |
| 28 | 3.920 | .2551 | .0171 | .0671 | 58.402 | 14.898 | 10.411 | 155.110 | 28 | |
| 29 | 4.116 | .2429 | .0160 | .0660 | 62.323 | 15.141 | 10.694 | 161.912 | 29 | |
| 30 | 4.322 | .2314 | .0151 | .0651 | 66.439 | 15.372 | 10.969 | 168.622 | 30 | |
| 31 | 4.538 | .2204 | .0141 | .0641 | 70.761 | 15.593 | 11.238 | 175.233 | 31 | |
| 32 | 4.765 | .2099 | .0133 | .0633 | 75.299 | 15.803 | 11.501 | 181.739 | 32 | |
| 33 | 5.003 | .1999 | .0125 | .0625 | 80.063 | 16.003 | 11.757 | 188.135 | 33 | |
| 34 | 5.253 | .1904 | .0118 | .0618 | 85.067 | 16.193 | 12.006 | 194.416 | 34 | |
| 35 | 5.516 | .1813 | .0111 | .0611 | 90.320 | 16.374 | 12.250 | 200.580 | 35 | |
| 40 | 7.040 | .1420 | .00828 | .0583 | 120.799 | 17.159 | 13.377 | 229.545 | 40 | |
| 45 | 8.985 | .1113 | .00626 | .0563 | 159.699 | 17.774 | 14.364 | 255.314 | 45 | |
| 50 | 11.467 | .0872 | .00478 | .0548 | 209.347 | 18.256 | 15.223 | 277.914 | 50 | |
| 55 | 14.636 | .0683 | .00367 | .0537 | 272.711 | 18.633 | 15.966 | 297.510 | 55 | |
| 60 | 18.679 | .0535 | .00283 | .0528 | 353.582 | 18.929 | 16.606 | 314.343 | 60 | |
| 65 | 23.840 | .0419 | .00219 | .0522 | 456.795 | 19.161 | 17.154 | 328.691 | 65 | |
| 70 | 30.426 | .0329 | .00170 | .0517 | 588.525 | 19.343 | 17.621 | 340.841 | 70 | |
| 75 | 38.832 | .0258 | .00132 | .0513 | 756.649 | 19.485 | 18.018 | 351.072 | 75 | |
| 80 | 49.561 | .0202 | .00103 | .0510 | 971.222 | 19.596 | 18.353 | 359.646 | 80 | |
| 85 | 63.254 | .0158 | .00080 | .0508 | 1 245.1 | 19.684 | 18.635 | 366.800 | 85 | |
| 90 | 80.730 | .0124 | .00063 | .0506 | 1 594.6 | 19.752 | 18.871 | 372.749 | 90 | |
| 95 | 103.034 | .00971 | .00049 | .0505 | 2 040.7 | 19.806 | 19.069 | 377.677 | 95 | |
| 100 | 131.500 | .00760 | .00038 | .0504 | 2 610.0 | 19.848 | 19.234 | 381.749 | 100 | |

| 6% | | | | Compound Interest Factors | | | | | | 6% |
|---|---|---|---|---|---|---|---|---|---|---|
| | Single Payment | | Uniform Payment Series | | | | Arithmetic Gradient | | | |
| | Compound Amount Factor Find F Given P | Present Worth Factor Find P Given F | Sinking Fund Factor Find A Given F | Capital Recovery Factor Find A Given P | Compound Amount Factor Find F Given A | Present Worth Factor Find P Given A | Gradient Uniform Series Find A Given G | Gradient Present Worth Find P Given G | |
| n | F/P | P/F | A/F | A/P | F/A | P/A | A/G | P/G | n |
| 1 | 1.060 | .9434 | 1.0000 | 1.0600 | 1.000 | 0.943 | 0 | 0 | 1 |
| 2 | 1.124 | .8900 | .4854 | .5454 | 2.060 | 1.833 | 0.485 | 0.890 | 2 |
| 3 | 1.191 | .8396 | .3141 | .3741 | 3.184 | 2.673 | 0.961 | 2.569 | 3 |
| 4 | 1.262 | .7921 | .2286 | .2886 | 4.375 | 3.465 | 1.427 | 4.945 | 4 |
| 5 | 1.338 | .7473 | .1774 | .2374 | 5.637 | 4.212 | 1.884 | 7.934 | 5 |
| 6 | 1.419 | .7050 | .1434 | .2034 | 6.975 | 4.917 | 2.330 | 11.459 | 6 |
| 7 | 1.504 | .6651 | .1191 | .1791 | 8.394 | 5.582 | 2.768 | 15.450 | 7 |
| 8 | 1.594 | .6274 | .1010 | .1610 | 9.897 | 6.210 | 3.195 | 19.841 | 8 |
| 9 | 1.689 | .5919 | .0870 | .1470 | 11.491 | 6.802 | 3.613 | 24.577 | 9 |
| 10 | 1.791 | .5584 | .0759 | .1359 | 13.181 | 7.360 | 4.022 | 29.602 | 10 |
| 11 | 1.898 | .5268 | .0668 | .1268 | 14.972 | 7.887 | 4.421 | 34.870 | 11 |
| 12 | 2.012 | .4970 | .0593 | .1193 | 16.870 | 8.384 | 4.811 | 40.337 | 12 |
| 13 | 2.133 | .4688 | .0530 | .1130 | 18.882 | 8.853 | 5.192 | 45.963 | 13 |
| 14 | 2.261 | .4423 | .0476 | .1076 | 21.015 | 9.295 | 5.564 | 51.713 | 14 |
| 15 | 2.397 | .4173 | .0430 | .1030 | 23.276 | 9.712 | 5.926 | 57.554 | 15 |
| 16 | 2.540 | .3936 | .0390 | .0990 | 25.672 | 10.106 | 6.279 | 63.459 | 16 |
| 17 | 2.693 | .3714 | .0354 | .0954 | 28.213 | 10.477 | 6.624 | 69.401 | 17 |
| 18 | 2.854 | .3503 | .0324 | .0924 | 30.906 | 10.828 | 6.960 | 75.357 | 18 |
| 19 | 3.026 | .3305 | .0296 | .0896 | 33.760 | 11.158 | 7.287 | 81.306 | 19 |
| 20 | 3.207 | .3118 | .0272 | .0872 | 36.786 | 11.470 | 7.605 | 87.230 | 20 |
| 21 | 3.400 | .2942 | .0250 | .0850 | 39.993 | 11.764 | 7.915 | 93.113 | 21 |
| 22 | 3.604 | .2775 | .0230 | .0830 | 43.392 | 12.042 | 8.217 | 98.941 | 22 |
| 23 | 3.820 | .2618 | .0213 | .0813 | 46.996 | 12.303 | 8.510 | 104.700 | 23 |
| 24 | 4.049 | .2470 | .0197 | .0797 | 50.815 | 12.550 | 8.795 | 110.381 | 24 |
| 25 | 4.292 | .2330 | .0182 | .0782 | 54.864 | 12.783 | 9.072 | 115.973 | 25 |
| 26 | 4.549 | .2198 | .0169 | .0769 | 59.156 | 13.003 | 9.341 | 121.468 | 26 |
| 27 | 4.822 | .2074 | .0157 | .0757 | 63.706 | 13.211 | 9.603 | 126.860 | 27 |
| 28 | 5.112 | .1956 | .0146 | .0746 | 68.528 | 13.406 | 9.857 | 132.142 | 28 |
| 29 | 5.418 | .1846 | .0136 | .0736 | 73.640 | 13.591 | 10.103 | 137.309 | 29 |
| 30 | 5.743 | .1741 | .0126 | .0726 | 79.058 | 13.765 | 10.342 | 142.359 | 30 |
| 31 | 6.088 | .1643 | .0118 | .0718 | 84.801 | 13.929 | 10.574 | 147.286 | 31 |
| 32 | 6.453 | .1550 | .0110 | .0710 | 90.890 | 14.084 | 10.799 | 152.090 | 32 |
| 33 | 6.841 | .1462 | .0103 | .0703 | 97.343 | 14.230 | 11.017 | 156.768 | 33 |
| 34 | 7.251 | .1379 | .00960 | .0696 | 104.184 | 14.368 | 11.228 | 161.319 | 34 |
| 35 | 7.686 | .1301 | .00897 | .0690 | 111.435 | 14.498 | 11.432 | 165.743 | 35 |
| 40 | 10.286 | .0972 | .00646 | .0665 | 154.762 | 15.046 | 12.359 | 185.957 | 40 |
| 45 | 13.765 | .0727 | .00470 | .0647 | 212.743 | 15.456 | 13.141 | 203.109 | 45 |
| 50 | 18.420 | .0543 | .00344 | .0634 | 290.335 | 15.762 | 13.796 | 217.457 | 50 |
| 55 | 24.650 | .0406 | .00254 | .0625 | 394.171 | 15.991 | 14.341 | 229.322 | 55 |
| 60 | 32.988 | .0303 | .00188 | .0619 | 533.126 | 16.161 | 14.791 | 239.043 | 60 |
| 65 | 44.145 | .0227 | .00139 | .0614 | 719.080 | 16.289 | 15.160 | 246.945 | 65 |
| 70 | 59.076 | .0169 | .00103 | .0610 | 967.928 | 16.385 | 15.461 | 253.327 | 70 |
| 75 | 79.057 | .0126 | .00077 | .0608 | 1 300.9 | 16.456 | 15.706 | 258.453 | 75 |
| 80 | 105.796 | .00945 | .00057 | .0606 | 1 746.6 | 16.509 | 15.903 | 262.549 | 80 |
| 85 | 141.578 | .00706 | .00043 | .0604 | 2 343.0 | 16.549 | 16.062 | 265.810 | 85 |
| 90 | 189.464 | .00528 | .00032 | .0603 | 3 141.1 | 16.579 | 16.189 | 268.395 | 90 |
| 95 | 253.545 | .00394 | .00024 | .0602 | 4 209.1 | 16.601 | 16.290 | 270.437 | 95 |
| 100 | 339.300 | .00295 | .00018 | .0602 | 5 638.3 | 16.618 | 16.371 | 272.047 | 100 |

| 7% | | | | Compound Interest Factors | | | | | | 7% |

| | Single Payment | | Uniform Payment Series | | | | Arithmetic Gradient | | | |
|---|---|---|---|---|---|---|---|---|---|---|
| | Compound Amount Factor Find F Given P F/P | Present Worth Factor Find P Given F P/F | Sinking Fund Factor Find A Given F A/F | Capital Recovery Factor Find A Given P A/P | Compound Amount Factor Find F Given A F/A | Present Worth Factor Find P Given A P/A | Gradient Uniform Series Find A Given G A/G | Gradient Present Worth Find P Given G P/G | | |
| n | F/P | P/F | A/F | A/P | F/A | P/A | A/G | P/G | | n |
| 1 | 1.070 | .9346 | 1.0000 | 1.0700 | 1.000 | 0.935 | 0 | 0 | | 1 |
| 2 | 1.145 | .8734 | .4831 | .5531 | 2.070 | 1.808 | 0.483 | 0.873 | | 2 |
| 3 | 1.225 | .8163 | .3111 | .3811 | 3.215 | 2.624 | 0.955 | 2.506 | | 3 |
| 4 | 1.311 | .7629 | .2252 | .2952 | 4.440 | 3.387 | 1.416 | 4.795 | | 4 |
| 5 | 1.403 | .7130 | .1739 | .2439 | 5.751 | 4.100 | 1.865 | 7.647 | | 5 |
| 6 | 1.501 | .6663 | .1398 | .2098 | 7.153 | 4.767 | 2.303 | 10.978 | | 6 |
| 7 | 1.606 | .6227 | .1156 | .1856 | 8.654 | 5.389 | 2.730 | 14.715 | | 7 |
| 8 | 1.718 | .5820 | .0975 | .1675 | 10.260 | 5.971 | 3.147 | 18.789 | | 8 |
| 9 | 1.838 | .5439 | .0835 | .1535 | 11.978 | 6.515 | 3.552 | 23.140 | | 9 |
| 10 | 1.967 | .5083 | .0724 | .1424 | 13.816 | 7.024 | 3.946 | 27.716 | | 10 |
| 11 | 2.105 | .4751 | .0634 | .1334 | 15.784 | 7.499 | 4.330 | 32.467 | | 11 |
| 12 | 2.252 | .4440 | .0559 | .1259 | 17.888 | 7.943 | 4.703 | 37.351 | | 12 |
| 13 | 2.410 | .4150 | .0497 | .1197 | 20.141 | 8.358 | 5.065 | 42.330 | | 13 |
| 14 | 2.579 | .3878 | .0443 | .1143 | 22.551 | 8.745 | 5.417 | 47.372 | | 14 |
| 15 | 2.759 | .3624 | .0398 | .1098 | 25.129 | 9.108 | 5.758 | 52.446 | | 15 |
| 16 | 2.952 | .3387 | .0359 | .1059 | 27.888 | 9.447 | 6.090 | 57.527 | | 16 |
| 17 | 3.159 | .3166 | .0324 | .1024 | 30.840 | 9.763 | 6.411 | 62.592 | | 17 |
| 18 | 3.380 | .2959 | .0294 | .0994 | 33.999 | 10.059 | 6.722 | 67.622 | | 18 |
| 19 | 3.617 | .2765 | .0268 | .0968 | 37.379 | 10.336 | 7.024 | 72.599 | | 19 |
| 20 | 3.870 | .2584 | .0244 | .0944 | 40.996 | 10.594 | 7.316 | 77.509 | | 20 |
| 21 | 4.141 | .2415 | .0223 | .0923 | 44.865 | 10.836 | 7.599 | 82.339 | | 21 |
| 22 | 4.430 | .2257 | .0204 | .0904 | 49.006 | 11.061 | 7.872 | 87.079 | | 22 |
| 23 | 4.741 | .2109 | .0187 | .0887 | 53.436 | 11.272 | 8.137 | 91.720 | | 23 |
| 24 | 5.072 | .1971 | .0172 | .0872 | 58.177 | 11.469 | 8.392 | 96.255 | | 24 |
| 25 | 5.427 | .1842 | .0158 | .0858 | 63.249 | 11.654 | 8.639 | 100.677 | | 25 |
| 26 | 5.807 | .1722 | .0146 | .0846 | 68.677 | 11.826 | 8.877 | 104.981 | | 26 |
| 27 | 6.214 | .1609 | .0134 | .0834 | 74.484 | 11.987 | 9.107 | 109.166 | | 27 |
| 28 | 6.649 | .1504 | .0124 | .0824 | 80.698 | 12.137 | 9.329 | 113.227 | | 28 |
| 29 | 7.114 | .1406 | .0114 | .0814 | 87.347 | 12.278 | 9.543 | 117.162 | | 29 |
| 30 | 7.612 | .1314 | .0106 | .0806 | 94.461 | 12.409 | 9.749 | 120.972 | | 30 |
| 31 | 8.145 | .1228 | .00980 | .0798 | 102.073 | 12.532 | 9.947 | 124.655 | | 31 |
| 32 | 8.715 | .1147 | .00907 | .0791 | 110.218 | 12.647 | 10.138 | 128.212 | | 32 |
| 33 | 9.325 | .1072 | .00841 | .0784 | 118.934 | 12.754 | 10.322 | 131.644 | | 33 |
| 34 | 9.978 | .1002 | .00780 | .0778 | 128.259 | 12.854 | 10.499 | 134.951 | | 34 |
| 35 | 10.677 | .0937 | .00723 | .0772 | 138.237 | 12.948 | 10.669 | 138.135 | | 35 |
| 40 | 14.974 | .0668 | .00501 | .0750 | 199.636 | 13.332 | 11.423 | 152.293 | | 40 |
| 45 | 21.002 | .0476 | .00350 | .0735 | 285.750 | 13.606 | 12.036 | 163.756 | | 45 |
| 50 | 29.457 | .0339 | .00246 | .0725 | 406.530 | 13.801 | 12.529 | 172.905 | | 50 |
| 55 | 41.315 | .0242 | .00174 | .0717 | 575.930 | 13.940 | 12.921 | 180.124 | | 55 |
| 60 | 57.947 | .0173 | .00123 | .0712 | 813.523 | 14.039 | 13.232 | 185.768 | | 60 |
| 65 | 81.273 | .0123 | .00087 | .0709 | 1 146.8 | 14.110 | 13.476 | 190.145 | | 65 |
| 70 | 113.990 | .00877 | .00062 | .0706 | 1 614.1 | 14.160 | 13.666 | 193.519 | | 70 |
| 75 | 159.877 | .00625 | .00044 | .0704 | 2 269.7 | 14.196 | 13.814 | 196.104 | | 75 |
| 80 | 224.235 | .00446 | .00031 | .0703 | 3 189.1 | 14.222 | 13.927 | 198.075 | | 80 |
| 85 | 314.502 | .00318 | .00022 | .0702 | 4 478.6 | 14.240 | 14.015 | 199.572 | | 85 |
| 90 | 441.105 | .00227 | .00016 | .0702 | 6 287.2 | 14.253 | 14.081 | 200.704 | | 90 |
| 95 | 618.673 | .00162 | .00011 | .0701 | 8 823.9 | 14.263 | 14.132 | 201.558 | | 95 |
| 100 | 867.720 | .00115 | .00008 | .0701 | 12 381.7 | 14.269 | 14.170 | 202.200 | | 100 |

| | | | | | | | | | |
|---|---|---|---|---|---|---|---|---|---|
| **8%** | | | | Compound Interest Factors | | | | | **8%** |
| | **Single Payment** | | **Uniform Payment Series** | | | | **Arithmetic Gradient** | | |
| | Compound Amount Factor Find F Given P | Present Worth Factor Find P Given F | Sinking Fund Factor Find A Given F | Capital Recovery Factor Find A Given P | Compound Amount Factor Find F Given A | Present Worth Factor Find P Given A | Gradient Uniform Series Find A Given G | Gradient Present Worth Find P Given G | |
| *n* | *F/P* | *P/F* | *A/F* | *A/P* | *F/A* | *P/A* | *A/G* | *P/G* | *n* |
| 1 | 1.080 | .9259 | 1.0000 | 1.0800 | 1.000 | 0.926 | 0 | 0 | 1 |
| 2 | 1.166 | .8573 | .4808 | .5608 | 2.080 | 1.783 | 0.481 | 0.857 | 2 |
| 3 | 1.260 | .7938 | .3080 | .3880 | 3.246 | 2.577 | 0.949 | 2.445 | 3 |
| 4 | 1.360 | .7350 | .2219 | .3019 | 4.506 | 3.312 | 1.404 | 4.650 | 4 |
| 5 | 1.469 | .6806 | .1705 | .2505 | 5.867 | 3.993 | 1.846 | 7.372 | 5 |
| 6 | 1.587 | .6302 | .1363 | .2163 | 7.336 | 4.623 | 2.276 | 10.523 | 6 |
| 7 | 1.714 | .5835 | .1121 | .1921 | 8.923 | 5.206 | 2.694 | 14.024 | 7 |
| 8 | 1.851 | .5403 | .0940 | .1740 | 10.637 | 5.747 | 3.099 | 17.806 | 8 |
| 9 | 1.999 | .5002 | .0801 | .1601 | 12.488 | 6.247 | 3.491 | 21.808 | 9 |
| 10 | 2.159 | .4632 | .0690 | .1490 | 14.487 | 6.710 | 3.871 | 25.977 | 10 |
| 11 | 2.332 | .4289 | .0601 | .1401 | 16.645 | 7.139 | 4.240 | 30.266 | 11 |
| 12 | 2.518 | .3971 | .0527 | .1327 | 18.977 | 7.536 | 4.596 | 34.634 | 12 |
| 13 | 2.720 | .3677 | .0465 | .1265 | 21.495 | 7.904 | 4.940 | 39.046 | 13 |
| 14 | 2.937 | .3405 | .0413 | .1213 | 24.215 | 8.244 | 5.273 | 43.472 | 14 |
| 15 | 3.172 | .3152 | .0368 | .1168 | 27.152 | 8.559 | 5.594 | 47.886 | 15 |
| 16 | 3.426 | .2919 | .0330 | .1130 | 30.324 | 8.851 | 5.905 | 52.264 | 16 |
| 17 | 3.700 | .2703 | .0296 | .1096 | 33.750 | 9.122 | 6.204 | 56.588 | 17 |
| 18 | 3.996 | .2502 | .0267 | .1067 | 37.450 | 9.372 | 6.492 | 60.843 | 18 |
| 19 | 4.316 | .2317 | .0241 | .1041 | 41.446 | 9.604 | 6.770 | 65.013 | 19 |
| 20 | 4.661 | .2145 | .0219 | .1019 | 45.762 | 9.818 | 7.037 | 69.090 | 20 |
| 21 | 5.034 | .1987 | .0198 | .0998 | 50.423 | 10.017 | 7.294 | 73.063 | 21 |
| 22 | 5.437 | .1839 | .0180 | .0980 | 55.457 | 10.201 | 7.541 | 76.926 | 22 |
| 23 | 5.871 | .1703 | .0164 | .0964 | 60.893 | 10.371 | 7.779 | 80.673 | 23 |
| 24 | 6.341 | .1577 | .0150 | .0950 | 66.765 | 10.529 | 8.007 | 84.300 | 24 |
| 25 | 6.848 | .1460 | .0137 | .0937 | 73.106 | 10.675 | 8.225 | 87.804 | 25 |
| 26 | 7.396 | .1352 | .0125 | .0925 | 79.954 | 10.810 | 8.435 | 91.184 | 26 |
| 27 | 7.988 | .1252 | .0114 | .0914 | 87.351 | 10.935 | 8.636 | 94.439 | 27 |
| 28 | 8.627 | .1159 | .0105 | .0905 | 95.339 | 11.051 | 8.829 | 97.569 | 28 |
| 29 | 9.317 | .1073 | .00962 | .0896 | 103.966 | 11.158 | 9.013 | 100.574 | 29 |
| 30 | 10.063 | .0994 | .00883 | .0888 | 113.283 | 11.258 | 9.190 | 103.456 | 30 |
| 31 | 10.868 | .0920 | .00811 | .0881 | 123.346 | 11.350 | 9.358 | 106.216 | 31 |
| 32 | 11.737 | .0852 | .00745 | .0875 | 134.214 | 11.435 | 9.520 | 108.858 | 32 |
| 33 | 12.676 | .0789 | .00685 | .0869 | 145.951 | 11.514 | 9.674 | 111.382 | 33 |
| 34 | 13.690 | .0730 | .00630 | .0863 | 158.627 | 11.587 | 9.821 | 113.792 | 34 |
| 35 | 14.785 | .0676 | .00580 | .0858 | 172.317 | 11.655 | 9.961 | 116.092 | 35 |
| 40 | 21.725 | .0460 | .00386 | .0839 | 259.057 | 11.925 | 10.570 | 126.042 | 40 |
| 45 | 31.920 | .0313 | .00259 | .0826 | 386.506 | 12.108 | 11.045 | 133.733 | 45 |
| 50 | 46.902 | .0213 | .00174 | .0817 | 573.771 | 12.233 | 11.411 | 139.593 | 50 |
| 55 | 68.914 | .0145 | .00118 | .0812 | 848.925 | 12.319 | 11.690 | 144.006 | 55 |
| 60 | 101.257 | .00988 | .00080 | .0808 | 1 253.2 | 12.377 | 11.902 | 147.300 | 60 |
| 65 | 148.780 | .00672 | .00054 | .0805 | 1 847.3 | 12.416 | 12.060 | 149.739 | 65 |
| 70 | 218.607 | .00457 | .00037 | .0804 | 2 720.1 | 12.443 | 12.178 | 151.533 | 70 |
| 75 | 321.205 | .00311 | .00025 | .0802 | 4 002.6 | 12.461 | 12.266 | 152.845 | 75 |
| 80 | 471.956 | .00212 | .00017 | .0802 | 5 887.0 | 12.474 | 12.330 | 153.800 | 80 |
| 85 | 693.458 | .00144 | .00012 | .0801 | 8 655.7 | 12.482 | 12.377 | 154.492 | 85 |
| 90 | 1 018.9 | .00098 | .00008 | .0801 | 12 724.0 | 12.488 | 12.412 | 154.993 | 90 |
| 95 | 1 497.1 | .00067 | .00005 | .0801 | 18 701.6 | 12.492 | 12.437 | 155.352 | 95 |
| 100 | 2 199.8 | .00045 | .00004 | .0800 | 27 484.6 | 12.494 | 12.455 | 155.611 | 100 |

**9%** Compound Interest Factors **9%**

| | Single Payment | | Uniform Payment Series | | | | Arithmetic Gradient | | |
|---|---|---|---|---|---|---|---|---|---|
| | Compound Amount Factor Find F Given P F/P | Present Worth Factor Find P Given F P/F | Sinking Fund Factor Find A Given F A/F | Capital Recovery Factor Find A Given P A/P | Compound Amount Factor Find F Given A F/A | Present Worth Factor Find P Given A P/A | Gradient Uniform Series Find A Given G A/G | Gradient Present Worth Find P Given G P/G | |
| n | | | | | | | | | n |
| 1 | 1.090 | .9174 | 1.0000 | 1.0900 | 1.000 | 0.917 | 0 | 0 | 1 |
| 2 | 1.188 | .8417 | .4785 | .5685 | 2.090 | 1.759 | 0.478 | 0.842 | 2 |
| 3 | 1.295 | .7722 | .3051 | .3951 | 3.278 | 2.531 | 0.943 | 2.386 | 3 |
| 4 | 1.412 | .7084 | .2187 | .3087 | 4.573 | 3.240 | 1.393 | 4.511 | 4 |
| 5 | 1.539 | .6499 | .1671 | .2571 | 5.985 | 3.890 | 1.828 | 7.111 | 5 |
| 6 | 1.677 | .5963 | .1329 | .2229 | 7.523 | 4.486 | 2.250 | 10.092 | 6 |
| 7 | 1.828 | .5470 | .1087 | .1987 | 9.200 | 5.033 | 2.657 | 13.375 | 7 |
| 8 | 1.993 | .5019 | .0907 | .1807 | 11.028 | 5.535 | 3.051 | 16.888 | 8 |
| 9 | 2.172 | .4604 | .0768 | .1668 | 13.021 | 5.995 | 3.431 | 20.571 | 9 |
| 10 | 2.367 | .4224 | .0658 | .1558 | 15.193 | 6.418 | 3.798 | 24.373 | 10 |
| 11 | 2.580 | .3875 | .0569 | .1469 | 17.560 | 6.805 | 4.151 | 28.248 | 11 |
| 12 | 2.813 | .3555 | .0497 | .1397 | 20.141 | 7.161 | 4.491 | 32.159 | 12 |
| 13 | 3.066 | .3262 | .0436 | .1336 | 22.953 | 7.487 | 4.818 | 36.073 | 13 |
| 14 | 3.342 | .2992 | .0384 | .1284 | 26.019 | 7.786 | 5.133 | 39.963 | 14 |
| 15 | 3.642 | .2745 | .0341 | .1241 | 29.361 | 8.061 | 5.435 | 43.807 | 15 |
| 16 | 3.970 | .2519 | .0303 | .1203 | 33.003 | 8.313 | 5.724 | 47.585 | 16 |
| 17 | 4.328 | .2311 | .0270 | .1170 | 36.974 | 8.544 | 6.002 | 51.282 | 17 |
| 18 | 4.717 | .2120 | .0242 | .1142 | 41.301 | 8.756 | 6.269 | 54.886 | 18 |
| 19 | 5.142 | .1945 | .0217 | .1117 | 46.019 | 8.950 | 6.524 | 58.387 | 19 |
| 20 | 5.604 | .1784 | .0195 | .1095 | 51.160 | 9.129 | 6.767 | 61.777 | 20 |
| 21 | 6.109 | .1637 | .0176 | .1076 | 56.765 | 9.292 | 7.001 | 65.051 | 21 |
| 22 | 6.659 | .1502 | .0159 | .1059 | 62.873 | 9.442 | 7.223 | 68.205 | 22 |
| 23 | 7.258 | .1378 | .0144 | .1044 | 69.532 | 9.580 | 7.436 | 71.236 | 23 |
| 24 | 7.911 | .1264 | .0130 | .1030 | 76.790 | 9.707 | 7.638 | 74.143 | 24 |
| 25 | 8.623 | .1160 | .0118 | .1018 | 84.701 | 9.823 | 7.832 | 76.927 | 25 |
| 26 | 9.399 | .1064 | .0107 | .1007 | 93.324 | 9.929 | 8.016 | 79.586 | 26 |
| 27 | 10.245 | .0976 | .00973 | .0997 | 102.723 | 10.027 | 8.191 | 82.124 | 27 |
| 28 | 11.167 | .0895 | .00885 | .0989 | 112.968 | 10.116 | 8.357 | 84.542 | 28 |
| 29 | 12.172 | .0822 | .00806 | .0981 | 124.136 | 10.198 | 8.515 | 86.842 | 29 |
| 30 | 13.268 | .0754 | .00734 | .0973 | 136.308 | 10.274 | 8.666 | 89.028 | 30 |
| 31 | 14.462 | .0691 | .00669 | .0967 | 149.575 | 10.343 | 8.808 | 91.102 | 31 |
| 32 | 15.763 | .0634 | .00610 | .0961 | 164.037 | 10.406 | 8.944 | 93.069 | 32 |
| 33 | 17.182 | .0582 | .00556 | .0956 | 179.801 | 10.464 | 9.072 | 94.931 | 33 |
| 34 | 18.728 | .0534 | .00508 | .0951 | 196.983 | 10.518 | 9.193 | 96.693 | 34 |
| 35 | 20.414 | .0490 | .00464 | .0946 | 215.711 | 10.567 | 9.308 | 98.359 | 35 |
| 40 | 31.409 | .0318 | .00296 | .0930 | 337.883 | 10.757 | 9.796 | 105.376 | 40 |
| 45 | 48.327 | .0207 | .00190 | .0919 | 525.860 | 10.881 | 10.160 | 110.556 | 45 |
| 50 | 74.358 | .0134 | .00123 | .0912 | 815.085 | 10.962 | 10.430 | 114.325 | 50 |
| 55 | 114.409 | .00874 | .00079 | .0908 | 1 260.1 | 11.014 | 10.626 | 117.036 | 55 |
| 60 | 176.032 | .00568 | .00051 | .0905 | 1 944.8 | 11.048 | 10.768 | 118.968 | 60 |
| 65 | 270.847 | .00369 | .00033 | .0903 | 2 998.3 | 11.070 | 10.870 | 120.334 | 65 |
| 70 | 416.731 | .00240 | .00022 | .0902 | 4 619.2 | 11.084 | 10.943 | 121.294 | 70 |
| 75 | 641.193 | .00156 | .00014 | .0901 | 7 113.3 | 11.094 | 10.994 | 121.965 | 75 |
| 80 | 986.555 | .00101 | .00009 | .0901 | 10 950.6 | 11.100 | 11.030 | 122.431 | 80 |
| 85 | 1 517.9 | .00066 | .00006 | .0901 | 16 854.9 | 11.104 | 11.055 | 122.753 | 85 |
| 90 | 2 335.5 | .00043 | .00004 | .0900 | 25 939.3 | 11.106 | 11.073 | 122.976 | 90 |
| 95 | 3 593.5 | .00028 | .00003 | .0900 | 39 916.8 | 11.108 | 11.085 | 123.129 | 95 |
| 100 | 5 529.1 | .00018 | .00002 | .0900 | 61 422.9 | 11.109 | 11.093 | 123.233 | 100 |

**10%**            Compound Interest Factors            **10%**

| | Single Payment | | Uniform Payment Series | | | | Arithmetic Gradient | | |
| | Compound Amount Factor Find F Given P F/P | Present Worth Factor Find P Given F P/F | Sinking Fund Factor Find A Given F A/F | Capital Recovery Factor Find A Given P A/P | Compound Amount Factor Find F Given A F/A | Present Worth Factor Find P Given A P/A | Gradient Uniform Series Find A Given G A/G | Gradient Present Worth Find P Given G P/G | |
| n | | | | | | | | | n |
|---|---|---|---|---|---|---|---|---|---|
| 1 | 1.100 | .9091 | 1.0000 | 1.1000 | 1.000 | 0.909 | 0 | 0 | 1 |
| 2 | 1.210 | .8264 | .4762 | .5762 | 2.100 | 1.736 | 0.476 | 0.826 | 2 |
| 3 | 1.331 | .7513 | .3021 | .4021 | 3.310 | 2.487 | 0.937 | 2.329 | 3 |
| 4 | 1.464 | .6830 | .2155 | .3155 | 4.641 | 3.170 | 1.381 | 4.378 | 4 |
| 5 | 1.611 | .6209 | .1638 | .2638 | 6.105 | 3.791 | 1.810 | 6.862 | 5 |
| 6 | 1.772 | .5645 | .1296 | .2296 | 7.716 | 4.355 | 2.224 | 9.684 | 6 |
| 7 | 1.949 | .5132 | .1054 | .2054 | 9.487 | 4.868 | 2.622 | 12.763 | 7 |
| 8 | 2.144 | .4665 | .0874 | .1874 | 11.436 | 5.335 | 3.004 | 16.029 | 8 |
| 9 | 2.358 | .4241 | .0736 | .1736 | 13.579 | 5.759 | 3.372 | 19.421 | 9 |
| 10 | 2.594 | .3855 | .0627 | .1627 | 15.937 | 6.145 | 3.725 | 22.891 | 10 |
| 11 | 2.853 | .3505 | .0540 | .1540 | 18.531 | 6.495 | 4.064 | 26.396 | 11 |
| 12 | 3.138 | .3186 | .0468 | .1468 | 21.384 | 6.814 | 4.388 | 29.901 | 12 |
| 13 | 3.452 | .2897 | .0408 | .1408 | 24.523 | 7.103 | 4.699 | 33.377 | 13 |
| 14 | 3.797 | .2633 | .0357 | .1357 | 27.975 | 7.367 | 4.996 | 36.801 | 14 |
| 15 | 4.177 | .2394 | .0315 | .1315 | 31.772 | 7.606 | 5.279 | 40.152 | 15 |
| 16 | 4.595 | .2176 | .0278 | .1278 | 35.950 | 7.824 | 5.549 | 43.416 | 16 |
| 17 | 5.054 | .1978 | .0247 | .1247 | 40.545 | 8.022 | 5.807 | 46.582 | 17 |
| 18 | 5.560 | .1799 | .0219 | .1219 | 45.599 | 8.201 | 6.053 | 49.640 | 18 |
| 19 | 6.116 | .1635 | .0195 | .1195 | 51.159 | 8.365 | 6.286 | 52.583 | 19 |
| 20 | 6.728 | .1486 | .0175 | .1175 | 57.275 | 8.514 | 6.508 | 55.407 | 20 |
| 21 | 7.400 | .1351 | .0156 | .1156 | 64.003 | 8.649 | 6.719 | 58.110 | 21 |
| 22 | 8.140 | .1228 | .0140 | .1140 | 71.403 | 8.772 | 6.919 | 60.689 | 22 |
| 23 | 8.954 | .1117 | .0126 | .1126 | 79.543 | 8.883 | 7.108 | 63.146 | 23 |
| 24 | 9.850 | .1015 | .0113 | .1113 | 88.497 | 8.985 | 7.288 | 65.481 | 24 |
| 25 | 10.835 | .0923 | .0102 | .1102 | 98.347 | 9.077 | 7.458 | 67.696 | 25 |
| 26 | 11.918 | .0839 | .00916 | .1092 | 109.182 | 9.161 | 7.619 | 69.794 | 26 |
| 27 | 13.110 | .0763 | .00826 | .1083 | 121.100 | 9.237 | 7.770 | 71.777 | 27 |
| 28 | 14.421 | .0693 | .00745 | .1075 | 134.210 | 9.307 | 7.914 | 73.650 | 28 |
| 29 | 15.863 | .0630 | .00673 | .1067 | 148.631 | 9.370 | 8.049 | 75.415 | 29 |
| 30 | 17.449 | .0573 | .00608 | .1061 | 164.494 | 9.427 | 8.176 | 77.077 | 30 |
| 31 | 19.194 | .0521 | .00550 | .1055 | 181.944 | 9.479 | 8.296 | 78.640 | 31 |
| 32 | 21.114 | .0474 | .00497 | .1050 | 201.138 | 9.526 | 8.409 | 80.108 | 32 |
| 33 | 23.225 | .0431 | .00450 | .1045 | 222.252 | 9.569 | 8.515 | 81.486 | 33 |
| 34 | 25.548 | .0391 | .00407 | .1041 | 245.477 | 9.609 | 8.615 | 82.777 | 34 |
| 35 | 28.102 | .0356 | .00369 | .1037 | 271.025 | 9.644 | 8.709 | 83.987 | 35 |
| 40 | 45.259 | .0221 | .00226 | .1023 | 442.593 | 9.779 | 9.096 | 88.953 | 40 |
| 45 | 72.891 | .0137 | .00139 | .1014 | 718.905 | 9.863 | 9.374 | 92.454 | 45 |
| 50 | 117.391 | .00852 | .00086 | .1009 | 1 163.9 | 9.915 | 9.570 | 94.889 | 50 |
| 55 | 189.059 | .00529 | .00053 | .1005 | 1 880.6 | 9.947 | 9.708 | 96.562 | 55 |
| 60 | 304.482 | .00328 | .00033 | .1003 | 3 034.8 | 9.967 | 9.802 | 97.701 | 60 |
| 65 | 490.371 | .00204 | .00020 | .1002 | 4 893.7 | 9.980 | 9.867 | 98.471 | 65 |
| 70 | 789.748 | .00127 | .00013 | .1001 | 7 887.5 | 9.987 | 9.911 | 98.987 | 70 |
| 75 | 1 271.9 | .00079 | .00008 | .1001 | 12 709.0 | 9.992 | 9.941 | 99.332 | 75 |
| 80 | 2 048.4 | .00049 | .00005 | .1000 | 20 474.0 | 9.995 | 9.961 | 99.561 | 80 |
| 85 | 3 299.0 | .00030 | .00003 | .1000 | 32 979.7 | 9.997 | 9.974 | 99.712 | 85 |
| 90 | 5 313.0 | .00019 | .00002 | .1000 | 53 120.3 | 9.998 | 9.983 | 99.812 | 90 |
| 95 | 8 556.7 | .00012 | .00001 | .1000 | 85 556.9 | 9.999 | 9.989 | 99.877 | 95 |
| 100 | 13 780.6 | .00007 | .00001 | .1000 | 137 796.3 | 9.999 | 9.993 | 99.920 | 100 |

**12%**             Compound Interest Factors             **12%**

| | Single Payment | | Uniform Payment Series | | | | Arithmetic Gradient | | |
| | Compound Amount Factor Find F Given P F/P | Present Worth Factor Find P Given F P/F | Sinking Fund Factor Find A Given F A/F | Capital Recovery Factor Find A Given P A/P | Compound Amount Factor Find F Given A F/A | Present Worth Factor Find P Given A P/A | Gradient Uniform Series Find A Given G A/G | Gradient Present Worth Find P Given G P/G | |
| n | F/P | P/F | A/F | A/P | F/A | P/A | A/G | P/G | n |
| 1 | 1.120 | .8929 | 1.0000 | 1.1200 | 1.000 | 0.893 | 0 | 0 | 1 |
| 2 | 1.254 | .7972 | .4717 | .5917 | 2.120 | 1.690 | 0.472 | 0.797 | 2 |
| 3 | 1.405 | .7118 | .2963 | .4163 | 3.374 | 2.402 | 0.925 | 2.221 | 3 |
| 4 | 1.574 | .6355 | .2092 | .3292 | 4.779 | 3.037 | 1.359 | 4.127 | 4 |
| 5 | 1.762 | .5674 | .1574 | .2774 | 6.353 | 3.605 | 1.775 | 6.397 | 5 |
| 6 | 1.974 | .5066 | .1232 | .2432 | 8.115 | 4.111 | 2.172 | 8.930 | 6 |
| 7 | 2.211 | .4523 | .0991 | .2191 | 10.089 | 4.564 | 2.551 | 11.644 | 7 |
| 8 | 2.476 | .4039 | .0813 | .2013 | 12.300 | 4.968 | 2.913 | 14.471 | 8 |
| 9 | 2.773 | .3606 | .0677 | .1877 | 14.776 | 5.328 | 3.257 | 17.356 | 9 |
| 10 | 3.106 | .3220 | .0570 | .1770 | 17.549 | 5.650 | 3.585 | 20.254 | 10 |
| 11 | 3.479 | .2875 | .0484 | .1684 | 20.655 | 5.938 | 3.895 | 23.129 | 11 |
| 12 | 3.896 | .2567 | .0414 | .1614 | 24.133 | 6.194 | 4.190 | 25.952 | 12 |
| 13 | 4.363 | .2292 | .0357 | .1557 | 28.029 | 6.424 | 4.468 | 28.702 | 13 |
| 14 | 4.887 | .2046 | .0309 | .1509 | 32.393 | 6.628 | 4.732 | 31.362 | 14 |
| 15 | 5.474 | .1827 | .0268 | .1468 | 37.280 | 6.811 | 4.980 | 33.920 | 15 |
| 16 | 6.130 | .1631 | .0234 | .1434 | 42.753 | 6.974 | 5.215 | 36.367 | 16 |
| 17 | 6.866 | .1456 | .0205 | .1405 | 48.884 | 7.120 | 5.435 | 38.697 | 17 |
| 18 | 7.690 | .1300 | .0179 | .1379 | 55.750 | 7.250 | 5.643 | 40.908 | 18 |
| 19 | 8.613 | .1161 | .0158 | .1358 | 63.440 | 7.366 | 5.838 | 42.998 | 19 |
| 20 | 9.646 | .1037 | .0139 | .1339 | 72.052 | 7.469 | 6.020 | 44.968 | 20 |
| 21 | 10.804 | .0926 | .0122 | .1322 | 81.699 | 7.562 | 6.191 | 46.819 | 21 |
| 22 | 12.100 | .0826 | .0108 | .1308 | 92.503 | 7.645 | 6.351 | 48.554 | 22 |
| 23 | 13.552 | .0738 | .00956 | .1296 | 104.603 | 7.718 | 6.501 | 50.178 | 23 |
| 24 | 15.179 | .0659 | .00846 | .1285 | 118.155 | 7.784 | 6.641 | 51.693 | 24 |
| 25 | 17.000 | .0588 | .00750 | .1275 | 133.334 | 7.843 | 6.771 | 53.105 | 25 |
| 26 | 19.040 | .0525 | .00665 | .1267 | 150.334 | 7.896 | 6.892 | 54.418 | 26 |
| 27 | 21.325 | .0469 | .00590 | .1259 | 169.374 | 7.943 | 7.005 | 55.637 | 27 |
| 28 | 23.884 | .0419 | .00524 | .1252 | 190.699 | 7.984 | 7.110 | 56.767 | 28 |
| 29 | 26.750 | .0374 | .00466 | .1247 | 214.583 | 8.022 | 7.207 | 57.814 | 29 |
| 30 | 29.960 | .0334 | .00414 | .1241 | 241.333 | 8.055 | 7.297 | 58.782 | 30 |
| 31 | 33.555 | .0298 | .00369 | .1237 | 271.293 | 8.085 | 7.381 | 59.676 | 31 |
| 32 | 37.582 | .0266 | .00328 | .1233 | 304.848 | 8.112 | 7.459 | 60.501 | 32 |
| 33 | 42.092 | .0238 | .00292 | .1229 | 342.429 | 8.135 | 7.530 | 61.261 | 33 |
| 34 | 47.143 | .0212 | .00260 | .1226 | 384.521 | 8.157 | 7.596 | 61.961 | 34 |
| 35 | 52.800 | .0189 | .00232 | .1223 | 431.663 | 8.176 | 7.658 | 62.605 | 35 |
| 40 | 93.051 | .0107 | .00130 | .1213 | 767.091 | 8.244 | 7.899 | 65.116 | 40 |
| 45 | 163.988 | .00610 | .00074 | .1207 | 1358.2 | 8.283 | 8.057 | 66.734 | 45 |
| 50 | 289.002 | .00346 | .00042 | .1204 | 2400.0 | 8.304 | 8.160 | 67.762 | 50 |
| 55 | 509.321 | .00196 | .00024 | .1202 | 4236.0 | 8.317 | 8.225 | 68.408 | 55 |
| 60 | 897.597 | .00111 | .00013 | .1201 | 7471.6 | 8.324 | 8.266 | 68.810 | 60 |
| 65 | 1581.9 | .00063 | .00008 | .1201 | 13173.9 | 8.328 | 8.292 | 69.058 | 65 |
| 70 | 2787.8 | .00036 | .00004 | .1200 | 23223.3 | 8.330 | 8.308 | 69.210 | 70 |
| 75 | 4913.1 | .00020 | .00002 | .1200 | 40933.8 | 8.332 | 8.318 | 69.303 | 75 |
| 80 | 8658.5 | .00012 | .00001 | .1200 | 72145.7 | 8.332 | 8.324 | 69.359 | 80 |
| 85 | 15259.2 | .00007 | .00001 | .1200 | 127151.7 | 8.333 | 8.328 | 69.393 | 85 |
| 90 | 26891.9 | .00004 | | .1200 | 224091.1 | 8.333 | 8.330 | 69.414 | 90 |
| 95 | 47392.8 | .00002 | | .1200 | 394931.4 | 8.333 | 8.331 | 69.426 | 95 |
| 100 | 83522.3 | .00001 | | .1200 | 696010.5 | 8.333 | 8.332 | 69.434 | 100 |

15% | Compound Interest Factors | 15%

| | Single Payment | | Uniform Payment Series | | | | Arithmetic Gradient | | |
|---|---|---|---|---|---|---|---|---|---|
| | Compound Amount Factor Find F Given P F/P | Present Worth Factor Find P Given F P/F | Sinking Fund Factor Find A Given F A/F | Capital Recovery Factor Find A Given P A/P | Compound Amount Factor Find F Given A F/A | Present Worth Factor Find P Given A P/A | Gradient Uniform Series Find A Given G A/G | Gradient Present Worth Find P Given G P/G | |
| n | | | | | | | | | n |
| 1 | 1.150 | .8696 | 1.0000 | 1.1500 | 1.000 | 0.870 | 0 | 0 | 1 |
| 2 | 1.322 | .7561 | .4651 | .6151 | 2.150 | 1.626 | 0.465 | 0.756 | 2 |
| 3 | 1.521 | .6575 | .2880 | .4380 | 3.472 | 2.283 | 0.907 | 2.071 | 3 |
| 4 | 1.749 | .5718 | .2003 | .3503 | 4.993 | 2.855 | 1.326 | 3.786 | 4 |
| 5 | 2.011 | .4972 | .1483 | .2983 | 6.742 | 3.352 | 1.723 | 5.775 | 5 |
| 6 | 2.313 | .4323 | .1142 | .2642 | 8.754 | 3.784 | 2.097 | 7.937 | 6 |
| 7 | 2.660 | .3759 | .0904 | .2404 | 11.067 | 4.160 | 2.450 | 10.192 | 7 |
| 8 | 3.059 | .3269 | .0729 | .2229 | 13.727 | 4.487 | 2.781 | 12.481 | 8 |
| 9 | 3.518 | .2843 | .0596 | .2096 | 16.786 | 4.772 | 3.092 | 14.755 | 9 |
| 10 | 4.046 | .2472 | .0493 | .1993 | 20.304 | 5.019 | 3.383 | 16.979 | 10 |
| 11 | 4.652 | .2149 | .0411 | .1911 | 24.349 | 5.234 | 3.655 | 19.129 | 11 |
| 12 | 5.350 | .1869 | .0345 | .1845 | 29.002 | 5.421 | 3.908 | 21.185 | 12 |
| 13 | 6.153 | .1625 | .0291 | .1791 | 34.352 | 5.583 | 4.144 | 23.135 | 13 |
| 14 | 7.076 | .1413 | .0247 | .1747 | 40.505 | 5.724 | 4.362 | 24.972 | 14 |
| 15 | 8.137 | .1229 | .0210 | .1710 | 47.580 | 5.847 | 4.565 | 26.693 | 15 |
| 16 | 9.358 | .1069 | .0179 | .1679 | 55.717 | 5.954 | 4.752 | 28.296 | 16 |
| 17 | 10.761 | .0929 | .0154 | .1654 | 65.075 | 6.047 | 4.925 | 29.783 | 17 |
| 18 | 12.375 | .0808 | .0132 | .1632 | 75.836 | 6.128 | 5.084 | 31.156 | 18 |
| 19 | 14.232 | .0703 | .0113 | .1613 | 88.212 | 6.198 | 5.231 | 32.421 | 19 |
| 20 | 16.367 | .0611 | .00976 | .1598 | 102.444 | 6.259 | 5.365 | 33.582 | 20 |
| 21 | 18.822 | .0531 | .00842 | .1584 | 118.810 | 6.312 | 5.488 | 34.645 | 21 |
| 22 | 21.645 | .0462 | .00727 | .1573 | 137.632 | 6.359 | 5.601 | 35.615 | 22 |
| 23 | 24.891 | .0402 | .00628 | .1563 | 159.276 | 6.399 | 5.704 | 36.499 | 23 |
| 24 | 28.625 | .0349 | .00543 | .1554 | 184.168 | 6.434 | 5.798 | 37.302 | 24 |
| 25 | 32.919 | .0304 | .00470 | .1547 | 212.793 | 6.464 | 5.883 | 38.031 | 25 |
| 26 | 37.857 | .0264 | .00407 | .1541 | 245.712 | 6.491 | 5.961 | 38.692 | 26 |
| 27 | 43.535 | .0230 | .00353 | .1535 | 283.569 | 6.514 | 6.032 | 39.289 | 27 |
| 28 | 50.066 | .0200 | .00306 | .1531 | 327.104 | 6.534 | 6.096 | 39.828 | 28 |
| 29 | 57.575 | .0174 | .00265 | .1527 | 377.170 | 6.551 | 6.154 | 40.315 | 29 |
| 30 | 66.212 | .0151 | .00230 | .1523 | 434.745 | 6.566 | 6.207 | 40.753 | 30 |
| 31 | 76.144 | .0131 | .00200 | .1520 | 500.957 | 6.579 | 6.254 | 41.147 | 31 |
| 32 | 87.565 | .0114 | .00173 | .1517 | 577.100 | 6.591 | 6.297 | 41.501 | 32 |
| 33 | 100.700 | .00993 | .00150 | .1515 | 664.666 | 6.600 | 6.336 | 41.818 | 33 |
| 34 | 115.805 | .00864 | .00131 | .1513 | 765.365 | 6.609 | 6.371 | 42.103 | 34 |
| 35 | 133.176 | .00751 | .00113 | .1511 | 881.170 | 6.617 | 6.402 | 42.359 | 35 |
| 40 | 267.864 | .00373 | .00056 | .1506 | 1 779.1 | 6.642 | 6.517 | 43.283 | 40 |
| 45 | 538.769 | .00186 | .00028 | .1503 | 3 585.1 | 6.654 | 6.583 | 43.805 | 45 |
| 50 | 1 083.7 | .00092 | .00014 | .1501 | 7 217.7 | 6.661 | 6.620 | 44.096 | 50 |
| 55 | 2 179.6 | .00046 | .00007 | .1501 | 14 524.1 | 6.664 | 6.641 | 44.256 | 55 |
| 60 | 4 384.0 | .00023 | .00003 | .1500 | 29 220.0 | 6.665 | 6.653 | 44.343 | 60 |
| 65 | 8 817.8 | .00011 | .00002 | .1500 | 58 778.6 | 6.666 | 6.659 | 44.390 | 65 |
| 70 | 17 735.7 | .00006 | .00001 | .1500 | 118 231.5 | 6.666 | 6.663 | 44.416 | 70 |
| 75 | 35 672.9 | .00003 | | .1500 | 237 812.5 | 6.666 | 6.665 | 44.429 | 75 |
| 80 | 71 750.9 | .00001 | | .1500 | 478 332.6 | 6.667 | 6.666 | 44.436 | 80 |
| 85 | 144 316.7 | .00001 | | .1500 | 962 104.4 | 6.667 | 6.666 | 44.440 | 85 |

18%                                        Compound Interest Factors                                          18%

| | Single Payment | | Uniform Payment Series | | | | Arithmetic Gradient | | |
|---|---|---|---|---|---|---|---|---|---|
| | Compound Amount Factor Find F Given P F/P | Present Worth Factor Find P Given F P/F | Sinking Fund Factor Find A Given F A/F | Capital Recovery Factor Find A Given P A/P | Compound Amount Factor Find F Given A F/A | Present Worth Factor Find P Given A P/A | Gradient Uniform Series Find A Given G A/G | Gradient Present Worth Find P Given G P/G | |
| n | | | | | | | | | n |
| 1 | 1.180 | .8475 | 1.0000 | 1.1800 | 1.000 | 0.847 | 0 | 0 | 1 |
| 2 | 1.392 | .7182 | .4587 | .6387 | 2.180 | 1.566 | 0.459 | 0.718 | 2 |
| 3 | 1.643 | .6086 | .2799 | .4599 | 3.572 | 2.174 | 0.890 | 1.935 | 3 |
| 4 | 1.939 | .5158 | .1917 | .3717 | 5.215 | 2.690 | 1.295 | 3.483 | 4 |
| 5 | 2.288 | .4371 | .1398 | .3198 | 7.154 | 3.127 | 1.673 | 5.231 | 5 |
| 6 | 2.700 | .3704 | .1059 | .2859 | 9.442 | 3.498 | 2.025 | 7.083 | 6 |
| 7 | 3.185 | .3139 | .0824 | .2624 | 12.142 | 3.812 | 2.353 | 8.967 | 7 |
| 8 | 3.759 | .2660 | .0652 | .2452 | 15.327 | 4.078 | 2.656 | 10.829 | 8 |
| 9 | 4.435 | .2255 | .0524 | .2324 | 19.086 | 4.303 | 2.936 | 12.633 | 9 |
| 10 | 5.234 | .1911 | .0425 | .2225 | 23.521 | 4.494 | 3.194 | 14.352 | 10 |
| 11 | 6.176 | .1619 | .0348 | .2148 | 28.755 | 4.656 | 3.430 | 15.972 | 11 |
| 12 | 7.288 | .1372 | .0286 | .2086 | 34.931 | 4.793 | 3.647 | 17.481 | 12 |
| 13 | 8.599 | .1163 | .0237 | .2037 | 42.219 | 4.910 | 3.845 | 18.877 | 13 |
| 14 | 10.147 | .0985 | .0197 | .1997 | 50.818 | 5.008 | 4.025 | 20.158 | 14 |
| 15 | 11.974 | .0835 | .0164 | .1964 | 60.965 | 5.092 | 4.189 | 21.327 | 15 |
| 16 | 14.129 | .0708 | .0137 | .1937 | 72.939 | 5.162 | 4.337 | 22.389 | 16 |
| 17 | 16.672 | .0600 | .0115 | .1915 | 87.068 | 5.222 | 4.471 | 23.348 | 17 |
| 18 | 19.673 | .0508 | .00964 | .1896 | 103.740 | 5.273 | 4.592 | 24.212 | 18 |
| 19 | 23.214 | .0431 | .00810 | .1881 | 123.413 | 5.316 | 4.700 | 24.988 | 19 |
| 20 | 27.393 | .0365 | .00682 | .1868 | 146.628 | 5.353 | 4.798 | 25.681 | 20 |
| 21 | 32.324 | .0309 | .00575 | .1857 | 174.021 | 5.384 | 4.885 | 26.300 | 21 |
| 22 | 38.142 | .0262 | .00485 | .1848 | 206.345 | 5.410 | 4.963 | 26.851 | 22 |
| 23 | 45.008 | .0222 | .00409 | .1841 | 244.487 | 5.432 | 5.033 | 27.339 | 23 |
| 24 | 53.109 | .0188 | .00345 | .1835 | 289.494 | 5.451 | 5.095 | 27.772 | 24 |
| 25 | 62.669 | .0160 | .00292 | .1829 | 342.603 | 5.467 | 5.150 | 28.155 | 25 |
| 26 | 73.949 | .0135 | .00247 | .1825 | 405.272 | 5.480 | 5.199 | 28.494 | 26 |
| 27 | 87.260 | .0115 | .00209 | .1821 | 479.221 | 5.492 | 5.243 | 28.791 | 27 |
| 28 | 102.966 | .00971 | .00177 | .1818 | 566.480 | 5.502 | 5.281 | 29.054 | 28 |
| 29 | 121.500 | .00823 | .00149 | .1815 | 669.447 | 5.510 | 5.315 | 29.284 | 29 |
| 30 | 143.370 | .00697 | .00126 | .1813 | 790.947 | 5.517 | 5.345 | 29.486 | 30 |
| 31 | 169.177 | .00591 | .00107 | .1811 | 934.317 | 5.523 | 5.371 | 29.664 | 31 |
| 32 | 199.629 | .00501 | .00091 | .1809 | 1 103.5 | 5.528 | 5.394 | 29.819 | 32 |
| 33 | 235.562 | .00425 | .00077 | .1808 | 1 303.1 | 5.532 | 5.415 | 29.955 | 33 |
| 34 | 277.963 | .00360 | .00065 | .1806 | 1 538.7 | 5.536 | 5.433 | 30.074 | 34 |
| 35 | 327.997 | .00305 | .00055 | .1806 | 1 816.6 | 5.539 | 5.449 | 30.177 | 35 |
| 40 | 750.377 | .00133 | .00024 | .1802 | 4 163.2 | 5.548 | 5.502 | 30.527 | 40 |
| 45 | 1 716.7 | .00058 | .00010 | .1801 | 9 531.6 | 5.552 | 5.529 | 30.701 | 45 |
| 50 | 3 927.3 | .00025 | .00005 | .1800 | 21 813.0 | 5.554 | 5.543 | 30.786 | 50 |
| 55 | 8 984.8 | .00011 | .00002 | .1800 | 49 910.1 | 5.555 | 5.549 | 30.827 | 55 |
| 60 | 20 555.1 | .00005 | .00001 | .1800 | 114 189.4 | 5.555 | 5.553 | 30.846 | 60 |
| 65 | 47 025.1 | .00002 | | .1800 | 261 244.7 | 5.555 | 5.554 | 30.856 | 65 |
| 70 | 107 581.9 | .00001 | | .1800 | 597 671.7 | 5.556 | 5.555 | 30.860 | 70 |

**20%** Compound Interest Factors **20%**

| | Single Payment | | Uniform Payment Series | | | | Arithmetic Gradient | | |
|---|---|---|---|---|---|---|---|---|---|
| | Compound Amount Factor Find F Given P F/P | Present Worth Factor Find P Given F P/F | Sinking Fund Factor Find A Given F A/F | Capital Recovery Factor Find A Given P A/P | Compound Amount Factor Find F Given A F/A | Present Worth Factor Find P Given A P/A | Gradient Uniform Series Find A Given G A/G | Gradient Present Worth Find P Given G P/G | |
| n | | | | | | | | | n |
| 1 | 1.200 | .8333 | 1.0000 | 1.2000 | 1.000 | 0.833 | 0 | 0 | 1 |
| 2 | 1.440 | .6944 | .4545 | .6545 | 2.200 | 1.528 | 0.455 | 0.694 | 2 |
| 3 | 1.728 | .5787 | .2747 | .4747 | 3.640 | 2.106 | 0.879 | 1.852 | 3 |
| 4 | 2.074 | .4823 | .1863 | .3863 | 5.368 | 2.589 | 1.274 | 3.299 | 4 |
| 5 | 2.488 | .4019 | .1344 | .3344 | 7.442 | 2.991 | 1.641 | 4.906 | 5 |
| 6 | 2.986 | .3349 | .1007 | .3007 | 9.930 | 3.326 | 1.979 | 6.581 | 6 |
| 7 | 3.583 | .2791 | .0774 | .2774 | 12.916 | 3.605 | 2.290 | 8.255 | 7 |
| 8 | 4.300 | .2326 | .0606 | .2606 | 16.499 | 3.837 | 2.576 | 9.883 | 8 |
| 9 | 5.160 | .1938 | .0481 | .2481 | 20.799 | 4.031 | 2.836 | 11.434 | 9 |
| 10 | 6.192 | .1615 | .0385 | .2385 | 25.959 | 4.192 | 3.074 | 12.887 | 10 |
| 11 | 7.430 | .1346 | .0311 | .2311 | 32.150 | 4.327 | 3.289 | 14.233 | 11 |
| 12 | 8.916 | .1122 | .0253 | .2253 | 39.581 | 4.439 | 3.484 | 15.467 | 12 |
| 13 | 10.699 | .0935 | .0206 | .2206 | 48.497 | 4.533 | 3.660 | 16.588 | 13 |
| 14 | 12.839 | .0779 | .0169 | .2169 | 59.196 | 4.611 | 3.817 | 17.601 | 14 |
| 15 | 15.407 | .0649 | .0139 | .2139 | 72.035 | 4.675 | 3.959 | 18.509 | 15 |
| 16 | 18.488 | .0541 | .0114 | .2114 | 87.442 | 4.730 | 4.085 | 19.321 | 16 |
| 17 | 22.186 | .0451 | .00944 | .2094 | 105.931 | 4.775 | 4.198 | 20.042 | 17 |
| 18 | 26.623 | .0376 | .00781 | .2078 | 128.117 | 4.812 | 4.298 | 20.680 | 18 |
| 19 | 31.948 | .0313 | .00646 | .2065 | 154.740 | 4.843 | 4.386 | 21.244 | 19 |
| 20 | 38.338 | .0261 | .00536 | .2054 | 186.688 | 4.870 | 4.464 | 21.739 | 20 |
| 21 | 46.005 | .0217 | .00444 | .2044 | 225.026 | 4.891 | 4.533 | 22.174 | 21 |
| 22 | 55.206 | .0181 | .00369 | .2037 | 271.031 | 4.909 | 4.594 | 22.555 | 22 |
| 23 | 66.247 | .0151 | .00307 | .2031 | 326.237 | 4.925 | 4.647 | 22.887 | 23 |
| 24 | 79.497 | .0126 | .00255 | .2025 | 392.484 | 4.937 | 4.694 | 23.176 | 24 |
| 25 | 95.396 | .0105 | .00212 | .2021 | 471.981 | 4.948 | 4.735 | 23.428 | 25 |
| 26 | 114.475 | .00874 | .00176 | .2018 | 567.377 | 4.956 | 4.771 | 23.646 | 26 |
| 27 | 137.371 | .00728 | .00147 | .2015 | 681.853 | 4.964 | 4.802 | 23.835 | 27 |
| 28 | 164.845 | .00607 | .00122 | .2012 | 819.223 | 4.970 | 4.829 | 23.999 | 28 |
| 29 | 197.814 | .00506 | .00102 | .2010 | 984.068 | 4.975 | 4.853 | 24.141 | 29 |
| 30 | 237.376 | .00421 | .00085 | .2008 | 1 181.9 | 4.979 | 4.873 | 24.263 | 30 |
| 31 | 284.852 | .00351 | .00070 | .2007 | 1 419.3 | 4.982 | 4.891 | 24.368 | 31 |
| 32 | 341.822 | .00293 | .00059 | .2006 | 1 704.1 | 4.985 | 4.906 | 24.459 | 32 |
| 33 | 410.186 | .00244 | .00049 | .2005 | 2 045.9 | 4.988 | 4.919 | 24.537 | 33 |
| 34 | 492.224 | .00203 | .00041 | .2004 | 2 456.1 | 4.990 | 4.931 | 24.604 | 34 |
| 35 | 590.668 | .00169 | .00034 | .2003 | 2 948.3 | 4.992 | 4.941 | 24.661 | 35 |
| 40 | 1 469.8 | .00068 | .00014 | .2001 | 7 343.9 | 4.997 | 4.973 | 24.847 | 40 |
| 45 | 3 657.3 | .00027 | .00005 | .2001 | 18 281.3 | 4.999 | 4.988 | 24.932 | 45 |
| 50 | 9 100.4 | .00011 | .00002 | .2000 | 45 497.2 | 4.999 | 4.995 | 24.970 | 50 |
| 55 | 22 644.8 | .00004 | .00001 | .2000 | 113 219.0 | 5.000 | 4.998 | 24.987 | 55 |
| 60 | 56 347.5 | .00002 | | .2000 | 281 732.6 | 5.000 | 4.999 | 24.994 | 60 |

25%                                    Compound Interest Factors                                    25%

|   | Single Payment | | Uniform Payment Series | | | | Arithmetic Gradient | | |
|---|---|---|---|---|---|---|---|---|---|
| | Compound Amount Factor Find F Given P | Present Worth Factor Find P Given F | Sinking Fund Factor Find A Given F | Capital Recovery Factor Find A Given P | Compound Amount Factor Find F Given A | Present Worth Factor Find P Given A | Gradient Uniform Series Find A Given G | Gradient Present Worth Find P Given G | |
| n | F/P | P/F | A/F | A/P | F/A | P/A | A/G | P/G | n |
| 1 | 1.250 | .8000 | 1.0000 | 1.2500 | 1.000 | 0.800 | 0 | 0 | 1 |
| 2 | 1.563 | .6400 | .4444 | .6944 | 2.250 | 1.440 | 0.444 | 0.640 | 2 |
| 3 | 1.953 | .5120 | .2623 | .5123 | 3.813 | 1.952 | 0.852 | 1.664 | 3 |
| 4 | 2.441 | .4096 | .1734 | .4234 | 5.766 | 2.362 | 1.225 | 2.893 | 4 |
| 5 | 3.052 | .3277 | .1218 | .3718 | 8.207 | 2.689 | 1.563 | 4.204 | 5 |
| 6 | 3.815 | .2621 | .0888 | .3388 | 11.259 | 2.951 | 1.868 | 5.514 | 6 |
| 7 | 4.768 | .2097 | .0663 | .3163 | 15.073 | 3.161 | 2.142 | 6.773 | 7 |
| 8 | 5.960 | .1678 | .0504 | .3004 | 19.842 | 3.329 | 2.387 | 7.947 | 8 |
| 9 | 7.451 | .1342 | .0388 | .2888 | 25.802 | 3.463 | 2.605 | 9.021 | 9 |
| 10 | 9.313 | .1074 | .0301 | .2801 | 33.253 | 3.571 | 2.797 | 9.987 | 10 |
| 11 | 11.642 | .0859 | .0235 | .2735 | 42.566 | 3.656 | 2.966 | 10.846 | 11 |
| 12 | 14.552 | .0687 | .0184 | .2684 | 54.208 | 3.725 | 3.115 | 11.602 | 12 |
| 13 | 18.190 | .0550 | .0145 | .2645 | 68.760 | 3.780 | 3.244 | 12.262 | 13 |
| 14 | 22.737 | .0440 | .0115 | .2615 | 86.949 | 3.824 | 3.356 | 12.833 | 14 |
| 15 | 28.422 | .0352 | .00912 | .2591 | 109.687 | 3.859 | 3.453 | 13.326 | 15 |
| 16 | 35.527 | .0281 | .00724 | .2572 | 138.109 | 3.887 | 3.537 | 13.748 | 16 |
| 17 | 44.409 | .0225 | .00576 | .2558 | 173.636 | 3.910 | 3.608 | 14.108 | 17 |
| 18 | 55.511 | .0180 | .00459 | .2546 | 218.045 | 3.928 | 3.670 | 14.415 | 18 |
| 19 | 69.389 | .0144 | .00366 | .2537 | 273.556 | 3.942 | 3.722 | 14.674 | 19 |
| 20 | 86.736 | .0115 | .00292 | .2529 | 342.945 | 3.954 | 3.767 | 14.893 | 20 |
| 21 | 108.420 | .00922 | .00233 | .2523 | 429.681 | 3.963 | 3.805 | 15.078 | 21 |
| 22 | 135.525 | .00738 | .00186 | .2519 | 538.101 | 3.970 | 3.836 | 15.233 | 22 |
| 23 | 169.407 | .00590 | .00148 | .2515 | 673.626 | 3.976 | 3.863 | 15.362 | 23 |
| 24 | 211.758 | .00472 | .00119 | .2512 | 843.033 | 3.981 | 3.886 | 15.471 | 24 |
| 25 | 264.698 | .00378 | .00095 | .2509 | 1 054.8 | 3.985 | 3.905 | 15.562 | 25 |
| 26 | 330.872 | .00302 | .00076 | .2508 | 1 319.5 | 3.988 | 3.921 | 15.637 | 26 |
| 27 | 413.590 | .00242 | .00061 | .2506 | 1 650.4 | 3.990 | 3.935 | 15.700 | 27 |
| 28 | 516.988 | .00193 | .00048 | .2505 | 2 064.0 | 3.992 | 3.946 | 15.752 | 28 |
| 29 | 646.235 | .00155 | .00039 | .2504 | 2 580.9 | 3.994 | 3.955 | 15.796 | 29 |
| 30 | 807.794 | .00124 | .00031 | .2503 | 3 227.2 | 3.995 | 3.963 | 15.832 | 30 |
| 31 | 1 009.7 | .00099 | .00025 | .2502 | 4 035.0 | 3.996 | 3.969 | 15.861 | 31 |
| 32 | 1 262.2 | .00079 | .00020 | .2502 | 5 044.7 | 3.997 | 3.975 | 15.886 | 32 |
| 33 | 1 577.7 | .00063 | .00016 | .2502 | 6 306.9 | 3.997 | 3.979 | 15.906 | 33 |
| 34 | 1 972.2 | .00051 | .00013 | .2501 | 7 884.6 | 3.998 | 3.983 | 15.923 | 34 |
| 35 | 2 465.2 | .00041 | .00010 | .2501 | 9 856.8 | 3.998 | 3.986 | 15.937 | 35 |
| 40 | 7 523.2 | .00013 | .00003 | .2500 | 30 088.7 | 3.999 | 3.995 | 15.977 | 40 |
| 45 | 22 958.9 | .00004 | .00001 | .2500 | 91 831.5 | 4.000 | 3.998 | 15.991 | 45 |
| 50 | 70 064.9 | .00001 | | .2500 | 280 255.7 | 4.000 | 3.999 | 15.997 | 50 |
| 55 | 213 821.2 | | | .2500 | 855 280.7 | 4.000 | 4.000 | 15.999 | 55 |

**30%**                                      Compound Interest Factors                                      **30%**

| | Single Payment | | Uniform Payment Series | | | | Arithmetic Gradient | | |
|---|---|---|---|---|---|---|---|---|---|
| | Compound Amount Factor Find F Given P | Present Worth Factor Find P Given F | Sinking Fund Factor Find A Given F | Capital Recovery Factor Find A Given P | Compound Amount Factor Find F Given A | Present Worth Factor Find P Given A | Gradient Uniform Series Find A Given G | Gradient Present Worth Find P Given G | |
| n | F/P | P/F | A/F | A/P | F/A | P/A | A/G | P/G | n |
| 1 | 1.300 | .7692 | 1.0000 | 1.3000 | 1.000 | 0.769 | 0 | 0 | 1 |
| 2 | 1.690 | .5917 | .4348 | .7348 | 2.300 | 1.361 | 0.435 | 0.592 | 2 |
| 3 | 2.197 | .4552 | .2506 | .5506 | 3.990 | 1.816 | 0.827 | 1.502 | 3 |
| 4 | 2.856 | .3501 | .1616 | .4616 | 6.187 | 2.166 | 1.178 | 2.552 | 4 |
| 5 | 3.713 | .2693 | .1106 | .4106 | 9.043 | 2.436 | 1.490 | 3.630 | 5 |
| 6 | 4.827 | .2072 | .0784 | .3784 | 12.756 | 2.643 | 1.765 | 4.666 | 6 |
| 7 | 6.275 | .1594 | .0569 | .3569 | 17.583 | 2.802 | 2.006 | 5.622 | 7 |
| 8 | 8.157 | .1226 | .0419 | .3419 | 23.858 | 2.925 | 2.216 | 6.480 | 8 |
| 9 | 10.604 | .0943 | .0312 | .3312 | 32.015 | 3.019 | 2.396 | 7.234 | 9 |
| 10 | 13.786 | .0725 | .0235 | .3235 | 42.619 | 3.092 | 2.551 | 7.887 | 10 |
| 11 | 17.922 | .0558 | .0177 | .3177 | 56.405 | 3.147 | 2.683 | 8.445 | 11 |
| 12 | 23.298 | .0429 | .0135 | .3135 | 74.327 | 3.190 | 2.795 | 8.917 | 12 |
| 13 | 30.287 | .0330 | .0102 | .3102 | 97.625 | 3.223 | 2.889 | 9.314 | 13 |
| 14 | 39.374 | .0254 | .00782 | .3078 | 127.912 | 3.249 | 2.969 | 9.644 | 14 |
| 15 | 51.186 | .0195 | .00598 | .3060 | 167.286 | 3.268 | 3.034 | 9.917 | 15 |
| 16 | 66.542 | .0150 | .00458 | .3046 | 218.472 | 3.283 | 3.089 | 10.143 | 16 |
| 17 | 86.504 | .0116 | .00351 | .3035 | 285.014 | 3.295 | 3.135 | 10.328 | 17 |
| 18 | 112.455 | .00889 | .00269 | .3027 | 371.518 | 3.304 | 3.172 | 10.479 | 18 |
| 19 | 146.192 | .00684 | .00207 | .3021 | 483.973 | 3.311 | 3.202 | 10.602 | 19 |
| 20 | 190.049 | .00526 | .00159 | .3016 | 630.165 | 3.316 | 3.228 | 10.702 | 20 |
| 21 | 247.064 | .00405 | .00122 | .3012 | 820.214 | 3.320 | 3.248 | 10.783 | 21 |
| 22 | 321.184 | .00311 | .00094 | .3009 | 1 067.3 | 3.323 | 3.265 | 10.848 | 22 |
| 23 | 417.539 | .00239 | .00072 | .3007 | 1 388.5 | 3.325 | 3.278 | 10.901 | 23 |
| 24 | 542.800 | .00184 | .00055 | .3006 | 1 806.0 | 3.327 | 3.289 | 10.943 | 24 |
| 25 | 705.640 | .00142 | .00043 | .3004 | 2 348.8 | 3.329 | 3.298 | 10.977 | 25 |
| 26 | 917.332 | .00109 | .00033 | .3003 | 3 054.4 | 3.330 | 3.305 | 11.005 | 26 |
| 27 | 1 192.5 | .00084 | .00025 | .3003 | 3 971.8 | 3.331 | 3.311 | 11.026 | 27 |
| 28 | 1 550.3 | .00065 | .00019 | .3002 | 5 164.3 | 3.331 | 3.315 | 11.044 | 28 |
| 29 | 2 015.4 | .00050 | .00015 | .3001 | 6 714.6 | 3.332 | 3.319 | 11.058 | 29 |
| 30 | 2 620.0 | .00038 | .00011 | .3001 | 8 730.0 | 3.332 | 3.322 | 11.069 | 30 |
| 31 | 3 406.0 | .00029 | .00009 | .3001 | 11 350.0 | 3.332 | 3.324 | 11.078 | 31 |
| 32 | 4 427.8 | .00023 | .00007 | .3001 | 14 756.0 | 3.333 | 3.326 | 11.085 | 32 |
| 33 | 5 756.1 | .00017 | .00005 | .3001 | 19 183.7 | 3.333 | 3.328 | 11.090 | 33 |
| 34 | 7 483.0 | .00013 | .00004 | .3000 | 24 939.9 | 3.333 | 3.329 | 11.094 | 34 |
| 35 | 9 727.8 | .00010 | .00003 | .3000 | 32 422.8 | 3.333 | 3.330 | 11.098 | 35 |
| 40 | 36 118.8 | .00003 | .00001 | .3000 | 120 392.6 | 3.333 | 3.332 | 11.107 | 40 |
| 45 | 134 106.5 | .00001 | | .3000 | 447 018.3 | 3.333 | 3.333 | 11.110 | 45 |

**35%**                                          Compound Interest Factors                                          **35%**

| | Single Payment | | Uniform Payment Series | | | | Arithmetic Gradient | | |
|---|---|---|---|---|---|---|---|---|---|
| | Compound Amount Factor Find F Given P | Present Worth Factor Find P Given F | Sinking Fund Factor Find A Given F | Capital Recovery Factor Find A Given P | Compound Amount Factor Find F Given A | Present Worth Factor Find P Given A | Gradient Uniform Series Find A Given G | Gradient Present Worth Find P Given G | |
| n | F/P | P/F | A/F | A/P | F/A | P/A | A/G | P/G | n |
| 1 | 1.350 | .7407 | 1.0000 | 1.3500 | 1.000 | 0.741 | 0 | 0 | 1 |
| 2 | 1.822 | .5487 | .4255 | .7755 | 2.350 | 1.289 | 0.426 | 0.549 | 2 |
| 3 | 2.460 | .4064 | .2397 | .5897 | 4.173 | 1.696 | 0.803 | 1.362 | 3 |
| 4 | 3.322 | .3011 | .1508 | .5008 | 6.633 | 1.997 | 1.134 | 2.265 | 4 |
| 5 | 4.484 | .2230 | .1005 | .4505 | 9.954 | 2.220 | 1.422 | 3.157 | 5 |
| 6 | 6.053 | .1652 | .0693 | .4193 | 14.438 | 2.385 | 1.670 | 3.983 | 6 |
| 7 | 8.172 | .1224 | .0488 | .3988 | 20.492 | 2.508 | 1.881 | 4.717 | 7 |
| 8 | 11.032 | .0906 | .0349 | .3849 | 28.664 | 2.598 | 2.060 | 5.352 | 8 |
| 9 | 14.894 | .0671 | .0252 | .3752 | 39.696 | 2.665 | 2.209 | 5.889 | 9 |
| 10 | 20.107 | .0497 | .0183 | .3683 | 54.590 | 2.715 | 2.334 | 6.336 | 10 |
| 11 | 27.144 | .0368 | .0134 | .3634 | 74.697 | 2.752 | 2.436 | 6.705 | 11 |
| 12 | 36.644 | .0273 | .00982 | .3598 | 101.841 | 2.779 | 2.520 | 7.005 | 12 |
| 13 | 49.470 | .0202 | .00722 | .3572 | 138.485 | 2.799 | 2.589 | 7.247 | 13 |
| 14 | 66.784 | .0150 | .00532 | .3553 | 187.954 | 2.814 | 2.644 | 7.442 | 14 |
| 15 | 90.158 | .0111 | .00393 | .3539 | 254.739 | 2.825 | 2.689 | 7.597 | 15 |
| 16 | 121.714 | .00822 | .00290 | .3529 | 344.897 | 2.834 | 2.725 | 7.721 | 16 |
| 17 | 164.314 | .00609 | .00214 | .3521 | 466.611 | 2.840 | 2.753 | 7.818 | 17 |
| 18 | 221.824 | .00451 | .00158 | .3516 | 630.925 | 2.844 | 2.776 | 7.895 | 18 |
| 19 | 299.462 | .00334 | .00117 | .3512 | 852.748 | 2.848 | 2.793 | 7.955 | 19 |
| 20 | 404.274 | .00247 | .00087 | .3509 | 1 152.2 | 2.850 | 2.808 | 8.002 | 20 |
| 21 | 545.769 | .00183 | .00064 | .3506 | 1 556.5 | 2.852 | 2.819 | 8.038 | 21 |
| 22 | 736.789 | .00136 | .00048 | .3505 | 2 102.3 | 2.853 | 2.827 | 8.067 | 22 |
| 23 | 994.665 | .00101 | .00035 | .3504 | 2 839.0 | 2.854 | 2.834 | 8.089 | 23 |
| 24 | 1 342.8 | .00074 | .00026 | .3503 | 3 833.7 | 2.855 | 2.839 | 8.106 | 24 |
| 25 | 1 812.8 | .00055 | .00019 | .3502 | 5 176.5 | 2.856 | 2.843 | 8.119 | 25 |
| 26 | 2 447.2 | .00041 | .00014 | .3501 | 6 989.3 | 2.856 | 2.847 | 8.130 | 26 |
| 27 | 3 303.8 | .00030 | .00011 | .3501 | 9 436.5 | 2.856 | 2.849 | 8.137 | 27 |
| 28 | 4 460.1 | .00022 | .00008 | .3501 | 12 740.3 | 2.857 | 2.851 | 8.143 | 28 |
| 29 | 6 021.1 | .00017 | .00006 | .3501 | 17 200.4 | 2.857 | 2.852 | 8.148 | 29 |
| 30 | 8 128.5 | .00012 | .00004 | .3500 | 23 221.6 | 2.857 | 2.853 | 8.152 | 30 |
| 31 | 10 973.5 | .00009 | .00003 | .3500 | 31 350.1 | 2.857 | 2.854 | 8.154 | 31 |
| 32 | 14 814.3 | .00007 | .00002 | .3500 | 42 323.7 | 2.857 | 2.855 | 8.157 | 32 |
| 33 | 19 999.3 | .00005 | .00002 | .3500 | 57 137.9 | 2.857 | 2.855 | 8.158 | 33 |
| 34 | 26 999.0 | .00004 | .00001 | .3500 | 77 137.2 | 2.857 | 2.856 | 8.159 | 34 |
| 35 | 36 448.7 | .00003 | .00001 | .3500 | 104 136.3 | 2.857 | 2.856 | 8.160 | 35 |

**40%** Compound Interest Factors **40%**

| | Single Payment | | Uniform Payment Series | | | | Arithmetic Gradient | | |
|---|---|---|---|---|---|---|---|---|---|
| | Compound Amount Factor Find *F* Given *P* | Present Worth Factor Find *P* Given *F* | Sinking Fund Factor Find *A* Given *F* | Capital Recovery Factor Find *A* Given *P* | Compound Amount Factor Find *F* Given *A* | Present Worth Factor Find *P* Given *A* | Gradient Uniform Series Find *A* Given *G* | Gradient Present Worth Find *P* Given *G* | |
| *n* | *F/P* | *P/F* | *A/F* | *A/P* | *F/A* | *P/A* | *A/G* | *P/G* | *n* |
| 1 | 1.400 | .7143 | 1.0000 | 1.4000 | 1.000 | 0.714 | 0 | 0 | 1 |
| 2 | 1.960 | .5102 | .4167 | .8167 | 2.400 | 1.224 | 0.417 | 0.510 | 2 |
| 3 | 2.744 | .3644 | .2294 | .6294 | 4.360 | 1.589 | 0.780 | 1.239 | 3 |
| 4 | 3.842 | .2603 | .1408 | .5408 | 7.104 | 1.849 | 1.092 | 2.020 | 4 |
| 5 | 5.378 | .1859 | .0914 | .4914 | 10.946 | 2.035 | 1.358 | 2.764 | 5 |
| 6 | 7.530 | .1328 | .0613 | .4613 | 16.324 | 2.168 | 1.581 | 3.428 | 6 |
| 7 | 10.541 | .0949 | .0419 | .4419 | 23.853 | 2.263 | 1.766 | 3.997 | 7 |
| 8 | 14.758 | .0678 | .0291 | .4291 | 34.395 | 2.331 | 1.919 | 4.471 | 8 |
| 9 | 20.661 | .0484 | .0203 | .4203 | 49.153 | 2.379 | 2.042 | 4.858 | 9 |
| 10 | 28.925 | .0346 | .0143 | .4143 | 69.814 | 2.414 | 2.142 | 5.170 | 10 |
| 11 | 40.496 | .0247 | .0101 | .4101 | 98.739 | 2.438 | 2.221 | 5.417 | 11 |
| 12 | 56.694 | .0176 | .00718 | .4072 | 139.235 | 2.456 | 2.285 | 5.611 | 12 |
| 13 | 79.371 | .0126 | .00510 | .4051 | 195.929 | 2.469 | 2.334 | 5.762 | 13 |
| 14 | 111.120 | .00900 | .00363 | .4036 | 275.300 | 2.478 | 2.373 | 5.879 | 14 |
| 15 | 155.568 | .00643 | .00259 | .4026 | 386.420 | 2.484 | 2.403 | 5.969 | 15 |
| 16 | 217.795 | .00459 | .00185 | .4018 | 541.988 | 2.489 | 2.426 | 6.038 | 16 |
| 17 | 304.913 | .00328 | .00132 | .4013 | 759.783 | 2.492 | 2.444 | 6.090 | 17 |
| 18 | 426.879 | .00234 | .00094 | .4009 | 1 064.7 | 2.494 | 2.458 | 6.130 | 18 |
| 19 | 597.630 | .00167 | .00067 | .4007 | 1 419.6 | 2.496 | 2.468 | 6.160 | 19 |
| 20 | 836.682 | .00120 | .00048 | .4005 | 2 089.2 | 2.497 | 2.476 | 6.183 | 20 |
| 21 | 1 171.4 | .00085 | .00034 | .4003 | 2 925.9 | 2.498 | 2.482 | 6.200 | 21 |
| 22 | 1 639.9 | .00061 | .00024 | .4002 | 4 097.2 | 2.498 | 2.487 | 6.213 | 22 |
| 23 | 2 295.9 | .00044 | .00017 | .4002 | 5 737.1 | 2.499 | 2.490 | 6.222 | 23 |
| 24 | 3 214.2 | .00031 | .00012 | .4001 | 8 033.0 | 2.499 | 2.493 | 6.229 | 24 |
| 25 | 4 499.9 | .00022 | .00009 | .4001 | 11 247.2 | 2.499 | 2.494 | 6.235 | 25 |
| 26 | 6 299.8 | .00016 | .00006 | .4001 | 15 747.1 | 2.500 | 2.496 | 6.239 | 26 |
| 27 | 8 819.8 | .00011 | .00005 | .4000 | 22 046.9 | 2.500 | 2.497 | 6.242 | 27 |
| 28 | 12 347.7 | .00008 | .00003 | .4000 | 30 866.7 | 2.500 | 2.498 | 6.244 | 28 |
| 29 | 17 286.7 | .00006 | .00002 | .4000 | 43 214.3 | 2.500 | 2.498 | 6.245 | 29 |
| 30 | 24 201.4 | .00004 | .00002 | .4000 | 60 501.0 | 2.500 | 2.499 | 6.247 | 30 |
| 31 | 33 882.0 | .00003 | .00001 | .4000 | 84 702.5 | 2.500 | 2.499 | 6.248 | 31 |
| 32 | 47 434.8 | .00002 | .00001 | .4000 | 118 584.4 | 2.500 | 2.499 | 6.248 | 32 |
| 33 | 66 408.7 | .00002 | .00001 | .4000 | 166 019.2 | 2.500 | 2.500 | 6.249 | 33 |
| 34 | 92 972.1 | .00001 | | .4000 | 232 427.9 | 2.500 | 2.500 | 6.249 | 34 |
| 35 | 130 161.0 | .00001 | | .4000 | 325 400.0 | 2.500 | 2.500 | 6.249 | 35 |

| 45% | | | | Compound Interest Factors | | | | | | 45% |
|---|---|---|---|---|---|---|---|---|---|---|
| | Single Payment | | Uniform Payment Series | | | | Arithmetic Gradient | | | |
| | Compound Amount Factor Find F Given P | Present Worth Factor Find P Given F | Sinking Fund Factor Find A Given F | Capital Recovery Factor Find A Given P | Compound Amount Factor Find F Given A | Present Worth Factor Find P Given A | Gradient Uniform Series Find A Given G | Gradient Present Worth Find P Given G | |
| n | F/P | P/F | A/F | A/P | F/A | P/A | A/G | P/G | n |
| 1 | 1.450 | .6897 | 1.0000 | 1.4500 | 1.000 | 0.690 | 0 | 0 | 1 |
| 2 | 2.103 | .4756 | .4082 | .8582 | 2.450 | 1.165 | 0.408 | 0.476 | 2 |
| 3 | 3.049 | .3280 | .2197 | .6697 | 4.553 | 1.493 | 0.758 | 1.132 | 3 |
| 4 | 4.421 | .2262 | .1316 | .5816 | 7.601 | 1.720 | 1.053 | 1.810 | 4 |
| 5 | 6.410 | .1560 | .0832 | .5332 | 12.022 | 1.876 | 1.298 | 2.434 | 5 |
| 6 | 9.294 | .1076 | .0543 | .5043 | 18.431 | 1.983 | 1.499 | 2.972 | 6 |
| 7 | 13.476 | .0742 | .0361 | .4861 | 27.725 | 2.057 | 1.661 | 3.418 | 7 |
| 8 | 19.541 | .0512 | .0243 | .4743 | 41.202 | 2.109 | 1.791 | 3.776 | 8 |
| 9 | 28.334 | .0353 | .0165 | .4665 | 60.743 | 2.144 | 1.893 | 4.058 | 9 |
| 10 | 41.085 | .0243 | .0112 | .4612 | 89.077 | 2.168 | 1.973 | 4.277 | 10 |
| 11 | 59.573 | .0168 | .00768 | .4577 | 130.162 | 2.185 | 2.034 | 4.445 | 11 |
| 12 | 86.381 | .0116 | .00527 | .4553 | 189.735 | 2.196 | 2.082 | 4.572 | 12 |
| 13 | 125.252 | .00798 | .00362 | .4536 | 276.115 | 2.204 | 2.118 | 4.668 | 13 |
| 14 | 181.615 | .00551 | .00249 | .4525 | 401.367 | 2.210 | 2.145 | 4.740 | 14 |
| 15 | 263.342 | .00380 | .00172 | .4517 | 582.982 | 2.214 | 2.165 | 4.793 | 15 |
| 16 | 381.846 | .00262 | .00118 | .4512 | 846.325 | 2.216 | 2.180 | 4.832 | 16 |
| 17 | 553.677 | .00181 | .00081 | .4508 | 1 228.2 | 2.218 | 2.191 | 4.861 | 17 |
| 18 | 802.831 | .00125 | .00056 | .4506 | 1 781.8 | 2.219 | 2.200 | 4.882 | 18 |
| 19 | 1 164.1 | .00086 | .00039 | .4504 | 2 584.7 | 2.220 | 2.206 | 4.898 | 19 |
| 20 | 1 688.0 | .00059 | .00027 | .4503 | 3 748.8 | 2.221 | 2.210 | 4.909 | 20 |
| 21 | 2 447.5 | .00041 | .00018 | .4502 | 5 436.7 | 2.221 | 2.214 | 4.917 | 21 |
| 22 | 3 548.9 | .00028 | .00013 | .4501 | 7 884.3 | 2.222 | 2.216 | 4.923 | 22 |
| 23 | 5 145.9 | .00019 | .00009 | .4501 | 11 433.2 | 2.222 | 2.218 | 4.927 | 23 |
| 24 | 7 461.6 | .00013 | .00006 | .4501 | 16 579.1 | 2.222 | 2.219 | 4.930 | 24 |
| 25 | 10 819.3 | .00009 | .00004 | .4500 | 24 040.7 | 2.222 | 2.220 | 4.933 | 25 |
| 26 | 15 688.0 | .00006 | .00003 | .4500 | 34 860.1 | 2.222 | 2.221 | 4.934 | 26 |
| 27 | 22 747.7 | .00004 | .00002 | .4500 | 50 548.1 | 2.222 | 2.221 | 4.935 | 27 |
| 28 | 32 984.1 | .00003 | .00001 | .4500 | 73 295.8 | 2.222 | 2.221 | 4.936 | 28 |
| 29 | 47 826.9 | .00002 | .00001 | .4500 | 106 279.9 | 2.222 | 2.222 | 4.937 | 29 |
| 30 | 69 349.1 | .00001 | .00001 | .4500 | 154 106.8 | 2.222 | 2.222 | 4.937 | 30 |
| 31 | 100 556.1 | .00001 | | .4500 | 223 455.9 | 2.222 | 2.222 | 4.938 | 31 |
| 32 | 145 806.4 | .00001 | | .4500 | 324 012.0 | 2.222 | 2.222 | 4.938 | 32 |
| 33 | 211 419.3 | | | .4500 | 469 818.5 | 2.222 | 2.222 | 4.938 | 33 |
| 34 | 306 558.0 | | | .4500 | 681 237.8 | 2.222 | 2.222 | 4.938 | 34 |
| 35 | 444 509.2 | | | .4500 | 987 795.9 | 2.222 | 2.222 | 4.938 | 35 |

50% | Compound Interest Factors | 50%

| | Single Payment | | Uniform Payment Series | | | | Arithmetic Gradient | | |
|---|---|---|---|---|---|---|---|---|---|
| | Compound Amount Factor Find F Given P F/P | Present Worth Factor Find P Given F P/F | Sinking Fund Factor Find A Given F A/F | Capital Recovery Factor Find A Given P A/P | Compound Amount Factor Find F Given A F/A | Present Worth Factor Find P Given A P/A | Gradient Uniform Series Find A Given G A/G | Gradient Present Worth Find P Given G P/G | |
| n | | | | | | | | | n |
| 1 | 1.500 | .6667 | 1.0000 | 1.5000 | 1.000 | 0.667 | 0 | 0 | 1 |
| 2 | 2.250 | .4444 | .4000 | .9000 | 2.500 | 1.111 | 0.400 | 0.444 | 2 |
| 3 | 3.375 | .2963 | .2105 | .7105 | 4.750 | 1.407 | 0.737 | 1.037 | 3 |
| 4 | 5.063 | .1975 | .1231 | .6231 | 8.125 | 1.605 | 1.015 | 1.630 | 4 |
| 5 | 7.594 | .1317 | .0758 | .5758 | 13.188 | 1.737 | 1.242 | 2.156 | 5 |
| 6 | 11.391 | .0878 | .0481 | .5481 | 20.781 | 1.824 | 1.423 | 2.595 | 6 |
| 7 | 17.086 | .0585 | .0311 | .5311 | 32.172 | 1.883 | 1.565 | 2.947 | 7 |
| 8 | 25.629 | .0390 | .0203 | .5203 | 49.258 | 1.922 | 1.675 | 3.220 | 8 |
| 9 | 38.443 | .0260 | .0134 | .5134 | 74.887 | 1.948 | 1.760 | 3.428 | 9 |
| 10 | 57.665 | .0173 | .00882 | .5088 | 113.330 | 1.965 | 1.824 | 3.584 | 10 |
| 11 | 86.498 | .0116 | .00585 | .5058 | 170.995 | 1.977 | 1.871 | 3.699 | 11 |
| 12 | 129.746 | .00771 | .00388 | .5039 | 257.493 | 1.985 | 1.907 | 3.784 | 12 |
| 13 | 194.620 | .00514 | .00258 | .5026 | 387.239 | 1.990 | 1.933 | 3.846 | 13 |
| 14 | 291.929 | .00343 | .00172 | .5017 | 581.859 | 1.993 | 1.952 | 3.890 | 14 |
| 15 | 437.894 | .00228 | .00114 | .5011 | 873.788 | 1.995 | 1.966 | 3.922 | 15 |
| 16 | 656.814 | .00152 | .00076 | .5008 | 1 311.7 | 1.997 | 1.976 | 3.945 | 16 |
| 17 | 985.261 | .00101 | .00051 | .5005 | 1 968.5 | 1.998 | 1.983 | 3.961 | 17 |
| 18 | 1 477.9 | .00068 | .00034 | .5003 | 2 953.8 | 1.999 | 1.988 | 3.973 | 18 |
| 19 | 2 216.8 | .00045 | .00023 | .5002 | 4 431.7 | 1.999 | 1.991 | 3.981 | 19 |
| 20 | 3 325.3 | .00030 | .00015 | .5002 | 6 648.5 | 1.999 | 1.994 | 3.987 | 20 |
| 21 | 4 987.9 | .00020 | .00010 | .5001 | 9 973.8 | 2.000 | 1.996 | 3.991 | 21 |
| 22 | 7 481.8 | .00013 | .00007 | .5001 | 14 961.7 | 2.000 | 1.997 | 3.994 | 22 |
| 23 | 11 222.7 | .00009 | .00004 | .5000 | 22 443.5 | 2.000 | 1.998 | 3.996 | 23 |
| 24 | 16 834.1 | .00006 | .00003 | .5000 | 33 666.2 | 2.000 | 1.999 | 3.997 | 24 |
| 25 | 25 251.2 | .00004 | .00002 | .5000 | 50 500.3 | 2.000 | 1.999 | 3.998 | 25 |
| 26 | 37 876.8 | .00003 | .00001 | .5000 | 75 751.5 | 2.000 | 1.999 | 3.999 | 26 |
| 27 | 56 815.1 | .00002 | .00001 | .5000 | 113 628.3 | 2.000 | 2.000 | 3.999 | 27 |
| 28 | 85 222.7 | .00001 | .00001 | .5000 | 170 443.4 | 2.000 | 2.000 | 3.999 | 28 |
| 29 | 127 834.0 | .00001 | | .5000 | 255 666.1 | 2.000 | 2.000 | 4.000 | 29 |
| 30 | 191 751.1 | .00001 | | .5000 | 383 500.1 | 2.000 | 2.000 | 4.000 | 30 |
| 31 | 287 626.6 | | | .5000 | 575 251.2 | 2.000 | 2.000 | 4.000 | 31 |
| 32 | 431 439.9 | | | .5000 | 862 877.8 | 2.000 | 2.000 | 4.000 | 32 |

| 60% | | | | Compound Interest Factors | | | | | 60% |

| | Single Payment | | Uniform Payment Series | | | | Arithmetic Gradient | | |
|---|---|---|---|---|---|---|---|---|---|
| | Compound Amount Factor Find F Given P F/P | Present Worth Factor Find P Given F P/F | Sinking Fund Factor Find A Given F A/F | Capital Recovery Factor Find A Given P A/P | Compound Amount Factor Find F Given A F/A | Present Worth Factor Find P Given A P/A | Gradient Uniform Series Find A Given G A/G | Gradient Present Worth Find P Given G P/G | n |
| 1 | 1.600 | .6250 | 1.0000 | 1.6000 | 1.000 | 0.625 | 0 | 0 | 1 |
| 2 | 2.560 | .3906 | .3846 | .9846 | 2.600 | 1.016 | 0.385 | 0.391 | 2 |
| 3 | 4.096 | .2441 | .1938 | .7938 | 5.160 | 1.260 | 0.698 | 0.879 | 3 |
| 4 | 6.554 | .1526 | .1080 | .7080 | 9.256 | 1.412 | 0.946 | 1.337 | 4 |
| 5 | 10.486 | .0954 | .0633 | .6633 | 15.810 | 1.508 | 1.140 | 1.718 | 5 |
| 6 | 16.777 | .0596 | .0380 | .6380 | 26.295 | 1.567 | 1.286 | 2.016 | 6 |
| 7 | 26.844 | .0373 | .0232 | .6232 | 43.073 | 1.605 | 1.396 | 2.240 | 7 |
| 8 | 42.950 | .0233 | .0143 | .6143 | 69.916 | 1.628 | 1.476 | 2.403 | 8 |
| 9 | 68.719 | .0146 | .00886 | .6089 | 112.866 | 1.642 | 1.534 | 2.519 | 9 |
| 10 | 109.951 | .00909 | .00551 | .6055 | 181.585 | 1.652 | 1.575 | 2.601 | 10 |
| 11 | 175.922 | .00568 | .00343 | .6034 | 291.536 | 1.657 | 1.604 | 2.658 | 11 |
| 12 | 281.475 | .00355 | .00214 | .6021 | 467.458 | 1.661 | 1.624 | 2.697 | 12 |
| 13 | 450.360 | .00222 | .00134 | .6013 | 748.933 | 1.663 | 1.638 | 2.724 | 13 |
| 14 | 720.576 | .00139 | .00083 | .6008 | 1 199.3 | 1.664 | 1.647 | 2.742 | 14 |
| 15 | 1 152.9 | .00087 | .00052 | .6005 | 1 919.9 | 1.665 | 1.654 | 2.754 | 15 |
| 16 | 1 844.7 | .00054 | .00033 | .6003 | 3 072.8 | 1.666 | 1.658 | 2.762 | 16 |
| 17 | 2 951.5 | .00034 | .00020 | .6002 | 4 917.5 | 1.666 | 1.661 | 2.767 | 17 |
| 18 | 4 722.4 | .00021 | .00013 | .6001 | 7 868.9 | 1.666 | 1.663 | 2.771 | 18 |
| 19 | 7 555.8 | .00013 | .00008 | .6011 | 12 591.3 | 1.666 | 1.664 | 2.773 | 19 |
| 20 | 12 089.3 | .00008 | .00005 | .6000 | 20 147.1 | 1.667 | 1.665 | 2.775 | 20 |
| 21 | 19 342.8 | .00005 | .00003 | .6000 | 32 236.3 | 1.667 | 1.666 | 2.776 | 21 |
| 22 | 30 948.5 | .00003 | .00002 | .6000 | 51 579.2 | 1.667 | 1.666 | 2.777 | 22 |
| 23 | 49 517.6 | .00002 | .00001 | .6000 | 82 527.6 | 1.667 | 1.666 | 2.777 | 23 |
| 24 | 79 228.1 | .00001 | .00001 | .6000 | 132 045.2 | 1.667 | 1.666 | 2.777 | 24 |
| 25 | 126 765.0 | .00001 | | .6000 | 211 273.4 | 1.667 | 1.666 | 2.777 | 25 |
| 26 | 202 824.0 | | | .6000 | 338 038.4 | 1.667 | 1.667 | 2.778 | 26 |
| 27 | 324 518.4 | | | .6000 | 540 862.4 | 1.667 | 1.667 | 2.778 | 27 |
| 28 | 519 229.5 | | | .6000 | 865 380.9 | 1.667 | 1.667 | 2.778 | 28 |

Continuous Compounding—Single Payment Factors

| rn | Compound Amount Factor $e^{rn}$ Find F Given P F/P | Present Worth Factor $e^{-rn}$ Find P Find F P/F | rn | Compound Amount Factor $e^{rn}$ Find F Given P F/P | Present Worth Factor $e^{-rn}$ Find P Given F P/F |
|---|---|---|---|---|---|
| .01 | 1.0101 | .9900 | .51 | 1.6653 | .6005 |
| .02 | 1.0202 | .9802 | .52 | 1.6820 | .5945 |
| .03 | 1.0305 | .9704 | .53 | 1.6989 | .5886 |
| .04 | 1.0408 | .9608 | .54 | 1.7160 | .5827 |
| .05 | 1.0513 | .9512 | .55 | 1.7333 | .5769 |
| .06 | 1.0618 | .9418 | .56 | 1.7507 | .5712 |
| .07 | 1.0725 | .9324 | .57 | 1.7683 | .5655 |
| .08 | 1.0833 | .9231 | .58 | 1.7860 | .5599 |
| .09 | 1.0942 | .9139 | .59 | 1.8040 | .5543 |
| .10 | 1.1052 | .9048 | .60 | 1.8221 | .5488 |
| .11 | 1.1163 | .8958 | .61 | 1.8404 | .5434 |
| .12 | 1.1275 | .8869 | .62 | 1.8589 | .5379 |
| .13 | 1.1388 | .8781 | .63 | 1.8776 | .5326 |
| .14 | 1.1503 | .8694 | .64 | 1.8965 | .5273 |
| .15 | 1.1618 | .8607 | .65 | 1.9155 | .5220 |
| .16 | 1.1735 | .8521 | .66 | 1.9348 | .5169 |
| .17 | 1.1853 | .8437 | .67 | 1.9542 | .5117 |
| .18 | 1.1972 | .8353 | .68 | 1.9739 | .5066 |
| .19 | 1.2092 | .8270 | .69 | 1.9937 | .5016 |
| .20 | 1.2214 | .8187 | .70 | 2.0138 | .4966 |
| .21 | 1.2337 | .8106 | .71 | 2.0340 | .4916 |
| .22 | 1.2461 | .8025 | .72 | 2.0544 | .4868 |
| .23 | 1.2586 | .7945 | .73 | 2.0751 | .4819 |
| .24 | 1.2712 | .7866 | .74 | 2.0959 | .4771 |
| .25 | 1.2840 | .7788 | .75 | 2.1170 | .4724 |
| .26 | 1.2969 | .7711 | .76 | 2.1383 | .4677 |
| .27 | 1.3100 | .7634 | .77 | 2.1598 | .4630 |
| .28 | 1.3231 | .7558 | .78 | 2.1815 | .4584 |
| .29 | 1.3364 | .7483 | .79 | 2.2034 | .4538 |
| .30 | 1.3499 | .7408 | .80 | 2.2255 | .4493 |
| .31 | 1.3634 | .7334 | .81 | 2.2479 | .4449 |
| .32 | 1.3771 | .7261 | .82 | 2.2705 | .4404 |
| .33 | 1.3910 | .7189 | .83 | 2.2933 | .4360 |
| .34 | 1.4049 | .7118 | .84 | 2.3164 | .4317 |
| .35 | 1.4191 | .7047 | .85 | 2.3396 | .4274 |
| .36 | 1.4333 | .6977 | .86 | 2.3632 | .4232 |
| .37 | 1.4477 | .6907 | .87 | 2.3869 | .4190 |
| .38 | 1.4623 | .6839 | .88 | 2.4109 | .4148 |
| .39 | 1.4770 | .6771 | .89 | 2.4351 | .4107 |
| .40 | 1.4918 | .6703 | .90 | 2.4596 | .4066 |
| .41 | 1.5068 | .6637 | .91 | 2.4843 | .4025 |
| .42 | 1.5220 | .6570 | .92 | 2.5093 | .3985 |
| .43 | 1.5373 | .6505 | .93 | 2.5345 | .3946 |
| .44 | 1.5527 | .6440 | .94 | 2.5600 | .3906 |
| .45 | 1.5683 | .6376 | .95 | 2.5857 | .3867 |
| .46 | 1.5841 | .6313 | .96 | 2.6117 | .3829 |
| .47 | 1.6000 | .6250 | .97 | 2.6379 | .3791 |
| .48 | 1.6161 | .6188 | .98 | 2.6645 | .3753 |
| .49 | 1.6323 | .6126 | .99 | 2.6912 | .3716 |
| .50 | 1.6487 | .6065 | 1.00 | 2.7183 | .3679 |

# FUNDAMENTALS OF ENGINEERING (FE) EXAM PRACTICE PROBLEMS

This set of homework problems has been developed in the multiple-choice style of the Fundamentals of Engineering (FE) examination. Passing the FE is an important step toward achieving professional licensure through the National Council of Examiners for Engineering and Surveying (NCEES). Problems have been divided by type and chapter to assist the student in learning and mastering the material in preparation for the examination. Bold italic signifies the topic covered in this book.

*I. Engineering economy* topic problems are found in the general morning session of the FE exam, as well as in several of the discipline-specific afternoon session exams. Engineering economy topics constitute 8% of the morning exam, and from 1 to 15% of some afternoon exams. Afternoon exams that include engineering economy topics include Chemical, Civil, Industrial, and Other Disciplines. The following listing of *engineering economy topics* found in the FE exam is from NCEES publications for 2010 exams.

*Morning Session:* (all examinees complete this section; there are 120 questions in 12 topical areas)

**Topic VI. Engineering Economics (8% of questions)**

A. *Discounted cash flow (e.g., equivalence, PW, equivalent annual FW, rate of return)*

B. *Cost (e.g., incremental, average, sunk, estimating)*

C. *Analyses (e.g., breakeven, benefit–cost)*

D. *Uncertainty (e.g., expected value and risk)*

*Afternoon Session:* (examinees chose one of seven possible modules: Chemical, Civil, Electrical, Environmental, Industrial, Mechanical, and Other Disciplines)

Chemical Engineering Session (60 questions in 11 topical areas)
### Topic VIII. Process Design and Economic Optimization (10% of questions)

**A.** Process flow diagrams (PFD)

**B.** Piping and instrumentation diagrams (P&ID)

**C.** Scale-up

**D.** *Comparison of economic alternatives (e.g., net present value, discounted cash flow, rate of return)*

**E.** *Cost estimation*

Civil Engineering Session (60 questions in 9 topical areas)
### Topic VIII. Construction Management (10% of questions)

**A.** Procurement methods (e.g., design–build, design–bid–build, qualifications based)

**B.** Allocation of resources (e.g., labor, equipment, materials, money, time)

**C.** Contracts/contract law

**D.** Project scheduling (e.g., CPM, PERT)

**E.** *Engineering economics*

**F.** Project management (e.g., owner/contractor/client relations, safety)

**G.** Construction estimating

Industrial Engineering Session (60 questions in 8 topical areas)
### Topic I. Engineering Economics (15% of questions)

**A.** *Discounted cash flows (equivalence, PW, EAC, FW, IRR, loan amortization)*

**B.** *Types and breakdown of costs (e.g., fixed, variable, direct and indirect labor, material, capitalized)*

**C.** *Analyses (e.g., benefit–cost, breakeven, minimum cost, overhead, risk, incremental, life cycle)*

**D.** *Accounting (financial statements and overhead cost allocation)*

**E.** *Cost estimating*

**F.** *Depreciation and taxes*

**G.** *Capital budgeting*

### Topic V. Manufacturing and Production Systems (13% of questions)

**A.** Manufacturing systems (e.g., cellular, group technology, flexible, lean)

**B.** Process design (e.g., number of machines/people, equipment selection, line balancing)

**C.** Inventory analysis (e.g., EOQ, safety stock)

**D.** Forecasting

**E.** Scheduling (e.g., sequencing, cycle time, material control)

**F.** Aggregate planning (e.g., JIT, MRP, MRPII, ERP)

**G.** Concurrent engineering and design for manufacturing

**H.** Automation concepts (e.g., robotics, CIM)

**I.** *Economics (e.g., profits and costs under various demand rates, machine selection)*

Other Disciplines Engineering Session (60 questions in 9 topical areas)

### Topic IV. Engineering Economics (10% of questions)

A. *Cost estimating*

B. *Project selection*

C. *Lease/buy/make*

D. *Replacement analysis (e.g., optimal economic life)*

*II. Engineering ethics* topic problems are found only in the general morning session of the FE exam. These questions constitute about 7% of the morning exam. The following list of engineering ethics topics found in the FE exam is from NCEES publications for 2010 exams.

*Morning Session:* (all examinees complete this section; there are 120 questions in 12 topical areas)

### Topic V. Ethics and Business Practices (7% of questions)

A. *Code of ethics (professional and technical societies)*

B. Agreements and contracts

C. *Ethical versus legal*

D. Professional liability

E. Public protection issues (e.g., licensing boards)

## PROBLEMS

### Chapter 1

*Decision Making*

D-1  Engineering economic analysis provides useful input in all of the following situations except which one?

(*a*) Determining how much we should pay for a machine that will provide a savings.

(*b*) Determining the priority of investing our company's retained earnings.

(*c*) Illustrating the economic advantages of one alternative over other feasible choices.

(*d*) Convincing management that one person should be hired over another.

D-2  Engineering economic analysis can be described by the following statement:

(*a*) Involves a systematic analysis of relevant costs and benefits.

(*b*) Involves a comparison of competing alternatives.

(*c*) Supports a rational economic decision-making objective.

(*d*) All of the above.

D-3  To which of the following questions does an engineering economy analysis provide useful input?

(*a*) Has the mechanical or electrical engineer chosen the most economical motor size given functional requirements?

(*b*) Has the civil or mechanical engineer chosen the best thickness for insulating a building?

(*c*) Has the biomedical engineer chosen the best materials for the company's artificial knee product?

(*d*) All of the above

D-4  Which of the following job functions potentially conducts and utilizes engineering economic analysis in decision making?

(*a*) Senior technical design engineer.

(*b*) Midlevel manager of business and finance.

(*c*) Senior management for new product development.

(*d*) All of the above.

D-5  Engineering economic analysis can be described by the following statement:

(*a*) Involves a systematic analysis of relevant costs and benefits.

(*b*) Involves a comparison of competing alternatives.

(*c*) Supports a rational economic decision-making objective.

(*d*) All of the above.

### Engineering Ethics

**D-6**  Engineers are acting in the most ethical way in which of these situations?

(*a*) They are making sure the company's interests are protected.

(*b*) They feel good about the decisions that they've made.

(*c*) They act to protect the interests of society in general.

(*d*) They ensure that their own best interest is protected.

**D-7**  To act as an ethical engineer, you should accept fees for engineering work in which situation?

(*a*) If you need the money to keep your business open and thriving.

(*b*) If you are competent to complete all aspects of the job.

(*c*) If the contract is a cost plus contract.

(*d*) If there were no other engineers who bid on the job.

**D-8**  A registered professional engineer (PE) has as a primary obligation to protect which of the following entities?

(*a*) The government

(*b*) The PE's company

(*c*) The PE's country

(*d*) The general public

**D-9**  Engineers should act in ethical ways for what reason?

(*a*) It creates a feel-good situation for them.

(*b*) The engineering code of conduct requires it.

(*c*) They may be considered for a raise because of it.

(*d*) They may be violating the law if they don't.

**D-10**  A design engineer is responsible for an important subelement of a large project at a firm. The project has fallen behind schedule, and the important client is very angry and threatening to sue. The boss is expecting the engineer's design review to go well so that the project can be shipped by the end of the week. The engineer notices during final design review that an element of the design is wrong and will create a major safety issue for the entire system. To rework that portion of the design will take several months. The engineer should do which of the following?

(*a*) Sign off on the drawings because of the threatened lawsuit and because the project is so far behind.

(*b*) Do not sign off on the drawings, and let the boss know what is found at the final design review.

(*c*) Tell the boss that to sign off on the design now, the engineer must have an immediate raise.

(*d*) Sign off and keep an eye on how construction goes. Maybe the engineer can correct the safety issues before the project is fully operational.

**D-11**  As team leader for your unit, you function as both engineer and manager. One of your roles is to approve major purchases, and you have been contacted by a new supplier in your area. The new company has invited you to expensive dinners, has offered trips to vacation spots to attend "product shows," and has recently been sending to your private address personal items such as collectible art, coins, sports tickets, and golf club memberships. You are unsure how to handle this situation. You should do which of the following?

(*a*) Accept all of the gifts because you know that everyone else is doing it and that this is your chance to get a share of the action.

(*b*) Decline the gifts and other offers that would be considered outside the scope of ordinary business or professional contact.

(*c*) Knowing that the gifts will not influence your purchasing decisions, accept the items with no guilt.

(*d*) Accept the gifts but make sure that your boss and others on your team share in the bounty.

### Cost Problems

**D-12**  A company produces several product lines. One of those lines generates the following annual cost and production data:

| | |
|---|---|
| Manufacturing/Materials costs | $200,000 |
| General/Administrative costs | 50,000 |
| Direct-labor costs | 170,000 |
| Other overhead costs | 60,000 |
| Annual production demand | 10,000 units |

The company adds 40% to its production cost in selling to the retailer. The retailer in turn adds a 50%

profit margin when selling to its customers. How much would it cost a retailer to buy 100 units of the product?

(a) $4800

(b) $6720

(c) $7200

(d) $1008

**D-13** An agribusiness is deciding what crop to plant in this area for the next growing season. The local agricultural extension office has provided the following data (in $100):

| Crop | Cost per acre | Income/acre at 100% Yield | Estimated Yield(%) |
|------|------|------|------|
| A | $30 | $45 | 80 |
| B | 45 | 75 | 65 |
| C | 15 | 25 | 90 |

Using a 200-acre plot as an example (subtract 10% for unusable areas of the field), which crop should be planted this year, and what is the total profit?

(a) A; $108,000

(b) C; $135,000

(c) C; $150,000

(d) B; $540,000

**Use the data below for Problems D-14 to D-17.**
A textbook publisher produces a textbook for $25 per book and sells a lot of 160 to the Campus Bookstore for $50 per unit. The bookstore sells the textbook new for $75 and used for $60. This edition of the book is used for 2 years (4 semesters). The bookstore sells all textbooks that it has at the beginning each semester, and it repurchases 50% of those at the end of each semester for $30.

**D-14** What is the total cost to the textbook publisher?

(a) $25

(b) $50

(c) $4000

(d) $8000

**D-15** What is the total cost to all parties (publisher, bookstore, students) over the life of this 160-unit lot of textbooks?

(a) $12,000

(b) $20,400

(c) $32,400

(d) $36,900

**D-16** What is the net profit for the bookstore (sales − costs) over the life of this 160-unit lot of textbooks?

(a) $7900

(b) $11,600

(c) $11,900

(d) $14,000

**D-17** What is the net profit for the publisher (sales − costs) over the life of this 160-unit lot of textbooks?

(a) $4000

(b) $8000

(c) $11,900

(d) $12,000

## Chapter 2

### Cost Concepts

**D-18** A firm bought a used machine 2 years ago for $1500. When new, the machine cost $8000. Today it could be sold for $500. Which of the following statements is true?

(a) The fixed cost for operating the machine can be ignored in any analysis.

(b) The $8000 purchase price is not included in the analysis.

(c) The $1500 paid 2 years ago is included in the analysis.

(d) The variable cost of ownership is the difference between what was paid and what the machine is now worth ($1500 − $500 = $1000).

**D-19** When considering two alternatives that are described only in terms of the cost of ownership, the breakeven point cannot be described by which of the following statements?

(a) The difference in initial cost between the two alternatives.

(b) The level of production (or activity) of each alternative under consideration is equivalent.

(c) Fixed plus variable costs of each alternative are equivalent.

(d) A rational decision maker should be indifferent between the two alternatives.

**D-20** If JMJ Industries realizes a profit of $4.00 per unit sold, what is the fixed-cost portion of their production costs? Their variable costs are $1.50 per unit, and they sell 1000 units per year at a price of $6.00 per unit.

(a) There are no fixed costs in this type of problem.

(b) $250

(c) $500

(d) $2000

**D-21** Consider the following production data for Alternatives *A* and *B* in a firm that uses a 10% interest rate.

|  | Alt. A | Alt. B |
| --- | --- | --- |
| Annual fixed cost per unit | $2 million | $3.5 million |
| Annual variable cost per unit | 850 | 250 |

If the company is going to produce 4000 units annually, which alternative should be chosen?

(*a*) Neither alternative should be chosen because the negative cash flows are greater than the positive cash flows for both alternatives.

(*b*) This problem cannot be solved because there is not enough data given.

(*c*) Alt. *A*

(*d*) Alt. *B*

**D-22** A company has annual fixed costs of $2,500,000 and variable costs of 0.15¢ per unit produced. For the firm to break even if they charge $1.85 for their product, the level of annual production is nearest to what value?

(*a*) 375,000 units

(*b*) 1,315,789 units

(*c*) 1,351,351 units

(*d*) 1,562,500 units

*Estimating in Engineering Economy*

**D-23** Which statement is not true with respect to estimating the economic impacts of proposed engineering projects?

(*a*) Order-of-magnitude estimates are used for high-level planning.

(*b*) Order-of-magnitude estimates are the most accurate type at about −3 to 5%.

(*c*) Increasing the accuracy of estimates requires added time and resources.

(*d*) Estimators tend to underestimate the magnitude of costs and to overestimate benefits.

**D-24** The Department of Transportation is accepting bids for materials to provide "signage and safety" for a new 25-mile section of a 4-lane highway. DOT estimates are:

| Lane paint | $500 per mile per lane |
| --- | --- |
| Reflective lane markers | $6 per 25 feet |
| Mile markers | $18 per unit |
| Flexible roadside delineators | $32 per unit; spaced at 1000 feet |
| Emergency boxes | $500 per unit spaced at 2.5 miles |
| Signage (various messages) | $1000 per mile; based on historical costs |

The bids that DOT receives should be in what range of costs?

(*a*) Less than $100,000

(*b*) Between $100,000 and $200,000

(*c*) Between $200,000 and $250,000

(*d*) Greater than $250,000

**D-25** A 250-gallon reactor cost $780,000 when it was constructed 20 years ago. What would a 750-gallon model cost today if the power-sizing exponent is 0.56 and the construction cost index for such facilities has increased from 141 to 556 over the last 20 years? Choose the closest value.

(*a*) $0.37 million

(*b*) $1.66 million

(*c*) $1.44 million

(*d*) $5.69 million

**D-26** A half-million-square-foot warehouse facility is being considered by your company for a location in Kansas City, KS. You have a bid for a similar type 25,000-ft$^2$ facility in Washington, DC, at $4,375,000. If the warehouse construction cost index value for KS is 0.75 and for DC it is 1.34, what range should the KS bid fall into if you assume that construction costs are linear across size?

(*a*) Less than $30 million

(*b*) Between $30 million and $60 million

(*c*) Between $60 million and $100 million

(*d*) Greater than $100 million

**D-27** A company has major clients in all 50 states. Fifteen (15) of the states have 4 clients, 10 have 3 clients, 20 have 6 clients and 5 have 10. The total number of clients the firm has is closest to what number?

(*a*) 23

(*b*) 50

(*c*) 250

(*d*) 260

# Chapter 3

*Simple Interest*

**D-28** Which of the following statements is not true?

(*a*) Simple interest is to be used only in simple decision situations.

(*b*) Compounded interest involves computing interest on top of interest.

(*c*) Simple interest is rare in practical situations of borrowing and loaning.

(*d*) If the interest is not stated as being simple or compounded, we assume the latter.

**D-29**  If you borrow $1000 from the bank at 5% simple interest per month due back in 2 years, what is the size of your monthly payments?

(*a*) $25

(*b*) $50

(*c*) $500

(*d*) $1200

**D-30**  Your quarterly payments on a loan are $500 and the interest that you are paying is 1% per quarter simple interest. The size of the principal that you have borrowed is closest to which value?

(*a*) $5000

(*b*) $12,500

(*c*) $20,000

(*d*) $50,000

**D-31**  The principal that you borrowed for a recent purchase was $15,000. You will pay the purchase off through a simple interest loan at 8% per year due in 3 years. The amount that is due at the end of 3 years is closest to what value?

(*a*) $1200

(*b*) $3600

(*c*) $16,200

(*d*) $18,900

**D-32**  If $10,000 is borrowed today at 5% simple interest, how much is due at the end of 10 years?

(*a*) $5000

(*b*) $10,000

(*c*) $15,000

(*d*) $16,289

*Compound Interest, Single Cash Flows*

**D-33**  A savings account's value today is $150 and it earns interest at 1% per month. How much will be in the account one year from today? Which of the following is correct to solve for the unknown value?

(*a*) $P = 150(1 + 0.01)^{12}$

(*b*) $150 = F(F/P, 12\%, 1)$

(*c*) $F = 150(1.12)^{-12}$

(*d*) $F = 150(F/P, 1\%, 12)$

**D-34**  To calculate how many years (*n*) an investment (*P*) must be kept in an account that earns interest at *i*%, in order to triple in amount, which of the following expressions should be used?

(*a*) $n = -P + F(P/F, i\%, n)$

(*b*) $n = \log(F/P)]/\log(1 + i\%)$

(*c*) $n = \ln(-P + F)]/\ln i\%$

(*d*) $n = -F(1 + i\%)^{P}$

**D-35**  If you invest $40,000 in a stock whose value grows at 2% per year, your investment is nearest what value after 5 years?

(*a*) $40,800

(*b*) $43,296

(*c*) $44,164

(*d*) $64,420

**D-36**  An account pays interest at 1.5% per month. If you deposit $5000 at the beginning of this year, how much could you withdraw at the end of next year?

(*a*) $5151

(*b*) $7148

(*c*) $49,249

(*d*) $265,545

**D-37**  A machine will need to be replaced 15 years from today for $10,000. How much must be deposited now into an account that earns 5% per year to cover the replacement cost?

(*a*) $1486

(*b*) $4810

(*c*) $6139

(*d*) $10,000

**D-38**  You deposit $5000 into an account that will grow to $14,930 in 6 years. Your rate of return on this investment is closest to what value?

(*a*) 18%

(*b*) 20%

(*c*) 22%

(*d*) 25%

**D-39**  An investment company owns land now worth $500,000 that is increasing in value each year. If the land value doubles in 7 years, what is the yearly rate of return nearest to?

(*a*) 0.0%

(*b*) 2.0%

(*c*) 7.0%

(*d*) 10.5%

**D-40**  Your friend withdrew $630,315 from an account into which she had invested $350,000. If the account paid interest at 4% per year, she kept her money in the account for how many years?

(*a*) 1.8 years

(*b*) 6.5 years

(*c*) 12.5 years

(*d*) 15 years

**D-41**  Annual revenues in our company are $1.5 million this year. If they are expected to grow at a compounded rate of 20% per year, what will they be 10 years from now?

(a) $3.89 million

(b) $9.29 million

(c) $10.9 million

(d) $57.51 million

**D-42** A student inherits $50,000 and invests it in government bonds that will average 3% annual interest. What is the value of the investment after 50 years?

(a) $67,195

(b) $219,195

(c) $355,335

(d) $5,869,545

**D-43** A zero-coupon bond has a face (par) value of $10,000. The bond is sold at a discount for $6500 and held for 3 years, at which time it is sold. If an annual rate of 10% is earned over that 3-year period, how much was the bond sold for?

(a) $8652

(b) $13,310

(c) $16,859

(d) $25,937

### Nominal and Effective Interest Rates

**D-44** A rate of 2% per quarter compounded quarterly is closest to what annual compounded interest rate?

(a) 2.00%

(b) 8.00%

(c) 8.24%

(d) 24.00%

**D-45** A nominal interest rate of "12% per year compounded yearly" is closest to:

(a) 1% per month effective rate

(b) 3% per quarter effective rate

(c) 6% per 6 months effective rate

(d) 12% per year effective rate

**D-46** An interest rate expressed as "1.5% per month" is exactly the same as:

(a) 4.5% per quarter effective interest

(b) 18% effective interest per year

(c) 18% per year compounded monthly

(d) None of the above

**D-47** A deposit of $50,000 is made into an account that pays 10% compounded semiannually. How much would be in the account after 10 years?

(a) $81,445

(b) $129,685

(c) $132,665

(d) $336,375

**D-48** A mining firm must deposit funds in a "reclamation" account each quarter. The account must have $25 million on deposit when a project reaches its horizon in 10 years. The account pays interest at a rate of 2% per quarter. How much is the quarterly deposit?

(a) $41,500

(b) $172,575

(c) $228,325

(d) $414,000

**D-49** A deposit of $1000 compounds to $2500 in 5 years. The interest on this account compounds quarterly. What is the closest nominal annual rate of return?

(a) 2.50%

(b) 4.70%

(c) 18.75%

(d) 20.11%

**D-50** If your local bank indicates that it pays interest on passbook savings accounts at a rate of 2.25%, the nominal and effective interest rates are nearest which the following?

(a) 2.25%, 2.25%

(b) 2.25%, 2.28%

(c) 2.25%, 27%

(d) 27%, 2.25%

### Continuous Compounding Interest Rate

**D-51** How long will it take money to triple in an account if it pays interest expressed as 8% nominal annual compounded continuously?

(a) 13.73 years

(b) 14.27 years

(c) Your money will never triple.

(d) None of the above

**D-52** A deposit of $500 per year (beginning of year) is made for a period of 2 years in an account that earns 6% nominal interest compounded continuously. How much is in the account after 2 years?

(a) $917

(b) $1062

(c) $1092

(d) $1095

**D-53** A firm offers to sell a zero-coupon bond (no semi-annual payments) to you today. When it matures in 5 years you will receive the par value of $10,000. If the firm pays interest at 15% compounded continuously on the bond how much would you pay for it today?

(a) $4724

(b) $4972

(c) $10,000

(d) $21,170

**D-54** If $1000 is invested annually at 6% continuous compounding for each of 10 years, how much is in the account after the last deposit?

(a) $1822

(b) $10,000

(c) $13,181

(d) $13,295

## Chapter 4

*Uniform Cash Flow Series*

**D-55** You place $100 per month into an account that earns 1% per month. Which of the following expressions can be used to calculate the account's value after 3 years?

(a) $P = 100(P/A, 1\%, 3)$

(b) $F = 100(P/A, 1\%, 36)(F/P, 1\%, 36)$

(c) $F = 100(1 + 0.01)^n - 1/0.01$

(d) $F = 100(F/A, 12.68\%, 3)$

**D-56** A machine must be replaced in 7 years at a cost of $7500. How much must be deposited at the end of each year into an account that earns 5% in order to have accumulated enough to pay for the replacement?

(a) $471

(b) $596

(c) $791

(d) $921

**D-57** Winners of the PowerState Lottery can take $30 million now or payments of $2.5 million per year for the next 15 years. These are equivalent at what annual interest rate? The answer is closest to what value?

(a) 1%

(b) 2%

(c) 3%

(d) 5%

**D-58** You deposit $1000 in a retirement investment account today that earns 1% per month. In addition, you deposit $50 at the end of every month starting this month and continue to do so for 30 years. The amount that has accumulated in this account at the end of 30 years is nearest to

(a) $35,949

(b) $42,027

(c) $174,748

(d) $210,698

**D-59** Suppose $10,000 is deposited into an account that earns 10% per year for 5 years. At that point in time, uniform end-of-year withdrawals are made such that the account is emptied after the 15th withdrawal. The size of these annual withdrawals is closest to what value?

(a) $2118

(b) $2621

(c) $3410

(d) $16,105

**D-60** A manufacturer borrows $85,000 for machinery. The loan is for 10 years at 12% per year. What is the annual payment on the machinery?

(a) $4843

(b) $8500

(c) $13,834

(d) $15,045

**D-61** How many years would you have to put $100 per year into an account that earns 15% annually to accumulate $6508?

(a) 17 years

(b) 21 years

(c) 30 years

(d) 65 years

**D-62** A $10,000 face value municipal bond pays $1000 interest at the end of every year. If there are 12 more years of payments, at what price today would the bond yield 18% over the next 12 years?

(a) $1372

(b) $4793

(c) $6165

(d) $10,000

*Gradient Cash Flows*

**D-63** Today $5000 is deposited in an account that earns 2.5% per quarter. Additional deposits are made at the end of every quarter for the next 20 years. The deposits start at $100 and increase by $50 each year thereafter. The amount that has accumulated in this account at the end of 20 years can be expressed as follows.

(a) $= 5000(P/F, 2.5\%, 20) + 100(F/A, 2.5\%, 20\%) + 50$

$(P/G, 2.5\%, 20 - 1)$

$(b) = 5000(F/P, 10\%, 80) + 150(P/G, 10\%, 80)$
$(F/P = 10\%, 80)$

$(c) = 5000(P/F, 10.38\%, 20) + 100(P/A, 10.38\%, 20)$
$+50(P/G, 10.38\%, 20)(P/F, 10.38\%, 1)$

$(d) = 5000(F/P, 10.38\%, 80) + 100(F/A, 2.5\%, 80)$
$+50(P/G, 2.5\%, 80)(F/P, 2.5\%, 80)$

**D-64** An investment returns the following end-of-year cash flows: Year 1, $0; Year 2, $1500; Year 3, $3000; Year 4, $4500; and Year 5, $6000. Given a 10% interest rate, what is the present worth?

(a) $5970

(b) $6597

(c) $9357

(d) $10,293

**D-65** A project's annual revenues will be $50,000 the first year and will decrease by $1500 per year over its 20-year life. If the firm's interest rate is 12%, what is the project's present worth?

(a) $305,998

(b) $373,450

(c) $384,654

(d) $440,902

**D-66** A cash flow series is described by the following: $10,000 + $250(t), where t is the number of compounding periods. The present worth of this series at the end of five periods, where interest is 2% per t, is nearest what value?

(a) $11,250

(b) $50,620

(c) $56,432

(d) $60,620

**D-67** Revenue from sales of a training video for the first year are estimated to be $350,000. In addition, revenue is expected to decrease by $25,000 per year over the life of the video (which is 10 years). If interest is 10%, the present worth of the revenue over the life of the video is nearest what value?

(a) $100,000

(b) $125,000

(c) $1,578,475

(d) $2,723,025

## Chapter 5
### Present Worth Analysis

**D-68** A project is being considered that has a first cost of $12,500, creates $5000 in annual cost savings, requires $3000 in annual operating costs, and has a salvage value of $2000 after a project life of 3 years. If interest is 10% per year, which formula calculates the project's present worth?

(a) $PW = 12,500(P/F, 10\%, 1) + (-5000 + 3000)$
$(P/A, 10\%, 3) - 2000(F/P, 10\%, 3)$

(b) $PW = -12,500 + (5000 - 3000)(P/A, 10\%, 3)$
$-2000(P/F, 10\%, 3)$

(c) $PW = 12,500(F/P, 10\%, 3) + (5000 - 3000)$
$(F/A, 10\%, 3) + 2000$

(d) $PW = -12,500 + 5000(P/A, 10\%, 3) - 3000$
$(P/A, 10\%, 3) + 2000(P/F, 10\%, 3)$

**D-69** A new packing machine will cost $57,000. The existing machine can be sold for $5000 now and the new machine for $7500 after its 10-year useful life. If the new machine reduces annual expenses by $5000, what is the present worth at 25% of this investment?

(a) −$18,388

(b) −$33,340

(c) −$34,145

(d) −$38,340

**D-70** A vendor is offering an extended repair contract on a machine. The firm's experience is that this will cover repair costs over the next 4 years of $200, $200, $400, and $500. At 6%, what is the extended repair contract worth now?

(a) $1040

(b) $1089

(c) $1099

(d) $1300

**D-71** Annual disbursements for maintenance of critical heavily used equipment will be $25,000 for the next 10 years, and then $35,000 into infinity. What is the present worth of the maintenance cost cash flow stream if interest is 15%?

(a) $166,667

(b) $183,147

(c) $192,367

(d) $233,334

**D-72** Manufacturing costs from a scrapped poor-quality product are $6000 per year. An investment in an employee training program can reduce this cost.

Program A reduces the cost by 75% and requires an investment of $12,000. Program B reduces the cost by 95% and will cost $20,000. Based on low turnover at the plant, either program should be effective for the next 5 years. If interest is 20%, the present worth of the two programs is nearest what values? (Consider cost reduction a positive cash flow.)

(a) A: −$25,460; B: −$37,049

(b) A: $1460; B: −$2951

(c) A: $5060; B: $1609

(d) A: $13,460; B: $17,049

## Chapter 6
### Annual Worth Analysis

**D-73** New product tracking equipment costs $120,000 and will have a $10,000 salvage value when disposed of in 10 years. Annual repair costs begin at $5000 in the fifth year and increase by $500 per year thereafter until disposed of. If interest is 10%, what is the closest equivalent annual cost of ownership?

(a) $21,505

(b) $21,766

(c) $21,844

(d) $23,109

**D-74** Your company is considering two alternatives:

|  | Alt. I | Alt. II |
|---|---|---|
| First cost | $42,500 | $70,000 |
| Annual maintenance | 6,000 | 4,000 |
| Annual savings if implemented | 18,500 | 20,000 |
| Salvage value | 12,000 | 25,000 |
| Useful life of alternative | 3 years | 6 years |

What is the annual dollar advantage of Alt. II over Alt. I at an interest rate of 15%?

(a) Alt. II has no annual advantage over Alt I.

(b) $3020

(c) $3500

(d) $7436

**D-75** Specialized bits (costing $50,000) used in the mining industry have a useful life of 5000 hours of operation and can be traded in when a new bit is purchased for 10% of first cost. The drilling machine that uses the bit is used 1000 hours per year. What is the equivalent uniform annual cost of these bits at 2.5%?

(a) $8559

(b) $9510

(c) $9828

(d) $10,920

**D-76** A new chemical remediation tank is needed. Current technology tanks, which cost $150,000, must be drained and treated every 2 years at a cost of $30,000; the tanks will last 10 years, and each will have a salvage value of 5% of first cost. A tank with new technology has just come on the market. There are no periodic maintenance costs, and a tank will last 20 years. If the new tanks cost $325,000, what minimum salvage value, as a percentage of first cost, would be required for this technology to be a better option? Use a 12% interest rate.

(a) 10%

(b) 36%

(c) 57%

(d) 72%

**D-77** A beautiful bridge is being built over the river that runs through a major city in your state. The cost of the bridge is estimated at $600 million. Annual costs of the bridge will be $200,000, and the bridge is estimated to last a very long time. If accountants in city hall use 3% as the interest rate for analysis, what is the annualized cost of the bridge project?

(a) $18 million

(b) $18.2 million

(c) $20,000 million

(d) $219,500 million

## Chapter 7
### Rate of Return Analysis

**D-78** Which of the following equations can be used to find the internal rate of return ($i$) for a project that has initial investment of $P$, net annual cash flows of $A$, and salvage value of S after $n$ years?

(a) $0 = -P + A(P/A, i\%, n) + S(P/A, i\%, n)$

(b) $(P - A)(P/A, i\%, n) = S(P/F, i\%, n)$

(c) $-A = -P(A/P, i\%, n) - S(A/F, i\%, n - 1)$

(d) $0 = -P(F/P, i\%, n) + A(F/A, i\%, n) + S$

**D-79** The rate of return on an investment of $1500 that doubles in value over a 4-year period, and produces a $300 annual cash flow, is nearest to which value?

(a) 15%

(b) 20%

(c) 25%

(d) 30%

**D-80** The interest rate that makes Alternatives $A$ and $B$ equivalent is in what range?

| Year | Alt. A | Alt. B |
|------|--------|--------|
| 0 | −$1000 | −$3000 |
| 1 | 100 | 500 |
| 2 | 100 | 550 |
| 3 | 100 | 600 |
| 4 | 200 | 650 |
| 5 | 200 | 700 |

(a) Less than 2%

(b) Between 2 and 5%

(c) Greater than 5%

(d) There is no interest rate that equates these two cash flow series.

**D-81** A corporate bond with a face (par) value of $10,000 will mature 7 years from today (it was issued 3 years ago). The bond just after the $6^{\text{th}}$ interest payment is being sold for $6950. The bond's interest rate is 4% nominal annual, payable semiannually. The yield of the bond if held to maturity is in what range?

(a) Less than 4%

(b) Between 4 and 6%

(c) Between 6 and 10%

(d) Greater than 10%

**D-82** A firm borrowed $50,000 from a mortgage bank. The terms of the loan specify quarterly payments for a 10-year period. If payments to the bank are $3750 per quarter, what effective annual interest rate is the firm paying?

(a) Less than 1%

(b) 7%

(c) 28%

(d) 31%

## Chapter 8
### Incremental Cash Flows and Analysis

**D-83** You are given the cash flow series for two projects, Alt. $A$ and Alt. $B$.

| Year | 0 | 1 | 2 | 3 | 4 | 5 | 6 |
|------|---|---|---|---|---|---|---|
| Alt. A ($) | −I1 | X | X | X | X | X | X+S1 |
| Alt. B ($) | −I2 | Y | Y | Y | Y | Y | Y+S2 |

Assume I2 > I1 and X, Y, S1, and S2 are positive; the incremental rate of return ($i$) on the additional investment in Alt. $B$ can be calculated with the following expression.

(a) $0 = -I2 + Y(P/A, i\%, 6) + S2(P/F, i\%, 6)$

(b) $0 = -(I2 - I1) + (Y - X)(P/A, i\%, 5) + (S2 - S1)(P/F, i\%, 6)$

(c) $0 = -(I2 - I1)(F/P, i\%, 6) + (Y - X)(F/A, i\%, 6) + (S2 - S1)$

(d) $0 = -(I2 - I1) + (Y - X) + (Y + S2) - (X + S1)$

**D-84** A firm is considering two mutually exclusive alternatives ($i = 8\%$):

| | Alt. Alpha | Alt. Omega |
|---|---|---|
| First cost | $10,000 | $30,000 |
| Annual maintenance | 2,800 | 2,000 |
| Annual savings if implemented | 5,000 | 6,500 |
| Salvage value | 2,000 | 5,000 |
| Useful life of alternative | 4 years | 8 years |

If Alt. Alpha will be replaced with a "like alternative" at the end of 4 years, what is the present worth of the incremental cash flows associated with going from an investment in Alpha to an investment in Omega?

(a) −$6201

(b) −$5942

(c) −$5028

(d) $852

**D-85** Project 1 requires an initial investment on $50,000 and has an internal rate of return (IRR) of 18%. A mutually exclusive alternative, Project 2, requires an investment of $70,000 and has an IRR of 23%. Which of the following statements is true concerning the rate of return on the incremental $20,000 investment?

(a) It is less than 18%.

(b) It is between 18 and 23%.

(c) It is greater than 23%.

(d) It cannot be determined from the data given.

**D-86** Alternative Uno has a first cost of $10,000 and annual expenses of $3000, whereas Alternative Dos has a first cost of $35,000, annual expenses of $2000, and a recurring cost of $5000 every 10 years. If both alternatives have an infinite life, which of the following equations can be used to solve for the rate of return on the incremental investment?

(a) $0 = -\$25,000 + \$1000/i - \$5000(A/F, i, 10)$

(b) $0 = -\$25,000 + \$1000/i + \$5000(A/P, i, 10)/i$

(c) $0 = -\$25,000 + \$1000/i - \$5000(A/F, i, 10)/i$

(d) $0 = +\$25,000 - \$1000/i + \$5000(A/F, i, 10)$

**D-87** Compare two competing, mutually exclusive new machines that have only cost data given and tell which of the following statements is true regarding the present worth of the incremental investment at your investment interest rate.

(*a*) If it is greater than zero, we chose the alternative with the largest initial investment expense.

(*b*) The internal rate of return will always be equal to the investment rate of return.

(*c*) Neither machine is chosen if there is only cost data and the present worth is less than zero.

(*d*) If it is less then zero, we chose the alternative with the smallest initial investment expense.

## Chapter 9

### Future Worth Analysis

**D-88** The future worth of a project with initial cost $P$, positive annual cash flows of $A$, salvage value S, and interest rate of $i$ over a life of $n$ years can be calculated using which statement?

(*a*) $FW = -P(F/P, i\%, n) + A(F/A, i\%, n) + S(F/P, i\%, n)$

(*b*) $FW = P(F/P, i\%, n) + A(F/A, i\%, n) + S$

(*c*) $FW = -P(P/F, i\%, n) + A(F/A, i\%, n) - A(P/A, i\%, n) + S$

(*d*) $FW = -P(F/P, i\%, n) + A(F/A, i\%, n) + S$

**D-89** A firm has been investing retained earnings to establish a building fund. The firm has retained $1.2 million, $1.0 million, and $950,000, respectively, 3, 2, and 1 year ago. This year the firm has $1.8 million to invest. If the firm earns 18% on invested funds, what is the value of the project that can be undertaken using the funds as a 25% down payment?

(*a*) $6.28 million

(*b*) $7.42 million

(*c*) $25.1 million

(*d*) $29.7 million

**D-90** A firm is considering the purchase of a software analysis package that costs $450,000. Annual licensing fees are $25,000 (payable at the beginning of each year, starting in Year 1). The firm is bidding on a large 4-year government project where the new software will be used. If the firm uses an interest rate of 20%, what value for software costs should be put in the bid?

(*a*) $514,725

(*b*) $527,650

(*c*) $1,067,540

(*d*) $1,094,346

**D-91** Which of the following is a true statement regarding the future worth of a single investment alternative?

(*a*) It will be equal to both the present worth and the annual worth if the same discounting interest rate is used.

(*b*) Choose to invest if the calculated amount is less than zero at the investment rate of return.

(*c*) It will yield a recommendation consistent with the present worth and annual worth methods if the same discounting interest rate is used.

(*d*) It cannot be used to evaluate single investment alternatives.

**D-92** Using the data for Uno and Dos from Problem 86, where the lives of both alternatives is 10 years, give the future worth on the incremental investment if the interest rate used is 10%.

(*a*) −$20,000

(*b*) −$20,783

(*c*) −$43,910

(*d*) −$53,910

### Benefit–Cost Ratio Analysis

**D-93** The annual benefits associated with construction of an outer belt highway are estimated at $10.5 million by a local planning commission. The initial construction costs will be $400 million, and the project's useful life is 50 years. Annual maintenance costs are $500,000 with periodic rebuilding costs of $10 million every 10 years. If interest is 2%, the benefit–cost ratio is closest to what value?

(*a*) 0.25

(*b*) 0.75

(*c*) 1.11

(*d*) 1.35

**D-94** A city needs a new pedestrian bridge over a local stream. The city uses an interest rate of 5%, and the project life is 30 years. The following data (in millions of dollars) summarizes the bids that were received.

| | Bid A | Bid B |
|---|---|---|
| Construction materials costs | $4.20 | $6.20 |
| Construction labor costs | 0.60 | 0.70 |
| Construction overhead costs | 0.35 | 0.03 |
| Initial administrative and legal costs | 0.60 | 0.01 |
| Annual operating costs | 0.05 | 0.075 |
| Annual revenue from operation | Unknown | 0.40 |
| Other annual benefits to the city | 0.22 | 0.25 |

What would the annual revenue of Bid *A* have to be for the two projects to be equivalent? Choose the closest value.

(a) 0.10

(b) 0.20

(c) 0.30

(d) 0.40

**D-95** When using the benefit–cost method of analyzing a project, which of the following is a true statement?

(a) It will always produce a recommendation consistent with the simple payback period method.

(b) It will always produce a recommendation consistent with present worth, future worth, and annual worth methods.

(c) It can be used only to evaluate projects from the public sector (such as bridges and roadways).

(d) None of the above.

**D-96** Project A has a first cost of $950,000 and will produce a $50,000 net annual benefit over its 50-year life. Project B costs $1,250,000 and produces an $85,000 net annual benefit. If interest is 3% per year, the benefit–cost ratios of Projects A and B are nearest what values?

(a) 0.52, 0.67

(b) 0.74, 0.57

(c) 1.35, 1.75

(d) 2.63, 3.40

**D-97** Using the data for Projects A and B in Problem 96, the benefit-cost ratio on the incremental investment is nearest what value?

(a) 0.17

(b) 0.33

(c) 3.00

(d) 5.83

### Sensitivity and Breakeven Analysis

**D-98** BVM manufactured and sold 25,000 small statues this past year. At that volume, the firm was exactly in a breakeven situation in terms of profitability. BVM's unit costs are expected to increase by 30% next year. What additional information is needed to determine how much the production volume/sales would have to increase next year to just break even in terms of profitability?

(a) Costs per unit

(b) Sales price per unit and costs per unit

(c) Total fixed costs, sales price per unit, and costs per unit

(d) No data is needed, the volume increase is $25,000 + 25,000(0.30) = 32,500$ units.

**D-99** Process A has fixed costs of $10,000 and unit costs of $4.50 each, and Process B has fixed costs of $25,000 and unit costs of $1.50 each. At what level of annual production would the two processes have the same cost?

(a) 50 units

(b) 500 units

(c) 5000 units

(d) 50,000 units

**D-100** A seasonal bus tour firm has 5 buses with a capacity of 60 people each. Each customer pays $25 for a one-day tour. Records show $360,000 in fixed costs per season, incremental costs of $5 per customer, and an average daily occupancy of 80%. The number of days of operation necessary each season to break even is closest to which value?

(a) 50 days

(b) 75 days

(c) 100 days

(d) 120 days

**D-101** Alternative I has a first cost of $50,000, will produce an $18,000 net annual benefit over its 10-year life and be salvaged for $5,000. Alternative II costs $150,000 and has a salvage value of $50,000 after its 10-year useful life. If interest is 15%, what is the minimum amount of annual benefit that Alternative II must produce to make it the preferred choice?

(a) This value can not be determined from the data given.

(b) $23,500

(c) $31,450

(d) $35,708

**D-102** Use the table to determine which project is best if it is known for sure that annual sales will be $7 million. All values are in millions (PW and Sales) of dollers.

| Ann. Sales | Estimated PW ($M) | | |
|---|---|---|---|
| | Proj. 1 | Proj. 2 | Proj. 3 |
| $ 0 | −75 | −10 | 0 |
| 5M | 125 | 15 | 150 |
| 10M | 325 | 40 | 300 |

(a) Project 1

(b) Project 2

(c) Project 3

(d) None, since Projects 1 and 2 have negative values.

## Chapter 10
### Uncertainty and Probability

**D-103** An interest rate of 15% is used to evaluate a new system that has a first cost of $212,400, annual operating and maintenance costs of $41,200, annual savings of $94,600, a life of 6 years, and a salvage value of $32,500. After initial evaluation, the firm receives word from the vendor that the first cost is 5% higher than originally quoted. The percentage error in the system's present worth from this is closest to what value?

(a) 5%

(b) 15%

(c) 100%

(d) 300%

**D-104** A machine has a first cost of $10,000 and annual costs of $3500. There is no salvage value, and interest is 10%. If the project's useful life is described by the following data, what is the annual worth of costs?

**Useful Life (years)**

|  | 4 | 5 | 6 | 7 |
|---|---|---|---|---|
| Prob. of life (%) | 5 | 22 | 41 | 32 |

(a) $3500

(b) $5127

(c) $5554

(d) $5796

**D-105** Three estimators have estimated a project with a 10-year life.

|  | Estimate 1 | Estimate 2 | Estimate 3 |
|---|---|---|---|
| First cost | $10,000 | $17,500 | $15,000 |
| Net annual cash flows | 7,500 | 8,000 | 6,000 |
| Salvage value | 3,500 | 0 | 10,000 |

Use an interest rate of 20% to determine the project's expected present worth. The value is closest to which of the following?

(a) $16,066

(b) $16,612

(c) $31,660

(d) $31,607

**D-106** The first cost (FC), life (n), and annual benefits (A) for a prospective project are uncertain. Optimistic (OP), most likely (ML), and pessimistic (PS) estimates are given. If the interest rate is 25%, what is the present worth difference between a total worst-case scenario and a total best-case scenario?

|  | Estimate | | |
|---|---|---|---|
| | | **Most** | |
| Parameter | Pessimistic | Likely | Optimistic |
| First cost | $150,000 | $100,000 | $80,000 |
| Annual cash flows | 25,000 | 45,000 | 50,000 |
| Project life, in years | 5 | 7 | 10 |

(a) $15.8

(b) $42.2

(c) $181.3

(d) $282.5

**D-107** Which of the following statements, related to the use of decision tree analyses to model a problem and recommend a solution, is not true?

(a) In modeling the decision, the sequence flows from left to right, with later outcomes and decisions shown to the right of earlier decision and outcomes.

(b) Branches at a decision point are "pruned off" if they maximize the benefit to the decision maker relative to other choices.

(c) In analyzing the best path, sequence flows from right to left as inferior branches are pruned at decision points.

(d) Expected value at outcome points is calculated by multiplying the effect of each branch by the probability of that branch event.

## Chapter 11
### Depreciation of Capital Assets

**D-108** The correct percentage to use to calculate the depreciation allowance for a MACRS 3-year property for Year 2 is which of the following?

(a) 14.4%

(b) 32.0%

(c) 33.3%

(d) 44.5%

**D-109** With reference to the straight-line depreciation method, which statement is false?

(a) An equal amount of depreciation is allocated in each year.

(b) The book value of the asset decrements by a fixed amount each year.

(c) The depreciation life (n) is set based on the MACRS property classes.

(d) The asset is depreciated down to a book value equal to the salvage value.

**D-110** A 7-year MACRS property has a cost basis for depreciation of $50,000. The estimated salvage (market) value is $10,000 after its 10-year useful life. The depreciation charge for the 4$^{th}$ year and book value of the asset after the 4$^{th}$ year of depreciation are closest to what values?

(a) $5760; $44,240

(b) $6245; $12,496

(c) $6245; $15,620

(d) $6245; $43,755

**D-111** A $100,000 asset has a $20,000 salvage value after its 10-year useful life. The depreciation allowance using straight-line depreciation is closest to what value?

(a) $2000

(b) $8000

(c) $10,000

(d) $12,000

**D-112** The book value of an asset that is listed as a 10-year MACRS property is $49,500 after the first year. If the asset's estimated salvage (market) value is $5000 after its 15-year useful life, what was the asset's original cost basis?

(a) $50,000

(b) $52,105

(c) $55,000

(d) $61,875

## Chapter 12
*Calculating Income Taxes*

**D-113** Which of the following is true?

(a) Tax rates are based on two flat-rate schedules, one for individuals and one for businesses.

(b) When businesses subtract expenses, they always include capital costs.

(c) For businesses, taxable income is total income less depreciation and ordinary expenses.

(d) When quantifying depreciation allowance, one must always divide first cost by MACRS 3-year life.

**D-114** If a corporation has taxable income of $60,000, which of the following expressions is used to calculate the federal tax owed?

(a) Flat 15% of taxable income

(b) Flat 25% of taxable income

(c) $7,500 + 25% over $50,000

(d) $13,750 + 34% over $75,000

**D-115** This past year CLL Industries had income from operations of $8.2 million and expenses of $1.8 million. Allowances for depreciated capital expenses were $400,000. What is CLL's taxable income and federal taxes owed for operations last year? Choose the closest values.

(a) $6.0 million; $1.93 million

(b) $6.0 million; $2.04 million

(c) $6.4 million; $1.93 million

(d) $6.4 million; $2.04 million

**D-116** Annual data for a firm for this tax year are:

| | |
|---|---|
| Revenues | $45 million |
| Operating and maintenance costs | 7 million |
| Labor/Salary costs | 15 million |
| Overhead and administrative costs | 3 million |
| Depreciation allowance | 8 million |

Next year the firm can increase revenue by 20% and costs will increase by 2%. If the depreciation allowance stays the same, what will be the change in firm's after-tax net profit? Choose the closest answer.

(a) $5.2 million

(b) $5.4 million

(c) $7.9 million

(d) $13.3 million

**D-117** Widget Industries erected a facility costing $1.56 million on land bought for $1 million. The firm used straight-line depreciation over a 39-year period; it installed $2.5 million worth of plant and office equipment (all classified as MACRS 7-year property). Gross income from the first year of operations (excluding capital expenditures) was $8.2 million, and $5.8 million was spent on labor and materials. How much did Widget pay in federal income taxes for the first year of operation?

(a) $680,935

(b) $1,002,750

(c) $1,321,815

(d) $2,788,000

## Chapter 13
*Replacement Analysis*

**D-118** Which of the following is a valid reason to consider replacing an existing asset?

(a) It has become obsolete and does not perform its intended function.

(b) There is a newer asset available that is technologically superior.

(c) It has become very costly to keep the asset in working order.

(d) All of the above.

**D-119** The replacement of a typical existing asset (defender) that is beyond its minimum economic life with a new asset (challenger) should be done when?

(*a*) It should never be done because the existing asset is working.

(*b*) The defender's marginal cost is greater than the challenger's equivalent annual cost.

(*c*) The challenger's average future cost becomes less than the existing asset.

(*d*) The defender's equivalent annual cost equals the challenger's equivalent annual cost.

**D-120** A factory asset has a first cost of $100,000, annual costs of $15,000 the first year and increasing by 7.5% per year, and a salvage (market) value that decreases by 20% per year over its 5-year life. The minimum cost economic life of the asset is what value? Interest is 10%.

(*a*) 2 years
(*b*) 3 years
(*c*) 4 years
(*d*) 5 years

**D-121** The minimum cost life of a new replacement machine is 6 years with a minimum equivalent annual cost of $6000. Given the existing machine's marginal cost data for the next 4 years, when should the existing machine be replaced?

| Year | Total Marginal Cost |
|------|---------------------|
| 1 | $5400 |
| 2 | 5800 |
| 3 | 6200 |
| 4 | 8000 |

(*a*) After Year 1
(*b*) After Year 2
(*c*) After Year 3
(*d*) After 6 years

**D-122** A material handling system was purchased 3 years ago for $120,000. Two years ago it required substantial upgrading at a cost of $15,000. It once again is requiring an upgrading cost of $25,000. Alternately, a new system can be purchased today at a cost of $200,000. The existing machine could be sold today for $50,000. In an economic analysis, what first cost should be assigned to the existing system?

(*a*) $50,000
(*b*) $65,000
(*c*) $75,000
(*d*) $80,000

## Chapter 14
*Inflation Effects*

**D-123** If the real growth of money interest rate for the past year has been 4% and the general inflation has been 2.5%, the combined (market) interest rate is closest to what value?

(*a*) 1.5%
(*b*) 6.5%
(*c*) 6.6%
(*d*) 10.0%

**D-124** To convert actual (inflated) dollars to constant purchasing power dollars (where $n$ = difference in time between today and purchasing base) that occur at the same point in time, one must:

(*a*) Multiply by $(P/F$, inflation rate %, $n)$.
(*b*) Multiply by $(F/P$, inflation rate %, $n)$.
(*c*) Multiply by $(P/F$, combined rate %, $n)$.
(*d*) Divide by $(P/F$, real interest rate %, $n)$.

**D-125** The cost of a material was $2.00 per ounce 5 years ago. If prices have increased (inflated) at an average rate of 4% per year, what is the cost per ounce now?

(*a*) $2.04
(*b*) $2.08
(*c*) $2.43
(*d*) $8.00

**D-126** A deposit of $1000 is made into an account that promises a minimum of 2% per year increase in purchasing power. If general price inflation is 3, 1, and 5%, respectively, over the next 3 years, the minimum value that will be in the account at Year 3 is closest to what amount?

(*a*) $1061
(*b*) $1092
(*c*) $1157
(*d*) $1177

**D-127** The cost of materials was $1000 per lot when the cost index is 145. Today the cost index is 210. What is the cost per lot?

(*a*) $690
(*b*) $1000
(*c*) $1448
(*d*) Cannot be determined with the data given.

## Chapter 15
*Capital Budgeting*

**D-128** Which of these statements can be said of projects considered to be mutually exclusive?

(*a*) The projects are equivalent or mutual in terms of their cash flows.

(*b*) Neither alternative should be chosen.

(c) All projects can be chosen as long as they meet minimum economic criterion.

(d) The selection of one in the set eliminates selection of others in that same set.

**Use the following data for Problems 129–131**
A firm has identified four projects for possible funding in the next budget cycle.

| Project | First Cost ($M) | PW($M) |
|---------|-----------------|--------|
| A | 12 | 225 |
| B | 18 | 250 |
| C | 24 | 320 |
| D | 30 | 400 |

**D-129** The projects are independent. What is the total number of possible investment strategies (combinations) that the firm can use?

(a) 4

(b) 6

(c) 10

(d) 15

**D-130** If the projects have the following contingencies, what is the total number of possible investment strategies (combinations) that the firm can use? Project A can be invested in only by itself; Project D is chosen only if Project B is chosen.

(a) 0

(b) 2

(c) 4

(d) 6

**D-131** If the projects are independent, and the project budget (capital limit) is $32 million, what investment combination maximizes PW?

(a) A and C

(b) D alone

(c) Invest in all projects to maximize PW.

(d) A and B

**D-132** A firm is considering the "make vs. buy" question for a subcomponent. If the part is made in-house, the production data would be: first cost = $350,000; annual costs for operation = $45,000; salvage value = $15,000; project life = 5 years; interest = 10%; and material cost per unit = $8.50. If annual production is 10,000 units, the maximum amount that the firm should be willing to pay to an outside vendor for the subcomponent is nearest?

(a) $10 per unit

(b) $16 per unit

(c) $22 per unit

(d) $28 per unit

**D-133** A firm is considering whether to buy specialized equipment that would cost $200,000 and have annual costs of $15,000. After 5 years of operation, the equipment would have no salvage value. The same equipment can be leased for $50,000 per year (annual costs included in the lease), payable at the beginning of each year. If the firm uses an interest rate of 5% per year, the annual cost advantage of leasing over purchasing is nearest what value?

(a) $2494

(b) $8694

(c) $11,200

(d) $12,758

## Chapter 17
### Accounting

**D-134** Which of the following summarizes a firm's revenues and expenses over a month, quarter, or year?

(a) Balance sheet

(b) Statement of assets and liabilities

(c) Income statement

(d) None of the above

**D-135** Which of the following financial ratios provides insight into a firm's solvency over the short term by indicating its ability to cover current liabilities?

(a) Acid-test ratio

(b) Quick ratio

(c) Current ratio

(d) Net profit ratio

**D-136** Which statement is most accurate related to an activity-based costing (ABC) system?

(a) ABC spreads the firm's indirect costs based on volume-based activities.

(b) ABC seeks to assign costs to the activities that drive those costs.

(c) ABC gives an inaccurate view of a firm's costs and should never be used.

(d) ABC is called ABC because it is an easy method to use.

**D-137** A firm's balance sheet has the following data:

| | |
|---|---|
| Cash on hand | $450,000 |
| Market securities | 25,000 |
| Net accounts and notes receivable | 125,000 |
| Retailers' inventories | 560,000 |
| Prepaid expenses | 48,500 |
| Accounts and notes payable (short term) | 700,000 |
| Accumulated liabilities | 120,000 |

The firm's current ratio and acid-test ratio are closest to what values?

(*a*)  1.42; 0.73

(*b*)  1.42; 0.79

(*c*)  1.47; 0.73

(*d*)  1.47; 0.79

**D-138** A firm's income statement has the following data (in $10,000):

| | |
|---|---|
| Total operating revenues | $1200 |
| Total nonoperating revenues | 500 |
| Total operating expenses | 925 |
| Total nonoperating expenses | 125 |

If the firm's incomes taxes were $60,000 what was the net profit (loss)?

(*a*)  All necessary data is not given, one cannot calculate net profit (loss).

(*b*)  −$53,500

(*c*)  $1,150,000

(*d*)  $6,440,000

**D-139** Annual manufacturing cost data (1000s) for four product lines are as follows:

| Data | Line 1 | Line 2 | Line 3 | Line 4 |
|---|---|---|---|---|
| Annual production | 4000 | 3500 | 9800 | 675 |
| Cost of direct materials | $800 | $650 | $1200 | $2500 |
| Cost of direct labor | $3500 | $3750 | $600 | $320 |

Rank the product lines from lowest to highest in terms of manufacturing cost per unit. Total indirect costs of $10.8 million are allocated based on total direct cost.

(*a*)  1-2-3-4

(*b*)  3-1-2-4

(*c*)  3-2-1-4

(*d*)  3-4-1-2

AASHTO. *A Manual on User Benefit Analysis of Highway and Bus-Transit Improvements.* Washington, DC: American Association of State Highway and Transportation Officials, 1978.

American Telephone and Telegraph Co. *Engineering Economy, $3^{rd}$.* McGraw-Hill, 1977.

Baasel, W. D. *Preliminary Chemical Engineering Plant Design, $2^{nd}$.* Van Nostrand Reinhold, 1990.

Benford, H. A. *Naval Architect's Introduction to Engineering Economics.* Ann Arbor, MI: Department of Naval Architecture and Marine Engineering Report 282, 1983.

Bernhard, R. H. "A Comprehensive Comparison and Critique of Discounting Indices Proposed for Capital Investment Evaluation," *The Engineering Economist.* Vol. 16, No. 3, 1971, pp. 157–186.

Block, S. "Are There Differences in Capital Budgeting Procedures between Industries? An Empirical Study," *The Engineering Economist.* Vol. 50, No. 1, 2005, pp. 55–67.

Elizandro, D. W., and J. O. Matson. "Taking a Moment to Teach Engineering Economy," *The Engineering Economist.* Vol. 52, No. 2, 2007, pp. 97–116.

*The Engineering Economist.* A quarterly journal of the Engineering Economy Divisions of ASEE and IIE.

Eschenbach, T. *Engineering Economy: Applying Theory to Practice, $3^{rd}$.* Oxford University Press, 2011.

Eschenbach, T. G., E. R. Baker, and J. D. Whittaker. "Characterizing The Real Roots for $P$, $A$, and $F$ with Applications to Environmental Remediation and Home Buying Problems," *The Engineering Economist*, Volume 52, Number 1, 2007, pp. 41–65.

Eschenbach, T. G., and J. P. Lavelle. "Technical Note: MACRS Depreciation with a Spreadsheet Function: A Teaching and Practice Note," *The Engineering Economist.* Vol. 46, No. 2, 2001, pp. 153–161.

Eschenbach, T. G., and J. P. Lavelle. "How Risk and Uncertainty Are/Could/Should Be Presented in Engineering Economy," *Proceedings of the $11^{th}$ Industrial Engineering Research Conference.* IIE, Orlando, May 2002, CD.

Eschenbach, T. G., N. A. Lewis, J. C. Hartman, and L. E. Bussey (with chapter author H. L. Nachtmann). *The Economic Analysis of Industrial Projects $3^{rd}$*, Oxford University Press, 2012.

Eschenbach, T. G., N. A. Lewis, and J. C. Hartman. "Technical Note: Waiting Cost Models for Real Options." *The Engineering Economist*, Vol. 54 No. 1, 2009, pp. 1–21.

Fish, J. C. L. *Engineering Economics.* McGraw-Hill, 1915.

Jones, B.W. *Inflation in Engineering Economic Analysis.* John Wiley & Sons, 1982.

Lavelle, J. P., H. R. Liggett, and H. R. Parsaei, editors. *Economic Evaluation of Advanced Technologies: Techniques and Case Studies.* Taylor & Francis, 2002.

Lewis, N. A., T. G. Eschenbach, and J. C. Hartman, "Can We Capture the Value of Option Volatility?" *The Engineering Economist*, Vol. 53 No. 3, July – September 2008, pp. 230–258, winner of Grant Award for best article in volume 53.

Liggett, H. R., J. Trevino, and J. P. Lavelle. "Activity-Based Cost Management Systems in an Advanced Manufacturing Environment," *Economic and Financial Justification of Advanced*

*Manufacturing Technologies.* H. R. Parsaei, W. G. Sullivan, and T. R. Hanley, editors. Elsevier Science, 1992.

Lorie, J. H., and L. J. Savage. "Three Problems in Rationing Capital," *The Journal of Business.* Vol. 28, No. 4, 1955, pp. 229–239.

Newnan, D. G. "Determining Rate of Return by Means of Payback Period and Useful Life," *The Engineering Economist.* Vol. 15, No. 1, 1969, pp. 29–39.

Perry, H. P., D. W. Green, and J. O. Maloney. *Perry's Chemical Engineers' Handbook,* $7^{th}$. McGraw-Hill, 1997.

Peters, M. S., K. D.Timmerhaus, M. Peters, and R. E.West. *Plant Design and Economics for Chemical Engineers,* $5^{th}$. McGraw-Hill, 2002.

Sundaram, M., editor. *Engineering Economy Exam File.* Oxford University Press, 2003.

Steiner, H. M. *Public and Private Investments: Socioeconomic Analysis.* John Wiley & Sons, 1980.

Terborgh, G. *Business Investment Management.* Machinery and Allied Products Institute, 1967.

Tippet, D. D., and P. Hoekstra. "Activity-Based Costing: A Manufacturing Management Decision-Making Aid," *Engineering Management Journal.* Vol. 5, No. 2, June 1993, American Society for Engineering Management, pp. 37–42.

Wellington, A. M. *The Economic Theory of Railway Location.* JohnWiley & Sons, 1887.

# INDEX